GLACIAL DEPOSITS IN GREAT BRITAIN AND IRELAND

GLACIAL DEPOSITS
IN
GREAT BRITAIN AND IRELAND

Edited by

JÜRGEN EHLERS
Geologisches Landesamt, Hamburg

PHILIP L. GIBBARD
University of Cambridge

JIM ROSE
Royal Holloway and Bedford New College, Egham

CRC Press
Taylor & Francis Group
Boca Raton London New York

CRC Press is an imprint of the
Taylor & Francis Group, an **informa** business

CRC Press
Taylor & Francis Group
6000 Broken Sound Parkway NW, Suite 300
Boca Raton, FL 33487-2742

© 1991 by Taylor & Francis Group, LLC
CRC Press is an imprint of Taylor & Francis Group, an Informa business

No claim to original U.S. Government works

Visit the Taylor & Francis Web site at
http://www.taylorandfrancis.com

and the CRC Press Web site at
http://www.crcpress.com

Contents

Foreword

RICHARD G. WEST

Gone are the days, I hope, when, according to Woodward (1907), superficial or drift deposits were described as *extraneous rubbish*. The basis for this hope is seen clearly in the following accounts of the glacial deposits of Great Britain and Ireland.

Glacial deposits are the most extensive of Quaternary deposits in Great Britain and Ireland. They are classically the characteristic sediments of the Quaternary, and were the original source of ideas of climatic change and of glacial episodes. Their study lead at an early date to classifications of Quaternary time and to widely-used schemes of stratigraphy. Such schemes, in north-west Europe at least, must still have a skeleton of glaciations, though the evidence for climatic change no longer relies most importantly on glacial deposits. Evidence from periglacial phenomena and palaeontology has resulted in a much wider knowledge of cold stage environments and climates, extending beyond the apparently simpler concept of glaciation. However, studies of present ice sheets and their related glacial and periglacial sediments has lead to a much better understanding of the complexity of processes related to past glaciations, belying this apparent simplicity.

No detailed survey of glacial deposits in the British Isles has been made before on the scale attempted in this book. The classical accounts of Carvill Lewis (1894), Wright (1937) and Charlesworth (1957) were made many years ago, before the vast development of knowledge of processes and sediments associated with ice movement and ice melting, a development which has taken place in the last few decades. The place of glaciation in Quaternary stratigraphy is more fully understood and the advent of radiocarbon dating has made possible much firmer resolution of the details of the last major ice advance in the last cold stage. The improvement of Quaternary stratigraphy has made it possible to segregate ice advances into particular cold stages, so further understanding the nature of the sequence of glaciations.

This improvement in our knowledge of glaciation and glacial deposits is certainly needed when we come to try and understand the glaciation of the British Isles. Although a relatively small area on the continental shelf of western Europe, the geographical background is complex. The ice caps were largely, perhaps even entirely, independent from the Scandinavian ice sheet, and more than one developed from centres of high ground in this oceanic situation, covering an expanse of latitude. Many areas in the British Isles must have been close to the centres of ice dispersion, making for a complicated mixture of erosion and deposition. Such arrangements contrast with the more regular sequence of sheets of glacial sediment found in the north European plain and resulting from the great expansion of the Scandinavian ice sheet.

Studies of British and Irish glacial deposits have probably profited from the development of a comparatively well-founded stratigraphy later than that of continental Europe, though many major problems remain, as will be seen in the following accounts. These are segregated into a section dealing with glacial events which have been widely recognised, a section containing detailed regional surveys, a section discussing particular topics involved in the study of glacial deposits, and a final overview. The extent of the contributions and the detail given is most timely in the development of Quaternary studies in Britain and Ireland, and will be seen to present an invaluable analysis and synthesis of the subject.

Glacial events

Time and space in the glacial sediment systems of the British Isles

D.Q. BOWEN

Three data sets from North Atlantic Ocean deep-sea cores are first order indicators of the volume of ice on surrounding continents: (1) variability in oxygen isotope ratios, controlled by global ice volume (Shackleton & Opdyke, 1973), (2) ice-rafted sediments (and $CaCO_3$ variability) (Ruddiman, 1977), controlled by the volume of ice on adjacent land masses, and (3) sea surface temperature (SST) estimates, based on planktonic fossils (Imbrie & Kipp, 1971). These data and climate modelling have shown that ice-sheets bordering the North Atlantic Ocean north of 45-50°N reduced temperatures in the ocean at the 41,000-year and 100,000-year orbital rhythms. Both these signals were 'transferred from the ice-sheets to the ocean via the atmosphere with little or no lag' (Ruddiman, 1987).

The first indication of moderately-sized ice sheets consists of indications of ice-rafting activity, dated to 2.4 million years B.P., in the Deep-Sea Drilling Project (DSDP) site at Rockall (Shackleton *et al.*, 1984). During the Matuyama Chron Northern Hemisphere ice sheets responded to the 41,000-year orbital rhythm (Ruddiman *et al.*, 1986; Ruddiman & Raymo, 1988). After 900,000 B.P. the climatic signals differ from those of the Matuyama Chron, and they suggest that Northern Hemisphere ice sheets grew to much larger maximum volumes, with the first $\delta^{18}O$ maximum occurring in Oxygen Isotope Stage 22. SSTs, % $CaCO_3$, and $\delta^{18}O$ indicators all show increasing variance after 900,000 B.P. at periods of 100,000 years (eccentricity), 23,000 years and 19,000 years (precession). They all show that the 100,000-year rhythm became dominant after 450,000 B.P. (Imbrie, 1985; Ruddiman & Raymo, 1988).

Although the North American ice sheet controlled one-half or more of the ice volume signal, it is unlikely that the American and British ice sheets were out of phase. Indeed the location of the British Isles on the margin of the northeast Atlantic Ocean, directly influenced by movements of the polar front,

and of the Gulf Stream (Ruddiman & McIntyre, 1976; 1981a) resulted in high ice sheet sensitivity to external forcing. Furthermore, the combination of a land mass chilled by orbital forcing alongside an ocean surface still warmed by the Gulf Stream in the northeast Atlantic Ocean would have been (Ruddiman *et al.*, 1980) an ideal juxtaposition leading to rapid and extensive ice growth. Critical to this process would be the availability of effective precipitation. Any spatial variability in precipitation would account for differences in the location, strength and extent of British ice centres at different times. If the variability of British ice centres can be established in time and space, and related to the larger Scandinavian ice sheet (Bowen & Sykes, 1988), it could provide an index of atmospheric variability for climate modelling.

Events in Scotland during the Devensian Stage are instructive because they allow inferences to be drawn on the importance of spatial variability in effective precipitation for the location and strength of ice centres. The rapid growth of the Loch Lomond ice sheet has been shown by radiocarbon dating and pollen analytical investigations (Sutherland, 1984a). This has been attributed to the rapid southward movement of the polar front in the North Atlantic Ocean about 11,000-10,000 B.P. (Sissons, 1979e). The coincidence of the two events in time is reasonably established, but what is not clear is exactly how the ocean surface is coupled with the atmosphere in such a way as to localise storm tracks to deliver the effective precipitation necessary for rapid ice growth. Using data from cirques occupied by Loch Lomond ice, Sissons & Sutherland (1976) suggested that precipitation was delivered mainly by occluded depressions. A further argument for spatial variability in precipitation at this time is the comparatively poor development of ice in North and Northwest Scotland.

A similar, although more extended sequence of events can be inferred for Late Devensian glaciation in Scotland. Ice from the Highlands expanded across the Central Lowlands of Scotland on to the Southern Uplands. Later, the Southern Uplands ice centre grew in strength and expanded into ground formerly covered by the Highlands ice (Geikie, 1894; Sissons, 1967a). The change in the relative strengths of the Highland ice and the Southern Uplands ice, has been attributed to the southward movement of the zone of maximum effective precipitation. This hypothesis is supported by evidence of ice-free areas in Northeast Scotland, in the Orkney Islands and in Caithness, as well as by maximum extension of the British ice sheet to the West Midlands of England, Southeast Ireland, and Southwest Wales (Sutherland, 1984a; Bowen *et al.*, 1986).

On the other hand, Caithness and the Orkney Islands were glaciated during the Early Devensian (Bowen & Sykes, 1988), while farther south in the British Isles any Early Devensian deposits have either been destroyed or overrun by the Late Devensian ice sheet. Thus it is clear that Early and Late Devensian ice sheets were not in spatial phase, a circumstance that can be attributed to variability in effective precipitation.

Glacial and glacial-related sediment systems

A glacial or glacial-related sediment system consists of: *a body of sediment either deposited directly by ice or processes associated with its wastage, or proglacial and para-glacial sediments*. The geographical extent of such systems is defined by ice sheet margins, ice flow lines, and the distribution of proglacial and paraglacial (for example, fluvial deposits with an outwash component) sediments which can carry the influence of the ice far beyond its margins. Glacial sediment systems are not necessarily mutually exclusive. The major sediment systems were active throughout most of the Pleistocene, although with different tempo in space and time. Some sub-systems were active for only part of the time, and may have been modified by catchment changes, or changes in the strength of different ice centres.

While a considerable variety of subsystems can be recognised on a detailed scale, it is only appropriate here to identify the major systems. These are: (1) The North Sea Basin Glacial Sediment System, which includes eastern Scotland, the north and east of England, with some input from Wales. (2) The Irish Sea Basin Glacial Sediment System, which includes glacier sub-systems from the North of England, mainland Ireland, Wales, with some input from Scotland. (3) The Hebridean Seas Glacial Sediment System.

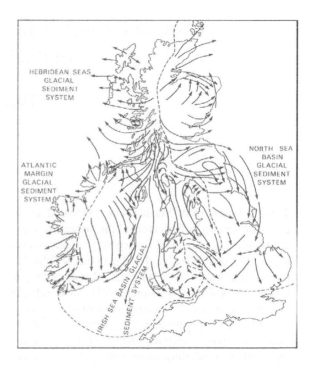

Figure 1: British Glacial Sediment Systems. Ice movement based, with additions, on Charlesworth (1957).

(4) The Atlantic Margin Glacial Sediment System (fig. 1).

Glacial sediment systems in time

The 1973 classification and correlation of Quaternary deposits in the British Isles (Mitchell *et al.*, 1973) recognised temperate and cold stages of chronostratigraphic status. Temperate ('interglacial') stages were defined on the basis of their characteristic vegetational history, and cold ('glacial') stages were defined on lithological criteria. Except for the later Pleistocene within the range of radiocarbon dating, no means of geochronometric dating was available. The succession and its correlation relied almost entirely on the identification of 'fixed points' provided by the vegetationally-defined 'interglacials'. On this basis, the numerous and incomplete rock successions throughout the British Isles were correlated and integrated into a model of earth history. Unfortunately the temperate stages ('interglacials') are not 'fixed' in time by first appearances or extinctions of fossil taxa. Their characteristic vegetational development is controlled by climate and not time. Such facies floras are inherently unreliable and unsuitable for dating and subdivision, and any classification based on them relies primarily on inferences about climate rather than on the actualities of the rocks (George, 1970; Bowen, 1978).

Proposals for the modification of the 1973 classification have been made (see, for example, table 1). Principally, these include additional stages based on a variety of evidence. But the case of the 'Wolstonian', is different and merits some discussion. It was originally defined as the 'Gipping Glaciation' which was thought to be intermediate in age between the Hoxnian and Ipswichian Stages in East Anglia. But after Bristow & Cox (1973) showed that the Gipping Till did not exist as a glacial lithostratigraphic unit separate from the Lowestoft Till (Anglian), a new type site for the cold interval between the Hoxnian and Ipswichian Stages was proposed at Wolston in the Midlands (Mitchell *et al.*, 1973). Lithological (Perrin *et al.*, 1979) and stratigraphical evidence (Rose, 1987), however, shows that the Wolstonian glacial sequence at its type site and elsewhere may be of Anglian age. Nevertheless, the case for retaining the label 'Wolstonian', because it refers

Table 1. Revised classification of the Quaternary of the British Isles. Amino acid ratios from Bowen *et al.* (1986), Bowen & Sykes (1988), Bowen *et al.* (unpublished). Ice margin positions shown by broken lines. Early Devensian glacial deposits shown by triangles. Upland ice centres shown by crosses. Kesgrave Formation shown by stipple.

STANDARD STAGES & events to be defined	Δ^{18}O Stages	Amino Acid Epimerization Marine Moderate	Marine Slow	Non-marine	Glaciation	
DEVENSIAN — Late	2	0.05 / 0.07		0.04	Loch Lomond / Late Devensian	
DEVENSIAN — Middle	3	0.09				
DEVENSIAN — Early	4				Early Devensian	
	5			0.07		
IPSWICHIAN	5e	0.16	0.1	0.1		
	6					
Stanton Harcourt	7	0.2	0.16	0.16		
	8				Paviland	
HOXNIAN	9	0.28	0.22	0.24		
	10					
Swanscombe	11	0.38	0.28	0.3		
ANGLIAN	12				Anglian	
CROMERIAN	13			0.34		
Waverley Wood	15			0.38	KESGRAVE FORMATION Up to 4 upland glaciations	
—0.73 Ma— Brunhes (approximate position) / Matuyama					A G B	
BAVENTIAN Norwich Crag		0.8			?Upland Ice	
Red Crag	1.09 / 2.4 Ma				Ice Rafting	

to post-Hoxnian, but pre-Ipswichian time, has been argued strongly (Gibbard & Turner, 1987). It is, however, difficult to see what purpose this would serve because it is not defined by rocks at a stratotype.

Because of the inherently homotaxial nature of Quaternary rock successions, and the inadequacy of a palaeontologically based system of dating and subdivision, it appears reasonable to conclude that only subdivision and correlation based on multiple dating methods is appropriate. Radiocarbon dating appears largely adequate for subdividing and correlating Late Devensian events (Rose, 1985). Uranium-series, Thermoluminescence and Electron Resonance Dating, have all been applied with some success, but have not yet been widely applied for a variety of problematical reasons. On the other hand, an amino acid geochronology based on fossil molluscs offers the only existing means of dating which is extensive in both time and space (Bowen *et al.*, 1986; Hughes, 1987; Bowen & Sykes, 1988; Bowen *et al.*, 1989). A provisional integration of both marine and non-marine amino acid data is assembled in table 1, which is used to subdivide the sequence.

The North Sea Basin Glacial Sediment System

The ice centres of the North Sea Basin Sediment System were in the Highlands and Southern Uplands of Scotland, northern and eastern England, together with a Scandinavian influence. For some considerable time the system was also fed by ice from the mountains of North Wales.

Heavy minerals from the Baventian Stage sediments of East Anglia, and remanie glacial deposits possibly of the same age on the Yorkshire Wolds (Catt, 1982b), represent the earliest known evidence for glaciation, although nothing is known about its extent. A maximum age is placed on the Baventian by amino acid ratios in *Macoma* from the Norwich Crag Formation beds at Wangford, which show that it is early Matuyama in age (table 1).

The Kesgrave Sediment Subsystem operated when the mountains of North Wales were part of the larger system. Rose (in Bowen *et al.*, 1986) has shown that input from North Wales probably ceased early during the Anglian Stage. It is represented by the Northern Drift on the dip-slope of the Cotswold Hills which is partly glacial and partly (decalcified) fluvial in origin. Sediments of this subsystem continue through the Middle Thames Valley, the Vale of St.Albans, into East Anglia (Rose & Allen, 1977), where they mark

the former courses of the Thames before its diversion. These deposits, mainly gravels, contain erratic clasts from North Wales (Gwynedd, Clwyd and Powys), and Kidderminster (Bunter) Conglomerate from the Northwest Midlands (Hey, 1965; 1986; Hey & Brenchley, 1977; Green *et al.*, 1980; Gibbard, 1985). The subsystem is named after the Kesgrave Formation of East Anglia (Rose & Allen, 1977; Rose *et al.*, 1976), which is continued offshore by the Yarmouth Roads Formation of the southern North Sea (Balson & Cameron, 1985), correlated with the Sterksel Formation of the Netherlands (Zagwijn, 1986).

The erratic clasts in the Kesgrave subsystem have been used to infer glaciation in the former headwaters of this drainage system. Rose (in Bowen *et al.*, 1986) has inferred at least four separate glaciations based on four major stratigraphic units: the Stoke Row Gravels, Westland Green Gravels, Beaconsfield Gravels ('Higher Gravel Train' of Wooldridge, 1938), and Gerrards Cross Gravels (Gibbard, 1985).

The age of the Kesgrave subsystem may be established within broad limits. It is younger than at least part of the Norwich Crag Formation, and is older than the Cromerian Stage (West Runton). Amino acid ratios on *Macoma* from Norwich Crag Formation sediments at Wangford, are of early Matuyama age. While on the basis of an epimerization model for non-marine mollusca, the Cromerian Stage (West Runton) is ascribed to Oxygen Isotope Stage 13 (Bowen *et al.*, 1989).

The Baginton-Lillington Gravels and the Baginton Sand (Baginton- Lillington Beds) of the Midlands are correlated with the Kesgrave Formation (Rose, 1987). Both units are overlain by glacial deposits of the Anglian Stage glaciation, an event which terminated the operation of the Kesgrave subsystem. Because the Anglian Stage is ascribed to Stage 12 (table 1) this fixes a close minimum age for the Kesgrave subsystem. Intraformational organic beds at Waverley Wood in the Midlands contain molluscs with amino acid ages ascribed to Stage 15 (table 1). A maximum age is probably best established by reference to some of its continental correlatives, namely the Sterksel Formation and the higher terrace gravels of the lower Rhine around Cologne. These show that sedimentation commenced during the Matuyama Chron. One of the main requirements in the British Isles at present is to fix the position of the Matuyama-Brunhes reversal within the rock sequence.

The top of the Aberdeen Ground Beds Formation in the area of the Forth approaches consists of glacial and glaciomarine sediments related to a Scottish ice-sheet. They occur at a stratigraphic level just above the Matuyama-Brunhes palaeomagnetic boundary. They could, therefore, be correlated with 'Glacial A' of the Dutch Cromerian 'Complex' (Stoker & Bent, 1985).

In the Midlands an early ice sheet from the North of England may be indicated by the Bubbenhall Clay (Shotton, 1953). If this is a till, it shows glaciation during the Matuyama Chron because it underlies the Baginton-Lillington beds (below).

The type section of the Anglian Stage at Corton in Suffolk (Mitchell *et al.*, 1973) includes the Cromer Till (Gunton Stadial), Corton Sands, and the Lowestoft Till. The Cromer Till is correlated with deposits of the North Sea Drift Glaciation, and with the Norwich Brickearth of northeast Norfolk. Lithologically these deposits are similar (Perrin *et al.*, 1979), and their far-travelled erratics and minerals show a Scandinavian influence. What is not clearly established, however, is whether these sediments were deposited as lodgement tills, or as glaciomarine or glaciolacustrine deposits, because it has been argued that many of the 'tills' exposed along the north Norfolk coast, including the 'contorted drift', were deposited subaqueously (Gibbard, 1980; Eyles & McCabe, 1989). On this basis, for example, the Corton Sands could be reinterpreted as shallow-water glaciomarine sediments.

The Lowestoft Till, however, is a land-based glacial deposit which occurs widely throughout East Anglia and the East Midlands (Perrin *et al.*, 1979). On its southern margin the Lowestoft Till ice-sheet blocked the course of the Kesgrave Thames (Gibbard, 1985). In the East Midlands it is coextensive with glacial deposits formerly classified as 'Wolstonian' (Perrin *et al.*, 1979; Rose, 1987). These also include the Triassic-rich Thrussington Till of northerly derivation, which shows that Pennine glaciers were extensive at that time. The Woodston Beds, which occur on the western margin of the Wash Basin, contain molluscs which amino acid geochronology has shown are of Hoxnian age. Because they are not covered by glacial deposits, this shows that no post-Hoxnian glaciation of this and adjacent areas occurred.

A maximum age for the Lowestoft Till Glaciation of Anglian age is fixed by the Cromerian Stage deposits of West Runton ascribed to Stage 13 of the Oxygen Isotope framework (table 1). The Hoxnian Stage, however, does not follow the Anglian because the minimum age for the Lowestoft Glaciation is fixed by the Nar Valley Beds of Norfolk: amino acid ratios in shells from both the lower freshwater deposits and upper marine clay are ascribed to Stage 11 of the Oxygen Isotope framework (table 1). Amino acid ratios also allow correlation of the Nar Valley Beds with the Swanscombe deposits of the Lower Thames Valley. These are also of immediately post-Lowestoft Till Glaciation age and represent early post-diversion Thames deposits. On this basis, therefore, the Lowestoft Till and the Anglian are ascribed to Stage 12 (Bowen *et al.*, 1989).

Farther west, in the Birmingham area, the sequences at Quinton and Nechells remain controversial. On the one hand, it has been argued that the Nechells lake beds are Hoxnian in age (Kelly, 1964), which would make the underlying till Anglian in age. But on the other hand, Sumbler (1983a) believes that the upper till at Nechells and Quinton is the same age as the 'Wolstonian' sequence farther east, now known to be of Anglian age. No independent means of dating, such as amino acid geochronology, are available, other than on the Bushley Green Terrace of the Severn, which is, at least in part, ascribed to Stage 9. The Nechells and Quinton organic beds could equally be Stage 9 or Stage 11.

On the northeast coast of England Scandinavian erratics occur in the Warren House Till of Durham (Francis, 1970) and the Basement Till of East Yorkshire (Catt & Penny, 1966). How they were introduced is unknown, and the question of deposition by ice or glaciomarine agencies remains open. Provisional amino acid data from the tills of East Yorkshire show that considerable complexity may occur, and that the term 'basement till' may apply to tills of diverse ages. If they are not Anglian in age, then on the basis of available information, they could be ascribed to Oxygen Isotope Stage 8 because: (1) only tills of the Late Devensian overlie Stage 7 marine deposits at Easington (Brown & Sykes, 1988), and (2) the IGCP on '*Glaciations in the Northern Hemisphere*' concluded that no extensive Stage 10 glaciation occurred in Europe (Šibrava *et al.*, 1986). It is, of course, possible that the British Isles, because of their oceanic position, did support extensive Stage 10 glaciers, but no evidence is known to support this.

Pressing questions also arise about material of Scandinavian provenance in eastern Scotland. Was the Scandinavian ice sheet ever so extensive? Or were these introduced in glaciomarine environments?

The age of pre-Devensian glaciations in Scotland is also uncertain. *Arctica* shells from the Benholm Burn Clay deposit in eastern Scotland have given isoleucine ratios of 0.36. If the clays are glaciomarine, then they date to the deglaciation of the

Anglian Stage ice sheet (Stage 12). But if the clay represents a till, then it is younger than Stage 11. The ages of the multiple glacial units at Kirkhill, in northeast Scotland are unknown (Sutherland, 1984a) but it is likely that one of them is of Anglian age.

In Caithness and the Orkney Islands, erratics in a lower till indicate glaciation from the mountains of Sutherland. Fabric analysis shows ice-movement towards the northeast or north-northeast in Caithness (Omand, 1973), but towards the north or north-northwest in Orkney (Rae, 1976). These tills are overlain without a break by a deposit containing shells, traditionally interpreted as shelly till. Moreover, the shelly tills are taken to indicate ice-flow out of the Moray Firth, because the Scottish ice was being deflected by Scandinavian ice (Croll, 1870; Peach & Horne, 1881). Amino acid ratios from the shelly deposits both in Caithness and Orkney are characterised by a youngest molluscan faunal element ascribed to Oxygen Isotope Substage 5e, unlike other shelly deposits in Scotland which also contain faunal elements of Middle and Late Devensian Substage ages (Bowen & Sykes, 1988). It would appear, therefore, that the shelly deposits are of Early Devensian Substage age. If, as seems likely, they are of the same glaciomarine origin as similar deposits in the Irish Sea, this would dispense with the need for the presence of Scandinavian ice to deflect the Scottish Moray Firth ice across Caithness and Orkney. The age of the sediments also shows an expansion of Early Devensian ice in the far north of Scotland across terrain unglaciated during the Late Devensian Substage maximum advance.

The extent of a composite ice-sheet of Late Devensian Substage age has been established through the work of numerous workers (Bowen *et al.*, 1986), and its timing constrained by radiocarbon dating to between 26,000 B.P. and c. 14,000 B.P. In the north of Scotland shelly deposits contain molluscs with amino acid ages of Middle Devensian Substage age, and sometimes of Late Devensian Substage age. These have not yet been investigated to detect glaciomarine influences.

Till units with characteristic properties, and of Late Devensian age, can be traced from northeast England to Norfolk. This glaciation has been named the Dimlington Stadial (Rose, 1985) based on a stratotype at Dimlington Cliff, Holderness (Catt & Penny, 1966). The Withernsea and Skipsea Tills overlie the Dimlington Silts with radiocarbon dates of 18,500 and 18,240 B.P. (Penny *et al.*, 1969). Such continuous lithological similarity of the sediments, including the Withernsea Till over some 200 km

offshore (Derbyshire *et al.*, 1984) makes a glaciomarine origin possible. The age of the Hunstanton esker of Northwest Norfolk, however, is fixed by amino acid ratios on *Macoma* fossils from its deposits which confirm its Late Devensian age.

The Irish Sea Basin Glacial Sediment System

Glacial sources include the Southern Uplands of Scotland, the Lake District, Ulster, Central and Western Ireland, and the Welsh mountains. Inland from the present-day coastal hinterland glacial deposits are characteristically dominated by local and regional lithologies, and erratics, and glacial landforms indicate the various ice flow directions (fig. 1). But around the margins of the Irish Sea an 'Irish Sea Till' lithology is ubiquitous. This is a massive clay deposit, usually red or dark blue in colour, which contains shells or shell fragments. Traditionally interpreted as till, it is increasingly being described as a glaciomarine mud (Eyles & Eyles, 1984; McCabe, 1987; Eyles & McCabe, 1989a).

The oldest known glacial deposits are the Kenn Gravels and Kenn Pier Till of the Bristol district (Andrews *et al.*, 1984). These underlie fluvial deposits at Kenn Pier and Yew Tree Farm, which accumulated close to sea-level, and which could involve reworking of estuarine and/or rocky shore deposits. The deposits contain shells of *Corbicula fluminalis* with isoleucine epimerization ratios of 0.39 ± 0.014 (6 shells) (Andrews *et al.*, 1984). These compare with 0.35 and 0.38 at Purfleet in the Lower Thames (Miller *et al.*, 1979; Andrews *et al.*, 1984). These beds, therefore, are Cromerian (table 1) or older, and the glacial deposits are pre-Cromerian (West Runton). If these relate to the most extensive glaciation in western Britain, then they may be the same age as the glacial gravels on Lundy Island in the Bristol Channel (Mitchell, 1968), the Crousa Common Gravels in Cornwall (Bowen *et al.*, 1986), and the glacial gravels of the northernmost Isles of Scilly (Mitchell & Orme, 1967), which have been redistributed by solifluction on possibly several occasions (Bowen, 1973 b). By the same token, the Irish Sea Glaciation of Gower and Southwest Wales which crossed Carmarthen Bay (although denied by Battiau-Quenny, 1981, despite well-documented erratic trains, Bowen, 1970), is the same age.

In the West Midlands the oldest known glacial sediments are the tills around Tryssull (Morgan, 1973). At Tryssull the till is overlain by organic beds containing operculae of *Bythinia tentaculata*, which by comparison with other amino acid data allow ascription of these 'interglacial' beds to Stage 9. The underlying till is, therefore, probably of Stage 12 age, and can be correlated with either the upper and lower tills at Nechells and Quinton.

No other unequivocal evidence of an Irish Sea ice sheet occurs before the Devensian Stage, unless the multiple sequence of deposits proved in the Point of Ayre borehole in the Isle of Man contains such information (Thomas, 1976). In Gower, South Wales, however, an advance of Welsh ice from the north, deposited the Paviland Moraine. Local field relations suggest that this is older than the Minchin Hole Sea-level raised beaches of Stage 7 age. On this basis the Paviland Glaciation is ascribed to Stage 8 (Bowen and Sykes, 1988). Given such an extensive expansion of Welsh ice, more or less beyond its Late Devensian Substage extent, it is unlikely that there was no Irish Sea ice-sheet at that time.

At Derryoge and Loughshinny, County Down, Ulster, amino acid analysis has shown that the youngest faunal element in shelly glacial deposits is of Substage 5e age of the Oxygen Isotope framework, in contrast to Late Devensian Substage age beds which contain Middle Devensian Substage and Late Devensian Substage faunas. The deposits are, therefore, of Early Devensian age.

During the Late Devensian Substage a large Irish Sea ice-sheet, composed of several ice-centres, pushed into the Cheshire-Shropshire-Staffordshire Plain of Northwest England to Four Ashes (stratotype of the Devensian Stage) and Tryssull (Morgan, 1973). Amino acid ratios on shells from this area confirm its Late Devensian age. In the Isle of Man, all the lithostratigraphic units currently exposed above sea-level are also of Late Devensian Substage age on the basis of the amino acid signatures of their contained marine molluscs. Along the Irish coast Late Devensian sediments outcrop as far south as the Screen Hills landform complex. On the Welsh side, Irish Sea deposits of Late Devensian age occur in Anglesey, the Llŷn Peninsula and in southwest Dyfed, notably at Abermawr. The southern margin of the Irish Sea ice-sheet appears to run from the Screen Hills of County Wexford to just north of Milford Haven in southwest Wales, as is suggested by a thickening of glacial deposits on the sea floor. At the maximum extent of the ice, it was coeval with land-based ice from North Wales and from Midland Ireland.

Following the pioneer work of Eyles & Eyles (1984) when they suggested that glacial deposits on the coastal fringe of the Isle of Man were glacio-marine, and not tills of various origins, such sediments have been recognised widely (e.g. McCabe, 1986a; 1987; McCabe *et al.*, 1986). The implications for widespread glaciomarine deposition in the Irish Sea Basin are considerable. Most, if not all, sediments formerly referred to as 'Irish Sea Till' are probably subaqueous muds, deposited from sediment plumes in proximal, but mostly distal tidewater environments. Amino acid dates from marine molluscs in these deposits provide close time constraints on the time of deposition. These include the 'Irish Sea Till' of Ballycotton Bay, in County Cork (Bowen & Vernon, unpublished), which is now reinterpreted as a distal glaciomarine mud. The extent of such glaciomarine deposition is widespread and the implications for isostatic depression and recovery considerable (Eyles & McCabe, 1989a). Deglaciation was probably complete in the region by 14,460 B.P.

Ice of Younger Dryas (Loch Lomond) age grew in the mountains of the Lake District (Sissons, 1980), the Wicklow Mountains (Colhoun & Synge, 1980), and in North (Seddon, 1957; Godwin, 1955) and South Wales (Walker, 1982). Outside the glaciated areas in Younger Dryas time periglaciation was both extensive and highly effective (Bowen, 1974).

The Hebridean Glacial Sediment System

This system includes the Seas of the Hebrides as well as the Malin Sea, off Northern Ireland. On occasions it overlapped in space with the Atlantic and Irish Sea Basin Glacial Sediment Systems. The general pattern of ice movement in this system is shown on fig. 1. In the northwest, however, it is possible that a radically different pattern obtained during Middle Pleistocene glaciations. The pattern shown for this area on fig. 1 is based mainly on Late Devensian ice movements.

Signs of an early and extensive glaciation of indeterminate age are inferred from erratics found on the Flannan Isles and the island of North Rona which lie off Northest Scotland, as well as from St. Kilda, west of the Hebrides (Sutherland, 1984a). Seismic stratigraphical investigations in the Sea of the Hebrides and the Malin Sea have shown a thick sequence of deposits with glaciomarine sediments and tills (Davies *et al.*, 1984), but of indeterminate age.

In Northwest Lewis (Outer Hebrides) Sutherland & Walker (1984) described a peat, tentatively ascribed to the Ipswichian Stage, which had not been overrun by ice. The peat is also overlain by a raised beach deposit, the Galson Beach (Sutherland & Walker, 1984; Peacock, 1984). The Galson Beach also overlies a till deposit which could, therefore, be of pre-Ipswichian age. It is not known if this till relates to the glaciation of the offshore islands (above).

It has been proposed that the high-level shell beds of Scotland, including those at Cleongart, in Kintyre, were emplaced when sea-level was relatively high during the growth of Early Devensian Substage ice (Sutherland, 1981b). This was supported by Sissons (1981b, 1982b), who explained high-level rock platforms in Western Scotland on the same basis. Amino acid analysis of *Arctica islandica* from the Cleongart Clays give ratios which are intermediate in age between Oxygen Isotope Stages 9 and 7 (Bowen & Sykes, 1988). The clays could well have been emplaced by such a mechanism as suggested by Sutherland (1981b), but in this case the age of the glaciation was probably time-equivalent to Stage 8.

It has been suggested that ice of the Late Devensian Substage completely covered all of Northwest and Western Scotland (Boulton *et al.*, 1977; Andersen, 1981). The discovery of an unglaciated enclave in Northwest Lewis (Sutherland & Walker, 1984), and the demonstration that an ice-cap independent of the main Scottish ice sheet occurred over the Outer Hebrides (Flinn, 1978b), showed that the northernmost expansion of the Scottish ice sheet was limited. As such it parallels the evidence for Caithness and Orkney (see *North Sea System*). The age of the Late Devensian glaciation is established on Lewis by the Tolsta Head site, where till overlies organic beds dated to 27,000 B.P. Amino acid ratios from the shelly drifts in Northwest Lewis show that glaciation occurred more or less about 17,000 B.P. (Bowen, unpublished). The extent to which Highland ice from the mainland crossed the northern part of The Minch is debatable. If the shelly drifts in northern Lewis are glaciomarine in origin, as seems possible at least around Port of Ness, then The Minch was not crossed by Highland ice. On the mainland, the Wester Ross Moraine marks an important ice margin, the significance of which is not yet established. All of the islands of the Inner Hebrides were ice-covered during the Late Devensian as well as the Malin Sea area (Davies *et al.*, 1984). During deglaciation of the Late Devensian ice sheet a series of shorelines developed in Western Scotland (see Sutherland, 1984a).

The Loch Lomond Glaciation is time-equivalent to the Younger Dryas (Gray & Lowe, 1977; Sutherland, 1984a). Glaciers developed on the high ground, the largest being located in the Western Grampians. Radiocarbon ages on shells in deposits either overridden or incorporated within deposits of the Loch Lomond Glaciation have shown that the maximum occurred after 10,900 B.P. Amino acid ratios on shells related to the glaciation can be incorporated into the British Isles aminostratigraphic framework (table 1).

The Atlantic Glacial Sediment System

Westward flowing ice from the mountains of Western Ireland comprise this system. Important individual ice centres were: the mountains of Donegal, Mayo, Galway and Kerry, with a substantial contribution from central Ireland via County Clare. The extent of this ice on to the continental shelf is shown by glacier ploughmarks on the outer shelf (Belderson *et al.*, 1973).

This critical region, on the very margin of the Northeast Atlantic Ocean, has not received sufficient attention by research workers. Except for limited areas the influence of Late Midlandian (Devensian) Substage ice is predominant, but there is little in the way of geochronology for correlation purposes.

Three major glaciations are recognised. The earliest, Munsterian, is pre-Oxygen Isotope Substage 5e in age, and it is generally agreed that all of Ireland was ice-covered at this time (McCabe, 1987). In Southwest Ireland for example, it has been shown that granite erratics from Galway were deposited on the Dingle Peninsula (Lewis, 1974). Interglacial deposits of Gortian age overlie these older glacial deposits at Gort, Kildromin and Baggotstown. But although early work correlated the Gortian of Ireland with the Hoxnian of East Anglia (Watts, 1964; 1985; Mitchell, 1960; 1972), this has been challanged, notably by Warren (1979, 1985), who maintains they are of 'Last Interglacial' (Substage 5e) age. At present there is no means of establishing the precise age or correlation of the Gortian.

Between Fenit and Spa, Tralee Bay, Mitchell (1970), described a marine deposit overlain by silt, peat and muds, lower and upper head deposits, and, at Tawlaght, till and outwash gravel. Whether the glacial beds are *in situ* is debatable. This has a bearing on the age of the underlying marine beds because the locality lies outside the extent of Late Midlandian (Devensian) ice.

Early Midlandian (Devensian) Substage glaciation has been described in the north of Ireland (McCabe, 1987). Fermanagh Stadial lodgement tills underlie organic beds at Aghnadarragh and Derryvree which date from the later part of the Early and Middle Midlandian Substages (Colhoun *et al.*, 1972; McCabe *et al.*, 1978; McCabe *et al.*, 1987). At its maximum extent this ice covered most of Ulster and fed the Donegal Bay glacier. It has been suggested that some tills in Southern Ireland may also be the same age (Devoy, 1983; Warren, 1985).

The extent of Late Midlandian (Devensian) ice shows that large ice-free areas existed in Inishowen, South West Donegal and in West Mayo (McCabe, 1987), which given their west coast location, is not entirely consistent with the considerable expansion of ice in the south and southeast at this time. These areas merit detailed re-examination. The 'Drumlin Phase' represents an almost 'continuous drumlin presence' (McCabe, 1987) within the Drumlin Readvance Moraine (Synge, 1969). This has been dated using a combination of amino acid ratios and radiocarbon dating on an *in situ* assemblage of *Macoma calcarea* from glaciomarine deposits in north County Mayo, to 17,000 B.P. (McCabe *et al.*, 1986; Bowen & Sykes, 1988). In Southwest Limerick, the Fedamore Moraine, of the Drumlin Phase, oversteps Late Midlandian glacial deposits on to Munsterian drifts (Synge, 1969). This is interpreted as an ice-marginal oscillation related to substrate conditions and glaciological parameters rather than caused by climatic forcing (McCabe, 1987). During deglaciation some raised beaches were formed, but as McCabe (1987) has pointed out these form only part of the uplift story because they are cut into glaciomarine beds. During the Nahanagan Stadial (Younger Dryas) some ice probably occupied the high-level cirques in Donegal, Mayo and Galway.

Conclusions

In the study of glacial sediment systems in space and time there can be little doubt of the importance of multiple dating systems to overcome the inherently homotaxial nature of the rock record. Subdivision by classical palaeontological means is now clearly inadequate for the task. Hitherto, the only such method capable of widespread application, both in time and space, is amino acid dating: specifically, using time-dependent isoleucine epimerization in molluscs. To this may be added dating by other means: for example, uranium-series, TL and ESR dating. Dates by these means, however, are not yet sufficiently widespread in space or in time to provide discrete frameworks, and for the present, should be tied into that provided by the amino acid geochronology work. This has shown a complexity of development in glacial systems comparable with the oxygen isotope record of oceanic variability (Bowen & Sykes, 1988; Bowen *et al.*, unpublished; see also table 1).

By constraining ice sheets in time and space valu-

able insight can be obtained about the temporal and spatial variability of precipitation effective enough to promote ice sheet growth and influence its relative strength. In this way an important contribution can be made towards establishing the interaction of atmosphere, oceans (not least through sea-level change) and the earth's crust in space and time.

Several outstanding problems require further evaluation: (1) Lithological investigations are urgently required on some sediments, for example, the glacial deposits of the Bristol district, and those of the Birmingham area, to establish provenance and ice-flow patterns. (2) An outstanding problem on the timing and extent of earlier glaciations concerns the ice advance dated to pre-Cromerian (West Runton) time in the west of England in the Bristol district (above), which appears to be the same age as the extensive advance recorded in Denmark in pre-Esbjerg aminozone time which was ascribed to an 'Elster I' advance of Stage 14 or older (possibly Stage 16) age (Bowen & Sykes, 1988). How extensive was this glaciation in the British Isles? And what were its effects? (3) The role of subaqueous sedimentation in glaciomarine tidewater environments requires evaluation throughout the coastal margins of the British Isles. This is important not only for establishing ice-marginal positions, notably in Northwest Scotland, Eastern England, and for evaluating the significance of Scandinavian erratics in Eastern England, but also as a means of estimating isostatic depression and subsequent recovery. For example, because obvious marine shorelines are frequently cut into glaciomarine deposits it is possible that the amount of isostatic depression and subsequent recovery has been seriously underestimated. This seems to be the case in the Irish Sea Basin where it is possible the ice was thicker and isostatic depression greater than in Scotland, a possibility entirely consistent with the major expansion of the Late Devensian ice sheet to the south, and not the north, of the British Isles.

A prescription for future work could include proposals for establishing relationships between glacial sediment subsystems, more multiple-dating programmes, re-evaluation of 'glacial' deposits in a glacio-marine paradigm, implications for isostatic movements and neotectonics, and the role of sea-level in the build-up, extent and decay of ice-sheets in and around the British Isles. It is a glorious irony that well over a century after their formulation some of the ideas of early workers on Quaternary drift deposits may have considerable relevance for fundamentals of glaciation in the British Isles and the British Seas.

Pre-Anglian glacial deposits and glaciations in Britain

RICHARD W. HEY

Sediments of late Pliocene or early Pleistocene age are widespread in East Anglia and the Thames drainage basin. Elsewhere in Britain, the only deposits that can be confidently placed within this interval are the marine St. Erth Beds, in Cornwall, and fossiliferous infillings of caves and fissures in SW England, Derbyshire and County Durham. In parts of southern England are unfossiliferous sediments lying on plateaux or interfluves, and many of these may indeed be accepted as pre-Anglian. None, however, can be precisely dated and some may even be pre-Pliocene.

East Anglia

The Pliocene and early Pleistocene deposits of East Anglia consist largely of shallow-marine sediments, the Crags, with non-marine sands and gravels towards the top of the sequence. The lowest unit, the Coralline Crag, has a temperate marine fauna of late Pliocene age. The succeeding Red Crag, probably also Pliocene (Funnell, 1977), contains boreal foraminifera and rare fragments of rocks from the Yorkshire coast (Double, 1924). The latter may perhaps have been carried southwards by icebergs in the North Sea.

The sequence between the Red Crag and the lowest Anglian deposits has been divided into nine climatic stages, of which four contain cold floras or faunas. These four, from oldest to youngest, are the Thurnian, Baventian, 'Pre-Pastonian' and Beestonian (West, 1961; 1980a).

At Easton Bavents (TM 518787) on the Suffolk coast, the marine Baventian, almost certainly Pleistocene, contains pollen and foraminifera suggesting that 'glaciers were present not far from East Anglia' (Funnell & West, 1962: 129). With them are small pebbles of chert from the Yorkshire Jurassic, in large enough numbers to leave little doubt that icebergs were now present in the North Sea (Hey, 1976).

At Sidestrand (TG 255405) on the north coast of

Norfolk, marine sands assigned by West (1980a) to Substage 'a' of the Pre-Pastonian Stage contain quartz grains with surface-textures indicating an early glacial history (Krinsley & Funnell, 1965). 9 km to the WNW, at Beeston Regis (TG 169434), a marine conglomerate of the same age contains abundant quartz and quartzite pebbles from the Bunter Pebble Beds (Lower Triassic, recently re-named Kidderminster Conglomerates) of the Midlands (fig. 2), together with some Lower Cretaceous chert from the Weald of SE England. Such assemblages, of which this is the earliest known example in East Anglia, are typical of the older gravels of the middle Thames. This suggests that an ancestral Thames may have supplied the sand grains of glacial origin found at Sidestrand (Hey, 1980).

Supporting evidence is provided by the fluviatile Kesgrave Sands and Gravels (Rose & Allen, 1977), which overlie the true Crags in many parts of East Anglia, are predominantly pre-Anglian, and are thought to have been laid down by the Thames itself during several successive aggradations. Their oldest subdivision, the Westland Green Member, contains glaciated sand grains, and its pebble content suggests that it may be a fluviatile equivalent of the Pre-Pastonian marine conglomerate at Beeston (Hey, 1980).

Lower and middle Thames basin

The Kesgrave Formation is widespread in Essex, between East Anglia proper and the Thames Estuary. Pre-Anglian gravels of Kesgrave type also occur discontinuously in the basin of the middle Thames, for the most part to the north of the modern river. They were evidently laid down by a river which traversed the Upper Cretaceous Chalk outcrop by a valley ancestral to the present Goring Gap, and which then turned ENE along the broad valley known as the Vale of St. Albans. It is believed that the Kesgrave

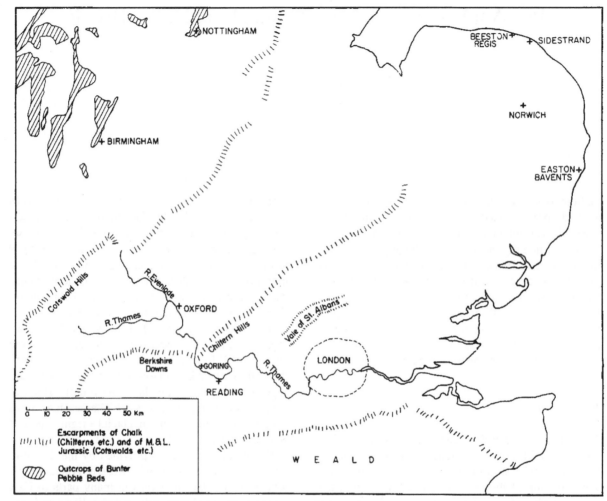

Figure 2: Map showing localities and topographical features mentioned in the text.

Formation of Essex and Suffolk comprises downstream equivalents of all but the oldest of these gravels (Hey, 1980; Green *et al.*, 1982).

The pre-Anglian sediments of the middle Thames valley, downstream from the Goring Gap, were divided by Gibbard (1985) into six units, named, from oldest to youngest, the Nettlebed, Stoke Row, Westland Green, Satwell, Beaconsfield and Gerrards Cross Gravels. The Nettlebed Gravels consist largely of locally derived flints, but the other five units all contain abundant quartz and quartzite pebbles, mostly from the Bunter. The Stoke Row Gravels thus record an important northward extension of the Thames catchment area, beyond the Chalk outcrop of the Chiltern Hills and Berkshire Downs.

The Westland Green Member of the Kesgrave Formation is the downstream continuation of the Westland Green Gravels of the middle Thames, and contains, as already mentioned, glaciated sand

grains. In the Beaconsfield Gravels ('Higher Gravel Train' of Wooldridge, 1938) of the middle Thames valley and the Vale of St. Albans, Green & McGregor (1978) noted an increase in the ratio of quartzite and sandstone pebbles to quartz pebbles. This they attributed to a further introduction of glacial material into the Thames basin. In the Gerrards Cross Gravels ('Lower Gravel Train') of the same area they noted an increase in proportions of all far-travelled material, including volcanic rocks, and suggested that this might indicate a third introduction of sediment of glacial origin.

On the Chilterns, and on the low hills to the south of the Vale of St. Albans, are patches of clay containing Bunter pebbles. This, the Chiltern Drift of Wooldridge (1938), has in the past been interpreted as a till of pre-Anglian age, but recent authors (e.g. Green & McGregor, 1983) are inclined to ascribe it to nonglacial processes such as solifluction, the Bunter pebbles being derived from older Thames gravels.

Upper Thames basin and Cotswold Hills

At Sugworth, 5 km south of Oxford, are fluviatile interglacial deposits considered by Shotton *et al.* (1980) to be Cromerian, although this has been recently questioned. Pleistocene deposits older than these occur discontinuously in and near the valley of the Evenlode, a northern tributary of the upper Thames, and on either side of the Thames between the Evenlode confluence and the Goring Gap. They consist largely of diamictons, with abundant Bunter pebbles in a matrix of sand and clay. Clasts from the local Jurassic limestones are notably scarce, bedding is seldom visible, and the deposits carry few obvious remnants of terraces. In addition to these coherent deposits, scattered Bunter pebbles and occasional flints have been found at many points on the Cotswolds, at altitudes up to 320 m O.D. Both the scattered pebbles and the coherent deposits were termed Northern Drift by Hull (1855).

Until recently it was believed (e.g. Dines, 1946; Arkell, 1947) that almost the whole of the Northern Drift was of glacial origin, a belief supported by the fact that the diamictons strongly resemble till and locally even contain striated clasts (Sandford, 1926). Shotton *et al.* (1980) considered that the high-level scattered pebbles might indeed be the last surviving remnants of a pre-Anglian till, perhaps Baventian, laid down by an ice-sheet advancing from the north. They concluded, however, that the coherent Northern Drift, all of it below 200 m O.D., consisted of decalcified river gravels and slope deposits, derived from the high-level Cotswold till but themselves deposited by non-glacial processes. Hey (1986) further suggested that the coherent Northern Drift might comprise upstream equivalents of the four youngest pre-Anglian fluviatile units of the middle Thames.

The hypothesis of Shotton *et al.* (1980) accounts both for the striated clasts in the Northern Drift, which could have been carried on ice-floes, and for the glaciated sand grains in the oldest beds of the Kesgrave Formation. By itself, however, it does not explain the variations of composition observed in the Northern Drift (Hey, 1986) and its supposed downstream equivalents (Green & McGregor, 1978; Gibbard, 1985). Nor, as was recognised by Shotton *et al.* (1980: 85), does it explain the fact that the Northern Drift contains very few clasts derived from any of the highly durable Palaeozoic formations of the Midlands.

A possible alternative hypothesis is that the coherent Northern Drift was laid down by a river issuing from an ice-front lying some distance to the north of the present Cotswold scarp. The upper course of such a river might well have been largely confined to Triassic outcrops, in which case its bedload might have come to be dominated by Bunter pebbles. The detailed composition of the bed-load might nevertheless have changed from time to time, in response to changes both in the glacier itself and in the non-glaciated areas drained by its outwash stream.

Other areas

Outside East Anglia and the Thames basin, there appears to be no deposit which can be securely dated as late Pliocene / early Pleistocene and which at the same time provides conclusive evidence for glaciation, whether contemporary or earlier.

Of the unfossiliferous high-level deposits that may be of this age, the most widespread is the Clay-with-flints which covers extensive areas on the Chalk dip-slopes of southern England. It is typically a diamicton containing both clay and abundant flints, the latter mostly fresh and unworn. Being unbedded it resembles a decalcified till and has been so regarded by many authors, from Sherlock & Noble (1912) onwards. This interpretation was adopted by Kellaway *et al.* (1975), who further suggested that the glaciation responsible might be pre-Cromerian, possibly Beestonian. Other recent authors, however, regard the Clay-with-flints as a mixture of weathering products from the Chalk and from the overlying Tertiary, virtually undisturbed except by periglacial processes (Jones, 1981: 92-98; Catt, 1986a).

Patches of unfossiliferous high-level gravel occur in many parts of southern England, especially towards the SW. None appears to carry any positive indications of a glacial origin, and some have been interpreted as normal river-gravels laid down at times when the topography was different from that of today (e.g. those near Bristol and Bath, discussed by Palmer, 1931).

Summary of conclusions

No pre-Anglian till has yet been definitely identified in Britain, though remnants of such a till may survive on the Cotswolds in the form of scattered pebbles.

There is, however, indirect evidence that a North Sea glacier approached the East Anglian coast in

Baventian times. There is also evidence that glacial sediment was being carried down the ancestral Thames during at least three pre-Anglian cold intervals, the earliest being the Pre-Pastonian (post-Bramertonian) Stage of West (1980a). The sediment could have been derived either from a still earlier till in the Cotswolds, or from a contemporary active glacier in the Midlands.

Anglian glacial deposits in Britain and the adjoining offshore regions

JÜRGEN EHLERS & PHILIP GIBBARD

The glaciation in the Anglian Stage was one of the most extensive of the Pleistocene. However, since for the greater part of its extent its deposits were later overriden by younger glaciations, much of the sedimentary record has been lost. There is therefore only fragmentary evidence for reconstructing the glacial and climatic history of the Anglian Stage.

For this reason the Anglian glacial sediments are best preserved beyond the margins of later ice advances, and Eastern England and the adjacent offshore region include extensive sequences of deposits and features formed during this event. This paper will therefore concentrate on discussion of these areas, but will also consider the implications of recent suggestions that glacial deposits, previously assigned to the younger Wolstonian (?Saalian) Stage, in the English Midlands may also be equivalent to the Anglian.

The dating of the Anglian glacial deposits in the type area is based on their relation to temperate stage sediments. In north Norfolk glacial deposits rest on Cromerian and throughout the region are overlain directly by Hoxnian Stage sediments. No evidence of interstadial or higher rank climatic oscillations has been identified intervening within the glacial sequence and therefore it is interpreted as representing a single complex glacial event. On the basis of strong lithological and detailed stratigraphical relationships the Anglian is equated with the continental Elsterian Stage, which is known to have included the most extensive Pleistocene ice sheets in eastern Europe.

East Anglia

The deterioration of climate that marked the beginning of the early Anglian Stage is recorded in fluviatile and pond sediments in the coastal exposures of Norfolk and Suffolk. The pond sediments have yielded fossil floras of typical cold climate affinities (West, 1980a) derived from herb-dominated communities.

Associated with these sediments are periglacial phenomena such as ice wedge casts (e.g. at West Runton) and polygonal patterned ground (e.g. at Corton) at various horizons. Throughout much of East Anglia periglacial structures and contemporary aeolian sediments are found during this time interval. The recognition that this implies a prolonged period of severe periglacial conditions led Rose et al. (1985) to propose that their Barham Arctic Structure Soil could be recognised across much of the region.

An important additional effect of the periglacial activity may have been the 'preparation' of the land surface, before arrival of the continental ice sheet in the region. These processes may be presumed to have generated a considerable quantity of regolith that readily became incorporated into the ice sheets' debris load.

The duration of the non-glacial cold interval of the early Anglian is not possible to determine, although by analogy with later cold stages it could have been lengthy. One indication that it may have been relatively rather short may be the apparent lack of extensive fluvio-periglacial deposition before arrival of the ice sheet in Britain and on the Continent. However, if the period was very arid, as thought by Rose et al. (1985) and since the deposits have been removed by the subsequent glaciation, the evidence available may be very misleading.

Subsequently the Anglian Glaciation began as the ice advanced from the NE and overrode the pre-existing sedimentary sequences throughout the region. This advance deposited a complex suite of interbedded tills and associated meltwater sediments, that are termed the North Sea Drift (Cromer Till) Formation. These sediments are best exposed on the northern and northeastern coasts of East Anglia, from Weybourne to Lowestoft (fig. 3). They can also be traced inland as far as Norwich and south towards Diss (Mathers et al., 1987).

In a series of papers Banham (1968, 1971, 1975, 1977) described and interpreted the coastal expo-

Figure 3: First Cromer Till of the 'North Sea Drift' Formation at Happisburgh, Norfolk, overlain by glaciolacustrine sands and silts (Photograph: Ehlers, 1986).

sures. He recognised three stratigraphically distinct till units deposited by oscillation of an ice lobe that entered the area from the southern North Sea basin. In places these sediments are considerably contorted and locally include huge rafts or 'schollen' of bedrock, some of these structures undoubtedly resulting from glaciotectonics (e.g. Banham, 1971). The exact nature of this advance is somewhat disputed, at the time of writing. Some interpret the sequence as representing deposition predominantly on land (Hart, 1988; Hart & Boulton, this volume), whilst others suggest deposition in water (Gibbard, 1980; 1988; Gibbard & Zalasiewicz, 1988; Eyles & McCabe, 1989) or most likely an interplay of the two processes (Lunkka, 1988). The precise genesis of the glacial sequences may be disputed, but there can be little doubt that the advance of an ice sheet, effectively out of a deep basin onto a relatively high ground area, would have given rise to a complex sedimentary

sequence such as that seen in this formation. (For further discussion of the palaeogeography see Ehlers *et al.*, this volume).

The ice that deposited the North Sea Drift sediments characteristically contains a suite of igneous and metamorphic erratics of southern Norwegian origin (fig. 4). Some of these erratics, specifically rhomb porphyry and larvikite, can be traced to exposures in the Oslo area and have therefore been generally accepted as evidence that the ice sheet originated in the area and crossed the North Sea basin. These erratics are found at all exposures of the North Sea Drift Cromer Tills. However, an indication that North Sea ice may have reached further south than previously thought is the present distribution of these Scandinavian erratics. Rhomb porphyries and larvikites can be found as far south as Bedford, Hitchin and Ipswich, a large number having been recognised from the area around Cambridge by Rastell &

Figure 4: Rhomb porphyry from the Oslo area, southern Norway, found at Cromer, North Norfolk (Photograph: Ehlers, 1986).

Romanes (1909). If the North Sea ice advance had been limited only to the area in which North Sea Drift (Cromer Till) is presently found, it would have been difficult to transport the boulders to their present positions, because all subsequent ice movement directions in Norfolk ranged only towards the northeast to southeast (Ehlers *et al.*, 1987).

Since the North Sea ice crossed mainly Tertiary and Pleistocene fluvial and deltaic deposits in the North Sea basin, it carried relatively sand-rich debris. Sandy tills below 'normal' Anglian tills (Lowestoft Till) have been reported from several areas (cf. Harmer, 1928). One occurrence of such till was found at Ingham, north of Bury St. Edmunds (Ehlers *et al.*, 1987). Evans (1976: 17) reports several finds of a sandy brown and red till (in boreholes at Ashwicken TF 189698 and Holt House Farm TF 186678 and exposed northwest of Ashwicken TF 195692) in northwestern Norfolk. These finds are assumed by Evans (*op. cit.*) to be from till older than the Lowestoft Till at Bawsey (see Straw, this volume). According to Turner (personal communication), three tills have been found in a borehole at Hockwold cum Wilton, the lowest being sand-rich. Some, although perhaps not all of these sand-rich tills may represent local till of no major stratigraphical significance, however.

During the next phase, ice advanced through the Vale of York and Lincolnshire into central and western East Anglia. The ice crossed the Fenland Basin, possibly exploiting a pre-existing gap in the Lincolnshire-Norfolk Chalk ridge. During this advance the ice eroded older, sandy North Sea Drift and apparently quarried a major portion of the relatively

soft Mesozoic clays, the latter providing much of the material for the till matrix. The Fenland Basin may in part owe its origin to this major erosional event.

On the basis of lithology (Perrin *et al.*, 1979), erratic content (Baden-Powell, 1948) and detailed fabric measurements (West & Donner, 1956; Ehlers *et al.*, 1987) it can be shown that the ice radiated outwards from the Fenland Basin in a fan-like pattern. These paths would have caused the ice to overtop the relatively high ground surrounding Fenland and therefore indicates the considerable energy available during this advance. Indeed it is probable that interaction of this 'British' ice with the North Sea ice was responsible for this unusual pattern of ice movement. Moreover, in the eastern part of the area the 'British' ice seems to have expanded close to its maximal position and replaced the North Sea ice. The latter apparently became less active during this phase. As the decay of the North Sea ice continued, the 'Great Eastern Glacier' of Harmer (1904) expanded progressively towards the east and northeast, reaching Lowestoft. The contemporary retreat of the North Sea ice, however, outpaced the advance of the ice from the west and southwest, so that an ice-free corridor existed between the two ice lobes for a period. It is in this ice-free zone that the Corton Sands were deposited.

The till deposited by the 'British' ice over a large part of East Anglia is the so-called Lowestoft Till. It has a grey to blue-grey clay matrix (up to 45%) and is rich in flint and subrounded chalk clasts; hence the old name 'chalky boulder clay' (plate 3). It frequently also includes Mesozoic fossil fragments, Jurassic limestone, Triassic clasts and less frequently Palaeozoic material, particularly chert. Igneous and metamorphic clasts, generally of unknown provenance, are rare, but do occur. Grain-size distribution is remarkably uniform across the region, minor changes being attributed to incorporation of local materials, such as North Sea Drift Formation sediments etc. (Perrin *et al.*, 1979).

The deposition of the Corton Sands has been the subject of much investigation in recent years. On the basis of the stratigraphy of the type section they were thought to represent interstadial conditions (cf. Mitchell *et al.*, 1973). However, an investigation of plant remains within silt bands in the sands by West & Wilson (1968) showed that they were of 'full-glacial' affinity. More recently Banham (1971), Pointon (1978) and Bridge & Hopson (1985) have refined the stratigraphy. Lithological studies and palaeocurrent analyses demonstrate that the sands were derived from the North Sea Drift ice to the N and NW (Bridge

& Hopson, 1985; Bridge, 1988). They were deposited by streams flowing in a west to east direction approximately parallel to the present River Waveney with an outlet close to Great Yarmouth. Here the streams may have entered a substantial lake in the southern North Sea basin (Gibbard, 1988). Subsequent advance of the Lowestoft Till ice from the west caused the sequence to be overridden.

Stratigraphically, therefore the Corton Sands are genetically related to the North Sea ice and therefore should properly be included as a member of the North Sea Drift (Cromer Till) Formation.

With the continued withdrawal of the North Sea ice, the Lincolnshire coast and adjacent offshore area became open for the expanding 'British' ice advancing southeastwards into Norfolk. Once more the Wash and the Breckland Gap played an important role in directing ice streams. Ice that flowed over the relatively high ground of the Chalk escarpment and down into the Fenland Basin led to deposition of the chalk-rich 'Marly Drift' till facies onto more typical Lowestoft Till in the vicinity of Bawsey (see plate 19 in Straw, this volume). This advance laid down what has been termed 'Gipping Till' by Baden-Powell (1948) and West & Donner (1956). Straw (this volume) interprets the 'Marly Drift' as a Wolstonian till. As a result of investigations by the present authors, however, it is not regarded as representing a separate glaciation but simply a facies deposited during a later phase of the Anglian.

The abundance of local material within this 'Marly Drift' till facies presumably arises because of strong incorporation of chalk following the stripping of covering materials by earlier advances, as well as the ice simply advancing over chalk for long distances.

During the 'Marly Drift' phase chalk-rich till was transported into much of East Anglia, including the Gipping Valley in Suffolk, where it overlies the Jurassic clay-rich tills (Lowestoft facies) of the preceding Lowestoft advance (fig. 5). By this time, however, the ice no longer reached its maximum limits. In East Anglia, the area beyond the Cromer Ridge was no longer covered by ice. As the ice retreated vast meltwater formations were laid down as sands and gravels, for example those between East Dereham, Swaffham and Fakenham. Here the gravels rest on the chalk-rich till of the last Anglian ice advance. Similarly in the area around the Glaven Valley in north Norfolk kames, dead ice topography and an esker are associated with the retreat of the chalk-rich till ice (Sparks & West, 1964).

Throughout East Anglia a series of deep, steep-sided valleys have been found cutting through the Chalk and associated bedrock. These valleys form a radiating pattern, broadly parallel to the dip of the Chalk i.e. they trend east-west in the north and north-south in the south of the region (fig. 6). Detailed study of these 'tunnel valleys', by Woodland (1970) and Cox (1985b) among others, indicates that they are normally filled with glacial sediments, predominantly meltwater sands, gravels or fines. Tills also occur but are less frequent. Closely comparable in size to

Figure 5: Various facies of Anglian tills, including pale, chalk-rich 'Marly Drift', overlying coarse outwash and Chalk bedrock at Newton, Norfolk (TF 838160; Photograph: Ehlers, 1986).

Figure 6: Buried tunnel valleys of East Anglia (after Woodland, 1971).

the *rinnen* of Denmark, Northern Germany, Poland etc., they are undoubtedly of glacial origin and probably result from subglacial drainage under high hydrostatic head (Ehlers *et al.*, 1984). This explains the commonly recorded reverse gradient and steep-sided form of these features. In places such valleys can be seen to continue downstream as normal channel-fill formations close to the surface, for example the Nar Valley and its apparent continuation towards Swaffham and East Dereham. Smaller scale 'tunnel valleys' have also been identified in Suffolk close to the maximum ice front position by Zalasiewicz *et al.* (this volume) and Cornwell & Carruthers (1986).

During deglaciation meltwater seems to have become concentrated into large channel-like valleys that formed in a generally radiating pattern again parallel to the dip of the Chalk and Lower Tertiary bedrock. In many places this alignment is determined by the course of the subglacial 'tunnel valleys'. Valleys such as the Stort, the Gipping, the Waveney and the Wensum are all of this type. However, in northern Norfolk substantial outwash sandur plains developed particularly at Kelling and Salthouse Heaths. The radial pattern of meltwater discharge continued until the ice withdrew beyond the Chalk escarpment, northwest of which drainage became directed by the regional slope into the Fenland Basin. The modern drainage system of the region was initiated during

this deglacial phase (fig. 7; cf. Rose *et al.*, 1985).

In the intervening areas between the meltwater channels the ice seems to have stagnated and produced a dead-ice topography. Hollows in this surface formed by melting of buried ice bodies formed kettle-like hollows many of which became filled with water. In some localities evidence for proglacial flow till, meltout, collapse faulting and related phenomena have been observed. Many of the hollows remained unfilled with sediment by the end of the Anglian and persisted as lakes into the subsequent Hoxnian temperate Stage.

In the marginal areas, particularly at Corton, the ice front retreated in standing water (Bridge & Hopson, 1985; Hopson & Bridge, 1987; Bridge, 1988). A similar phenomenon is found in the Nar Valley (Ventris, 1986) where the 'tunnel valley' deepens towards the west i.e. in the direction of the ice retreat. A proglacial ice-dammed lake seems to have formed in the valley and caused the thinning ice initially to float, depositing subaquatic till, and later to retreat completely leaving laminated clays to accumulate. Subsequent deglaciation gave rise to drainage of the lake and establishment of fluvial sedimentation.

Southeast England and the Thames Valley

Advance of the Lowestoft Till ice into Essex and Hertfordshire brought it into the region influenced by the River Thames and its tributaries. A substantial series of Thames deposits termed the Kesgrave Formation are aligned WSW-ENE to W-E across central Essex and Suffolk and represent the preglacial course of the river (see Hey, this volume). These deposits are rich in quartz and quartzite clasts. The ice overrode these terrace-like accumulations in all but the southernmost part of East Anglia. A characteristic palaeosol, the Valley Farm Soil, developed on these terrace surfaces was also buried by the till and forms an important marker horizon (Kemp, 1985; Rose & Allen, 1977; Rose *et al.*, 1985; 1976).

Oscillations of the ice sheet margin have been repeatedly identified in the southern marginal zone, particularly in the Ipswich, Chelmsford and Hertfordshire areas (Allen *et al.*, this volume). Similar frontal movements almost certainly also occurred in the north and east London areas (Gibbard, 1979 and unpublished). These oscillations had profound effects on the River Thames since the Lowestoft ice advanced into and overrode the river's contemporary valley in Hertfordshire; the till units interdigitating with the fluvial sediments (plates 1 and 2, figs. 8 and

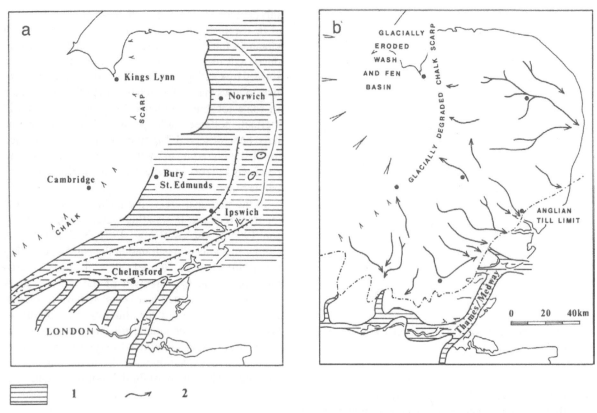

Figure 7: The landscape of eastern England (a) at the time of the Barham Soil and (b) after the Anglian glaciation; 1 = Thames River terraces and associated marine deposits, 2 = Glacially initiated river courses (after Rose *et al.*, 1985).

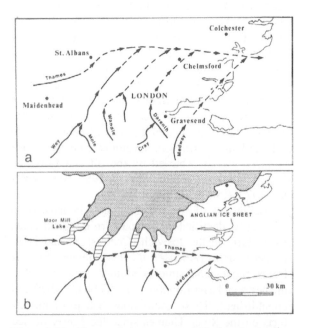

Figure 8: The Thames drainage system (a) before the Anglian glaciation and (b) during the maximum Anglian glaciation (based on Gibbard, 1983).

9). The ice not only advanced up the valley of the Thames itself, but also those of southern tributaries. This dammed the rivers and the resulting proglacial lakes progressively overspilled until the water reached an unglaciated valley, the Medway, via which the river could return to its earlier course in easternmost Essex (Gibbard, 1977; 1979; 1985; 1988; Bridgland, 1980; 1983; Bridgland *et al.*, in press). The present course of the Thames through London was therefore established during the latter part of this phase of the Anglian glaciation.

During the course of this glaciation the connection with the source of quartz-rich material seems to have been severed in the western Midlands, the late Anglian Black Park Gravel and all subsequent units are poor in quartz in the Middle and Lower Thames (Gibbard, 1985). This accords well with the Upper Thames, where the post-Northern Drift Formation gravels are similarly poor in quartz (Briggs & Gilbertson, 1973; 1980; Gibbard, 1988).

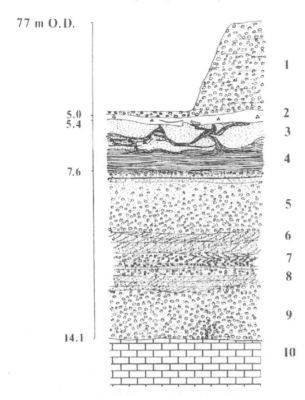

Figure 9: Moor Mill Quarry (TL 143025); Lowestoft Till overlying large blocks of sandy silt resting within strongly contorted Moor Mill laminated clays overturned at the top by the Anglian ice advance; 1 = Medium to coarse gravel with some slightly distorted current bedding, 2 = Brown partially waterlain till, 3 = Buff sandy silt blocks resting in contorted laminated clays, 4 = Laminated silty clays, 5 = Coarse to medium gravel, 6 = Trough crossbedded sand and fine gravel, 7 = Coarse to fine gravel, 8 = Tabular crossbedded sand, 9 = Coarse to medium gravel with frost wedge, 10 = Chalk (after Gibbard, 1974).

The English Midlands

The glacial sequence of the English Midlands in the area around Coventry and Birmingham has been traditionally considered to be of Wolstonian age. This followed from the classic work of Shotton (1953, 1971, 1983a, b). Broadly speaking the sequence comprises a series of glacial sediments identified over a large area in the west and central Midlands, as follows (see also Rice & Douglas, this volume, for a fuller discussion of the Midlands deposits):

1. Dunsmore Gravel
2. Upper Wolston Clay ⎫
3. Wolston Sand and Gravel ⎬ Oadby Till
4. Lower Wolston Clay ⎭ Thrussington Till
5. Baginton Sands

6. Baginton-Lillington Gravels
7. Bubbenhall Clay

Of these, numbers 1-6 provide the major evidence for extensive glaciation of the region as far south as Morton-in-Marsh in Gloucestershire. Unit 7 was thought by Shotton (1953) to represent localised remains of an earlier ice-dammed lake of possible Anglian age, analogous to the later Lake Harrison (represented by the Wolston Clays). However, there has been much discussion over whether the Bubbenhall Clay (7) is a true glaciolacustrine deposit or just remobilised Triassic bedrock, mainly by Sumbler (1983a, b). The same author has questioned the basis for the dating of the Wolston Formation sequence, pointing to the apparent continuity with the Anglian deposits further to the east and therefore concluding that the Midlands' sequence should also be assigned to the Anglian Stage.

Debate over the age as well as the association of the Wolston Formation members has increased recently as a result of the suggestion by Rose (1987) that the Baginton-Lillington Gravel might represent a longer time period than previously thought. This author has stated that it is possible to trace these gravels through the East Midlands, across the Fenland Basin (where they are now absent) and into East Anglia. Gravels derived from the East Midlands are indeed present in East Anglia particularly in the area of Ingham near Bury St. Edmunds, where they have been recognised by Hey (1980) and Clarke & Auton (1982) as evidence of a tributary to the preglacial Thames. Opinion is divided over whether these Ingham Gravels represent a possible preglacial River Trent or Rose's West Midland river. If the latter is correct the implication is that the Wolston Formation, which most agree represents a broadly continuous sequence, must be of Anglian age. There are, however, several complexities arising from these arguments which are beyond the scope of this paper. Suffice to say that at present the evidence for equating the East Anglian and Midlands sequences, as Rose (1987) suggests, remains equivocal. In spite of continual discussion and recent discoveries at four localities of fossiliferous sediments associated with the Baginton-Lillington Gravel Member the question of the true age of the glacial and associated sediments remains unresolved.

Nevertheless, undoubted Anglian glacial sediments do indeed occur in the west Birmingham area. Here two lake basin infillings, the form and situation of which closely resemble kettle holes, have yielded characteristic Hoxnian pollen sequences. Both the Nechells (Kelly, 1965) and Quinton (Horton, 1974) deposits occur interstratified in glacial and associated sediments and therefore potentially hold the key to answering some of the Midlands questions. However, their occurrence overlying glacial sediments, including the Nurseries Till derived from the NW, filling channel-like depressions in the underlying bedrock (Horton, 1974), emphasises that Anglian events in the Midlands may have been somewhat similar to those in eastern England.

North Sea area

Most of the marginal areas of the Anglian glaciation of the British ice sheet lie not on land but at the bottom of the sea. Therefore it might be expected that a sedimentary basin like the North Sea would provide an excellent sediment trap in which the sediments of the Anglian glaciation would be nicely preserved. However, this is not the case. The occurrence of Anglian tills is limited to relatively small areas in the southern part of the North Sea (south of Dogger Bank). Further to the north, the whole sedimentary sequence representing the Pleistocene glaciations consists of marine and glaciomarine deposits.

On the floor of the southern North Sea, as well as further north, many buried channels occur, the oldest of which have been shown to be of Anglian age (Balson & Cameron, 1985; Cameron *et al.*, 1987). They contain tills and meltwater sediments assigned to the Swarte Bank Formation. These channels are remarkably similar in form and scale to the 'tunnel valleys' of East Anglia or the *rinnen* of the adjacent Continent, already discussed.

If the channels at the bottom of the North Sea are of the same origin as those on the land, it must be concluded that the sedimentary record gives a very incomplete picture of the glacial history of the region. Two widely divergent interpretations of this history exist. Whilst some workers interpret the incisions as 'tunnel valleys' (e.g. Dingle, 1971; Wingfield, unpublished), the majority of workers at present do not think that the North Sea was covered by an ice sheet during either the Devensian or the Wolstonian. Indeed even in the Anglian an ice sheet is only thought to have covered the southern North Sea (Cameron *et al.*, 1987).

This is, however, in conflict with geomorphological considerations:

i. During the Pleistocene glaciations sea level was lowered eustatically by about 150 m during the Devensian, and by about 200 m during the Wolsto-

nian and Anglian glaciations. Average water depth in the central North Sea is at present about 90 m and in the northern North Sea 140 m. If glaciomarine sediments had been deposited during all three glaciations, as can be demonstrated from boreholes and seismic records, this implies that the floor of the North Sea must have been depressed isostatically. This, however, means that the region must have been covered initially at least by a grounding ice sheet.

ii. Along the English and Scottish east coasts tills are found that contain numerous Scandinavian erratics (Catt, this volume; Hall & Connell, this volume; Ehlers, 1988; see also above). These can only have been emplaced by glacial transport. Notwithstanding the possibilities for reworking, drift ice transport would require even greater isostatic depression than that mentioned above (i), which would again require a pre-existing grounding ice sheet.

iii. The buried channels beneath the floor of the North Sea do not resemble river systems or landforms created by tidal scour, similar to those of tidal inlets. The size, shape and irregular thalwegs of the channels resemble those of the well-known 'tunnel valleys' and it therefore seems most probable that the sea floor features are of the same origin.

Closing observations

It is now generally accepted that the Middle and Late Pleistocene cold stages appear to begin with a prolonged period of slow climatic deterioration, punctuated by interstadials, which become progressively rarer until the latter part of the stage when a major ice sheet may develop and extend into temperate regions. This pattern of development is clearly reflected in the deep sea record, which as a rule shows a gradual increase in $^{18}O/^{16}O$ ratio until the glacial maximum, which is followed by a sudden decrease towards the onset of the subsequent temperate (interglacial) event. However, one must be aware that such 'sudden' events in reality spanned several thousand years and may have included stadial and interstadial type oscillations.

Unlike the Devensian (Weichselian) and continental Saalian Stages, no such interstadial-stadial patterns have been identified in the Anglian so far. However, the possibly prolonged non-glacial phase before arrival of the ice sheet, together with the apparent abrupt transition from the late Anglian glacial into the following Hoxnian interglacial climate bears comparison with patterns of the later stages.

Wolstonian glacial deposits and glaciation in Britain

R. JOHN RICE & TERRY DOUGLAS

In southern Britain glacigenic sediments from as far west as the Scilly Isles (Mitchell & Orme, 1967; Coque-delhuille & Veyret, 1984) and as far east as the Norfolk coast (Catt, 1981; Straw, 1983) have been ascribed to the Wolstonian glaciation. However, in these coastal extremities any direct correlation remains very tentative, and this paper will therefore concentrate on the type area for the Wolstonian Stage in Britain, namely the central and eastern Midlands.

The stage name derives from that of a small village on the river Avon in north Warwickshire (fig. 10) where Shotton (1953) first described the deposits that were later to be designated as the stratotype (Mitchell et al., 1973). Particular attention will be given to the often contentious issues of the nature of proglacial Lake Harrison and the relationships of the Midland Wolstonian deposits to those of neighbouring areas.

It should be noted that alternative views on the stratigraphic status of these deposits have been expressed by Sumbler (1983a, b), Bowen et al. (1986) and Rose (1987).

Stratigraphic sequence

For present purposes the nomenclature advocated by Sumbler (1983a) will be employed:

Dunsmore Gravel
Upper Wolston Clay
Wolston Sand and Gravel ⎱ Oadby Till
Lower Wolston Clay ⎰
Thrussington Till
Baginton Sand and Gravel

The Wolston type area

Baginton Sand and Gravel. This earliest member of the succession, locally attaining a thickness of some ten metres, consists of a basal gravel overlain by a fine-to-medium cross-bedded red sand. The deposit occupies the floor of a broad northeast-southwest trending valley excavated in the Mercia Mudstone. The gravel is confined to the lowest part of the valley, with the sand along each margin overstepping on to the bedrock. In the vicinity of Wolston over 80% of the gravel is composed of Bunter quartz and quartzites with most of the rest potentially from the same conglomeratic sources. As Shotton (1953) originally recognised, the Baginton Sand and Gravel was laid down by water flowing in a northeasterly direction, totally at variance with the modern southwesterly draining Avon. The direction of flow is attested both by the regional slope of the deposit and also by internal cross-bedding structures. Since he was able to demonstrate that the bedrock relief continues to decline northeastwards into the catchment of the modern Soar in Leicestershire, Shotton adopted the name Proto-Soar for the river believed to be responsible for the Baginton Sand and Gravel.

Thrussington Till. Overlying the sand above a sharp and generally flat-lying boundary is a matrix-dominated till that is normally 3 to 5 m thick. The red matrix comes from the Mercia Mudstone and among the more common erratics are green sandstone and siltstone fragments from the same bedrock source, together with coal and Bunter pebbles. Less common clasts include Carboniferous limestone, Bromsgrove sandstone, and south Leicestershire igneous rocks, the whole assemblage implying transport from the north or northwest. The relatively structureless nature of the deposit suggests that it is a lodgement till, and the undeformed basal contact may be explained by permafrost strengthening the underlying Baginton Sand (see below).

Lower Wolston Clay. This division of the glacial sequence consists predominantly of brown and reddish brown silts and silty clays that commonly attain an aggregate thickness of some 8–10 m. The deposit is virtually always bedded and in places is conspicuously laminated. In small exposures the material may appear stoneless, but in laterally extensive sec-

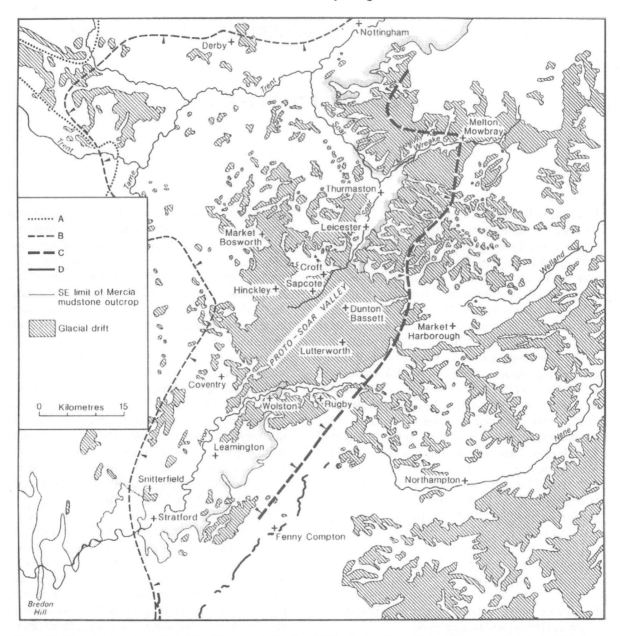

Figure 10: Map showing the extent of the drift cover in the East Midlands, together with localities and selected glacial features discussed in the text. Most of the drift is till, but as used here the term also encompasses lacustrine, glaciolacustrine and glaciofluvial sediments. The last are especially prominent in the area between Hinckley, Rugby and Coventry. A = Devensian ice limit; B = minimum position reached by 'Oadby-till ice'; C = minimum position reached by 'Thrussington-till ice' (compare with limit of Mercia mudstone outcrop); D = Lake Harrison shoreline bench as mapped by Dury (1951).

tions horizons with stones are almost invariably present and in some of these the abundance of clasts makes them difficult to differentiate from till. The provenance of the earliest clasts is akin to that of the debris in the Thrussington Till, but there is a noteworthy upward increase in material from northeasterly sources and a tendency for the clays themselves to become more grey in colour. Shotton regarded the whole deposit as accumulating in what he termed Lake Harrison, a large body of water originally visualised as impounded between advancing ice on the northern and western sides and the higher ground of the Jurassic escarpments on the southeastern side, although with recognition of the glacigenic origin of the underlying material it became necessary to envisage a period of melting to initiate the lacustrine phase.

Wolston Sand and Gravel. At least one widespread stratum of sand, with local development of gravel, interrupts the lacustrine sedimentation represented by the Wolston Clay. Even after due allowance for the effects of compaction, some of this coarser material lies at altitudes well below those attained by the highest of the stillwater deposits, and since internal sedimentary structures attest to deposition by flowing water there must presumably have been one or more periods when the lake level fell sufficiently to permit accumulation of proglacial *sandar*; further witness to subaerial conditions is provided by occasional ice-wedge casts observed in the sand and gravel. The material generally becomes finer southwards and eventually may even grade into slightly coarser horizons within the main body of silts and clays. Detailed mapping indicates that north of the Avon valley two separate horizons of sand and gravel may occur (Sumbler, 1983a), with the upper one in particular containing so much flint and Jurassic limestone as to demonstrate that, by this stage, glacial transport from the northeast was firmly established.

Upper Wolston Clay. In the vicinity of Wolston the sand and gravel just described is overlain by a further 5 to 10 m of bedded silts and clays whose presence indicates the continuing existence of Lake Harrison. The material now accumulating was predominantly of northeastern origin and includes occasional lenses of chalky till. The maximum altitude reached by this Upper Wolston Clay is close to 125 m O.D., a level that may have been determined by overflow cols across the Jurassic scarplands to the southeast.

Oadby Till. North of the Avon valley the youngest lacustrine sediments are overlain by a bluish grey lodgement till, up to 20 m thick, that contains an abundance of chalk, flint and Jurassic limestones; such glacigenic material of northeasterly provenance was named Oadby Till by Rice (1968). More recently Sumbler (1983a) has argued that, a short distance further east, identical material occurs as the lateral equivalent of beds as early as the Lower Wolston Clay, implying a sharply diachronous boundary to the base of the till. South and southwest of Wolston the Oadby Till is only intermittently preserved, but the patches that do remain indicate that the northeastern ice eventually overrode the whole district under review and was thus responsible for extinguishing Lake Harrison as it had previously existed.

Dunsmore Gravel. This youngest member of the glacial succession comprises a relatively thin but extensive sheet of sandy and clayey gravel, with both flint and Bunter pebbles as common constituents. Below a decalcified upper horizon the deposit in

places also contains chalk and Jurassic limestone. The deposit slopes towards the southwest, and this disposition, allied to the nature of the material, suggests that the Dunsmore Gravel represents the outwash sandur from the decaying 'Oadby Till' ice. There may well have been significant local erosion of both the Oadby Till and the Upper Wolston Clay prior to accumulation of the Dunsmore Gravel, and when traced further down the Avon valley the gravel was once held to grade into the highest of the local river terraces, namely the Avon No.5 of Tomlinson (1925).

Regional significance of glacial deposits in the east and central Midlands region

In this part of the review four themes that have already been introduced in discussing the Wolston area will be elaborated in a broader regional context. A less formal stratigraphic framework will be adopted, and it is worth emphasising that one reason for this is that any rigorous litho- or chronostratigraphic regional analysis would demand a much more elaborate classificatory scheme than has either been proposed or indeed is currently feasible.

(i) *The Proto-Soar drainage system.* At the onset of the Wolstonian Glaciation the watershed at the head of the Proto-Soar valley probably lay west of Stratford. Gravels composed largely of Bunter quartz and quartzite pebbles, together with an admixture of local sandstones, are found widely along the eastern side of this watershed at elevations between 90 and 105 m O.D. (Cannell & Crofts, 1984). These appear to be among the most westerly representatives of the Baginton Sand and Gravel. Around Snitterfield the gravels also contain as much as 25% limestone, suggesting that there were significant feeders flowing across the modern Avon valley from the Jurassic outcrops to the south. Close to Leamington another right-bank tributary supplying substantial quantities of Jurassic limestone has been identified (Shotton, 1953), but across most of north Warwickshire analyses of the Baginton Gravel reveal over 80% of the clasts to be of 'Bunter' derivation (Crofts, 1982). Where the sand and gravel emerges in Leicestershire from beneath the thick drifts of the Severn-Trent watershed it still has a very similar composition although its culminating height has now declined to about 80 m O.D. Along the modern Soar valley in central Leicestershire remnants of the gravel are widely preserved at around 70 m O.D. as far north as Thurmaston at the confluence of the Wreake. It is a curious fact that further downstream along the Soar valley no occurrence of the Baginton Sand and Gravel has ever been detected, and yet there is increasing evidence for such material along the lower Wreake valley to within a few kilometres of Melton Mowbray. The presence of this waterlain deposit of western provenance in a valley carrying a stream from the east at least hints at the possibility of drainage reversal. Such a hypothesis receives support from an earlier reconstruction of a large drift-filled valley extending from Melton Mowbray eastwards to the Fenland margin and debouching at just over 40 m O.D. near Thurlby (Rice, 1965; Wyatt, 1971). Only near Thurlby is the sand and gravel that floors the old valley exposed at the surface and here the clasts in the gravel are found to consist of 30% 'Bunter' quartz and quartzite, followed by 25% ironstone and 17% limestone (Booth, 1983). This composition certainly appears consistent with the Proto-Soar at one time having fed 'western' material into the head of the buried valley near Melton Mowbray. On the other hand it should be pointed out that, on current evidence, there seems to be a bedrock sill just west of Melton Mowbray that precludes any simple reconstruction of a Proto-Soar valley declining smoothly and regularly from Thurmaston to Thurlby. Clearly more information is required before the precise history of the drainage system in this northern area can be written.

To date biotic finds from the Baginton Sand and Gravel have been so sparse as to make the exact specification of the contemporaneous climate rather difficult. The most productive sites have been in north and central Warwickshire where the gravels have yielded a fauna that includes *Coelodonta antiquitatis, Mammuthus primigenius, Palaeoloxodon antiquus, Equus caballus* and *Rangifer tarandus* (for full list see Shotton, 1983a). The one site that has also yielded floral remains suggests a moderately cold climate with open birch woodland, but the presence of a few thermophilous species led Kelly (1968) to question whether the vegetation was in complete equilibrium with the climate. At several locations the overlying sands have been observed to contain ice wedges, suggesting that there may have been a significant refrigeration of the climate as the aggradation proceeded.

(ii) *Advance of the 'Thrussington Till' ice.* With very few exceptions the material overlying the Baginton Sand and Gravel between central Warwickshire and north Leicestershire is a red Trias-derived till with most of its erratics transported from northerly or northwesterly sources. The southeastward ice movement is attested not only by the provenance of the materials, but also by fabric studies (e.g. Rice,

1981a; Shotton, 1983a) and most recently by striations on a newly bared igneous rock surface at Croft; the quoted fabric studies yield directions of 120° and 155° respectively, and the striations a direction of 125°. Two critical questions concerning this 'Thrussington Till' ice are the limits that it eventually reached and its relationship to ice advancing from the northeastern quadrant. Unfortunately neither of these questions can yet be answered with any great precision. Fig. 10 incorporates a preliminary attempt to depict the boundary of the area over which Triassic-derived till forms the earliest of the glacigenic sediments. As might be anticipated the distinctive red till extends southeastwards to rest on the Lower Lias bedrock. Yet within a few kilometres it either disappears, or at least becomes much more patchy. There can be no presumption, however, that the change marks an identifiable ice limit since the Trias-derived material becomes interleaved with a grey Lias-rich till which rapidly assumes local dominance. Information about this transition zone is inevitably sparse since there are almost no sections and the data come mainly from augering and boreholes. From the parallelism between the transition zone and the outcrop pattern of the basal Lias the probability seems to be that once the 'Thrussington Till' ice had crossed on to the Jurassic rocks it began incorporating the local bedrock into its ground moraine. If correct this interpretation means that the limit reached by the 'Thrussington Till' ice is likely to be very difficult to detect, since the northeastern ice proceeding along the strike of the Mesozoic rocks also picked up immense quantities of local material. The point may be illustrated by reference to Northamptonshire where an early widespread diamict composed mainly of local debris clearly pre-dates the major spread of chalky till. In the writers' view it would be premature to equate this deposit with the Thrussington Till, although it is worth noting that there are erratics in Northamptonshire that require an overall transport from the northwest at some stage during the Pleistocene (Sabine, 1949). For the present it is probably better to adopt a more cautious interpretation and simply argue that the 'Thrussington Till' ice covered the whole width of the Proto-Soar valley and advanced at least on to the fringes of the Jurassic scarplands further to the southeast.

It is only near Stratford upon Avon in the far southwest that the Baginton Sand and Gravel is succeeded by material other than till. Exposures around Snitterfield have displayed the uppermost sands grading into several metres of red laminated silt and clay. This evidence of lacustrine sedimentation indi-

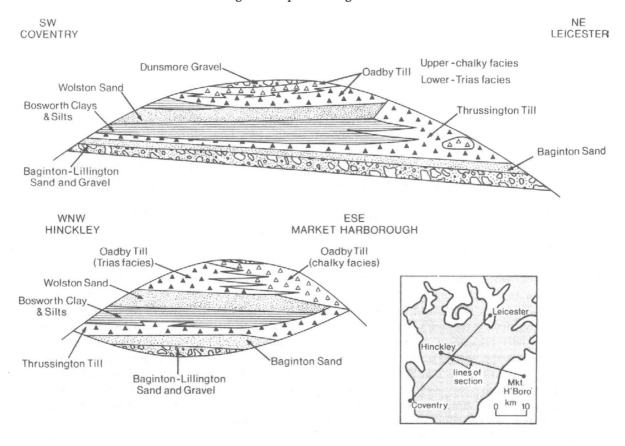

Figure 11: A simplified model illustrating the stratigraphic relationships of the Wolstonian beds in the Leicester-Coventry-Hinckley area.

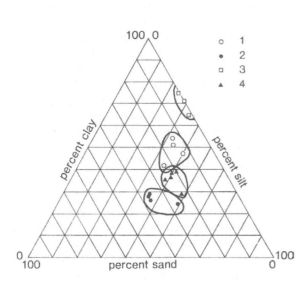

Figure 12: Grain-size analyses. 1 = Oadby Till (chalky facies); 2 = Oadby Till (Trias facies); 3 = Bosworth Clay; 4 = Thrussington Till.

cates either that the initial advance of the 'Thrussington Till' ice failed to reach the Stratford area, or at the very least that a phase of ponding supervened before the locality was overrun by the ice.

(iii) *Basis and history of proglacial Lake Harrison.* Following Shotton's recognition of Lake Harrison (1953), several other workers have mapped parts of the lake basin and within this area the regional correlation is reasonably secure (Bishop, 1958; Rice, 1968; 1981; Douglas, 1980). Based on the stratigraphy revealed in these investigations, a sequence of events can be devised, which is complicated only by the undoubted fact that many of the boundaries between lithostratigraphical members are diachronous. The nomenclature derived by Sumbler (1983a) shows this clearly as the Oadby Till member is seen as a time equivalent to a range of other beds. A diagrammatic representation of this can be seen in fig. 11, which shows a simplified model of the arrangement of Wolstonian beds in the Leicester-Coventry-Hinckley area. The grain-size characteristics of the different members are given in fig. 12.

Shotton's original view of Lake Harrison has been questioned by some recent authors. Firstly, Ambrose & Brewster (1982) pointed to the results of their geophysical surveys in southeast Warwickshire where they contended that for part of its length the lake bench identified by Dury (1951) is a feature formed by the slightly more resistant '70' Marker Horizon within the Lower Lias. If this explanation were to be accepted, then the evidence for Lake Harrison would depend on its sedimentary record rather than its geomorphological imprint as a shoreline feature. Secondly, Sumbler (1983a) in discussing the type Wolstonian glacial deposits 'accepted that lacustrine deposition occurred over a large area', but concluded that 'the widespread occurrence of till within the Wolston Clay implies the close proximity of an ice sheet, which suggests that the concept of Lake Harrison as a single semi-permanent body of water may be an oversimplification'. It is appropriate, therefore, to reconsider the idea of Lake Harrison, not only because its existence has been challenged, but also because its deposits have played an important role in the elucidation of the glacial history of the Midlands.

The extent of the lacustrine members is quite well known from augering and boreholes. Relatively few exposures however have been opened in the lake clays and silts which have allowed detailed sedimentological analysis. Lake clays regarded as being deposited in Lake Harrison have been described from two stratigraphic horizons. Firstly, between the Thrussington Till and the Wolston Sand where they are most widespread and secondly, particularly south of Coventry, they are found above the horizon of the Wolston Sand. The lacustrine beds thicken northwards with the increasing depth of water available as the Proto-Soar valley descends to the north-northeast. Indeed it is in this area that evidence of an extensive Lake Harrison along the lines of Shotton's 1953 model must be demonstrated, for in the south of the Lake Harrison basin in the vicinity of Warwick and Stratford much of the drift succession is absent, presumably eroded by the incision of the Warwickshire Avon. It is therefore difficult to use these fragments of lacustrine material to refute the view that the occurrence of widespread lacustrine deposits could be the product of a 'succession of transient ice-marginal lakes' (Sumbler, 1983a).

As long ago as 1898, Harrison identified lake clays in the Hinckley area belonging to what he termed Lake Bosworth, the forerunner of Shotton's Lake Harrison. This area was mapped by Douglas (1974, 1976, 1980) and the name given to the lake deposits,

the Bosworth Clays and Silts, was used by Shotton (1976) as a member of his amplified Wolstonian nomenclature. The distribution of the Bosworth Clays and Silts is indicated in fig. 13. Where these are seen in section, they are frequently laminated and often varved. In an unweathered state, the clays are dark red-brown towards the base of the member and often grey-brown in the upper part. The latter facies is usually associated with several calcareous concretions locally known as 'race'. The difference in colour and carbonate levels may imply that the ice impounding the proglacial lake during the deposition of the Bosworth Clays and Silts shifted from a dominantly northwestern Trias-rich phase to the chalk-rich eastern phase; a contrast noted in the associated tills.

Arguably the best exposure of the Bosworth Clays and Silts was a temporary pit near Aston Flamville which was excavated in 1975. A series of brown, laminated silts and clays with overlying grey silts, in total up to 10 m in thickness rested on a till with a reddish-brown matrix which contained only very few Jurassic or Cretaceous erratics. The silts and clays contained isolated narrow bands of sand. The laminations were best developed in the brown silts and on a small scale showed a number of slump structures. This lacustrine sequence showed discrete but not infrequent very small inclusions of a reddish till probably dropped from floating bergs. The junction between the underlying Thrussington Till and the clays was usually a sharp one, but interfingering of the till and bedded material convincingly demonstrated that no length of time elapsed between the deposition of the two and that the lake became established on the withdrawal of the ice. However, this site has been shown by Rice (1981) to be close to the northeastern limit of the Bosworth Clays and Silts within the Proto-Soar valley, thus the frequency of dropstones and the occurrence of sand probably reflects a proximal position in relation to the ice dam. More distal facies of the Bosworth Clays and Silts show a preponderance of clay and very low frequencies of dropstones or till inclusions.

Not only are the Bosworth Clays and Silts a persistent member of the succession in the Hinckley area, being found consistently between the largely Trias-rich Thrussington Till and the overlying Wolston Sand and Gravel, but they constitute the thickest member of the Wolstonian sequence in western Leicestershire. It is significant that the surface onto which the lake clays were deposited is a broad valley opening towards the south-southeast. This is the Market Bosworth arm of Lake Harrison (Shotton, 1953), or the Hinckley Valley of Douglas (1980). The

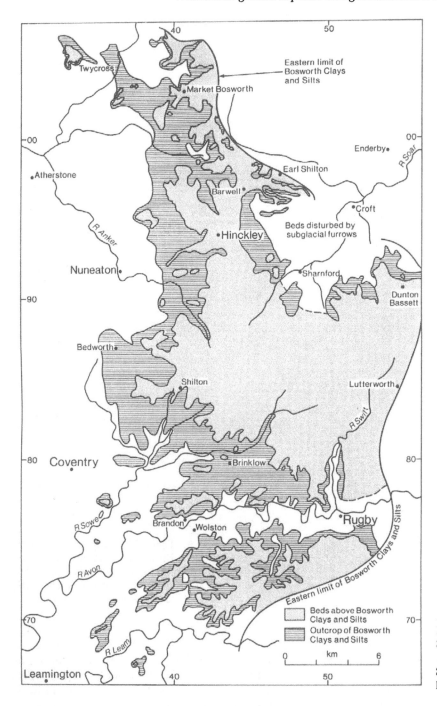

Figure 13: The distribution of the Bosworth Clays and Silts (Sources: Shotton, 1953; Douglas, 1980; Rice, 1981a; Sumbler, 1983a; Geological Survey sheets and field mapping by Douglas).

member is thickest in the centre of this valley and thins to a feather-edge against rising ground to the east (fig. 14). The corresponding feather-edge on the western side has invariably been removed by erosion as the present drainage of the Anker is normal to the trend of the Hinckley valley; however at Twycross, the Bosworth Clays and Silts thin to about 4 m (fig. 14). Thicknesses up to 35 m have been recorded for the Bosworth Clays and Silts in the southern part of the Hinckley valley and up to 50 m can be estimated on the basis of augered hillslope traverses. The height attained by the lake deposits can best be judged by noting the altitude of the feather-edge of the lake clays as they thin against the rising bedrock or Thrussington Till surface. This limit is between 110 and 119 m O.D. and is remarkably consistent consid-

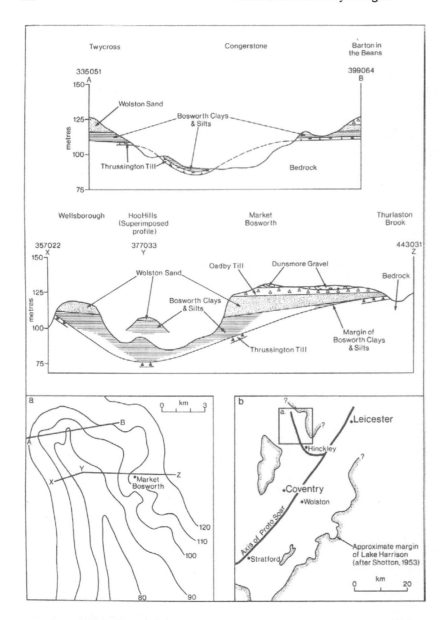

Figure 14: a) Sections through the Bosworth Clays and Silts near Market Bosworth; b) the form of the sub-drift valley and its relationship to Lake Harrison.

ering the glacio-isostatic factors which are likely to have influenced the proglacial environment. Away from the margin of the lake clays, the altitude reached by this member is largely controlled by post-depositional compaction. Douglas (1980) showed through trend surface analysis, that the top of the Bosworth Clays and Silts described a basin almost exactly coincident with the shape of the underlying Hinckley valley. On the assumption that the lake deposits were built up almost to the maximum surface level of the lake, compaction factors of about 30 - 40% have been calculated.

The continuity of the Bosworth Clays and Silts in western Leicestershire carries certain implications. The Hinckley valley is a tributary to the proto-Soar and in order that lacustrine sedimentation could occur in the Market Bosworth area, it seems that similar conditions must have existed throughout most of what Shotton regarded as Lake Harrison. The only other way in which so extensive a series of lake clays could accumulate would be in the improbable circumstance that the Hinckley valley was blocked at its southern end by ice. As the earlier red-brown facies of lake sediments shows what may be expected to be a northwestern provenance, this would seem to be out of the question. In other words, widespread lacustrine sedimentation in the Hinckley valley of western Leicestershire is the clearest evidence in support of a Lake Harrison of similar dimensions to Shotton's 1953 model.

It is possible to trace the Bosworth Clays and Silts southwards from the Market Bosworth area, by way of boreholes near Hinckley and Nuneaton and along the line of the M69 (Shotton, 1976) to the Coventry area where it has been mapped by Shotton (1953) and Sumbler (1983a). Similarly, it is possible to trace the beds eastwards to the area in south Leicestershire mapped by Rice (1981), where the increasing occurrence of till testifies to the presence of an ice front during the lacustrine phase (fig. 13). The extent of these beds and their relationship to the basin in which Lake Harrison was ponded indicate that it is possible to consider that the Bosworth Clays and Silts not only have a lithostratigraphic value, but can also be used in a chronostratigraphic context to refer to that phase of Lake Harrison before the deposition of the Wolston Sand and Gravel sandur.

The presence of an ice front restricting drainage in the Croft- Leicester area is indicated by the distribution of the Wolstonian members (figs. 11, 13) as well as the lack of any evidence that there was a phase of erosion during or immediately following the deposition of the Bosworth Clays and Silts. Both Shotton (1976) and Douglas (1980) on the basis of varved sediments have demonstrated that the Bosworth Clays and Silts represented a lengthy period of deposition, possibly of the order of 10,000 years. The present authors, however, would not argue that the lengthy withdrawal of ice from the maximum of the Thrussington Till advance constituted an interstadial as Shotton (1976) has suggested. It seems unnecessary to adopt the term 'interstadial' for a phase in which the ice stood well to the south of its position at the maximum of the Dimlington Stadial of the Devensian Stage and for a phase when no biological indicators of interstadial character have been recorded in the Midlands.

The picture of Lake Harrison confirmed above does not account for all the occurrences of lake beds in the pre-Devensian Quaternary deposits of the Midlands. It is important however, to reserve the name Bosworth Clays and Silts for the most widespread manifestation of Lake Harrison which took place before the Wolston Sand had been deposited. Shotton (1953) and more recently Sumbler (1983a) showed that in the Coventry and Wolston areas renewed ponding took place after the Wolston Sand and Gravel had been deposited as evidenced by the Upper Wolston Clay. Near Leicester, Rice (1968) has iden-

tified the Glen Parva and Rotherby Clays which represent more local ponding.

(iv) *Advance of the 'Oadby Till' ice*. Over a remarkably wide area the uppermost glacigenic sediment is a till sheet, in places over 30 m thick, in which chalk and flint are among the commonest erratics. Much of the material possesses the character of a lodgement till, and its approximate western limit is shown on fig. 10. In addition to these depositional effects, the advance of the 'Oadby Till' ice locally had important disruptive consequences whereby earlier superficial deposits were subject to severe dislocation. One result is that whereas the drift succession around Wolston may be regarded as still lying in reasonably orderly fashion, further north any such presumption needs to be treated with caution. A minimum of two important disruptive processes can be distinguished, namely glaciotectonic disturbance and the erosion of elongated subglacial furrows; in addition the possibility of areal scouring of the earlier Wolstonian sediments needs to be borne in mind.

Glaciotectonics can be most readily demonstrated at sites near Dunton Bassett and Sapcote (Rice, 1981a). At the former locality a now disused pit exhibited severe disruption of the Wolston Sand and Gravel. Large-scale thrust faults, overturned isoclinal folds and low-angle shears all testified to compressive forces from the north or slightly east of north. *Décollement* generally occurred on the Lower Wolston Clay just below the former sandur, while strata up to and including the base of the overlying till were affected: the latter fact seems to imply that the ice edge had already advanced beyond the site of the pit when the major movements took place. Near Sapcote sections in the M69 Motorway revealed displaced masses not only of Thrussington Till and Lower Wolston Clay but also of Mercia Mudstone. Huge 'rafts' of the bedrock, up to 10 m thick and 100 m across have been thrust over the local drift sequence. Beds incorporating significant quantities of chalk and flint are involved, although there is no overall cover of Oadby Till. At neither Dunton Bassett nor Sapcote do the glaciotectonic structures possess any surface expression, and at the latter site near the floor of the Soar valley post-glacial erosion has probably stripped away substantial thicknesses of Oadby Till. The absence of any topographic expression makes instances of glaciotectonics very difficult to identify without deep sections; there are, however, several suspected but unproven cases in a broad zone extending from east of Dunton Bassett to beyond Sapcote in the west.

The reconstructed Proto-Soar valley with its continuous spread of Baginton Sand and Gravel provides an invaluable datum against which to assess concentrated subglacial erosion. The best example is provided by the distribution of chalky tills and other sediments in the area just southwest of Leicester. Here the bedrock floor lay at 60 m and the culmination of the fluvial aggradation at 72 m O.D. Yet boreholes show grey chalky sediments within an elongated furrow carved through the Baginton Sand and Gravel and into the underlying Mercia Mudstone to descend below 45 m O.D. The fill comprises tills, sands, gravels and laminated clays, with the majority of the strata containing clasts of chalk and flint. The stratigraphy is extremely disordered and even where there is a dense network of boreholes the absence of deep sections renders correlation of individual beds difficult and often virtually impossible. Nevertheless it is clear that a substantial proportion of the material is waterlain, with indications of both high-velocity and quiescent conditions. The sides of the furrow are locally very steep; in one of the few available exposures the Baginton Sand and Gravel was seen to be almost vertically truncated by the wall of the furrow, while only a few metres away the normal upward progression into the Thrussington Till lay undisturbed. Although the precise conditions responsible for the excavation and infilling of the furrow remain problematical, it is tempting to invoke a change in thermal conditions at the base of the ice as one contributory factor; while the Baginton Sand and Gravel was accumulating permafrost conditions prevailed, but by the time the furrow was being infilled there was patently a very aqueous environment.

Numerous other instances of concentrated linear erosion might be quoted, ranging from near Derby in the north to near Northampton in the south. A further indication of the power of the 'Oadby Till' ice comes from the displacement of huge 'rafts' of Middle Lias marlstone and Lincolnshire limestone in northeastern Leicestershire and adjacent parts of Lincolnshire (Straw, 1979a). Less obvious but equally pertinent to the present discussion is the possibility of areal scouring of the earlier Wolstonian deposits. One indication that this has occurred may be found in the diversity of sediments sometimes accompanying the more characteristic chalky till. For example in cuttings along the M1 Motorway near Lutterworth the grey chalky till was observed to be interleaved with large lenticular masses of red or pinkish brown till; in some instances there were even deposits with a Mercia Mudstone matrix containing an abundance of chalk fragments (Poole *et al.*, 1968). A possible further sign of reactivation of northwesterly-derived till by pass-

age of the 'Oadby Till' ice may come from the widespread occurrence of banded tills (see, for example, Douglas, 1980; Rice, 1981b) although the precise mechanism involved in the production of these curious features remains enigmatic.

The foregoing account illustrates the very active nature of the 'Oadby Till' ice as it passed across the Midlands. There have been conflicting views about the direction assumed by this ice sheet (e.g. Perrin *et al.*, 1979), but the local evidence as opposed to broader regional correlations points unequivocally to a southerly or southwesterly movement. This statement may be substantiated by reference to fabric studies (e.g. West & Donner, 1956; Shotton, 1983a), by the vector of compressive stresses in glaciotectonic areas (Rice, 1981a), by the orientation of erratic trains (e.g. Straw, 1979a), and by slickensides observed on a low angle thrust at Croft (Rice, 1981c). The last-mentioned is of particular interest because, at the one site, striations beneath the Thrussington Till have an orientation of 125°, and slickensides at the Thrussington Till/Oadby Till interface have an orientation of 215°.

In summary, the advance of the 'Oadby Till' ice was from the north or northeast. In Lincolnshire, where the movement was broadly along the strike of the rocks, distinctive tills lie in contiguous parallel belts (Straw, 1979a). Further south the ice seems to have fanned out, with a stronger westerly component to its movement carrying Cretaceous debris into the heart of the Midlands. Where there are no widely recognisable marker horizons to act as a guide, the vagaries of local till stratigraphy make it extremely difficult to decipher the precise age relationships of the Thrussington and Oadby Tills. The great majority of workers undertaking detailed local investigations (e.g. Bishop, 1958; Bridger, 1981; Douglas, 1980; Rice, 1981a, c; Sumbler, 1983a) have found evidence for an early input of northeastern erratics, and the most appropriate model seems to envisage a persistent composite ice sheet, with the junction between northeasterly and northwesterly ice streams gradually migrating from east to west. Even when the 'Oadby Till' ice reached its maximum extent it was still flanked along its western margin by ice of more westerly origin.

Stratigraphic significance of the Wolstonian glacigenic units

The nature of the Wolstonian tills has been fully discussed above. It is, however, worthy of mention here that the extensive lake clays and the overlying outwash are important regional lithostratigraphical markers which allow the tills to be allocated to the earlier advance and the subsequent readvance: the Thrussington and Oadby Tills respectively. The pattern of provenance and stratigraphic position is extremely complex as the area in question lies at the western limit of the chalky till province. Only the earliest Wolstonian deposit of the area, the Baginton-Lillington Gravels seems to be devoid of Cretaceous material in the Midlands. Rice (1981a) has shown that the Thrussington Till in south Leicestershire although consisting predominantly of a Trias facies, locally contains northeastern material with Cretaceous erratics. Naturally, this latter facies is more frequently found in the east of the region. Similarly, in the Oadby Till, both northwestern and northeastern provenances are represented with a regional trend suggesting that the latter eventually gained dominance, so that the uppermost till facies is commonly a chalky till.

It is clear however, that the earlier schemes which linked each till lithology with a separate glacial advance can no longer be sustained (cf. West & Donner, 1956). The transition between the two distinct till facies is invariably complex and often takes the character of a banded till. This zone is typically 1–2 m in thickness and consists of alternating bands of predominantly chalky and red (Trias-rich) till. One measured section displayed 15 alternations within 1.5 m. Many of these bands are laterally persistent over 15 m, eventually thinning to become imperceptible streaks (plate 13).

What is apparent from the foregoing is that the internal consistency of the correlations within the Midlands is largely the result of lateral persistence of the lacustrine member of the succession. The associated tills are often extensive, but provenance and character are at best unreliable guides to stratigraphic position. The problem is exacerbated when attempts are made to extend the correlations outside the region, for as Wills observed as long ago as 1937, 'the weakest links in the chain of evidence are those that connect the Midlands with other glaciated areas'. The ice of the Late Devensian Dimlington Stadial reached the peripheries of the Midlands at Wolverhampton in the west and into Lincolnshire to the east. It is the correlations between the pre-Devensian deposits, however, which have caused considerable controversy.

Shotton (1983a, b) has summarised the opposing arguments for the age of the Wolstonian and has argued strongly that the Wolston stratotype should remain as the representative of the penultimate major glaciation of the British Isles. At the Wolston type site the glacial beds cannot be related directly to interglacial deposits. Correlation over longer distances means straying from the clear lithostratigraphy imposed by the Lake Harrison sediments. Much debate has centred on the age and interpretation of the chalky tills. Perrin *et al.* (1979) advanced the argument that the 'Chalky Boulder Clay' was the product of a single glaciation on the basis of its consistent matrix character. As deposits of Hoxnian age overlie this till in East Anglia, a post-Hoxnian i.e. Wolstonian age for it in the Midlands clearly presents problems. Shotton (1983a) has re-stated the evidence for the conventional chronology and has argued for the *status quo* on the grounds of the contained fauna of the Baginton-Lillington Gravels, the links between the Lake Harrison overflow and the River Thames terraces and the stratigraphy in the Birmingham area as revealed by boreholes. Recent evidence shows that the fluvial Baginton-Lillington Gravels and the Baginton Sand which immediately precede the Wolstonian glacigenic deposits in the Midlands can be traced into East Anglia where they underlie deposits of Anglian age (but see also Ehlers & Gibbard, this volume). This further weakens the argument for retaining the term Wolstonian to refer to deposits of the penultimate glaciation (Rose, 1987). What is clear, however, is that within a large part of the English Midlands, Quaternary beds can be successfully correlated with the Wolston stratotype. This is largely due to the establishment of a lithostratigraphy throughout the Lake Harrison basin. The variable provenance of the tills which sandwich these beds highlights the difficulties in assigning ages to glacigenic sediments.

Pre-Midlandian glacial deposits in Ireland

PETER G. HOARE

Some Quaternary workers have shown little hesitation in recognising signs of pre-Midlandian* glaciation in Ireland: "The first occasion on which great masses of ice actually formed ... probably took place about 200,000 years ago..." (Mitchell, 1976: 34). Although others have been more cautious (see McCabe & Hoare, 1978, for example), most publications on the glacial stratigraphy of the country distinguish Munsterian and Midlandian drift (see Mitchell, 1957; 1972; 1976; Synge & Stephens, 1960; Stephens & Synge, 1965; Synge, 1968; 1970a; 1979a; 1980; Colhoun, 1970; 1971a; McCabe, 1972; 1986b; Mitchell *et al.*, 1973; Stephens *et al.*, 1975, and Finch, 1977, for example). Whilst there are numerous references to glaciations of pre-Midlandian age, and to some of Early and Middle Midlandian date, few of these can be substantiated. Synge & Stephens (1960) considered that there was no clear evidence of pre-Munsterian glaciation.

An extremely confused picture of glacial and other Quaternary events in Ireland has emerged for a variety of reasons. Geochronometric dates and marker horizons, such as interstratified organic beds, are not generally available. In addition, sequences have often been dealt with in a cavalier manner, and workers may be seen as stratigraphic 'lumpers' and 'splitters'. Unnecessarily complex models have been devised to explain phenomena which may be accommodated within existing schemes, and attempts have been made to relate the Irish glacial stratigraphy to what is known about the extent and timing of continental European glaciations (Synge, 1981). The origin of certain critical beds is not fully understood as little detailed sedimentological work has been undertaken, either in the field or in the laboratory.

Till of possible pre-Midlandian age underlies clastic marine sediments in St. George's Channel, the Celtic Sea and the Bristol Channel; the overlying material may be Ipswichian (Garrard, 1977) or may belong to a Devensian interstadial (Devoy, 1983). Examination of the glacigenic material and its correlation with drift exposed on the Irish mainland is at a preliminary stage and these subjects are not discussed further here (but see Huddart, 1981a and b, and Hoare, this volume). Pre-Quaternary glaciofluvial gravels such as those of (?)Ordovician age that crop out in the west of Ireland (Williams, 1981) are also excluded from this selective review.

The positions of sites referred to in the text are shown on fig. 15.

The start of the Quaternary record in Ireland

The oldest known Quaternary material in Ireland may be represented by the cold-loving, marine molluscs, similar to those of the Waltonian Crag of East Anglia, that were transported within glacial erratic 'rafts' of Crag sediment (Mitchell, 1972) or were reworked into outwash of (?)Midlandian (=Devensian?) age (Mitchell *et al.*, 1973). They have been reported from a number of sites in County Wexford (Griffith, 1835; Bell, 1888; 1889; 1890; 1915; 1919b; Cole & Hallissy, 1914; Harmer, 1922), from Killincarrig, County Wicklow (O 2910) (McMillan, 1938; 1964; Mitchell, 1972; Mitchell *et al.*, 1973), from Kill-o'-the-Grange, County Dublin (O 231265) (Sollas & Praeger, 1895), and from other locations. The more coherent part of the stratigraphic record, however, is usually regarded as beginning with deposits of Gortian age; it represents, therefore, only a fragment of Quaternary time.

---

*The Irish Quaternary stages are considered by many to be as follows (from older to younger): pre-Gortian (which may be equivalent to the Baggotstownian of Watts, 1964), Gortian, Munsterian, an un-named warm stage, Midlandian and Littletonian. Synge (1979a, 1981) employed the term Munsterian to refer to a particular glacial episode as well as to a Quaternary cold stage. Warren (1985) suggested the name Fenitian for the last cold stage, and Ballybunnionian for the penultimate (or an earlier) cold stage.

However, thick, organic-rich and laminated lacustrine sediment of temperate character lies below till in Ballyline Townland, County Kilkenny (S 391467) (Coxon & Flegg, 1985). The deposit probably belongs to the middle part of the Quaternary but can only provide a maximum date for the younger glacial material. Late Pliocene or Early Quaternary temperate sediments occur at Pollnahallia, County Galway (M 3347) (Coxon, 1986b). Further early Quaternary deposits will almost inevitably be discovered, particularly as a result of the investigation of enclosed depressions and caves in the extensive limestone areas of Ireland.

The stratigraphic position of the Gortian Stage

The position of the Gortian Stage is of considerable relevance to the search for pre-Midlandian glacigenic sediments. Drift rests upon Gortian organic material at Newtown, County Waterford (S 7007) (Mitchell, 1948; 1962; 1970; Watts, 1959a; 1967), at Boleyneendorrish, County Galway (M 5306) (see below), and at a limited number of other sites (Mitchell, 1948; Watts, 1959a; 1970; 1985). Elsewhere, it is often assumed that all or most of the glacial sequence post-dates this temperate episode. There is one principal school of thought on the stratigraphic setting of the Gortian Stage, but two other proposals require consideration:

(i) Watts (1985) re-stated the majority view which is based on palynological evidence. He believed that all Quaternary 'interglacial' material belongs to the Gortian Stage, and that this is equivalent to the Hoxnian of Britain. According to this scheme, either part of the post-Gortian glacial sequence is of Munsterian (=Wolstonian?) age or the whole of it must be accommodated within the Midlandian (as the most recent drift dates from the Midlandian; Hoare, this volume). It is assumed that Ipswichian sediments remain to be discovered (but see Watts, 1964; Colhoun & Mitchell, 1971, and below).

(ii) Warren (1979a, 1985) and Synge (1980, 1981) believed that stratigraphic and palynological considerations indicated that Gortian material is of the same age as Ipswichian deposits in the remainder of the British Isles; all post-Gortian glacial drift therefore belongs to the cold Midlandian Stage. Although the stratigraphic position of the Gortian cannot be determined by a count-from-the-top approach, the possible correlation should not be overlooked.

(iii) Synge (1977a, 1979a, 1980, 1981) suggested that a number of the glacial events that have been documented in Ireland may pre-date the Gortian Stage. He followed Kinahan (1874) in believing that solifluction may have caused older sediments to descend onto younger ones in coastal cliff exposures. As a result, 'glacialoid' till (Kinahan, 1874) came to rest on Gortian material at Newtown, County Waterford, and on raised beach deposits elsewhere (Synge, 1977a; 1981). Detailed fieldwork is required to test Synge's claim that a fundamental misunderstanding of the glacial history of Ireland has arisen as a consequence of the failure to recognise this phenomenon.

In view of these uncertainties, the writer feels compelled largely to discount glacial events that are placed without good reason in the Munsterian or in earlier cold stages; these episodes are outlined below and some are discussed in detail by Hoare (this volume). Instead, drift that is clearly older than the Gortian Stage is identified in order to demonstrate pre-Midlandian glaciation.

Indications of a pre-Gortian cold stage

The Gortian Stage stratotype occurs at Boleyneendorrish, approximately 9 km northeast of Gort, County Galway (Kinahan, 1865; 1878). Fine-grained beds with thin layers of plant debris lie below the Gortian material (Jessen *et al.*, 1959). Pollen and macroscopic plant remains from this lower unit record the arrival of arctic and subarctic pioneer species at the close of the preceding cold stage, but do not provide evidence of glacial conditions.

Plates 1–3

1. Harper Lane Quarry (TL 164019); Westmill Gravel of the Proto-Thames aligned northeastwards through the Vale of St. Albans, overlain by Moor Mill laminated clays of the Anglian ice-dammed lake formed between the Lowestoft Till ice advance and the Thames Valley. The clay grades upwards into Lowestoft Till (Photograph: P. Gibbard, 1971).
2. Laminated clays with dropstones underlying till at Bunker Hill Quarry, Hertfordshire (TL 301095), deposited marginal to the Anglian ice sheet (Photograph: P. Gibbard, 1971).
3. Lowestoft Till at the coastal cliff of Kessingland (Photograph: J.A. Zalasiewicz, 1985).

Sub-Gortian glacial deposits

An important Quaternary sequence was uncovered during the digging of two wells at Baggotstown, County Limerick (R 6635) (Watts, 1964). Organic lacustrine deposits with strong Gortian affinities rest on a "Stony clay with erratics..." (Watts, 1964: 169) in Well 1 (table 2); Watts considered that the older material is the first Elsterian (=Anglian/pre-Gortian?) till to be found in Ireland. It underlies temperate sediments that cannot be younger than Ipswichian, whatever the stratigraphic position of the Gortian. The lower till, if that is what it is, may be assigned with some confidence to a glaciation within the 'Baggotstownian Stage' (Watts, 1964). The site is located inside the Southern Irish End-Moraine (Charlesworth, 1928a), a feature widely regarded as marking the maximum extent of the Midlandian ice-sheet (but see Hoare, this volume), and it was assumed that the upper till represents two post-Gortian glaciations.

Organic-rich material of Gortian aspect lies at depth in ·Well 2 at Baggotstown (table 3); the gyttja which occurs above it may be of Ipswichian age (Watts, 1964). The stratigraphically higher organic unit is unlikely to have been derived from the lower one (Watts, 1964; Mitchell, 1976). The till that separates them may date from a Munsterian glaciation.

Lacustrine material with Gortian affinities was

Table 2. The stratigraphy of Well 1, Baggotstown, County Limerick (Watts, 1964); thicknesses are in metres.

Upper till	4.90
Organic material	1.80
Modified top of the underlying deposit	2.10
Lower till	not known

Table 3. The stratigraphy of Well 2, Baggotstown, County Limerick (Watts, 1964); thicknesses are in metres.

?Upper till	1.35
Gyttja	0.03
Till	3.82
Sandy silt with small organic fragments towards the base	2.10
Gravel	not known

Table 4. The stratigraphy at Kildromin, County Limerick (Watts, 1967); thicknesses are in metres.

Upper till	4.25
Disturbed top of the underlying material	0.30
Organic sediment	3.75
Varved, proglacial lake sediment	1.80
Lower till	not known

also recorded in a borehole at Kildromin, County Limerick (R 7141), 7 km northeast of Baggotstown (Watts, 1967) (table 4). The "Reddish-brown, tenacious, unweathered boulder-clay..." (Watts, 1967: 170) at the base of the sequence (the lower till in table 4) would seem to represent a pre-Midlandian glaciation. Only one till rests upon the interglacial material, although it is generally thought that the area was overwhelmed by ice on at least two occasions since the Gortian Stage.

Although the lower till in Well 1 at Baggotstown and at Kildromin occupy a similar stratigraphic position, they may not belong to the same glaciation. Even if they are contemporaneous, it is not possible to reconstruct a pattern of ice-movement or to determine the maximum extent of the event. An ice-mass of considerable proportions was involved if the distance to the nearest glaciated uplands is taken into account, but the position of the centre of dispersion is not known. The lower till in Well 2 at Baggotstown appears to signify a later, but nonetheless pre-Midlandian, glaciation. The nature of the gravel at the bottom of the succession in Well 2 was not established.

The sequences at Baggotstown and Kildromin provide evidence for two pre-Midlandian glacial episodes; their considerable significance demands a re-investigation of the sites and of the surrounding areas.

Further possible evidence of pre-Midlandian glaciation

The following section contains examples of some of the less than satisfactory arguments that have been advanced to demonstrate that glaciation took place

Plates 4–5

4. The wave-cut platform at Clogga Head, County Wicklow (Photograph: Peter G. Hoare, 1969).
5. Glacial erratics resting on limestone pavement, the Burren, County Clare: free boulders or *remanié* deposits? (Photograph: Peter G. Hoare, 1970).

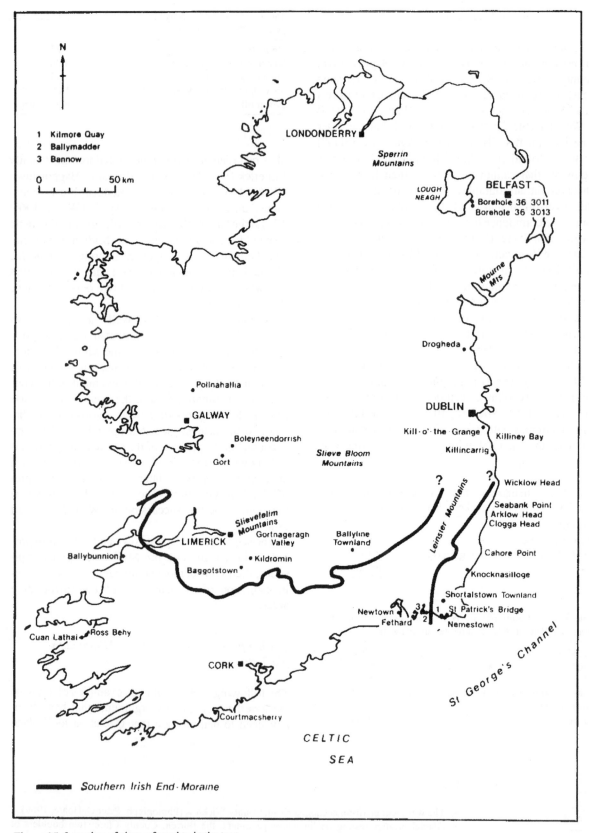

Figure 15: Location of sites referred to in the text.

before the Midlandian Stage.

Many regional Quaternary successions in Ireland have been arranged into two sets of deposits, one of Munsterian and the other of Midlandian age. The distinction has been made on lithostratigraphic and morphostratigraphic grounds and is summarised by the terms 'Older Drift' and 'Newer Drift' (the Older and Younger Series of Wright, 1937). The former lies at the surface outside the Southern Irish End-Moraine and is described as deeply altered, much dissected, displaying smooth slope profiles and generally lacking features such as drumlins, kames, eskers, moraines and kettle-holes which are assumed to have been destroyed by subaerial processes during succeeding cold and warm episodes (see Synge, 1970a, for example). 'Newer Drift' occurs within this glacial limit and sometimes rests upon 'Older Drift'; it is said to be characterised by unweathered till and by 'fresh', hummocky, constructional forms. Bowen (1973b) and Warren (1981, 1985) were unable to accept this approach. In addition, periglacial phenomena are reported to be as common inside the Southern Irish End-Moraine as they are outside it (Warren, 1981; Lewis, 1985). However, the density of these features must reflect the texture of the material and the hydrology of the site, as well as the length of time the deposit was exposed to frost action. These points are discussed in greater detail by Hoare (this volume).

The following episodes are conventionally placed in one or more pre-Midlandian cold stages: the Greater Cork-Kerry Glaciation of southwest Ireland (Farrington, 1954); early glacial events in the Sperrin Mountains (Colhoun, 1970; 1971a) and the Moneydorragh More/Ballymartin Glaciation of the Mourne Mountains (Stephens *et al.*, 1975); the Drogheda Glaciation of County Louth (McCabe, 1972; 1973; Colhoun & McCabe, 1973); the Enniskerry (Farrington, 1934; 1944) and Brittas (Farrington, 1942; 1944; 1949) Glaciations of the Leinster Mountains; the rather more widespread Eastern General/'Irish Sea' Glaciation of the eastern and southern coasts of the country (Farrington, 1944; Stephens *et al.*, 1975); and the Munster General Glaciation of western and southern Ireland (Mitchell, 1957; 1976). Material associated with these episodes makes up the 'Older Drift' and is usually placed in the Munsterian Stage although Synge (1979a) suggested that it may either pre-date the Gortian Stage or belong to the early Midlandian. The Eastern General Glaciation of northeastern and eastern Ireland was regarded as Midlandian in age by Hill & Prior (1968) and by Synge (1981). In the absence of interstratified deposits of Ipswichian age, the distinction is based on less

than convincing, and sometimes on extremely flimsy, evidence.

The tills, sands and gravels of the Killiney Bay exposure, County Dublin (O 260234 - 267195), have been subdivided, without good reason, according to this convention. Early workers identified a Lower Boulder Clay, a bed of shelly sand and gravel (the Middle Drift), and an Upper Boulder Clay (see Harkness, 1869; Hull, 1871, and Lamplugh *et al.*, 1903, for example). This interpretation was accepted by Synge (1963b, 1977a), Bowen (1973b) and Mitchell *et al.* (1973) who proposed that the tills are of Munsterian and Midlandian age, respectively. Hoare (1977a and b) suggested that the entire succession was deposited during a single but complex episode of uncertain date associated with ice emerging from the Irish Sea Basin. He believed that a later glacial advance from the Irish midlands failed to remove all of this drift and to lay down an 'upper till'. McCabe (1972, 1973) presented a similar picture for eastern Counties Louth and Meath.

Farrington (1944, 1957b) recorded weathered drift of Scottish/Irish Sea Basin provenance at un-named sites in the Dublin area; features in the Killiney Bay exposure might be interpreted in this way. Tills are oxidised to a red-brown from an original grey-brown colour and are decalcified to depths of up to 4.3 m in Counties Dublin, Wicklow and Wexford (Synge, 1964; Davies & Stephens, 1978). By contrast, decalcification has only taken place to the relatively shallow depth of 1.5 m in what is assumed to be the corresponding deposit in northeast Ireland (Hill & Prior, 1968). The so-called 'weathering' in the Dublin area may have occurred during a Midlandian interstadial (Farrington, 1949; Synge, 1979a), an unspecified warm stage (Farrington, 1957b), the Ipswichian (Mitchell, 1960; Davies & Stephens, 1978) or during "...a minor break in the succession." (Synge, 1977a: 201).

Hoare (1977a and b) suggested that the 'weathering profiles' in the Killiney Bay exposure are due to the alteration of the upper part of tills by percolating water held up at the base of the overlying, freely-draining, outwash gravels. The time at which this sub-surface process began cannot be established, but modern seepage lines illustrate that the mechanism is still taking place. A final, if unscientific, point might also be contemplated: the modifications described above occur at two stratigraphic levels in the sequence; it would be of no little significance if evidence of two pre-Midlandian temperate episodes was present at one site in a region which otherwise has none!

These features in the Killiney Bay exposure have not been examined in any scientific detail. Numerous sequences throughout the island have been dealt with in an equally superficial manner. A great deal of further work is required if Munsterian and Midlandian drift is to be distinguished.

A well-developed wave-cut platform and beach occurs at approximately −1.0 to +7.6 ± 0.5 m O.D. (B) (Farrington, 1966b)* along the south coast of the country; the 'Main Platform' lies at 5.2 ± 1.0 m O.D. (B) (Wright & Muff, 1904). The type-exposure of both features is at Courtmacsherry, County Cork (W 5643) (Mitchell, 1957). They are sealed beneath a variety of Quaternary sediments throughout the area. A thin layer of Gortian peat appears to overlie them at Newtown, County Waterford (Mitchell, 1948; 1962; Watts, 1959a), and was radiocarbon 'dated' to > 38,000 B.P. (Mitchell, 1970). A similar platform is located between 9.0 m and 20 m O.D. (B) at sites on the east coast (plate 4). Synge (1964) considered it is of a similar age to that on the south coast, but subsequently (Synge, 1981) suggested that it is older.

There is lack of agreement over the timing and climate of the period or periods during which the Courtmacsherry Platform and Beach developed. They may date from the 'pre-glacial' (Wright & Muff, 1904), an unspecified cold episode (Wright & Muff, 1904; Bryant, 1966; Farrington, 1966b; Stephens, 1970; Synge, 1977a; McCabe, 1986b), an unspecified 'interglacial' (Mitchell, 1948), a pre-Hoxnian episode (West, 1977b) or from the Elster-Saale (=Gortian?) temperate stage (Mitchell, 1957). The platform may be 'pre-glacial' (Synge & Stephens, 1960), 'composite' (Mitchell, 1972), "...of very considerable age." (Mitchell *et al.*, 1973: 67), Cromerian (Mitchell, 1960; 1962) or Hoxnian (Synge & Stephens, 1960). The beach may date from the Hoxnian (Mitchell, 1960; 1962; 1976; West, 1977b), from the Ipswichian (Bowen, 1973b; Synge, 1977a; 1980; 1981; 1985; Warren, 1979a; 1985; Devoy, 1983), vary in age from place to place (Davies & Stephens, 1978) or date from a Midlandian cold or interstadial interval (Synge, 1977a; 1981; 1985;

Devoy, 1983), although oceanic oxygen isotope evidence suggests a low sea-level at this time (Shackleton & Opdyke, 1973; Shackleton, 1977).

Doubt has been expressed over the true identity of the beach-like material at Fethard, County Waterford (S 7905) (Carter & Orford, 1981; Devoy, 1983), and at a number of other locations. Geographical variations in the nature of the sediment may mirror facies changes seen in modern beaches, but some deposits may have been wrongly assigned to this unit.

The Courtmacsherry Beach contains certain small clasts that have been interpreted as erratics of local origin (Wright & Muff, 1904; Synge & Stephens, 1960; Synge, 1964) and as far-travelled erratics of northeasterly provenance (Mitchell, 1957; 1960; 1962; Mitchell *et al.*, 1973; Synge, 1970a; 1977a; 1980). They may have been emplaced by icebergs or masses of pack-ice (Wright & Muff, 1904; Synge, 1964) during the Late Hoxnian or Early Wolstonian (Mitchell, 1972), or by mats of seaweed (Wright & Muff, 1904). Alternatively, they may have been derived from an early till of unspecified age (Mitchell, 1957), from glacigenic sediment of Anglian age, none of which has survived (Mitchell, 1960), from the Clogga Till of eastern Counties Wicklow and Wexford (Synge, 1970a), or from the Bannow Till (see Synge, 1964) of County Wexford and the Ballyvoyle Till (see Watts, 1959a) of County Waterford (Synge, 1980). Yet the Bannow and Ballyvoyle Tills were originally thought to post-date the beach (Watts, 1959a; Synge, 1964). There is, within this extraordinarily confused picture, the possibility that some clasts in the Courtmacsherry Beach represent glacial material of pre-Midlandian age.

At Ballybunnion, County Kerry (Q 8641), till is overlain by gravel which is sealed beneath soliflucted (?geliflucted) till (Synge, 1981; Warren, 1981; 1985). The gravel, which was equated by Synge and by Warren with the Courtmacsherry Beach (=Ipswichian?) of the south coast, rests on a platform cut across the till. Warren (1985) placed the lower till in a cold Ballybunnionian (=Wolstonian?) Stage. The gravel at Ballybunnion, and at the other coastal locations from which it has been reported, requires detailed examination in an attempt to distinguish cold and warm stage marine deposits from glacial outwash. Similarly, a proposal that till exposed between Ross Behy (V 644909) and Cuan Lathaí (V 622899), County Kerry, represents a pre-Midlandian glaciation (Warren, 1977; 1981) cannot be considered until further details are provided.

Synge 1979a, 1981 assigned the "...fairly featureless..." (1979a: 8,) drift of northern County Kerry, a

---

*Quaternary workers have measured the heights of features against a variety of datums. This is of little consequence when mountain summits are concerned, but is of considerable importance in connection with raised beaches and related phenomena. Elevations of marine features given in this contribution relate to Ordnance Datum at Belfast (O.D.(B)) and are taken from Devoy (1983) who has discussed the problem in some detail.

drift-sheet with denuded eskers, moraines and kames in northwest County Mayo, the carriage of Galway Granite erratics to the Slieve Bloom Mountains in Counties Offaly and Laois, and certain other phenomena to a cold Connachtian Stage (=Saalian of continental Europe?). This evidence may be accommodated within existing models of glaciation (McCabe, 1986b) and Synge's interpretation was seen as "...stratigraphic speculation..." by Devoy (1983: 240). Deposits that are traditionally regarded as Munsterian in age (see below) were placed in the Elsterian Stage.

A green (Synge, 1964), olive or ochre silt with rounded quartz "...grains and pebbles..." (Synge, 1977a: 204) underlies till of Scottish/Irish Sea Basin provenance at Nemestown, County Wexford (S 9703). Similar material crops out at St. Patrick's Bridge (S 976035) and at Ballymadder (S 861073), County Wexford (Synge, 1977a). The gravelly silt rests on the Courtmacsherry Beach and Platform at Bannow (S 8307) and Kilmore Quay (S 9603), County Wexford; the overlying (?)Elsterian Bannow Till at these sites may not be *in situ* (Synge, 1981). The Nemestown bed may be a weathered till (Cole & Hallissy, 1914; Mitchell, 1951; Woldstedt, 1958; Huddart, 1976; 1977b), a 'gumbotil' (Farrington, 1939) or a periglacial deposit (Mitchell, 1960; 1962; Synge, 1964). It has been regarded as Anglian (Synge, 1977a), Early Wolstonian (Mitchell, 1960) and Munsterian (Huddart, 1976; 1977b) in age. De-

tailed investigation of the timing and nature of the event that led to the accumulation of the Nemestown deposit is clearly required.

A till on the floor of the Gortnageragh Valley, County Limerick (R 8553), may have been associated with ice situated over the Slievefelim Mountains during the Elsterian Stage (Synge, 1966b). No details were given of the location or character of the deposit, nor did Synge seek to justify the proposed timing of the glacial episode.

Farrington (1954) and Synge (1964, 1970a, 1973, 1975b, 1977a, 1981) described a lower, olive-grey, gravelly till from Clogga Head, County Wicklow (T 252692) (fig. 16) and from other coastal exposures in Counties Wicklow and Wexford. The lithology of the material and the underlying bedrock striae suggested a northwesterly provenance. The till was described as weathered, eroded and solifluction (Synge, 1964; 1973) and desiccation and/or frost cracks penetrate the upper part of the bed (Davies & Stephens, 1978). Synge (1970a) denied that it was weathered; he later speculated that a weathered layer might have been removed by marine action (Synge, 1975b). A marine-abrasion platform, the Clogga Rock Platform (Synge, 1981), which is excavated in the adjacent bedrock and rises to 12 - 14 m O.D.(B) at Wicklow Head (T 3492) and Arklow Head (T 2070), County Wicklow, cuts across the Clogga Till which partially fills a small valley-like feature (Synge, 1977a; 1981). The till was considered to date from the Munsterian (Wol-

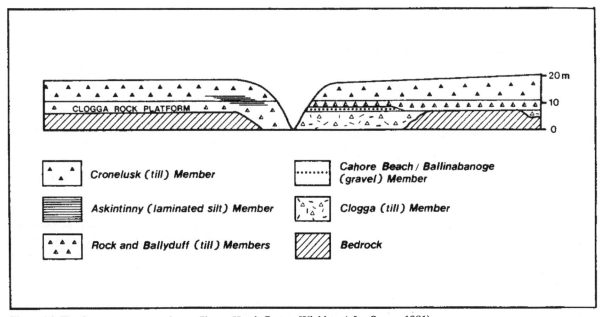

Figure 16: The Quaternary succession at Clogga Head, County Wicklow (after Synge, 1981).

stonian?) Stage (Mitchell, 1972; Synge, 1973; 1977a; 1981:310) or from the Elsterian (=Anglian?) (Synge, 1981:307). The rock platform may be Ipswichian (Synge, 1977a) or Holsteinian (=Hoxnian?) (Synge, 1981) in age.

A bed of poorly-sorted gravel, the Cahore Beach (Synge, 1964) or the Ballinabanoge Member (Synge, 1977a), overlies the Clogga Till at Clogga Head and is sealed by till of Scottish/Irish Sea Basin provenance (Synge, 1964). A similar sequence of deposits may be seen at Seabank Point, County Wicklow (T 2675), and in the most southerly exposure of the Clogga Till at Cahore Point, County Wexford (T 2247) (Synge, 1964; 1977a). The gravel may be a glaciofluvial deposit (Farrington, 1954; Martin, 1955; Davies & Stephens, 1978; Huddart, 1977b; 1981a) or an 'interglacial' beach (Synge, 1964; 1970a; 1975b; Mitchell, 1972; 1976) which is equivalent to the Courtmacsherry Beach of the south coast (Synge, 1977a). Warren (1985) appeared to agree with both interpretations. The weathering (*sic*) of the Clogga Till, the excavation of the Clogga Rock Platform and the accumulation of the Ballinabanoge Member has been associated with an unspecified temperate stage (Synge, 1964) and with the Ipswichian (Synge, 1975b). The presence of shell fragments and of flint and limestone clasts in the gravel unit at Seabank Point suggested that it is connected with the overlying till of Scottish/Irish Sea Basin provenance at this exposure (Synge, 1977a), although Synge (1964) described it as "...non-calcareous and ... devoid of shells..." (p. 76), and Mitchell (1972) suggested it was a beach of Ipswichian age.

Synge (1977a), Synge & Huddart (1977) and Huddart (1981a) added considerable detail to earlier descriptions of the Clogga Head exposure (table 5).

Table 5. The Quaternary sequence at Clogga Head, County Wicklow (Synge, 1977a; Synge & Huddart, 1977; Huddart, 1981a); thicknesses are in metres.

Cronelusk Member	'Irish Sea' till	> 4.00
Askintinny Member	Marine or glacio-fluvial?	1.13
Ballyduff Member	Clay-rich 'Irish Sea' till	2.00
Rock Member	Gravelly 'Irish Sea' till	2.00
Ballinabanoge Member	Gravel	not given
– – – – – – Wave-cut platform – – – – – –		
Clogga Member	Inland till	not given

Huddart (1981a) described foraminifera and ostracods from the laminated silts of the Askintinny Member; he suggested a marine origin and a possible Ipswichian age with which Synge (1981) appeared to concur. However, the fauna may be contemporaneous with the enclosing sediment, derived, or both, and the significance of the deposit and of its possible correlative at Knocknasilloge, County Wexford (T 1433) (Huddart, 1981a and b) has not been established. Thomas & Summers (1981) regarded the Knocknasilloge Member as a glaciolacustrine deposit but Huddart (1981b) was unwilling to accept this proposal.

The Clogga Head exposure may contain evidence of two temperate stages and two pre-Midlandian glacial events; however, much sedimentological and petrographic work is required before such a statement can be substantiated. This sequence is discussed further by Hoare (this volume).

A mass of beach gravel and estuarine shelly sand, measuring approximately 200 m in length, was temporarily exposed in February 1968 in Shortalstown Townland, County Wexford (T 0214) (Colhoun & Mitchell, 1971). The site lies approximately 1.5 km inside the limit of the (?)Late Midlandian ice-sheet and 6 km from the present coastline (see Hoare, this volume). The material is both underlain and overlain by till, but the sequence has been greatly disturbed by thrusting caused by glacial movements. The marine deposits contain pollen corresponding to the warmest part of a temperate stage, and elm pollen are more common than in any material that has been associated with the Gortian Stage. Colhoun & Mitchell (1971) considered that it might be of Ipswichian age, a proposal that Synge (1977a) did not accept. Colhoun & Mitchell (1971) placed the tills in separate cold stages, and Mitchell (1972) and Davies & Stephens (1978) considered the lower till is Munsterian and the upper one Midlandian in age. Nevertheless, the extent to which the stratigraphy has been disturbed or repeated is uncertain, and the sands and gravels may have been transported as a glacial 'raft'. The presence of a rather distinctive warm stage pollen assemblage, however, suggests that a satisfactory 'Ipswichian'

Table 6. The stratigraphy in the Institute of Geological Sciences (now British Geological Survey) boreholes 36/3011 (J 088658) and 36/3013 (J 118682), County Antrim, Northern Ireland (Baztey, 1978); thicknesses are in metres.

George's Island Till	< 34
Sallow Island Sands and Clays	< 30
Oglis Till	< 10

site may yet be discovered in Ireland (but see above for a discussion of the Kildromin site).

Results from exploration drilling in the Lough Neagh Basin, County Antrim, Northern Ireland, indicate that temperate deposits may separate two glacial sediments in this area (Bazley, 1978) (table 6). The Sallow Island Sands and Clays occupy an area of at least 2 km². They have yielded spores of uncertain environmental significance.

Summary and conclusions

Until the stratigraphic position of the Gortian Stage is resolved, it is necessary to take the precaution of identifying drift that is older than this event in order to demonstrate pre-Midlandian glaciation. If the Gortian is equivalent to the Hoxnian Stage, all younger glacial sediments may belong to the Midlandian Stage, although this is highly unlikely. It will not be possible to show that post-Gortian drift belongs to two cold stages rather than to one until interstratified 'Ipswichian' deposits are found or geochronometric dates are obtained on a number of beds.

However, a succession of glacial events of considerable complexity would fall within the Midlandian Stage if Gortian material represents the Ipswichian Stage in Ireland. Even greater congestion results if glaciation did not occur until the Late Midlandian. Synge (1977a) suggested that many of the glacial episodes occurred during the last 20,000 years of the Midlandian Stage, although Munsterian and pre-Gortian glaciations were also recognised by him. The lower part of the record is placed in the Munsterian Stage by those who believe that a considerable period of time was required for this complex stratigraphy to accumulate (see McCabe, 1972, and Stephens *et al.*, 1975, for example). This approach may have been responsible for delaying a serious review of the position of the Gortian Stage and of the way in which some of the glacial sequences were emplaced (McCabe & Hoare, 1978; McCabe, 1986b).

Whatever fundamental changes are made to the Irish Quaternary stratigraphic column, it will still be possible to demonstrate pre-Midlandian glaciation. The small number of sites and the lack of detailed information about the deposits prevents a comprehensive reconstruction of the events concerned.

Acknowledgements

The author is most grateful to Dr S.J.Gale for suggesting a number of improvements to this paper and to Ms R.H.A. Smith for drawing the figures.

Possible early Devensian glacial deposits in the British Isles

PETER WORSLEY

Whether or not early Devensian glaciation can be undisputedly demonstrated in Great Britain and Ireland is a problem which has as yet no clear solution. Whilst few would deny the probability that during parts of the pre Late Devensian, glacial systems were present in at least highland Britain, curiously no stratigraphic proof for glaciation in the interval between the Ipswichian Interglacial and the Dimlington Stadial, acceptable to the consensus view, has been forthcoming. However, a number of workers have taken a more positive attitude towards the somewhat ambiguous evidence and concluded in favour of the proposition. The debateable evidence bearing upon this issue will be reviewed in the following regionally based discussion.

Lincolnshire area

Eastern England is the focus of the most persistent advocacy of an early Devensian glacial event which, at its maximum extent, covered a greater area than the Dimlington Stadial glaciation. Straw (1979a) has promoted the hypothesis that early Devensian glacigenic sediments and allied landforms crop out in a relatively narrow north-south oriented belt lying immediately west of the Dimlington Stadial ice advance limit. In the south he envisages the later limit to run off-shore near to Skegness. Hence the bulge forming the most southerly Devensian glacigenic sediments on the English east coast - those covering The Wash and northwest Norfolk – are regarded as being of early Devensian age. Northwards, near to Scarborough, this outer older advance limit is thought to be overstepped by the later Dimlington Stadial limit, thus creating a diachronous limit to Devensian glaciation in this region. Further westwards in the southern Vale of York, a similarly more extensive earlier limit is postulated beyond the classical arcuate Escrick Moraine (fig. 17).

Straw's bipartite (multistadial) model of eastern England Devensian glacial history was originally formulated in east Lincolnshire (Straw, 1961a). From this area his interpretive model was extended into north Norfolk and Yorkshire. The three principal factors identified by him which led to the early Devensian concept are:

(i) two terminal morainic belts linked to a bipartite glacial stratigraphy;

(ii) prominent contrasts in the form and pattern of meltwater channels along the eastern margins of the Lincolnshire Wolds;

(iii) glacial terrain with differing morphological expressions, one being dissected and weathered, the other more continuous with fresher looking constructional landforms.

That these mainly geomorphologically-based subdivisions are observable is difficult to deny and hence the crux of the debate revolves around the chronological significance of the landform contrasts. In Straw's judgement, the above criteria cannot be satisfactorily explained by arguing that the inferred time difference is of a few thousand years only, and hence can be accomodated within a single stadial.

Morphological contrasts alone do not, of course, give quantitative values on the time differential and recourse is necessary to data which involved other than morphostratigraphic indices. The key, according to Straw, lies in the record of the proglacial lakes which were thought to have arisen in two separate stadials as a consequence of ice blocking eastward directed drainage. As is evident from fig. 17 the more extensive Devensian ice of phase 1 blocked both the Humber and Wash gaps to create the high-level proglacial lake Humber/Fenland. Altimetric considerations of the lake perimeter suggest that a col on the south shore permitted overspill into the River Waveney valley in central East Anglia. Phase 2 ice left the Wash unobstructed and a lower-level lake resulted. Since all modern workers agree that the latter lake is of Dimlington Stadial age it need be considered no further here.

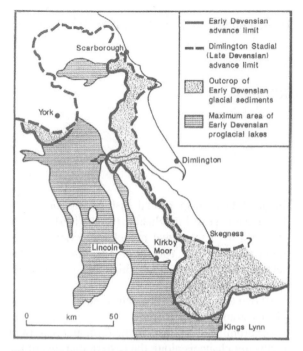

Figure 17: Early and Late Devensian ice limits in eastern England as postulated by Straw (1979) and modified from his fig. 3.1. The possible outcrop of early Devensian glacial deposits is indicated together with the area inundated by the phase 1 proglacial lake at its maximum level.

Unfortunately, as yet the "history (of the lakes) have never been fully elucidated" (Straw, 1979a: 30) and consequently reliance has to be placed on generalised relationships. Two important localities for placing stratigraphical limits on the postulated Lake Fenland are Wretton in Norfolk and Tattershall in Lincolnshire where respectively the Rivers Wissey and Bain enter the modern Fens. At Wretton (20 km south of Kings Lynn) a fluvial sequence forming a terrace at 6 m O.D. overlies undisputed Ipswichian deposits. The intraformational biostratigraphy associated with the gravels indicates an early Devensian age (West *et al.*, 1974), and Straw (1979a) suggests that the uppermost unfossiliferous metre of gravel immediately below the terrace surface relates to the Lake Fenland submergence episode but gives no substantiation. Similarly in the Tattershall area a fluvial gravel forming a low terrace of the Bain overlies Ipswichian sediments. Biogenic-rich horizons in the post-Ipswichian gravels are associated with radiocarbon dates spanning 28,000 - 46,000 B.P. and no intraformational lacustrine sediments have been identified. Hence it is necessary to combine data from both sites in order to postulate that any lake episode must antedate 46,000 B.P. and postdate the early Deven-

sian as defined on biostratigraphy, effectively bracketing the 50,000 - 60,000 timespan.

The Tattershall area is crucial to Straw's interpretation since here he recognises a delta (now dissected) which he envisages as forming in a high level (30 m O.D.) Lake Fenland coeval with the maximum extent of Devensian ice on the east coast. This lake would have flooded the lower Bain Valley. Quarry exposures in the alleged delta at Kirby Moor reveal some 6 m of near horizontally stratified sands with occasional angular flint clasts resting on pre-Devensian Wragby Till. Sedimentologically, the sands are indicative of an extensive low-relief fluvial depositional environment as typified by a distal sandur. No evidence for standing water is present and therefore the character of the exposed sequence is not compatible with deltaic deposition. An additional problem which confronts the delta hypothesis is that early Devensian sedimentation at Kirby Moor would necessitate the prior total infill of the lower Bain valley and its later removal some time before 46,000 B.P. when the biostratigraphy indicates valley bottom aggradation. Indeed the recent sand and gravel resource assessment survey (Power & Wild, 1982) has demonstrated that the Kirby Moor succession is part of an extensive spread of sands which appear to be related to the last deglaciation of the area (pre-Ipswichian). On the basis of present knowledge it has to be concluded that the landforms and sediments in the Tattershall area lend no support to the palaeolake Fenland concept.

A pointer to the likely ultimate fate of the early Devensian lake hypothesis can be derived from Gaunt's work in the southern Vale of York - Humberhead area. Gaunt has investigated the Devensian ice limits and lake relationships in this key region and supports the high- and low-level lake model. However, in his most recent synthesis (Gaunt, 1980) he made no mention or comment on a possible early Devensian glacial event and took the view that the "glacial activity in the region was confined to only a few millenia" (p. 122) and thus considered the two lake levels to be broadly contemporaneous. Accordingly, some mechanism other than at least a 30,000 years time differential would appear to be necessary to account for the terrain contrasts which so impressed Straw. The case for an early Devensian glaciation in eastern England remains unproven.

Cheshire

At Oakwood Quarry near to Chelford in east Cheshire, extensive exploitation of the Chelford Sands

Formation has exposed the normally buried unconformity beneath the formation. This reveals that it
takes the form of a shallow valley trending and falling
from the SE to NW. Although cut primarily in Triassic
Mercia Mudstones pre-Chelford Sands Pleistocene
sediments mantle most if not all of the bed-rock
surface. Immediately below the sands a lag of erratic
materials is common and these frequently exhibit
evidence of wind abrasion. The lag rests on a diamict
which has informally been called Oakwood till
(Worsley *et al.*, 1983). This unit is considered to be
glacigenic but, like the till associated with the Stockport Formation which overlies the Chelford Sands, it
is of variable character being laminated in parts and at
one locality contained a ventifact which must have
been derived. At least a degree of resedimentation is
suspected but very restricted exposure makes assessment difficult. Since there is undisputed evidence that
the Oakwood till antedates the pre-sumed early Devensian Chelford Interstadial lying within the Chelford
Sands, there is great interest in its age but for the
moment this must necessarily be speculative.

A single trial pit dug into the quarry floor above the
palaeovalley axis revealed that the Oakwood till
overlay a thin fluvial gravel succession intervening
between it and little weathered Mercia Mudstone.
Although the gravels contained biogenics they were
not particularly diagnostic for chronological purposes and simply indicated a former harsh open
environment. Amino raceminisation analyses on
Lymnaea peregia from the biota yielded D:L ratios
consistent with a general early Devensian age but
"the data (...) cannot be regarded as definitive"
(Bowen *et al.*, 1988). However, the latter workers do
show some favour towards the notion that the Oakwood till is of early Devensian age and "adds to the
considerable body of evidence now accumulating for
early Devensian glaciation in the British Isles". Yet,
Worsley *et al.* (1983) considered this possibility with
respect to the Oakwood evidence but concluded that
it was more likely to be pre-Ipswichian in age. This
judgement was without knowledge of the amino-acid
data but even if the molluscan fauna in the gravels is
ultimately shown to have an early Devensian age, the
Oakwood till overlying them at the trial pit site is not
necessarily Devensian since it may well have been
derived by mass-wasting processes from a source
higher on the palaeovalley slopes. A number of separate till units are known from beneath Chelford
Sands (Worsley, 1985) but none can be attributed to a
specific stadial or stage. It is here suggested that on
balance there is currently no good evidence for an
early Devensian glaciation in Cheshire.

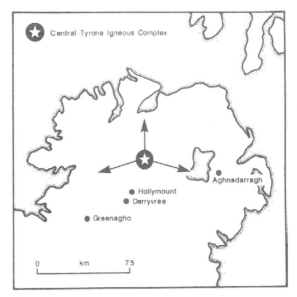

Figure 18: Location of some key localities relating to the early
Devensian in Northern Ireland.

Ireland

At present three Irish localities (see fig. 18) are
known where it is believed that pre-Late Midlandian
(Devensian) biogenic sequences lie sandwiched between two tills. Generalised logs from these sites are
shown in fig. 19 together with possible correlations.
For the past decade, the lower till unit – the Fermanagh Stadial tills – have normally been accepted by
Irish workers as the products of an early Midlandian
glacial event.

The first inklings of such a concept in an Irish
context appear to arise from the study of Colhoun *et
al.* (1972) of exposures at Derryvree, near Maguiresbridge in County Fermanagh. A road cut through a
drumlin revealed the presence of two distinct till units
separated by fine sands and organic silts containing a
fossil biota indicative of 'full glacial aspect'. A finite
radiocarbon date of 31,000 B. P. on the organics suggested a late Middle Devensian age for at least part of
the ice-free interval between deposition of the two
tills. An important observation was that the upper
surface of the lower till was unweathered; nevertheless the investigators were not inhibited by their
conclusion that the lower till was probably pre-
Ipswichian and unlikely to be early Midlandian.

A second site with a broadly comparable lithostratigraphy was later discovered at Hollymount just
over 1 km to the northwest of Derryvree. However,
this latter site, a natural river cut, did not expose the
lower part of the succession. The base of the inter-till
biogenic-rich silty sediments and upper part of the

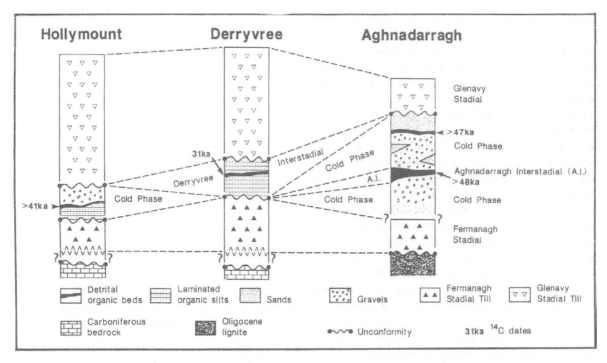

Figure 19: Outline logs of the stratigraphies at the three localities in Ulster known to have inter-till biogenic sequences, with an emphasis on the potential early and middle Devensian (adapted from McCabe, 1987: fig. 11). Note that the base of the Pleistocene is not known at Derryvree and Hollymount and the contact between the lower till and overlying sequence at Aghnadarragh is of doubtful status.

lower till had to be proved in a borehole. McCabe *et al.* (1978: 81) on the basis of samples from the borehole have concluded "that the surface of the till was unweathered and that there was no evidence of a hiatus between the till and the overlying laminated series". A radiocarbon date on the detrital organic material was infinite – 42,000 B.P. and palaeoecological study of the contained fossils suggested a 'tundra' environmental attribution. In their discussion, McCabe *et al.* considered that the lower till was probably lateral equivalent to that at Derryvree but the possibility of the Hollymount lower till antedating that at Derryvree was not discounted. As to the age status of these lower tills, they were non-committal recognising that either a pre-last interglacial or early Midlandian date was possible.

However, one of the authors, had previously been more emphatic for Mitchell (1977) offered the view that the lower part of the Hollymount succession which he gave as till, varved clay, lacustrine deposit with vegetable debris was "deposited without interruption" giving no opportunity for the till either to be subaerially weathered, or to be disturbed by frost action before the arrival of the vegetable debris. Despite the infinite radiocarbon date he rejected the

possibility that the tundra episode might be older than the last interglacial in order to declare that the lower till "must be of Devensian age" and noted that in Britain there is no accepted record of a till so early in the Devensian. He speculated that the age of the glaciation was early Middle Devensian c. 45,000 - 50,000 years B.P. and followed a major permafrost episode after the Chelford Interstadial of England. Later in Mitchell (1981) he stated "both of these sites appear to give strong indications that in Ireland ice did form during the early part of the Midlandian Cold Stage". Here we should note that both lower tills were implicitly regarded as being directly related and the lack of weathering at the top of the Derryvree lower till (recalling the +10,000 years age hiatus) was not regarded as a problem requiring comment.

A spectacular 17 m thick Pleistocene succession has recently been discovered at Aghnadarragh, close by the eastern shore of Lough Neagh. The exposure cropped out in the walls of a large trial pit excavated to evaluate the commercial potential of the underlying Oligocene lignite. This has been fully described by McCabe *et al.* (1987) as "the most complete (Devensian) sequence known in Ireland". The latter paper was preceded by preliminary reports by Mc-

Cabe *et al.* (1986) and summary accounts in Bowen *et al.* (1986) and McCabe (1987). Reacting to the first two papers published – McCabe *et al.* (1986) and Bowen *et al.* (1986), Worsley (1988) challanged the chronological conclusions concerning the lower part of the succession, in particular he rejected the claim for an early Devensian glaciation in the light of comparisons with the British, French and North Atlantic successions. In the event, this alternative opinion did not require amendment in the light of the data given in the two subsequently published reports.

Immediately above the lignite lies a tectonised lower till containing a small component of Tyrone Igneous Complex material which encouraged correlation with the lower tills of Derryvree and Hollymount. Just like the situation at the latter two sites, the Aghnadarragh lower till was seen to have a gradational and conformable passage upwards into nonglacigenic sediments and consequently was interpreted as lacking a hiatus. In the overlying sequence allochthonous peats and wood enabled a palaeoenvironmental reconstruction of a boreal woodland dominated by birch, pine and spruce (Aghnadarragh Interstadial). In contrast an alternative interpretation of the Aghnadarragh flora has been proposed by Mitchell (1986a) giving greater credence to the presence of *Taxus* seeds and *Corylus* nuts. As a result he believes that the flora represents the initial deterioration of a warm stage climax woodland but does not specify which other than stating that it must be Upper Pleistocene because of the presence of *Mammuthus* in the underlying sediments.

Infinite radiocarbon dates and the biota characteristics clearly pointed towards a pre-Middle Midlandian age and a broad similarity with the Chelford Interstadial flora and fauna persuaded McCabe *et al.* (1987) that they were correlates. It follows from this conclusion that the underlying lower till was assigned to an early Midlandian age in view of the conformable succession without any evidence for an intervening sub-aerial weathering phase. This line of reasoning is, of course, the same as that finally adopted at the Derryvree and Hollymount sites and the chronological non sequitar which arises from this has been noted above. The result is that if the reasonable assumed equivalence of the three lower tills is accepted and dated as pre-Chelford then this demands the presence of an even greater time hiatus at Derryvree. If the arguments for assigning an age of some 100,000 years B. P. to the Chelford Interstadial given by Worsley (1986, 1988) are accepted then this might be of the order of 70,000 years.

As all three sites require differing temporal values

to the 'gradational passage' from till up into nonglacigenic sediments, we have grounds for suspecting that even at Aghnadarragh a hidden hiatus might be present. This arises since uncertainty persists with regard to both the status of the stadial - interstadial transition and the validity of the correlation between the Aghnadarragh and Chelford Interstadials. By way of example we may note McCabe *et al.*'s (1987) reasoning that the Aghnadarragh Interstadial *per se* must have persisted for a sufficiently long period for otherwise there could not have been a migration of *Picea* into Ulster. This inference arises since the acceptance of a continuous sequence from the preceding stadial neccessitates time for the build-up and decay of a considerable ice mass over Ulster, otherwise the lower till could not have been emplaced during the 'Fermanagh Stadial'. During this time forest trees must have been displaced southwards. If, however, the till is pre-Devensian in age this specific problem of tree migration rates is eliminated.

It is difficult to avoid the conclusion that the 'reinforcement syndrome' has exerted an unwarranted influence on the perceived chronological status of the lower till (also note the erroneous inclusion of Greenagho as a fourth Ulster site with a till-biogenic-till sequence in McCabe (1987: fig. 11; cf. Dardis *et al.*, 1985). Perhaps the comment by McCabe *et al.* (1987) that at Aghnadarragh "it is recognised than nonsequences are present and represent substantial periods of geologic time" has a greater applicability than those workers envisioned. Thus, we may suggest that no unambiguous evidence is currently known in Ulster which confirms an early Devensian glaciation in Ireland. Assigning the 'Fermanagh Stadial' to an earlier cold stage alleviates the stratigraphical anomaly.

Scotland

Unfortunately, at present, there is a paucity of convincing stratigraphic data favouring the identification of early Devensian glacial deposits. A number of workers have alluded to their occurrence and generated a number of speculative hypothesis. However, a critical assessment of the evidence concurs wholly with the conclusion in Bowen *et al.* (1987: 322) "no glacial sediments have been directly dated to the Early Devensian in Scotland and the concept of an Early Devensian glaciation remains unproven".

Note

PRIS Contribution nr 47.

Late Devensian glacial deposits and glaciation in Scotland and the adjacent offshore region

DONALD G. SUTHERLAND

In the Middle and Late Pleistocene Scotland and the adjacent offshore areas were glaciated by ice sheets on at least four and possibly five distinct occasions (Bowen et al., 1986). The most recent of these ice sheets is of Late Devensian age and it is to this period of glaciation that the great majority of the glacial deposits of the Scottish Lowlands and much of the Scottish Highlands can be attributed. Widespread glacial and glaciomarine deposits of this age also occur on the neighbouring shelves. The Late Devensian, as conventionally defined (26,000 - 10,000 B.P.; Mitchell et al., 1973), encompasses two phases of glaciation in Scotland - the last ice sheet (the main build-up and retreat of which broadly accords in time with the Dimlington Stadial of Rose, 1985) and the more minor Loch Lomond Readvance which broadly coincides with the Loch Lomond Stadial during the last c. 1000 years of the Late Devensian. As the Loch Lomond Readvance is discussed elsewhere (Gray & Coxon, this volume) the present chapter will only consider the last mainland ice sheet and the contemporaneous island ice caps.

Chronology

There are only a limited number of radiometric dates that place direct constraints on the chronology of the last ice sheet. The relevant dates are given in table 7. None of these dated samples can be directly linked to the presence of the ice sheet. However, these dates on materials (organic debris, marine shells, speleothems) that accumulated in a non-glacial environment provide evidence of ice-free conditions and hence limiting ages on glacial events. A further relevant date is that from borehole 74/7 in the Marr Bank Formation which, if correct, supports the attribution of that formation to the Late Devensian which implies that the co-eval Wee Bankie Formation represents the limit of the last ice sheet in the west central North Sea (Sutherland, 1984a; Stoker et al., 1985b).

Other important sites are Crossbrae Farm (Hall, 1984) and Toa Galson (Sutherland & Walker, 1984) (fig. 20) the evidence from both of which indicates that neither of these localities was glaciated during the Late Devensian.

Stratigraphic evidence from the Southern Uplands and the Central Lowlands indicates that the initial expansion of the ice sheet was centred in the western Highlands and that this ice advanced to within c. 20 km of the Southern Uplands ice centre. As the glaciation proceeded, the Southern Uplands ice grew in strength ultimately excluding the Highland ice from large areas it once occupied in the Central Lowlands.

The retreat of the ice sheet around the coasts has been traced in greatest detail through the sequences of raised shorelines that formed progressively on deglaciation. The most detailed sequence of such shorelines occurs in SE Scotland (Sissons et al., 1966; Sissons, 1983) and an approximate chronology for this period of ice retreat can be calculated on the basis of shoreline gradient against age relationships (Andrews & Dugdale, 1970; Sissons, 1976c: 120). Such a calculation (using the latest information on the ages of younger shorelines) suggests that the five shorelines identified by Cullingford & Smith (1980) along the east coast of Central Scotland formed progressively between c. 16,000 B.P. and c. 14,000 B.P. (fig. 20). These shorelines (and the later Main Perth Shoreline) were formed as the high-arctic Errol Beds and their offshore equivalent, the St. Abbs Formation, were being deposited. These sediments postdate the Marr Bank Formation and hence, if the radiocarbon sample previously mentioned from borehole 74/7 is correct, started to form sometime after c. 17,700 B.P. The arctic sediments ceased to be deposited on the general warming at the opening of the Lateglacial Interstadial at approximately 13,000 - 13,500 B.P. and hence the sediment and shoreline chronologies broadly agree on the timing of this phase of deglaciation. A similar shoreline gradient against age calculation but based on less well estab-

Table 7. Radiometric dates on samples providing limiting age control on the Late Devensian ice sheet.

Locality	Lab. No.	Material	Age
(a) Samples pre-dating the ice sheet			
Bishopbriggs	GX-0597	bone	$27,550^{+1370}_{-1680}$
Inchnadamph	SRR-2103	bone	$25,360^{+810}_{-740}$
	SRR-2104	bone	$24,590^{+790}_{-720}$
Tolsta Head	SRR-87	peat	$27,333 \pm 250$
Garrabost	SRR-2367	marine shell	$26,300 \pm 320$(o)
			$23,000 \pm 230$(i)
Assynt	SU1-80A	speleothem	$26,000 \pm 3000$
Assynt	AU2-80	speleothem	$30,000 \pm 4000$
Assynt	SU12-80B	speleothem	$26,000 \pm 2000$
(b) Samples most closely post-dating ice sheet			
Hawthornhill	SRR-2009	marine shell	$13,775 \pm 135$
Shiells	SRR-391	organic debris	$13,640 \pm 130$
Inchinnan	SRR-925	marine shell	$13,095 \pm 265$
Loch Ette-ridge	SRR-304	lacustrine	$13,150 \pm 390$
Loch an t' Suidhe	SRR-1805	lacustrine	$13,140 \pm 100$
Cam Loch	SRR-253	lacustrine	$12,940 \pm 240$
Roberthill	Q-643	wood	$12,940 \pm 250$

The speleothem samples have been dated by U-series decay, the other samples by radiocarbon. References to the samples are: Rolfe (1966), Lawson (1984), von Weymarn & Edwards (1973), Sutherland & Walker (1984), Atkinson *et al.* (1986), Browne *et al.* (1983), Aitken & Ross (1982), Browne *et al.* (1977), Sissons & Walker (1974), Lowe & Walker (1986), Pennington (1975), Bishop & Coope (1977). The reliability of many of the radiocarbon dates is discussed in Sutherland (1980, 1986).

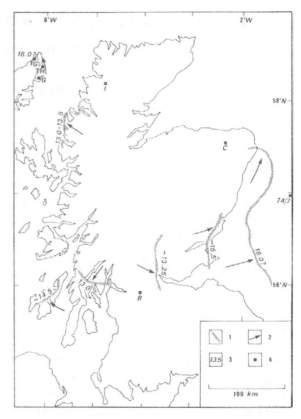

Figure 20: Estimated ages of various ice-marginal positions of the Late Devensian ice sheet in Scotland. 1 = approximate position of ice margin at indicated time; 2 = general direction of ice flow with respect to indicated ice margins; 3 = estimated ages in thousands of years B. P.; 4 = sites mentioned in the text as follows: TG = Toa Galson; TH = Tolsta Head; G = Garrabost; I = Inchnadamph, Assynt; C = Crossbrae Farm; B = Bishopbriggs.

lished data can be made for certain shorelines on the west coast which suggests that Shoreline L1 of Dawson (1982) formed approximately 14,500 B.P. (fig. 20). This indicates that parts of western Jura and north-western Islay were deglaciated by this time.

For the greater part of the deglacial phase presently available evidence suggests that the ice sheet retreated progressively with no major stillstands or readvances. An exception occurs in Wester Ross where an extensive end moraine system has been mapped (Robinson & Ballantyne, 1979; Sissons & Dawson, 1981). This is as yet undated but pre-dates the radiocarbon date of $12,810 \pm 155$ B. P. (Q-457) at Loch Droma (Kirk & Godwin, 1963) and has been associated with a shoreline with a gradient of be-

Plates 6–7

6. Streamlined forms and striae on peridotite. Note open fractures caused by stress relief (glacial unloading?). Locality (NM 373990) west of Loch Bealach Bhic Neill, Rhum (Photograph: J.D. Peacock, 1975).

7. Section 6 m high in weakly bedded brown lodgement till with locally imbricated clasts. Bed of gravelly till by hammer. Holm (NB 453307) near Stornoway, north Lewis (Photograph: J.D. Peacock, 1977).

6
―――
7

tween 0.33 and 0.39 m/km⁻¹. This shoreline is partly represented by bedrock erosional features which has been interpreted as implying formation in a severe climate (Sissons, 1982b), and the abandonment of this shoreline and the corresponding moraines may be tentatively dated to 13,000 - 13,500 B. P. (fig. 20). Possibly correlative end moraines also occur in Easter Ross (Sissons, 1982a; Sutherland, 1984a). Along the west coast stillstands during ice sheet retreat have been interpreted from breaks in the marine limit near the mouths of many of the sea lochs (Peacock, 1970; Sutherland, 1981a) and in the SW Highlands such a stillstand or minor readvance has been dated to approximately 13,000 B. P. (Sutherland, 1981a). These breaks in the marine limit may be the result of climatic change influencing the retreat of the ice sheet or they may solely relate to topographic control on ice flow near the mouths of sea lochs (Sutherland, 1984a).

The chronology of the last ice sheet glaciation may therefore be summarised as follows. Somewhat prior to 25,000 B. P. much of Scotland was ice-free, though the climate was severe (cf. Birnie, 1983) and glaciation of part of the Highlands may reasonably be inferred. The ice sheet centred in the western Highlands expanded after this and as the glaciation proceded the more southerly ice centres gained in strength in comparison to the more northerly while ice-free areas persisted on the northern fringes of the ice sheet. There is thus a possibility that the limits of the ice sheet were not reached synchronously on its northern and southern margins. At no place, however, is the maximum of the ice sheet dated. Along the eastern side of the ice sheet deglaciation may be inferred to be underway between 16,000 and 17,000 B. P. and despite a very severe prevailing climate progressive deglaciation of almost all lowland areas continued until, at around 13,000 B. P., the ice mass was largely confined to the western Highlands (fig. 20). At this time, either due to topographic control

and/or climatic change, stillstands or minor readvances occurred on various margins of the ice sheet. The extent of deglaciation during the Lateglacial Interstadial cannot be established as the ice everywhere retreated to within the limits attained during the subsequent Loch Lomond Readvance. The great majority of ice sheet build-up and disintegration thus appears to have taken place in no more than 12,000 years.

The maximum extent of the ice sheet

Dated sites in the Western Isles, North East Scotland and the North Sea Basin have provided limiting evidence on the extent of the Late Devensian ice sheet in Scotland. However, the precise position of the ice margin reached at the maximum is only known for a relatively small part of the ice sheet. The longest established ice margin is that of the North Sea Basin (fig. 21) at the junction of the Marr Bank Formation and the Wee Bankie Formation (Thomson & Eden, 1977; Sutherland, 1984a; Cameron *et al.*, 1987). This limit very probably corresponds with the limit of the 'Red Series' drift in eastern Aberdeenshire (Jamieson, 1882; Synge, 1956; Hall, 1984) but there is no direct dating control that confirms this correlation. An ice margin dated to the Late Devensian has also been established for the Outer Hebrides ice cap in the north of the Isle of Lewis (Sutherland & Walker, 1984) and this also places constraints on the margin of the mainland ice sheet in the same region (fig. 21). The offshore limit of the ice sheet along most of the western continental shelf has yet to be established, however.

No clear limits have been reported for the ice lobe that occupied the Moray Firth area (fig. 21). On the southern margin of this lobe it is not known whether the 'Blue-Grey Series' glacial deposits are of Late Devensian age (Synge, 1956; Hall, 1984) or older

Plates 8–10

8. Till with angular blocks of grey gneiss. Samala (NF 800622), Baleshare, North Uist (British Geological Survey Photograph D2898, 1979).

9. Kirkhill Quarry, SE face. Upper till resting unconformably on grey gelifluctate complex 3. Beneath is the lower till which in turn overlies gelifluctate complex 2, containing many blocks of angular felsite. The prominent grey band is the lower buried soil, developed on the lower sands and gravels which, in turn, rest on felsite bedrock (Photograph: 1979).

10. Kirkhill Quarry, lower buried soil, E face. The base of gelifluctate complex 2 rests on a thin transported organic horizon. Below this is the truncated lower buried soil, with an upper bleached horizon and lower iron-rich horizon, with iron pan, which is developed on the lower sands and gravels (Photograph: 1979).

8
---
9 | 10

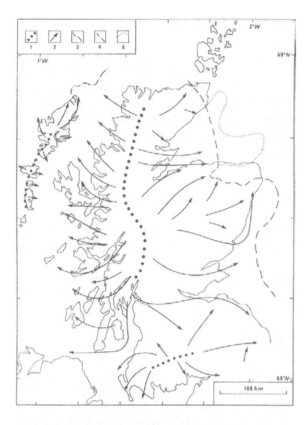

Figure 21: Ice flow directions, major ice sheds and approximate ice-marginal positions at maximum of Late Devensian
ice sheet. 1 = ice shed; 2 = ice flow direction; 3 = margin
according to Sutherland (1984a); 4 = margin according to
Sutherland (unpublished); 5 = alternative margin in Moray
Firth area according to Cameron et al. (1987).

(Sutherland, 1984a) whilst on the northern side of the
lobe the question of the age of the last glaciation of
much of Caithness and Orkney is yet to be resolved
(Sutherland, this volume). Similarly, while it is very
probable that the last glaciation of Shetland was of
Late Devensian age, the extent of the local ice cap on
these islands has not been established (Flinn, 1978a;
Cameron et al., 1987; Sutherland, this volume).

The distribution of erratics in the glacial deposits
attributed to the Late Devensian ice sheet indicates
clearly that the ice mass was not a simple dome but
was polycentric with a number of initially independent ice bodies which subsequently coalesced. The
major centre of ice dispersal was the western Highlands where an elongate ice shed extended from
Sutherland in the north to the Cowal Peninsula in the
south (fig. 21). In the west, the ice flowing outwards
from this ice shed abutted but could not overwhelm
independent ice centres on the islands of Skye, Mull

and Arran as well as the ice cap that built up on the
Outer Hebrides (Sissons, 1967a; 1976c; Sutherland,
1984a). The second largest ice mass became established in the western Southern Uplands with an ice
shed orientated approximately east-west from the
Merrick across the Loch Doon Basin towards the
Lowther Hills (Cornish, 1982) (fig. 21). In the eastern
part of the country more minor ice centres appear to
have existed in the Cheviot Hills, the South East
Grampians and the Cairngorms.

Despite the complex nature of the ice body and the
often irregular topography on which the ice sheet
flowed, ice-flow patterns were relatively simple (fig.
21) which implies that the ice, at its maximum extent,
was thick enough to flow largely independently of the
topography. Striations and erratics as high as 1,100 m
O.D. in the Ben Nevis and Glen Coe areas and
Rannoch granite erratics at over 1,000 m O.D. in the
Central Grampians have been considered to support
the idea that the last ice sheet overtopped all the
mountains in the SW Grampians. However, the production of these features by the last ice sheet, although probable, is not proven. Similarly, the occurrence of Loch Doon granite erratics on the summit of
the Merrick (842 m) in the Southern Uplands
strongly suggests that the ice sheet here over-topped
these mountains.

The evidence for a restricted northern and eastern
margin to the ice sheet makes it probable, however, in
contrast to earlier models based on an ice sheet
extending to the edge of the continental shelf (Boulton et al., 1977; Andersen, 1981) that some of the
mountain summits away from the central areas of ice
dispersal would have stood above the ice sheet as
nunataks. Thus, although the altitudinal distribution
of periglacial features resulting from macrogeliva-
tion on mountain summits in the relevant areas still
remains to be mapped in detail (Ballantyne, 1984)
certain authors have recognised marked contrasts
across narrow altitudinal bands between ice-moulded
and periglacially modified rock surfaces that may be
explicable in terms of an upper limit of an ice sheet.
Thus Godard (1965: 605) has suggested such 'periglacial trim lines' occur, for example, at 950 m on An
Teallach, 870 m on Ben More Assynt and 680 - 700 m
on Ben Loyal - Ben Hope. As with the distribution of
high-level erratics and striations there is no direct
method of dating these features and their attribution
to the last ice sheet cannot be unequivocally established.

Much work remains to be done before the extent of
the Late Devensian ice sheet is known in detail but
the recent evidence is indicative of a much smaller

ice mass than was considered likely even a few years ago (Boulton *et al.*, 1977; Andersen, 1981; Price, 1983). The main ice body probably had a maximum ice thickness of little more than 1,300 m in the ice shed area of the SW Grampians and a flow-line distance of less than 200 km to the ice margin in the west-central North Sea Basin. Similar figures of the northwestern margin of the ice-sheet are a maximum ice thickness of c. 880 m and a flow-line distance to an ice margin to the north of the Outer Hebrides of c. 120 km. Glaciological and palaeoclimatic models must reconcile these figures with the known extent of the southern margin of the ice mass in South Wales, the Midlands and the East of England.

Glacial deposits

The great majority of the glacial deposits on the Scottish land area and significant thicknesses of Quaternary sediments offshore to both east and west can be attributed to the Late Devensian ice sheet. There are, however, relatively few sedimentological studies of these deposits that would assist in the development of glaciological models of the nature of ice-sheet advance or retreat.

A number of studies have been published in which the quantitative variations in erratic content of tills ascribed to the Late Devensian ice sheet have been established. The most detailed such study was that of Kirby (1968) working in the Midlothian Basin to the south of Edinburgh who concluded, on the basis of such evidence combined with analyses of the fabrics of the tills (Kirby, 1969a), that an early advance of Highland ice into the Midlothian Basin was succeeded without any hiatus by Southern Upland ice from the SW. During deglaciation, Kirby considered that an uppermost till (the Roslin Till) represented a readvance of Highland ice into the area vacated by the retreating Southern Uplands ice but detailed analyses of the facies variations in the Roslin Till led Martin (1981) to conclude that the Roslin Till was comprised of various flow till units deposited as part of a single deglacial sequence. This interpretation is in accord with certain aspects of the outwash terrace sequences that developed between the two ice masses during deglaciation (Kirby, 1969b; Sutherland, 1984a).

Elsewhere in the south Central Lowlands, McLellan (1969) and Holden (1977) have provided pebble count and fabric evidence from tills that has supplemented and made more precise the details of the interaction of the Highland and Southern Uplands ice

masses which had previously been deduced from the qualitative observations on the distribution of particular erratics as recorded in the relevant Geological Survey Memoirs. In the Rhins of Galloway, Kerr (1982) on the basis of erratic content, fabric, particle-size analyses and colour has differentiated two tills, the lower of which he has ascribed to an initial period of glaciation by Highland ice and the upper, deposited without any break in sedimentation, related to Southern Upland ice.

Within the area of influence of Southern Uplands ice, Cornish (1982, 1983) has studied the transport of erratics in the ice-shed zone. He found that an initial movement of locally nourished glaciers into the valleys of the ice-shed area was succeeded, as the ice built up to inundate these valleys, by an outwards flow to both north and south that, in the vicinity of the ice shed itself, produced relatively little redistribution of erratic material. On the eastern margin of the Southern Uplands ice in the Tweed valley, Kerr (1978) was able to demonstrate much less restricted ice flow to the east on the basis of the transport of specific erratic types.

The influence of Highland ice in the Central Lowlands has been studied in detail by Menzies (1981) in the Glasgow area. On the basis of variation in erratic content relative to particular rock outcrops and the particle size of tills, Menzies argued that two till units, previously distinguished by colour and ascribed to separate glacial episodes, were different facies of the same till deposited during the Late Devensian. A detailed study of erratic transport related to a single source has been reported by Shakesby (1978) who demonstrated the diminution down-ice in size and frequency of erratics from the Lennoxtown essexite outcrop to the east of Glasgow.

In North East Scotland the complex drift sequence to the north of and inland from Aberdeen has been studied by Murdoch (1977; and in Munro, 1986). On the basis of colour variations, particle size analyses, fabric and erratic content as well as field relations, three separate tills were recognised and all attributed to various ice mass movements during the Late Devensian. A lack of dating control leaves the attribution of all these tills to the Late Devensian open to dispute (cf. Hall, 1984; Sutherland, 1984a).

By analogy with other glaciated areas it might be presumed that glaciotectonic deformation of drift and underlying bedrock would be relatively common in Scotland but there are few examples in the literature. In East Lothian to the east of Edinburgh, open-cast coal mining revealed large anticlinal structures with allied faulting and thrusting which has affected the

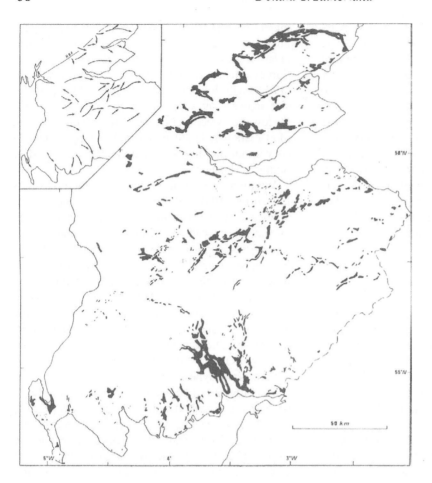

Figure 22: Distribution of glacio-fluvial sands and gravels in southern Scotland. Numerous sources. Inset: meltwater routes during ice sheet retreat as indicated by distribution of sands and gravels and meltwater channels (see Sutherland, 1984a). Late reversal of meltwater drainage in lower Clyde valley represented by broken arrow.

Carboniferous strata to a maximum depth of 35 m (McAdam & Tulloch, 1985). To the south and east of this area a number of particularly large erratics of Carboniferous strata have been located, including one measuring over 500 m by 400 m in plan (Clough *et al.*, 1910) and the glaciotectonic features seen in the open-cast workings may be considered a stage in the process of detaching such large erratics from bedrock. Much smaller scale glaciotectonic structures involving only drift have been described from the Glasgow area by McMillan & Browne (1983).

The deglacial phase of the Late Devensian ice sheet is particularly notable for the landforms and sediments produced by meltwater. Fig. 22 shows the distribution of glaciofluvial sands and gravels in southern Scotland as well as the major directions of meltwater drainage that may be deduced from the distribution and internal characteristics of these deposits as well as associated meltwater channels (cf. Sutherland, 1984a: fig. 9). A considerable research effort has gone into mapping the geomorphological expression of these deposits (e.g. Sissons, 1958b;

Kirby, 1969b; Sugden, 1970; Young, 1974) and into establishing their dimensions as a basis for aggregate resource assessment (e.g. Goodlet, 1970; Paterson, 1977; Laxton & Nickless, 1980; Aitken *et al.*, 1984). However, relatively few studies have considered the sedimentology of these deposits and consequently the processes involved in their formation, which have important consequences for understanding the pattern and mode of deglaciation, are only poorly known.

An example of the wide variation of interpretation is that of the Carstairs Kames. These were described last century (Geikie, 1863) and subsequently have been variously interpreted as moraines formed at the margin of Highland ice (Charlesworth, 1926), as meltwater deposits laid down within a dead-ice mass between active Highland and Southern Upland ice masses to the north and south respectively (Macgregor, 1927), as subglacial eskers draining to the east during the Perth Readvance (Sissons, 1961b; 1964a) or deglaciation of the Southern Uplands ice sheet (McLellan, 1969), as being partly a terminal moraine and partly kames formed within dead ice related to the retreat of the Southern Uplands ice (Goodlet, 1964), or as supraglacial fluvial deposits that accumulated around ice-cored moraines (Boulton, 1972). Each of these interpretations has its own implications for the mode of deglaciation of central Scotland and more detailed sedimentological studies are necessary in order to understand the genesis of this and other, similar, deposits elsewhere in Scotland.

The limited amount of published sedimentological work on the stratified sediments produced during the deglacial phase of the last ice sheet emphasises the diversity of sedimentary environments relating to this period. In North East Scotland, Thomas (1984c) and Thomas & Connell (1985) have described the facies variations produced where a subglacial esker discharged subaquatically into an ice-dammed lake. In that example the glaciolacustrine sediments were dominantly clays and the coarse-grained esker sediments were largely deposited very close to the ice margin. This contrasts with the large glaciodeltaic sequence in the upper Avon Valley in Central Scotland analysed by Martin (1980). In this latter area the apparently much greater supply of sand-sized and coarser sediments resulted in these being the dominant sediment (up to 35 m thick) of the glaciolacustrine environment with only a minor contribution of silts and clays at the base of the sequence. One other example of esker sedimentation has been published by Terwindt & Augustinus (1985) who concluded that the bulk of the sediment in the esker had accumulated laterally rather than longitudinally and partly in sub-aerial and partly in subaquatic conditions.

In the coastal and offshore areas during deglaciation glaciomarine deposits accumulated. At greater distances from the ice margins fine-grained laminated sediments with dropstones containing low-diversity high-arctic marine fauna have been widely reported (Peacock, 1975b; Thomson & Eden, 1977; Paterson *et al.*, 1981; Stoker *et al.*, 1985b). There are few detailed descriptions of the sediments deposited in the more complex proximal glaciomarine environment. Rose (1980a) has described a glaciomarine till, interpreted to have been deposited from beneath a floating ice margin in the Clyde Estuary. It is underlain by a lodgement till and overlain by laminated glaciomarine silts and clays. In the estuarine environments of the Tay and Forth valleys of eastern Scotland, coarsening-upwards sequences of clays, sands and gravels have been described and related to the progradation of large glacier-fed deltas formed during deglaciation (Sissons, 1969; Paterson *et al.*, 1981).

Conclusion

In recent years there has been a considerable development of the glaciological theory of how ice sheets develop and flow and models of varying degrees of sophistication have been constructed for the former northern hemispheric ice sheets (Paterson, 1981; Andrews, 1982). Such models have been used in various research programmes such as investigating the links between climatic change, ice-sheet growth and decay and sea level variations (e.g. Paterson, 1972; Budd & Smith, 1979), explaining the distribution of glacial landforms and sediments in areas affected by the last ice sheets (e.g. Boulton *et al.*, 1977; Sugden, 1977), and predicting the dispersal of erratics of interest to mineral exploration programmes (Boulton, 1984).

The observational basis for the accurate construction and critical testing of such models is, however, considerable. It is necessary to know the dimensions of the ice mass both in terms of plan position and altitude, preferrably at various times during both advance and retreat as well as ice maximum. The chronology of advance and retreat that matches knowledge of the dimensions is also necessary. The distribution of landforms and sediments needs to be known as well as assurance that these do indeed relate to the ice sheet being modelled as well as an understanding as to whether they were formed during ice advance or ice retreat. From the review of the evidence presented in this chapter, it is apparent that knowledge of the Late Devensian Scottish ice sheet is at present inadequate to allow the construction of sophisticated glaciological models and a considerable field effort is needed in order to acquire the necessary information.

Late Devensian glacial deposits and glaciations in eastern England and the adjoining offshore region

JOHN A. CATT

Much of eastern England from the Scottish border to the north Norfolk coast and large areas of the adjacent North Sea floor (fig. 23) are covered by tills, glaciofluvial and glaciolacustrine deposits resulting from the major glaciation that occurred during the period of the late Devensian time defined by Rose (1985) as the Dimlington Stadial (26,000 to 13,000 B. P.). During the later Loch Lomond Stadial (11,000 to 10,000 B. P.) glaciers reformed in parts of Scotland and northwest England, but never reached eastern England. An early Devensian glaciation of Lincolnshire was suggested by Straw (1961a, 1969), based on geomorphological evidence and the recognition of gravels separating two lithologically similar tills east of a morainic feature, but this was disputed by Madgett & Catt (1978). As there is no other evidence for an early Devensian glaciation elsewhere in eastern Britain or from the North Sea succession (Stoker et al., 1985), we must assume for the moment that there was only the one (Dimlington Stadial) glaciation of eastern England during the Devensian.

By analogy with the well-dated glacial successions elsewhere in northern Europe and in North America, and with the generalised climatic history deduced from deep ocean sediments, it is now assumed that the Dimlington Stadial ice sheet reached its maximum in eastern England at or soon after 18,000 B. P. This is indeed supported by ^{14}C dates of $18,500 \pm 400$ and $18,240 \pm 250$ years B. P. for moss from the Dimlington Silts, which underlie the Devensian tills in southeast Holderness (Penny et al., 1969). However, Holderness and other areas south of the North Yorkshire Moors seem to have had a fairly simple Devensian glacial history, related to a single coastal 'surge' (Boulton et al., 1977). Further north there is often evidence for two or more phases of different ice movements, and in the absence of such precise dating as Holderness has provided, these suggest that the Dimlington Stadial glaciation started earlier and consequently lasted longer than in Holderness. For example, Holmes (1977) suggested from dating of channels in parts of the North Sea east of Scotland that there was an extensive glacial cover there before 25,000 B. P. However, it is also possible (though on present evidence unlikely) that some of the earlier Devensian ice movements pre-date the Dimlington Stadial.

Northern England is an area of quite variable bedrock type, and tills usually vary in composition over short distances, depending on the direction of ice movement. Lateral correlation of deposits is consequently difficult, except in the coastal cliff sections and other areas of almost continuous exposure. However, in the region south of the North Yorkshire Moors till lithology is much less variable, and characteristics such as erratic suites, heavy minerals, till matrix colour and particle size distribution have been used for lithostratigraphic subdivision and correlation. Correlation of offshore successions with those on land is also problematic, mainly because the seafloor deposits are known principally from their geophysical characteristics with limited sampling. Another problem with seafloor successions is that they have been truncated to an unknown extent by marine erosion during the Flandrian.

Ice movement directions on land

Ice movement directions on land (fig. 24) are based mainly on striae, erratics, till macrofabric and drumlin orientations. On the east coast south of the Tees Estuary the surge lobe moved generally southwards, but there was a strong tendency for a southwestward movement inland into low-lying areas, such as the Esk Valley, Robin Hood's Bay and Vale of Pickering (Kendall, 1902), Holderness (Penny & Catt, 1967), and The Wash and Lincolnshire Marsh (Straw, 1969). This has usually been attributed to the westward pressure of ice from Scandinavia. However, at present there is little evidence that Scandinavian ice did extend across southern parts of the North Sea during

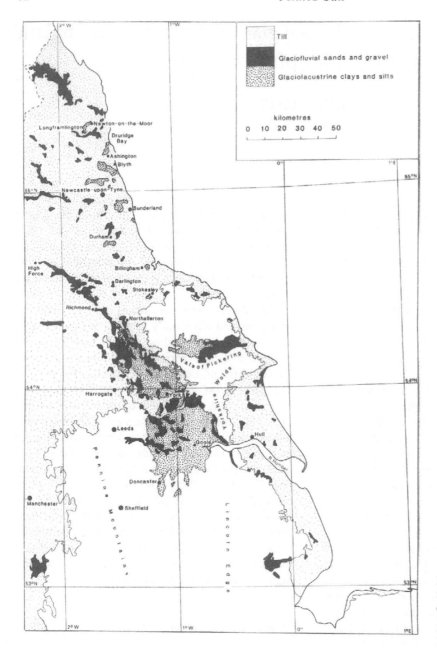

Figure 23: Outcrop of tills, glacio-fluvial and glaciolacustrine deposits of the Dimlington Stadial in eastern England, based upon Soil Survey of England and Wales (1983).

the Devensian (Jansen *et al.*, 1979; Cameron *et al.*, 1987). Norwegian erratics (larvikite and rhomb porphyries) occur in coastal Devensian tills of eastern England, but they could have been derived from pre-Devensian glacial deposits or from marine sediments containing rocks dropped by icebergs of Scandinavian origin. The landward movement probably resulted only from lateral spreading of the surge lobe, which was at least 150 km wide.

In the Vale of York the main ice movement was NNW-SSE, controlled by high ground to east and

west. As on the east coast, the initial advance was a surge, though here it was into the water of Lake Humber (Gaunt, 1981). Ice fronts were later stabilised slightly to the north of the surge limit, to form the York and Escrick Moraines (see Catt, this volume). On its western margin the Vale of York glacier was joined by ice streams flowing E or SE down Swaledale and Wensleydale (Raistrick, 1926), Nidderdale (Tillotson, 1934) and Wharfedale (Raistrick, 1931a). In Airedale, the next valley to the south, the ice terminated west of Leeds and never joined the

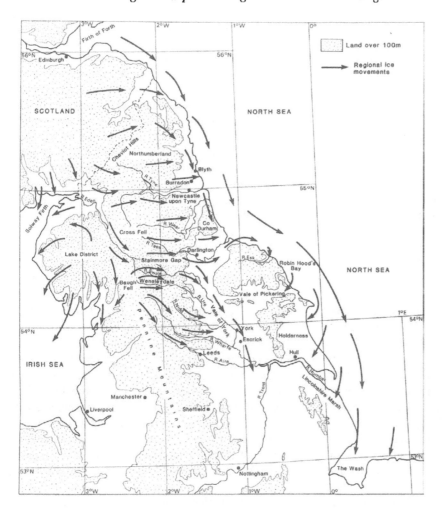

Figure 24: Main ice movements in eastern England during the Dimlington Stadial (Late Devensian).

trunk glacier in the Vale of York. All these tributary glaciers originated from a local ice cap, which formed on higher parts of the Pennines, such as Baugh Fell. Indeed, at the maximum of the Dimlington Stadial glaciation it is likely that only a few of the highest Pennine peaks, such as Ingleborough, projected through this ice cap as nunataks (Raistrick, 1934).

In areas adjacent to upper Swaledale and upper Wensleydale, high level striations and the few indicator erratics suggest a radial ice flow from Baugh Fell which was almost independent of topography (Raistrick, 1926). But striations, drumlins and drift tails (King, 1976) in slightly lower parts of these and other dales all indicate ice movement along the valleys, closely controlled by relief. This conflict of evidence is conventionally resolved by postulating that the flow was originally independent of topography because the ice was thick enough to cover the entire landscape, and became confined to valleys

during only the later stages of deglaciation (Rose, 1980c). However, the extent to which, even at the glacial maximum, the ice covered the interfluve moorlands between the dales on the eastern slopes of the Pennines remains uncertain. Beneath the blanket peat these areas are extensively covered by loamy drifts containing stones of various local Palaeozoic rocks, but these could be partly periglacial deposits or pre-Devensian glacial sediments. Undoubted till is confined to the valley floors and side slopes, for example in parts of Wensleydale above Aysgarth. The final stages of deglaciation in the dales produced sequences of retreat moraines as the ice-front receded up valley (Raistrick, 1934); many of these moraines have valley trains of outwash gravel downstream and lake flats upstream.

North of Swaledale, the local Pennine ice was deflected eastwards and then southeastwards into Arkengarthdale (and thus back into lower Swaledale) by a strong ice stream moving eastwards through the

Stainmore Gap (Wells, 1954; Mills & Hull, 1976). Lake District erratics (Shap Granite, Borrowdale Volcanics and Permian Brockram) are common on and east of Stainmore, indicating that the ice stream was forced from the eastern Lake District up the Vale of Eden and through the Stainmore Gap in the Pennines by very large volumes of ice in southwest Scotland and the Solway Firth. On the eastern side of Stainmore the Lake District ice invaded lower Teesdale, displacing northwards into County Durham the ice from upper Teesdale, which had probably originated on the Pennines in the Cross Fell region. The Lake District ice then passed through the Vale of Mowbray south of Darlington to become the Vale of York trunk glacier, thus explaining the presence of Lake District erratics in the Vale of York.

Lake District ice also crossed the Pennines further north through the Tyne valley. Lake District indicator erratics from this stream occur as far north as Blyth (Land, 1974), beyond which they are replaced by erratics from southern Scotland and the Cheviot Hills, which were deposited by ice still flowing essentially eastwards towards the North Sea basin. Between the Tyne and Teesdale streams, central parts of County Durham were invaded by Pennine ice also flowing eastwards down Weardale. Woolacott (1907) and most later workers have suggested that, as in areas further south, the westward pressure of Scandinavian ice in the North Sea basin deflected all these eastward-flowing ice streams southwards parallel to the present coast. However this explanation is no more acceptable here than in Yorkshire and Lincolnshire; any ice occupying the North Sea basin off Northumberland and County Durham is more likely to have come from the mountains of northeast Scotland and the Firth of Forth. Evidence for an earlier eastward flow and a later SE or SSE movement, such as the superimposed striations at Burradon (National Grid Reference NZ 277730) reported by Smythe (1912), can be explained if ice continued to accumulate in southeast Scotland and the Cheviot Hills later than in other areas, and flowed southwards down the coast after the originally more powerful western ice streams had largely melted (Carruthers *et al.*, 1930; Smith, 1981).

Distribution of surface deposits

The 1:250,000 soil map of England and Wales (Soil Survey of England and Wales, 1983) provides the most recent comprehensive small-scale remapping of the country's superficial deposits. The distribution of till, glaciofluvial sands and gravels and glaciolacus-trine clays and silts in northeast England shown in fig. 23 is based on this work. It shows that till is much more widespread throughout the region than either glaciofluvial or glaciolacustrine deposits. For the sake of clarity, areas of soils developed in coversand of Loch Lomond Stadial age and in Flandrian deposits (peat, alluvium, etc.) overlying the Dimlington Stadial glacial sediments have been omitted. The areas of glaciolacustrine clays and silts do not indicate the full extent of glacial lakes, mainly because these deposits are partly obscured by glaciofluvial sediments and till, for example in the Sunderland / Newcastle-upon-Tyne area and central parts of the Vale of Pickering.

Glaciofluvial sands and gravels occur mainly in valleys, such as the eastern Vale of Pickering, the Aln and Till Valleys in Northumberland, the North and South Tyne Valleys, the Wear Valley north of Durham, the Tees Valley between High Force and Darlington, the Gilling Beck Valley north of Richmond, the Swale Valley between Richmond and Northallerton, parts of Wensleydale and Bishopdale, the Aire Valley below Leeds, and the Ouse Valley between York and Goole. Many of these are probably valley trains formed during a late stage in deglaciation, when the last remnants of melting ice were confined to the major valleys. In addition there are numerous patches of glaciofluvial deposits scattered throughout the region, though most are too small to show in fig. 23.

Glaciolacustrine deposits are most extensive in southern and western parts of the Vale of York, where they were deposited in Lake Humber, mainly at its late (8 m) stage (Gaunt, 1981). This lake was impounded by ice fronts at the York and Escrick lines to the north, by the high ground of the Pennines to the west, Lincoln Edge to the south and the Yorkshire Wolds to the east, and by ice and later by a moraine blocking the Humber Gap. Other important areas of similar deposits are in the western Vale of Pickering, in the Leven Valley west of Stokesley (North Yorkshire), in the lower Tees Valley east of Stockton, in the Wear Valley south of Durham, also west of Sunderland, north of Newcastle-upon-Tyne, and in parts of eastern Northumberland south and west of Blyth, between Ashington and Druridge Bay, and between Newton-on-the-Moor and Longframlington. All of these are within 45 km of the east coast, and most were probably deposited in lakes formed at various times between the margin of the coastal ice and higher ground to the west.

The laminated clays in the Sunderland and Newcastle areas probably accumulated in a single large

Figure 25: Laminated glaciolacustrine clay (Tyne-Wear Complex) overlying Lower Boulder Clay, A 19 road cutting, Wellfield, County Durham (Photograph: D. B. Smith).

Figure 26: Upper Boulder Clay overlying Ryhope Sands, cliffs near Salterfen, near Sunderland, Tyne and Wear (Photograph: D.B. Smith).

lake, termed Lake Wear by Raistrick (1931b), which was fed by water from the retreating remnants of western ice in both Tynedale and Weardale. The lake deposits also include brown sands with lenses of stony clay, generally overlying the laminated clays, all grouped as the Tyne-Wear Complex (Smith, 1981). The deposits of this complex reach a maximum height of 132 m O.D., implying a lake level in excess of this, though most of the deposits are below 90 m O.D.

In addition to Lake Pickering in the western part of the Vale of Pickering, Kendall (1902) proposed a series of smaller proglacial lakes in the Cleveland Hills between the Vale of Pickering and the Tees estuary. The largest of these were in Eskdale, Kildale, Glaisdale, Wheeldale, Harwooddale and the Hackness Valley. Each was thought to be delimited by marginal drainage channels, strandlines and lake floor and deltaic deposits. But many of the channels and deposits have been reinterpreted as subglacial (Gregory, 1962; 1965), and this casts doubt upon the existence of many of the lakes, especially as the 'strandlines' often coincide with lithological boundaries in the gently folded Jurassic bedrock.

Till stratigraphy

Throughout most of the upland areas of northeastern England, such as the Pennine and Cheviot Hills, till is patchy and very variable in both lithology and thickness. It is usually thicker in the valleys draining these uplands, and there it is often divisible into a lower clay-rich lodgement or deformation till and an upper sandier and less compact flow till or supraglacial melt-out till. These two layers are often separated by a thin bed of sand and gravel. The lateral variation in the upland tills, which results from differences in bedrock lithology and ice flow directions, makes correlation from area to area very uncertain, and few of the deposits are designated as stratigraphic units. A Dimlington Stadial age is usually assumed from the apparently weak dissection and shallow, weakly to moderately developed soils, but direct dating evidence is rare.

In the lowland areas of eastern England the Dimlington Stadial tills are less variable in lithology and in many areas can be consistently divided into two or more distinct units. Throughout Northumberland a lower greyish till with many large stones from local sources is overlain by a reddish-brown till with fewer and smaller far-travelled stones (Carruthers *et al.*, 1930, 1932). As in the upland valleys, this sequence may often have resulted from a single ice advance, the lower till being a local lodgement till from basal layers of the glacier, and the upper till a flow or melt-out till from higher layers containing mainly far-travelled stones. However, in some areas both have the characteristics of compact lodgement tills, and the difference in erratic suites may indicate a change in the source of the ice. Eyles & Sladen (1981) suggested that the upper reddish-brown till in coastal Northumberland is a deep Flandrian weathering profile on a grey sulphide-rich lodgement till. This implies that locally there may be only one Dimlington Stadial till, but this need not be true throughout Northumberland.

The Tyne-Wear Complex further south in Durham overlies the eroded surface of a greyish-brown till deposited by the earlier western ice (fig. 25). This is usually termed the Lower Boulder Clay (Smith, 1981), though two varieties, one with abundant Carboniferous erratics and the other with Permian material, were named the Wear and Blackhall Tills respectively by Francis (1970). The far-travelled erratics in the Lower Boulder Clay include Lake District volcanic rocks and granites, Scottish greywackes and granites, and red sandstones of Devonian and Triassic age. In the coastal areas of Tyne and Wear,

County Durham and north Cleveland, the Lower Boulder Clay is overlain by sands, the Peterlee Sands of Francis (1970) and Ryhope Sands of Smith (1981), and an Upper Boulder Clay (fig. 26), which Francis (1970) named the Horden Till. The upper till contains Cheviot erratics and is lithologically similar to much of the upper till in Northumberland. It was probably deposited by the later ice stream flowing southwards down the coast from the Tweed-Cheviot area, and Smith (1981) suggested that it was responsible for impounding Lake Wear.

The youngest widespread till-like deposit in County Durham and adjacent areas is a reddish-brown silty clay with a variable stone content, which has been named the Pelaw Till (Francis, 1970) or Pelaw Clay and Prismatic Clay (Smith, 1981) in County Durham and Tyne and Wear, and the Teesside Till (Francis, 1970) in Cleveland. Francis (1970) believed that much of this deposit, which is usually 1 - 2 m thick (locally up to 9 m) and often contains thin contorted beds of sand, is a flow till. However, a readvance to deposit such a thin till over almost all parts of the area up to about 130 m O.D. seems unlikely, and Smith (1981) suggested that "it is a congeliturbate formed by periglacial modification and redistribution of all pre-existing deposits (but especially of the plastic laminated clays, hence the coincidence of highest occurrences at 130 - 132 m) following the drainage of Lake Wear".

In the Vale of York, the Dimlington Stadial till is mainly reddish-brown clay, with abundant erratics of Magnesian Limestone and Triassic sandstones, as well as some Lake District, Carboniferous and Lower Palaeozoic rocks. In western parts of the Vale, where tributary glaciers from the dales joined the trunk glacier, the till is usually greyer, sometimes more sandy, and consistently richer in Carboniferous limestone, sandstone and chert erratics from the Pennines (Lovell, 1982). The single till layer present even in northern parts of the Vale of York contrasts with the multiple till sequences further north, especially in parts of Cleveland, where three tills occur south of Hartlepool (James, 1982), and a sequence of four tills is exposed at Rockliffe Scar (NZ 313086) on the bank of the River Tees about 6 km SSE of Darlington. According to Francis (1970: 148), the lowest till at Rockliffe Scar is equivalent to the Wear and Blackhall Tills, and the uppermost is the Teesside Till. The two intermediate tills are reddish-brown, and Francis (1970) stated that they wedge out on the northern side of the lower Tees Valley (i.e. towards Hartlepool), where they are overlain by a southern extension of the Horden Till. No similar reddish tills are known further south in the Vale of York, so it seems that the two intermediate tills were probably deposited by an ice stream confined to the lower Tees Valley at a time between deposition of the Durham Lower and Upper Boulder Clays.

Cliff exposures on the Yorkshire coast show that there are two distinct lodgement tills south of the Tees Estuary. The lower of these, the Drab Clay of Bisat (1940) or Skipsea Till of Madgett & Catt (1978), is greyish-brown. At Dimlington Farm in southeast Holderness it overlies the Dimlington Silts dated to approximately 18,500 B.P. This till extends to the Devensian glacial limit in Lincolnshire and north Norfolk, though where it is > 5 m thick it is reddened throughout by oxidation; this led earlier workers to distinguish it erroneously as a separate deposit, which was named the Hessle Till, Hunstanton Brown Boulder Clay (Suggate & West, 1959) or Lincolnshire Marsh Till (Straw, 1969). The upper of the two tills on the Yorkshire coast is more restricted in extent than the lower: in Holderness it has a narrow arcuate outcrop adjacent to the coast between Mappleton and Easington, and it is absent from Lincolnshire and Norfolk. It is reddish-brown, even in unweathered form, and has abundant Triassic sandstone and shale erratics, also many Liassic, Permian and Carboniferous rocks and occasional igneous erratics from the Lake District. Previously known in Holderness and Filey Bays as the Purple Clay, it was renamed the Withernsea Till by Madgett & Catt (1978). Both the Skipsea and Withernsea Tills contain pyrite and siderite, and have been oxidized during the Flandrian to depths of 5 m or more.

On the basis of matrix colour and erratic suites, the Skipsea Till is correlated by all recent workers with the Lower Boulder Clay of County Durham. The Withernsea Till could be equivalent to either the Durham Upper Boulder Clay (the Horden Till) or to one or both of the intermediate tills at Rockliffe Scar, all of which are reddish-brown. However, the presence of Lake District rather than Cheviot erratics in the Withernsea Till suggests that it should be correlated with either or both of the intermediate tills at Rockliffe Scar, which were probably deposited by part of the Stainmore ice stream.

Madgett & Catt (1978) suggested that superimposition of Stainmore ice from the lower Tees valley onto the coastal (Lower Boulder Clay) ice brought about the southward surge that resulted in simultaneous deposition of the Skipsea and Withernsea Tills from a single, two-tiered ice sheet. However, Francis (in discussion of Madgett & Catt, 1978) disagreed.

He suggested that both the Withernsea Till and the main Vale of York till were deposited by the southward-flowing Tweed-Cheviot ice stream, which deposited the Upper Boulder Clay in County Durham, and then divided into two streams passing east and west of the Cleveland Hills. As explained by Catt (this volume), Francis' reconstruction of ice movements fails to account for the existence of Lake Humber in southern parts of the Vale of York, because at its early high-level stages this lake was impounded by the Vale of York glacier to the north and by ice in the Humber Gap which deposited the Skipsea Till. If there was an ice-free interval between deposition of the Lower and Upper Boulder Clays in County Durham, as Smith (1981) suggested, then the ice that deposited Skipsea Till in the Humber Gap would have disappeared before ice invaded the Vale of York, and water would not have been impounded to create a high-level Lake Humber.

Consequently it is likely that the Upper Boulder Clay or Horden Till was deposited by a southward ice advance reaching only to the northern side of the lower Tees Valley, and extending inland as far as a moraine between Easington and Elwick. This occurred at a somewhat later stage in the Dimlington Stadial than the main invasion of eastern England by ice that had crossed the Pennines through the Tyne and Stainmore Gaps and then flowed southwards into the Vale of York, eastern Yorkshire, east Lincolnshire and north Norfolk. However, it must be admitted that more work is required to clarify the correlation of tills between Yorkshire and County Durham and thus improve our knowledge of ice movements at different times in northern England during the Dimlington Stadial. The most critical area requiring detailed study is clearly the lower Tees Valley, where it would be useful to have several fully cored boreholes through the Quaternary succession.

Offshore deposits

Many of the problems associated with the Dimlington Stadial events in northeast England might be resolved if more were known about the nature and distribution of glacial deposits on the floor of the adjacent parts of the North Sea. Because it is less important in petroleum exploration, the glacial history of this area is in fact much less well known than parts of the North Sea north of 56°N, where the Dimlington Stadial glacial deposits are known as the Wee Bankie Formation (Stoker *et al.*, 1985b).

Dingle (1970) reported that 20 - 45 km east of the

North Yorkshire coast, tills similar to those of Holderness overlie Mesozoic rocks in a NNW-SSE belt. However, most of the sea floor between this belt and the coast has lost any till cover it may once have had (Dingle, 1971a). Holderness-type tills occur east and south of Flamborough Head, forming an almost continuous cover as far as the Dogger Bank and Norfolk coast (Donovan & Dingle, 1965; Balson & Cameron, 1985). According to Donovan (1973) most of the till exposed on the sea bed is Withernsea Till, and the Skipsea Till has only limited outcrops close to southeast Holderness and further east near the Silver and Sole Pits. These pits and other enclosed elongate hollows on the sea floor are usually attributed to erosion beneath the ice sheet, though they may have

been modified by tidal scour in the early Flandrian when the sea level was much lower than at present (Donovan, 1973).

In the area immediately offshore from the Tees Estuary and North Yorkshire coast, which is critical for correlation between County Durham and areas further south, any deposits left by the Dimlington Stadial ice sheet or sheets have therefore been almost entirely removed. This was probably the result of marine erosion during the early Flandrian eustatic rise of sea level. Where till is extensively preserved on the sea floor, as in areas further east and south, it seems to be of British rather than Scandinavian origin, but has not been sufficiently well studied to clarify patterns of ice movement in northeast England.

PREAMBLE

The following two papers concern the events during the most recent glaciation of Ireland. Both are included since they illustrate a contrasting approach and stratigraphical interpretation: The first represents the established view of Hoare and the second the recently proposed ideas of Warren. This inclusion of both papers is thought appropriate since the editors felt readers would welcome the opportunity to judge the evidence for themselves.

Late Midlandian glacial deposits and glaciation in Ireland and the adjacent offshore regions

PETER G. HOARE

Although there is considerable controversy over the timing of glacial events in Ireland, it is widely believed that Midlandian* ice-masses did not develop until towards the end of the cold stage. However, a small ice-sheet may have formed in central County Tyrone during a short-lived Early Midlandian (?pre-Chelford Interstadial) Fermanagh Stadial (McCabe, 1986b; see also Mitchell, 1976; 1977, and Warren, 1985); and Synge (1981) held the unorthodox view that the country had been influenced by Early, Middle and Late Midlandian glaciations. The more recent publications on the subject are reviewed here, with particular emphasis on evidence from the key area of eastern Ireland. Those glacial events that are thought to have preceded the formation of Late Midlandian ice-masses are described by Hoare (this volume).

The pattern of movement, timing and maximum extent of the Late Midlandian ice-sheets

Most of the sites and features referred to in the text are located on fig. 27; the remainder are shown on figs. 28–31.

An Irish ice-sheet spread out from a curved axis of dispersion which extended northeastwards from the west-central part of the low-lying midlands to the Lough Neagh region (J 0070) (Close, 1867; Lamplugh et al., 1903; Farrington, 1934; 1939; Synge & Stephens, 1960; Hill & Prior, 1968; Synge, 1969; 1970a; Colhoun, 1970; Davies & Stephens, 1978). The ice did not accumulate over the Atlantic seaboard as Charlesworth (1963b) suggested. Drumlinised till deposited by this ice-mass overlies organic-rich

material of Derryvree 'Interstadial' age (McCabe, 1986b) in Derryvree Townland, County Fermanagh (H 361390) (Colhoun et al., 1972). A radiocarbon date of 30,500 B.P. was obtained on the interstadial sediment and thus provides a maximum age for the 'Drumlin Readvance' phase (see below) of the Late Midlandian glaciation of the area. The connection between the Derryvree till and glacigenic sediments at other sites in the country cannot be demonstrated, but near-contemporaneity is often assumed.

The greater part of the drift sequence in eastern Ireland is of Late Midlandian age if a radiocarbon date of 24,050 B.P. that was determined on shells reworked by ice into the older of two tills at Glastry, County Down (J 6463) (Hill & Prior, 1968), is adopted. However, shells of such apparent antiquity may provide unreliable dates (see Shotton, 1966; Synge, 1970a; Mitchell, 1972; Mitchell et al., 1973; Stephens et al., 1975, and Davies & Stephens, 1978, for example), although some have accepted the Glastry result (Hill & Prior, 1968; Warren, 1985). Synge (1977a, 1981) also considered that most of the glacial events to have affected eastern Ireland occurred during the Late Midlandian. Such unorthodox views are not necessarily incorrect, of course.

An upper, diachronous, limit to the age of the Late Midlandian glaciation of the lowland areas of eastern Ireland is indicated by radiocarbon dates from organic deposits that overlie the drift. A selection of these dates may be found in Mitchell (1957), McAulay & Watts (1961), Colhoun & Mitchell (1971) and Watts (1977).

The maximum extent of the ice-sheet may be marked by a discontinuous moraine that occurs along the southern edge of the limestone lowlands; the feature was discovered by Lewis (1894) and mapped in detail by Charlesworth (1928) who called it the Southern Irish End-Moraine. Whilst this limit has been adopted by most workers (see Mitchell, 1972; Synge, 1973; McCabe, 1985, and Watts, 1985, for example), a small number have questioned its valid-

---

*It is assumed that the Midlandian Stage of Ireland is equivalent to the Devensian of the rest of the British Isles; the previous cold stage is known as the Munsterian (Mitchell et al., 1973). The term Devensian is used here when it has been applied to material which lies on the bed of the Irish Sea.

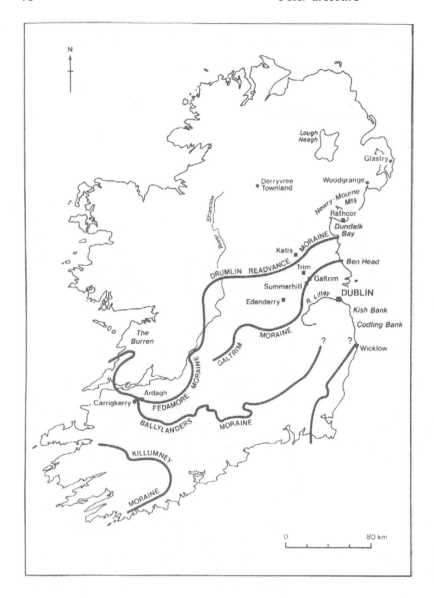

Figure 27: The location of sites and features mentioned in the text (others are shown on figs. 28-31).

ity (see Bowen, 1973b, and Warren, 1979a; 1985, for example). It is generally believed that the south of Ireland was free of ice during the Late Midlandian glaciation, whereas almost the whole of the country was inundated by Munsterian ice-masses. The configuration of the Southern Irish End-Moraine indicates that the Late Midlandian ice reached its maximum extent in the north and west somewhat later than in the south and east (Synge, 1969; 1970a; 1979a; 1981). The Ballylanders (County Tipperary, R 7625) or Tipperary Moraine (see Mitchell, 1957) in the south and southeast of the country may date from 20,000 (Synge, 1969: 89; 1973) or 23,000 years ago (Synge, 1969: 92; McCabe, 1985; 1986b). The south-western part of the limit is known as the Fedamore

(County Limerick, R 5944) or Drumlin Readvance Moraine (see below).

The maximum extent of the ice that deposited the so-called 'Newer Drift' has been established in a variety of ways in areas where the Southern Irish End-Moraine did not develop or has not survived. These include the distribution of topographically 'fresh', hummocky, constructional forms such as moraines, kames and eskers with kettle holes, the presence of drumlins, the occurrence of surface tills showing a lack of deep weathering and of certain soil-types (see Mitchell, 1957; 1972; Synge & Stephens, 1960; Synge, 1966b; 1969; 1970a and b; 1979a; Synge & Finch, 1966, and Finch, 1971, for example). It is widely assumed that the more widespread 'Older Drift' is of Munsterian age and that its subdued surface expression and smooth slope profiles and the rearrangement of its fabric to depths of approximately 3 m (Synge, 1970a) are the result of prolonged modification by temperate and periglacial processes (Synge & Stephens, 1960; Stephens & Synge, 1965; Synge, 1968; 1969; 1970a). However, some glacial sediments may date from the early part of the last cold stage, and the distinction between 'subdued' Munsterian landforms and 'fresh' Midlandian examples may not be a reliable one (Synge, 1979a). In addition, the use of periglacial phenomena to separate the two drift-sheets has been questioned by Lewis (1985).

Although views on the general position of the limit of this ice-sheet have remained essentially unaltered since Charlesworth made his proposals, many relatively minor modifications have been suggested. Even where a well-developed moraine does exist, the ice appears in places to have extended beyond it. Thus the Southern Irish End-Moraine may be seen as a complex feature which locally records a stillstand during the retreat of the ice. Variations in the lithology of surface tills formed the basis for making adjustments in Counties Tipperary (Farrington, 1945) and Limerick (Synge, 1966b; 1969; 1970b; 1979a), although these differences may be explained in other ways (Warren, 1979; 1985). Finch (1971)

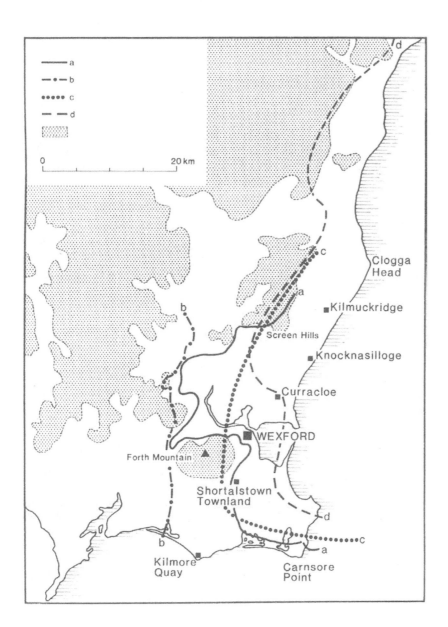

Figure 28: The Late Midlandian glacial limit in southeast Ireland: a) after Colhoun & Mitchell (1971); b) after Culleton (1978a and b); c) after Davies & Stephens (1978); d) after Synge (1981).

suggested a southerly extension of the limit in south-east County Tipperary and southwest County Kilkenny following a survey of the soils of the area. Changes have also been indicated for eastern Ireland (see Farrington, 1942; 1957b; Martin, 1955, and Synge, 1964; 1977a, for example) and for the south-eastern part of the country (see Gardiner & Ryan, 1964; Colhoun & Mitchell, 1971; Culleton, 1978a and b; Davies & Stephens, 1978, and Synge, 1981, for example) (fig. 28). Further modifications will almost inevitably follow.

The Irish ice-sheet pressed against the northwest-ern flanks of the Leinster Mountains but did not override them. The farthest extent of the ice may be represented by the Hacketstown (County Carlow, S 9780) Line (Synge, 1973: 564) (fig. 29), although this moraine may date from the Munsterian (Synge, 1973: 562) or from the Middle Midlandian (Synge, 1981). Synge (1973) suggested that the Hacketstown Line is somewhat older than the Ballylanders Moraine in County Tipperary (see above). The maximum elevation of the Late Midlandian glaciation is recorded by a low mound of limestone-rich gravel at Killakee, County Dublin (O 122225), at 380 m O. D.

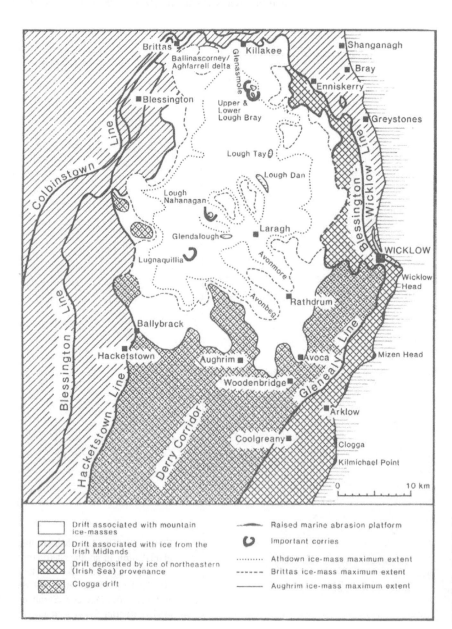

Figure 29: The distribution of glacial material associated with the Clogga/Enniskerry Glaciation in Counties Wicklow and Wexford and the extent of various glacial episodes at the northern end of the Leinster Mountains (after Synge, 1981).

(Synge & Stephens, 1960; Mitchell *et al.*, 1973), and not at 336 m (Synge, 1970a) or 490 m (Synge, 1979b).

An unweathered drift-sheet, consisting of limestone-rich till and ice-marginal sands and gravels, occurs on the eastern side of the Leinster Mountains (Farrington, 1934; 1944; Martin, 1955; Synge, 1964; 1970a; 1973; 1981). It is not clear whether these are the deposits of a Scottish or of an Irish ice-sheet, although they are said to be of 'Irish Sea Basin' provenance. Their upper limit, which is marked by the Glenealy Line (Synge, 1973; 1977a; 1981) (fig. 29), coincides with the proposed maximum extent of the (?)Munsterian 'Eastern General' ice-sheet (Farrington, 1944) near Enniskerry, County Wicklow (O 2217). This material is never found above 200 m O.D. in eastern County Wicklow (Synge, 1973), and it lies at approximately 100 m O.D. between Wicklow Head, County Wicklow (T 3492), and Forth Mountain, County Wexford (S 9718) (Synge, 1977a) (fig. 28), a distance of about 80 km. The lack of an appreciable gradient may indicate that the drift was deposited by "...a glacier of shelf ice..." (Synge, 1977a: 208). The Glenealy Line may correspond with the Hacketstown Line on the western side of the Leinster Mountains (Synge, 1973: 568; Davies & Stephens, 1978) or post-date it (Synge, 1981). Synge (1981) equated the Glenealy Line with the Screen Hills kame-and-kettle moraine in County Wexford (see below), and suggested that they are of Late Midlandian age.

Material deposited by the Irish ice-sheet is patchily distributed in the eastern parts of Counties Louth, Meath and Dublin; considerable areas are drift-free or bear only a thin mantle of glacial debris (McCabe, 1972; 1973; Hoare, 1975; 1977a and b; and this volume; McCabe & Hoare, 1978). This outcrop pattern has led to controversy over the extent of Scottish ice in the area at this time (see below and Hoare, this volume). Bedrock is exposed over substantial parts of many upland regions, and there has been speculation about the original extent of drift on the highly karstified limestone outcrop of the Burren, County Clare (plate 5; cf. Farrington, 1965).

The withdrawal of the Late Midlandian ice-sheet

Moraines mark the stillstands and readvances which interrupted the retreat of the ice (Charlesworth, 1928; 1973; Synge, 1950; McCabe, 1971; 1972; 1973; Hoare, 1975; McCabe & Hoare, 1978). Synge (1977a) was not correct in suggesting that none of

these landforms exists in County Dublin (see Hoare, 1975, and McCabe & Hoare, 1978, for example). McCabe (1971) described twenty-one recession and readvance moraines between Gormanstown, County Meath (O 1667), and Dromiskin, County Louth (O 0598). Eskers, esker-systems and related features have also been documented in some detail (see Charlesworth, 1928; Synge, 1950; 1970a; Synge & Stephens, 1960; Farrington & Synge, 1970, and McCabe, 1985, for example). Charlesworth (1963b) mistakenly identified eskers as moraines in Counties Mayo, Galway and Westmeath (Synge, 1970a). The wide distribution of moraines and the possible 'beaded' nature of the eskers near Trim, County Meath (T 8056) (Synge, 1950; 1979a), indicate that the ice did not become entirely stagnant and melt *in situ* as Flint (1930) and Mitchell (1976) claimed (but see below).

The ice withdrew northwestwards from the Hacketstown Line on the western flanks of the Leinster Mountains and then readvanced, possibly by as much as 34 km (Synge, 1977a) or 50 km (Synge, 1981), to the Blessington Line (Synge, 1977a; 1979a; 1981) (fig. 29). An interstadial may have separated the two events (Synge, 1979a). This younger limit may be equivalent to the Wicklow Line on the eastern side of the mountains (fig. 29) and to the Ballylanders Moraine of County Tipperary (Davies & Stephens, 1978; Synge, 1979a), and may date from 18,000–19,000 years ago (Synge, 1977a) (other possible figures are given above).

Only a few of the many lakes that Charlesworth (1928, 1937) considered were dammed by the Irish ice-sheet within the northern Leinster Mountains have been substantiated. The largest and most complex of these occupied the upper part of the King's River and River Liffey Valleys in Counties Dublin and Wicklow (Farrington, 1934; 1957a; Synge, 1977a; 1981; Davies & Stephens, 1978). At its fullest development, Glacial Lake Blessington measured approximately 6 km by 21 km and rose to 280–283 m O.D. (Farrington, 1957a) or to 306–309 m O.D. (Synge, 1981). A large delta-moraine accumulated near Blessington, County Wicklow (N 980142) (Cohen, 1976; 1979), and the Ballinascorney-Aghfarrell delta formed in a northern arm of the lake (Farrington, 1942; 1957a; Hoare, 1977c and d). A smaller proglacial lake was impounded at the head of Glenasmole, County Dublin (O 104210) (Synge, 1971; Hoare, 1975). Varved sediments accumulated on the bed of Glacial Lakes Blessington (Cohen, 1976; 1979) and Glenasmole (Synge, 1971).

The second major stillstand during the withdrawal of the ice from the northwestern side of the Leinster

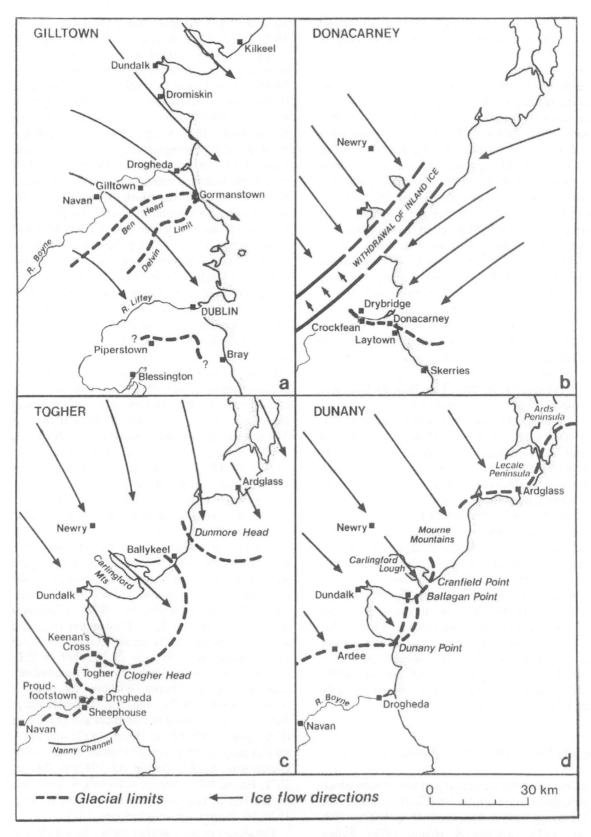

Figure 30: Late Midlandian glacial events in the northern part of eastern Ireland (after McCabe & Hoare, 1978).

Mountains is represented by the Colbinstown (County Wicklow, S 8398) Line (Synge, 1981) (fig. 29).

The Galtrim Moraine (fig. 27) extends from Edenderry, County Offaly (N 6332), via Galtrim, County Meath (N 8652), to Ben Head, County Louth (O 1869), a distance of 70 km (Synge, 1950; 1970a; Synge & Stephens, 1960; McCabe, 1971). That part of the moraine that lies to the southwest of Summerhill, County Meath (N 8448), may delimit a minor oscillation of the ice (Synge, 1950). The spacing of (?)annual beads in two northwest-southeast aligned eskers within the complex of twelve eskers or eskersystems near Trim, County Meath, suggests that the ice withdrew from this moraine at 76 m and 130 m a^{-1} (Synge, 1950).

The retreat of the ice was again interrupted during the episode known as the Kells (County Meath; Synge, 1952), Drumlin (Synge & Stephens, 1960; Synge, 1969), Dunleer (County Louth; McCabe, 1972; 1973) or Dunany (County Louth; McCabe & Hoare, 1978) Readvance. A line of moraine extends from Kells (N 7475) to Dunany Point, County Louth (O 1591), and to a similar feature at Rathcor, County Louth (J 1804), on the opposite side of Dundalk Bay (Synge, 1970a; McCabe, 1971; 1972; 1973). The limit may be traced southwestwards from County Westmeath to the mouth of the River Shannon (Synge, 1981). It is represented in County Limerick by the Fedamore Moraine (R 5944) which crosses the older Ballylanders Moraine between Ardagh (R 2738) and Carrigkerry (R 2238) (Synge, 1969; 1970a). The Ballyardel (= Drumlin) Readvance limit lies at the mouth of Carlingford Lough, on the boundary between Counties Down and Louth (Stephens *et al.*, 1975). It extends northeastwards from the southern coast of the Lecale peninsula (J 5637) to the tip of the Ards peninsula (J 6245), County Down (Hill & Prior, 1968) (fig. 30d).

Most of the extensive drumlin swarms are thought to have developed during this major episode (Synge, 1952; 1969; 1970a), and orderly withdrawal of the ice-sheet may then have been replaced by wholesale stagnation (Synge & Stephens, 1960). However, the drumlin fields may have formed at different times, and the moraines which lie at their outer edges may not be contemporaneous (McCabe, 1985; Warren, 1985).

The Drumlin Readvance line marks the maximum extent of Irish ice in the western and northern parts of the island during the Late Midlandian. It was associated with a westward or northwestward shift in the position of the axis of ice-dispersal (Synge,

1969; 1981; Stephens *et al.*, 1975); Synge (1970a) suggested an earlier eastward movement had taken place. This phase of renewed activity may have been due to a climatic deterioration (McCabe, 1972; 1973; Davies & Stephens, 1978) or to a surge (Dardis *et al.*, 1984; McCabe, 1986b). McCabe & Hoare (1978) proposed that rapid advances might account for some of the complexities of Irish glacial stratigraphy. Synge (1969) showed remarkable foresight when he suggested a date of 17,000 B.P. for this event (and not 19,000 B.P., McCabe, 1973). Glaciomarine muds deposited penecontemporaneously with marine delta-moraines at the limit of the Drumlin Readvance in northern County Mayo were radiocarbon dated to 16,940 and 17,300 B.P. (McCabe *et al.*, 1986b).

The large Screen Hills kame-and-kettle moraine lies between Kilmuckridge (T 1640) and Curracloe (T 0927), County Wexford (fig. 28). It may mark the maximum extent of Late Midlandian ice in southeast Ireland (Charlesworth, 1928; Synge, 1981), although the limit is generally regarded as being farther to the west and south (see Colhoun & Mitchell, 1971; Culleton, 1978a and b, and Davies & Stephens, 1978, for example) (fig. 28). At least four phases of deposition may be distinguished within the morainic complex (Huddart, 1976; 1977b; 1981a); each is recorded by proglacial sandur sediments and an overlying basal (?lodgement) readvance till which accumulated during small oscillations of the ice-sheet during its withdrawal.

The gravels and tills of the Screen Hills moraine overlie the Knocknasilloge Silts and Sands which consist of density underflows, suspension sediments, flow tills and dropstones, and contain foraminifera and ostracods (Huddart, 1976; 1977b; 1981a and b). They may be of glaciomarine origin and indicate that the ice was floating at the time the kame-and-kettle complex accumulated (Huddart, 1976; 1977b; 1981b). However, Huddart (1981a) also suggested a marine origin and a possible Ipswichian age, despite the signs of glaciation. He correlated them with the Askintinny Member of the Clogga Head succession, County Wicklow (T 252692) (see Hoare, this volume), with the polleniferous estuarine shelly sand which lies between two tills in Shortalstown Townland, County Wexford (T 0214) (Colhoun & Mitchell, 1971), and with the marine sediments that separate two tills on the floor of St. George's Channel (Garrard, 1977). Thomas & Summers (1981) argued for the glaciolacustrine nature of the Knocknasilloge Silts and Sands.

The influence of Scottish ice during the Late Midlandian

Irish ice emerging from the Lough Neagh lowlands was obstructed by Scottish ice at the height of the Late Midlandian glaciation (Vernon, 1966; Hill, 1968; Hill & Prior, 1968). However, the complex flow patterns reflect not only the conflict between these two ice-sheets, but also considerable topographic control and a change in the position of the centre of dispersal of the Irish ice-mass. The Irish ice was forced to flow northwestwards along the eastern coast and then westwards near Benmore (Fair Head) (D 1743) (Hill & Prior, 1968). The Scottish ice-sheet may not have gained access to the Irish mainland at this time (but see below).

McCabe & Hoare (1978) described a complex set of events involving both Irish and Scottish ice and affecting Counties Louth, Meath and Dublin (fig. 30); there is no evidence that the ice from Scotland extended south of Donacarney (O 1374), County Meath (fig. 30b). However, Davies & Stephens (1978) suggested that the latter ice-mass covered a narrow coastal strip of the central part of this region during the Late Midlandian. An alternative explanation of the distribution of glacial material in this area has been provided by McCabe (1972, 1973), Hoare (1975, 1977c and d, and this volume) and McCabe & Hoare (1978).

Although the Antrim Coastal Readvance (Dwerryhouse, 1923; Charlesworth, 1939; 1963b; 1973) does not appear to have taken place (see Synge & Stephens, 1960; Hill, 1968; Hill & Prior, 1968, and Stephens *et al.*, 1975, for example), there was a limited incursion by Scottish ice southwestwards into County Antrim towards the end of the Midlandian Stage. This event has been called the Armoy Readvance (Dwerryhouse, 1923), and the Armoy Stage (Stephens *et al.*, 1975) and it led to the deposition of the North Antrim Scottish Till (Hill & Prior, 1968). The limit of the episode is marked by a prominent end moraine which extends from Ballymoney (C 9425) in

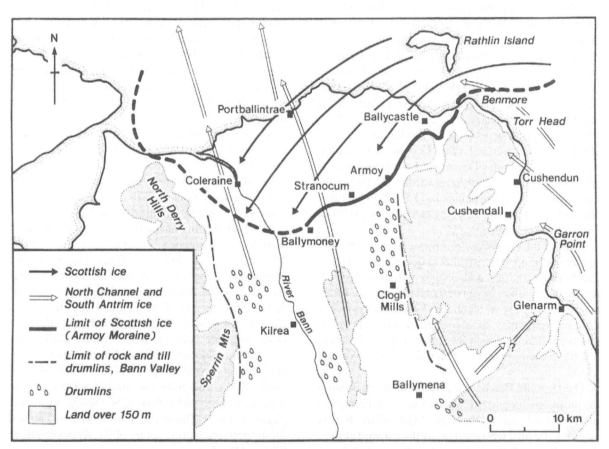

Figure 31: The extent of Late Midlandian Scottish ice and other glacial features in County Antrim, Northern Ireland (after Stephens, Synge, Prior & Creighton in Stephens *et al.*, 1975).

the west to Ballycastle (D 1140) in the east (Prior, 1968; Hill & Prior, 1968; Stephens *et al.*, 1975) (fig. 31). The Armoy Moraine defines the northern limit of drumlins in the Bann valley (Charlesworth, 1939; 1963b; Prior, 1968) and may be equivalent, therefore, to the Drumlin Readvance Moraine farther south (see above). Scottish and Irish ice-masses came into contact in northeast County Antrim (Hill & Prior, 1968; Creighton, 1974; Stephens *et al.*, 1975), and this moraine may mark their point of contact in the Bann valley (Creighton, 1974; Stephens *et al.*, 1975).

Late Midlandian ice-masses elsewhere in Ireland

Considerably smaller ice-masses also developed over parts of the mountainous rim of the country during the latter part of the Midlandian Stage. The most substantial of these was the Lesser Kerry-Cork (later Cork-Kerry) (Farrington, 1954; 1959) or Killumney ice-cap (Mitchell, 1957) which occupied an area of approximately 8,500 km² in the southwest (fig. 27). Although this event may have coincided with the Drumlin Readvance phase of the Irish ice-sheet (Synge, 1981), the two sets of deposits do not come into contact and are not chronometrically dated.

The Athdown ice-cap formed over the Leinster Mountains during the Late Midlandian (Charlesworth, 1928; 1937; Farrington, 1934; Synge, 1971; 1981; Hoare, 1975; McCabe & Hoare, 1978) (fig. 29). This event has been described in some detail by Hoare (this volume).

The final phase of Late Midlandian glacial activity took place more recently than 11,500 years ago and is represented by the Nahanagan Stadial glaciers (Farrington, 1957b; 1966a; Colhoun & Synge, 1980), a discovery anticipated by Mitchell (1957) (fig. 29). It followed a brief period of relative warmth known as the Woodgrange (County Down) Interstadial (see Morrison & Stephens, 1965; Singh, 1970; Mitchell, 1976, and Watts, 1977, for example).

Late Midlandian/Late Devensian drift in the offshore zone

The topography of the bed of the Irish Sea, including the possible existence of morainic or other 'land-bridges' at the close of the last cold stage, has figured prominently in discussions of the way plants and animals may have entered Ireland as the ice-masses retreated (see Mitchell, 1976; Synge, 1985; Stuart & van Wijngaarden-Bakker, 1985, and Preece *et al.*, 1986, for example). Mitchell (1960, 1963) made an

early attempt to interpret the geomorphology of the floor of the basin, but bedrock underlies some of the landforms he identified as glacial in origin (Dobson *et al.*, 1971; 1973). Subsequently, Mitchell (1972) suggested that marine action had removed most of the drift from the sea-bed. More recently, drilling, coring and remote sensing devices have demonstrated the considerable thickness, widespread distribution and stratigraphy of the glacial material. Glacigenic sediments that now lie on the sea-bed may have suffered little solifluction or subaerial erosion before they were covered by the rise of sea-level, but marine erosion and reworking is likely to have taken place. Many of the descriptions and conclusions reached so far are of a rather generalised nature, and much remains to be learnt of the extent of grounded and floating ice-masses and of the precise nature and age of the deposits. Correlation with sediments that crop out on the Irish mainland cannot as yet be made with any confidence.

Drift rests on marine material of (?)Ipswichian age in the southern part of the basin (Garrard, 1977), but no further refinement of the timing of the glaciation is possible. A younger till lies to the west of the Lleyn peninsula, north Wales, and may record a readvance of an ice-sheet to a point south of Bardsey Island (Garrard, 1977). This upper glacial unit was said to be of 'Irish Sea Basin provenance', which serves little purpose when the sediment is located in the central part of the basin.

Pantin (1977) described the Quaternary sediments of the northern part of the Irish Sea Basin, but only those to the northwest, west and southwest of the Isle of Man are discussed here. Bedrock is overlain by an almost ubiquitous reddish-orange till. The thickness of this material was not stated, although it amounts to several metres in the northeastern part of the basin. The till is a stiff, mainly silty clay, sediment with occasional gravel clasts that was assumed to be of Devensian age. The absence of pre-Devensian till, which occurs in the southern part of the basin (Garrard, 1977), may be due to greater erosion by Devensian ice closer to its Scottish source.

A proglacial sediment associated with the retreat of a Late Devensian ice-sheet is preserved in a few depressions in the western part of the northern Irish Sea Basin where it rests upon (?)Devensian till. It consists of a well-bedded, reddish-orange mud with sand or coarse silt laminae; a till-like facies and dropstones, including those composed of material resembling till, also occur. Foraminifera and other microfossils indicate deposition in water connected to the open sea, but the salinity may have been low as there are few macrofauna. The dropstones indicate the presence of floating ice which may have restricted water circulation. The beds appear to pass laterally into marine or glaciomarine sediments.

Whittington (1977) examined the nature of the floor of the west-central part of the basin, principally by the use of shallow seismic profiling. Considerable bedrock topography indicates the arrangement of older, more obdurate rocks and younger, less resistant, lithologies; glacially-moulded bedrock suggests a north-south movement. The overlying till was considered to be Devensian in age. Morainic deposits were tentatively identified to the east of Wicklow (T 3193) and the Codling Bank (fig. 27). The hummocky surface which underlies deeper water to the east of the Kish Bank (fig. 27) may represent a drumlin field. Confirmation of this proposal would have interesting implications for the timing and extent of the drumlin-forming phase of the Late Midlandian glaciation described above. To the northwest, estuarine sediments rest on 'periglacial fluvial gravels', which may have been laid down by meltwater flowing along the Liffey valley and its tributaries, and this in turn overlies 'Midland General' (=Late Midlandian?) till at the mouth of the River Liffey, County Dublin (Naylor, 1965).

Devensian drift underlies the greater part of the southern Irish Sea Basin, with the notable exception of the eastern part of Cardigan Bay (Garrard & Dobson, 1974; Garrard, 1977). It is made up largely of an unweathered, mature, highly calcareous, shelly and sandy till; this texture may be contrasted with the muddy nature of the unit in the northern part of the basin. The ice, which may have originated in Scotland (Garrard, 1977), transported felsite and granite erratics from the Newry-Mourne Mountain igneous complex to the Codling Bank (Dobson *et al.*, 1971); it is equally possible that the ice advanced from the Lough Neagh centre of dispersal (see above).

Devensian drift is up to 70 m thick in the southern part of the Irish Sea Basin and averages 30–40 m. The material thins southwards to an approximate thickness of 10 m off the west coast of Dyfed, south Wales. It appears as a lobe of sediment extending into the Celtic Sea and ending due west of St Gowan's Head (latitude 51°34'); this feather-edge of drift possibly marks the maximum extent of the ice-sheet (Garrard, 1977).

Lodgement till may be the most common form of drift, although laminated flow or melt-out till is also thought to occur, together with some glaciofluvial silt, sand and gravel (Garrard & Dobson, 1974; Garrard, 1977). Garrard (1977) described the colour, particle-size distribution, lithological composition and total carbonate content of the tills, although it is not clear how their origins were determined. Eyles & Eyles (1984), Eyles *et al.* (1985b) and McCabe (1986b) suggested that some material within the Irish Sea Basin may have been too readily interpreted as basal or lodgement till. They considered that 'diamictons' and 'diamicton assemblages' within the Quaternary sequence at the northern end of the Isle of Man were of marine and glaciomarine origin and that other deposits around the shores of the basin may have had a similar origin. Thomas (1976, 1977, 1984a) and Thomas *et al.* (1985) considered the material in the Isle of Man was deposited by land-based ice, and Thomas & Dackombe (1985) defended this interpretation of these sediments (see Dackombe & Thomas, this volume).

Summary and conclusion

Although it has been claimed that the history of the last glaciation in Ireland is much clearer than that of earlier glacial episodes (Synge, 1979a), certain fundamental problems still require clarification. Thus, the number, precise timing and limits of the Late Midlandian ice-masses are not known with any degree of certainty. The relatively few radiocarbon dates that are available do not permit a detailed reconstruction of the retreat of the Irish ice-sheet. McCabe (1985) has outlined some of the ways in which progress may be made. It is clearly rather too early to attempt to correlate events in Ireland with those preserved in the deep-sea oxygen isotope record (cf. Warren, 1985).

Acknowledgement

The author is most grateful to Ms R. H. A. Smith for drawing the figures.

Fenitian (Midlandian) glacial deposits and glaciation in Ireland and the adjacent offshore regions

WILLIAM P. WARREN

The term 'Midlandian' for the last cold stage of the Quaternary in Ireland has been superseded by 'Fenitian' (Warren, 1985). The extent of 'Midlandian' deposits had been based on a very poorly constructed and largely morphostratigraphic model (Mitchell *et al.*, 1973) which the application of standard stratigraphic procedures rendered redundant (Bowen, 1973; Warren, 1979a; 1985). Since most, but not all, of the deposits previously ascribed to the penultimate cold stage seem best regarded as last glaciation in age, and the stratotypes relating to that stage have been incorporated into the Fenitian, the term 'Munsterian' no longer has a useful stratigraphic function. Deposits of a pre-Fenitian cold stage, probably penultimate, are recognised at Ballybunnion, County Kerry. This cold stage is termed the Ballybunnionian (Warren, 1985).

Bowen *et al.* (1986) have questioned this stratigraphic framework and prefer to rely on the stratigraphically dubious concept (Hedberg, 1976; Bowen, 1978) of morphostratigraphic regions as the basis of their stratigraphy. But it would seem that some of the difficulties which Bowen *et al.* (1986) have with Warren's (1985) stratigraphic model, particularly with regard to the subdivision of the Fenitian Stage, are more apparent than real and are based on a misunderstanding of stratigraphic principles (see below). It is nevertheless necessary to emphasise where the basis of the differences in approach lie.

Bowen (1973) was the first in modern times to draw up a stratigraphic model which was based on the available lithostratigraphic evidence and which demanded that either the Southern Irish End Moraine of Lewis (1894) and Charlesworth (1928a) be abandoned as the southern limit of last glaciation or be justified on the basis of normal stratigraphic criteria. A close examination of the morphostratigraphic basis upon which the moraine was established (Charlesworth, 1928a) and upon which it is maintained (Bowen *et al.*, 1986) reveals that even upon the criteria used, and irrespective of the more normal

stratigraphic criteria, it does not withstand scrutiny. Bowen *et al.* (1986) summarised the characteristics of the Munsterian glacigenic sediments, as 'subdued morphologies, deep weathering profiles and a distinct lack of fresh glacial bedforms' in contrast to the absence of such characteristics inside the limits of the moraine. Their proposition fails for the following reasons:

1. As has been demonstrated elsewhere (Warren, 1978; 1979a; 1980; 1981; 1985), the generalisations are not a true reflection of conditions south of the moraine. Indeed the 'Midlandian' limit has been shifted south of the moraine in an attempt to exclude 'unweathered' tills from the 'Munsterian' (Finch & Synge, 1966; Finch, 1971). In addition, deposits which were once quite unequivocally placed in the 'Munsterian' area specifically with reference to these precise criteria are now included within the Midlandian sediments (cf. Farrington, 1936; 1954; 1959; Synge & Stephens, 1960; Synge, 1970a; 1979; McCabe, 1985).

2. If the generalisations were true they do not suggest, much less prove, the intervention of interglacial conditions between the deposition of the two suites of deposits (see Bowen, 1973; Warren, 1985; Warren, this volume).

Extent of glacial deposits

The problems in delineating the extent of the deposits of the last glaciation in Ireland are underlined by the absence of distinct dateable marker horizons. It has been argued that the Courtmacsherry (Raised Beach) Formation of the south and southwest coasts is essentially an interglacial deposit (Mitchell *et al.*, 1973; Warren, 1985 and this volume). Bowen (1973) and Warren (1979a, 1985) have shown that stratigraphically this is likely to be the most recent interglacial, and Synge (1977a, 1981) has suggested that it is early glacial. It may be best to regard it as the deposit of the high sea level that immediately preceded the last cold stage (see Orme, 1966) and, as the base of the Fenitian Stage is marked by the base of the periglacial silts overlying the raised beach deposits at Fenit (Warren, 1985), the beach by definition belongs to the previous, Gortian Stage. Whether others interpret it as last interglacial or early Fenitian, it is clear that any *in situ* overlying glacial deposits belong to the Fenitian. On this principle the tills represented by the Bannow Formation, which overlies the Courtmacsherry Formation indicate that inland ice of the last glaciation extended beyond the south coast, and only a small area of north Kerry and possibly small parts of southwest Limerick and north Cork, remained unglaciated during this period (Warren, 1980 and this volume; Warren *et al.*, 1986).

It is probable that the so-called Southern Irish End Moraine represents an important series of ice limits across the southern Irish midlands. It has been shown to be diachronous (Synge, 1970a) but it is not known to what degree. The moraine is a composite feature composed of the Hacketstown Moraine, the Ballylanders Moraine and the Fedamore Moraine (fig. 32) and it is clear that Synge (1977b) envisaged a significant time lapse between the deposition of the Hacketstown Moraine and the Fedamore Moraine. Whether these represent halt stages or readvances of the ice front as it receded from a position off the present south coast is a moot point. McCabe *et al.* (1986) insisted that the areas of drumlinisation reflect a single glacial readvance that, for unstated reasons, produced these bedforms simultaneously in northeast, northcentral, northwest, west and southwest Ireland (presumably including the Bantry Bay and

Figure 32: The morphological expression of glacigenic deposits in Ireland.

Figure 33: Interpretations of the 'Drumlin Readvance Moraine'.

Kenmare River drumlins). As has been stated elsewhere (Warren, 1985) there is absolutely no known reason why such a time constraint should be placed on the formation of drumlins. Less still is there any basis for the corollary use of drumlins as a morphostratigraphic marker horizon as is evident has been done to justify some of the proposed changes in stratigraphic nomenclature of some of the subunits of the Fenitian Stage (Bowen *et al.*, 1986).

The Fenitian glacial deposits

The Fenitian glacial sediments are particularly varied both in their sedimentary structures and their surface expression. There are areas of thin till cover with very little surface expression, for example most of County Wexford (except the Screen Hills), County Waterford, north County Cork, north County Dublin, east County Meath, south County Louth and west County Mayo. There are also areas of little or no glacial sediment cover, particularly in Connemara (west County Galway) and the Burren of County Clare. The deglacial sediments of the Irish midlands form a most striking association of glacial and glaciofluvial sediments. Here interbedded glaciofluvial, glaciolacustrine and direct glacial deposits are commonly in excess of 50 m thick and form such striking features as the Curragh of County Kildare and the very striking esker chains of Counties Westmeath, Offaly and Galway. The eskers of the midlands and those of Galway/Mayo clearly reflect a continuous process of deglaciation from east to west across the midlands and from north to south from the northwest coastal area. Synge (1970a) specifically included the main groups of eskers within his 'Drumlin Readvance Moraine' which joined Charlesworth's (1939) Carlingford Moraine to Synge's Fedamore Moraine in Limerick. Charlesworth (1973) adopted this line for his Carlingford Readvance Moraine which he had previously (1963) drawn between Carlingford and Galway. Synge (1977b, 1979e), followed by McCabe *et al.* (1986), later redrew this line, without explanation, to exclude most of the midland eskers (fig. 33). The confusion regarding the precise location of this 'moraine' is emphasised by Charlesworth's (1973) mistaken belief that his line includes the Galtrim Moraine of Synge (1950).

The other striking group of features is the drumlin swarms that extend from Antrim and Down across the north midlands and up into south Donegal, and the smaller swarms in Counties Galway and Mayo, Clare and Limerick, north Donegal, and south Kerry. These are composed, by and large, of till, but are nevertheless quite varied in composition (McCabe, 1985). The three main groups of drumlins are separated by the midland glacial and glaciofluvial sediments and more particularly by the two main groups of eskers. As there is no indication of a disruption in the esker systems (each esker group appears to represent a continuous period of deglaciation), any readvance moraine associated with drumlin formation would of necessity include within it the major groups of esker chains as was recognised by Synge (1970a). However, neither the Galtrim Moraine (Synge, 1950) nor Carlingford Moraine (Charlesworth, 1939: 270-272) shows definitive evidence of readvance, as distinct from ice front stand-still during retreat, or change in direction of ice movement. It would be more rational therefore to regard any of the other moraines that form the composite Southern Irish End Moraine (see below) as a 'Drumlin Readvance Moraine' than to base the supposed readvance on such an ephemeral feature.

Early and Late Fenitian glacial deposits

We are constrained by the fact that whereas all the good dateable stratigraphic sequences belonging to the Fenitian Stage occur in the north of the country, the sediments which have given rise to most debate about their age are in the extreme south. Yet the dated deposits at Derryvree (Colhoun *et al.*, 1972), Hollymount (McCabe *et al.*, 1978) and Aghnadarragh

(McCabe & Hirons, 1986; Bowen *et al.*, 1986) do offer valuable insights in our interpretation of the Fenitian sediments. Of these, that at Derryvree, which contains organic silts between two tills, offers the most useful finite date. The silts which contain mainly grass and sedge pollen, indicative of open tundra conditions, are dated at 30,500 ± 1170/1030 B.P. by radiocarbon determination (Colhoun *et al.*, 1972). The dates at the other sites are not finite and are therefore less useful in providing minimum or maximum dates for the other sediments in the sequence, particularly the subjacent tills. The lower till at Derryvree (Derryvree Formation) is probably early Fenitian as there is no evidence of interglacial conditions intervening between it and the overlying organic silts.

At Hollymount and Aghnadarragh the organic deposits are also sealed between two tills. The organic freshwater muds at Hollymount reflect in their macrofossils, which have produced a radiocarbon date of > 41,500 B.P. (McCabe *et al.*, 1978), an open countryside with vegetation of a northern type. At Aghnadarragh the organic sediments are more complex. They seem to represent full interstadial conditions with *Pinus*, *Picea* and *Abies* present. The deposit is a woody detritus peat which, in addition to identifiable *Picea* and *Abies* fragments, contains unidentified hardwood (McCabe & Hirons, 1986). A date > 48,180 B.P. is quoted by Bowen *et al.* (1986). This is presumably a radiocarbon date and may be questionable on the basis of the very high potential for contamination from subjacent Tertiary lignite. This is particularly true since there is evidence that the woody detritus was transported some distance before deposition and is probably a reworked deposit. Further information with regard to this date is awaited.

Early Fenitian glaciation

The Derryvree Formation represents an early Fenitian glacial episode, the Derryvree Substage, the ice of which had vacated the north of the country by 41,000 B.P. and possibly as early as 50,000 B.P. The extent of this event is unknown. It could possibly have been responsible for the glacigenic sediments south of the Southern Irish End Moraine complex. The contention that this event must have been short-lived and of limited extent (Bowen *et al.*, 1986; McCabe, 1987) is questionable. There is as yet insufficient evidence to determine either its extent or duration.

Late Fenitian glaciation

The tills overlying the organic sediments at Derryvree represent a later glacial event which affected that area sometime after 30,500 B.P. Once more the precise extent of this event is not known. Fenitian ice might, for example, have reached its maximum extent at this time or it might only have reached any one of the main moraines that go to form the Southern Irish End Moraine (Synge, 1970a). McCabe *et al.* (1986) suggest that an ice margin lay close to the north coast of County Mayo by about 17,000 B.P. They regarded this as within the limit of an undefined 'Drumlin Substage' which seems to relate to the 1979 limit of Synge's (1970a, 1979e) Drumlin Readvance Moraine.

The concept of a Drumlin Readvance Moraine was proposed by Synge (1970a) but has not found general acceptance as an isochronous unit (Hill, 1970; Sugden & John, 1976). The feature has never been subjected to close scrutiny. It is based on the occurrence of a number of broad spreads of dead-ice topography in counties Meath and Westmeath, west of the Carlingford (Synge, 1970a) or Dunany Moraine (Synge, 1979e) close to the margin of the drumlin field to the north. These are projected about 180 km to the southwest to join a similar feature, in County Limerick (the Fedamore Moraine), some distance south of the drumlins that occur along the Shannon estuary. The position of the northern moraine is a matter of some conjecture for it has been placed in distinctly different localities on different maps (Synge, 1970a; 1979e). The morphological and sedimentary associations of the midlands indicate a very close relationship between the ice that deposited the drumlin swarms and that associated with the esker chains, and raise strong doubts as to any significant advance or readvance associated with drumlinisation (cf. Hill, 1970). It is very difficult to envisage anything but a more-or-less continuous process of deglaciation following the deposition of the Ballylanders Moraine, as is suggested in the models of Sugden & John (1976) and Boulton *et al.* (1977). A tentative conclusion that the minimum extent of late Fenitian (Maguiresbridge Substage) ice is marked by the Ballylanders Moraine (largely the old Southern Irish End Moraine) is proposed, but it is thought likely that it extended much further. If not, then the tills to the south represent the Fermanagh Substage.

The concept of a drumlin morphostratigraphic region relating singularly to the 'Drumlin Readvance phase' has led Bowen *et al.* (1986) to suggest that so-called 'drumlin-forming till' cannot represent the deposits of the 'main glacial phase', which in this conceptual framework relates to a different 'morphostratigraphic region'. These concepts are quite baffling as they imply that the glacial event which produced the drumlins of the northwest could not have deposited the sediments outside the drumlin belt, and they belie the distinct sedimentary continuity that extends in places from one of the so-called morphostratigraphic regions to another. This sort of application of morphostratigraphic units is generally unacceptable in stratigraphic interpretation (Hedberg, 1976; Bowen, 1978) and in this case it is based on the unfounded premise of a 'Drumlin Readvance'. Thus, whereas the term 'geomorphological region' would have some meaning when applied to the drumlin swarms, the term 'morphostratigraphic region' is misapplied.

Boulton *et al.* (1977) proposed that final drumlinisation occurred close to the ice margin, during deglaciation, in areas where there had been sufficient time for thick lodgement till and drumlins to develop in zones of low ice velocity. It is axiomatic that the distribution of drumlins is directly related to process and that they may have formed at any time provided that the parameters that controlled the process were in operation. To impose a time constraint by insisting, without any corroborating evidence, that the Irish drumlin fields all relate to a specific drumlin readvance (McCabe *et al.*, 1986; Bowen *et al.*, 1986) is to challenge, without evidence, the principle of uniformitarianism, Thus the Maguiresbridge Formation, a till which overlies the Derryvree organic deposits, represents the period of the late Fenitian glacial event, the Maguiresbridge Substage, which followed the Derryvree Substage, and the objections of Bowen *et al.* (1986) are without foundation.

Direction of ice movement

It is difficult to model the flow patterns of Fenitian ice sheets as there is some evidence of shifting ice sheds or centres of dispersion (Synge, 1970a). However, the outline pattern of ice movement can be reconstructed from the distribution of erratics and other ice direction indicators (fig. 34). This is not to suggest that direction of ice movement did not change or that there was no reworking of sediments in the course of glaciation. Three major ice bodies affected the Irish area: a general Irish ice sheet that related to an ice shed in the north midlands, an ice sheet that occupied the Irish Sea basin and was fed partly through the Firth of Clyde and an extended icecap centred on the mountains of Kerry and Cork.

William P. Warren

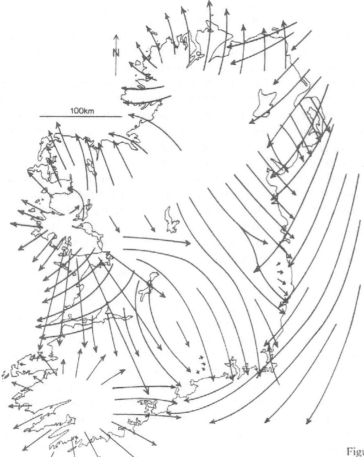

Figure 34: The pattern of ice movement in Ireland (from Geological Survey of Ireland records).

The Irish ice sheet

Galway Granite provides perhaps the clearest and most widespread indicator erratic that throws light on the direction of ice movement over an extended period of time. It extends east-southeast to Slieve Bloom and its distribution fan forms an arc that reaches deep into southeast Tipperary and as far south as Cork City and reaches Fahamore in Kerry to the southwest (fig. 35). Its distribution has been ascribed to the 'Munsterian' glaciation with the suggestion that its occurrence in 'Midlandian' deposits is a result of reworking (cf. Charlesworth, 1963b). Current work in the midlands indicates that in Counties Offaly, Laois and east Galway the direction of ice movement, at least during deglaciation, is entirely consistent with direct transport of erratics from the area of Galway Granite to Slieve Bloom and southeast into Tipperary. Although Galway Granite erratics are widespread in County Clare, the striae

pattern in north and west of the county is indicative of a northeast-southwest ice movement, as is drumlin orientation, and is not consistent with the erratic distribution. In the southeast of the county, however, and in County Limerick, striae are oriented northwest-southeast and are more consistent with the pattern of erratic distribution. Farrington (1965) has suggested that the granite erratics in north Clare have been reworked by the northeast to southwest ice movement. The alternative, that the boulders were deposited after the striae had been cut and the drumlins formed, must also be borne in mind. This could have occurred during deglaciation as the ice margin receded northward over County Clare. Hill (1970) has suggested that north-south oriented drumlins on the north of the Ards peninsula near Bangor in County Down were overridden by northwest-southeast drumlin-forming ice which did not significantly alter or reorient them but deposited a thin cover of till over them. The striae in the

Bangor area also trend north-south (Hill & Prior, 1968).

In the northeast, erratics of Ailsa Craig microgranite occur in a lower shelly till that extends as far southwest as Lough Neagh, illustrating that ice from Scotland pushed across the North Channel. This was later replaced by Irish ice of northwestern provenance which deposited an upper till. Stephens *et al.* (1975) suggest the axis of the Irish ice was aligned northeast-southwest across Lough Neagh to the south Antrim coast. It is likely that the northeast end of this axis waxed and waned across Antrim and Down during the Fenitian (see above).

This ice axis may be extended southwest to County Galway as outlined by Hull (1878). A number of centres of ice dispersal which formed a rough axis along this line would probably best explain the distri-

bution of direction indicators which, south of the line, show a west-east pattern converging with a northwest-southeast pattern and then running almost north-south through the south midlands and across the south coast. The migration of these centres and their waxing and waning would have produced both the broad Galway Granite erratic fan and the contradictory evidence of ice movement direction in north Clare and elsewhere.

During deglaciation the ice front seems to have retreated to two main centres, one in County Galway as indicated by the convergence of the north Connaught esker chain and the main midland esker chain towards Connemara, and the other in Fermanagh/Tyrone and the Lough Neagh basin (Colhoun, 1970), on which the ice margins to the north and those of the northeast and the north midlands converged.

Figure 35: The distribution of Galway Granite erratics. The source area is indicated in black.

Figure 36: The distribution of shelly tills and shells associated with glacial deposits. The heavy line indicates the extent of the coastal area in which the shelly tills and associated shelly deposits occur (from Geological Survey of Ireland records).

Irish Sea Basin ice

A clay/silt-rich shelly till with Scottish and Irish Sea Basin erratics is known mainly from coastal sites and inland sites in coastal counties (Wright, 1914). The erratic suite in the shelly till (the Ballycroneen Formation of Warren, 1985) of eastern and southern Ireland demonstrates an ice movement out of the Firth of Clyde, down the Irish Sea basin and along the south coast as far as Ballycroneen just east of the mouth of Cork Harbour (Wright & Muff, 1904). This ice pushed onshore covering much of Counties Antrim and Down and north Derry. It also pushed inland along the east coast to cover parts of Counties Louth, Meath, Dublin, Wicklow and Wexford (Synge, 1977a) and it may have extended much further inland (fig. 36). Gerrard (1977) indicates that ice occupied

the Irish Sea basin during the last glaciation and possibly during a previous one. The extent of Fenitian ice in this basin has long been a matter of dispute (see Synge, 1977a) as has the nature of the ice sheet. Synge (1977a, 1981) has suggested that the ice sheet in the Irish Sea basin projected an ice shelf into the Celtic Sea as far south as Cornwall in England and as far west as Cork Harbour. Warren (1985) envisaged terrestrial ice and Bowen *et al.* (1986) questioned whether some of the Ballycroneen Formation deposits might not be glaciomarine. McCabe (1987) suggested it is a subaqueous flow deposit associated with a tidewater glacier. Both shelf and glaciomarine deposition would require high relative sea levels. Synge acknowledged this and suggested a very rapid early Fenitian ice advance down the Irish Sea basin before eustatic level had fallen significantly from its interglacial level. No method of deposition was suggested but, as no rise in relative sea level was proposed, subaquatic sedimentation was probably not intended, for Synge (1985) later suggested an early surge into the Celtic Sea without any mention of shelf ice. The suggestion by Bowen *et al.* (1986) and McCabe (1987) that the Ballycroneen Till may be glacio-marine would presuppose a relative sea level somewhat higher than the present shoreline. McCabe (1987) invoked a complex isostatic mechanism for which there is no independent support. Whereas there are undoubted subaquatic facies associated with the Ballycroneen Formation it is inaccurate to suggest, as have Bowen *et al.* (1986) and McCabe (1987), that it is characteristically a subaqueous deposit. It is characteristically a massive clay/silt-rich diamicton with shell fragments and few phenoclasts. It has few sorted lenses or beds but has some shear structures and, on the east coast, a frequently well-striated substrate. Associated laminated silts occur in places and indicate an occurrence of standing water, and subaquatic sequences have been recorded in south County Wexford (Thomas & Summers, 1982). But these are to be expected particularly in deglacial facies.

The shelly till extends to an altitude of 300 m O.D. in south County Dublin and County Wicklow where it is clearly not a subaqueous deposit. Wright & Muff (1904) remarked on the lack of associated striated rock surfaces on the south coast where it is usually underlain by earlier Quaternary sediments. It is likely, whether or not it was associated with high relative sea level, that the till here was deposited from a laterally expanding glacial margin without much erosive power (Synge, 1985).

In the offshore area beyond the south coast between Cork and west Waterford, according to Geological Survey of Ireland records, Quaternary sediments are sparse (R. Keary, personal communication). According to Garrard (1977) there is no till south of the northern margin of the Celtic Sea. However, Pantin & Evans (1984) have recorded till deposits as far south as the central and southern part of the Celtic Sea. They interpret these as ice-rafted deposits but at one site at about 50°N, 8°20'W, close to the latitude of the Isles of Scilly, they sampled a till which is reported to have the geotechnical properties consistent with a lodgement till (see also Scourse *et al.*, this volume). The whole Celtic Sea area is anomalous (Pantin & Evans, 1984). In the southern and central area there is an hiatus between late Pliocene/early Pleistocene and late Pleistocene/early Flandrian (Holocene) deposits (Pantin & Evans, 1984). The

occurrence of isolated till patches and mounds in the south and central Celtic Sea, interpreted as iceberg deposits, lack a corroborating, thickening and more extensive glaciomarine cover to the north (Garrard, 1977). Within the Irish Sea basin massive tills are extensive while in the northern part (from Dublin north), according to Geological Survey of Ireland records, laminated muds with interbedded diamictons seem to prevail (M. Geoghegan, personal communication; cf. Pantin, 1977). This seems to reflect an initial low relative sea level followed by a rapid eustatic rise during deglaciation, but Synge's model (1985) might suggest a large proglacial lake in the central Irish Sea during deglaciation.

The Kerry/Cork ice sheet

Glacial deposits in counties Kerry and Cork largely reflect a Fenitian ice cap that extended from an ice dome centred on the area south of the MacGillycuddy's Reeks. The stratigraphic sequence on the south coast and direction indicators indicate that it extended eastward beyond Cork Harbour. The kame and kettle deposits between Cork City and Midleton, and the Cloyne esker (Lamplugh *et al.*, 1905), were deposited by this ice, but it is difficult to draw the eastern limit of its deposits owing to the lack of distinctive tracer erratics. In the Youghal area, however, striae, which to the west are generally oriented west-northwest, begin to show a stronger northern component. East of Youghal the striae tend to fall northwest to north suggesting a source to the north rather than the west (see Warren, this volume). It is likely that ice from the north merged with the Kerry/Cork ice in the Youghal area. The northern boundary in County Cork is even more difficult to draw, but erratics in the Cork City area along with a change in striae pattern around the northeast shore of Cork Harbour may indicate a confluence of the two ice bodies in that area. In County Kerry felsite erratics from the Lough Guitane area (east of Killarney) extend northwards in an arc with a radius of about 30 km. There is some question as to whether all of these erratics relate to the Fenitian Stage. The erratic distribution, together with the evidence of sedimentary sequences on the northern side of Sliabh Mis, suggest that the ice sheet, probably added to by local glaciers, passed over the lower cols of Sliabh Mis and extended as far as Fahamore (Warren *et al.*, 1986 and this volume).

The pattern of deglacial sediments associated with this ice sheet is very different to that associated with the ice sheet centred on the midlands. While there are two small drumlin fields in Bantry Bay and Kenmare River, large spreads of kame deposits and extensive esker chains are absent. A significant halt stage of the eastern ice front associated with a belt of gravel moraine and extensive outwash gravels occurs at Killumney near Cork City (Lewis, 1894; Farrington, 1959). Intermittent spreads of gravel and till occur west of this moraine. North of the MacGillycuddy's Reeks deglacial deposits include extensive kame terraces, gravel and till moraines, often remarkably fresh and continuous, as well as ice marginal and interlobate glaciolacustrine deposits. Small spreads of kame and kettle gravels also occur (Warren, 1978).

Cirque and valley ice

Most of the larger mountain groups that were not overrun by extraneous ice contain cirque and valley moraines. The Wicklow Mountains held a small ice cap that extended to the margins of the Irish Sea ice sheets to the east and the midland ice sheet to the west (Synge, 1977a). There is evidence, in the Clogga Till on the east coast (Synge, 1977a) and in erratic boulders to the west, of an earlier and more extensive ice cap. The age of this event is not known. If gravels overlying the Clogga Till relate to the Courtmacsherry (raised beach) Formation it is probably pre-Fenitian (Synge, 1977a). If not then it is likely to be Fenitian in age (Warren, 1985).

Comment

The lack of agreement as to the extent of glaciation during the last cold stage (cf. Warren, 1985; Bowen *et al.*, 1986) reflects to some extent the inadequacy of the available biostratigraphic evidence. It also reflects shortcomings in the principles upon which Quaternary biostratigraphic correlations are made (cf. Birks *et al.*, 1975; Warren, 1979a; Gennard, 1984). However, most important, it reflects the differences between the standard stratigraphic approach of geologists (Bowen, 1973; Warren, 1979a; 1985) and the morphostratigraphic approach of physical geographers (Bowen *et al.*, 1986) upon which, in the last analysis, the traditional interpretation of Irish interglacial biostratigraphy rests (cf. Watts, 1985: 162-163).

The Loch Lomond Stadial glaciation in Britain and Ireland

J. MURRAY GRAY & PETER COXON

In the absence of any unequivocal evidence that glaciers existed in the British Isles during the Little Ice Age or earlier in the Flandrian, it appears that the last glaciation of Britain and Ireland occurred during the Loch Lomond (Younger Dryas) Stadial, c. 11,000 - 10,000 years B.P. Although the existence of a late phase of valley glaciers had long been recognized in Scotland, the term Loch Lomond Readvance was first used by Simpson (1933) on the basis of field evidence at Loch Lomond itself and at Menteith in the Upper Forth valley. Sissons (1977a), believing that the glaciers built up from scratch rather than resulting from a readvance of the last ice sheet, suggested that the term Loch Lomond Advance would be more appropriate, but this proposal has been challenged, since there is no direct evidence of complete ice sheet decay (for example Sutherland, 1984a). The title of this paper avoids both terms, though it should be stressed that in Ireland the name Nahanagan Stadial is used in preference (Colhoun & Synge, 1980).

Tracing the glacial limits

With some important exceptions, little research has been carried out on the stratigraphy and sedimentology of the stadial glacial deposits. Over the last 15 years the main research thrust has been devoted to defining the extent of the stadial glaciers, mainly on the basis of geomorphological mapping by J.B. Sissons and his research students. The main geomorphological evidence (fig. 37) used in delimiting the extent of glaciers has been:

(i) *Hummocky moraines*. These steep-sided, sharp-crested mounds and ridges, frequently strewn with boulders, are common in many Highland valleys where they have long been associated with a valley glacier phase (fig. 38). Sometimes the hummocky moraines are closely associated with large-scale fluted moraines. Sissons' mapping demonstrated that the hummocky moraines often have a distinct and abrupt down-valley limit that he interpreted as representing the maximal extent of glaciers (fig. 37a). Good examples from northern Scotland are shown in fig. 39. However, the association of hummocky moraine with the Loch Lomond Stadial has been disputed (for example Sugden, 1970) and is difficult to understand on glaciological grounds. Indeed there are several localities where it has been demonstrated that steep-sided hummocks predate the Loch Lomond Stadial glaciation (for example Clapperton *et al.*, 1975; Sissons, 1982a). Thus care must be taken to interpret the distribution of hummocky moraines in the context of landform assemblages over a wide area.

(ii) *End and lateral moraines*. These are common in some areas but apparently rare in others. For example of 35 glacial limits mapped in Snowdonia, 30 utilise end or lateral moraines (Gray, 1982; see fig. 40), whereas in the Gaick area of the Scottish Highlands Sissons (1974a) found that end moraines were very poorly developed. In some places recessional moraines occur up-valley of the end moraines (fig. 37b), the most impressive series occurring at Lough Nahanagan in Ireland.

(iii) *Boulder and drift limits*. Boulder limits occur where there is an abrupt limit to an arcuate spread of boulders, contrasting with either a boulder-free area outside the limit (fig. 37c) or with very different boulder lithologies. The best published descriptions are from northwest Scotland where Sissons (1977a) reported spreads of white Cambrian quartzite boulders terminating abruptly and contrasting sharply with the red Torridonian Sandstone bedrock beyond. Drift limits are similar but as well as a contrast in the concentration of boulders there is also a sharp change in drift thickness (fig. 37d). A good example occurs at Cwm Ffynnon Llugwy in Snowdonia. There are also more complex situations where limits are partly defined by end moraines and partly by drift and boulder limits.

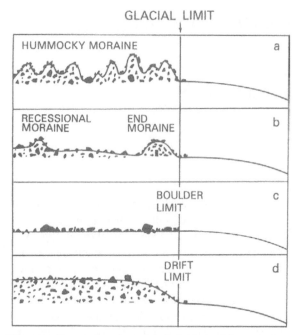

Figure 37: Some types of geomorphological evidence used to delimit the extent of Loch Lomond Stadial glaciers; a) limit of hummocky moraine, b) end moraine, c) boulder limit, d) drift limit.

(iv) *Weathering contrasts.* As Thorp (1981, 1986) pointed out, most research has focussed on delimiting the snouts of the Loch Lomond Stadial glaciers and relatively little attention has been paid to reconstructions of the vertical extent of the glaciers in the accumulation zones. Working in the western Grampians, he was able to map a trimline whose altitude varies consistently and which could be traced down into the glacial termini mapped by other workers. The main evidence used was the contrast between smooth, ice-moulded bedrock below, and severely periglaciated surfaces above, the proposed former glacier margins (fig. 41). Several other workers have used contrasts in periglacial weathering to aid in the reconstruction of the stadial glaciers (for example Sissons, 1974a; Ballantyne & Wain-Hobson, 1980). Recently, J. Rose (personal communication) has been attempting to determine whether the Schmidt hammer can be used to pick up contrasts in rock strength that might be interpreted in terms of differential weathering on either side of the glacier limits. Earlier, Shaw (1976) had rather limited success in using other relative dating methods based on boulder weathering in the Lochnagar area. On the other hand J. Bibby (personal communication) has found that the Lussa Soil Association on the Isle of Mull corresponds closely with the Loch Lomond limits mapped independently from geomorphological evidence (Gray & Brooks, 1972), thus demonstrating a potentially important mapping technique.

(v) *Bog stratigraphy contrasts.* This is a further important method used in tracing the glacial limits. The principle is described in the dating section below.

Using this evidence for snout and lateral limits it has been possible to reconstruct the outlines of the stadial glaciers, though this has often involved a large degree of extrapolation into accumulation/headwall areas, and in areas where there is limited evidence of the types described above.

Figure 38: Hummocky moraine, Glen Fuaron, Isle of Mull (Photograph: J. M. Gray).

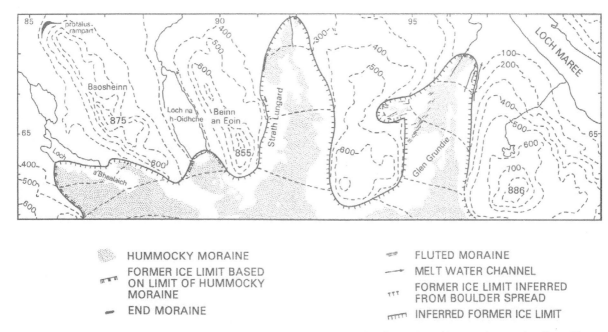

HUMMOCKY MORAINE

FORMER ICE LIMIT BASED ON LIMIT OF HUMMOCKY MORAINE

END MORAINE

FLUTED MORAINE

MELT WATER CHANNEL

FORMER ICE LIMIT INFERRED FROM BOULDER SPREAD

INFERRED FORMER ICE LIMIT

Figure 39: Former glacier limits in part of NW Scotland, based largely on the distribution of hummocky moraine (from Sissons, 1977a).

Figure 40: End moraine at Cwm Drws-y-Coed, Snowdonia, North Wales (Photograph: J.M. Gray).

1   Frost-riven bedrock
2   Fossil scree
3   Stone-banked solifluction lobes
4   Tor-like summit
5   Solifluction sheet
6   Turf-banked solifluction terraces
7   Blockfield
8   Debris-strewn slope
9   Thick gullied till
10 Boulder spread
11 Hummocky moraine
12 Roches moutonées
13 Till
14 Ice-moulded bedrock

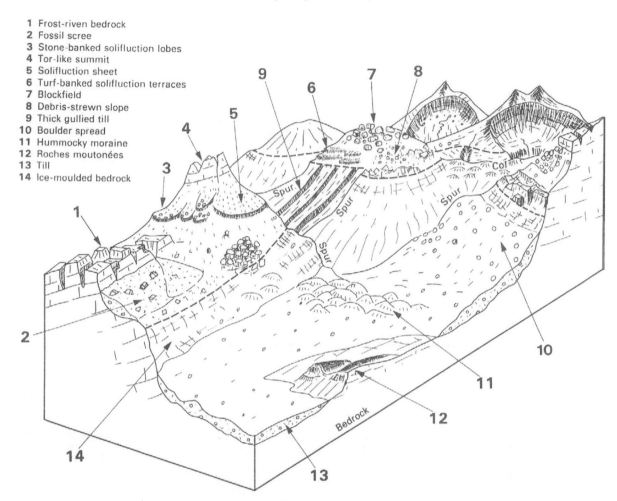

Figure 41: Evidence used by Thorp (1986) in delimiting stadial trimlines in the western Scottish Highlands.

Other geomorphological features

As well as the geomorphological evidence described above used in the reconstruction of the Loch Lomond glacial limits, several other interesting geomorphological features are related to the glaciation and include:

(i) *Protalus ramparts*. These are particularly common in areas marginal for glacier formation during the stadial, for example Lake District, Snowdonia, Cader Idris, Brecon Beacons (fig. 42), though they occur in Scotland as well. They indicate the presence of semi-permanent snow-beds, though there is a clear transition through to arcuate ridges that are glacial in origin (Gray, 1982). The 'remarkable' example described by Sissons (1976a) has been partly re-interpreted by Ballantyne (1986) as a lateral moraine.

(ii) *Rock glaciers*. A few features interpreted as fossil rock glaciers have been assigned to the Loch

Lomond Stadial. Examples occur on the Isle of Jura (Dawson, 1977, see fig. 43) and at Moelwyn Mawr and Cader Idris in N. Wales (S. Lowe, personal communication). The last example grades laterally into a normal glacier.

(iii) *Other periglacial features*. Among the other periglacial features related or probably related to the stadial are pingos, solifluction sheets, fossil screes (fig. 42), ice wedges, blown sand and river gravels, and these come from sites widely scattered over the British Isles (see, for example, Sissons, 1979e for brief descriptions and references).

(iv) *Drumlins*. South of Loch Lomond the glaciers moulded the till into a series of drumlins (fig. 44) that are significantly smaller than those produced during the Dimlington Stadial. Rose & Letzer (1977) suggested that this was probably due to lower glacier stress/bed resistance levels during the Loch Lomond Stadial.

Figure 42: Protalus rampart at the head of the Tal-y-llyn valley, mid-Wales. Bedded talus exposed in the foreground (Photograph: J.M. Gray).

Figure 43: Rock glacier, Isle of Jura, Scotland (Photograph: J.M. Gray).

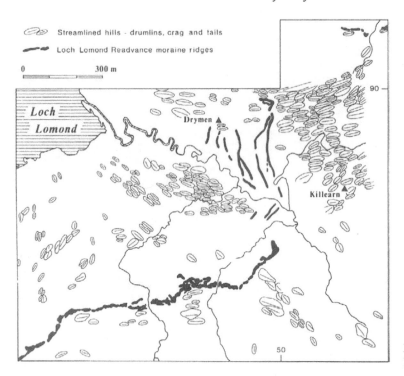

Figure 44: Drumlins associated with the Loch Lomond glacier (from Rose, 1981).

(v) *Glaciofluvial features.* Hundreds of small (1-4 m deep) meltwater channels have been mapped within the limits of the Gaick ice-cap in the central Grampians, while the most impressive glaciofluvial deposits are the outwash spreads and related deposits formed as the stadial glaciers began to retreat in western Scotland, for example at Loch Etive, where there is also an impressive kame terrace and outwash system (Gray, 1975, and this volume).

(vi) *Ice-dammed lakes.* The famous Parallel Roads of Glen Roy are associated with lakes dammed up by Loch Lomond Stadial glaciers advancing up the valleys north and east of Spean Bridge. A considerable amount of research has enabled reconstruction of the history of lake drainage including details of river terraces, fluvial adjustment and jökulhlaups discharging northwards via Loch Ness (for example Sissons, 1978; 1979c, f; Sissons & Cornish, 1982). The lake shorelines were shown to be at least partly erosional, probably due to periglacial processes acting on the lake shore (Sissons, 1978). Ice-dammed lakes were also formed in Glen Moriston (Sissons, 1977c) and around Loch Tulla (Ballantyne, 1979).

(vii) *Coastline.* Some workers have also argued for periglacial erosion of the sea coast at this time and have assigned the development of a major erosional shoreline - the Main Lateglacial Shoreline - to this period. The shoreline occurs as a buried or sub-merged platform in eastern Scotland (Sissons, 1969; 1976b), but it is best seen in the area around Oban where it is clearly much better developed outside the proposed limits of Loch Lomond Stadial glaciers than within their limits. It is glacio- isostatically tilted from c. 11 m O.D. north of Oban to below sea-level in Southern Kintyre (Gray, 1974; 1978a). There is increasing evidence that the stadial ice caused a redepression of the crust (Firth, 1986).

Deposits and their genesis

Only a few studies have analysed the glacial sediments related to the stadial glaciers, but there has been considerable controversy as to the nature and origin of the deposits, particularly in relation to the genesis of the hummocky moraine. Sugden (1970) argued that in the Cairngorm valleys most of the landforms are fluvioglacial and most of the sediments are sands and gravels. He believed that the 'hummocky moraine' was formed as a result of stagnation of patches of ice isolated at the heads of glens by general downwastage of the last ice-sheet. Sissons (1979a) disputed many of Sugden's observations and interpretations related to the Cairngorms. He believed the features to be hummocky moraines, and though sections are rare, those that do exist indicate a sandy till composition. This accords with many other observations of sections in hummocky moraines from various parts of Scotland. Furthermore, the general absence of recessional moraines formed during the wastage of the Loch Lomond Stadial glaciers has led some workers to conclude that the glaciers probably stagnated *in situ* as a response to rapid climatic amelioration, rather than undergoing active snout retreat (Sissons, 1974a; 1979e; Gray & Lowe, 1977; Gray, 1982).

Eyles (1979, 1983a, c, 1984) agreed that most sections in hummocky moraines reveal a coarse-grained diamict, but he argued that the moraines were the result of supraglacial sedimentation at the snouts of glaciers that remained active during their retreat, the debris having been mainly derived from severe periglacial weathering of exposed valley sides. He presented detailed observations of these processes operating on modern Icelandic glaciers and argued that these provide a modern analogue for the deposition of the Scottish hummocky moraine.

The most detailed sedimentological analysis of hummocky moraine sediments was carried out on samples from northwest Scotland by Hodgson (1982), including particle size distributions, fabrics, clast shape and clast composition. Evidence from sections and pits dug into the moraines confirmed that they are predominantly composed of a bouldery till with a sandy matrix. He also demonstrated that in the hummocky moraines he studied, the composition of the till changed rapidly immediately down-ice from changes in bedrock lithology. On the other hand, there is a distinct lack of correspondence between till composition and the bedrock lithology of the valley sides and headwalls. This led Hodgson to the conclusion that most of the sediment was picked up from the valley floors and carried only a short distance by the Loch Lomond Stadial glaciers. He

suggested that the material was at least in part bulldozed by the advancing glaciers and subsequently overridden and partly reworked by the ice. This subglacial modification is supported by the common association between hummocky moraines and fluted moraines (see fig. 45) since the latter are clearly of subglacial origin. This work is described in more detail by Hodgson (1986).

Reworking of previously deposited drift is also particularly evident in coastal localities where the tills contain large proportions of shelly marine clays and silts either mixed in with the glacial debris to produce a till with a fine-grained shelly matrix, or incorporated with the glacial debris in complex glaciotectonic structures. Examples occur at Rhu Point (Rose, 1980b), Kinlochspelve on Mull (Gray & Brooks, 1972) and Loch Creran (Peacock, 1971c). In other places end moraines contain pushed-up pre-existing lake sediments, as at Lough Nahanagan (Colhoun & Synge, 1980), or sheared ice-dammed lake sediments, as at Gartness south of Loch Lomond (Rose, 1980b), or are dominated by glaciofluvial sediments, as at Loch Don on Mull (Gray, 1975).

Dating

That the glacial limits defined by the methods described above were produced during the Loch Lomond Stadial has been clearly demonstrated by 'limiting' radiocarbon dates. On the one hand, dates that provide a *maximal* limit are obtained from marine shells or organic lake silts within deposits that were either over-ridden or transported by the glaciers. These are listed in table 8 and indicate that the ice advance culminated after 10,900 B. P.

On the other hand, *minimal* limits are provided by dates for the earliest organic sediments at the base of

Table 8. Maximal radiocarbon dates for culmination of glaciation, modified after Sutherland (1984a).

Location	Laboratory no.	Radiocarbon age yr BP (1 Range)	Material	Reference
Drymen	I-2235	11,530–11,870	Shell	Sissons (1967)
Menteith	I-2234	11,630–11,970	Shell	Sissons (1967)
South Shian	IGS-C14/16	11,320–11,740	Shell	Peacock (1971)
South Shian	IGS-C14/17	11,630–11,990	Shell	Peacock (1971)
South Shian	IGS-C14/18	11,210–11,650	Shell	Peacock (1971)
Kinlochspelve	I-5308	11,160–11,500	Shell	Gray & Brooks (1972)
Lough Nahanagan	Birm-321	11,340–11,860	Organic silt	Colhoun & Synge (1980)
Lough Nahanagan	Birm-320	10,950–12,050	Organic silt	Colhoun & Synge (1980)
Rhu	HAR-931	10,860–11,370	Shell	Rose (1980)
Loch Goil	T-1456	12,110–12,410	Shell	Sutherland (1981)
Balloch	SRR-1530	10,780–11,050	Shell	Browne & Graham (1981)

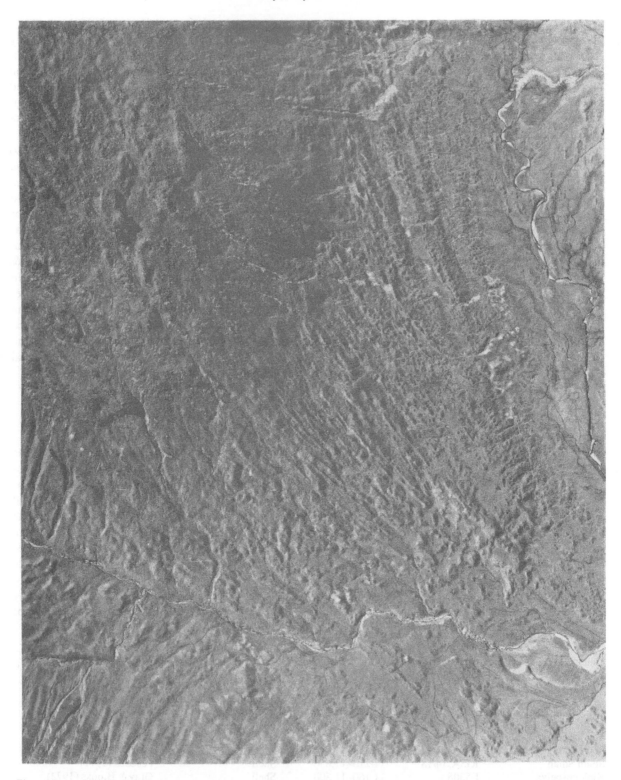

Figure 45: Aerial photograph of fluted/hummocky moraine, Glen Forsa, Isle of Mull (Crown Copyright: RAF photograph).

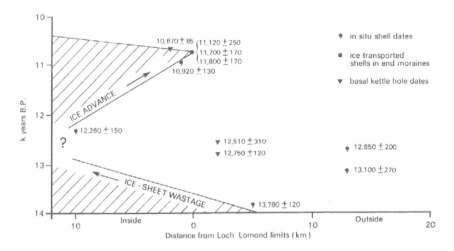

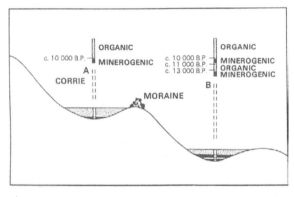

Figure 46: Lateglacial chronology, north Clyde area.

Figure 47: Schematic diagram to illustrate the different bog stratigraphies on either side of Loch Lomond Stadial glacier limits.

surface kettle holes located within the glacial limits. The large number of available dates indicate that widespread stagnation of the ice masses had occurred by 10,200 B.P. (See for example Gray & Lowe, 1977; Lowe & Walker, 1980; Walker & Lowe, 1982).

The fullest sequence of dates occurs in the southwest Scottish Highlands north of Glasgow where the chronology depicted in fig. 46 has been established. The presence of overridden Lateglacial marine sediments (Clyde Beds) at the Loch Goil site is significant in indicating an ice advance of at least 10 km. A glacial maximum around 10,750 B.P. is indicated, though there would have been significant local and regional variations around Britain. For example, in south Wales and southeast Ireland the full effects of the stadial were not felt until about 10,600 B.P. (Craig, 1978; Walker, 1980) and in these southern parts of the British Isles the stadial was probably relatively short.

The kettle hole sites at which the oldest basal sediments are from the Devensian/Flandrian transition have also been helpful in defining the extent of stadial glacies (fig. 47). The principle is based on the presence of Lateglacial Interstadial sediments in kettle holes unglaciated during the stadial and absence of them in sites glaciated at this time (for example Sissons *et al.*, 1973). The method has been used effectively in various parts of Britain (for example Walker, 1980), though it should be noted that the real or apparent absence of Lateglacial Interstadial sediments from a kettle hole can arise in other ways (Gray, 1975).

Distribution of glaciers

By far the largest ice mass of the stadial glaciation was situated over the western Scottish Highlands (fig. 48), stretching from Loch Torridon in the north to Loch Lomond in the south (175 km), and from Loch Shiel in the west to Loch Rannoch in the east (100 km). However, being a mountain ice cap there were numerous nunataks projecting through the ice, Thorp (1986) having mapped over 60 in the central portion alone. Trimline mapping on these and other valley sides showed that the ice surface exceeded 700 m O.D. in places, while in the west several glacier snouts descended to below present sea level. Ice thicknesses in excess of 400 m occurred in Rannoch Moor and the Great Glen. The ice cap was markedly asymmetric with a large area over 600 m O.D. situated towards its eastern margin (Thorp, 1984; 1986).

Over 200 other independent ice masses have been mapped in the Scottish Highlands and Inner Hebrides (Sutherland, 1984a). Three of these were smaller

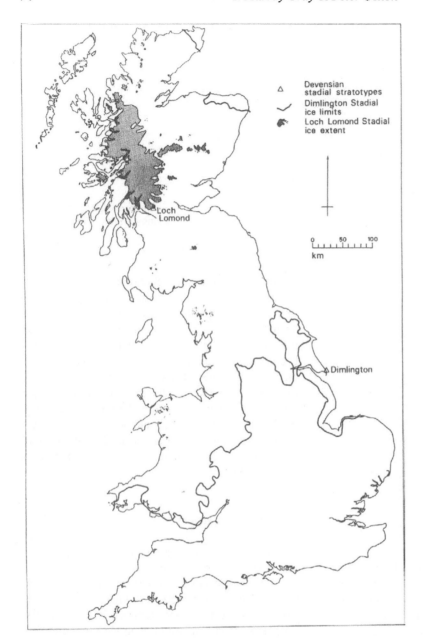

Figure 48: Distribution of Loch Lomond Stadial glaciers in Britain.

ice-caps - in the Gaick (c. 45 × 10 km) and Glen Mark (c. 10 × 10 km) areas of the eastern Highlands (Sissons, 1972; 1974a), and on the Isle of Mull (c. 20 × 16 km, Gray & Brooks, 1972). Most of the remainder were valley or corrie glaciers, with major groups having been mapped in the north-west Highlands (Sissons, 1977a; Robinson, 1977), Skye (Sissons, 1977b), Rhum (Ballantyne & Wain-Hobson, 1980), Mull (Gray & Brooks, 1972), Arran (D.G. Sutherland personal communication), the Cairngorms (Sissons, 1979a), and the southeast Grampians (Sissons, 1972). Glaciers also developed on Harris and southwest Lewis in the Outer Hebrides during the stadial (Sutherland, 1984a), while on St. Kilda two protalus ramparts have been mapped (Sutherland *et al.*, 1984).

In the western Southern Uplands a group of 11 small glaciers has been mapped by Cornish (1981) around Merrick and the Rhinns of Kells, while in the central section Price (1983) describes a small plateau ice-cap between Broad Law and Hart Fell. May (1981) mapped three small glaciers southeast of Green Lowther and Sissons (1979e) mentions an isolated glacier on the Cheviot.

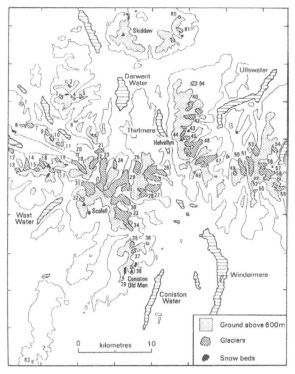

Figure 49: Loch Lomond Stadial glaciers in the Lake District as numbered by Sissons (from Sissons, 1980).

In the Lake District a total of 64 glaciers was mapped by Sissons (1980) though the largest, north east of Scaffell, was only 5 km long (fig. 49). These glaciers covered an area of only 55 km² and their combined volume was only 3.3 km³. In the upper Eden valley of the north Pennines, Rowell & Turner (1952) mapped 8 small glaciers between Mallerstang and Ravenstonedale commons, though Letzer (1978) argued that the morphology was more akin to rock glacier tongues or rotational landslips. At one of the sites - Swarth Fell - there is a good example of an end moraine/protalus rampart but further research is underway into the origin of the features (W. Mitchell, personal communication). Other possible glacier sites in northern England are mentioned by Manley (1959).

Wales was also marginal for glaciation during the stadial, this being reflected in the small size of the glaciers and the numerous protalus ramparts. For example, in Snowdonia Gray (1982) mapped 35 glaciers with a total area of only 17.5 km² and 16 protalus ramparts. The largest glacier, in Cwm Llydaw east of Snowdon, was only 4 km long (fig. 50). Farther south S. Lowe (personal communication) has mapped glaciers, protalus ramparts or rock glaciers in

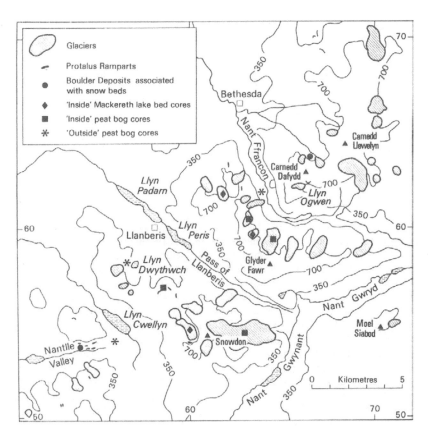

Figure 50: Loch Lomond Stadial glaciers in Snowdonia (from Gray, 1982).

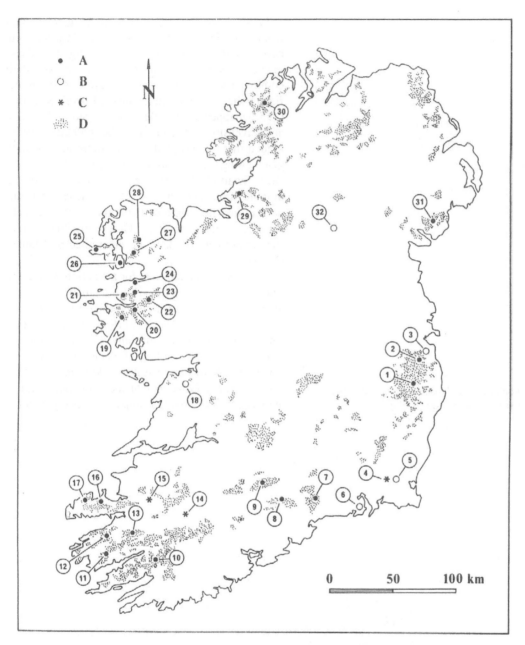

Figure 51: Location map of sites and areas containing moraines and protalus ramparts of *possible* Nahanagan Stadial age, dated pingo remnants and some of the more important biostratigraphical sites containing Nahanagan Stadial material. An (×) denotes radiocarbon dates available at that site. N.B. only site 1 has dated glacial features. 1 = L. Nahanagan (×), 2 = L. Bray (and other Wicklow Mt. areas), 3 = Ballybetagh (×), 4 = Camaross, 5 = Coolteen (×), 6 = Belle Lake (×), 7 = Comeragh Mts., 8 = Knockmealdown Mts., 9 = Galty Mts., 10 = Shehy Mts., 11 = Cahernageeha, 12 = Teermoyle Mt., 13 = Macgillcuddy's Reeks, 14 = Meenskeha (×), 15 = Castleisland, 16 = Slievanea, 17 = Brandon Mt., 18= Gortalecka, 19 = Twelve Pins, 20 = Maumturk Mts., 21 = Mweelrea Mts., 22 = Partry Mts., 23 = Sheefry Hills, 24 = Croagh Patrick, 25 = Achill, 26 = Corraun, 27 = L. Anaffrin and Nephin Beg Mts., 28 = Corslieve, 29 = Glennif and Glenade, 30 = Donegal Mts., 31 = Mourne Mts., and 32 = Drumurcher (×). The construction of this map involved many personal observations and the use of published records of Nahanagan Stadial features not all of which are necessarily accepted by this author (PC). The map can be used as a first approximation of potential sites of Nahanagan age. Data sources include: Browne, 1986; Colhoun, 1971, 1981; Colhoun & Synge, 1980; Coope *et al.*, 1979; Coudé, 1983; Coxon, 1985, in press; Craig, 1978; Kenyon, 1982; Mitchell, 1973, 1976, 1977; Synge, 1963, 1968, 1978; Warren, 1970, 1979 and Watts, 1977, 1985. A = Cirques or upland areas containing cirques with possible Nahanagan moraines or protalus ramparts; B = Important biostratigraphic sites (not included in above category) that contain deposits of Nahanagan age; C = Biostratigraphically dated pingo remains; D = Land above 300 meters.

a number of localities including the Cader Idris, Arenig and Aran areas. In the Brecon Beacons and neighbouring Forest Fawr and Mynydd Du areas of south Wales Ellis-Gruffydd (1977) described small stadial moraines and ramparts in 25 localities, though recently this area has been remapped (D. Robertson, personal communication).

Ireland

The Loch Lomond Stadial in Ireland has been named informally by Mitchell (1976) as the Nahanagan Stadial. This term stems from work by Colhoun & Synge (1980) at the cirque lake of Lough Nahanagan (fig. 51). The cirque, eroded in granite of the Leinster mountain chain, lies between 420-520 m O.D. and was studied in 1968-69 when the cirque lake was lowered by just under 40 m during the construction of a pump-storage scheme. 60% of the bed of the cirque lake was exposed, revealing a series of moraines and laminated and ice-pushed clay deposits. The ice-pushed lake clays from a moraine contained abundant pollen of *Gramineae* and *Cyperaceae* with *Juniperus*, *Rumex* and *Empetrum* also present. This pollen assemblage (*Juniperus-Empetrum*) and the radiocarbon dates obtained from the lake clays incorporated within moraine B (11,600 ± 260 a B.P. and 11,500 ± 550 a B.P.) indicate that the clays were deposited during the Woodgrange Interstadial and subsequently deformed by ice-push. A core from behind this moraine contained pollen of the *Artemisia* phase at the base showing that the renewal of glaciation at Lough Nahanagan did not cover the entire *Artemisia* phase and was probably of short duration (in the order of 4-500 years, Watts, 1977). On this basis, the ice activity in Lough Nahanagan is believed to have occurred after 11,500 B.P. probably during the period between c. 11,000-10,500 B.P. (Colhoun & Synge, 1980).

This site does not provide particularly good stratigraphic control, for example it lacks radiocarbon dates bracketing the glacial activity that produced the inner cirque moraines. However, it is the only site in Ireland that unequivocally shows renewed glaciation in the Lateglacial Stadial through lithostratigraphical analysis, biostratigraphical information and radiocarbon dating.

The Nahanagan Stadial in the rest of Ireland is well represented lithostratigraphically and biostratigraphically at a number of sites as an inorganic (usually clayey and often containing coarse sand or pebbles) sediment sequence and an *Artemisia* pollen assembl-

age respectively (Watts, 1985). Numerous radiocarbon dates are available from this period (e.g. Craig, 1978) but no published work, other than that detailed above, proves glacial activity during the Nahanagan Stadial.

Many small cirque moraines around Ireland have been assigned to the Nahanagan Stadial (referred to in older literature as Lateglacial Zone III) on the basis of their location, well within larger moraines and their 'fresh' appearance but no absolute dates exist for these features. The need for further research into dating Irish cirque moraines is clear although the problem is not an easy one to solve (Browne, 1986). Warren (1979c) has commented on the problem of a lack of dating for cirque moraines in the MacGillycuddy's Reeks area (County Kerry) and concluded that although Nahanagan Stadial moraines are probably present, there is simply no evidence for the age of morainic features.

One area that has been the subject of recent research is the Nephin Beg mountain range in County Mayo. Here Synge (1968) suggested that only three out of twelve cirques in the area contained moraines of Nahanagan age. He had originally commented (Synge, 1963c) that Younger Dryas moraines were small and limited to cirque walls in a few cirques except at one site (Corranabinna) where a small valley glacier may have existed. On the other hand Kenyon (1982, 1986), who carried out detailed geomorphological mapping of the Nephin Beg area, put forward that glaciers were much more extensive during the Nahanagan Stadial than had been previously thought. However, when reconstructing the glaciers he still only postulated small-scale cirque glaciation (with glaciers reaching between 0.05 - 0.77 km^2 in area) producing small moraines in some of the higher cirques as at Corslieve 2, 3 and 4 and Corranabinna (fig. 52a). Coudé (1983, 1985) also states that Nahanagan Stadial (Tardiglacial) features (including moraines and/or protalus ramparts - the genetic origin not always being clear) are widespread, occurring in 41 out of 139 cirques in Iar Connaught. The features Coudé has mapped are small morainic arcs very near to the upper reaches of the cirque walls. This figure includes cirques in the Nephin Beg mountains (Corslieve 1, 2, 3 and 4, Lough Annafrin and Corryloughaphuill are amongst those mapped by him), Sheefry Hills and Partry Mountains. Coudé (1983) noted that the bigger cirques whose volume was greater than 10 × 10^6 m^3 contained moraines thought to be this age while only 24 out of 117 cirques of lesser volume possess this characteristic.

Work by Browne (1986) on cores taken from

cirque lakes in the Nephin Begs has shown that no Nahanagan Stadial glaciation occurred in Lough Annafrin (184 m O.D.) and if a glacier did exist in Corslieve 1 (320 m O.D.) then it was probably very small. Browne has also suggested, on the basis of palynological work, that conditions during the Nahanagan Stadial, even at high altitude in the Nephin Begs were not particularly severe. So even in an area that has been worked over by numerous researchers, using a variety of approaches, we have little idea of the extent of Nahanagan Stadial glaciation and more seriously no dated evidence of glacial activity. One conclusion that can be drawn here is that glaciation in the Irish mountain cirques was only of limited extent and did not occur on the scale found in northwestern Britain.

Like the glacial features, dated periglacial features in Ireland are rare. Even the detailed work of Colhoun (1971b) in the Sperrin Mountains could not satisfactorily distinguish between features of Lateglacial Zones I and III (Nahanagan Stadial) age. Many periglacial features exist (Lewis, 1985) but to date only the organic infilling of pingo remnants (dated to the end of the Nahanagan Stadial) and solifluction deposits have been radiocarbon-dated. The sediments in pingo remnants suggest that the pingos themselves formed, or were reactivated, during the Nahanagan Stadial itself (Mitchell, 1973; Lewis, 1985; Coxon, 1986a). Even these dates are not entirely satisfactory as they do not necessarily date the pingo's existence. Solifluction of inorganic debris into lake sediments has been dated to the Nahanagan Stadial at numerous sites and at Drumurcher in County Monaghan, a solifluction deposit of sandy silt was dated to 10,515 ± 195 B.P. (Coope et al., 1979). On the basis of 'freshness of form' both Colhoun (1981) and Coxon (1985b) have put forward a Nahanagan Stadial age for large protalus ramparts (fig. 52b) but such an age is only conjecture. Indeed Kinahan (1894) observed protalus ramparts - or clocha sneachta snow stones in gaelic - being added to in County Wicklow in the Winter of 1876-77 (Warren, 1979c; Edwards & Warren, 1985), perhaps indicating that smaller-scale features may be more recent.

Many authors have thought the Nahanagan Stadial to have been a severe climatic period (for example Mitchell, 1976) and Watts (1977; 1985) has pointed out the importance of Ireland's proximity to the movements of the Polar Front (Ruddiman & McIntyre, 1981b) in controlling Younger Dryas vegetation response to climate. If the climate of Ireland was severe during the Nahanagan Stadial and if Ireland's climate was especially responsive because of the Polar Front then clearly Ireland is an area where much more work on the chronology, biostratigraphy and geomorphology of the Lateglacial and the Nahanagan Stadial in particular should be done. So far, the limited evidence available suggests that during the Nahanagan Stadial glacial activity was very limited in extent, producing only small inner cirque moraines; but it must be emphasised that much work remains to be done on Ireland's glacial history. However, periglacial processes, especially solifluction, were widespread and the vegetational record provides a distinctive picture of a deterioration in plant cover.

Palaeoclimatic inferences

Having reconstructed the stadial glaciers as described above, it becomes possible to attempt palaeoclimatic reconstruction. Early attempts were made by Manley (1959), but following the detailed geomorphological mapping of the stadial glaciers, a more comprehensive approach has been developed by Sissons & Sutherland (1976), Sissons (1979b) and Sutherland (1984b). The first step is to contour the glacier surfaces using modern glaciers as analogues and for each ice mass calculate the glacier surface areas within successive altitudinal bands. These values are then substituted in an equation developed by Sissons (1974a) to give the altitude of the equilibrium line (ELA):

$$ELA = \sum A_i \times h_i : \sum A_i$$

where A_i = area of the glacier surface at contour interval i (in km^2), h_i = altitude of midpoint of contour interval i.

These calculations have been carried out for numerous Scottish glaciers and the resulting spatial distribution of ELAs is shown on fig. 53. A clear rise from values below 400 m in the west to over 1000 m in the east is evident, a trend that is mainly explicable in terms of an eastward decrease in precipitation similar to that of the present day. Support for the idea of stadial aridity in the eastern Grampians comes from the high Artemisia values in the area at the time (Birks & Mathewes, 1978). Sissons (1979b) suggests that stadial precipitation in the area may have been as low as 200-300 mm in places. In the Outer Hebrides, ELAs of between 125 and 300 m have been calculated (D.G. Sutherland, personal communication).

More surprising is the southeast to northwest rise away from the Highland edge in the southeast Gram-

Figure 52a: Moraines at Corslieve 2, Nephin Beg Mountains of possible Nahanagan Stadial age (Photograph: P. Coxon).

Figure 52b: Protalus rampart, Gleniff, County Sligo (Photograph: P. Coxon).

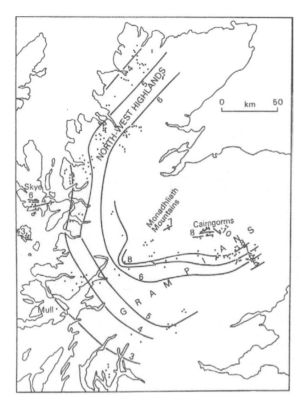

Figure 53: Contoured equilibrium line altitude map for stadial glaciers in Scotland (from Sissons, 1979e).

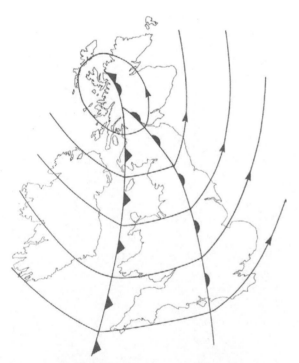

Figure 54: Suggested synoptic situation that would give stadial snowfall associated with southerly winds preceding warm fronts over Scotland.

pians. Sissons & Sutherland (1976) argued that this was a consequence of the main snow-bearing winds coming from south or southeasterly directions rather than as expected from the southwest. Evidence from elsewhere in the Scottish Highlands and Islands supports this suggestion. In many individual mountain groups the glaciers occur at lower altitudes on the south-facing sides compared with the northern sides, and south-facing glaciers are often larger than north-facing ones. Since both observations oppose the influence of direct insolation, Sissons concluded that the most likely factor over-riding this was snow-bearing airstreams from a southerly direction. Specifically, he believed that the typical synoptic situation during the stadial was as shown in fig. 54 when "warm or occluded fronts approached from the west or southwest, the major snowfalls occurring with south to southeasterly winds before the fronts arrived" (Sissons, 1979b, p.519). Ruddiman et al. (1977) demonstrated that the oceanic polar front was situated at or south of the latitude of southwest Ireland during the stadial, and Sissons (1979b) has argued that this resulted in many more depressions following more southerly tracks than normal at the present day. The above relationships have been developed for the Scottish Highlands and it is probably significant that in the Southern Uplands, Lake District and Snowdonia apparently there was a less dominant southerly snowfall influence, though southerly winds were still important for deflation (Sissons, 1980; Cornish, 1981; Gray, 1982). Mean ELAs of 495 m for the southwestern Southern Uplands, 540 m for the Lake District and 600 m for Snowdonia, show the expected southward rise and may be compared to the pattern shown in fig. 54. Still further south, in the Brecon Beacons area, D. Robertson (personal communication) suggests that without snowdrifting from plateau summits, there may have been insufficient snowfall to allow glaciers to develop.

Sissons (1979b) also pointed out the contrast between the small size of the Loch Lomond Stadial glaciers in the Southern Uplands, Lake District and Wales in comparison with the major centres of ice accumulation that developed in these areas during the Dimlington Stadial. According to Sissons the explanation lies in the southward migration of polar waters at the start of each glaciation and accompanying southward migration of the zone of maximum precipitation. Under this interpretation, during the Loch Lomond Stadial the zone of maximum precipitation did not extend south of the Scottish Highlands before the polar waters retreated again, whereas during the

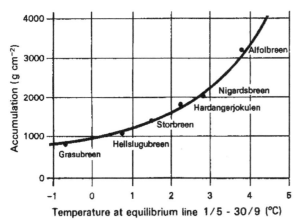

Figure 55: Curve produced by O. Liestøl showing the relationship between accumulation and summer temperatures at the equilibrium line for some present day Norwegian glaciers (from Sissons, 1979e).

Dimlington Stadial the zone of maximum precipitation progressively reached and passed the Southern Uplands, Lake District and Wales.

Attempts have also been made to calculate stadial July temperatures by employing a relationship developed by O. Liestøl (see Sissons, 1979b) for modern Norwegian glaciers between accumulation values and summer temperatures at the equilibrium line (fig. 55). By using estimates of stadial precipitation, summer stadial temperatures at ELAs can be derived from Liestøl's curve and reduced to sea level using an environmental lapse rate of 0.6° C/100 m. For the southwest Highlands with stadial ELAs of about 400 m and present annual precipitation of 3000-4000 mm, a stadial mean July temperature of 7° C has been calculated (Sissons, 1979b), a value that accords with independent calculations based upon coleopteran assemblages (Coope, 1977). Given the evidence of ice wedge development during the stadial and thus a mean annual temperature at least as low as -1° C, a stadial mean January temperature below -9° C is indicated for the southwest Highlands. Clearly the climate was considerably more continental than at present (Sissons, 1979b), but within a few hundred years Britain and Ireland had rapidly emerged from the rigours of the Loch Lomond Stadial glaciation, into the balmier conditions of the Flandrian.

Acknowledgements

We thank numerous colleagues for allowing us to quote unpublished work, and to various institutions and publishers for permission to reproduce figures. Mrs. Eileen Russell (Geography, Trinity College, Dublin) kindly translated parts of Dr. Coudé's work.

Critical regions

Glacial deposits of the Hebridean region

J. DOUGLAS PEACOCK

The Hebridean region is conventionally subdivided into the *Inner Hebrides*, which are the islands adjacent to the Scottish west coast, and the *Outer Hebrides*, which lie some 50 km to the west (figs. 56 and 57). With the exception of the extreme north of Lewis (fig. 56), all were glaciated during the Late Devensian maximum glaciation and most of the glacial deposits are of this age. The long held view that the Outer Hebrides were crossed by Scottish mainland ice during the last glaciation (Geikie, 1873; 1878) has only recently been largely discounted (Coward, 1977; von Weymarn, 1974) and it is now accepted that these islands supported a local ice sheet at the Late Devensian maximum (Flinn, 1978b; Peacock, 1981, 1984; Peacock & Ross, 1978; Sutherland & Walker, 1984; von Weymarn, 1979), though mainland ice may have impinged on the north of Lewis. In contrast, the Inner Hebrides were entirely glaciated by mainland ice, apart from the Cuillin Hills of Skye and the mountains of Mull (Bailey *et al.*, 1924; Peach *et al.*, 1910). The retreat of this ice in the Inner Hebrides was accompanied by the formation of raised beaches (up to 40 m O.D. in places), but Lateglacial beaches are not found in the Outer Hebrides, which were further from the major centres of ice loading. Here, however, a probably pre-Late Devensian raised beach has been recorded (fig. 56, table 9).

The solid geology of the region is diverse. The Outer Hebrides are underlain chiefly by coarse-grained grey Lewisian gneiss (Archaean) with intercalations of amphibolite and metasedimentary rocks and plutons of partly metamorphosed intermediate and basic igneous rocks (Fettes *et al.*, in press). Large areas of the Inner Hebrides are composed of early Tertiary basaltic lavas and intrusions and, in Islay and Jura, of gneisses and chiefly quartzitic metasediments (Johnstone, 1966; Richey, 1961). Triassic to Cretaceous sediments are extensive offshore between the Scottish mainland and the Outer Hebrides; they are exposed on land in a few parts of the Inner Hebrides and in the Stornoway area in the Outer Hebrides.

In this chapter the greater attention given to the Outer Hebrides partly reflects the writer's experience and partly the fact that there are few detailed descriptions of glacial *sediments* in the Inner Hebrides. For St. Kilda, which lies west of the Outer Hebrides, the glacial deposits are described by Sutherland *et al.* (1984).

Table 9. Summary of Pleistocene history, Outer Hebrides.

Glacial events	Extra-glacial events	Suggested correlations
Valley glaciation of Lewis and Harris (in part?)	Formation of boulder lobes on mountains	Loch Lomond Readvance (Late Devensian) 11,000–10,000 BP
Valley glaciation of Lewis and Harris (in part?)		Wester Ross Readvance (Late Devensian) ?13,500 BP
Local ice-sheet, external ice in north Lewis	Formation of periglacial features in north Lewis	Late Devensian maximum glaciation 18,000 BP
	Formation of interstadial and interlacial deposits in north Lewis	
	Formation of raised beach in Lewis and Barra	Pre-Late Devensian
Glaciation by mainland ice		Pre-Late Devensian

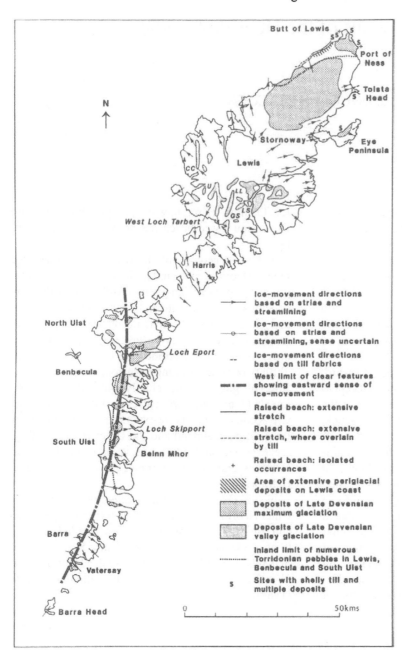

Figure 56: Quaternary deposits, land-forms and ice movement directions in the Outer Hebrides (modified from Peacock, 1984). CC = Abhainn Cheann Chuisil; GS = Glen Skeaudale; LL = Loch Langa-vat; LS = Loch Seaforth; U = Glen Ul-ladale.

OUTER HEBRIDES

Pre-Devensian glacial deposits

In the extreme north of Lewis the pre-Late Devensian raised beach is locally underlain by brown, red, and grey till (figs. 58 and 59) and bedded sand and gravel, the last including sparse pebbles of red sandstone (see below). The relationship of these deposits to a possibly interglacial peat which also occurs below

the raised beach (Sutherland & Walker, 1984) is not known.

Deposits of the Late Devensian maximum glaciation

Till. Till is only patchily distributed, a continuous cover being confined to the peat-covered area of northern Lewis (fig. 56). Some of this is normal lodgement till which forms level or gently undulating sheets. Characteristic sections north of Stornoway

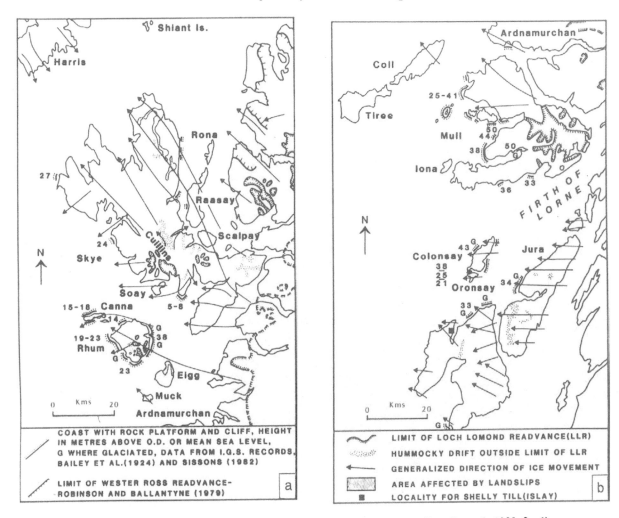

Figure 57: Quaternary features of the Inner Hebrides; (a) northern part, (b) southern part (from Peacock, 1983; fig. 1).

Figure 58: Pre-Late Devensian raised beach gravel about 1 m thick lying on brown till and overlain by soliflucted beach gravel and till. Cunndal, north Lewis (NB 513655) (Photograph: J.D. Peacock, 1977).

Figure 59: Pre-Late Devensian raised beach gravel overlain by locally derived bouldery till. Cliad, Barra (NF 673048) (Photograph: J.D. Peacock, 1978).

Figure 60: Part of rock-cored drumlin with fractured rock passing upward into till with angular blocks. Quarry (NF 950723), Lochpartain, North Uist (Photograph: J.D. Peacock, 1977).

Figure 61: Section of mound showing locally derived bouldery till, south Harris (NF 088960) (Photograph: J.D. Peacock, 1977).

Figure 62: Shattered granitic gneiss (light) apparently sheared over till. South Harris (NF 089959) (Photograph: J.D. Peacock, 1977).

show it to be up to 4 m thick, soft to hard, grey and brown, with subangular clasts of local rocks up to about 1.5 m across. This is locally underlain by a more gravelly deposit with angular clasts and overlain by a layer up to 1 m thick of 'washed' till with water-sorted sand and numerous boulders up to 2 m across. The gneiss below the till is commonly broken, but striated surfaces are preserved locally. An excellent section can be seen at Holm (plate 7). Shelly till and diamicton occurs at several localities in north Lewis (fig. 56), particularly in association with multiple deposits (see next). Small areas of lodgement till

similar to that in north Lewis are known as far south as Vatersay. The red and brown coloration in some of the north Lewis tills is related in the Eye Peninsula to the underlying Triassic conglomerate and, elsewhere in north Lewis, probably to Triassic sandstone and shale offshore. However, the red sandstone erratics so far examined onshore are identical to the siliceous Precambrian Torridonian sandstone of the mainland rather than the usually calcareous and friable Triassic sandstone offshore. The siliceous sandstone erratics occur in the raised beach and underlying till, these probably being the source for such erratics in the Late

Figure 63: Mounds formed of locally derived gneiss boulders up to 2 m across. Loch Laxdale, south Harris (NF 113963) (Photograph: J. D. Peacock, 1977).

Figure 64: Interbedded diamicton (flow till) and glaciofluvial gravel, Port of Ness (NB 537637) (Photograph: J. D. Peacock, 1977).

Devensian tills (Peacock, 1984; Sutherland, 1985).

In some places there are areas in which the till forms mounds, the most extensive being in North Uist. At some localities the mounds, which are usually 2 - 5 m high, are aligned in the direction of ice movement, as at Diraclett (NG 150900) and Loch Carran (NG 086960) in south Harris. At Baleshare in North Uist, ridges in the Samala area (NF 800620) are apparently aligned at an angle both to the foliation in the country rock and to the direction of ice move-

ment further east, whereas south of Howmore (NF 760362) in South Uist, till ridges are parallel to the NNW to SSE foliation and at an angle to the easterly direction of ice movement. Both these localities are near the ice-shed. At other localities, such as those described below in North Uist, there is little correlation with other linear features and the mounds themselves are randomly orientated.

Sections suggest that the mounds consist largely of angular gneiss fragments set in a sparse sandy matrix (figs. 60 and 61; plate 8). This material grades into true till full of angular gravel. In south Harris, cross-sections of mounds up to 3 m high at a locality (NG 088960) show bands of debris derived from differing rock lithologies (gneiss and dolerite) passing downward into angular material and bedrock. A nearby section is shown in fig. 62. At Greaneclett (NF 927720), exposures in mounds 2 to 3 m high show little-transported angular fragments which are in places apparently in continuity with the underlying disturbed bedrock. Similar sections are to be seen at nearby localities. Elsewhere, as in South Uist (NF 757305), the angular rock debris overlies striated bedrock, from which it is separated by a very thin silty deposit possibly marking a shear plane. In most of the above sections, the angular fragments range up to boulder size, but are rarely more than a metre across. Blocks larger than this are commonly shattered (fig. 62). Exceptionally, where the granitic gneiss is massive as in the area about Loch Laxdale (NG 108962) in south Harris, mounds are found consisting essentially of collections of large blocks (fig. 63).

In view of the low relief of the moundy areas it is suggested that the hummocks are formed of material derived entirely from the bedrock below the ice, probably as the result of shear stress at the ice/rock interface acting together with a variable hydraulic pressure on joints and other discontinuities (cf. Broster *et al.*, 1979). Stress relief due to glacial

unloading could also have played a part (plate 6). The material was transported in some cases a short distance before being deposited by a meltout or lodgement process. This mechanism of till formation was clearly widespread in the Outer Hebrides and could account for the occurrence of some of the head-like deposits at the base of lodgement till discussed earlier. The presence of such mounds in the ice-shed area suggests that this was traversed by moving ice at some earlier stage.

Multiple deposits. Shelly tills and diamictons with Torridonian erratics, associated in places with beds of silt, sand and gravel, occur at several localities in northern Lewis (fig. 56). Though it has been suggested that the multiple deposits comprise more than one glacial phase (Baden-Powell, 1938; Geikie, 1878) subsequent work suggests that many if not all of the multiple deposits can be attributed to one depositional episode (von Weymarn, 1974). It should be noted, however, that none of the successions has yet been studied in detail.

At Garrabost (east of Stornoway, fig. 56) a moraine ridge extending NNW through the village is formed of interbedded diamicton, sand, clay and gravel, all of which contain shell fragments. These have yielded a radiocarbon age of roughly 23,000 B.P. (Sutherland & Walker, 1984), which would support a Late Devensian age for the deposit. Similar sediments occur farther north at Tolsta (fig. 56). At Port of Ness the section in the cliffs south of the village shows inter-

Figure 65: Flow tills with sand lenses and pockets of gravel. Cliff south of Traigh Sands (NB 513645), north Lewis (Photograph: J.D. Peacock, 1977).

Figure 66: Asymmetrical transverse moraine 4 - 6 m high and 100 m long with flutes 0.2 to 0.5 m high on its proximal (right) side aligned in the direction of ice-movement. Glen Skeaudale (NB 153031), north Harris (Photograph: J.D. Peacock, 1978).

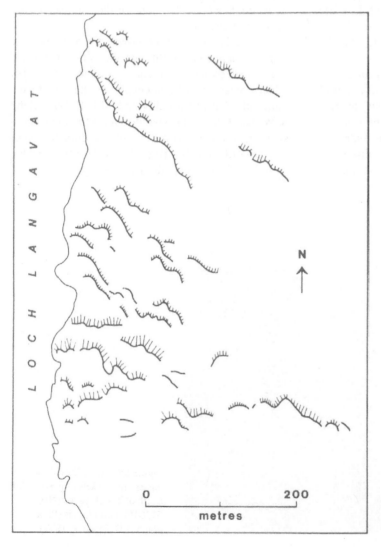

Figure 67: Asymmetrical transverse moraines on the east side of Loch Langavat (NB 1614 and 1615).

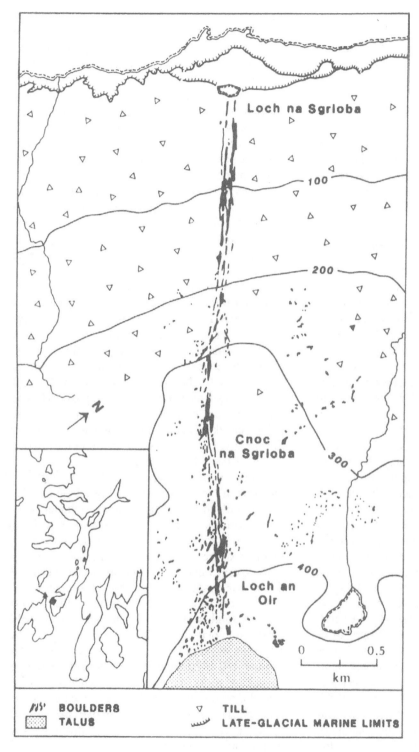

Figure 68: Medial moraine complex in west Jura (after Dawson, 1979; fig. 1).

bedded diamicton, gravel and sand, which apparently pass northwards into subaerial outwash gravels (fig. 64). On the NW coast the multiple deposits which form the cliffs backing the sandy beaches from 1.5 km west of the Butt of Lewis reach up to 25 m in thickness. They comprise water-sorted sand and silt with isolated pebbles, finely laminated silts and diamictons, including flow tills (fig. 65). Till crowded with angular clasts occurs in places at the base. The multiple deposits here have yielded a largely fragmentary molluscan fauna including both warm and cold water species (Baden-Powell, 1938). Deposition during the Late Devensian is suggested by finite radiocarbon ages of more than 35,000 B.P. and unpublished amino-acid determinations (Sutherland & Walker, 1984).

Other meltwater deposits. Deposits and landforms associated with glacial meltwater, other than those interbedded with the multiple deposits referred to above, are of small extent and virtually restricted to Lewis and north Harris (Peacock, 1984).

Deposits of Late Devensian valley glaciations

In SW Lewis, north Harris and very locally farther south the retreat of the main Late Devensian glaciers was followed by an as yet undated vigorous valley glaciation (fig. 56). This resembles the Loch Lomond Readvance of the Scottish mainland in the presence of landforms such as fluted drift, but the relationship to periglacial features suggests it may be older, at least in part (Peacock, 1984). The deposits include till, which is often bouldery, together with local silt and sand.

Hummocky moraines. The chief component of many hummocks is a very poorly sorted diamicton composed of a matrix of clayey or sandy silt, charged with subrounded to angular clasts up to boulder size. Most of the rock fragments, of which only a few are striated, are derived from the local gneiss. The diamicton, in which the character of the clasts may vary greatly from bed to bed, may contain pockets and wisps of sand and silt, and be interbedded or interlaminated with a variable proportion of silt and sand which occasionally shows traces of cross-bedding. In places, the diamicton and associated deposits are contorted and dip at high angles, otherwise the bedding is usually gently dipping, either conforming to, or being truncated by, the surface of the mound. It is suggested that most of the diamicton in the hummocks is flow till originating from debris (perhaps pre-existing glacial deposits; see Hodgson, 1982) incorporated at the base of the ice and deposited near the ice-front. Morainic hummocks formed mainly of rock fragments with little or no matrix occur in the vicinity of Tarbert in south Harris. Such deposits resemble the moundy till described above and may likewise have been formed by deformation, shearing and lodgement.

Fluted drift. Drift fluted by the ice in the direction of movement is common in the valleys of Harris west of Loch Seaforth and is very well developed locally in SW Lewis. The individual flutes, which are generally formed of till or gravel, are up to 300 m long (though generally much shorter), 6 m high and 30 m across. They tend to be best developed adjacent to steep slopes or crags. In places the ridges are broken into chains of mounds. Good examples can be seen in Glen Skeaudale (fig. 66) and Glen Ulladale (NB 074134) where they lie on the proximal sides of asymmetrical transverse moraines. In the valley of the Abhainn Cheann Chuisil in SW Lewis (fig. 56) the flutes locally cross earlier-formed drumlinoid morainic mounds up to 5 m high (NB 032223). The flutes are similar to those described by Hodgson (1982), some being ascribed to reworking of drift already deposited and some to the deposition of new material.

Asymmetrical transverse moraines. These occur at many localities in the area of valley glaciation in north Harris (Peacock, 1984), but three examples will suffice. Numerous small moraines which occur on the valley side east of Loch Langavat (fig. 67) consist of arcuate ridges 1 - 5 m high, slightly to strongly convex down valley (north), with steep (20 - 30°) north-facing distal slopes and gentle proximal slopes. The ridges are highest at their central, most distal points and the distance along the chord joining the low proximal tips is 10 - 40 m. In places the proximal slopes of the ridges display traces of parallel flutes 0.5 - 1.0 m high, aligned apparently in the northward direction of the ice movement. The lobate moraines tend to unite to form ridges running obliquely down slope, perhaps defining former ice front positions. Apart from scattered boulders on the surface, the lithology of the mounds here is concealed by vegetation and peat, but is likely to be till. In Glen Skeaudale the moraine ridge shown in fig. 66 becomes more symmetrical to the north. Near here (NB 153031) there is another ridge, 1 - 2 m high and concave upstream in plan, with a fluted proximal face. These ridges are apparently formed of till, but others in the vicinity (NB 147029), which are parallel to striations on nearby rock, are formed of boulders. Possible annual moraines in Glen Meavaig, which are spaced at intervals of 30 to 150 m, seem to have been formed by active ice which was receding northwards across the present-day watershed.

Other morainic ridges. The valley glaciers of SW Lewis locally ended at bouldery terminal moraines. Arcuate moraines which occur on the south shore of Loch a'Ghlinne (NB 023126) in north Harris mark the termination of a small corrie glacier. Possible annual moraines have been reported in north Harris.

INNER HEBRIDES

Deposits of the Late Devensian maximum glaciation

During the glacial maximum mainland ice flowed generally westwards across the area (fig. 57), though Sissons (1982b) believes that the ice-front may have been situated over parts of the islands for considerable periods. Detailed descriptions of the associated lodgement till are lacking, but on Islay it is "a tough argillaceous or arenaceous clay, varying from bright brick-red to light brown or grey ... plentifully charged with well rounded and striated boulders" (Wilkinson, 1907). It is extensive on Islay, but sparse-ly distributed elsewhere. A local deposit of shelly till in northern Islay contains well preserved fragments of the delicate arctic bivalve *Palliolum groenlandicum* (Peacock, 1974). In Jura Dawson (1979) has described a remarkably straight fossil medial moraine complex some 3.5 km long (fig. 68) consisting dominantly of local quartzite blocks. The ridges trend in the general direction of ice movement. Glaciofluvial deposits are generally scarce, but are locally more common in central Islay where terraced gravels and eskers have been partly reworked by the Lateglacial sea.

Deposits of Late Devensian valley glaciations

The retreat of the mainland ice was interrupted by readvances such as the Wester Ross Readvance (fig. 56) of Robinson & Ballantyne (1979). A suggested readvance limit in central Islay is marked by a moraine belt (fig. 57, Dawson, 1982; 1983). Outside the limit of the Loch Lomond Advance or Readvance (LLR) in central and eastern Skye the orientation of drumlinoid mounds suggests the presence of active valley glaciers emanating from local centres of glaciation (Donner & West, 1955; Peach *et al.*, 1910).

The LLR in the Inner Hebrides is clearly defined and its limits are now well known (Ballantyne & Wain-Hobson, 1980; Gray & Brooks, 1972; Sissons, 1977e). Within the terminal moraines there are belts of hummocky moraines and fluted drift which are similar to those in the Outer Hebrides. In places the LLR glaciers reached the sea and on Mull they ploughed up marine clay and sand formed during the previous Windermere (Lateglacial) Interstadial. The largest glaciers were in eastern Mull and the 'Black' Cuillin Hills of Skye. In Rhum, eleven sites of small glaciers have been identified. On the western side of Eigg some of the landslips seem to have been associated with the development of LLR glacier ice (Peacock, 1975a).

The glaciation of the Shetland and Orkney Islands

DONALD G. SUTHERLAND

As the implications of the theory of glaciation and the possible extent of former ice sheets were being assimilated by geologists working in Scotland during the last century, it became apparent that the Orkney and Shetland islands occupied a crucial position with respect to the extents and respective strengths of the Scottish and Scandinavian ice sheets. James Croll (1870), in particular, in one of his many insights into the effects and causes of glaciation (cf. Croll, 1875), envisaged periods of maximum glaciation in which the northern North Sea was inundated by ice derived in part from Scotland and in part from Scandinavia and that the two ice masses coalesced and then flowed to the NW across Orkney and Shetland. Croll (1875: 444-452) suggested that Shetland and perhaps the north of Orkney had been glaciated by Scandinavian ice whilst Caithness and the southern part of Orkney at this period of maximum glaciation were overridden by Scottish ice that had been forced to turn to the NW by the presence of Scandinavian ice to the north and east. Although, as discussed below, much more evidence related to glaciation has become available during the last century, both from the land areas of the islands and the neighbouring sea beds, all subsequent discussion of the glaciation of these is land groups has been in the terms originally proposed by Croll.

In recent years studies of palaeoclimate based on terrestrial deposits and offshore cores have emphasised the importance of the oceanic polar front in the North Atlantic as a control on the glaciation of the neighbouring continental areas (e.g. Ruddiman & McIntyre, 1979; 1981a). This concept further emphasises the significance of the position of Orkney and Shetland for the entry of North Atlantic Drift water to both the Norwegian and the North Seas would be past these islands (fig. 69).

Early glaciation and interglacial deposits on Shetland

Under the influence of Croll's ideas of the likely nature of glaciation of Shetland, Peach & Horne (1879) conducted a brief field programme (during their annual leave from the Geological Survey) which apparently confirmed the hypothesis that Scandinavian ice had indeed passed over the Shetlands. Following this 'primary' glaciation Peach & Horne recorded that there was a period of local glaciation in which ice nourished in the Shetlands flowed outwards 'in all directions' (p. 810). It was not until the work of Chapelhowe (1965) that this concept of a two-phase glaciation of the Shetlands was amended and evidence was produced, at the Fugla Ness site, that indicated that there was a much longer Quaternary sequence in these islands.

The lithostratigraphy of the Fugla Ness site has still to be described in detail. Table 10 summarises Chapelhowe's (1965) description from which it can be seen that a lowermost till unit is separated from two upper till units by a peat bed of interglacial character (Birks & Ransom, 1969). On the basis of the pollen content and macrofossil remains, Birks & Ransom (1969) suggested that the Fugla Ness peat may date from the Hoxnian Stage but the basis for

Table 10. Stratigraphy of Fugla Ness site (after Chapelhowe, 1965).

Sediment	Depth cm
Sandy, slightly organic soil	0–14
Reddish till with granite pebbles	14–219
Grey-brown till, horizontally stratified. Pebbles mainly local with some granite	219–536
Compact structureless peat with pine cones. Silty and stony lenses common	536–586
Compact structureless peat with some wood between 586–630 cm. Large stones at 606 cm	586–691
Grey, cemented till, somewhat similar to upper grey-brown till	> 691

this correlation is not secure (Lowe, 1984). Accepting that certain aspects of the pollen record imply a Middle Pleistocene age and that the thermophilous nature of the macrofossils indicates correlation with a period of very mild oceanic climate lead Sutherland (1984a) to suggest that the Fugla Ness peat may correlate with the event recorded in deep-sea cores at c. 380 ka.

Whilst the age of the Fugla Ness peat is important in understanding the glacial sequence at this locality the lack of detailed information on the till units results in two distinct interpretations as to the glacial history of this site. This is because the existing description fails to distinguish between the till units that occur above and below the peat. It is possible that they relate to the same glaciation with the peat ocurring as an erratic (cf. Birks & Ransom, 1969). If this is so, then the site indicates only two periods of glaciation both post-dating the peat. Alternatively, if these till units relate to separate glacial events then three periods of glaciation are recorded at this site the earliest of which pre-dates the interglacial peat and hence may be Middle or Early Pleistocene in age. As is discussed later, much of the debate on the relative influences of Scandinavian or local ice has been related to indicators of the direction of ice movement. It is therefore unfortunate that no information has been published that relates to the direction of flow of the ice that deposited the lower till or tills at Fugla Ness. The uppermost till unit contains granite erratics and is apparently derived from the SE (Chapelhowe, 1965) in conformity with the last local ice cap glacia-

tion of probable Late Devensian age (see below).

A further peat bed underlying till has been reported from Shetland, at Sel Ayre (fig. 70) (Mykura & Phemister, 1976; Birks & Peglar, 1979). On the basis of its pollen content this peat has been tentatively assigned to the last (Ipswichian) interglacial Stage (Birks & Peglar, 1979) but both its interglacial status and its correlation with the Ipswichian may be questioned (Lowe, 1984). A sample from the peat gave a radiocarbon age of 36,800 +1950/-1560 (SRR-60) (Mykura & Phemister, 1976) but this date, as with the finite dates from Fugla Ness (Page, 1972; Harkness & Wilson, 1974; Harkness, 1981), must be treated with caution (Sissons, 1981b) and may be quite misleading as to the true age of the deposit. The Sel Ayre peat is interbedded with sands and gravels but the provenance of the gravels has not been published and hence there is no information from this site as to glaciation prior to the deposition of the peat. The overlying till contains erratics of local sandstone as well as rare basic lavas which have been derived from the SE (Mykura & Phemister, 1976). The ice that deposited this till may therefore be correlated with the latest period of local ice cap glaciation. This site thus has potential for providing a limiting age for the latest local glaciation but the uncertainties as to the age of the peat limit the inferences that can presently be made.

Influence of Scandinavian ice

Since the paper by Peach & Horne (1879) the debate as to the nature of the last glaciation of Shetland and the relative importance of Scandinavian and local ice has been concerned almost exclusively with two types of evidence: striated and ice-moulded rock surfaces and transport of erratics. The evidence of these types that has been published in recent years has been compiled in fig. 70. As the encroachment of Scandinavian ice on the Shetlands has been considered by all workers to pre-date the latest local ice cap glaciation, the evidence related to this ice movement is considered first.

Flinn (1977, 1978a) considered that the only evidence for the glaciation of Shetland by Scandinavian ice was the transport of certain erratics in the southern part of the islands. The westward transport of certain rock types across the reconstructed ice shed of the later local ice cap at least as far as Foula (fig. 70) was interpreted by Flinn as relating to Scandinavian ice having crossed the southern islands during a previous glaciation or at an early phase of the last glaciation. It

Figure 69: Location of Orkney and Shetland.

is also in this area, at Dalsetter (fig. 70), that the only Scandinavian erratic that has been found in Shetland occurs (Finlay, 1926) although the history of transport of a single erratic may be complex (Hoppe, 1974). The westward transport of erratics has also been described from certain of the northern isles (fig. 70) and here too the influence of Scandinavian ice has been invoked (Mykura, 1976).

In assessing the influence of Scandinavian ice, Hoppe (1974) placed emphasis on an older set of striations occurring on and to the west of Bressay (fig. 70) which he considered to have been formed by ice flowing from the NE. The sense of direction of some of these striations has been contradicted by Flinn (1977: 141), however. The evidence for Scandinavian ice reaching the Shetlands is therefore rather weak and relates solely to the margins of the islands. In particular it may be noted that no shelly tills or North Sea Basin erratics have been reported from Shetland, a situation in marked contrast to the eastern Orkney Islands. It may be concluded that if Scandinavian ice ever crossed the Shetland islands the evidence for this has been largely obliterated by the subsequent local glaciation(s).

Local ice cap

The great majority of the evidence relating to transport of erratics and ice-moulded surfaces can be explained in terms of a local ice cap covering the Shetland islands with ice flowing outwards from an ice shed located along the long axis of the islands. This evidence is summarised in fig. 70. Since westerly directed ice moulding and erratic transport on the western side of the islands is compatible with local as well as Scandinavian glaciation, most significance has been given to the evidence for eastward movement of ice on the eastern side of the islands. Such evidence has been found, with the possible exception of the island of Unst in the far north, along the length of the eastern Mainland and islands. Such evidence includes, for example, the transport of erratics from Mainland as far east as the Out Skerries (fig. 70).

That Shetland has been glaciated by a local ice cap has been agreed by all workers since Peach and Horne (1879) but the age and extent of this ice cap have been the subject of debate. Although it is broadly agreed that the last phase of local ice cap glaciation dates from the Late Devensian, direct evidence for this is lacking (Sissons, 1981b). If the Sel Ayre peat is indeed of last interglacial age and the overlying till relates to the latest local ice cap, then this only indicates Devensian glaciation. A minimum age for the termination of the local ice cap is provided by radiocarbon dates on basal organic sediments resting on the glacial deposits. The oldest published such date, however, is only 12,090 ± 900 yr BP (St-1640) (Hoppe, 1974).

The extent of the local ice cap is also poorly defined. Hoppe (1974) considered that the ice cap was solely a remnant of the Scandinavian ice that had earlier (during the Late Devensian maximum) crossed the Shetland isles. He envisaged the ice mass having been isolated from the Scandinavian ice sheet during general deglaciation as a result of rapid ablation in the deeper waters between Shetland and Scandinavia. The local ice, on this hypothesis, was dynamically active but climatically dead, the exact pattern of ice flow being controlled by more rapid ablation in the deeper waters around the Shetlands, this giving rise to the radial pattern of striations which are frequently oriented normal to the coastline. In contrast, both Flinn (1977, 1978a) and Mykura (1976) have suggested that the ice cap built up *in situ* and hence was climatically healthy. Mykura differentiated two phases of ice cap glaciation, an earlier phase in which the ice cap was confluent with and deflected by Scandinavian ice (which may at this stage have overridden the northern and southern parts of Shetland) and a later phase in which the Scandinavian ice had withdrawn and the local ice flowed outwards unimpeded. Flinn (1978a) did not consider that there was any evidence for the local ice being in contact with Scandinavian ice and has recently (1983) suggested that the northern ice margin of the ice cap may have been located only a short distance to the north of Shetland.

Recent offshore investigations in the northern North Sea Basin have relevance to this debate. Current work (Long & Skinner, 1985; Stoker & Long, 1985) has suggested that during the Late Devensian ice from Shetland reached some 70 - 80 km to the east of the islands but that it was not confluent with Scandinavian ice which terminated on the western margin of the Norwegian Trench (see also Rise *et al.*, 1984). If these interpretations of the offshore sediments are correct, then the evidence for western transport of erratics in the northern and southern parts of Shetland must relate to either an earlier phase of glaciation or, as the relevant areas are relatively close to the reconstructed ice shed, to movements of that ice shed during the course of the glaciation.

Little information is available as to the mode and timing of retreat of the local ice cap. A 'moraine belt'

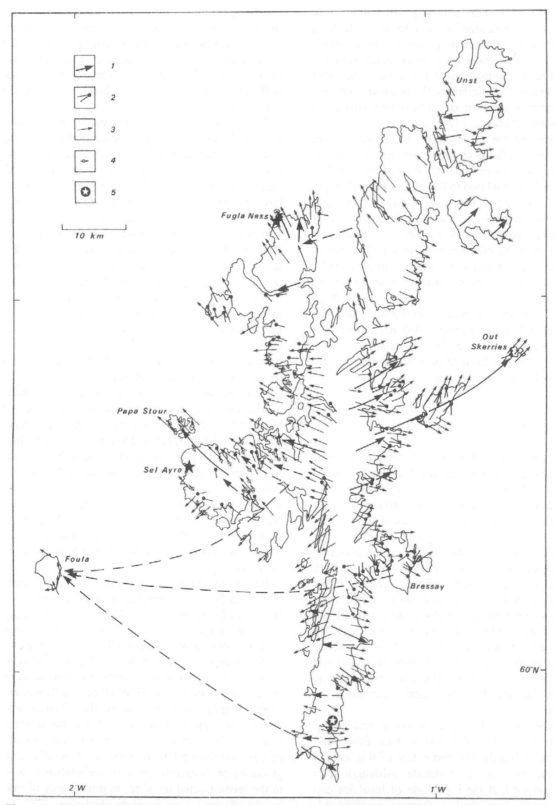

Figure 70: The glaciation of Shetland. 1 = Directions of transport of erratics (after Chapelhowe, 1965; Mykura, 1976; Mykura & Phemister, 1976; Flinn, 1978a); 2 = Striations from Hoppe (1974); 3 = General direction of ice moulding and striations from Flinn (1978a, 1983); 4 = Striations from Mykura & Phemister (1976); and 5 = Dalsetter erratic.

has been described from the western island of Papa Stour (Mykura & Phemister, 1976) but its significance is unclear. The absence of raised shore-lines limits the possibilities for studying ice cap retreat. Hoppe (1974), on the basis of striation patterns being a reflection of calving during deglaciation has suggested that the contemporaneous sea level was only 20 - 25 m lower than it is today but this figure must remain speculative. Minor moraines have been variously described on some of the hills (Charles-worth, 1955; Mykura, 1976; Mykura & Phemister, 1976; Flinn, 1977) and attributed to the Loch Lomond Stadial. Which of these features are moraines has been disputed among the authors quoted, Whichever interpretation is adopted, however, the Loch Lomond Readvance was of very limited extent on Shetland.

Orkney

By comparison with Shetland, there has been relatively little research into the glaciation of Orkney in recent years. Almost all the glacial deposits appear to relate to a single phase of glaciation and no organic interglacial deposits have been reported. The main influence on glaciation of the Orkneys has been an external one and the role of local ice is extremely limited. There is therefore a major contrast with Shetland despite their proximity and broad topographic similarities.

The stratigraphically earliest Quaternary deposit reported from Orkney is a raised cobble beach exposed in the north of Hoy (fig. 71). The beach occurs at 6 - 12 m above present sea level and rests on the inner margin of a marine abrasion ramp. No erratic material and no fossils have been observed in the beach which is overlain by a head deposit and then a till deposited during the last regional glaciation (Wilson *et al.*, 1935; Sutherland, unpubl.). The age of the beach is unknown.

Much of the low land of the Orkneys is covered by a thin sheet of glacial deposits that locally thickens to over 10 m in hollows. Two till units have been recognised by Rae (1976). Based on fabric analyses, erratic content and the orientation of underlying striations, Rae concluded that the lower till had been deposited by ice flowing towards the north or NNW. The upper till unit, which is that which has been most widely described in the literature, was deposited during the same glacial phase as the lower till as Rae could find no evidence for a break in deposition between the two. Fig. 71 shows the available evidence relating to ice movement during the deposition

of this upper till. As noted by Peach & Horne (1880) in the first paper synthesising data on the glaciation of the Orkneys, orientation of striations and the direction of carry of erratics is indicative of glaciation by external ice flowing broadly from the SE to the NW and being little influenced by local topography.

Particularly in the eastern islands this upper till is notable for the occurrence of far-travelled erratics from both the Scottish Mainland and the North Sea Basin. These erratics include metamorphic and granitic rocks from the Northern Highlands, Jurassic sediments from Sutherland and, in the northern islands rocks that have been assigned to the Grampians and even the Central Valley of Scotland (Peach & Horne, 1880; 1893). One boulder has been found (Saville: fig. 71) that has been thought to come from Scandinavia (Peach & Horne, 1880; Wilson *et al.*, 1935) although it may be derived from Sutherland (Rae, 1976). The farthest travelled erratics have been recorded from the northern isles and there is therefore the possibility that the ice that crossed the north of Orkney originated in the southern Highlands.

The occurrence of marine shell fragments in the till (fig. 71) has provided an opportunity to place a maximum date on the glaciation. An 'infinite' radiocarbon age from a sample of shell fragments was reported by Rae (1976) whilst amino-acid ratios on fragments of *Arctica islandica* have indicated that different shell populations, some quite old, are contained within the till (Sutherland, unpubl.). The youngest shell fragment dated is of Middle Devensian age which would imply that Late Devensian glaciation deposited the till but as this is a single fragment more analyses are needed before a conclusion can be reached. On geological and glaciological grounds it has been argued that Late Devensian ice did not cross Orkney (Sutherland, 1984a) for it is difficult to envisage the Scottish ice being diverted to the NW across a topographic barrier without Scandinavian ice being in contact with the Scottish ice in the North Sea Basin and the evidence from the central and western North Sea Basin appears to imply that the two ice masses were not in contact during the Late Devensian (Sutherland, 1984a; Cameron *et al.*, 1987). The date of the last regional glaciation of the Orkney islands therefore remains unknown and direct dating evidence is awaited.

There is no published information on the mode or direction of retreat of the last ice sheet to cover Orkney. Few morainic features have been reported and none interpreted. In northern Hoy small glaciers subsequently developed and deposited end moraines for example at Dwarfie Hamars and Enegars Corrie

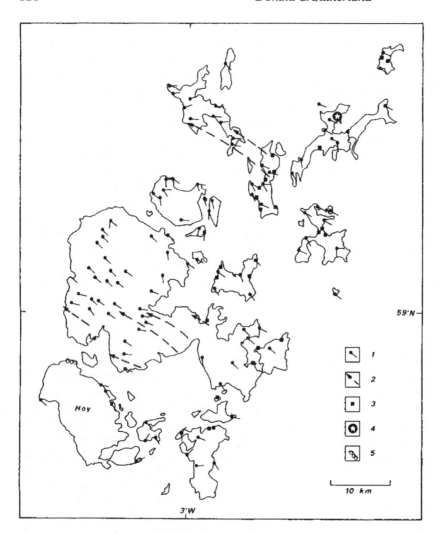

Figure 71: The last glaciation of Orkney. 1 = Striations from Wilson *et al.* (1935); 2 = Transport of erratics recorded in Peach & Horne (1880); 3 = Shelly till localities (after Peach & Horne, 1880; Rae, 1976; Sutherland, unpubl.); 4 = Saville erratic; 5 = Raised beach in the north of Hoy.

(Sutherland, unpubl.). Although not directly dated, these most probably relate to the Loch Lomond Readvance and represent the last minor phase of glaciation of Orkney. These small glaciers formed in favoured localities and have equilibrium line altitudes of c. 150 m.

Conclusion

Since the pioneering work of Peach & Horne that apparently confirmed Croll's hypothesis on the nature of glaciation of the Shetland and Orkney Isles the evidence that has accumulated has led to significant modifications of the original ideas but no similarly coherent, simple hypothesis can now be advanced to explain the glaciation of this region. Such evidence as exists for the glaciation of Shetland by Scandinavian ice is confined to the northern and southern parts of the islands and if such a glaciation took place it now seems unlikely to have occurred during the last glacial maximum. The last glaciation of Shetland was solely that of a local ice cap but the exact age of this event is not yet clarified. It is generally considered to have occurred during the Late Devensian. In contrast, the last major glaciation of the Orkney islands clearly relates to overriding of the islands by Scottish ice in which an initial phase of northerly ice flow was succeeded by a northwesterly flow. The age of this glacial event is not yet established. This northwesterly flow across Orkney would appear to confirm part of Croll's original hypothesis but if the glaciation was of Late Devensian age, then the principal element of that hypothesis, the northern North Sea being occupied by Scandinavian ice, would appear not to be realised. Direct dating of the various glacial events involved is needed before these issues can be resolved.

Both Orkney and Shetland are relatively low island groups with only very restricted areas extending above 300 m. The largest hill mass occurs in northern Hoy in Orkney but this attains only 480 m. That Shetland nourished a local ice cap apparently during the last glacial maximum is therefore of some palaeoclimatic significance and is comparable with the ice cap that developed on the Outer Hebrides at that time. An additional puzzle is therefore presented for the Orkneys for no evidence has yet been reported that suggests that those islands ever nurtured a local ice cap.

The glacial deposits of Buchan, northeast Scotland

ADRIAN M. HALL & E. RODGER CONNELL

The lowlands of northeastern Scotland are a key area for understanding the Quaternary history of Scotland. The region has been glaciated at various stages by three separate ice streams moving towards the North Sea Basin along the Moray Firth, out of the mountains and foothills of the eastern Grampians and northwards along Strathmore and the North Sea coast. In the east, Buchan has been an area of marginal glaciation, as shown by the preservation of preglacial deeply weathered bedrock (Hall, 1985) and the existence of interstratified sequences of glacial and periglacial deposits (Connell & Hall, 1987). These provide important evidence of pre-Devensian glacial events in Scotland (Hall, 1984). The extent and timing of Devensian glaciation in northeast Scotland remains controversial, with recent argument for widespread glaciation in the Early Devensian (Sutherland, 1981b; 1984a; Hall, 1984). A summary of the main Quaternary events in the region is given in table 11.

PRE-DEVENSIAN GLACIATIONS

The earliest indications of glacial climatic conditions in northeast Scotland are provided by early Middle Pleistocene glacial and glaciomarine sediments in the west central North Sea, deposited by an ice sheet with a limited offshore extent (fig. 72; Stoker & Bent, 1985). These early 'Cromerian Complex' deposits are part of the Early to Middle Pleistocene Aberdeen Ground Formation which locally approaches to within 20 km of the present coastline (Stoker et al., 1985a, b) and which probably provided the supposed 'Crag' shells found in the Late Devensian Kippet Hills esker (Sutherland, 1984a).

The Kirkhill sequence

On land the most complete sequence of pre-Devensian deposits occurs at Kirkhill (Connell et al., 1982; Hall, 1984; Connell & Hall, 1987; fig. 73; plate 9). Here an early Devensian(?) till of westerly provenance lies above an older till derived from the NW. The upper surface of the lower till shows a range of weathering changes, including mottling, clay translocation, alteration of clasts and down-profile changes in colour and clay mineralogy which indicate soil formation under humid temperate interglacial conditions. After formation the upper buried soil was truncated and disturbed by periglacial activity before deposition of the upper till.

The lower till is underlain by periglacial solifluction deposits and weakly organic sands which in turn rest on a distinctive marker horizon provided by the lower buried soil (plate 10). This was originally interpreted as a humid temperate podsol (Connell, Edwards & Hall, 1982) but micromorphology suggests affinities with 'cold water gley' soils of arctic environments (Romans, in Hall, 1984: 70). Superimposition of periglacial soil characteristics on an earlier interglacial profile seems likely, particularly since pollen assemblages contain elements of interglacial and interstadial character in the overlying organic deposits (cf. Lowe, 1984).

The lower buried soil is developed in sands and gravels. Boulder gravels appear to correlate with coarse gravels in nearby Leys quarry. The latter were deposited as part of a valley sandur which drained towards the North Ugie valley from an ice margin which lay east of Kirkhill (Hall & Connell, 1986).

The base of the Kirkhill sequence is formed by stratified sands and gravels underlain and locally interstratified with a blocky gelifluctate containing erratics of western provenance. The entire sequence rests in bedrock channels and basins up to 10 m deep probably cut by meltwater.

The ages of the deposits in the Kirkhill sequence are uncertain. The organic deposits above the lower buried soil have now been shown to be beyond the range of conventional radiocarbon dating (Hall,

Table 11. Quaternary stratigraphy of northeast Scotland. Litho- and chronostratigraphy of west-central North Sea from Stoker, Long & Fyfe (1985b). All other stratigraphic units are informal and assignment to specific stages is provisional. Gf = Glaciofluvial; Gl = Glaciolacustrine; Gm = Glaciomarine; T = Till; F = Fluvial; M = Marine. Numbers refer to key references to sites: 1 = Godwin & Willis (1959); 2 = Bremner (1934); 3 = FitzPatrick (1965); Sutherland (1984); 4 = Peacock (1966, 1971); 5 = Hall (1984), Connell & Hall (1987); 6 = Munro (1986); 7 = Synge (1963); 8 = Campbell (1934), Donner (1979); 9 = Stoker, Long & Fyfe (1986b).

Stage name		Environment	Coastal Moray and Banff, lower Spey valley	Buchan[5]	Aberdeen - Strathmore	West-central North Sea[9]
Late Devensian	Loch Lomond Stadial	Periglacial	Garral Hill gelifluctate[1]	Woodhead gelifluctate		
	Windermere Interstadial	Humid, temperate	Garral Hill peat[1]	Woodhead peat	Mill of Dyce peat[6]	
	Dimlington Stadial	Glacial - periglacial	Blue-grey Series (T, Gf, Gl) Tills from west and south-west[2]	Red Series (T, Gf, Gl) Blue-grey Series (T, Gf, Gl) Crossbrae gelifluctate	Red Series (T, Gf, Gl) Tills from west[6]	Wee Bankie Formation (T) Marr Bank Formation (Gm)
Middle Devensian		Humid, cold		Crossbrae peat	?Benholm peat[8]	
Early Devensian		Glacial - periglacial		?Inland series (T, Gf) Kirkhill gelifluctate complex 3		Coal Pit Formation (Gm, M)
Ipswichian		Humid, warm temperate	Teindland buried soil[3] Boyne Bay weathering profile[4]	Kirkhill upper buried soil Moreseat weathering profile	King's Cross weathering profile[7]	
Wolstonian		Glacial - periglacial	Teindland lower sand and gravel (?Gf)[3] Boyne Bay lower till[4]	Kirkhill lower till Moreseat till (?'Mesozoic drift') Kirkhill gelifluctate complex 2	King's Cross till[7] ?Benholm shelly till[8]	Fisher Formation (Gm, T)
Hoxnian		Humid, warm temperate becoming cold		Kirkhill lower buried soil and organic deposits		Ling Bank Formation (M)
Anglian		Glacial - periglacial		Kirkhill cryoturbate Kirkhill and Leys gravels (Gf) Kirkhill lower sands and gravels (F) Kirkhill gelifluctate complex 1		Aberdeen Ground Formations *pars* (Gm, M, T)

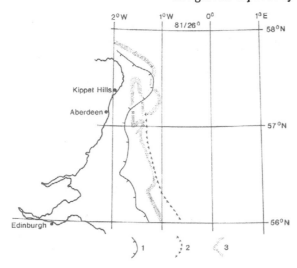

Figure 72: Drift limits in the west central North Sea. 1 = Eastern limit of the Wee Bankie Formation, marking the maximum extent of Scottish ice in the Late Devensian (Stoker, Long & Fyfe, 1985b). 2 = Maximum limit of Scottish ice during the early 'Cromerian Complex' glaciation (Stoker & Bent, 1985). 3 = Western limit of the Aberdeen Ground Formation (Stoker, Long & Fyfe, 1985b). Borehole 81/26: this borehole includes Wolstonian till of Scottish origin but lay beyond the Devensian glacial maximum (Sejrup *et al.,* 1984).

1984). The upper buried soil is of interglacial status and so the lower till must be Wolstonian in age or older. The lower buried soil represents an older interglacial stage (Hoxnian?) and so the two earliest glacial events inferred from glaciofluvial features at Kirkhill and Leys are probably Anglian or older (table 11).

Pre-Devensian weathered tills of inland derivation

The weathered lower till at Kirkhill may correlate with other weathered tills derived from inland at King's Cross, Aberdeen (Synge, 1963a), Moreseat (Hall & Connell, 1982) and Portsoy (Peacock, 1966) and with the buried interglacial soil at Teindland (FitzPatrick, 1965; Sutherland, 1984a). Except at Moreseat, the weathered tills, together with the Teindland soil, are buried in part by glacial deposits of probable Devensian age and weathering probably occurred during the last interglacial. Firm correlation, however, awaits detailed study of these sites, particularly of their pedogenic characteristics, but together these weathered tills point to at least one major advance of inland ice close to or beyond the present coastline prior to the Devensian.

Early glaciation of Buchan from the Moray Firth

Evidence also exists for onshore movement of Moray Firth ice prior to the Devensian in the form of dark grey-black clayey tills and erratics, loosely termed the 'Mesozoic Drift' (Hall, 1984), found beneath or incorporated into tills of inland derivation of probable Devensian age (fig. 74). These tills are spatially and lithologically distinct from the calcareous, shelly Late Devensian Blue-Grey Series surface till sheet (Synge, 1956) and are not to be correlated with these deposits (Sutherland, 1984a). In Buchan these tills incorporate Jurassic and Cretaceous macro- and microfossils (Jamieson, 1906; Hall & Connell, 1982). The absence of Jurassic outcrop in the central North Sea demonstrates that these tills have been carried by ice moving out of the Moray Firth. The ability of ice from the Moray Firth to cross Buchan at this stage indicates an ice stream of considerably greater power than its Late Devensian counterpart and this may well have been the same Wolstonian ice sheet responsible for the deposition of Scottish till of this age in borehole 81/26 in the outer Moray Firth (fig. 72; Sejrup *et al.,* 1984). At Moreseat a grey-black clayey till apparently underlies weathered till of inland derivation and possible pre- Devensian age (McMillan & Aitken, 1981) and may relate to an earlier incursion of Moray Firth ice (Hall, 1984).

Early glaciation from the south

Dark grey clayey tills also occur beneath Devensian tills from Aberdeen southwards to Burn of Benholm (Bremner, 1939; Campbell, 1934). These have been grouped with the dark, clayey tills around Ellon belonging to the Mesozoic Drift (Bremner, 1939). The shelly till at Benholm, however, contains reworked Upper Cretaceous and Tertiary dinoflagellate cysts and is derived from the E or SE and this till must represent movement of ice over the bed of the North Sea. Peat containing pollen of interstadial character incorporated into the base of the red till at Benholm (fig. 74) has given a date of > 42,000 years B.P. (Donner, 1979) and if this peat originally developed on the surface of the grey shelly till (Campbell, 1934) then this lower till must predate the Late Devensian.

Possible Scandinavian glaciation

Earlier workers (Bremner, 1939; 1943; Synge, 1956) attempted to relate the dark grey clayey tills along the

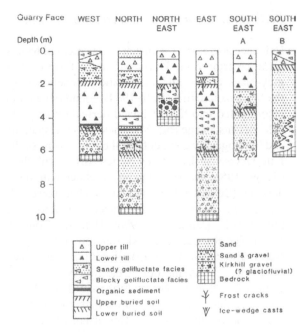

Figure 73: Kirkhill Quarry, schematic stratigraphy.

North Sea coast with the scatter of Scandinavian erratics in this area (Read *et al.*, 1925; Campbell, 1934; Bremner, 1939; 1943; Maclean, 1977) (fig. 74) and so infer invasion by Scandinavian ice. In fact, with one dubious exception (Bremner, 1943), no Scandinavian erratics are known from these grey shelly tills. Moreover, there is no evidence of Scandinavian ice crossing the North Sea to northeast Scotland during the Quaternary (Stoker *et al.*, 1985a); ice rafting followed by entrainment by Scottish ice masses remains the most likely mode of transport (Clapperton & Sugden, 1977). Finds of single Scandinavian erratics in early(?) Devensian tills at Kirkhill and Sandford Bay show, however, that some Scandinavian erratics, at least, were carried westwards from the North Sea coast during or before the Middle Pleistocene and then reworked into tills of inland derivation.

DEVENSIAN GLACIATIONS

During the Devensian northeast Scotland was glaciated by ice from the Moray Firth, Strathmore and the eastern Grampians. Each ice stream laid down a distinctive series of deposits (fig. 75).

(i) The Blue-Grey Series. Ice moving out of the Mesozoic basin of the Moray Firth deposited the Blue-Grey Series (Synge, 1956) along the coast and

as far SE as Peterhead. The tills of the Blue-Grey Series are typically calcareous, fine-grained and dark grey in colour and contain abundant reworked Late Jurassic and Early Cretaceous microfossils. Ice-worn shells are often present and at Clava and King Edward dark grey tills are found in association with beds with high shell contents which have long been regarded as *in situ* marine deposits (Jamieson, 1858;

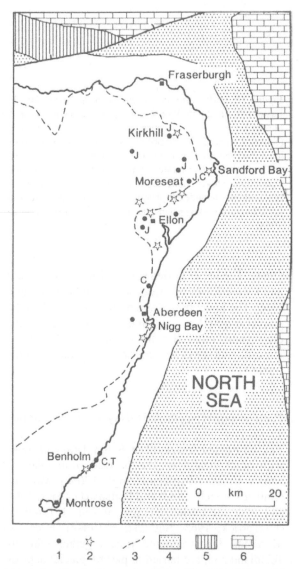

Figure 74: Sites with pre-Devensian tills derived from offshore and with erratics of Scandinavian origin. 1 = Site with pre-Devensian dark grey-black clayey till or erratic, derived from offshore. Contains Jurassic (J), Cretaceous (C) or Tertiary (T) macro-/micro-fossils. 2 = Site with Scandinavian erratic(s). 3 = Inland limit of Late Devensian Red Series and Blue-Grey Series drift sheets. 4 = Permo-Trias. 5 = Jurassic. 6 = Cretaceous.

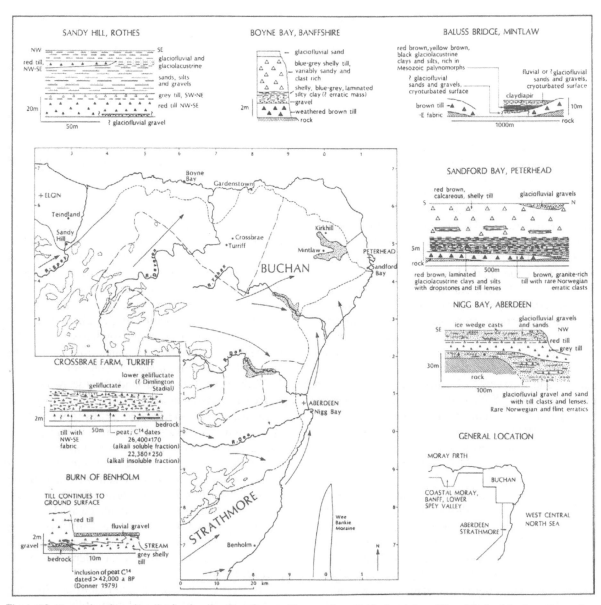

Figure 75: Devensian deposits: distribution, ice flow lines and key sections. Pecked and dotted line: Limit of the Late Devensian unglaciated area in north-east Scotland according to Synge (1956). Arrows: Ice flow lines.

1906; Horne *et al.*, 1893; Sutherland, 1981b), although transport as erratic masses seems more likely (Bell & Kendall, in Horne *et al.*, 1893; Read, 1923),

(ii) The Red Series. Ice moving north along the North Sea coast deposited the Red Series drift sheet (Jamieson, 1906). Red Series tills are typically a vivid red-brown colour, calcareous and contain materials derived from Devonian rocks in Strathmore and Permo-Triassic, Devonian and Cretaceous rocks offshore (Clapperton & Sugden, 1977). North of Aberdeen, however, the tills form only part of a

complex sequence of interstratified deposits in which basal lodgement tills are overlain by flow tills, subaquatic debris flows, thick glaciolacustrine muds and sands and ice-contact glaciofluvial sands and gravels (Merritt, 1981; Connell *et al.*, 1985).

(iii) Inland tills. Ice streams moving out of the eastern Grampians deposited sandy tills whose character strongly reflects the character of the local bedrock and which often incorporate significant proportions of reworked weathered rock. Thick inland tills are largely confined to the Dee valley and to valleys

south and west of Elgin. In Buchan inland tills are generally thin, discontinuous and have been disturbed by periglacial activity (Connell & Hall, 1987).

Stratigraphical sequence of the Devensian

Glacial sequences deposited by the three main ice streams are found interstratified around the coast of northeast Scotland and allow a relative sequence of Devensian events to be established (fig. 75).

Around Elgin grey shelly tills from the bed of the Moray Firth are found beneath reddish-brown sandy tills with erratics from the west and northwest (Peacock *et al.*, 1968; Aitken *et al.*, 1979). At the classic Sandy Hill section in the lower Spey valley at Rothes a red till which includes Old Red Sandstone and Jurassic erratics, derived from the NW is overlain by grey till with erratics from the south and southwest. These lower tills are overlain by bedded sands and an upper red till representing a final advance of ice up the Spey valley and, finally, thick stratified silts, sands and gravels of glaciolacustrine origin (Bremner, 1934; Sutherland, 1984a). The till stratigraphy in Moray indicates initial movement of ice from the NW Highlands across the Moray Firth and, probably without interruption, subsequent expansion of ice from the Great Glen, Monadliath and Cairngorms.

East of Elgin, Read (1923) recognised an upper till from the south and southwest overlying grey shelly tills from the Moray Firth. Work at the key section at Boyne Bay (Peacock, 1966; 1971b), however, indicates that an upper grey till with Huntly erratics is simply a facies of the underlying grey shelly till and there is now general agreement that the last ice movement along the outer Moray Firth coast was from the NW (Hall, 1984). Striae (Read, 1923) and till fabrics demonstrate an initial SW-NE movement of inland ice in northern Banff. That this inland ice had retreated from the coast during or before deglaciation is shown by the tongue of Blue-Grey Series till which extends south up the Deveron valley towards Turriff (fig. 75) and by the existence of large ice-marginal lakes ponded between the southern margin of the Moray Firth ice mass and the cliffed coastline into which the sands and gravels of the 'Coastal Deposits' were laid down (Read, 1923; Peacock, 1971b).

Along the North Sea coast between Ellon and Peterhead, a till with igneous and metamorphic erratics from the west is overlain unconformably by the Red Series (Jamieson, 1906; Hall, 1984). North of

Peterhead to St. Fergus, Jamieson (1906) recognised that the Red Series became intermingled with the Blue-Grey Series from the NW. Sutherland (1984a) has recently suggested that interstratification of these deposits was a result of glaciotectonic disruption of a pre-existing grey till by northward movement of the ice which deposited the Red Series. Sections and borehole evidence from the Peterhead area (McMillan & Aitken, 1981), however, leave no doubt that the Red and Blue-Grey Series ice masses were broadly contemporaneous (fig. 76).

Occupation of the Buchan coast by Red and Blue-Grey Series ice masses took place at a time when Inland Series ice had retreated to the west. Glaciolacustrine deposits, locally resting on inland tills, extend many kilometres up the Ugie and Ythan valleys and were deposited in lakes ponded by the margins of the coastal ice masses (Hall, 1984) (fig. 75).

In the Aberdeen area, notably at Nigg Bay (fig. 75), the local interdigitation of grey tills derived from the west with red tills from the south has been taken to indicate confluence of ice streams from the Dee and Don valleys and Strathmore (Clapperton & Sugden, 1977; Munro, 1986). Around Aberdeen, however, and even in Strathmore, tills of inland provenance again generally underlie Red Series tills (Bremner, 1915; Auton & Crofts, 1986). Initial expansion of inland ice into the Aberdeen area appears to have been followed by a period of weakening flow along the Dee and Don valleys which allowed incursion of ice from the south. Probably during this stage glaciolacustrine silts and sands accumulated in a lake in the Don valley west of Dyce (Auton & Crofts, 1986) and ice from Strathmore was able to penetrate into the mouths of the Angus glens and deposit red tills (Synge, 1956).

The stratigraphy for the area east of the Spey and north of Montrose shows that inland tills consistently underlie tills of the coastal ice masses. This regional stratigraphy indicates initial expansion of ice from the eastern Grampians to or beyond the present coastline prior to expansion of the Moray Firth and North Sea ice streams. In Buchan waning of inland masses caused a large ice-free area to develop but elsewhere the inland and coastal ice masses appear to have remained in contact. Final extension of coastal ice streams into Buchan was closely controlled by coastal topography, with ice forming lobes into valleys and avoiding higher ground, implying thin glaciers.

The maximum offshore extent of the Late Devensian ice sheet in the North Sea is marked by the eastern limit of the till of the Wee Bankie Formation (fig. 74). South of Stonehaven, this limit coincides

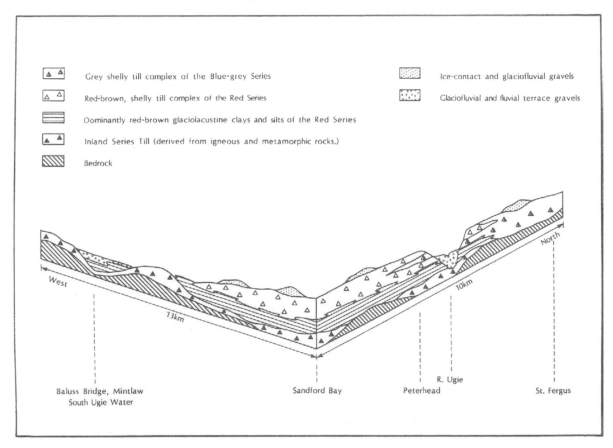

Figure 76: Schematic Devensian drift stratigraphy of the Peterhead area. N.B. Vertical scale greatly exaggerated.

with a prominent end moraine, the Wee Bankie Moraine. The Wee Bankie Formation is apparently the offshore extension of the Red and Blue-Grey Series deposits (Stoker *et al.*, 1985b).

Timing of Devensian glaciation

East of the occurrence of the Wee Bankie Formation, contemporaneous shell-bearing glaciomarine sands and silts, the Marr Bank Formation, are found. Shells from this formation are dated to the Late Devensian (Thomson & Eden, 1977; Holmes, 1977). The Wee Bankie Formation and the contiguous Red and Blue-Grey Series deposits are thus also of Late Devensian age (Stoker *et al.*, 1985b). Glaciation by the coastal ice streams must therefore date to the Dimlington Stadial (Rose, 1985).

The timing of the last glaciation of inland areas is more controversial. At Crossbrae, near Turriff, a peat containing pollen and macrofossils of interstadial character rests on till correlated with the Inland Series

and is overlain by up to 2.7 m of periglacial solifluction deposits. The peat has given radiocarbon dates of 26,000 ± 170 and 22,380 ± 250 years B.P. There is no evidence at this site for glaciation since peat formation close to the Middle/Late Devensian boundary. At Kirkhill Quarry, the upper till is correlated with the Inland Series and overlies the upper buried soil of probable Ipswichian age. These sites indicate that the last glaciation of inland areas of Buchan was during the Early or Middle Devensian and that the ice-free area and zone of enhanced periglacial activity found beyond the limits of the coastal ice streams date from the Dimlington Stadial (Connell & Hall, 1987). More dates are needed to confirm this sequence of events, however, particularly as the inland limit of Late Devensian ice in northeast Scotland has yet to be clearly demonstrated. It is too early to reject an alternative interpretation that the Inland Series represents an early stage of the Late Devensian glaciation and that the glacial lakes and periglacial features of Buchan merely reflect early deglaciation.

Conclusion

Significant advances have been made in the understanding of the Quaternary stratigraphy of northeast Scotland in recent years, with the discovery of pre-Devensian deposits at new sites and increased knowledge of the stratigraphy and depositional environments of Devensian glacial sequences. A regional litho- stratigraphy is emerging which includes deposits from at least three and probably four or more major glacial phases, together with multiple periglacial phases (Connell & Hall, 1987). Recent work in the Central North Sea also recognises four major episodes of glaciation (Cameron *et al.*, 1987) and the possibilities for land-sea correlation in the Buchan area are clear. Such correlations, however, will await firm dating of a far greater number of Quaternary deposits in northeast Scotland. Considerable chronostratigraphic potential does exist. Dateable organic materials occur in sediments at several sites, now buried, reported by early workers (Peacock, 1980; Edwards & Connell, 1981), most notably the 'indigo' shelly till from around Ellon (Jamieson, 1906). These sites need to be re-investigated by pitting or drilling. Amino-acid age determinations on shells from tills and shell beds are also required. Finally, the growing number of Pleistocene soils reported from the region suggests that soil stratigraphy will play a major role in future correlation.

It is sobering that after more than a century of Quaternary research the extent of the last glaciation in northeast Scotland should remain controversial. The existence of an ice-free area in the Late Devensian is demonstrated by the glaciolacustrine fills of the Ugie and Ythan valleys. Its eastern and northern limits are defined by the sharp drift limits of the Red and Blue-Grey Series but the western and southern margins of the ice-free area identified by Synge (1956) need to be re-evaluated, particularly in the light of new subsurface data from the Don valley (Auton & Crofts, 1986). The Crossbrae evidence indicates that an ice-free area existed in central Buchan throughout the Late Devensian. This site, however, provides the only firm dating control on the duration of the ice-free period and there is a pressing need for the withdrawal of inland ice from Buchan to be dated at other sites. The acceptance of the concept of an ice-free enclave in northeast Scotland in recent reviews (Sutherland, 1984a; Bowen *et al.*, 1986) is premature, particularly in view of the major implications that such an enclave carries for the limits of the last ice sheet in northern Scotland.

The glaciation and glacial deposits of the western Grampians

PETER W. THORP

Agassiz (1840) was the first scientist to recognise that glaciers once existed in Scotland. In a letter published in 'The Scotsman' on October 7th, 1840 Agassiz correctly interpreted the 'parallel roads' of Glen Roy as shorelines relating to ice-dammed lakes (Price, 1983). He also recognised that the polished rock surfaces and moraines at the foot of Ben Nevis provided clear evidence of former glacial activity. By the 1860's it was possible for Jamieson (1865) to publish inferred ice sheds and lines of movement of an ice sheet covering mainland Scotland. His map shows the western Grampians as the most important centre of ice accumulation in Scotland, with ice radiating out from Rannoch Moor. After 120 years of subsequent research, that basic ice flow pattern remains virtually unchanged (see for example Sutherland, 1984a; fig. 13). However, in the last two decades the pace of Quaternary research in the western Grampians has quickened and much more detailed information is now available, on deposits and landforms formed especially during the last ice advance (Loch Lomond Advance, c. 11,000 - 10,000 B.P.). Although much new information is available for areas such as Glen Roy, Glen Spean and the coastal area of Loch Linnhe, the mountain interior of the western Grampians has remained, until the last decade, a relatively neglected area for Quaternary research.

Topography and climate combine in the western Grampians to produce an upland environment that is likely to be quick to respond to a falling snowline and the onset of glaciation (Manley, 1949). Currently most of the region receives between c. 1,500 and 4,000 mm yr^{-1} of precipitation with amounts on most of the mountains greater than 2,500 mm yr^{-1} (fig. 77). Many peaks in the region exceed 1,000 m O.D., including Ben Nevis (1,344 m) which is the highest mountain in the British Isles. Dissection of the region by rivers and ice has been severe. It is a region of glacial erosional features *par excellence* with numerous low-level glacial breaches and cols, over 350 cirques, many classic U-shaped valleys and long tidewater lochs, such as the 30 km-long Loch Etive.

No deposits earlier than the Late Devensian (c. 26,000 B.P. to c. 10,000 B.P.) have been recorded in the western Grampians. During the 'Devensian Main Glaciation' the western Grampians formed the most important source area for an ice sheet that extended beyond the present coastline of Scotland (Sutherland, this volume). The patterns of dispersal of erratics and the ice flow directions inferred from striae show that the principal ice shed extended in a roughly N-S direction across the western Grampians, only 90 km from the Atlantic coast (fig. 77).

Widespread and abundant geological evidence of the former extent of a mountain icefield has been mapped in the western Grampians (summarised in Thorp, 1986). Radiocarbon-dated marine shells from beneath till at South Shian by Loch Creran (fig. 77) and from outside the study area in the Forth valley, on the Isle of Mull and by Loch Lomond (Sutherland, 1984a) securely date the icefield to the Loch Lomond Stadial (approximately equivalent to the Younger Dryas of Scandinavia; c. 10,800 - 10,300 B.P.). Further support for a Loch Lomond Stadial age for the icefield is provided by radiocarbon-dated sites on Rannoch Moor (Lowe & Walker, 1980) and by pollen-stratigraphic studies in glens Spean and Roy (Macpherson, 1980; Lowe, unpublished).

The Loch Lomond icefield covered about 65% of the area shown in fig. 78 so that for many areas only the deposits of the last and comparatively recent glacial event have survived. This account, therefore, is confined to the mainly qualitative description and analysis of the glacial, glaciofluvial and glaciolacustrine deposits and landforms associated with the glaciers of the Loch Lomond Advance.

The Loch Lomond Advance icefield

Charlesworth (1955), McCann (1966), Peacock (1971a, 1977) and Gray (1972, 1975) referred mainly

Figure 77: The main centre of ice accumulation in Scotland (modified after Sissons, 1976) and places mentioned in the text.

to proglacial glaciofluvial landforms in proposing various limits to former outlet glaciers that descended westwards to sea level from the Loch Lomond icefield in the western Grampians (fig. 78). Thompson (1972) mapped the limits of the icefield along its eastern margin in the Loch Rannoch and Glen Lyon areas. Sissons (1979a) was able to reconstruct in detail the glacial limits and surface form of the icefield from the Great Glen to the Strath of Ossian on the basis of the mapped distribution of numerous end and lateral moraines (fig. 78). The ice-dammed lakes and associated deposits and landforms in the Glen Roy area, in particular, were the focus of a series of

research projects undertaken by Sissons, and by Sissons & Cornish (summarised in Ballantyne & Gray, 1984). The results of extensive field mapping, especially in the former accumulation and ice shed areas of the Loch Lomond Advance icefield, were presented by Thorp (1981a, b; 1984; 1986). Former glacier limits, the majority of which have been termed trimlines by Thorp (1981b), were inferred from contrasting glacial and periglacially-derived deposits and landforms inside and outside the advance limits respectively (figs. 79-83). Such limits, based on the field evidence on 225 spurs, 84 cols and on numerous mountain ridges, complement the limits,

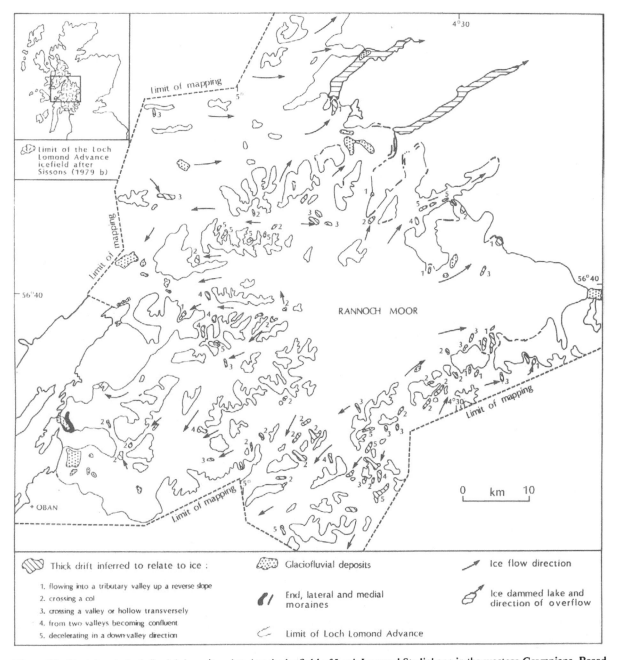

Figure 78: Glacial and glaciofluvial deposits related to the icefield of Loch Lomond Stadial age in the western Grampians. Based on published data in Peacock (1971), Gray (1975), Sissons (1979a) and Thorp (1986) and on unpublished data in Gray (1972), Thompson (1972) and Thorp (1984).

based on end, lateral and hummocky moraines, the upper limit of thick drift and the ice-contact slope of glaciofluvial landforms, in the ablation zones of the former glaciers. This field evidence enabled, for the first time, the direct correlation of the western and eastern margins of the icefield and the reconstruction of the altitudinal form, ice sheds and ice flow directions of the western Grampians icefield (fig. 84).

The geomorphological evidence

End and lateral moraines. The maximum extent of the Roy, Treig, Ossian, Alder and Rannoch outlet glaciers on the northeast side of the icefield (fig. 77) is delimited by well-developed end and lateral moraines (fig. 78) that total c. 35 km in length and that range from 200 to 670 m O.D. (Thompson, 1972;

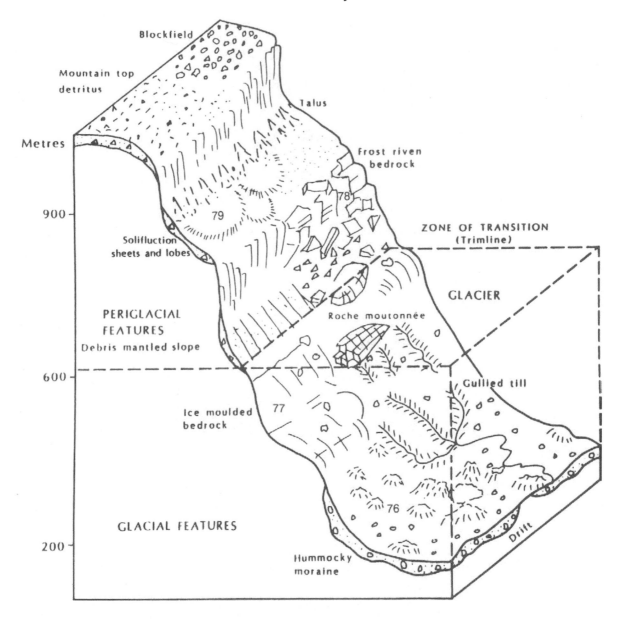

Figure 79: Diagrammatic representation of the glacial and periglacial features observed on a spur on the north side of Beinn Dubhchraig (NN 308255) south of Tyndrum. Location of figs. 80-83 shown.

Sissons, 1979a; Thorp, 1986). In contrast no unequivocal lateral moraine has been recorded in the southwest part of the icefield shown in fig. 78, west of the main ice shed (Thorp, 1984). Furthermore the only clearly-developed end moraines, that total only c. 3 km in length in the western part of the former icefield, are those delimiting the maximal extent of the Creran glacier (fig. 78). The reasons for such contrasts are not clear although they may relate to differences in topography, palaeoclimate and glacier dynamics

across the former icefield, a view supported by the following points.

First, the Creran, Etive, Linnhe and Duror outlet glaciers all terminated in tidal water ranging from only a few metres (Duror and Creran) to c. 80 m (Linnhe) in depth. The apparent lack of end moraines at the inferred limit of the Linnhe and Etive glaciers may either relate to their unidentified presence in deep water on the fjord floors or imply that the glaciers did not reach a steady-state position long

Figure 80: Hummocky moraine at a valley confluence near Tyndrum at c. 200 m O.D.

Figure 81: Ice-moulded schist bedrock at c. 450 m O.D.

Figure 82: Frost-riven bedrock and angular debris at c. 650 m O.D.

Figure 83: Solifluction lobes mantling the slope at c. 900 m O.D.

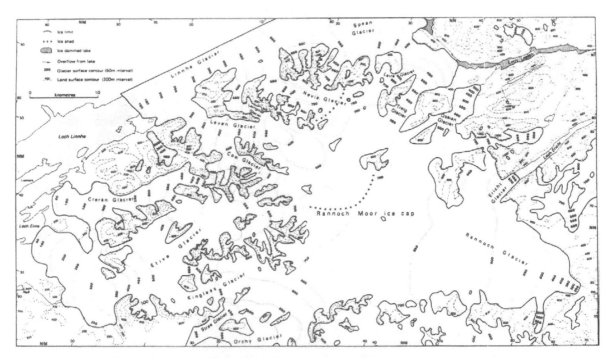

Figure 84: Extent and form of the western Grampian icefield (after Thorp, 1986).

enough for end moraines to form (Mercer, 1961; Paterson, 1981). Second, the steep sides of the fjords and glacial troughs in the southwestern part of the icefield may have inhibited the formation and preservation of lateral moraines (cf. Chinn, 1979). Third, the climate across the icefield would have differed considerably with heavy precipitation of 2,700 -

4,000 mm yr^{-1} on the mountains in the southwest (Sissons, 1979b) decreasing to less than 1,000 mm yr^{-1} on some of the mountains to the northeast of the icefield (Thorp, unpublished). Since the glaciers on the western side of the icefield mainly faced to the southwest and flowed for much of their length below c. 600 m O.D. they would have been subjected to higher rates of ablation than the outlet glaciers on the eastern side of the icefield (fig. 84). Such high-accumulation, high-ablation glaciers generate considerable quantities of meltwater (Flint, 1971; Gustavson & Boothroyd, 1982). It is suggested that much of the debris being transported supraglacially, englacially and subglacially by the glaciers may have been entrained by meltwater preventing the formation of clear end and lateral moraines.

Materials within the end and lateral moraines reflect the varying environments that existed in the western Grampians during the Loch Lomond Stadial. The end moraine at South Shian by Loch Creran (fig. 77) is composed principally of till formed from Late Devensian marine silt and silty clay (Peacock, 1977) bulldozed from the floor of the sea loch (fjord). Similar fine-grained lacustrine sediments mixed with till and glaciofluvial sands and gravels have been reported near the terminus of the Roy glacier (Sissons, 1977d), demonstrating that the glacier advanced northwards over lake-floor sediments. Conversely much of the material within the Ossian, Treig and Rannoch end and lateral moraines is composed of coarse 'rubbly' debris containing many large boulders > 1 m in length (Sissons, 1979a; Thorp, 1984) reflecting their origin especially from local outcrops of bedrock and from periglacially-weathered mountain slopes upglacier.

Hummocky moraine

The term hummocky moraine has been applied, in perhaps a general and indiscriminate fashion (cf. Lundqvist, 1977), to the irregular, moundy terrain that characterises many areas within the Scottish Highlands (Sissons, 1967a; Price, 1983; Eyles, 1983c). At one end of the spectrum the terrain comprises many high (> 20 m), steep-sided (> 30°) conical-shaped mounds and ridges studded with boulders (1 - 3 m long) and interspersed with deep kettles, as exemplified at Tyndrum (Lowe & Walker, 1981). At the other end of the spectrum the term has been applied to low mounds (2 - 3 m) separated by flat ground or by shallow, peat-filled depressions (Thompson, 1972; Thorp, 1984). Such differences in the size and steepness of the hummocks may relate, in some areas, to topographical factors. For example, it has been noted by some workers (Sissons & Grant, 1972; Thompson, 1972; Thorp, 1984) that hummocky moraine is much more subdued on valley sides, especially where the slope gradient is moderately steep, compared with the often highly pronounced hummocks to be found on valley floors and at valley confluences. Sections within the hummocks generally reveal a wide variety of sediments (fig. 85) that range from water-sorted silts, sands and gravels to very coarse rubbly 'till' with little or no matrix

Figure 85: Section in typical hummocky moraine.

(Sisson & Grant, 1972; Sissons, 1974a; Eyles, 1983c).

In the western Grampians the distribution of areas of hummocky moraine within the limits of the Loch Lomond Advance indicates important spatial patterns. Many areas of hummocky moraine appear to correlate with large inputs of rock debris onto the surfaces of former glaciers from cirque and valley-side rock walls by means of rock falls, ice and snow avalanches and other mass wasting processes (Sissons & Grant, 1972; Sissons, 1967a; Eyles, 1983c). In a number of areas the abrupt termination of large solifluction boulder lobes at glacier limits, inferred from other forms of evidence, suggests that much supraglacial rock debris was derived from snow-free slopes undergoing severe periglacial weathering (Sissons, 1976c; Ballantyne, 1984; Thorp, 1986).

Studies of the glacial sediments forming at the present time along the margins of actively retreating glaciers in Iceland led Eyles (1983c) to suggest that the hummocky moraine in the Scottish Highlands was largely formed by the dumping of supraglacial diamict during active retreat of the Loch Lomond Advance glaciers. Although such a depositional model may well apply to many areas of hummocky moraine in the western Grampians, the model is difficult to apply to the drift hummocks on Rannoch Moor. Here there are many thousands of hummocks extending for about 30 km from the western edge of the basin to the slopes above Loch Rannoch (Thorp, 1986). Some of the rock debris in the hummocks is volcanic in origin and is clearly derived from the high mountains to the west of the basin (Hinxman *et al.*, 1923). A good proportion of the volcanic debris is likely to have been transported supraglacially and englacially into the basin. However, most of the rock debris within the hummocks is local in origin (i.e. Rannoch Moor Granite) and derived from the floor of the basin and is, therefore, most likely to have been incorporated subglacially as the glacier tongues flowed from the mountain rim in the west and south towards Loch Rannoch (cf. Hinxman *et al.*, 1923). Since it is likely that 'hummocky moraine' is formed in different ways, it is perhaps premature to apply a supraglacial origin to all areas of hummocky moraine in the Scottish Highlands.

Thick drift. Many areas within the limits of the Loch Lomond Advance icefield comprise bare, ice-scoured bedrock, especially on the western side of the former icefield where slopes tend to be steeper and ice velocities are inferred to have been generally greater than on the eastern side (Thorp, 1984). Else-where within the limits of the icefield large areas of bedrock are concealed by hummocky moraine or by drift that varies in thickness from only a metre or less to a maximum estimated thickness of up to 80 m, as in Glen Roy (Sissons, 1979a). Extensive areas of drift exceeding 5 m in thickness can be identified in the field from the depth of sections formed by post-glacial streams cutting into the sediment (often as far as the underlying solid bedrock). In many localities stream dissection of the drift has created a badland-type topography with numerous deep gullies separated by sharp-edged ridges (cf. King, 1976: 114). The spatial arrangement of the thick sediment sequences within the limits of the icefield (fig. 78) suggests that five types of topographical traps (allied to an abundant supply of rock debris within the glacier) played an important part in their formation. Such thick sediment infills occur in valleys tributary to the main outlet glaciers (type 1), below cols crossed by tongues of ice (type 2) and in valleys transverse to the inferred ice-flow direction (type 3). Additional areas of thick sediment are frequently found at valley confluences (type 4) and, less commonly, in wider valley sections (type 5). The five types of topographical traps are shown schematically in fig. 86.

The two most common locations for a thick infill occur either where basal ice flowed up a reverse slope into an ice-free tributary valley or where ice flowed transfluently across a col (types 1 and 2, fig. 78). In the latter situation the thick drift is normally found on the upglacier side of the col. Such a spatial pattern is very apparent in upper Glen Lyon where congested ice found escape routes across high-level cols (c. 550 - 700 m O.D.) northwards into the Rannoch Basin and southwards into Strath Fillan. A hypothetical sequence of depositional processes is envisaged as occurring as a tongue of ice moves into an ice-free valley and eventually across a col at the head of the valley (stages 1 and 2, fig. 87). A hypothetical stage (stage 3) during deglaciation is also depicted.

As the icefield grew in size evidence provided by striae, friction cracks and *roches moutonnées* indicates that the ice flow direction in many areas was transverse to a number of small valleys and ravines. These depressions frequently contain thick sediment infills (type 3; > 5 m) whilst the bedrock surfaces of adjacent ridges are usually drift-free and ice-scoured (e.g. south of Loch Ericht, between upper Glen Etive and Glen Creran, and between upper Glen Nevis and Loch Treig; fig. 78). Thick valley infills of drift, orientated at an angle to the former inferred ice-flow direction, have been observed elsewhere, for exam-

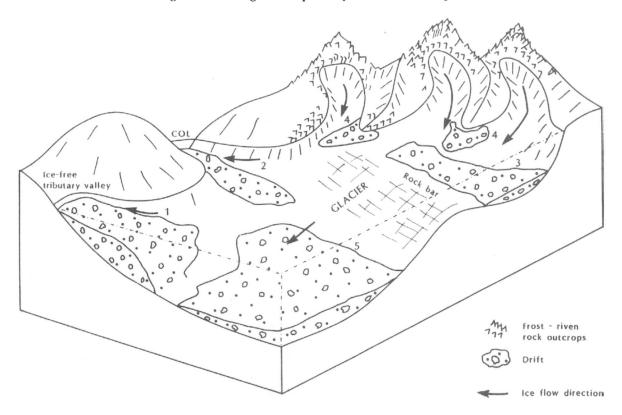

Figure 86: Schematic representation of the location of thick drift within the limits of the Loch Lomond Advance icefield in the western Grampians. For explanation of numbers see text and fig. 78.

ple western U.S.A. (Thorp, 1985), Norway (Garnes, 1979), northern England (King, 1976) and in other parts of Scotland (Barrow *et al.*, 1905; Sissons, 1967a; 1974a). The process associated with the deposition of thick drift in valleys transverse to the ice-flow direction (type 3) and also in wide valley sections (type 5) are probably similar since thickening of the ice occurs in both situations leading to a reduction in the basal ice velocity and a retardation of the debris in the basal layers of the ice. Lodgement and basal meltout tills would be expected in such locations since the thermal gradient in the ice would increase over the hollow (Nobles & Weertman, 1971), thus causing a faster rate of basal melting and hence a faster rate of deposition to occur. However, the detailed sedimentological and stratigraphical work to support such views has yet to be done.

Thick sequences of drift, sometimes associated with an irregular, moundy surface, or more rarely with a morainic ridge (i.e. medial moraine) occur at a number of valley confluences within the limits of the Loch Lomond Advance icefield (Thorp, 1986). For example Barrow *et al.* (1905) noted that in Glen Lyon

areas of thick drift were mostly located 'under the lee of hill spurs and in the side valleys'. Such deposition (type 4, fig. 78) is inferred to relate to the periglacial mass wasting of the valley walls and the erosion subglacially of the bedrock and the transport supraglacially and englacially of the rock debris to the point of confluence below a spur (Eyles & Rogerson, 1978) and across the exit of the tributary valley.

Glaciofluvial deposits and landforms. A wide variety of glaciofluvial landforms formed in the western Grampians at and within the maximal limits of the Loch Lomond Advance icefield. In the west large quantities of glaciofluvial sands and gravels, covering 21 km² in total, were deposited proglacially in a marine environment. Fluvioglacial deltas, each covering c. 4 km² formed close to the maximal limits of the Creran and Etive glaciers (Gray, 1975) in shallow tidal water probably less than 40 m deep. Deltas of similar origin occur well within the inferred maximal limits of the Linnhe glacier at Corran Ferry, North Ballachulish and at Annat by Loch Eil (figs. 77 and 78; McCann, 1966; Peacock, 1970). Minimum thicknesses of sediment of 25 and 42 m respectively

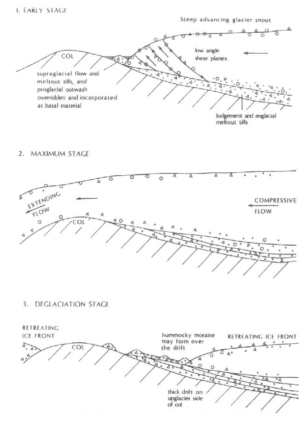

Figure 87: Theoretical depiction of the inferred stages (1, 2, 3) in the formation of thick drift by wet-based ice flowing into ice-free tributary valleys and across cols in the western Grampians.

are indicated for the Achnacree (Etive) and Corran Ferry glaciofluvial deltas by the depth of the kettles within the sediments, while a borehole in the Annat delta failed to reach rockhead at 60 m O.D. (Peacock, 1970). Four fan-shaped masses of gravel totalling c. 1 km² flank the edge of Loch Linnhe near Kentallen (fig. 78). These have been interpreted by Thorp (1984, 1986) as representing outwash fans formed laterally along a calving ice front in a manner analogous to outwash fans forming at the termini of some present day tidewater glaciers in Alaska (cf. Powell, 1981). Three of the fans have a steep ice-contact slope with a surface that slopes to the south-west from 10 - 11 m O.D. at the apex, at the foot of a raised cliffline. At the time of their formation the ice front in Loch Linnhe was between 3.5 and 5.5 km wide and it was calving into tidal water at least 80 m deep.

Sea level in Lochs Linnhe, Creran and Etive at the time of the formation of the features described above is considered to have been less than 13 m O.D. (Peacock, 1970; Sissons, 1967a; Gray, 1972). Gray

(1975) has described laminated clays close by Loch Etive occurring up to altitudes of 10 - 11 m O.D. He has suggested that if the clays accumulated in sea water they would provide a minimal level for the sea at that time. Such a suggestion accords with the 10 - 12 m altitude of the raised Main Rock Platform (Main Lateglacial Shoreline) in the vicinity (Gray, 1978a), that is believed to have formed mainly during the Loch Lomond Stadial (Sissons, 1974b).

Glaciofluvial sediments deposited proglacially are less extensive on the eastern side of the icefield and cover less than 7 km². The largest glaciofluvial landform (c. 3 km²) is the outwash valley train that slopes downvalley for 3 km from an ice-contact slope at the eastern end of Loch Rannoch. Another large sand and gravel mass, covering 1.5 km² and formed at the level of the 261 m O.D. ice-dammed lake in Glen Roy, is located at the junction of the Roy and Turret Glens (see below and fig. 89).

Surprisingly few eskers and kame terraces have been mapped within the limits of the western Grampians icefield, although this generalisation does not apply to kames which are more widespread. Most eskers are generally less than 3 m high and less than 300 m long and rarely occur in groups of more than 2 or 3. Exceptions include groups of eskers to the southwest of Loch Ericht (Thorp, 1984) and north and southwest of Loch Rannoch (Thompson, 1972; Thorp, 1984). Fine examples of kame terraces have been mapped and instrumentally-levelled along the edge of Loch Etive by Gray (1972). Sections in the terraces generally reveal poorly developed bedded sands and gravels with occasional lenses of laminated clays. Kame terraces also occur in the Loch Teig-Spean valley area where, in places, they form the 261 m O.D. 'road' (Sissons, 1977d). Small kame terraces (< 15 m wide) flank the edges of a shallow valley (NN 620470) southwest of Loch Ericht (Thorp, 1984) where they occur in association with kettles, subglacially engorged eskers and large flat-topped kames. The whole assemblage of landforms is believed to have formed during the downwasting of a stagnant lobe of ice. Only a few other kame terraces have been mapped elsewhere within the limits of the icefield, across an area exceeding 2,000 km².

There are two main implications of the spatial distribution of the glaciofluvial deposits and landforms described above. First, in the west the location of glaciofluvial deltas was controlled primarily by loch constrictions and rock thresholds (cf. Mercer, 1961; Peacock, 1971a). Initial retreat of the calving ice fronts in Lochs Linnhe, Creran and Etive from their maximum position may have been triggered by

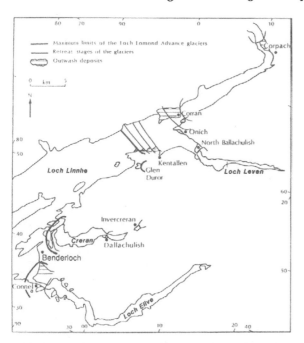

Figure 88: Glaciofluvial deposits and inferred retreat stages of the Loch Lomond Advance glaciers in Loch Linnhe and vicinity.

climatic change at or close to the end of the Loch Lomond Stadial. However, this view may not be correct since tidewater glaciers can undergo large-scale asynchronous advances and retreats that are not directly related to climatic changes (Brown *et al.*, 1982). The major controlling factors are the depth of the water and the width of the fjord, with constrictions and rock thresholds forming 'anchor points' for the glaciers. The retreat of the Linnhe glacier, for example, from its maximal position near Kentallen is likely to have been rapid (cf. Sutherland, 1984a for the retreat of the southwestern margin of the Late Devensian ice sheet) until it divided into two and stabilised at 'anchor points' at Corran Ferry and North Ballachulish. This enabled the development of the glaciofluvial deltas at these two locations. Similarly, other glaciofluvial deltas formed at loch narrows and rock thresholds in lochs Creran, Etive and Eil (fig. 88) during general deglaciation of the icefield. Second, the general lack of assemblages of ice-stagnation landforms (i.e. kames, eskers, kettles and kame terraces) in many areas in the western Grampians, that are topographically suitable for the development of downwasting masses of ice, suggests that the Loch Lomond Advance icefield did not undergo widespread stagnation (Sissons, 1976c) but retreated actively (cf. Eyles, 1983c; Sutherland, 1984a).

Glaciolacustrine deposits and landforms. During the Loch Lomond Stadial proglacial ice-dammed lakes were impounded in Glens Roy, Gloy and Spean (Jamieson, 1863; Sissons, 1979a), on the northeast side of the western Grampians icefield (fig. 78) to form the famous 'parallel roads' (lake shorelines). Ice from the high snow accumulation areas west of the Great Glen, and to a much lesser extent from the Ben Nevis Range, advanced up Glen Roy to form a rising sequence of lake levels at 261, 325 and 350 m O.D.. During deglaciation the levels were occupied in reverse order. In Glen Gloy there is only one well-developed shoreline (355 m O.D.) and only one in Glen Spean (261 m O.D.). All four lake levels correspond with outlet cols at approximately the same altitude as the shorelines.

Jamieson (1863) attributed the origin of the shorelines to wave action but recent detailed investigations by Sissons (1978) led him to propose severe frost action as the primary process with the rock debris carried away by ice floes during the spring melt of the lake ice. In support of this proposal he cited the angular, slabby nature of much of the debris on the shorelines and the irregular surface and considerable width in places (6 - 12 m) of the platform where cut across solid rock. Sissons argued that given the relatively short time that the lakes were likely to be in existence it was improbable that wave action alone could cut such wide rock platforms. As further evidence for severe frost action operating at the level of the lake surface Sissons (1977d) described sections on the lower ground in upper Glen Roy which comprise fine-grained lacustrine clays, silts and sands containing numerous angular stones with well-developed dropstone structures. Such stones he suggested were unlikely to be derived from the calving ice front because of their dominantly angular shape. Their source was most likely the lake shore.

Three large fans of coarse gravel occupy the floor of upper Glen Roy. The largest (c. 1.5 km²) is at the confluence of Glens Roy and Turret and for some time its origin aroused much controversy (see Gray, 1978b). Initially interpreted as a fluvial fan by Sissons (1977d) he later reinterpreted the features as an outwash fan associated with a glacier terminus in Glen Turret (Sissons & Cornish, 1983). Sections in the two fluvial fans show coarse gravels with clasts to 0.8 m long overlain by laminated silts and clays typically 0.5 - 1.3 m thick (Sissons & Cornish, 1983). These stratigraphical relationships, together with the bulk of the fan gravels occupying the valley sides and floor below the 261 m. O.D. lake level, led Sissons & Cornish to propose that the gravels were deposited

Figure 89: Dissected glaciofluvial fan at the confluence of glens Turret and Roy. Coarse sands and gravels overlain by lacustrine silts and clays are exposed in the section cut by the river Turret (centre).

subaquatically in the 261 m O.D. lake of the rising sequence by torrential spring snowmelt floods and were later covered by lacustrine sediments when the lake level stood at 325 and 350 m O.D.. Only thin sheets of gravel were deposited, mainly subaerially, over the lacustrine deposits and coarse gravels when the lake level descended to 261 m as the Roy glacier retreated downvalley. All three gravel fans have undergone considerable dissection by fluvial activity (fig. 89) since the drainage of the 261 m O.D. lake.

Additional deposits and landforms, associated with the 261 m O.D. lake, were created during ice

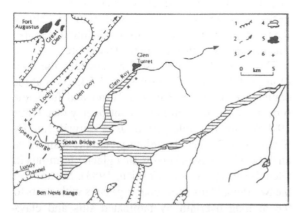

Figure 90: The 261 m O.D. ice-dammed lake in Glen Roy and adjacent valleys at its greatest extent during the wastage of the Loch Lomond Advance icefield (modified after Sissons, 1979c). 1 = Ice margin. 2 = Jökulhlaup route. 3 = End moraines. 4 = Ice-dammed lake. 5 = Outwash. 6 = Fluvial fan.

retreat in Glen Spean (fig. 90). At its maximal extent at Spean Bridge the 261 m O.D. lake had a surface area of 73 km², a volume of 5 km³ and it was 200 m deep near the ice dam (Sissons, 1979c). Sissons argued that drainage of such a large lake took place, as with many present-day ice-dammed lakes, in the form of a catastrophic subglacial *jökulhlaup* (or a series of *jökulhlaups*) via the 30 m deep Spean gorge. Such a flood he envisaged as raising the level of Loch Ness by 8.5 m and depositing the vast gravel spread near Fort Augustus (Sissons, 1979c). Later (1981a) Sissons ascribed a similar origin to the 40 m thick fan-shaped mass of gravel at Inverness overlying marine deposits of the Main Lateglacial Shoreline and underlying Flandrian deposits. He further proposed that five end moraines up to 5.5 km long in the Spean Bridge area were related to the discharge of lake waters southwestwards through the Lundy channel, that is some 70 - 80 m O.D. higher than the Spean gorge (fig. 90).

Accurate levelling by Sissons (1979f) of about 20 terraces in Glen Spean and lower Glen Roy has shown that a complex sequence of events occurred during further deglaciation of the Loch Lomond Advance icefield. Initially a small lake was impounded in mid-Glen Roy by an 80 m thick drift barrier, just inside the Loch Lomond Advance limit. A staircase sequence of terraces downstream of the drift barrier records the breaching of the barrier and the infilling of the final Spean ice-dammed lake with sand. Further retreat of the ice is recorded by the dissection of the sand infill by the Spean and by the creation of another series of terraces that slope down steeply to

Loch Lochy. A large fluvial fan built up against the ice from the Spean gorge exit and kame terraces (fig. 90) along the south side of Loch Lochy demonstrate that ice still occupied the Great Glen. Terrace terminations show that the level of the ice-dammed Loch Lochy was about 39 - 40 m O.D. (some 13 - 14 m above its present natural level) and that it overflowed to the northeast along the floor of the Great Glen. As the ice front retreated to the southwest the level of Loch Lochy fell to 34 m, during which time a large fan of sand relating to this level was deposited at the gorge exit. The fan merges into a large terrace that slopes to the southwest indicating that the present drainage of the Spean towards Loch Linnhe had finally become established.

Conclusions

Much of the basic field mapping of the geomorphological evidence relating to the Loch Lomond Advance icefield in the western Grampians has now been completed. Detailed coverage, however, has been uneven areally. Much information is available for the Glen Roy - Glen Spean area in terms of former ice-marginal landforms, ice-dammed lake levels, lake escape routes, fluvial fans and river terraces, that can be placed into an approximate chronological framework. Quaternary landforms in many other areas of the western Grampians have yet to be studied exhaustively, while in some areas no research on Quaternary landforms and sediments has been undertaken. For example examination of sediments and submerged landforms on the floors of the sea lochs (fig. 77) may prove rewarding, but these areas have been sadly neglected until recently (cf. Boulton *et al.*, 1981). Moreover, no detailed quantitative studies of Quaternary sediments (e.g. till fabrics, particle-size and shape analysis, compressibility tests and the analysis of glaciotectonic structures and sediment stratigraphy) have been published (as far as the writer is aware) for the western Grampians. There is considerable scope for such studies, especially for understanding the genesis of the hummocky moraine and thick drift sequences that occur within the limits of the Loch Lomond Advance icefield.

Another fruitful line of research would be to investigate the relationships between the thermal regimes, thicknesses, surface slopes and ice-flow velocities of the former Loch Lomond Advance glaciers and the palaeoclimatic controls operating at the time of their formation. Such information might help to reconcile the view that the icefield was entirely wet-based in the western Grampians (Thorp, unpublished) with the view that palaeoclimatic conditions at the time were believed to be severe enough for permafrost (Rose, 1975; Sissons, 1976c), with the mean annual sea level temperature possibly as low as -8° C to -10° C (Ballantyne, 1984). Finally, such glaciological information would be of great value in helping to determine the possible relationships that existed between the thermal regimes of the former glaciers and the sedimentary characteristics and spatial distribution of the glacial and glaciofluvial deposits in the western Grampians.

The glacial history and glacial deposits of the North and West Cumbrian lowlands

DAVID HUDDART

The lowlands of North and West Cumbria form a broad, west-east trending area stretching from the Solway Firth to the Pennines (fig. 91). This lowland decreases in width to the south but there is an extension down the Edenside valleys, between the Lake District and the Pennines. It is a critical area for the study of glacial deposits in Britain for two reasons. Firstly, there has been dispute over the importance of stagnant ice during deglaciation. Goodchild (1875, 1887) stressed the complexity of the deposits in Edenside and considered that practically all were formed subglacially/englacially during the melting of a stagnant ice sheet. However, Trotter (1929) and Hollingworth (1931) fitted the deposits into a model whereby the landforms and sediments were associated with a continuously retreating ice front and extra-glacial deposition. Subsequently Huddart (1967, 1970, 1981a, 1983) has reinterpreted the deglaciation of some lower areas of the Lake District valleys and Edenside in terms of downmelting and marginal stagnation. This has resulted in complex stratigraphic sequences with basal, melt-out and flow-till deposition; subglacial/englacial fluvioglacial deposition and erosion, with the creation of ice-walled lake plain, ice-walled stream trench, moulin kame, esker, ice-walled lake delta and marginal kame terrace environments.

However, in the Carlisle Plain, in the lower reaches of the Irthing and Eden valleys there is evidence of proglacial lacustrine deltaic and deep-water sediments associated with a westerly retreating, Devensian 'Main Glaciation' ice sheet (Huddart, 1970). Topography can cause changes in deglaciation style and this has happened in Cumbria with resultant changes in deglacial environment, landforms and sedimentary succession. Secondly, Cumbria is the type locality for the Scottish Readvance, which has been a recurrent theme in the discussion of the Devensian glaciation of the northern Irish Sea basin. It was first suggested by Trotter (1922) and occurred at a late stage in deglaciation. Whilst the concept has been re-evaluated and supported by Huddart (1970, 1977), Huddart & Tooley (1972) and Huddart et al. (1977) using stratigraphic, sedimentological and geomorphological methods, other workers have suggested either that this episode did not occur or that it is largely illusory. In this chapter the major concern will be to establish that there is evidence for the readvance and to discuss its regional significance.

The problem of the Scottish Readvance

A discussion of the glacial chronology in the Cumbria lowlands is found in Huddart (1971b), Huddart et al. (1977), Vincent (1985) and Gale (1985) where four glacial phases have been recognised. The latest or Scottish Readvance was established by the Geological Survey in the 1920's as a separate readvance of southern Scottish ice onto the Cumbria lowland after the 'Main Glaciation'. The principal effects of this readvance were thought to have been the damming of lakes in the Carlisle Plain and the deposition of an upper till. However, although a limit for this readvance in the Carlisle area was presented by Trotter (1929), the evidence was meagre and the author himself stated that unless there were underlying sands the Upper Boulder Clay could not be differentiated from the Lower Boulder Clay. Since the 1920's several authors have drawn generalised lines for the marginal limits of this readvance in northern Britain (see figures in Huddart & Tooley, 1977; Thomas, 1985b).

Trotter (1922) first recognised an Upper Boulder Clay resting on Middle Sands near Longtown. He stated that this upper till was deposited from a distinct readvance of ice from the Southern Uplands and was not a readvance of 'Main Glaciation' ice from the Irish Sea. This readvance was subsequently thought to have extended over the Carlisle Plain up to the 400 m contour in the Brampton region (Trotter, 1929). The deposits from it were said to occur as broad sheets and discontinuous patches, both of which ap-

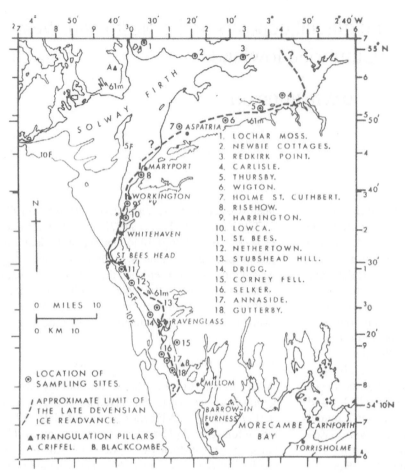

Figure 91: Location map and the approximate limit of the Late Devensian ice readvance.

peared to be plastered on the underlying deposits. No serious modification of the pre-existing landforms was reported, although drumlins with steeper and blunter ends to the west were developed. The absence of any end moraine and the progressive thinning of the Upper Boulder Clay, with the preservation of the original forms of the Middle Sands towards the outer limit suggested that this readvance was short and of no great intensity. The recessional deposits from this readvance were traced by Trotter & Hollingworth (1932). They recognised that the readvance impinged on the present day coast at St. Bees, where they thought its 'ground moraine' overlaid peats, silts and loams that were the deposits of an intervening warm period, and the ice may have reached as far as the Whicham valley.

Trotter's (1924) limits in West Cumbria, the Carlisle Plain and the north coast of Ireland were extended by Charlesworth's (1926) North-East Ireland - Isle of Man - Cumberland Moraine to the Bride Hills Moraine in the Isle of Man and to the Tweed

valley. In 1939 Charlesworth correlated the Antrim readvance with the Stranraer-Lammermuir moraines and the Solway and Bride Hills Moraine and with the Irish Carlingford Moraines.

A moraine traced by Synge (1952) from Brampton, south to St. Bees Head, passed out to sea and reappeared in the Bride Hills. This ice limit was correlated with the Irish Kells Moraine, as both marked the southern limit of drumlin belts. Mitchell (1960, 1963) recognised moraines on the Irish Sea bed and linked these with moraines at Gormanstown, the Bride Hills and St. Bees.

Penny (1964) suggested a line delimiting the Scottish Readvance in the Cumberland lowland which was merely the mapped feather-edge of the Upper Boulder Clay. Sissons (1964b) suggested that most of the drumlins were formed during the Lammermuir advance which occurred after the maximum of the 'Newer Drift' but before the Perth Readvance. He thought ice extended well south of the Lake District at this stage and linked the ice front with the Kells

Moraine. A limit for the Scottish Readvance further south was also suggested by Gresswell (1967) who correlated his Kirkham Moraine with the Bride/Kells Moraines. Saunders (1968) suggested that the limit of the Scottish Readvance should be brought even further south and equated with the Dinas-Trevor-Brynkir Moraine in Lleyn.

Walker (1966) discussed the significance of the Scottish Readvance glaciation. Although he accepted the work of the Geological Survey in delimiting its extent, he established from sites both inside and outside the supposed extent of the readvance that there were differences in sedimentation history. At Scaleby and Oulton Mosses, inside the supposed limit, the first accumulation was shown to occur in his zone C.5 (Alleröd). However, at Moorthwaite and Abbott Mosses, outside the limit, deposits pre-dating the Alleröd were found. He correlated his zones C.1 to C.3 with the Main Glaciation - Scottish Readvance interstadial and zone C.4 with the period of maximum effect of the Scottish Readvance which produced soliflucted sediments around the moss-basin margins. However, Pennington (1970) suggests that there is no convincing argument for a 'Cumbrian Interstadial' with tree birches and pine and postulates that the whole basal succession at Abbott and Moorthwaite Mosses could represent the early stages of Zone 1 (Devensian Late-Glacial), complicated by secondary pollen. As a result of this Pennington (1970) gave the view that the present interpretation of the glacial succession in the Vale of Eden/Carlisle Plain by the Geological Survey requires no readvance and that the so-called Upper Boulder Clay is not the result of a separate and distinct glaciation. This is the view expressed by Evans & Arthurton (1973). Nevertheless, Taylor *et al.* (1971) state "In a final phase of the glaciation in north-west England, a thin reddish till-like stony clay was deposited in low-lying parts of the Solway basin and along the coast of northwest Cumberland. The clay swathes pre-existing drumlins and outwash deposits, the tops of which are undisturbed. Its origin and age are uncertain, although it has been interpreted as the till of a minor readvance of Scottish ice. The Bride Hills Moraine ... has been taken as marking the limit of this supposed readvance but has no counterpart in the Solway Basin." Mitchell (1972) suggests that "many of our much fought over 'advances' and 'retreat-stages' are probably largely illusory, and depend as much on personal whim as on field evidence." He correlated the Bar Hill - Wrexham Moraine with the Bride Hills, suggesting both formed at around 18,000 B.P. There is no indication of any readvance limit in

the Cumberland lowland. Other workers, such as Sissons (1974c) have questioned the reality of this readvance, particularly as the major lakes are known to have been ice free since before 14,500 B.P. Thomas (1985b) suggests that the concept of a major Scottish Readvance during the Devensian glaciation is largely illusory, lacks stratigraphic or chronological foundation and cannot be used as substance for stratigraphic classifications based on an assumed response to climatic change. The present evidence suggested to him that the deglaciation was not punctuated by major readvance but was characterised by rapid retreat, accompanied by minor snout oscillation caused by essentially local controls.

With this uncertain background as to the Scottish Readvance limits in northern Britain and even as to whether there is any evidence for such a readvance, the criteria used to establish the validity of this glacial phase are presented. The approach was based on a study of both landforms and stratigraphy to establish the depositional environment of any particular sedimentary association. This work established proglacial environments. If these proglacial morphostratigraphic units can be shown to have been associated with an advancing ice sheet rather than a retreating one and their marginal limits mapped, then the validity and extent of the readvance can be established.

Several depositional environments, outlined below, are thought to be associated with a readvance of Irish Sea ice in the Cumbria lowlands. The limit for the readvance has been established by mapping the morphology of the sedimentary association. Thomas (1985b) warns that morphological evidence from moraines is unreliable as a large moraine may mark a readvance, a recessional stillstand or even a terminal maximum. However, at each locality the stratigraphy and lithology of the units was recorded and the depositional environment determined by analysing the grain-size, fabric and lateral and vertical sedimentary changes.

The following depositional environments associated with the Scottish Readvance were established

i) Proglacial lacustrine with overlying till in the eastern Carlisle Plain (Huddart, 1970).

ii) Subglacial in the western Carlisle Plain (Huddart, 1970).

iii) Proglacial sandur at Broomhills in the eastern Solway lowlands (Huddart, 1970).

iv) Subglacial esker at Thursby in the eastern Solway lowlands (Huddart, 1973).

v) Proglacial lacustrine at Holme St. Cuthbert in the Solway lowland (figs. 92 and 93) (Huddart &

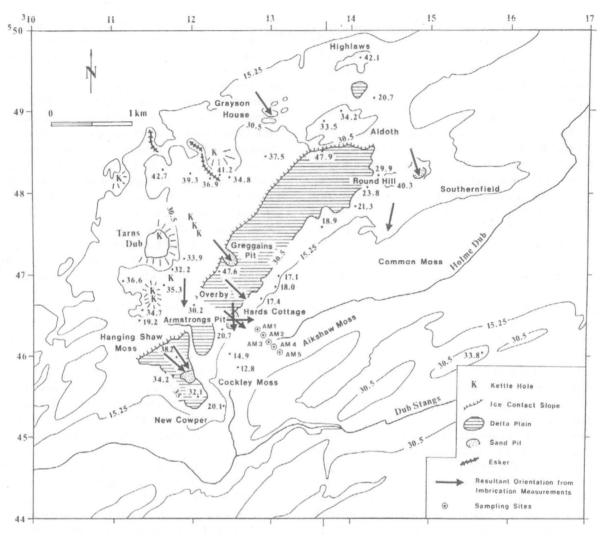

Figure 92: Proglacial lacustrine landforms and sediments, Holme St. Cuthbert; location map and landforms.

Tooley, 1972).

vi) Proglacial sandur at Harrington in West Cumbria (figs. 94 and 95) (Huddart & Tooley, 1972).

vii) Proglacial sandur and end moraine at St. Bees and between Nethertown and Seascale in West Cumbria (figs. 96, 97 and 98) (Huddart & Tooley, 1972; Huddart, 1977a).

viii) Proglacial lacustrine in lower Wasdale (Huddart, 1970).

ix) Proglacial sandur and end moraine along the Black Combe coastal lowland (Huddart & Tooley, 1972; Huddart, 1977a).

Examples of the morphostratigraphic approach to studying the Readvance landforms

a) *Glacial and fluvioglacial landforms, west of*

Black Combe (fig. 96). The glacial sequence is similar to the succession described from St. Bees by Huddart & Tooley (1972). Although the stratigraphy is more complex, the sediments are relatively undeformed. The lowest Selker till is interpreted as the basal till of the 'Main Glaciation'. During the deglaciation subglacial meltwater eroded the channels and deposited the subglacially-engorged eskers on Corney, Bootle and Little Fells (Smith, 1967; Huddart & Tooley, 1972). After an interval of unknown duration the Irish Sea ice readvanced, with proglacial, braided rivers producing the sandur sequence which Smith (1912) had attributed to the retreat of the 'Main Glaciation' ice. That the ice sheet was advancing is indicated by the vertical increase in grain-size and the change from lower to upper flow regime indicators (fig. 99). Proximal, proglacial deposition is indicated

a

b

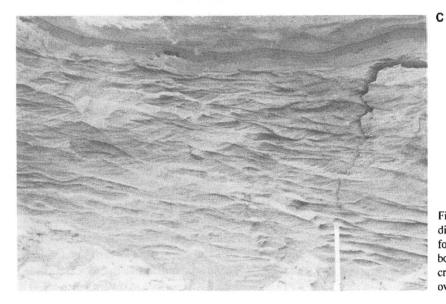

c

Figure 93: Proglacial lacustrine sediments, Holme St. Cuthbert; a = foreset / bottomset transition; b = bottomset sequence; c = type 'A' cross-stratification in bottomsets, overlain by parallel lamination.

by the pebble gravel units which have many of the characteristics of longitudinal sandur bars. The interbedded, thin tills are flow tills from the ice front. The imbricate gravels indicate deposition from the northwest and west (fig. 100). The overlying Gutterby Spa complex is interpreted as deposition from the basal layers of the Irish Sea ice which advanced over the sandur, with little disturbance of the underlying sediments. This ice advance did not reach the Black

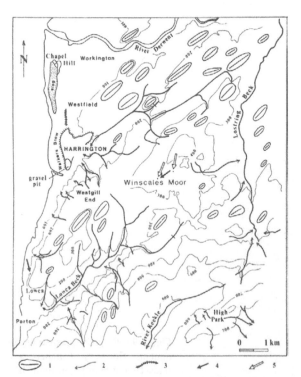

Figure 94: Glacial landforms in the Harrington area. 1 = Drumlin, 2 = Glacial drainage channel, 3 = Esker, 4 = Till macro fabric orientation, 5 = Till micro fabric orientation.

Combe foothills and till fabrics indicate ice movement from between 255° and 328° (fig. 101). The ice sheet margin decayed *in situ* at its maximum extent, producing a till - sandy clay - till complex and kettleholes.

b) *Evidence for proglacial Lake Wasdale.* The evidence presented by Smith (1912, 1931) for glacial lakes in lower Wasdale, Eskdale and Miterdale was based on supposed overflow channels, deltas, beaches and deep-water sediments. Several lakes were postulated which were brought about by the split of the Lake District and Irish Sea ice sheets and the appearance of Muncaster Fell as a nunatak. The sequence of lakes was thought to be between 36.6 m and 152.4 m O.D. with a channel system across Muncaster Fell acting as a spillway from the Wasdale lake into Lake Eskdale when the Eskdale Green gap was still blocked by ice. The level of Lake Eskdale was controlled by a series of marginal channels along the foothill flanks to the south but these have been reinterpreted as subglacial in origin by Smith (1967). The site of Lake Eskdale is thought to have been occupied by stagnant ice and similarly in Wasdale the evidence points to downwasting ice, with the 'lake beaches' forming as kame terraces at several levels. However, there is borehole evidence in lower Wasdale which suggests that lake sediments are present which could be referred to deposition in the 68.6 m O.D. lake. The location of the boreholes and the successions are given in figs. 102 and 103.

The main feature is the thickness of glacial sediments exposed above bedrock at Haggs Wood (86.6 m), Moorgate, Aikbank and Mainsgate (80.0 m when bedrock was not reached). The drift thins rapidly to the east. The lowest unit above bedrock is a sandy till (41.8 m thick) at Moorgate and Wardbarrow (30.8 m

a

b

c

Figure 95: Proglacial sandur sediments, Harrington; a = horizontal, coarsening-upwards sequence; b = large-scale cross- stratified gravels representing downstream end of longitudinal bar avalanching into deeper water, overlain by horizontal pebble gravel of the bar surface; c = cross-stratified sands representing more distal, transverse sand bars interbedded with gravel foresets of longitudinal bars.

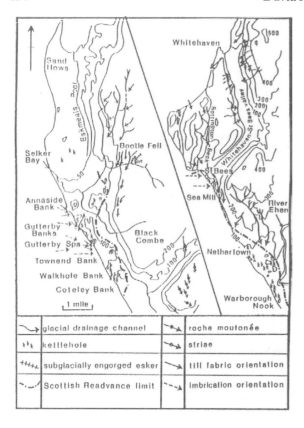

Figure 96: Location of the St. Bees and Black Combe Moraines.

thick). This is succeeded by sandy clay, clay and sand and/or gravel in the boreholes east of a line through Gubbergill. However, at Greenlands the basal unit is 21.5 m of gravel and at Mainsgate there are three tills interspersed in the clay. At Haggs Wood there are 72.5 m of clays but west of Gubbergill there is a predominance of sand and gravel. There is an upper till in the records at Addyhouse, Moorside and Gubbergill. This depositional sequence is interpreted as the result of two distinct phases. The lowest till and gravel units probably represent the 'Main Glaciation' deposits. The second phase is represented by three separate facies from west to east and is thought to be the result of a later Irish Sea readvance which dammed the drainage in lower Wasdale. To the east are lacustrine clays; deltaic sands/gravels are in the middle and to the west there is basal till and proglacial gravel. However, in the Mainsgate area the lacustrine facies is interrupted by three till units which suggests that stagnant ice may still have been present in the valleys. Deltas have been described by Smith (1931) and Trotter *et al.* (1937) which seem to have the typical deltaic lobate front and ice-contact slope but the sections are confined to Stubshead Hill where there are fluvial sediments similar to the topset facies at Holme St. Cuthbert. Sequences of horizontally stratified or tabular cross-stratified, medium to coarse sands are succeeded by channelled and imbricate

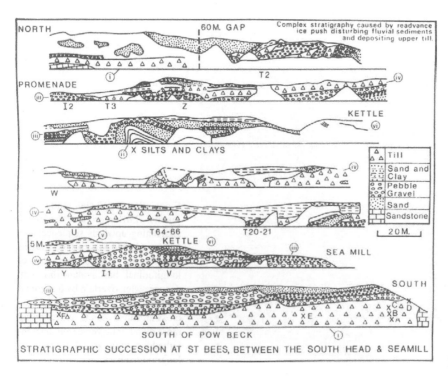

Figure 97: Stratigraphy at St. Bees.

granule to pebble gravels which indicate that deposition was from west to east (orientation: 252°). The westerly slope of the hill shows a fault pattern with normal faults, dipping at 60 degrees, with throws of a metre. Stubshead Hill is interpreted as a proximal delta plain with a western ice-contact slope. Delta foresets should lie laterally to the east.

The validity of the Upper Till as a basal, readvance till

The stratigraphic position of the upper till in the Cumbria lowlands was the fundamental difference between the 'Main Glaciation' basal till and the Scottish Readvance basal till of the Geological Survey workers and was the initial basis for assigning the two tills to distinct separate formation periods. However, these upper tills are not automatically of basal origin as was thought by these workers in the 1920's. Upper tills could have originated by ablation, flow or from an englacial deposition (Boulton, 1972) and that two till units are separated by Middle Sands does not necessarily mean that the sequence indicates a threefold, ice advance-retreat-advance sequence (Boulton, 1977b). All can be deposited from the decay of a complex ice sheet and the fact that there is an upper till in Cumbria will not stand by itself as evidence for a readvance. However, the upper till does have characteristics which indicate that it is of basal origin. The till, especially in the Carlisle Plain, shows incorporation of the underlying sediments which are disturbed. For example at Brunstock, the New Eden Bridge, Tower Farm, Scotby and the Rosehill interchange, the till, although only between 42 cm and 2 m thick, overlies laminated clays and silts, contains clasts of this sediment and in thin section shows silt and clay veins (Huddart, 1971b). There is an almost indistinguishable junction between the till and the clay/silt, even though the latter is contorted in the upper 50 cm. West of the New Eden Bridge, where the upper till overlies bedrock, the till shows an irregular, erosional contact with the white Kirklinton Sandstone and the lowest 60 cm of till has a sand matrix with white sandstone clasts, whereas above this lower zone the till has a silt/clay matrix derived from the laminated clay to the west (fig. 104a).

The point made by Thomas (1985b) is correct that even where evidence for override exists, as above and at St. Bees, this does not necessarily indicate a readvance of wide extent. However, because the upper till is of regional significance with a relatively uniform thickness, as in the Carlisle Plain and along the Black Combe coast, this indicates that the depositional processes are not just of local importance. In local areas the till fabric orientations give consistent results. The dominant orientation is W-E or NW-SE, with an imbrication indicating deposition from the west (Huddart, 1970). The till has a consistently fine grain-size which gives much lower values of sand/matrix ratio than the 'Main Glaciation' till and is visually much finer (figs. 104b-d) (Huddart, 1971a). Its stratigraphic position, especially in south and west Cumbria, where it is above a proglacial, sandur sequence indicating an advancing ice front and its lithological suite composed dominantly of southern Scottish erratics (especially Criffel Granite and Lower Calciferous Sandstone Conglomerate from Kirkcudbrightshire) together with lithologies which indicate some erosion of Main Glaciation sediments by an advancing ice sheet (Huddart, 1970) suggest a basal origin. It is realised that flow tills could be associated with an advancing ice front and could overlie proglacial sediments. This is the case along the Black Combe coast where thin tills up to 78 cm thick are occasionally associated with sandur sediments. These flow tills and gravels are overlain by much thicker, basal tills.

Readvance marginal limits in the Cumbria lowland

The precise ice-marginal limits of the readvance of the Irish Sea ice are difficult to establish in many parts of the Cumbria lowland (approximately indicated in figs. 91 and 96 and, in greater detail, in Huddart, 1970). This is especially the case in the Carlisle Plain where the readvance deposited only a thin till and did not form an end moraine. In the Carlisle area the ice is thought to have readvanced into the topographically low area, spreading out as a relatively thin lobe. The ice seems to have reworked the marginal slopes but did not override the Rosehill deltas which reach a maximum height of 31.8 m O.D. An upper till has been located as far east as Greenholme, near Corby Hill (Huddart, 1971b) which is the maximum distance east that the effects of the readvance ice have been traced. Trotter's (1929) readvance margin between Lanercost, Brampton and Cumwhitton is thought to be too far east. Similarly in Edenside, outside the Carlisle Plain, there is no evidence of an upper till. In the Petteril valley the upper till has not been traced further south than Harraby. Within the valley proper the glacigenic sediments all suggest that they were formed during the decay of the 'Main

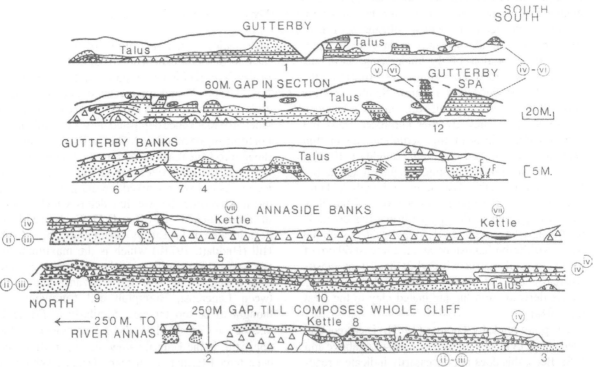

SAME SEQUENCE, CLIFFS LOWER, TILLS THIN, FLUVIAL SEDIMENTS THE SAME THICKNESS

Figure 99a: Stratigraphy, Black Combe Coast.

Figure 98: Sediments exposed in large-scale fold, St. Bees; A = St. Bees Till (upper readvance till); B = gravels (proximal sandur); C = sands (distal sandur); D = St. Bees Silts and Clays (proglacial lacustrine).

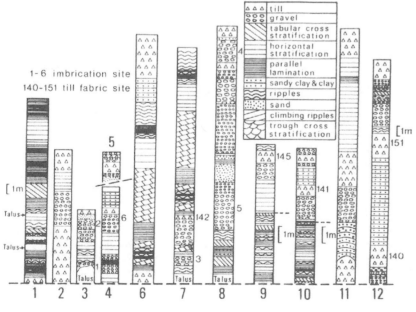

Figure 99b: Stratigraphy, Black Combe Coast.

David Huddart

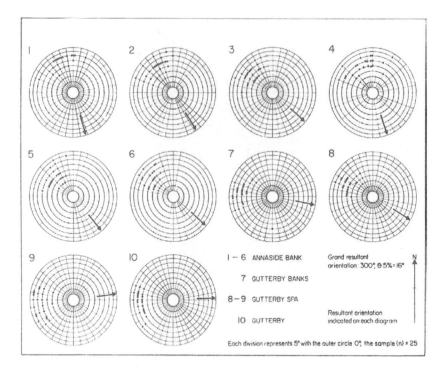

Figure 100: Palaeocurrent data from imbricated gravels, Black Combe Coast.

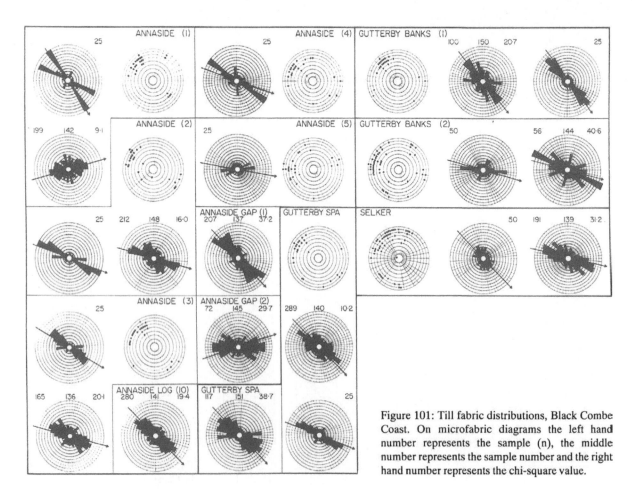

Figure 101: Till fabric distributions, Black Combe Coast. On microfabric diagrams the left hand number represents the sample (n), the middle number represents the sample number and the right hand number represents the chi-square value.

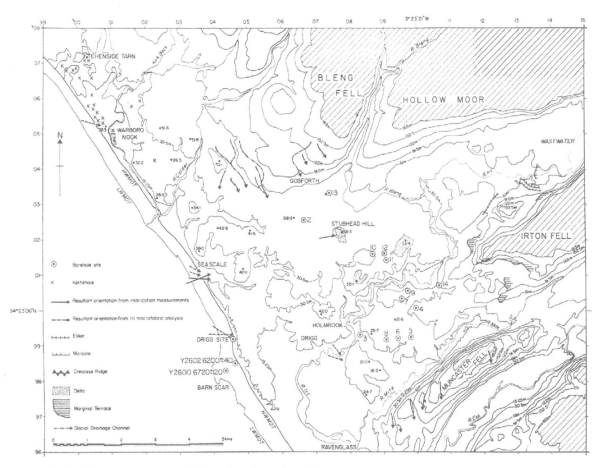

Figure 102: Glacial landforms and borehole locations, lower Wasdale.

Glaciation' ice sheet (Huddart, 1970; 1981; 1983). No upper till has been found in other valleys where topographically it might have been expected. It is suggested that these valleys were still occupied by stagnant, 'Main Glaciation'ice during the time period when the ice readvanced into the Carlisle Plain.

Southwest of the Carlisle Plain the readvance limit is difficult to trace, although sandur sediments at Broomhills and the esker system around Thursby were associated with the readvance (Huddart, 1973). The deltas around Wigton (Eastwood *et al.*, 1968) could have been associated with a lake ponded up by the readvance ice front. The Holme St. Cuthbert fluvioglacial complex marks the marginal position of the readvance ice but south of this the ice did not reach the present-day coast until the region of Workington, a possible exception being at Risehow, where there is a high percentage of Criffel Granite in the till and there is a till fabric orientation indicating

ice movement from the north-west (Huddart, 1971b). Here the sandur at Harrington was deposited in association with the readvancing icesheet.

South of Harrington the readvance ice did not override the present-day coast because the Whitehaven and St. Bees Sandstone cliffs reach between 60 - 138.6 m O.D. The next positive marginal position is at St. Bees where a proglacial sequence and upper till form the end moraine. South of St. Bees the readvance ice formed the 'kame and kettle' topography between Nethertown and Drigg. The stratigraphy reveals proglacial outwash overlain, in places, by an upper till. The marginal position of the readvance ice is represented by the deltas in lower Wasdale. Around and to the south of Ravenglass the readvance ice is not considered to have reached the Lake District foothills but an end moraine was formed along the Black Combe coast.

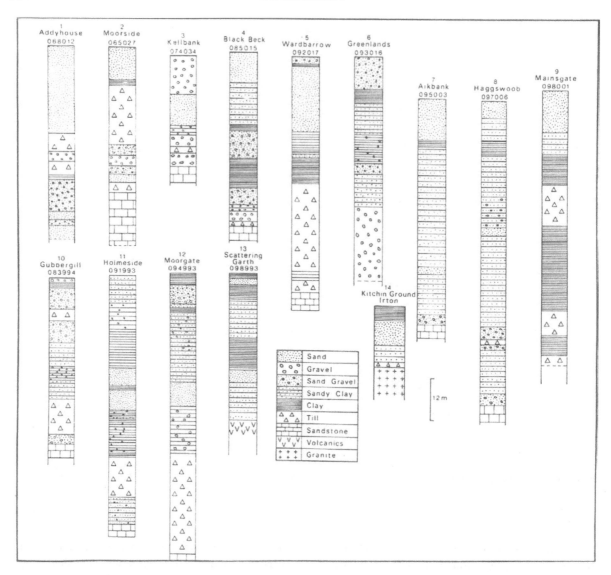

Figure 103: Stratigraphic succession, lower Wasdale.

Regional significance of the Scottish Readvance in the eastern Irish Sea basin and its place in the Devensian chronology

It seems likely that the maximum glacial phase of the Devensian Stage in Britain occurred between 25,000 and 18,000 B.P. According to Mitchell (1972) the ice reached its southern limit at the Shortalstown / Mathry / southern Lleyn / Wolverhampton line. During the retreat from this maximum, marginal ice readvances in the Irish Sea basin were common and have been claimed by various authors in mid-Cheshire, Kirkham (although this moraine has been re-evaluated by Longworth, 1985), the Isle of Man

(Thomas, 1976; 1977; 1984a; 1985b), Wexford (Huddart, 1977b; 1981b; Thomas & Summers, 1983; 1984), Low Furness (Huddart *et al.*, 1977) and in Cumbria. Thomas (1985b) suggests that in all cases these readvances represented short-distance, localised oscillations of the snout. He claims that they have no significant stratigraphic importance and do not mark ice sheet fluctuations driven by climatic change.

Very few dateable, organic sediments have been found associated with these postulated readvances and so at present they cannot be fitted into a chronological Devensian sequence. This is the case in the Cumbrian lowland. No organic material has been

Figure 104: Tills, Cumbria lowlands; a = Contact between Readvance till and Kirklinton Sandstone, Carlisle Plain, scale: 15 cm; b = 'Main Glaciation' till, Bullgill; c = Readvance till, Gutterby Spa.

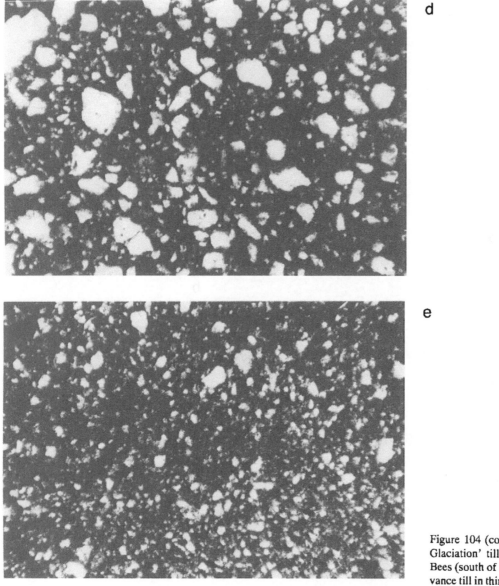

d

e

Figure 104 (continued): d = 'Main Glaciation' till in thin section, St. Bees (south of Sea Mill); e = Readvance till in thin section, St. Bees.

found with what are considered to be readvance sediments and there is no evidence to date this time period from extraglacial, biogenic sequences.

Nevertheless, the evidence is that readvances did occur and that they did produce stratigraphically important sequences. The detailed work undertaken in Cumbria suggests that there is evidence for various types of depositional environment which have been used for morphological correlation, even though there is no proof yet of time parallelism.

From evidence in the Isle of Man and in the Irish Sea basin between that island and Cumbria (Pantin,

1975; 1977; 1978), Thomas (1985b) suggested a tentative reconstruction of glacial conditions to the east of the Isle of Man at approximately 15,000 B.P. To the east a narrow arm of the sea extended northwards to abut directly an ice margin running from the Bride Moraine towards Cumbria. Both to the east and west where the ice margin rose against the bedrock highs of the Isle of Man and Cumbria the margin was terrestrial and large volumes of proglacial outwash accumulated. The Scottish Readvance depositional environments described in this chapter are the result of this terrestrial deposition and, although it was thought possible at one stage that the St. Bees Silts and Clays (fig. 98) were marine in origin, they have no included fauna and have been interpreted as lacustrine (Huddart & Tooley, 1972). They do not appear to be the equivalent of Thomas' Dog Mills Series. However, in Cumbria readvance ice built up proglacial sandur sequences, dammed drainage to form proglacial lakes and produced end moraines. There is no evidence for associated glaciomarine conditions as in the Isle of Man and in the Irish Sea basin. Contrary to many writers' views, the eastern Irish Sea ice sheet oscillated at least twice to give the Low Furness and Scottish Readvances. Whilst these two readvances may be synchronous there is no evidence for any time parallelism in the sequences produced. Yet it is believed that the sequence of depositional environments and associated significant stratigraphic sequences are broadly time-parallel and can be provisionally used for long-distance correlation in Cumbria. Thomas' mechanisms for local readvances are interesting but they have not (yet) been proven in Cumbria. What seems to be conclusive is that readvances did occur in these lowlands and that the Scottish Readvance problem should be no longer with us. Readvance landforms and sediments can be proven but the magnitude of the time gap, if any, between the readvances in local areas is still unknown.

The Geological Survey in the 1920's and Eastwood *et al.* (1968) considered that the Scottish Readvance ice reached as far as the 120 m contour but this contour limit is considered too high and too far east in the Cumbria lowland. Maximum heights of landforms associated with the readvance are from north to south: 27 m O.D. in the Carlisle Plain, 47.9 m O.D. in the Holme St. Cuthbert area, just over 30 m O.D. at Harrington and St. Bees, 68.3 m O.D. in lower Wasdale and 55.2 m O.D. at Gutterby Spa, near Black Combe. In conclusion, evidence has been given to suggest that the Scottish Readvance did occur in Cumbria. This has been based on a series of proglacial environments associated with an advancing ice front and landforms indicating a terminal position. The mapped extent of this readvance has been suggested which is not as far east or as high as was suggested by the Geological Survey in the 1920's. It is not considered that the whole northern Irish Sea basin was deglaciated during the interval between the 'Main Glaciation' and the readvance, as was suggested by Trotter (1929). There is no evidence for the mechanism for such readvances, whether there was a significant time gap between the 'Main Glaciation' and any readvance or whether all the various landform assemblages indicative of a readvance or marginal position are synchronous and time parallel.

Glacial deposits of Northumbria

TERRY DOUGLAS

Considering the significance of Northumbria in the history of glacial research in Britain, it is perhaps surprising that relatively little recent work has been published on the area. As long ago as 1840 William Buckland, on his famous tour, recognised 'enormous moraines' at the margins of the eastern valleys of the Cheviot Hills (Boylan, 1981). It was also in this area that Kendall's pioneering work on 'overflow channels' in North Yorkshire was extended to the Cheviot Hills, an area with arguably the most comprehensive and spectacular sequence of glaciofluvially eroded channels in England (Kendall & Muff, 1903). The significant shift in the way in which such channels were no longer interpreted as overflows from proglacial lakes on the Kendall model was established by work in southern Northumberland by Peel (1949) and most importantly by Sissons (1958c). Also worthy of note is Carruthers' monograph of 1953 entitled 'Glacial Drifts and the Undermelt Theory', which drew heavily on the experience of geological mapping in Northeast England and contained some of the ideas later to be enshrined as models of glacial deposition.

In part, the limited attention recently afforded the area, may be the result of the lack of a clear stratigraphic context for much of the glacial material in Northumbria. Interglacial material is largely absent and the complex till sheet is regarded as belonging to the Dimlington Stadial of the Devensian Stage (Rose, 1985), largely as a result of the constructive landforms which it carries and the correlation of the east coast tills with the type site for the Stadial at Dimlington in east Yorkshire. Several detailed stratigraphic tables have been prepared for parts of the Northumbria region (Smith & Francis, 1967; Francis, 1970; Smith et al., 1973; Lunn, 1980), but considerable difficulties have been encountered in assigning ages to any of the pre-Dimlington Stadial events and so a simplified stratigraphy is presented in table 12.

Recent approaches to glacial geology have stressed the importance of a landsystems approach (Eyles, 1983a). This is quite appropriate as a model for much of Northumbria where three distinct terrain types can be mapped:

(a) The lodgement till plain of the coastal lowlands and Tyne gap, often drumlinised.

(b) Widespread glaciofluvial landforms occurring in association with (a) and along major valleys and in piedmont locations.

(c) The glaciated valley terrain of the uplands which has invariably been extensively modified by periglacial mass wasting.

These terrain types are considered separately below.

The lowland tills

Tills form a nearly continuous cover over much of the lowland area from the Tweed to the Tees (fig. 105). The surface morphology of the till sheet has often been destroyed as a result of coal mining activities and some of the coastal sections have been obscured by the tipping of coal waste, including the only locality of the Scandinavian Drift (Warren House Till of Francis, 1970). The Scandinavian Drift was identified by Trenchman (1915) and has widely been regarded on the basis of erratic content and stratigraphic position, as representing a pre-Devensian glacial event. If this till is correlated with other pre-Devensian east coast tills - the Basement Till of east Yorkshire and the Welton Till of Lincolnshire (Bowen et al., 1986) - then it can be placed in the interval between the Hoxnian and Ipswichian Stages; but in the absence of a convincing regional stratigraphic framework, it is safer to regard this remnant of an earlier glacial stage as being simply pre-Devensian.

Until quite recently, there was considerable debate about the age and status of the extensive formations of tills, silts, sands and gravels which cover much of lowland Northumbria. The coastal sections of County Durham lent themselves to a so-called 'tripartite

Table 12. A simplified Quaternary stratigraphy of Northumbria.

Stage		Representative deposits
Flandrian		A variety of deposits, the most significant being upland blanket peats, fluvial deposits and coastal dunes
Devensian	Loch Lomond Stadial	Bizzle cirque moraine Solifluctates of uplands
	Windermere Interstadial	Local organic deposits in kettles etc.
	Dimlington Stadial	Tills and related deposits throughout the region, including widespread glaciofluvial and glaciolacustrine deposits
Pre-Devensian		Temperate Stage(s): Easington Raised Beach Gravel (>38,000 B.P.); Fissure fillings in Magnesian Limestone Cold Stage: Warren House Till (Scandinavian Drift)

classification' of Lower Boulder Clay, Middle Sands and Upper Boulder Clay, and work by Woolacott (1921) identified several glacial episodes to account for these drifts. Carruthers developed his monoglacialist views starting in 1939 and culminating with his monograph of 1953. Although many of the mechanisms and glaciological arguments that he advanced can no longer be sustained in their entirety, the essence of his message that a complex sequence of lacustrine clays, fluviatile sediments and tills can be produced by a single phase of glaciation was an idea ahead of its time. Smith & Francis (1967) showed that the evidence for the age of the Lower Boulder Clay of the Durham coast was conflicting, but on the basis of its relationship with the Easington raised beach gravels, it pre-dated this interglacial feature and was therefore the product of an earlier glacial stage.

North of the River Tyne, the relative persistence of the 'Middle Sands' is not so evident, although waterlain material is still frequent within the till sequence revealed in coastal exposures and in the sections afforded by opencast coal mining on the Northumberland coalfield. Detailed sedimentological work by Eyles & Sladen (1981) and Eyles, Sladen & Gilroy (1982) has advanced a model which demonstrates that the complex stratigraphy consists of crosscutting lodgement till units and explains colour differences between upper and lower till units as the result of the development of a post-depositional weathering profile. These authors uphold the now widely held view that the entire sequence is attributable to the Dimlington Stadial chronozone. The genetic implications of several superimposed facies of lodgement tills are supported by the surface morphology, which although not as clearly drumlinised as the Lower Tweed valley or the Tyne gap, is streamlined.

The rockhead surface on which the tills rest is also usually striated.

The weathering profile extends to 8 m in places and has been caused by oxidation which has reddened the upper units in comparison to the lower grey tills, decalcification and other systematic changes. The abrupt lower limit of the weathering profile which is often observed, frequently coincides with the occurrence of 'subglacial inclusions' which act as internal drainage blankets and control the depth of weathering (Eyles, Sladen & Gilroy, 1982). This model also offers an explanation of the observation that clast content decreases upwards through the till sequence, the 'lower till' often being distinguished not only on the grounds of matrix colour but also on clast frequency. Thus the number of completely weathered clasts increases upwards to give the appearance of a reduction in clast frequency and secondly, whereas the lowest lodgement till units will incorporate significant quantities of bedrock, successively higher units will be bounded at their base by shear planes in till and would have incorporated beds already deposited.

The pattern of glacial deposition outlined above, relies heavily on the interpretation of sections exposed along the coast (fig. 105). Inland, the tills are also widespread and the pioneering efforts of Raistrick (1931b) on the distribution of erratics in the tills is still widely used as an indicator of ice movement during what would now be termed the Dimlington Stadial. The pattern in east Northumberland is relatively straightforward with an ice stream moving parallel to the coast bringing Cheviot igneous lithologies with it. In southern Northumberland and much of County Durham, this stream was confluent with ice bearing Lake District erratics moving through the Tyne gap in an easterly direction as shown by the

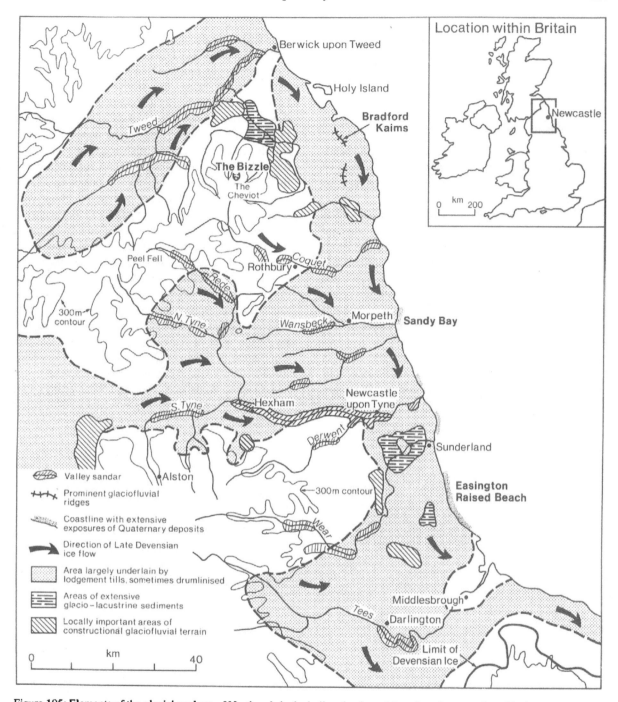

Figure 105: Elements of the glacial geology of Northumbria, including the sites of those locations mentioned in the text.

drumlin fields of western Northumberland (Frost & Holliday, 1980) and till fabrics from east County Durham (Beaumont, 1971).

Extensive glaciolacustrine beds have been described from several areas of Northumbria. Whereas the more localised lenses of silts and laminated clays can be incorporated in the subglacial model (Eyles & Sladen, 1981), widespread surface occurrences of lake clays can often be linked to geomorphological evidence for proglacial lakes during deglaciation. Thus in Durham, Lake Edder Acres and Lake Wear have been recognised (Smith & Francis,

1967; Smith, 1981) and in in north Northumberland around the margins of The Cheviot, several lakes were impounded (Clapperton, 1971a). These major spreads of glaciolacustrine material are located in fig. 105.

Glaciofluvial sediments

A wide variety of glaciofluvial sediments and landforms are encountered in Northumbria. For ease of description three environmental situations can be distinguished.

Firstly, relatively isolated constructive landforms of sand and gravel are associated with the lowland area of lodgement till deposition. Parsons (1966) mapped a system of esker ridges known as the Bradford Kaims near Bamburgh (fig. 105) for a distance of over 8 km. These ridges of sand and gravel are aligned in the direction of ice flow and represent subglacial channels draining the Devensian ice sheet.

Secondly, certain areas comprise terrain which is almost entirely the product of ice sheet wastage and consists of extensive ice-contact forms. These are

best developed on the margin of the Cheviot Hills near Wooler and to the south of Cornhill in the Tweed valley (fig. 105). The latter has been referred to as the Cornhill 'kettle moraine', and was at one time regarded as marking the limit of the Aberdeen-Lammermuir Readvance (Sissons, 1967a), although this readvance was subsequently rejected. Clapperton (1971b) mapped the meltwater deposits between Wooler and the River Breamish as an almost continuous belt which covers the piedmont zone below 275 m O.D. A complex network of ridges or eskers, kames, flat-topped terraces and kettle holes were interpreted as ice-contact deposits, predominantly of sand and gravel and the product of downwasting ice in the topographically controlled basins between the Cheviot Hills and the Carboniferous cuestas.

Thirdly, glaciofluvial deposits are found in Northumbria as outwash and sometimes ice-contact features aligned along the major river valleys. These benches are particularly widespread in the Tyne valley to the west of Newcastle where they have provided a major local source of aggregate. Elsewhere, spreads of gravels, often terraced, are found in the Aln, Coquet, Wansbeck, Tyne and Wear valleys. One

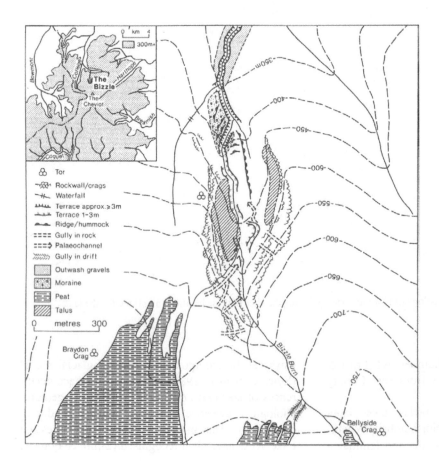

Figure 106: Geomorphological map of The Bizzle, Cheviot Hills, Northumberland, showing the location of the moraine attributed to the Loch Lomond Stadial. Source: field mapping by Mr. Ray Browning and the author.

example, characteristic of much of the area, has been described from the Derwent valley by Allen & Rose (1986). Here, deltaic and ice-contact sediments are associated with a meltwater channel system and a former proglacial lake. Analysis of these forms in their regional context again demonstrates the interplay of topography and the characteristics of the wasting Devensian ice sheet in controlling the geometry and location of glaciofluvial sedimentation.

The glacial deposits of the uplands

Whereas Raistrick (1931b) and other early workers had established the general pattern of ice flow over Northumbria, the detailed pattern, particularly of the upland drifts is still often poorly known. Clapperton (1970) established on the basis of the distribution of Cheviot and 'foreign' erratics, that the Cheviot Hills had supported an ice cap during the Devensian which deflected Scottish and Lake District ice around it. Similar arguments have been advanced for local ice occupying the Northern Pennine dales of south Tyne, Wear and Tees. Yet the occurrence of deposits and glacial landforms in these upland areas is patchy. A few drumlins and drift tails indicate that the direction of ice movement often followed the lines of major valleys, but the morainic topography which is widespread in parts of the English Lake District is largely absent. However, a moraine occupies the Bizzle cirque which is incised into the northern flanks of The Cheviot. The moraine is not very substantial but its position led Clapperton (1970) to consider that it

might represent the limit of a Lateglacial cirque glacier. Recent field mapping has confirmed this ridge as being a moraine. The north-facing aspect of the cirque and the altitude of the lip (about 500 m O.D.) point to this site as being the only likely representative of Loch Lomond Stadial ice in Northumbria (fig. 106).

Throughout the remainder of the uplands, a strong case can be made that any former glacial landforms have been substantially modified by periglacial processes, notably solifluction, after the wasting of the Dimlington Stadial ice sheet and during the cold climate of the Loch Lomond Stadial. Clark (1971) and Tufnell (1969) have described a range of periglacial features found in Northumbria and Douglas & Harrison (1987) have shown that these periglacial conditions have resulted in the formation of a landform-sediment association consisting of smooth slopes underlain by soliflucted debris. *In situ* till is rarely found at the surface in these upland valley locations. Douglas & Harrison (1985) described a 9 m high stream section near Linhope in the Breamish valley at 300 m O.D. (plates 11 and 12). The exposure is arcuate in plan and shows both dip and strike sections through bedded solifluction deposits which overlie till. The till shows geotechnical properties which are very different from the 6 m of overlying solifluctate, which here contains a high proportion of coarse-grained debris from upslope sources of deeply weathered granite. The solifluctate can be subdivided into several units of about 1 m in thickness, each of which parallels the smooth terrace-like surface of the slope foot. At Makendon at 330 m O.D., in the upper

Figure 107: Characteristic 'terrace feature' underlain by solifluction sheets and truncated by Flandrian stream incision, Makendon, Coquet valley, Northumberland.

reaches of the Coquet valley near the Scottish border, a section 8 m high has been exposed by a gully cutting through a solifluction sheet (fig. 107). *In situ* till is found at the base of the section with an overlying unit of soliflucted till and an upper clast-rich layer of soliflucted gelifractate derived from andesite outcrops upslope. The widespread occurrence of quite thick sequences of sediment reworked by solifluction is consistent with the absence of frost-susceptible glacial landforms in the uplands. The sharp topography of the glaciofluvial sediments in the Cheviot piedmont has been preserved as the constituent sandy gravels are not so susceptible to periglacial mass wasting.

Glacial deposits of the English Lake District

JOHN BOARDMAN

The English Lake District is a deeply dissected upland area with peaks reaching almost 1,000 m. The principal rock types are lavas and tuffs of the Borrowdale Volcanic Group and thinly cleaved mudstones of the Skiddaw Group, both of Ordovician age. There is abundant evidence for regional glaciation during which the landscape was submerged by ice sheets. There is also evidence for a recent phase of glaciation which was restricted to corries and valleys (Manley, 1959; Sissons, 1980). Many valley floors are occupied by lakes in rock-cut basins or dammed by glacial debris. Evidence regarding the scale of glacial erosion is equivocal and the Late Devensian glaciation appears to have had limited impact on the landscape (Boardman, 1980; 1988). On valley sides, periglacial slope deposits frequently overlie bedrock or till (Boardman, 1978). Mean annual precipitation exceeds 4,000 mm on the highest summits.

Despite early work of the Geological Survey (Ward, 1875; 1876) glacial deposits have been neglected over most of the Lake District. Detailed consideration will therefore be given to the northeastern part of the region where recent work has been carried out (Boardman, 1981). In particular, the glacial deposits of the Mosedale Beck area will be described (fig. 108).

There is no evidence in the Lake District for incursions of ice from elsewhere. Huddart (1971b) has shown that Scottish ice reached Gillcambon Beck in the extreme north, and to the east, Scottish ice undoubtedly moved south along the Vale of Eden at some time prior to the main Late Devensian glacial event (Letzer, 1981).

Stratigraphy

The formal stratigraphy proposed for the northeastern Lake District is shown in table 13 (Boardman, 1985). Three lithostratigraphic units of glacigenic origin are accorded formation status; each comprises glacial and glaciofluvial deposits. These are the Thornsgill, Threlkeld and Wolf Crags Formations. In the Mosedale Beck valley the relationship between the units can be clearly demonstrated (fig. 109).

Thornsgill Formation

The Thornsgill Formation outcrops at several localities in the Mosedale Beck valley between sites 'D' and 'F' (fig. 110), and at site 'N' in the Thornsgill Beck valley. It overlies slates of the Skiddaw Group and is overlain by younger Quaternary sediments.

The main component of the Thornsgill Formation is the Thornsgill Till. At all sites this unit is severely weathered, its original clast composition, texture and colour is difficult to establish. However, at site 'N' in Thornsgill, a basal horizon in unweathered till is succeeded by a partially weathered horizon which in turn grades into the severely weathered zone (fig. 111). This site is designated the type site of the Thornsgill Till and the zone of weathering, the Troutbeck Palaeosol (Boardman, 1985).

At site 'N', the clast composition of the till changes up the profile. In the unweathered zone the till is rich in slate, higher in the profile the dominant clasts are from the Borrowdale Volcanic Group and the Threlkeld Microgranite which outcrops 6 km to the west. Clasts up to 0.5 m in diameter are found in the severely weathered zone. The up-profile lithological change suggests that far-travelled material from high in the ice body was deposited above debris-rich basal ice which included sheared blocks of local bedrock. Glaciotectonic disturbance of bedrock, till and sands and gravels is in evidence close to the base of the profile (fig. 111). The texture and colour of the till is strongly influenced by weathering. Typical colours of the unweathered till are dark grey (N 3/0) whereas the severely weathered zone is characterised by yellowish brown (10 YR 5/8) colours. Clasts are often soft,

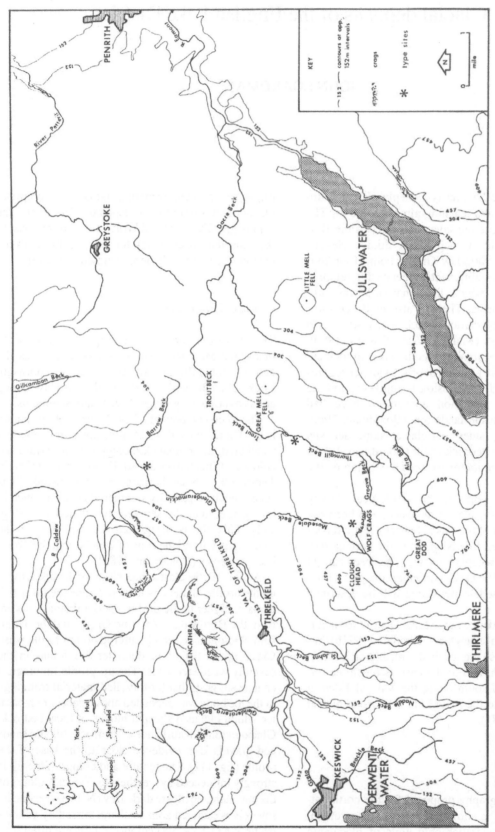

Figure 108: Topography and drainage of the northeastern Lake District.

Table 13. Formal stratigraphy proposed for the Quaternary of the northeastern Lake District (from Boardman, 1985: reprinted by permission of John Wiley & Sons Ltd).

Stage name	Sediments and soil properties	Lithostratigraphy Formation	Member	Soil stratigraphy	Environment
Flandrian	Colluvium alluvium, landslip debris, etc.		Not investigated		
	Rubification, silt and clay translocation, gleying			Laddray Wood Paleosol	Humid, warm temperate
Devensian	Scree	Millbeck	Dodd Wood Scree Latrigg Grèzes Litées (stratified scree)		Periglacial
	Gravel Till	Wolf Crags	Wolf Crags Gravel Wolf Crags Till		Glacial/glaciofluvial
	Till Sand, gravel and laminated beds Gravel	Threlkeld	Threlkeld Till Lobbs Sand and Gravel Mosedale Gravel		Glacial/glaciofluvial
Ipswichian and earlier	Clast decomposition, rubification, solution, oxidation, clay flowage, clay translocation, gleying			Troutbeck Paleosol	Humid, warm temperate
Unknown	Till, sand and gravel	Thornsgill	Thornsgill Till including sand and gravel bed		Glacial/glaciofluvial

Table 14. Site 'N', Thornsgill: particle size, lithological and weathering characteristics. Sampling points are shown on figure 111 (from Boardman, 1985: reprinted by permission of John Wiley & Sons Ltd).

Sample ref.no	Percentage total sample				Percentage sample < 2 mm			Percentage mudstone (n)	Percentage severely weathered (n)
	Gravel	Sand	Silt	Clay	Sand	Silt	Clay		
N9*					84.0	12.4	3.6		
N8	62.3	26.6	8.5	2.6	71.8	21.2	7.0		
N6	68.2	21.0	7.9	2.9	66.1	24.9	9.0		
N5	71.7	18.8	6.9	2.6	66.5	24.5	9.0	16.3 (1294)	100.0 (1074)
N4					18.9	52.2	28.9		
N7					18.7	46.0	35.3		
N3	35.3	25.4	28.2	11.1	39.2	43.7	17.1	91.1 (549)	83.6 (67)
N2	23.9	30.8	28.2	17.1	40.5	37.1	22.4		
N1	26.9	26.1	31.0	16.0	35.7	42.4	21.9	83.0 (1281)	23.1 (186)

*From laboratory crushing test
(n) = Number in sample

yellow and weathered throughout. At the base of the profile the till is rich in silt and clay (N1) whereas in the weathered zone sand predominates (N8, table 14). Weathered igneous clasts break down chiefly into sand-size materials (sample N9, table 14).

The characteristics of the severely weathered Thornsgill Till as seen at the type site, are repeated at many sites in the Mosedale Beck valley. At Caral Gully, the till is 14 m thick and the severe weathering occurs throughout its depth. Beneath the till, the slate bedrock is also weathered. The detailed character of the weathered Thornsgill Till is described elsewhere using micromorphological, X-Ray Diffraction (XRD) and Scanning Electron Microscope (SEM) analytical techniques (Boardman, 1983; 1985). Macrofabric analyses of clast orientation in the till suggest ice movement from the southwest and west (fig. 112). Although this assessment is based on only two statistically significant samples it is supported by analysis of clast lithology.

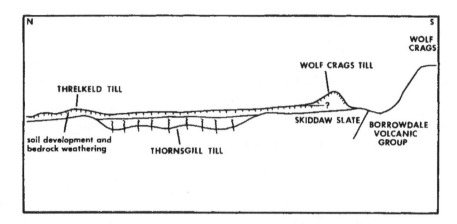

Figure 109: Diagrammatic section along Mosedale Beck valley to show the relationship of the three till units.

The Thornsgill Formation occurs in an area of about 8 km.[2] There are two possible explanations for the distribution (Boardman, 1984):

(i) Parts of the outcrop lie within a buried bedrock valley which runs at right angles to the direction of Devensian ice movement, it was therefore protected from subsequent erosion;

(ii) Devensian ice in the Vale of Threlkeld may have been cold-based and therefore non-erosive, at least until a late phase of lower ice velocities and sediment deposition.

Threlkeld Formation

The Threlkeld Formation comprises three units of which the Threlkeld Till is by far the most extensive. The till is found over most of the northeastern Lake District. It generally forms the ground surface unless overlain by periglacial or glaciofluvial deposits, or on steep slopes where bedrock may outcrop. In many valley-bottom locations drumlins are the characteristic landform of the till sheet. The position of the Threlkeld Till in relation to other deposits and landforms, and its geographical distribution, imply that it is the till of the last major regional glaciation to have affected the area.

The type site of the Threlkeld Till is at Barrow Beck (fig. 108) where about 5 m of till are exposed in a river cliff. It is an olive grey (10 Y 4/2) clayey till (sand 37%, silt 35% and clay 28%) with 66% Borrowdale Group clasts, 26% Skiddaw Group and 4% Threlkeld Microgranite. In the upper metre of the profile the till is oxidised and gleyed, of a yellowish grey (2.5 Y 5/6) colour with orange mottling. A macrofabric gives a typical lodgement till pattern with a resultant vector 151/351° and erratic types

implying movement from the south.

The Threlkeld Till in the northeastern Lake District is a variable deposit which differs considerably across the region. The extent and reasons for its variability will be summarised.

(i) Bedrock strongly influences till coloration, for example in the east near Penrith, Carboniferous sandstones give rise to reddish colours. The degree of weathering also influences till colour; it is not always possible to sample unoxidised till.

(ii) The percentage of fines in the till is related to three factors:

a) distance from source area;

b) distribution of easily comminuted rock types;

c) till type - ablation till has lower percentages than lodgement till.

(iii) The particle-size distribution of samples of Threlkeld Till is shown in fig. 112. However, gravel percentages are difficult to assess because 3 - 4 kg samples may not give statistically valid estimates of gravel fractions of stony tills. Alternatively, the < 2 mm fraction may be considered and Threlkeld Till samples cluster in the clay loam category (Hodgson, 1974) (fig. 113).

(iv) The clast lithology of the till is related to the position of the site in relation to bedrock outcrops, comminution of less resistant lithologies during transport, and deposition from ice rich in either local or far-travelled material. The clast lithology is also influenced by the particle-size class which is sampled, in this study, the 4 - 8 mm range was selected.

(v) Macrofabrics from the till indicate regional ice movement and supplement information from streamlined landforms and clast lithology (fig. 114).

(vi) The Threlkeld Till is often < 5 m thick but at many sites the full thickness is not apparent. In areas

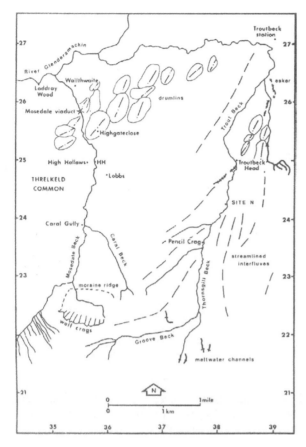

Figure 110: Mosedale and Thornsgill area (from Boardman, 1985; reprinted by permission of John Wiley & Sons, Ltd.).

of drumlin formation the till is thicker, for example around Burns Farm borehole evidence shows it to be over 25 m thick (Boardman, 1982).

(vii) At almost all sites the Threlkeld Till appears to be a lodgement till. At two sites, a bouldery upper component is probably an ablation till and at Caral Gully an upper unit occurs in association with sands, gravels and laminated beds which is likely to be a meltout or flow till.

Macrofabric analyses at several sites suggest that the till has been reworked by post-depositional processes. This is especially evident at the foot of long north-facing slopes where sampling of the upper 2 m of the till gave fabric patterns with the resultant vector close to the orientation of the slope. Macrofabrics from lower in the exposures had preferred clast orientations in the direction of regional ice movement as reconstructed using other evidence. Reworking is most likely to have occurred under periglacial conditions of the Loch Lomond Stadial.

There is some evidence for loss of fines during reworking.

The Lobbs Sand and Gravel Member (table 13) consists of glaciofluvial units including laminated clays, silts and fine sands, deposited in association with the Threlkeld Till from melting stagnant ice. Morphostratigraphically, this member is associated with kames, kettles and esker complexes in, for example, the area south of Keswick, the Vale of St. Johns, and Naddle (Boardman, 1982). In these areas sands and gravels are found within kames which occur alongside till hummocks. Lithologically, as might be expected from their wide distribution, the sands and gravels are variable. In contrast to sites where the Threlkeld Till forms drumlins, it is clear that in these areas the till was deposited at a late stage during the final melting of the ice.

Wolf Crags Formation

This formation comprises two members, a till and a gravel, which are the product of a phase of corrie glaciation. The position of the Wolf Crags Formation in relation to the Threlkeld Formation shows that it represents the most recent glacial event to have affected the northeastern Lake District (fig. 109).

The type site of the Wolf Crags Till Member is the stream-breached gap in the moraine ridge at Wolf Crags corrie (fig. 110). On the western side of the breach about 15 m of till are exposed. It is a dull yellowish brown (10 YR 4/3) sandy till (for example sand 61%, silt 28% and clay 11%). The till is oxidised throughout its exposed depth and for this reason it is informally defined. The Wolf Crags corrie basin straddles the boundary between Borrowdale and Skiddaw Group rocks but in addition to these lithologies within the till are found small numbers of Threlkeld Microgranite clasts. This implies reworking of previously deposited glacial sediments by the corrie glacier. The Wolf Crags Till is probably a product of two processes, sub-glacial erosion of the basin and frost action on the corrie backwall, material reaching the moraine ridge by sub-, en- and supraglacial routes. Deep oxidation of the till is a reflection of its sandy, well-drained character.

The Wolf Crags Till Member has only been examined in any detail at the type site. However, Sissons (1980) has shown that glaciers contemporaneous with that at Wolf Crags existed at 64 corrie and valley sites in the Lake District. At many sites the till occurs in fluted, hummocky and ridge forms and must therefore have been deposited from both active and stagnant ice.

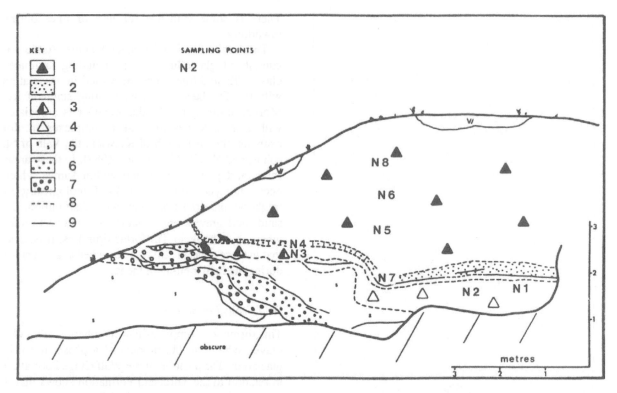

Figure 111: Thornsgill, site 'N' stratigraphy and sampling points (from Boardman, 1985; reprinted by permission of John Wiley & Sons, Ltd.). 1 = weathered till; 2 = silty clay band, gleyed in shaded areas; 3 = partially weathered till; 4 = unweathered till; 5 = deformed, brecciated Skiddaw Slate; 6 = sands and gravels; 7 = gravel; 8 = lithological boundaries; 9 = shears.

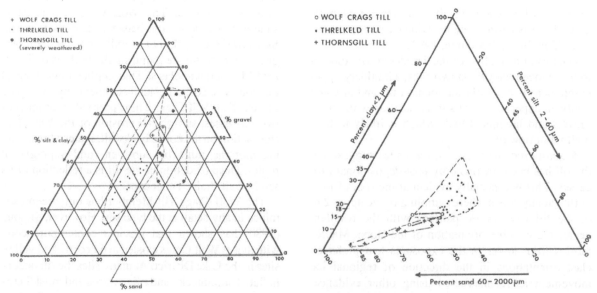

Figure 112: Particle size of tills in the northeastern Lake District.

Figure 113: Particle size of tills in the northeastern Lake District, < 2 mm fraction.

Table 15. Differentiation of Thornsgill and Threlkeld Tills.

Criteria	Thornsgill Till	Threlkeld Till
Distribution	Restricted to Mosedale and Thornsgill valleys	Extensive: most of northeastern Lake District except steep slopes
Relative stratigraphic position	Lower (never at ground surface)	Upper (usually forms ground surface)
Colour	Yellowish-brown, e.g. 10 YR 5/6	Varies with rock type, e.g. dark grey (N 3/0), dull reddish brown (2.5 YR 4/4)
Weathering	Severel weathered: rotten, friable clasts without original colour	Some oxidation but generally unweathered
Texture		
(i) total sample	Sandy gravel, few fines (< 25% silt and clay)	Rich in fines (30–65% silt and clay)
(ii) fine fraction (after Hodgson 1974)	Sandy loam	Clay loam or sandy silt loam
Macrofabric } Clast lithology }	No clear differences	

The Wolf Crags Gravel Member occurs in a series of terraces within the Mosedale Beck valley. It is a coarse gravel with inclusions of finer gravel and sandy sediments. Clasts are frequently up to 1 m in diameter and are concentrated into large bars with imbricate structure. Clasts are of the Borrowdale Volcanic Group and Threlkeld Microgranite rock types and are mainly derived from erosion of the Thornsgill Till. Transport and deposition of the gravel was a result of glacial meltwater and snow-melt discharges probably contemporaneous with deposition of the Wolf Crags Till. High velocities and discharges were necessary for gravel transport (Boardman, 1981; Rose & Boardman, 1983). Recent work on soil development in Mosedale suggests that the upper terraces are of late Devensian age and the lower ones late Flandrian, the latter resulting from a flood or floods in the last few hundred years (Smith & Boardman, 1989).

Differentiation of the till units

The two major till units in the northeastern Lake District are the product of regional glaciation though only remnants of the Thornsgill Till survive. Table 15 summarises the differences between the tills. The Wolf Crags Till is geographically and morphologically distinct in that it occurs in corries and high upland valleys usually within hummocks and ridges. Only in the Mosedale Beck valley are the three tills seen in close proximity.

Age of the glacial deposits

The Wolf Crags Formation results from a phase of restricted glaciation that occurred in western and northern Britain during the Loch Lomond Stadial, 11,000 - 10,000 B.P. (Sissons, 1979e). In the Lake District, the deterioration of climate during the stadial has been inferred from pollen evidence (Pennington, 1977) and the glacial landforms have been mapped by Sissons (1980). Pennington (1978) has shown that Wolf Crags corrie was occupied by ice at that time.

The Threlkeld Formation was deposited in association with ice during the Dimlington Stadial, 26,000 - 13,000 B.P. (Rose, 1985), during the last regional glaciation of northern Britain. Deposition of the formation pre-dates 14,623 ± 360 B.P., by which time the central Lake District was ice free (Pennington, 1978). The Threlkeld Formation may be correlated, on the basis of its age and relationship to the last major glacial event, with the Wear Till in Durham (Francis, 1970), the Stockport Formation in Cheshire (Worsley, 1967), and the Skipsea and Withernsea Tills of Holderness (Madgett & Catt, 1978).

The Thornsgill Formation cannot be directly dated. The Troutbeck Palaeosol, developed in the formation, represents a substantial period of humid temperate weathering. The duration of this period has been tentatively estimated at between 100,000 and 150,000 years (Boardman, 1985). On this basis the glaciation responsible for deposition of the Thornsgill Till occurred in Oxygen Isotope Stage 10 or 12. A peat bed above the Thornsgill Till has yielded a radiocarbon date of > 56,000 years and a Uranium series age estimation of about 90,000 (Boardman *et al.*, in preparation). There are very few occurrences of pre-Devensian tills in northern Britain and estimates of their ages are extremely speculative.

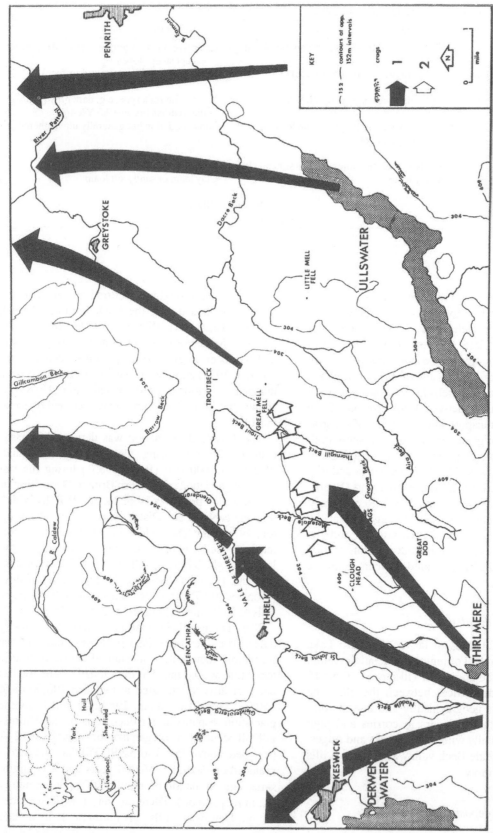

Figure 114: Direction of ice movement in the northeastern Lake District; (1) Dimlington Stadial: based on erratic distribution, macrofabrics in Threlkeld Till, and orientation of streamlined landforms; (2) Pre-Devensian: based on two macrofabrics in the Thornsgill Till.

Correlation of the Thornsgill Till with deposits beyond the northeastern Lake District is not at present possible.

Conclusions

The glacial deposits of the Lake District represent two styles of glaciation, the regional event when the mountains were inundated by an ice sheet (Boulton *et al.*, 1977), and the spatially limited corrie and upland-valley glaciation. The characteristic deposit of the active phase of regional glaciation is lodgement till with the phase of ice wastage being represented by ablation till and glaciofluvial deposits. In the Lake District, the glacial deposits of the Loch Lomond glaciation have not been investigated to the extent that generalisations can be made. In this area of steep slopes, reworking of the upper zones of tills after deposition, appears to have occurred at many sites.

The Quaternary history and glacial deposits of East Yorkshire

JOHN A. CATT

On the coast of East Yorkshire glacial deposits are exposed in several bays (mainly Holderness, Filey Bay and Robin Hood's Bay) bounded by headlands of Cretaceous and Jurassic bedrock. The best exposures are on the Holderness coast between Flamborough Head (Chalk) and the Humber estuary, where they are maintained by rapid coast erosion controlled by an unusual process of southward movement of beach sediment (Pringle, 1985). Most of the deposits are late Devensian in age, but the Wolstonian Basement Till is exposed locally between the Devensian Skipsea Till and the bedrock surface.

Coastal exposures of the Wolstonian Basement Till

The very dark grey to olive grey (Munsell Colour: 5Y 3/1 to 4/2) Basement Till is seen, as beach conditions allow, in the foreshore and cliff foot between Kilnsea Beacon (TA 412176) and Hompton (TA 376237) on the southeast Holderness coast. The surface is at 2 - 6 m O.D. and the base at -30 m to -35 m O.D. over Chalk. Boreholes drilled in 1985 on the foreshore near Dimlington Farm (fig. 115) showed that the till varies little in lithology throughout this thickness, and that there is probably no earlier till in Holderness (Catt & Digby, 1988). Erratics include chalk, flint, sandstones, Magnesian Limestone, Carboniferous Limestone, larvikite, rhomb porphyry and other igneous and metamorphic rocks from Scotland and Scandinavia. Also north of Dimlington Farm (fig. 115) there are large masses of fossiliferous grey clay derived from the floor of the North Sea. The fauna of this clay (the Sub-Basement Clay of Bisat, 1939) includes 180 species of molluscs, notably *Arctica islandica, Astarte semisulcata, Macoma balthica, Mya truncata, Dentalium entalis* and *Turritella tricarinata*, 11 species of marine ostracods, and 24 - 45 species of foraminifera (Reid, 1885; Bell, 1917; 1919; Catt & Penny, 1966). Many are coldwater forms. Stones (probably dropstones) within the Sub-Basement Clay are almost entirely Scottish and Scandinavian in origin. The macrofabric of the Basement Till usually indicates NE-SW ice movement, but has often been modified by the overriding Devensian glacier, suggesting that the surface of the till was pushed to its present level by the later ice advance (Penny & Catt, 1967). This explains why the till surface at Dimlington is above the level (c. 1 m O.D.) to which it was bevelled by the sea at Sewerby during the Ipswichian Stage (Catt & Penny, 1966).

The only other locality in Holderness where Basement Till is exposed, albeit rarely, is on the foreshore between Bridlington and Sewerby (Lamplugh, 1882; 1883; 1890a). Its surface level near Bridlington Harbour again suggests pushing by Devensian ice, but exposures here are too infrequent for this to have been verified by macrofabric studies. The last known occasion when beach sediments were temporarily removed to expose Basement Till east of Bridlington Harbour was in 1964 (de Boer *et al.*, 1965). In this area the Basement Till contains large erratic rafts of glauconitic marine sand yielding a similar fauna to the rafts of grey clay at Dimlington. These sandy inclusions have been described as the Bridlington Crag (Reid, 1885). Catt & Penny (1966) suggested that the rafts at Bridlington and Dimlington were derived from late Hoxnian marine sediments, because they contain a typical glacial assemblage of heavy minerals presumably derived from the earliest (Anglian) definite glacial sediments in the North Sea area. However, Baden-Powell (1956) thought that the molluscs were 'early glacial' (i.e. Anglian) in age, and Reid & Downie (1973) reported rich assemblages of pollen and dinoflagellate cysts indicating a Pastonian age. Perhaps the inclusions are derived from North Sea sediments of various ages.

North of Holderness, the Basement Till has been reported locally on Flamborough Head (Lamplugh, 1890b; 1891; 1892), on the foreshore near Reighton Gap (TA 142763) (Lamplugh, 1879; Catt & Penny, 1966), and above the Speeton Shell Bed on New

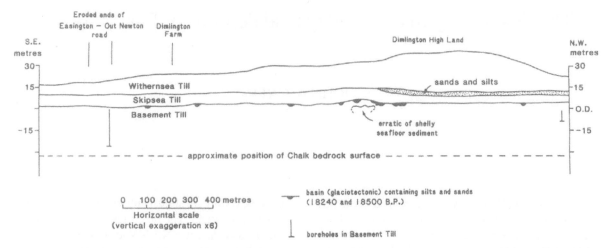

Figure 115: Measured cliff section near Dimlington Farm, S.E. Holderness, based mainly on observations 1956-64 (from Catt & Penny, 1966).

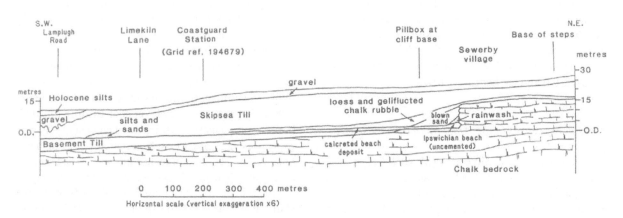

Figure 116: Cliff section at Sewerby, sketched from periodic cliff and foreshore observations 1960-64 (from Catt & Penny, 1966).

Closes Cliff (TA 147758) (Versey, 1938a; Catt & Penny, 1966). Further north a similar deposit, the Scandinavian Drift of Trechmann (1915), was previously exposed at Warren House Gill, County Durham. There was also a single exposure of Basement Till in east Lincolnshire at Welton-le-Wold; this was described as the Welton Till by Alabaster & Straw (1976). The Basement Till therefore has a very patchy distribution in coastal areas of northeast England. Presumably it originally formed a continuous sheet, which was strongly dissected during the Ipswichian and earlier parts of the Devensian.

The Wolstonian age of the Basement Till is indicated by: (a) it underlies the Ipswichian beach deposits at Sewerby (Catt & Penny, 1966) (see fig. 116), and (b) it overlies gravels containing derived Hox-

nian bones and artefacts at Welton-le-Wold (Alabaster & Straw, 1976). The relationship to the estuarine Speeton Shell Bed is less significant, because the till is thin and restricted in occurrence, suggesting it could be an erratic raft in the base of the Devensian till above. Also the various types of biostratigraphical evidence for the age of the Shell Bed are contradictory. West (1969) tentatively related pollen spectra from it to mixed oak forest assemblages in substage II (f) of the Ipswichian at some sites in southeast England, but the molluscan assemblage is identical to that of estuarine silts occurring at a similar height at Kirmington in north Lincolnshire (Reid, 1885), and the Kirmington silts contain typical Hoxnian pollen spectra (Watts, 1959b).

The Dimlington Stadial (Late Devensian) deposits of coastal areas

The coastal cliffs of Holderness, Filey Bay and other bays between Filey Brigg and the Tees estuary are cut mainly in tills deposited during the Dimlington Stadial of Rose (1985). The type site for this stadial is the cliff section at Dimlington Farm, southeast Holderness (figs. 115 and 117), where the Basement Till is overlain by two distinct tills, the Skipsea and Withernsea Tills of Madgett & Catt (1978). The lower (Skipsea) till is locally separated from the Basement Till beneath by silts containing arctic moss remains (fig. 115), which have given radiocarbon dates of 18,500 ± 400 B.P. (I-3372) and 18,240 ± 250 B.P. (Birm.-108) (Penny et al., 1969).

In Holderness the Skipsea Till is mainly very dark greyish-brown (10YR 3/2) and the overlying Withernsea Till dark brown (7.5YR 3/2), but the Skipsea Till is the more variable in colour and erratic suite, and was divided into five units by Bisat (1940). The distribution of these subdivisions of the Skipsea Till (= Drab Clay of Bisat) throughout the Holderness cliff sections was surveyed by Bisat between 1932 and 1951. His cliff section was reproduced by Catt & Madgett (1981), but the coastline has subsequently retreated by up to 60 m and the subdivisions are no longer easy to trace. The lithological variation within the Skipsea Till results from differences in the amounts of three main components: (a) chalk and flint, (b) red (Triassic) silt and sandstone, and (c) grey shale, sandstone, coal and limestone. Some parts are composed almost entirely of one of these components; for example, thin impersistent bands of white,

Figure 117: Till facies exposed at Dimlington on the East Yorkshire coast. The Basement Till forms the lowermost unit, separated by a marked shear plane from the overlying Skipsea Till. The section is capped by the Withernsea Till which at this locality is desiccated following transport down the front of the cliff as a mudflow (Photograph: Ehlers, 1987).

intensely chalky till occur in the Dimlington cliff section, and a dull red (Trias-rich) band has been traced for several kilometres south of Hornsea (Reid, 1885). Norwegian erratics (larvikite and rhomb porphyries) occur in lower parts of the Skipsea Till, and rocks from the Cheviots and Southern Scotland throughout. Chalk and flint become more abundant southwards.

The Withernsea Till (= Purple Clay of Bisat and earlier authors) has a more restricted distribution than the Skipsea Till. In the Holderness cliff sections it is seen above the Skipsea Till between Easington and Mappleton only, and inland it has an arcuate margin no more than 10 km from the coast (Madgett, 1975). Its erratics are mainly red Triassic sandstone and shale, grey Liassic and Carboniferous shales, chalk, Magnesian Limestone and Carboniferous Limestone; igneous rocks are rarer than in the Skipsea Till, and come mainly from the Lake District (e.g. Shap Granite).

The junction between the Withernsea and Skipsea Tills is usually sharp, with no evidence of weathering or disturbance, and only rare examples of incorporation of Skipsea Till within the basal Withernsea Till. The two tills are locally separated by stratified sands or gravels (fig. 115), but these do not indicate an ice-free interval, because they contain derived but no indigenous fossils, and because lenses of similar sands and gravels occur as commonly within either of the two tills and are obviously infillings of englacial meltwater channels. The Withernsea Till is overlain by deposits infilling a kettle-hole at The Bog, Roos (fig. 118), which have given a radiocarbon date of 13,045 ± 270 (Beckett, 1981); taken with the evidence of the radiocarbon dates for the moss in basins beneath the Skipsea Till at Dimlington Farm, this shows that both tills were deposited in 5000 years or less. An even shorter period of glaciation in east Yorkshire is indicated by the date of 16,713 ± 340 for moss from the base of a kettle-hole at Kildale Hall (NZ 609097), but this may be affected by a hardwater error (Jones, 1977). The two tills also have similar macrofabric characteristics, indicating ice movement from NNE to SSW throughout most of Holderness (Penny & Catt, 1967).

Many of these features suggest that the Skipsea and Withernsea Tills were deposited by a single ice sheet rather than two separate advances. A similar sequence of two tills can be traced northwards as far as the Tees estuary (though neither is as chalky north of Flamborough Head), but over most of County Durham there is no equivalent of the Withernsea Till; the Skipsea Till, however, probably correlates with

the Lower Boulder Clay of County Durham (Smith, 1981). Catt & Penny (1966), Madgett & Catt (1978) and Edwards (1981) have therefore supported in modified form the earlier suggestion of Carruthers (1953) that the glacier invading eastern Yorkshire was a composite ice sheet comprising superimposed tributary glaciers. This postulates that the Skipsea Till was deposited by ice which originated in Northumberland and southern Scotland and moved southwards along the coast, whereas the Withernsea Till came from a Tees valley ice-stream, which overrode the coastal (Skipsea Till) ice near the mouth of the Tees and was then carried southwards on its back into eastern Yorkshire. The erratic suites support these different ice movements, and the assemblage of derived pre-Quaternary palynomorphs in the Skipsea Till confirms that this part of the glacier moved southwards down the coast from Northumberland (Hunt *et al.*, 1984), though the microfossils in the Withernsea Till have yet to be studied. Carruthers attributed the Basement Till to a basal (Scandinavian ice) layer of the same multi-tiered glacier, but it is now clear that the Basement was deposited in an earlier and completely different glaciation.

This unusual hypothesis for simultaneous deposition of the Skipsea and Withernsea Tills by one ice advance explains several other features of the Dimlington Stadial glaciation of east Yorkshire:

i) At Robin Hood's Bay (NZ 953048) and other sites in northeast Yorkshire the Skipsea and Withernsea Tills are separated by a till of intermediate type, up to 8 m thick, which has mixed colour and erratic characteristics of both tills (Catt & Madgett, 1981). This is more easily explained by mixing of the two layers of the composite ice sheet than by incorporation of Skipsea Till into basal layers of a separate Withernsea Till glacier.

ii) Where Skipsea Till is locally incorporated into lower parts of the overlying Withernsea Till, it is often because of shearing where bedrock obstructions retarded basal ice movement (Edwards, 1981). The low-angle shear planes can often be traced from one till into the other, suggesting that the ice from which both tills were deposited was present at the time this deformation occurred.

iii) A computer-generated model of the British Devensian ice sheet based on known ice thicknesses and gradients (Boulton *et al.*, 1977) failed to explain why east coast ice reached south of the Tees estuary, unless there was some instability and surging. Overriding of the coastal ice by a Tees valley ice stream could have provided this instability, because the additional weight would have in-

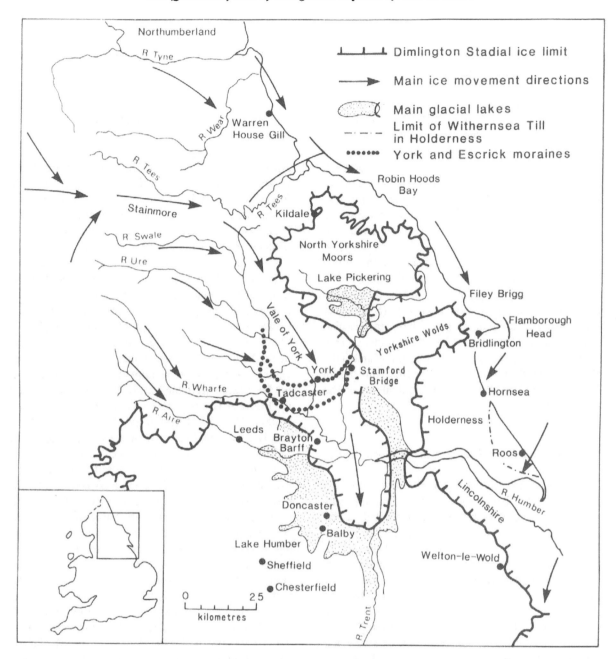

Figure 118: Late Devensian glacial features in east Yorkshire and surrounding areas, and main localities mentioned in the text.

creased basal melting, thus lubricating the postulated surge lobe.

iv) A large lake (Lake Humber) was impounded at various levels up to 33 m O.D. in southern parts of the Vale of York (fig. 118) by ice simultaneously blocking the Humber gap as well as northern parts of the Vale (Gaunt, 1981). The ice to the north came mainly from northwest England, crossing the Pennines via Stainmore and the Tees valley; near Darlington it divided into two streams, one passing southwards into the Vale of York, and the other continuing eastwards to the Tees estuary and eventually depositing the Withernsea Till along the east coast. However, the east coast ice that blocked the Humber gap deposited Skipsea Till, the surface of which reaches a maximum height of approximately 8 m O.D. in the gap itself; the Withernsea Till terminates 20 - 25 km east of the gap. The high levels of Lake Humber, dated to

21,835 ± 1600 B.P. or later (Gaunt, 1974), therefore imply that ice which deposited the Skipsea Till in the Humber gap was contemporaneous with ice in the Vale of York, which was part of the Stainmore stream that also deposited the Withernsea Till.

Catt & Penny (1966) and most earlier authors identified a third Devensian till in east Yorkshire, which was known as the 'Hessle Clay' or 'Upper Purple'. This was said to overlie the Purple Clay (= Withernsea Till) and to rest directly on the Drab Clay (= Skipsea Till) where the Withernsea Till is absent. However, its stated characteristics were mainly features resulting from post-depositional weathering and pedogenesis (reddish colours caused by oxidation, greyish faces on fissures caused by gleying, etc.), and Madgett & Catt (1978) showed that it is a Flandrian weathered mantle, decalcified to 70 - 90 cm and oxidised to about 5 m depth, formed on whichever of the two Devensian tills (Skipsea or Withernsea) occurs at the surface. Their analyses also showed that the Devensian tills in Lincolnshire (the Hessle Clay of Jukes-Browne, 1885, and Marsh Tills of Straw, 1969) and north Norfolk (the Hunstanton Boulder Clay) are all weathered equivalents of the Skipsea Till.

The Dimlington Stadial deposits in the Vale of York

The ice in the Vale of York, fed partly from Stainmore and partly from true valley glaciers flowing eastwards down the Yorkshire dales, deposited sands, gravels and grey or brown till of variable composition. The till is often as deeply oxidised at the surface as in Holderness, and extends approximately as far south as an arcuate ridge (the Escrick Moraine) extending from Stamford Bridge in the east to beyond Tadcaster in the west (fig. 118). This and a similar lobate ridge to the north (the York Moraine) have long been regarded as the positions of stable ice margins during the Devensian glaciation. However, Gaunt (1981) suggested that the ice initially surged much further south (as far as Wroot, SE 715030) into the waters of a high-level Lake Humber. The presence of ice south of the Escrick Moraine is indicated not by till, but by gravels containing Carboniferous and Permian erratics. These rest on a periglacial land surface, which predates the formation of Lake Humber, and extends northwards beneath the till.

The ice of this initial advance melted quickly in the high-level lake, and ice-fronts were established later at the York and Escrick lines, probably when the lake level had declined (Gaunt, 1981). Laminated clays deposited in the lake underlie and overlie the till and gravels forming the Escrick Moraine, showing clearly that the moraine was formed where the glacier terminated in the lake.

The 'Older Drifts' of the Vale of York, Wolds and Moors

Apart from the Basement Till of coastal areas, the clearest evidence for pre-Devensian glaciation in Yorkshire comes from southern parts of the Vale of York, especially the Pennine footslopes east of Leeds, Wakefield, Sheffield, and Chesterfield. In this area patches of grey till and gravel containing erratics of Carboniferous sandstone, limestone, chert and coal, vein quartz, Permian limestone, and occasional Lake District rocks were mapped by the British Geological Survey (Wray et al., 1930; Bromehead et al., 1933; Edwards et al., 1940; 1950; Mitchell et al., 1947; Stephens et al., 1953; Eden et al., 1957). The patches lie well outside even the extended Devensian limit of Gaunt (1981), and are usually at greater heights than the Devensian till in the Vale of York, reaching > 100 m O.D. near Leeds, > 200 m O.D. near Sheffield and > 300 m O.D. near Chesterfield. The till is often more deeply decalcified (2 m) than the Devensian till. In central and eastern parts of the southern Vale of York there are patches of similar till at Balby (SE 562004), Brayton Barff (SE 585305) and near Bawtry (Gaunt et al., 1972), again occupying slightly higher ground than the nearest Devensian till. Laminated clays are closely associated with the till at several sites, including Brayton Barff, suggesting that there was a pre-Devensian as well as a late Devensian Lake Humber. High-level gravels overlying Triassic and Jurassic rocks and containing Carboniferous erratics occur at Church Hill, Holme upon Spalding Moor (SE 8238) and other sites up to 70 m O.D. along the western foot of the Yorkshire Wolds (de Boer et al., 1958), and are probably related to the same glacial episode. It may also have been responsible for a large erratic of Cave Oolite over 100 m long and 4 m thick, which overlies chalky gravel with far-travelled erratics at 50 m O.D. near South Cave (Stather, 1922).

Near Doncaster several deep, narrow, steep-sided channels aligned NW-SE and closed at either end were probably cut as tunnel valleys beneath the pre-Devensian Vale of York glacier, as they are partly filled with over-consolidated clays and often covered by grey till (Gaunt, 1981). Reddish sands with Bunter pebbles derived from areas to the south, such as the Upper Trent Basin, overlie the till and channel deposits southeast of Doncaster. From their northward cross-bedding dips, these seem to have been deposited by glacial meltwater from the Midlands. They also contain a few flint pebbles, which suggest that the meltwater was from the glacier that deposited the Wolstonian chalky till of the Middle Trent Basin (Gaunt, 1981). If so, this means the 'Older Drift' of the Vale of York is also Wolstonian in age. The northward flow of meltwater implies considerable isostatic depression to the north at this time.

The western margin of the Devensian coastal ice in east Yorkshire is usually drawn at the feather edge of the weathered Skipsea Till on the rising bedrock surface of the Yorkshire Wolds and North Yorkshire Moors (fig. 118). On higher parts of these hills beyond this line there is a pattern of meltwater channels. In addition, small lakes were impounded by the Devensian ice in some valleys, often with interconnecting overflow channels (Kendall, 1902; de Boer, 1945; de Boer *et al.*, 1958). However, many of the discordant channels in these areas seem to be subglacial in origin (Gregory, 1962; 1965; Foster, 1986), suggesting that the Devensian ice extended beyond the till margin. There are also scattered deposits of sand or gravel with far-travelled rocks, often preserved in fissures (Sheppard, 1904; Mortimer, 1905;

Versey, 1938b). Bisat (1940) attributed these deposits to a pre-Devensian glaciation, which could even be as old as Baventian (Catt, 1982b). But Foster (1986) suggested that many of these 'Older Drifts' on the Wolds were deposited by the Devensian ice sheet, upper layers of which could have been too clean to deposit till. However, the Wolds have a thin but fairly continuous cover of loess (Catt *et al.*, 1974), which accumulated just before or during the Dimlington Stadial glaciation in Holderness (Wintle & Catt, 1985), and this would not have survived intact if it had been either deposited before ice advanced onto the higher Wolds or deposited originally on a glacier surface.

At present it therefore seems that higher parts of the Wolds and North Yorkshire Moors were glaciated at least once before the Devensian. Remnants of the resulting deposits are preserved locally, but some of the 'Older Drift' patches could be late Devensian, because the exact position of the Dimlington Stadial ice margin is still uncertain. The erratics and heavy minerals in deposits high on the Wolds do not suggest correlation with any of the tills or glacial gravels of Devensian or Wolstonian age in coastal regions of the Vale of York, so it is likely that they are pre-Wolstonian in age.

Quaternary history and glacial deposits of the Peak District

CYNTHIA V. BUREK

Only scattered remnants of glacial deposition are recorded in the Peak District, nevertheless the area forms an important link between Quaternary sediments formed from the Irish Sea and North Sea ice sheets. The Peak District, whilst not generating its own ice sheet, affected and modified the ice that crossed the region from the west, resulting in a series of complex sedimentary deposits. Lying astride the main watershed of England, the district separates the plains of East Anglia and Lincolnshire to the East, the drift-covered rolling hills of the South Midlands and the Triassic plains of Cheshire to the West and formed an important natural barrier between the Irish and North Sea ice sheets during the Pleistocene.

Geological background

The Peak District is broadly speaking a dome of Carboniferous Limestones over 300 m in height overlooked by Millstone Grit scarps on three sides (fig. 119). The dome forms the southern limit of the Pennines before the Carboniferous dips under the Triassic sediments of the Midlands. It represents an area of approximately 647.5 km² and consists of at least 1000 m of massive and bedded limestone intermixed with basal lavas, tuffs and dolerites which have weathered to form clay horizons in places ('clay wayboards'). The area is barren of any pre-glacial, post-Carboniferous deposition, weathering or erosional episodes with the exception of minor Mio-Pliocene fluviatile sand and clay deposits that dot the southern half of the dome filling solution depressions e.g. at Brassington (Boulter et al., 1971).

Quaternary history

The earliest evidence of Pleistocene activity in the Peak District occurs in high level caves such as Water Icicle Close Cave above the Lathkill valley. Here evidence of speleothem development associated with the Cromerian Stage is found. In Victoria Quarry Cave Close, Dove Holes (SK 07927695) very rare warm temperate faunas, normally assigned to the Lower Pleistocene (Villefranchian stage), have been recognised. Between the Cromerian deposits and the fluvial temperate Mio-Pliocene Brassington Sands, there appears to be an hiatus in sedimentary deposition for which no adequate explanation has yet been provided.

Above the Cromerian, evidence of Anglian ice passing over the Peak District can be seen in the high level tills and erratics at isolated sites (see figs. 120, 121 and 122).

In the succeeding Hoxnian Stage during the warm, moist conditions cemented screes and tufas formed and speleothem development was active in the high level caves at Castleton. Furthermore, the Hathersage Terrace, a notable landmark in the Peak District was formed during the extensive *in situ* weathering and soil formation of the interglacial. Ford (1985) has also provided evidence for substantial cave development during or before the Hoxnian through isotopic dating of Castleton speleothems.

The Wolstonian Glaciation was characterised by extensive till deposition in Stoney Middleton Dale and the valleys of the Wye and Lathkill (Burek, 1978; 1985; Aitkenhead et al., 1985). In addition, the Derwent river underwent a glacial diversion (Straw, 1968).

The temperate Ipswichian Stage brought further soil formation and *in situ* weathering resulting in the formation of extensive residual insoluble residue deposits particularly over limestones. This was probably a feature of the Hoxnian also, but the soils produced were eroded by glaciation in the Wolstonian.

In the early Devensian periglacial conditions prevailed during which loess was formed by deflation of pre-existing glacial debris. Aeolian material probably derived from the east, and possibly also from the west

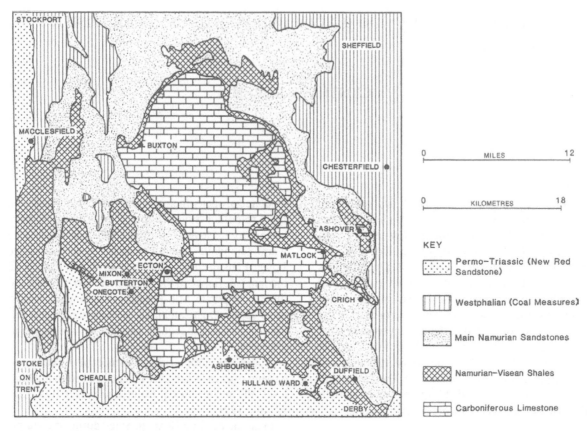

Figure 119: Distribution of the principal geological units around the Peak District.

KEY

Permo-Triassic (New Red Sandstone)

Westphalian (Coal Measures)

Main Namurian Sandstones

Namurian-Visean Shales

Carboniferous Limestone

may have also contributed to this silt source.

The climatic amelioration of the Upton Warren Interstadial resulted in clay translocation in soils (Cazalet, 1968). This was followed by a return to cold periglacial conditions that caused cryoturbation of the insoluble residue, left by *in situ* weathering from the previous warm period, and loess to produce a deposit locally termed 'silty drift' (Pigott, 1962). Also during this period, increased phreatic activity and valley incision through the Lower Carboniferous Shale cover of the Peak District, caused a lowering of the water table and gave rise to the development of an extensive dry valley system. The valley incision occurred not only during the periglacial conditions of the Devensian, but also during the temperate Ipswichian Stage. Prior to the Ipswichian and Devensian, a shale cover of the Peak District can be recognised from the close geochemical similarity in the fine fraction of Wolstonian tills and the Carboniferous shale.

Glacial deposits

Although the distribution of glacial deposits is limi-

ted in the Peak District, it has been firmly estab-lished that climatic and depositional conditions during the Quaternary were complex (Burek, 1977b; 1985). Much of the plateau surface is covered by non-glacial deposits, which have been classified in detail by Burek (1978). However, only the tills and the 'silty drift' will be discussed in detail below.

Fig. 120 shows the distribution of the tills on the limestone and the immediate surrounding area. Extensive spreads of till are recognised on both sides of the Wye valley between Bakewell and Rowsley and on the south side of the Lathkill valley. The best exposure is near Youlgreave in Shining Bank Quarry (SK 26 NW 35; figs. 121 and 122), where 9 m of till containing striated limestone boulders up to 0.9 m diameter as well as smaller sandstone pebbles, and basalt boulders is recognised. Granite, Borrowdale Volcanic and amygdaloidal basalt clasts from as far as the Lake District and the Highlands of Scotland are also recorded (Burek, 1978; 1985). Raper Pit (SK 217652) on the Lathkill, less than two kilometres away, has a similar matrix composition but does not show this variety in large erratic content.

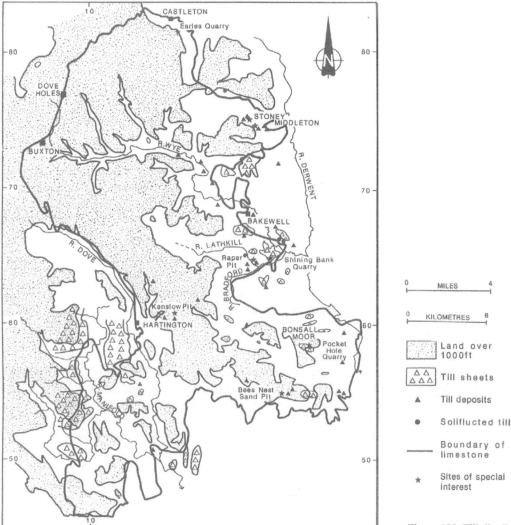

Figure 120: Till distribution.

This contrast can be attributed to glacial dynamics. Nye (1952), Boulton (1972, 1975a), Francis (1975), Sugden & John (1976) and Boulton *et al.* (1977) have shown from Spitsbergen and other areas that glaciers carry their debris in stress patterns which trend upwards in the accumulation zone and downwards in the ablation zone. If there is ever a minor change in the bedrock form, relief, or climate, the sensitive stress patterns may alter and deposition frequently results. For example, two streams of ice merging together could produce debris deposition because of the resulting compressive flow, subsequent thickening of the ice and pressure increase. Thus the nature and distribution of till particularly in an upland area with the highly variable relief, can alter substantially from one place to another within a

few hundred feet (Boothroyd *et al.*, 1977). This may account for the variable distribution and mineralogical/chemical nature of the Peak District tills. Thus a lobe of ice coming from the west down the former Lathkill valley amalgamated with the merged Wye and Derwent ice flows, thereby altering the glacial regime to such an extent that substantial deposition resulted. The lodgement till (*sensu* Boulton, 1972; Francis, 1975; INQUA Commission 2, 1977) of Shining Bank Quarry and Raper Pit must therefore have been de-posited under the altered ice flow conditions and the variability in erratic suites between the two sets produced by differences in depositional system or erratic suites in different shear stress lines.

Fresh Peak District till is typically a fine-grained, gritty, brown grey clay (Munsell colour 10YR4/2-3).

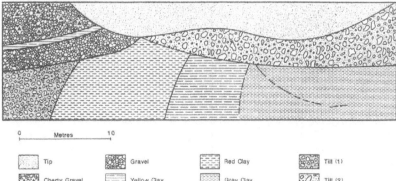

0   Metres   10

- ▫ Tip
- ▨ Cherty Gravel
- ▨ Gravel
- ▨ Yellow Clay
- ▨ Red Clay
- ▨ Grey Clay
- ▨ Till (1)
- ▨ Till (2)

Figure 121: Kenslow Pit (after Shotton, 1962; in Warwick, 1977).

For the most part its upper layers are leached and boulder-free, but beneath it contains striated boulders of both local and foreign material. Local limestone and basalt erratics appear to be larger, more striated and more numerous than well-travelled erratics such as Lake District volcanic and igneous rocks. The till overlies all rock types in the Peak District but is best preserved on shales possibly because it is more clayey and therefore more cohesive. Also, the apparent preference for this occurrence above shales in a predominantly limestone area is possibly a result of the loss of the clay fraction into joints and ultimately into the extensive cave systems present in the area, leaving only scattered erratics on the limestone surface.

High-level tills. Peak District tills can be divided into two units dependent upon altitude. The majority of the deposits lie below 244 m O.D.. However, highly weathered complex till deposits scattered over the Mio-Pliocene sand pits (Boulter *et al.*, 1971), at heights of 330-360 m O.D., lie unconformably over limestone collapse structures and are laterally limited. These are important in providing insight into pre-Wolstonian glacial events. Three solution hollow fills (Kenslow Pit, Friden Pit and Bees Nest sand pit) comprise complex Pleistocene deposits. At Kenslow Pit (fig. 121) the till contains chert fragments mainly of local origin and only minor quantities of limestone suggesting that the till was predominantly derived from limestone residue. This was presumably assimilated by the glaciers after the limestone had emerged from beneath the Triassic sandstone and Carboniferous shale cover and after a period of prolonged warm climate chemical weathering. Glaciation followed, as indicated by the occurrence of cherty till which rests unconformably on top of lower Pliocene clays. Above this till the following sequence is recorded:

  thin soil
  loess with few cherts ('silty drift')
  cherty gravel
  yellow clay
  gravel
  till

Here we can reasonably assume that the gravel is outwash or outwash-derived and the yellow clay is chemically weathered, indicating a mild climate. The chert gravel was derived as an insoluble residue from the limestone and reworked by running water, probably indicating solifluction or colluviation into the pit. The loess appears similar to that deposited over all the limestone and must have been subjected to cryoturbation to form 'silty drift'. Lastly, a thin soil has developed, presumably during Postglacial time. Depending on the interpretation of the cherty gravel, this would indicate at least one if not two periods of glacial or periglacial conditions since the last till was deposited. Fabric analysis performed on the basal till, showed a WSW orientation, possibly indicating influence of eastern ice or rotation on subsidence. Also, in Friden Pit, Shotton noticed two tills (Yorke, 1954), one containing chert fragments and no limestone, the other containing small pieces of limestone separated from the former by a waterlaid gravel. Fabric analysis of the cherty lower till indicated a similar possible eastern derivation. At Bees Nest sand pit at Brassington, a silty, roughly stratified, cherty deposit with scattered pebbles overlying the Mio-Pliocene sands and clays was interpreted as soliflucted till because of the presence of rare far-travelled erratics (Ford, 1972).

These high-level, older tills differ from the younger, low-level tills mineralogically and are more deeply weathered. They are devoid of fossils or datable materials. However, based on their geographical location and physical characteristics a tentative Anglian age has been assigned to them (Burek, 1978). High-level till is overlain by a loessic silt in a fissure at Earles Quarry, Hope (SK 161082). No other till has been observed near this locality probably because the till occurs below the general level of the plateau surface.

Scattered erratics lying up to a height of 396.5 m O.D. (Dale, 1900) suggest a previous glacial episode over the Peak District. The glaciation was possibly the Anglian since ice during this event reached as far south as North London. Furthermore, erratics from the Dove and Manifold areas have been found in the East Midlands in Anglian deposits. It would seem therefore that Anglian ice overode the Peak District and probably greatly modified the former topography, but deposits were only preserved in favoured locations such as fissures or over solution hollow infills.

Low-level tills. It has been suggested that, in contrast, low-level till deposits are of Wolstonian age and that Lake District ice moved southwards eventually coming into contact with the Irish Sea and Welsh ice sheets. Subsequently, the ice flow took the course of least resistence across the Cheshire Plain and thence over a col in the South Pennines at Dove Holes. The main tongue flowed down the Wye, Dove and Manifold valleys (the rivers at that time were flowing about 45 m higher than they are today) and joined the Derwent and Trent. Because it is doubtful that such a small and narrow col as Dove Holes could accomodate all the ice which overflowed the rest of the White Peak, two phases of Wolstonian glacial advance have been postulated (Waters & Johnson, 1958). To the south of the Peak District in the Wolstonian, the Chalky Boulder Clay ice and the Irish Sea ice met in the East Midlands and produced interdigitating deposits of the two tills (e.g. Huncote, Leicestershire; Rice, 1977).

The Peak District ice moving south was halted and then stagnated, resulting in few glacial erosion features and a general scatter of erratics (Dale, 1900; Jowett & Charlesworth, 1929; Dalton, 1958). It is hypothesised that relatively clean ice overrode the col at Dove Holes (fig. 122) (till lies at 305 m O.D. at Draglow), producing tills containing limited far-travelled erratics. Ice then travelled via Peak Dale where it scoured bedrock from a relatively flat area around Wormhill and Miller's Dale at about 305 m and deposited material below Monsal Head as a consequence of rapid change in valley configuration (fig. 120). The till, which appears to be lodgement in

the Bakewell area, was deposited by two ice lobes issuing from the Wye valley, Baslow and Rowsley at a height of c. 248 m O.D. bordered on the north by Longstone Edge and on the southwest by the high land from Sheldon to Over Haddon. This till lies as low as 60 m O.D. at Bakewell cemetery (that near Haddon Hall at 137 m O.D. is soliflucted), at 164.7 m O.D. in Shining Bank Quarry (SK 229652) and as high as 248 m O.D. at Rowland (SK 213725). The direction of ice movement in this area is supported by a northwest to southeast orientation on striated pavement at Shining Bank Quarry. Upstream from Monsal Head, only scattered terrace deposits high on the flanks of the Wye valley (SK 179722) are recognised.

Around Eyam and Stoney Middleton Dale, a striated limestone pavement giving a general west to east orientation overlain by patchy till occurs at 213-248 m O.D. consistent with the general level of deposition in the Bakewell area. In places equivalents of this patchy till lie on terraces of an earlier, possibly Wolstonian age. Because aggradation is common in the upper reaches of a river in periglacial times, till could have been deposited subsequently, for example in the early Devensian. The depth of subsequent river erosion and the isolation of the till patches, however, point to a greater age. The Stoney Middleton till is therefore assigned to the Wolstonian.

During the Wolstonian, ice flowed over the col at Dove Holes, following the previously established drainage system. This was less entrenched than at present so the ice may also have flowed via Peak Forest to Foolow, Eyam and Stoney Middleton as well as down the Wye or the Dove. Unfortunately the only evidence remaining of this route is the distribution of erratics (fig. 122). The less resistant tills have been removed by subaerial processes. The derived sediments have been redeposited below ground in cave systems or incorporated into other glacial deposits to the south of the Peak District.

Since the onset of any glaciation is preceded by climatic deterioration until periglacial conditions prevail, one would expect periglacial deposits to underlie tills. This does not appear to be the case in the Peak District. The tills for the most part lie directly on bedrock and the only evidence of periglacial activity can be seen at Raper Pit (SK 218653) where a small rotated block of clay, sand and gravel was found in the till (Burek, 1977b). It is thought that the sand was picked up by the advancing ice as a frozen block and moved as an erratic. This block has been rotated during transportation. In addition it is probable that the block had not been transported far

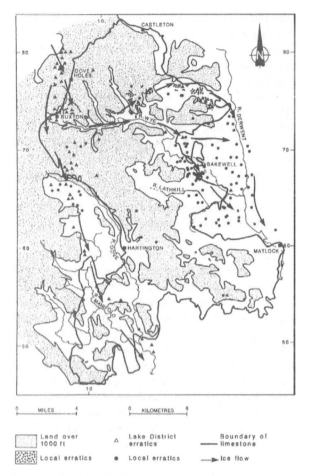

Figure 122: Ice flow direction and erratic distribution.

Mineralogy

Mineralogically, Peak District tills can be divided into three groups. Two groups represent tills deposited from the older glaciations, and the third from a younger glaciation. The general mineralogy of the older tills is twofold. The upper part of for example Pocket Hole Quarry on Bonsall Moor above Matlock (SK 261585), has been leached of calcite and contains only quartz and clay minerals. However, below 1.8 m calcite is present in the section. High clay mineral peaks from X-ray diffraction traces are significant in these tills (fig. 123). In Earles Quarry (SK 161816) till found in an enlarged joint contained quartz, feldspar, clay minerals and baryte, but again no calcite. Baryte is a typical vein mineral and indicates that the limestone must have been stripped of the shales and grits at the time of till deposition because the mineralisation is only present in the Lower Carboniferous limestone. Thus the tills must have originally contained calcite but this has been removed. The younger tills all contain calcite. Anatase (TiO_2) is present in the 'clay wayboards' of the area but none was detected in the tills or the loess samples.

The clays of the older tills are illite, montmorillonite (12 Å and 14 Å), chlorite, kaolinite and in Pocket Hole Quarry halloysite. In the leached upper part, mixed-layer clays - montmorillonite/illite and montmorillonite/chlorite occur but no halloysite. The latter evolves from a weathering of igneous rocks in the following sequence:

Montmorillonite → Halloysite → Metahalloysite → Kaolinite

The younger tills all contain substantial amounts of quartz, calcite, frequently feldspar, most often orthoclase, oligoclase, labradorite, andesine and bytownite and clay minerals. The predominant clay minerals in all younger tills are illite (+ muscovite), chlorite and kaolinite. Occasionally montmorillonite, mixed-layer clays, halloysite, chamosite and vermi-

because, even in a frozen state, the material would not have been resistant.

Occasionally, the till lies directly on gravels, but these are thought to be fluvial. Above the till, loess is the only deposit found extensively over the Peak District. Sands and gravels are not found overlying till in North Derbyshire apparently because the stagnation of ice severely restricted the development of outwash deposits.

Plates 11–12

11. View of sediments exposed at Linhope Burn, Northumberland. Upper part shows a dip section of bedded solifluctate; the darker band is a coarse sandy bed saturated with throughflow which does not drain through the impermeable till exposed in the bottom one third of the section. The form of the smooth bench is evident (Photograph: T. Douglas, 1985).
12. Detail of section shown in plate 1. The boundary between the compact lodgement till and the overlying solifluctate is evident (Photograph: T. Douglas, 1985).

11
—
12

PLATE 13

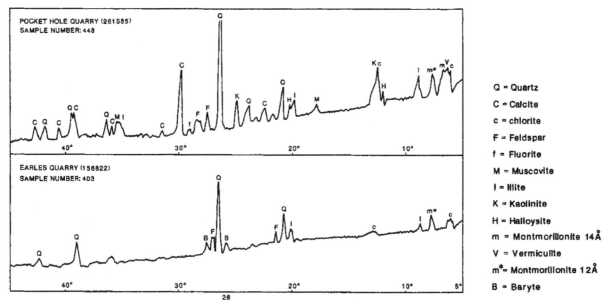

Figure 123: Typical older till X-ray diffraction trace.

culite are present. Dolomite, baryte and fluorite can also be present representing erratic breakdown due to weathering. Fig. 124 is a typical X-ray diffraction trace.

Clay mineralogy within the superficial deposits is immature with respect to climate. Many of the unstable clay minerals, such as halloysite, present in a leaching environment (such as the current moist, temperate climate) are identified. The clay mineralogy also supports the assumption that minerals from outside the area were introduced since some are not present in the underlying limestone. The widespread nature of the clay mineralogy supports the aeolian influence on the superficial sediments first recorded by Pigott (1962). The geochemistry of the 'silty drift' and loess is dealt with in detail elsewhere in this volume, in Burek (1978, 1985b) and in Burek & Cubitt (1979).

Loess and 'silty drift'

In the Peak District, the majority of the limestone plateau is covered with a thin bed of (< 0.3 m) loessic material. It is characteristically an orange-brown (Munsell: 5YR4/4), homogeneous and fine-grained siltloam with 85-88% silt and 20-25% sand. The presence of detrital quartz grains, a richer assemblage of heavy minerals than is present in the limestone and kaolinite and mica clay minerals suggests that the loess was derived from the Millstone Grit and shale areas surrounding the limestone (Kazi, 1972; Catt, 1977a, b).

The loessic deposit lies directly under present-day soils and was probably the most recently accumulated sediment. In places however, it has subsequently been mixed by cryoturbation with underlying insoluble residue derived from limestone bedrock to form the widespread 'silty drift' of Pigott (1962). In addition we can conclude from the nature of 'silty drift' that firstly any deposit formed intermediate between the loess and insoluble residue was eroded after deposition on the latter; secondly a period of chemical weathering preceded the deposition of the loess and formed the insoluble residue; and thirdly a period of intense cryoturbation followed loess deposition.

There is one problem with this simple picture, however. Loess could have been deposited during each of the cold periods of the Quaternary, under non-

Plate 13

13. Banded tills, Cadeby, western Leicestershire. Each band reveals distinct matrix type and erratic content (Photograph: T. Douglas, 1974).

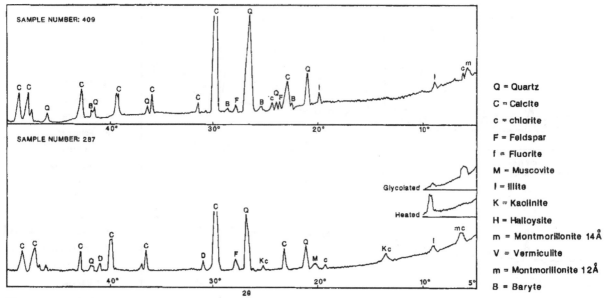

Figure 124: Typical younger till X-ray diffraction trace.

glacial conditions. For conclusive evidence of the age(s) of the loess in the Peak District, various techniques must be employed to distinguish the units, such as heavy mineral analysis for provenance, ratios of trace elements to establish weathering events (Burek, 1985b) and relative dates from other areas. In Lincolnshire, for example, loess occurs extensively above tills of Devensian age, and has been shown to be derived from them (Straw, 1963b; Catt *et al.*, 1974).

In some places, the loess has been leached to a depth of as much as 1.25 m, indicating a period of chemical weathering since deposition. Loess is also present in many joints, fractures and caves in the limestone, colouring them orange.

At Dove Holes, for example, the limestone generally has little cover but orange loessic material is seen in joints at great depths in the quarry faces suggesting that it was previously more widespread than today.

Mineralogy

European loess generally contains quartz, chlorite, occasionally vermiculite and well-crystallised mica.

In 1962, Pigott described the X-ray diffraction mineralogy of his 'silty drift' as containing quartz, (the principal peak) illite, muscovite and kaolinite, (peaks equal in height) and chlorite. The latter he attributed to the 'clay wayboards' because the insol-

uble residue of the limestone from Bradwell and Coombs Dale contained quartz, kaolinite and mica, while that from Cressbrook and Miller's Dale contained equal quantities of only quartz and kaolinite. There was no chlorite recorded in the limestone.

The typical mineralogy obtained from loess and 'silty drift' samples from the Peak District by Burek (1985a) is outlined in fig. 125. Mineralogy common to all samples is quartz, plagioclase, orthoclase, chlorite, kaolinite, and illite and/or muscovite. Occasionally vermiculite, chamosite, montmorillonite (both types) and mixed-layered illite/montmorillonite clay minerals are identified.

In contrast, Pigott does not record feldspar from the 'silty drift' and European loess is also not thought to contain this mineral. However there may be several reasons for the occurrence of feldspar in the loessic deposits of the Peak District. Millstone Grit surrounding the area does contain abundant feldspar especially sandstones below the Kinderscout Grit (i.e. Shale Grit and Mam Tor beds) which are predominantly subarkosic in character with substantially more feldspar than Namurian sandstones elsewhere in the Pennines (Gilligan, 1920; Eden *et al.*, 1957; Stevenson & Gaunt, 1971). This would make more feldspar available for deposition than normally found in loess. The absence of feldspar from Pigott's samples cannot be accounted for at present. The underlying limestones contain no feldspar in the insoluble residue only euhedral quartz, chert, silicified shells and occasional pyrite (Pigott, 1962; Cox &

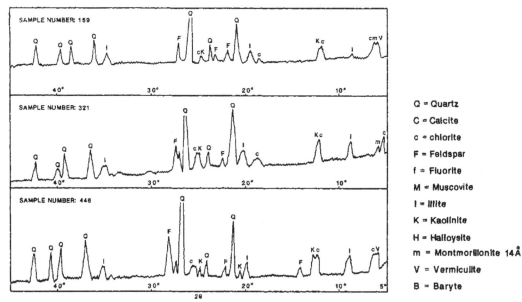

Figure 125: Typical loess and 'silty drift' X-ray diffraction trace.

Bridge, 1977; Harrison & Adlam, 1985), and thus this feldspar must have been derived from outside the area. As outlined by Catt *et al.* (1984), windblown deposits in Lincolnshire and East Yorkshire were carried a relatively short distance by winds blowing from approximately north or northeast off the ice sheet. It is interesting to note that the quartz and feldspar are the principal mineralogical constituents of the aeolian deposits in those two counties. Thus it is not inconceivable that the feldspars originated from the Namurian sandstones.

Conclusions

Glacial deposits are limited in the Peak District, but it has been possible using a variety of techniques to establish the sequences of climatic and depositional conditions during the glacial part of the Quaternary. Non-glacial deposits cover much of the plateau surface and these help to determine environmental conditions during the non-glacial Pleistocene. Thus, North Derbyshire has been directly affected by ice during two glacial episodes, the Anglian and the Wolstonian, and indirectly by one other, the Devensian. Prior to the Cromerian little evidence of deposition has been found. The geomorphology and sedimentology of the Peak District is a direct result of the alternation of climatic conditions during the past two million years. The complex nature of the deposits in the area reflects the location of the Peak District between the two major ice masses affecting England during the glacial episodes, the ice sheet issuing from the North Sea to the East and that rising out of the Irish Sea and the Welsh, Scottish and Lake District mountains to the north-west. A continuing assessment of their influence on the area can only enlighten discussion on events in the surrounding lowlands.

Acknowledgements

I would like to thank the secretarial and graphics departments of Poroperm-Geochem Ltd. for their assistance and my husband Dr. John M. Cubitt for critically reading the manuscript and his endless help and encouragement.

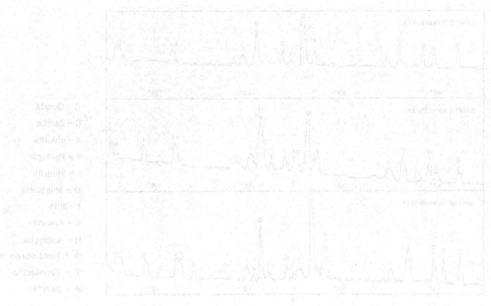

Q = Quartz
C = Calcite
I = Illite
F = Feldspar
M = Mica
K = Kaolin
H = Halloysite

Figure 2. X-ray diffraction traces.

Glacial deposits of the lowlands between the Mersey and Severn rivers

PETER WORSLEY

This area forms part of a tract of low ground which extends from the shores of the Irish Sea in the north, southwards to what was the main watershed of England prior to a major glacial diversion of drainage (see fig. 126). To the west, the area's border is formed by the abrupt limit of the Welsh massif and similarly in the east the Pennine uplands form a clear morphological feature especially in the northern and central sectors. Glaciers probably advanced into the lowlands during at least three glacial stages in the Pleistocene, with the main ingress being powered by ice originating in the north, particularly the Lake District and Southern Uplands of Scotland. Somewhat anomalously, local ice accumulation was restricted to the Welsh uplands, for there is little evidence to suggest that the southern Pennines were a significant area of ice accumulation.

The glacial deposits form an extensive but not complete cover over the low ground. Glacial morphological expression is variable and a classical range of ice contact, morainic, dead ice, lacustrine and proglacial landforms are present. In thickness the glacigenic sediments range from zero where the bedrock is exposed to extremes in excess of 100 m. For much of the area the bedrock consists of Permo-Triassic sediments, with mudstones, sandstones and evaporites predominating. These extend beneath the waters of the Irish Sea in an up-glacier direction and hence it is no surprise that the Permo-Triassic materials dominate the ultimate characteristics of the glacigenic sediments. Along the low ground margins in both the east and west, the occurrence of Carboniferous shales, coals and sandstones imparts a rather different aspect to the glacial deposits with the dull reds being replaced by dark browns and greys. Overall the tills have a fine-grained matrix and a representative series of sand-silt-clay determinations on the till matrix and are plotted in the ternary diagram forming fig. 127. Natural exposures of glacial deposits are rare and virtually all data pertaining to them are derived from studies of quarries and boreholes.

Historically, in terms of the development of glacial geological concepts, the area has played an important role. For almost 150 years studies of glacial sedimentary facies and stratigraphy have generated controversy over interpretation. A major influence on discussion has been the common occurrence of Pleistocene marine molluscan fossils in association with the glacigenic materials and this factor delayed the universal acceptance of the land ice hypothesis as the key to unravelling the glacial record in the region (Thompson & Worsley, 1966). Indeed the possibility of the high relative sea levels during the last ice retreat are currently the topic of debate. Until relatively recently the Geological Survey has tended to promote an interpretative model which has placed emphasis on the persistence and regularity of lithological units such that the glacial succession was classified into a tripartite lithological scheme utilising the terms 'Lower Boulder Clay', 'Middle Sands' and 'Upper Boulder Clay' which were then given chronostratigraphic status. As understanding improved this classification became redundant. A milestone was reached when Jackson et al. (1983) reported from the same area where the tripartite classification had previously been forcefully advocated (Poole & Whiteman, 1966) and stated "the glacial deposits ... comprise a complex and laterally variable which cannot be accomodated" in the tripartite framework. Shortly afterwards the long delayed Chester and Winsford memoir (Earp & Taylor, 1986) declared, p.64, "that most of the drift of the district and all that can be mapped at the surface, can be explained in terms of one Devensian glacial advance and retreat". Thus the potential complexities of sedimentation within a single glacial event were finally given due recognition.

Following an initial attempt to apply a more modern stratigraphic nomenclature (Worsley, 1967), progressively over the last two decades a more rational litho- and biostratigraphic ordering has emerged. Current practice is to group all those glacial deposits

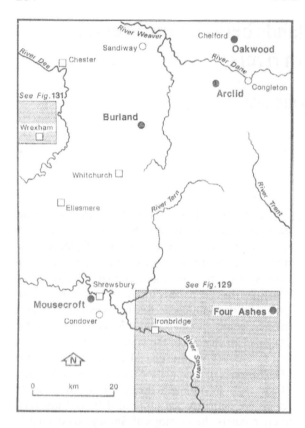

Figure 126: Location map showing the principle sites discussed in the text.

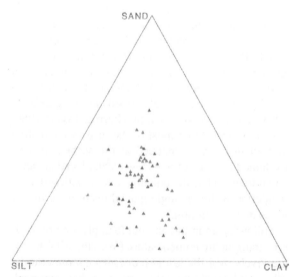

Figure 127: Plot of till matrix grain-size analyses for a range of tills from the area.

with a surface outcrop which were deposited by essentially southward moving ice, into the Stockport Formation. Hence the latter embraces the products of the last glaciation to move into the area from out of the Irish Sea Basin.

The Pleistocene succession

(a) *The Devensian Stage type site – Four Ashes.* Following the pioneer systematisation of the British Quaternary (Mitchell *et al.*, 1973) the type site selected for the last glacial stage was at Four Ashes close to the south east limit of the region under consideration (fig. 126). There an aggregate quarry worked a relatively thin gravel (maximum thickness of 4 m) which was patchily overlain by till. During the late 1960's the gravel contained a number of intraformational organic rich lenses and more rarely lenses occupying shallow channels cut in the Triassic bedrock at the base of the gravel. Studies of the biota associated with the organics, A. Morgan (1973) and Andrew & West (1977) have revealed a range of palaeoenvironments which, even at the time of writing, remain the most diverse last glacial assemblages known in Britain from such a small area. As a result of these studies the provisionally chosen type site at Chelford, in the northeast of the area, was dropped in favour of Four Ashes. Somewhat anomalously, the newly coined stage name – Devensian – was retained despite the fact that it's derivation was supposedly from a Romano British tribe which lived close to Chester (the Roman Deva) on the River Dee. The Four Ashes drainage actually ends in the North Sea!

The Four Ashes succession is illustrated diagrammatically in fig. 128 where its meagre thickness when compared to other important sequences in the area can be appreciated. As noted above, it is the fossil biota which are of paramount importance and it is remarkable that the some 50 or so discrete lenses, each of which appear to have retained their own integrity, survived in such a restricted body of sediment without reworking. An older age limit is imposed by two of the basal channels since these respectively contain a biota indicative of interglacial and interstadial conditions. Comparison with sites elsewhere strongly suggests that these are correlates with the Ipswichian Interglacial and Chelford Interstadials. A number of both finite and infinite radiocarbon dates in the context of the site stratigraphy point to the main gravel body having accumulated largely during the Middle Devensian. It follows that the till which unconformably overlies the Four Ashes Gra-

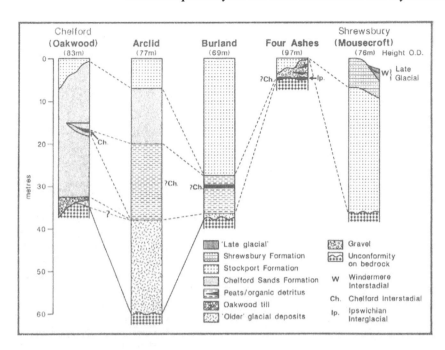

Figure 128: Outline logs to show the lithological successions at five key sites in the lowlands together with tentative correlations between them. Note that although each site has a similar height above sea level the logs are drawn as though they were at the same elevation. The Arclid and Burland logs are derived solely from borehole data.

vels is likely of Late Devensian age and presumably related to the Dimlington Stadial. As is apparent in fig. 128 the till at Four Ashes is regarded as part of the Stockport Formation. Currently the youngest finite radiometric date at Four Ashes is 30,000 B.P. and hence there is less precision in dating the ice advance than is possible on the Humberside coast of eastern England. However at the time of aquisition, the Four Ashes radiocarbon dates encouraged a major reassesment of the dating of the last ice advance into the north west Midlands (Shotton, 1968) and led to the general acceptance that the spread of 'Newer Drift' deposits (i.e. Stockport Formation) to the 'Newer Drift' limit were of Late Devensian age, confirming results obtained in the northern part of the area (Boulton & Worsley, 1965). Ice wedge casts at several horizons indicate recurrent Devensian permafrost.

(b) *The sequence about the southern limit of Devensian Glaciation.* The southern limit of the Stockport Formation corresponds with the maximum extent of the Late Devensian ice. Its position shown in fig. 129 is largely after Wills (1924) and the Geological Survey. The limit crosses the River Severn just south of Bridgnorth. North of the latter town, the Severn occupies a deeply incised valley which leads into the Ironbridge Gorge, a feature normally interpreted as a major drainage diversion associated with the Late Devensian glaciation. Previously the upper Severn discharged into the Irish

Sea. Unfortunately the evidence relating to gorge formation is meagre although it is likely to have been initiated by subglacial meltwater erosion and later adopted by the developing proglacial drainage network as the ice receded to the north.

Recent work has shown the glacial record in the area immediately east of the Ironbridge Gorge to be more complex than previously thought. First, geological mapping and borehole analysis connected with the expansion of Telford 'New Town' has revealed the presence of a totally buried channel system - the Lightmoor Channel - paralleling the modern Severn (Hamblin, 1986). The Lightmoor feature has an uneven floor, humped longitudinal long profile, steep-sided walls and in places a width of 1 km. Second, Hollis and Reed (1981) undertook a mapping investigation of the ground (adjoining Hamblin's study area), paying particular attention to exposures provided by a sand and gravel working in the downstream portion of the Lightmoor Channel. In contrast to Hamlin's view that the sediments south of the former watershed are best interpreted in terms of a unitary phase of subglacial sedimentation, Hollis & Reed invoke a two phase advance sequence with an intervening retreat phase. It may not be without significance that the latter workers (p. 102) state "it is tempting to equate the earlier ice advance with a Middle Devensian, pre Upton Warren glaciation". However, their views of proglacial lake drainage and vertically conformable passages led them to finally

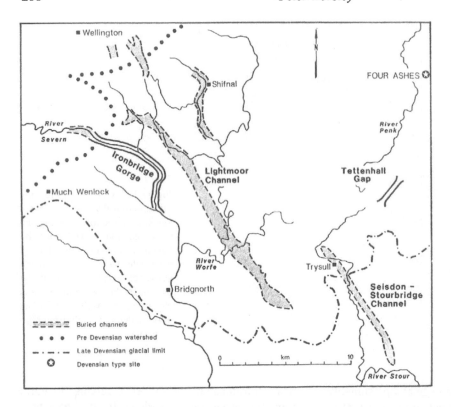

Figure 129: Map of the area about the southern limit of Late Devensian glaciation to show the relationships between the Ironbridge Gorge and the Lightmoor and Seisdon-Stourbridge buried channels. Note how the Ironbridge Gorge cuts across the pre-Devensian watershed and how its upstream portion leads into a buried channel below the level of the modern River Severn. Also the diachronous nature of the Seisdon-Stourbridge channel to the Devensian glacial limit.

favour "only a short interval between the two ice advances".

Neither of these two recent studies discuss the possible implications arising from A.V. Morgan's (1974) work on the Seisdon - Stourbridge Channel only 5 km to the northeast of the Lightmoor Channel (see fig. 129). In the Seisdon feature he identified, beneath an unconformable thin cover of Late Devensian till, a shallow basin fill on a 27 m thick proximal outwash succcession, the Trysull sands and gravels. On the basis of seemingly good biostratigraphy, fossiliferous lake basin sediments where correlated with part of the early Hoxnian Interglacial. Thus Morgan was able to postulate an Anglian Stage age for the thick outwash facies in the main channel fill at Trysull. This conclusion is consistent with the continuation of the channel feature across and beyond the Devensian glacial limit.

Since the Seisdon-Stourbridge channel extends well south of the Late Devensian glacial limit, there is proof, independent of the biostratigraphy, that earlier glacial deposits lie within the Late Devensian limit. From this the possibility arises that the erosion of the Lightmoor Channel may be related to an Anglian event and this dating could also be extended to include the lower part of the channel infill. This scenario more readily accounts for the problem of where the

sediments eroded to create the Lightmoor Channel were deposited, since a major Devensian aged sediment flux beyond the southern termination of the channel is unknown. It may also be pertinent to note that Hollis & Reed were able to recognise non-glacigenic local gravels beneath a veneer of Devensian till which they assumed to be of 'Devensian' age occurred only in bedrock hollows. The latter might well be composite in age as is demonstrably the case at Trysull. Hence there is reason to speculate that the presence of pre-Devensian glacial deposits might be more widespread than the conventional wisdom suggests and that the subglacial surface may have a more complex history of development than hitherto suspected.

(c) *Chelford Sands Formation*. In east Cheshire a distinctive sedimentary body lies beneath the basal Stockport Formation unconformity. This unit, the Chelford Sands Formation, is dominated by exceptionally well-sorted sands but additionally contains minor gravel, silt, and peat lenses plus occasional *in situ* tree stumps and wood clasts. Natural exposures are virtually non-existent and recourse is necessary to exposures in a small number of quarries and borehole data.

Erosion of the sands prior to and during the succeeding glaciation and their incorporation into the overlying Stockport Formation frequently makes the confident identification of the unconformity between the Chelford Sands and Stockport Formations difficult, especially away from the quarry exposures. A further problem is access to much of the borehole data since confidentiality restrictions limit its availability. However, the latter problem does not apply to the Geological Survey and Evans *et al.* (1968) have produced postulated contours on what they consider to be the terminal depositional surface of the formation prior to its dissection and burial by glacigenic materials. These show an east-west declining surface extending from the lower slopes of the sharp bedrock-controlled western limit of the southern Pennines. Overall the surface has an exponential decrease in gradient to the west and an overall average slope of 6 m/km over some 15 km. Owing to the irregular floor beneath the Chelford Sands there is no simple pattern of thinning away from the Pennine footslopes as broad low angle alluvial fans. In the Congleton area the sands are known to have attained almost 70 m in thickness. Westwards towards the central parts of the Cheshire Plain a valley network eroded into the pre-sands surface (bedrock and older glacial deposits) became progressively infilled with the sands over time and in many instances the earlier formed relief was ultimately overwhelmed such that sand deposition extended over the interfluves. An example of this is afforded by the Oakwood, Chelford exposures (see fig. 130).

This reconstruction of the broad palaeoenvironment is consistent with the sedimentology. The sedimentary structures are mainly those associated with ephemeral stream processes and indicate that the sands accumulated in a low relief landscape. A more enigmatic issue concerns the degree to which aeolian generated sedimentary structures are present. Towards the top of the sand sequence very extensive planar laminae with subtle repetitions of grain size frequently occur. These may be generated by adhesion of sand carpets moving over moist sand surfaces in a manner similar to that reported by Good & Bryant (1985). At various horizons throughout the sands sporadic rock clasts invariably show evidence of wind abrasion and this is manifest in a variety of ways from a high degree of polish to multiple faceting to form classical ventifacts. The influence of wind processes is also undisputed where dune-like infills with amplitudes of 4 m can be seen, building out from the margins of fluvially incised channels.

The characteristics noted above are consistent with a depositional environment associated with arid or semi-arid palaeoclimates but the temperature values are indeterminate. This deficiency can be remedied when it is recognised that within the sands cast structures generated by thermally induced contraction of permafrost are present. These structures, mainly ice wedge casts, signify mean annual temperatures of at least -6°C and hence indicate that the Chelford Sands are at least in part sediments which accumulated under cold semi-arid conditions.

(d) *Chelford Interstadial sediments.* Undoubtedly the most well-known aspect of the Chelford Sands are the organic components found in the Chelford

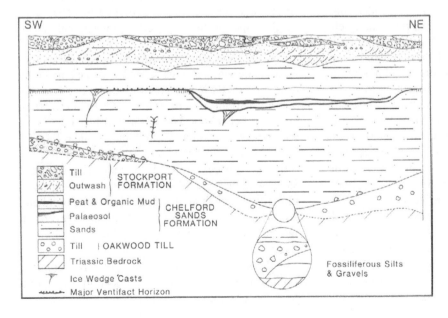

Figure 130: Schematic cross section of the Pleistocene succession at Oakwood, Chelford, showing the buried valley cut into the bedrock and the palaeochannel within the Chelford Sands Formation with its fill of interstadial sediments. It is likely that the main peat bed formerly filled the palaeochannel and has suffered postdepositional compaction. The Oakwood till sometimes appears as though it has been subject to resedimentation.

area. Indeed these include the type succession for the Chelford Interstadial at the old Farm Wood quarry, the biota of which are described by Simpson & West (1958) and Coope (1959, 1977) and the radiocarbon dating by Worsley (1980). Later quarry developments at Oakwood, Chelford enabled a better appreciation of the stratigraphic setting of the interstadial sequence (see fig. 130). There it was clear that the organic-rich interstadial deposits formed part of the infill sequence of a major palaeochannel system cut into the lower part of the sand succession (Worsley, 1985). The conventional wisdom is that the Chelford Interstadial, which is characterised by a *Betula – Picea – Pinus* tree flora, is of early Devensian age although in the Chelford region *per se* there is no good stratigraphic control on its age other than it is likely to relate to the pre-Stockport Formation Upper Pleistocene.

A thick succession of peats and organic detritus within Chelford Sands at Arclid are probably of the same age as those at Chelford (see figs. 126 and 130). Little is known about these organics except that macro fossil material pumped from below the water table appears identical to that at Chelford and one infinite radiocarbon age has been obtained. A recent reinvestigation by Bonny *et al.* (1986) of a sequence of fossiliferous silts and peats at Burland has yielded a virtually identical floristic record to that at Chelford and this strongly suggests that they are coeval. However, the Burland deposits are not within Chelford Sands but rather form part of a sequence of clays and silts which Bonny *et al.* suggest may have accumulated in a lake basin or abandoned river channel (see fig. 128). In view of the local setting, a lake may well have been induced by subsidence associated with solution of saliferous beds in the underlying Triassic bedrock. The occurrence of a potential Chelford Interstadial shallow channel fill overlying bedrock at Four Ashes has previously been noted.

(e) *Older glacial deposits.* As discussed in section (b) above, there are reasons for suspecting that two entirely separate suites of glacial deposits are present in the southern limits of the area. Since they have a common direction of derivation, i.e. from a northerly source via the Irish Sea basin, their lithological characters are similar and when superimposed without intervening non-glacial sediments or weathering phenomena their discrimination is problematic.

Recognition of older glacial deposits is facilitated by the presence of either overlying biogenic-rich sediments or distinctive lithologies of non-glacial genesis. Coincidentally, all the three Cheshire localities with significant Chelford Interstadial deposits – Arclid, Burland and Chelford (Oakwood) – display clear evidence for a glaciation antedating the interstadial. Dependence upon fragmentary core material derived from boreholes over 40 m below the ground surface at the first two sites inevitably means that little is known about the sediment body other than that till is present. Fortunately, at Oakwood Chelford, quarry workings have shown that widespread till, informally known as the Oakwood till, is present lying between the basal sand unconformity and the bedrock surface. At one locality on the quarry floor at the bottom of a palaeovalley cut into the bedrock an excavation revealed a thin sequence of fossiliferous silts, sands and gravels directly above the bedrock – see fig. 130 (Worsley *et al.*, 1983 and this volume). Unfortunately, the biota were non-distinctive apart from representing a harsh open environment and hence of little use in correlation. Despite this, it is certain that prior to quarrying the type Chelford Interstadial succession formerly overlay the site and hence the biota certainly antedate the interstadial. Similarly, the Oakwood till is older than the interstadial.

Besides, those localities which have undoubted biogenic materials above older glacial deposits, a number of boreholes strongly suggest the presence of glacigenic materials beneath what are thought to be Chelford Sands. Bearing in mind the problem of directly superimposed glacigenic units we can be certain that one glaciation, at least as extensive as that of the Late Devensian, has affected the area. Only in the Trysull area can a reasonably certain date for the event be suggested and there the biostratigraphic control rather remarkably indicates a likely Anglian Stage age. Therefore, if it is tentatively assumed that all the disparate 'Older glacial deposits' are of a similar age we are thus encouraged to propose an Anglian age for the pre-Chelford Interstadial glaciation of the area.

(f) *Late Devensian glacial deposits.* As noted previously, the sediments associated with ice emanating from a northerly bearing are grouped into the Stockport Formation. At Four Ashes, close to the southern limit of the Stockport Formation, an older age limit is imposed by the youngest radiocarbon date (30,000 B.P.) obtained from the Four Ashes Gravels. In Cheshire, as has been discussed, the underlying Chelford Sands with their included interstadial biostratigraphy, similarly establish an older age limit, although here, following the consensus view of an Early Devensian age for this interstadial, a looser lower age limit is given.

In the early 1960's, the presence of derived marine molluscan faunas within the Stockport Formation gave the possibility of establishing a radiocarbon-

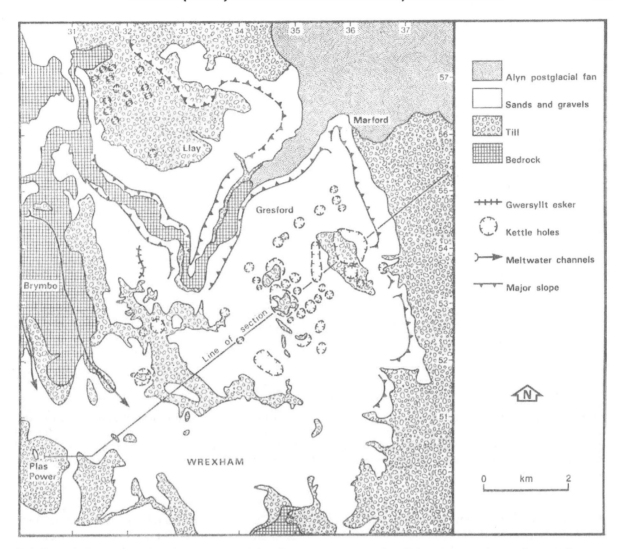

Figure 131: A glacial geological map of the Wrexham district in the Welsh border region where the northern and western ice sheets interacted.

dated control on the age of the formation. The availability of a recently discovered fauna from a sand pit at Sandiway, to the southwest of Northwich, led to the submission of two whole shells of *Nucella lapillus* for dating (Boulton & Worsley, 1965; Thompson & Worsley, 1966). The resultant age estimate of some 28,000 B.P. for the shells gave a fortuitously young lower age limit to the formation and assuming the validity of the result this demonstrated a Late Devensian age for the last glacial event in Cheshire. Subsequent dating of derived shells from similar contexts have given older ages. Recent unpublished amino acid D:L ratio determinations on the same faunal assemblage as that from Sandiway have confirmed the general accuracy of the pioneering radiocarbon assay.

Reference has already been made to the complexity of the glacial stratigraphy in certain parts of the area. A number of workers, including Paul (1983), Shaw (1971, 1972a), and Thomas (1985a) have focussed specifically on the glacial sedimentology. Others including Boulton (1972) have commented upon how observed field relationships have influenced their work on modern glacial environmental analogues, especially that of supraglacial sedimentation. The latest attempt to interpret the landforms and sediments is that of Thomas (1989) who has attempted to synthesise some 700 mineral assessment boreholes in conjunction with field reconnaissance plus map and air photograph interpretation in the area between Wrexham and Shrewsbury. Thomas claims that it is possible to identify spatially

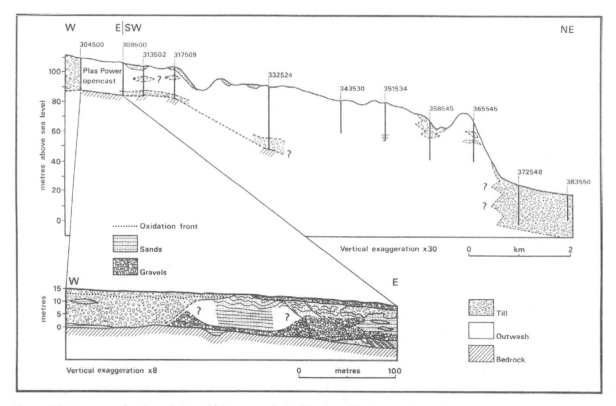

Figure 132: A cross section through the glacial sequence in the Wrexham district largely based upon borehole data. The amplified cross section from the Plas Power open cast coal mine shows an example of the complex stratigraphy generated by the two different ice sheets. The westernmost till body consisted of 'Welsh' materials whereas the eastern equivalent was composed of the tzpical Permo-Triassic till associated with ice from the Irish Sea basin. Hence this site corresponds to the contact between the Shrewsbury and Stockport Formations.

ordered patterns of landform – sediment assemblages based upon reconstructions of sediment body geometry, together with lithological and geomorphological character. He uses these to infer that the retreating ice was characterised by a marginal zone of rapidly changing lacustrine and supraglacial conditions with occasional frontal oscillations. As he freely admits, multiple-till sequences are difficult to correlate solely on the basis of borehole data and it was necessary to use intuition derived from experience elsewhere. An example of an area of complex landform-sediment relationships may be seen in fig. 131 and a corresponding cross-section in fig. 132 (Worsley, 1985).

Although the main direction of Devensian ice flow was from the north via the Irish Sea basin, it has long been appreciated that the western margins of the lowlands experienced glaciation from a western (Welsh) source. Along a belt extending from Shrewsbury northwestwards through Ellesmere into the Wrexham district the two different ice streams came into contact. The width of the zone of interaction is variable but generally lies within the range of 5-15

km. The type succession relating to the two ice flows is located at Mousecroft Lane in the southwest suburbs of Shrewsbury. A schematic representation of the overall stratigraphy is shown in fig. 128. Although it cannot be proven, it is assumed that the bulk of the main sequence above the bedrock is attributable to the Stockport Formation and hence to glacial activity derived from an Irish Sea basin source, a conclusion supported by the presence of comminuted marine shell material. Above a well-marked unconformity lie glaciofluvial sediments of totally different provenance being dominated by 'Welsh' greywackes and allied Lower Palaeozoic clasts. These constitute the lower part of a distinctive twofold sequence of coarsening upwards sandur-type gravels capped by till. The suite of 'Welsh' tills and outwash lithologies will be grouped into the Shrewsbury Formation.

Both formations in the Shrewsbury area have been affected by post-depositional deformation due to the later meltout of buried ice. At Mousecroft Lane, quarrying has bisected a kettle hole and the underlying sediments to expose above the Shrewsbury For-

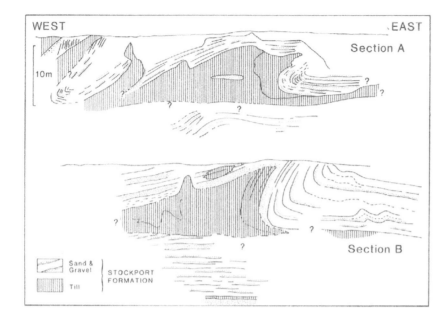

Figure 133: Glaciotectonic structures exposed at the Wood Lane Quarry, Ellesmere in 1964. The upper section A was drawn in May 1964 and the lower section B in September 1964 after the working face had retreated by some 50 m. The fold axes trended to 30° N having been produced by 'Welsh' ice which moved from the northwest and pushed into the Stockport Formation sediments. A 'Welsh' facies was previously visible to the west of the section. Note the flow till unit within the thick fluvial facies close to the base of section B.

mation – a fine Late Glacial sequence consisting of a solifluction mantle overlain by a Windermere Interstadial peat and in turn by a further solifluction layer of presumed Loch Lomond Stadial age. Some 8 km southwest of Mousecroft at Condover, quarrying of a kettle hole fill in 1986 yielded three nearly complete mammoth skeletons although in this instance the site, being beyond the outcrop of the Shrewsbury Formation, was solely within the Stockport Formation.

Although in the Shrewsbury area the Stockport and Shrewsbury Formations are simply superposed, northwestwards their relationships become more complex. It is evident that the two ice sheets were contemporaneous and the location of the medial junction between the two was subject to lateral shift over time. The net result was the generation of a complex zone of sedimentation the details of which can only be glimpsed since little is exposed and borehole interpretation is difficult. A large opencast coal mine at Plas Power, close to Wrexham, gave a rare opportunity to see some of the complexity at outcrop. A section from Plas Power forms part of fig. 132. Similarly aggregate working at Wood Lane to the southeast of Ellesmere has revealed the effects of glacial tectonics in determining the detailed stratigraphy in an area where the two ice sheets were in conflict (see fig. 133).

Conclusion

Much has been learnt concerning the processes of glacial deposition and landform generation in association with the Late Devensian glaciation of the lowlands between the Severn and Mersey rivers. Considerable scope, however, remains for learning more, in particular the details of the glacial sedimentology require further attention. Only a tantalisingly small insight is currently available relating to pre-Devensian glaciation. Where apparently unitary thick glacigenic successions occupy areas where the bedrock has been deeply eroded future detailed investigations may enable the recognition of multiple glaciation. The Chelford Sands remain somewhat enigmatic both with respect to their stratigraphy and ultimate source rocks. More than one boreal forest type of interstadial episode may have occurred and the thick sequence of biogenic-rich sediments at Arclid deserve closer inspection. We can be reasonably certain that the selective outline presented here will require major revision in the years to come.

Note

PRIS Contribution nr. 48.

Glacial deposits of Lincolnshire and adjoining areas

ALLAN STRAW

The region, everywhere below 160 m O.D., is part of the eastern scarplands of England underlain by broadly parallel outcrops of east-dipping Jurassic and Cretaceous rocks which, through the Tertiary, have controlled the formation and recession of a series of cuestas. By the onset of glaciation, the clay lowlands of the lower Trent vale, central Lincolnshire and the Fen basin probably stood at least 40 to 50 m above present sea-level (Straw, 1979a). The pre-glacial river pattern, to be disrupted by glaciation, comprised two master streams, the Humber and the Trent, which flowed eastwards across the various Mesozoic outcrops, and their tributaries.

Although fluvial and glacial processes have been largely responsible for shaping the region, marine action has also played a part. For some period(s) of pre-Devensian time the eastern edge of the Wolds and the northern margin of Norfolk were a cliffed high-energy coastline. By contrast easternmost Lincolnshire and much of the Fen basin were sufficiently low-lying to be inundated during the middle and late Flandrian transgression. Alluvial and estuary areas now render impossible any direct observation of connections between the glacial deposits of Lincolnshire and northwest Norfolk. An important question yet to be resolved is whether the two areas have or do not have comparable glacial successions.

Devensian

The most recent ice advances to reach Norfolk and Lincolnshire occurred in the Devensian Stage, when North Sea ice pressed onto the east side of the Wolds and penetrated a little way through the Humber and Wash gaps (fig. 134). The glacial deposits now pass beneath Flandrian alluvium inland of the modern coast so that in complete contrast to Holderness, cliff sections are largely absent.

In northwest Norfolk, reddish-brown sandy-clay tills underlie a narrow zone below 37 m O.D. along the foot of Chalk slopes that probably represent a degraded marine cliff (Straw, 1960). Poor exposures over the years have revealed variously weathered diamictons with clasts of chalk, flint, quartzite pebbles, and a wide range of erratics derived from coastal and offshore Mesozoic and Palaeozoic outcrops to the north, including Whin Sill dolerite and eastern Scottish igneous and metamorphic rocks. A weak esker survives in Hunstanton Park, and low morainic mounds elsewhere, but on their northern side the drifts, probably never more than 10 m thick, pass beneath Flandrian alluvium. They extend down the east side of the Wash (fig. 134) to Wolferton (Gallois, 1978).

Some 40 km to the northwest, Devensian drifts rise through Fen deposits to form the modest Stickney Moraine, and then extend northward to the Humber. Comprising tills and much glaciofluvial material they pass, with recognisable morainic features, into the lower reaches of Wold valleys and onto spurs to a maximum height of 114 m O.D. near Louth, closely associated with many meltwater channels (Straw, 1961a). East of the Wolds over the Lincolnshire 'Marsh' the drifts, averaging 20 to 30 m in thickness, rest on a plane abraded surface of Chalk and Lower Cretaceous rocks (Straw, 1961b).

Natural sections and quarry exposures are now poor and infrequent. Most descriptions have been of thin or weathered portions of the deposits. Stratigraphic knowledge relies mainly on log records of boreholes and wells, many hundreds of which have been sunk through the drifts to the underlying aquifers. These logs, used also to reconstruct the form of the basal platform (fig. 135), reveal compact tills disposed in several fairly continuous layers generally 3 to 10 m but frequently over 20 m thick. The tills are separated by lenses or persistent layers of silts, sands or gravels ranging from a few centimetres to 10 m thick. Logs rarely record more than four till layers and although it is impossible to differentiate genetic type, over the area around Grimsby and south

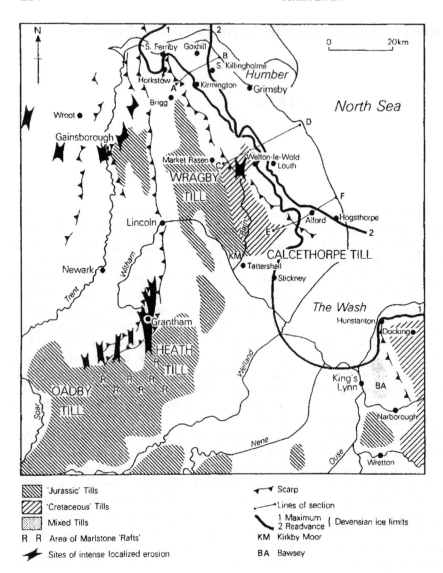

Figure 134: Devensian glacial limits, and distribution of pre-Devensian glacial drifts and sites of intense localised erosion in Lincolnshire and northwest Norfolk.

to Alford, two major layers are common, separated by sands and gravels. For example in South Killingholme parish (fig. 134) the sixteen available logs show an upper till mostly 15 to 20 m thick separated from 2 to 14 m of lower till by sands and gravels, in places silts, some 0.5 to 6 m thick. Because it rests on the Chalk platform, the lower till demonstrably has an irregular upper surface. The overlying deposits, especially the silts, seem to occupy lower parts of this surface. The upper till, from description, extent and

Plates 14–16

14. 2 m of upper flint gravel, view northwest, at Docking Common (TF 790357). Strong involutions involve albic horizon and organic matter of palaeosol. Spade 1 m (Photograph: A. Straw, 1969).

15. Palaeosol, view northeast, at Docking Common (TF 790357). Albic and spodic horizons pass upward into disturbed grey sand and organic layers. Upper flint gravel strongly involuted. 1.5 m of tape extended (Photograph: A. Straw, 1973).

16. Terrace gravels of the River Bain, view west, 3 km north-northeast of Tattershall (TF 229604). Mid-Devensian flint and chalk gravel with sands in planar and lenticular beds, transport to south. Ice-wedge cast 3 m wide near surface and at least 5 m deep (Photograph: A. Straw, 1977).

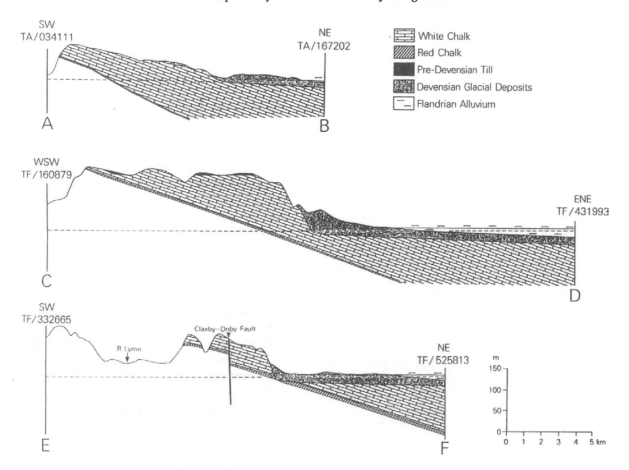

Figure 135: Sections across the Lincolnshire Wolds and 'Marsh'. Devensian drifts lie against and over a former marine cliff and platform.

thickness is at least in part a lodgement till.

By contrast twenty-two available logs for boreholes in Goxhill, Barrow-upon-Humber and Barton parishes (fig. 134) record one layer of till 3 to 13 m thick containing only thin and impersistent sand and gravel partings. A single tough purplish-brown till resting on an abraded Chalk surface is exposed in the Humber cliff at South Ferriby (fig. 136). In the south,

Plates 17–20

17. Calcethorpe Till, view north, at Welton-le-Wold (TF 280885). Flint and chalk clasts (Photograph: A. Straw, 1984).
18. Welton Till, view north, at Welton-le-Wold (TF 281883). Chalk and flint clasts (Photograph: A. Straw, 1959).
19. Chalky- and Jurassic-type tills, view southeast, at Bawsey (TF 680193). Lower till includes chalk clasts and many large pieces of shaley Kimmeridge Clay. Upper till occurs irregularly over the lower till, always with sharp contact, and includes much chalk and flint in chalky-clay matrix. Irregular weathering-front in upper till marks base of Flandrian soil, which is strongly leached in an overlying thin layer of blown sand, beneath quarry waste (Photograph: A. Straw, 1969).
20. Discrete sand body in the stratified Third Cromer Till. Note the original undisturbed bedding in the sand body and distorted laminae in the till beneath which suggest the sand was enplaced as a frozen 'clast' dropped into the till (Photograph: P. Gibbard, 1985).

17	19
18	20

from Alford to Stickney one till layer, normally 6 to 17 m thick, is generally recorded.

Many logs east of the Wolds record up to 10 m of chalk 'gravel' beneath the glacial deposits. Thickest near the Wold edge they owe much to periglacial gelifluction and niveofluvial processes.

The area underlain by the more complex sequences of drifts also has the more uneven terrain. Although relative relief is modest, ridges and mounds parallelled by gravel trains and in part by modern streams exist south of Grimsby and close to the Wold edge northwest and southeast of Louth. A broader ridge zone from Alford to the coast constitutes the Hogsthorpe Moraine (Straw, 1961a). Northwest of Grimsby the drifts thicken and the surface rises to form a north-south drainage divide (Killingholme Moraine) west of which lies a narrow north-graded outwash train; the gravels, like those against the

Figure 136: Cliff section in Devensian till, view south, at South Ferriby on the south side of the Humber estuary (SE 995222). Purplish-brown lodgement till with small chalk clasts and erratics becomes increasingly weathered towards the surface. Chalk bedrock lies 1.5 m below the base of the cliff (Photograph: A. Straw, 1959).

Hogsthorpe Moraine, contain derived marine molluscan shells. Southwest and west respectively of these moraines the drift sequences are simpler and the surfaces lack constructional forms except for the low ridges in the Fens (Stickney Moraine) and in the Ancholme valley (Horkstow Moraine).

Lithological descriptions of the drifts including those by Madgett & Catt (1978) of necessity have been mainly restricted to near-surface deposits. Sands consist mainly of quartz, flint, chalk and coal grains, and gravels normally are composed of erratic rocks typical of the tills and much chalk and flint especially when connected with the meltwater channel systems. Tills are silty or sandy clays with colour, from dark purplish-grey to light reddish-brown, betraying both a considerable content of Jurassic and Triassic material and various effects of weathering. All diamicts contain macro-fragments of chalk and flint (though amounts vary greatly), water-worn pebbles, and erratics of northern England, eastern Scotland and sometimes Scandinavian rocks. Borehole logs confirm the general characters, but with no deep exposures, caution should be observed in ascribing to the total drift complex the results of analysis of near-surface samples.

The maximum extent of Devensian ice in northwest Norfolk and Lincolnshire can be delimited with much confidence (Straw, 1960; 1961a), but controversy persists about the age, number and consequences of ice advances into the area.

Following Catt & Penny (1966), Madgett & Catt (1978) maintained that the Devensian deposits of Holderness (their Skipsea and Withernsea Tills) comprise a single group deposited, on evidence of the Dimlington ^{14}C date of about 18,250 years B.P. (Penny *et al.*, 1969), in the Late Devensian. They claim these tills to extend through east Lincolnshire to north Norfolk.

Straw (1961a, 1979a) prefers a multistadial interpretation of borehole stratigraphy, surface morphology and meltwater channel systems, identifying a second advance to the Hogsthorpe- Killingholme line (fig. 134). Whether multi- or single-phase, a Late Devensian age for all the drifts raises difficult questions (Straw, 1979b), not least concerning the existence of proglacial lakes in the Humberhead and Fen basins.

Gaunt (1976) following Edwards (1937) and Gaunt *et al.* (1972) provided evidence for the former presence of 'Lake Humber' to a maximum height of 33 m O.D. and for a lower shallow stage of the same lake during deposition of the '25-Foot Drift'. Because pre-Devensian quartzite gravels had been

Figure 137: Kirkby Moor Sands, view southeast, surface at 27 m O.D. (TF 228627). Up to 9 m of well-sorted sand in planar beds with occasional thin layers of flint gravel. Transport to south, and interpreted as delta-head material deposited by the River Bain into 'Lake Fenland' (? Early Devensian). Wragby Till underlies the sands just below the level of the quarry floor (Photograph: A. Straw, 1977).

Figure 138: Lower flint gravel, view northeast, at Docking Common (TF 790357). Involuted flint gravels containing lenses and layers of wind-blown silts and fine sands. Palaeosol horizons on left are overlain by upper flint gravel and are destroyed by cryoturbation as they rise towards the surface (Photograph: A. Straw, 1969).

Figure 139: Quarry section, view north, at Welton-le-Wold (TF 285883). 2.5–4 m of Devensian till overlies 3–5 m of Welton Till and 4 m of Welton Gravels. Fabric analysis indicates transport of Devensian till from east and of Welton Till from north. Welton Gravels accumulated under periglacial conditions and have yielded derived mammalian fossils and humanly-worked artefacts (plate 11) (Photograph: A Straw, 1963).

carried through the Lincoln gap by the Trent below 20 m O.D., Straw (1963a, 1979b, 1980a) has argued that the deeper Lake Humber must have been continuous and coeval with a proglacial lake in the Fen basin. The latter could only have been impounded when Devensian ice reached northwest Norfolk, yet no Late Devensian lacustrine deposits have been identified in the Fen basin to compare with the '25-Foot Drift' of the Humberhead area. Extensive sand deposits on Kirkby Moor (figs. 134 and 137) are demonstrably older than nearby Middle Devensian Bain gravels (plate 16), yet appear to be a strandline deposit built by the River Bain into a body of water more likely lacustrine than marine (Straw, 1966). On such evidence the higher Lake Humber, Lake Fenland and the ice advance to northwest Norfolk were Early Devensian phenomena. The later advance to the Hogsthorpe-Killingholme line more likely represents the Late Devensian glaciation, when Lake

Humber was re-impounded to the lower level. The debate continues, and more firmly-dated stratigraphic evidence is needed.

Circumstantial evidence bearing on Devensian cold episodes if not glaciations has been reported from near Docking in north-west Norfolk (Straw, 1980b). Here, in a valley-floor location, a palaeosol with well-defined horizons (plate 15) lies between units of involuted flinty sandy gravels containing seams of aeolian silt (plate 14 and fig. 138). Organic matter from the upper part of the palaeosol yielded ^{14}C dates of 19,300 ± 300 years B.P. (Birm-350) and 24,000 ± 550 years B.P. (Birm-412). The lower gravels confirm severe climatic conditions before a milder Middle (?) Devensian interlude when pedogenesis occurred.

A final point concerns the Devensian limit in northwest Lincolnshire. Palmer (1966) and Gaunt (1976) claimed that Vale of York ice advanced south to the latitude of Doncaster. Gaunt's claim rested entirely on interpretation of a discontinuous series of sands and gravels between East Cowick on the River Aire and Wroot as materials deposited along the edge of an ice lobe standing, or even floating, in the shallow waters of Lake Humber. But the stratigraphical position of these gravels is uncertain, their regular alignment and morphology are more typically subglacial and their constituents are dominantly Carboniferous and Permian. A plausible alternative is that these deposits are pre-Devensian and represent subglacial drainage south beneath stagnant ice. There are no Devensian glacigenic materials, and no unambiguous evidence for an ice advance, south of Selby.

Pre-Devensian

Pre-Devensian tills and associated deposits occur widely over the region west and south of the Devensian glacial limit (fig. 134). Because of later valley erosion they lie generally on interstream areas, and have been removed from steeper slopes perhaps mainly under Devensian periglacial conditions. Only at Welton-le-Wold in east Lincolnshire has Devensian till been observed directly superposed on an older till (fig. 139). The drift surface is smooth and no glacial terrain survives except for the possible subglacial ridge partly buried by Flandrian alluvium north of Wroot. Lodgement tills predominate, but a large sandur (Biscathorpe Gravels) containing lenses of flow till exists in the Bain valley. However it lies beneath tills (Calcethorpe and Belmont Tills) and its

deposits are exposed only as a consequence of deep valley erosion (Straw, 1966). Quartzite pebble gravels (Hilton Terrace of the River Trent) between Newark and Lincoln and bordering the Witham valley below Lincoln are essentially glaciofluvial deglacial materials (Straw, 1963a).

A three-fold division of the drift can be made in Lincolnshire (Straw, 1983). In the Wolds west of Louth a highly-chalky lodgement till (Calcethorpe Till; plate 17) with large quantities of grey tabular flint and some far-travelled dolerite and sandstone erratics occupies interfluve areas. Although normally less than 10 m it reaches 16 m thickness around the upper Bain valley in old valley depressions. West of the Chalk outcrop, the Calcethorpe Till incorporates much Lower Cretaceous material (Spilsby Sandstone, Tealby Limestone) and can be traced 27 km south to the Fen margin (fig. 134).

In central Lincolnshire particularly south of Market Rasen up to 10 m of grey lodgement till (Wragby Till) with many chalk clasts rests on a smooth eroded surface of Ampthill and Kimmeridge Clays. Along its eastern margin it incorporates much Spilsby Sandstone (Kelly & Rawson, 1983: Appendix) and around Tattershall it passes beneath Fen alluvium.

West of the Fen basin over the plateau-like interfluves of Lincoln Heath a broken sheet of lodgement till conceals a number of pre-glacial valleys over which it reaches 30 to 60 m thickness (Wyatt, 1971). This Heath Till is composed mainly of Lias and Oxford Clays with clasts of chalk and flint, Bunter pebbles, Keuper Marl and *Gryphaea* fossils.

Figure 140: Acheulean artefacts ('hand-axes') recovered from the upper division of the Welton Gravels, Welton-le-Wold (Photograph: A. Teed, 1975).

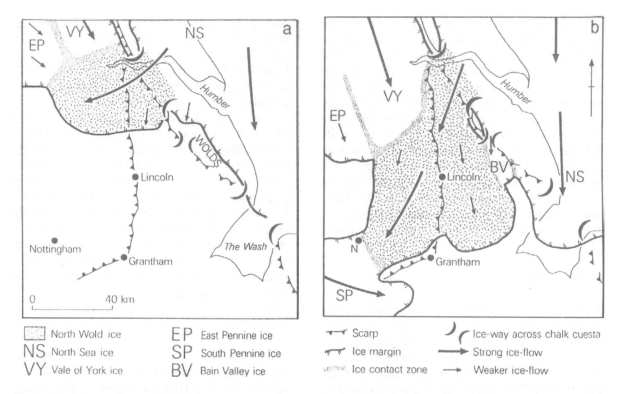

Figure 141: Maps to illustrate the initial westerly flow of a lobe of North Sea ice across the northern Wolds and northwest Lincolnshire, and its subsequent deflection to the southwest by Vale of York ice, during the Wolstonian glaciation. This 'North Wold' ice was forced deep into the Midlands as general southerly flow became established over Lincolnshire.

Notably southward of Grantham it encloses large slabs (rafts) of Middle Lias Marlstone, Northampton Ironstone and Lincolnshire Limestone, often well over 100 m across.

Other tills worthy of note are of restricted extent. West of the middle Bain valley at least 8 m of deformation till (Belmont Till) composed largely of masses of local Lower Cretaceous rocks underlies Calcethorpe Till. East of the upper Bain valley at Welton-le-Wold, at least 10 m of tough brownish-grey silty-clay lodgement till (Welton Till; plate 18) contains many striated clasts of chalk and flint, and erratics of dolerite, Carboniferous sandstones and Scottish igneous and metamorphic rocks. Covered partly by Calcethorpe Till, it overlies a valley-floor deposit of flinty gravels (Welton Gravels) (Alabaster & Straw, 1976).

These different facies of older drift are not the consequence of multiple glaciation but of a single phase of ice flow approximately parallel to the north-south outcrops of Jurassic and Cretaceous rocks. They rest on bedrock and are stratigraphically equivalent (Straw, 1983).

In Norfolk south of the Devensian limit at least two lithologically distinct tills occur. Over interfluves east and south from the Docking area lie a few metres of highly-chalky till, designated the Marly Drift (Straw, 1965) which in this western area contains chalk and flint clasts to large size, some of the latter being marine cobbles, and several types of hard sandstone, metamorphic and igneous rocks. It might be an outlier of this till observed in quarries at Bawsey east of King's Lynn (Evans, 1975; Straw, 1979c). Here masses of crushed chalk with chalk boulders, much flint and rare erratic pebbles rest with sharp contact (plate 19) on and in places incorporate a dark grey clayey lodgement till that is continuous over the Sandringham Sands in the quarry area. Composed largely of Kimmeridge Clay the latter may be related to other Jurassic-rich diamicts of the King's Lynn area (Gallois, 1978) which, curiously, survive inseveral linear depressions that seem to extend the west-draining valleys of the Lower Cretaceous outcrop over the floor of the Fen basin, and connect tenuously with the Bawsey tills and other interfluve patches. Clasts in the low-level diamicts are mainly local Upper Jurassic and Lower Cretaceous materials and although claimed as lodgement tills (Gallois, 1978) with ice movement from the northwest, they may in part be of periglacial origin.

The age of the pre-Devensian deposits has been the subject of much recent debate. Perrin *et al.* (1979) favoured an Anglian age and argued for a divergent pattern of flow by North Sea ice from the Wash gap. Straw (1965, 1979a, 1983) has long argued for ice flow generally from the north-northwest on the west side of the Chalk cuesta, and north-south over the Wolds and north-west Norfolk, ice thereby moving mainly *across* the Wash gap rather than *through* it to the southwest.

Indirect support for a Wolstonian age derives from the superposition of Ipswichian interglacial deposits on Wragby Till at Tattershall (Girling, 1974) and on Heath Till at Wing in Leicestershire (Hall, 1980), and from the rare survival of Acheulean artefacts in the region compared with their profusion in East Anglia and the Thames basin (Wymer & Straw, 1977). Unfortunately the stratigraphic relationships of interglacial deposits at Kirmington in northeast Lincolnshire are unclear because they may well have been displaced after deposition (Carruthers, 1947), they are located within the Devensian end moraine and there are no older drifts in the immediate vicinity. Truncated Hoxnian interglacial sediments lie in the Nar valley of northwest Norfolk (Stevens, 1959; Gallois, 1978; Ventris, 1984) above Jurassic-rich till, but Wolstonian ice might still have passed over the site or at least have been responsible for the introduction of large quantities of marine flint cobbles to the area such as in the Blackborough End gravels.

The best indication of Wolstonian age for the tills comes from Welton-le-Wold where Calcethorpe and Welton Tills overlie the flinty Welton Gravels that accumulated in a former dipslope valley of the Wolds (Straw, 1982). A meagre mammalian fauna and three Acheulean handaxes (fig. 140) recovered from an upper gravel unit indicate late Hoxnian or early Wolstonian aggradation. Loessic layers and cryoturbation within this gravel unit confirm the onset of cold conditions prior to the arrival of Welton / Calcethorpe Till ice (Alabaster & Straw, 1976).

Over seventy-five years ago Harmer (1910) claimed that ice had flowed generally south over Lincolnshire. Straw (1983) supported this in general but drew attention to the presence in the Midlands (Rice, 1968; Shotton, 1953) of Jurassic-type tills with chalk and flint clasts, the latter including tabular flints derived in all probability from Upper Chalk rocks in northeast Lincolnshire. It can be postulated therefore that Wolstonian ice crossed the north Lincolnshire Chalk outcrop and moved west into the lower Trent

valley (fig. 141a) before being deflected and pushed far to the south-southwest by a stronger flow of ice moving generally south from the Vale of York (fig. 141b).

Although most of the older tills may be Wolstonian there can be little doubt that the area was wholly glacierised within the Anglian Stage, because tills in south Norfolk, Suffolk and Essex incorporate Upper Jurassic clays, Oolite limestones, Bunter quartzites, Keuper Marl, and Coal Measure and other Carboniferous rocks confirming the flow of ice southeast over Lincolnshire and the Fen basin. In Lincolnshire some Anglian drifts may survive beneath the Kirmington interglacial deposits, and Wyatt *et al.* (1971) regard material within some of those southwestern valleys buried by Heath Till as possible vestiges of earlier glacial deposits. Their virtual absence is not exceptional considering the wealth of evidence for widespread erosion by Wolstonian ice (Straw, 1979d). Anglian deposits appear to survive more extensively in northwest Norfolk, but if this is the case why are there no representatives here of the Wolstonian tills of central Lincolnshire?

Conclusion

Lincolnshire and northwest Norfolk have been glaciated on probably four occasions:

(i) Within the Anglian Stage ice flowed southeast over the whole region but most deposits have been subsequently destroyed except perhaps in northwest Norfolk.

(ii) During the Wolstonian Stage, Lincolnshire was again glacierised but the ice limit in Norfolk is still undetermined. To the later phases of glaciation belong the different facies of chalk-bearing till and associated sands and gravels.

(iii) Ice entered only east Lincolnshire during the Devensian Stage. At its maximum extent it reached Norfolk and entered the Humber and Wash gaps. Deglaciation, followed by an advance to the more restricted Hogsthorpe-Killingholme line can be argued on stratigraphical and geomorphological evidence. Whether the earlier advance took place in the Early Devensian or whether both advances occurred in the Late Devensian is not yet certain. Protagonists of the latter situation should provide convincing explanations for circumstances surrounding proglacial lakes in the Fen and Humberhead basins, for the disposition of outwash and fluvial terrace materials and for changes in the courses of rivers.

The glacial deposits of northwestern Norfolk

JÜRGEN EHLERS, PHILIP GIBBARD & COLIN A. WHITEMAN

Whilst the cliff sections of northeastern Norfolk had from early times been the focus of glacial geological investigations (e.g. Lyell, 1840; Wood & Harmer, 1868; Woodward, 1887; see also Hart & Boulton, this volume), northwestern Norfolk was largely neglected. Interest was awakened by the discovery of interglacial marine clays of Hoxnian age in the Nar Valley (fig. 142), the so-called 'brickearths of the Nar' (Rose, 1865). However, detailed investigations of these deposits were only eventually undertaken by Stevens (1960), Evans (1976), and more recently by Ventris (1985).

In the last decades of the 19th century parts of East Anglia were mapped by the Geological Survey. In North Norfolk the area around Fakenham, Wells and Holt was surveyed in 1876-83. In the accompanying memoir Geikie summarised the results by the statement that "these (glacial) deposits are mainly a prolongation of the Contorted Drifts of Cromer..." (Woodward, 1884), which is true for the Holt area but does not hold further to the west.

During the Pleistocene, East Anglia was only twice affected by glaciations: during the Anglian and during the Devensian. During the Devensian the ice sheet only reached the coastal areas of northwest Norfolk (Madgett & Catt, 1978; Straw, this volume). The previous opinion that major parts of the area had also been glaciated during the Wolstonian (Baden-Powell, 1948; West & Donner, 1956; Straw, 1965; 1983) or exclusively at that time (Bristow & Cox, 1973; Cox & Nickless, 1972) is now discarded by most workers (West, 1977; Gladfelter, 1975; Perrin, Rose & Davies, 1979; Cox, 1981; Boulton et al., 1984).

During the Anglian the ice advanced as far south as the northern margin of London. Much as in the marginal areas of the Scandinavian ice sheet, the Anglian glaciation of East Anglia was characterised by intensive subglacial meltwater activity. The principal system of large tunnel valleys of East Anglia has been known since the fundamental work of Woodland (1970). In contrast to the glaciated parts of North Germany or Poland which are underlain by several hundred metres of unconsolidated Tertiary deposits, in East Anglia the courses of the major buried valleys appear to have been largely determined by pre-existing valley systems in the surface of the Upper Chalk. However, during recent years, British Geological Survey workers have also been able to identify numerous smaller tunnel valleys in the marginal areas of the glaciation (see Mathers et al., this volume).

The tills

During the Anglian three different types of tills were deposited in East Anglia which can be clearly distinguished: the Cromer Tills, the Lowestoft Till and the 'Marly Drift'.

The Cromer Tills. The Cromer Tills (Reid, 1882) are sandy tills, exposed in the coastal cliff sections of northern Norfolk and northern Suffolk between Lowestoft and Weybourne (plates 20 - 22). Originally, only two Cromer Tills were distinguished. However, Banham (1968) demonstrated that the till of the so-called 'Contorted Drift' rests on the two previously identified members and represents a third Cromer Till. The Cromer Tills have been correlated with the Norwich Brickearth, a similarly sandy till in the area around Norwich. In contrast to the Cromer Tills of the cliffs between Happisburgh and Weybourne, the Norwich Brickearth is largely decalcified, for example in the Scratby cliff section.

The lithological composition of the Cromer Tills differs from that of the other tills of North Norfolk. The Cromer Tills contain comparatively little chalk and flint, a relatively high proportion of quartz and quartzite (from reworked preglacial sands and gravels) as well as about 1% of far-travelled erratics, amongst which some Norwegian indicator rocks occur (rhomb porphyries and larvikites). These tills

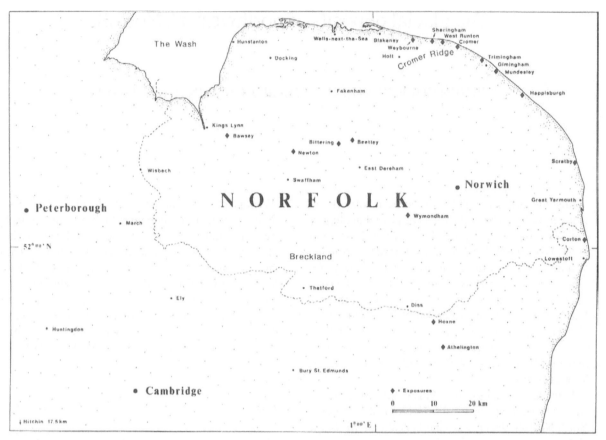

Figure 142: Location map.

are often referred to as the 'North Sea Drift'.

According to unpublished investigations by Gibbard, at least some parts of the Cromer Tills are not basal tills but represent waterlain till, interfingering with other glaciolacustrine deposits. However, the Cromer Tills are certainly not simply another facies of the Lowestoft Till (see below), but deposits of an ice advance from a different direction (NNE).

The Lowestoft Till. Since the 19th century it has been known that the most widespread till in East Anglia is a chalk-rich, clay-rich till. Harmer (1902) interpreted the deposition of this 'Chalky Boulder Clay' to be the result of an ice advance by his 'Great Eastern Glacier' from the northwest. The term 'Lowestoft Till' was introduced by Baden-Powell (1948). The Lowestoft Till occurs as a largely continuous till sheet especially in the more southerly areas. The dark colour results from the reworking of considerable quantities of Jurassic (Kimmeridge) clays, which strongly suggests westerly to northwesterly source areas. Other sedimentary rocks of English origin, especially Chalk, also occur. Norwegian erra-

tics are found occasionally, but they may be reworked from older deposits.

Originally, a more chalky variant of the Lowestoft Till was named 'Gipping Till' by Baden-Powell (1948), a terminology used subsequently by West & Donner (1956). The 'Gipping Till' was found in a number of places overlying the Lowestoft Till. The original concept, however, that this till was deposited during a separate glaciation (the Wolstonian) could not be maintained. In the Gipping Valley, the type area of the 'Gipping Till', it cannot be clearly distinguished from the Lowestoft Till by either structural characteristics or petrographical differences (Allen, 1984).

The 'Marly Drift'. Relatively early it was observed that in northwest Norfolk extremely chalk-rich tills occurred. The term 'Marly Drift' was first used for these deposits by Boswell (1914: 123; 1916: 87). For a long time it was unclear whether this till was a chalk-rich variety of the 'North Sea Drift' (Boswell) or chalk-rich 'Chalky Boulder Clay' (Harmer, 1928; West & Donner, 1956). As a result of their laboratory investigations, Banham, Davies & Perrin (1975: 255) came to the conclusion that there were four different types of Marly Drift:

    (i) reconstituted chalk
    (ii) Lowestoft Till enriched with chalk
    (iii) Cromer Till enriched with chalk
    (iv) a mixture of (ii) and (iii)

The cover of Pleistocene deposits is thin and discontinuous in northwest Norfolk. Where Marly Drift occurs, it directly overlies the bedrock (Chalk) in many cases. More often, however, we have found a thin layer of glaciofluvial sediments between the till and the underlying substratum. This layer, which often has a thickness of just a few centimetres, records the role of meltwater during the ice advance. The presence of this sand and gravel layer also demonstrates that glacial erosion of the underlying Chalk was not an areal process but was restricted to escarpments and protrusions.

The distribution of the Marly Drift has been variously delineated by different authors. Whilst Harmer (1909) and Banham, Davies & Perrin (1975) put the southern boundary on the Docking - Fakenham line, we would prefer to follow Straw's (1965) definition (cf. Straw, this volume) and include the Breckland tills. We consider that the Marly Drift represents a till which is characterised by an extremely high chalk and flint content and the almost complete absence of any other lithological components.

Stratigraphical questions

One of the main problems of the North Norfolk till stratigraphy results from the fact that the different till types rarely occur in superposition. In the distribution area of the Cromer Tills, Marly Drift and Lowestoft Till are absent, in the Marly Drift area there are no Cromer and Lowestoft Tills, and in the Lowestoft Till area the other two tills do not occur. There are, however, a few exceptions:

(i) At Corton, north of Lowestoft, a Cromer Till is overlain by two other tills, each separated by sands. Whilst the uppermost till, the so-called Pleasure Gardens Till, is only of local importance (interpreted as a flow till by Banham, 1971 and Pointon, 1978: 68), the middle till is of typical Lowestoft type. From this succession it is clear that the Lowestoft Till is younger than the Cromer Tills. This section has been designated the type locality for the Anglian Stage (Mitchell *et al.*, 1973).

(ii) One of the few sections in which the relationship of the Marly Drift with another till can be seen is the former sand pit at Bawsey (TF 683194; Evans,

1976; Ventris, 1985; cf. Straw, this volume). Here Marly Drift directly overlies the clay-rich, blackish-grey Lowestoft Till. This situation strongly argues against the supposition that the Marly Drift is just a chalky facies of the Cromer Tills.

The cliffs of North Norfolk are key sections for the stratigraphic relationship between Cromer Tills and Marly Drift. By detailed mapping of the cliff sections between Sheringham and Weybourne (fig. 143) we hoped to find some evidence bearing on the true stratigraphical relationship of the Marly Drift.

The cliffs between Sheringham and Weybourne

The sedimentary sequence of the Pleistocene deposits exposed in the coastal cliffs of North Norfolk is very complicated and so far has not been explained in detail. Just south of the Norfolk/Suffolk border, near Lowestoft, Lowestoft Till is found overlying a glacigenic deposit of the Cromer Till type. Further to the north, however, no evidence of the occurrence of Lowestoft Till is present. The cliff sections at Happisburgh show only three Cromer Tills, each separated by a few metres of meltwater deposits.

Towards the west this relatively simple sedimentary sequence becomes increasingly complicated. West of Mundesley the original bedding is disturbed by a sequence of large basins and diapirs. The layers of till, silt lenses and glaciolacustrine sands involved in these dislocations are hard to correlate with the undisturbed sequences. The strong diapirism is ex-plained by Banham (1975) as a result of plastic deformation of water-saturated, unconsolidated sediments under the overburden of an ice sheet.

The texture of the tills in these cliff sections is so strongly disturbed, that fabric measurements are of little value. Long axis orientation of clasts in the basal parts of the till diapirs between Sheringham and Weybourne confirmed the results published by Banham & Ranson (1965). Locally, however, considerable deviations occur. Fabric measurements at the base of the sand basins reflect the sideways squeezing-out of the till material. Chalk clasts within the remaining till layer several centimetres thick, have been drawn into cigar-shaped forms. The large alterations of the till fabric in these areas caution against over-interpretation of the measurements in other, apparently less disturbed parts of the sections. Diapirism has to be regarded as the ultimate formative process. The preglacial sands underlying the lowest Cromer Till, however, are only affected by slight distortions of the bedding.

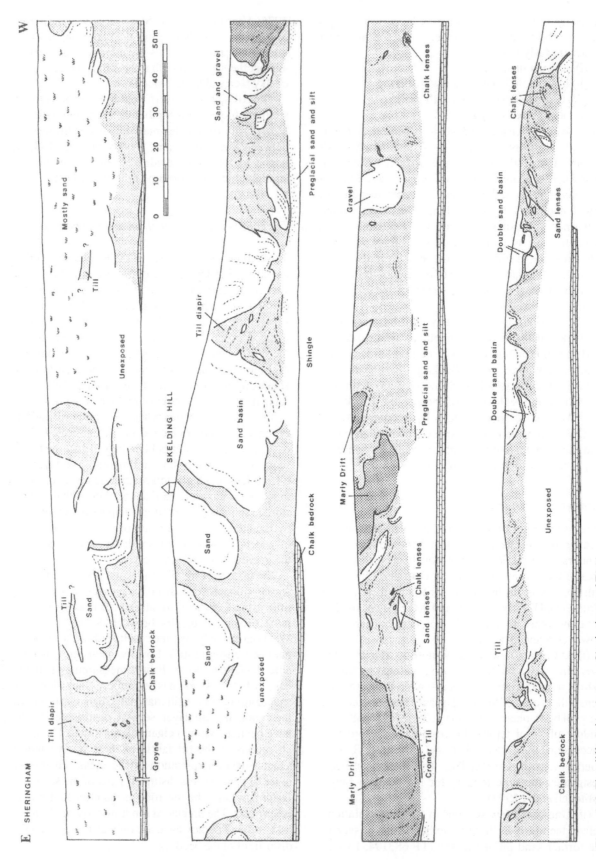

Figure 143: The cliff sections between Sheringham and Weybourne.

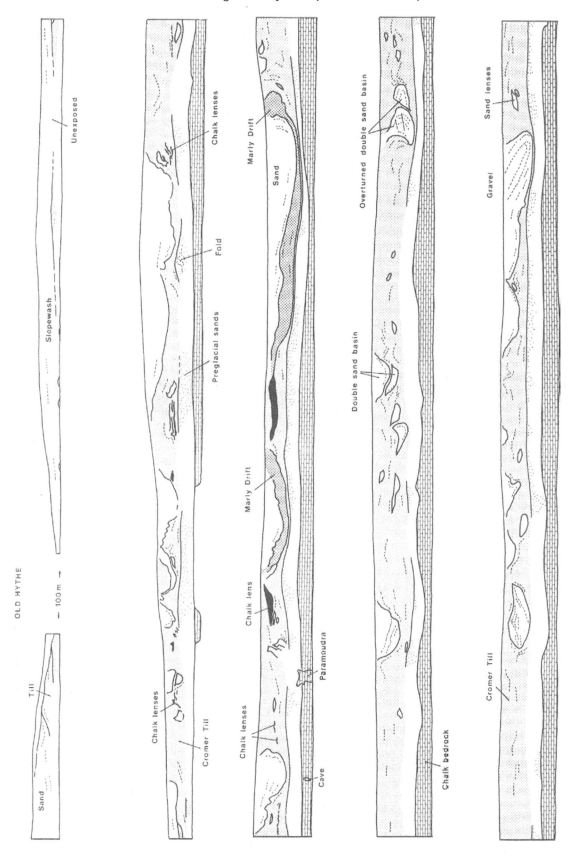

OLD HYTHE

← 100 m →

Sand

Till

Unexposed

Slopewash

Chalk lenses

Fold

Preglacial sands

Marly Drift

Sand

Chalk lenses

Cromer Till

Chalk lens

Marly Drift

Chalk lenses

Paramoudra

Cave

Overturned double sand basin

Double sand basin

Chalk bedrock

Gravel

Sand lenses

Cromer Till

Figure 143 (continued).

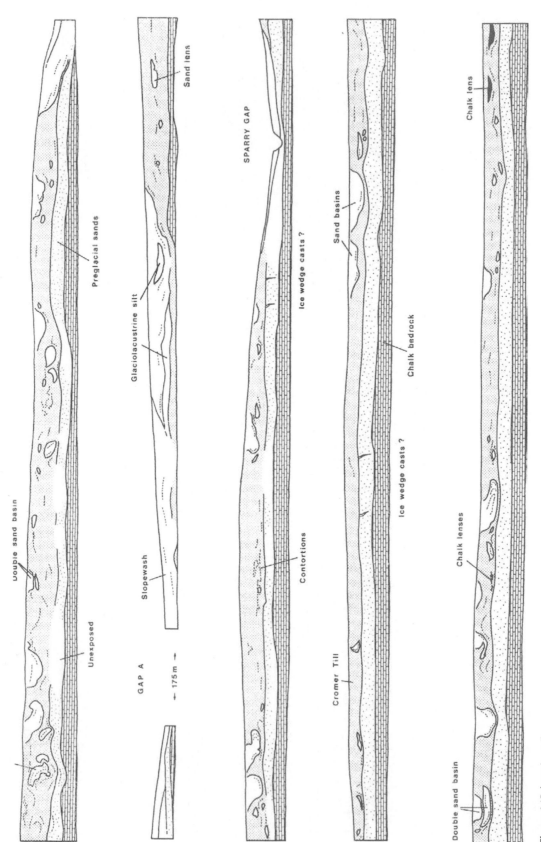

Figure 143 (continued).

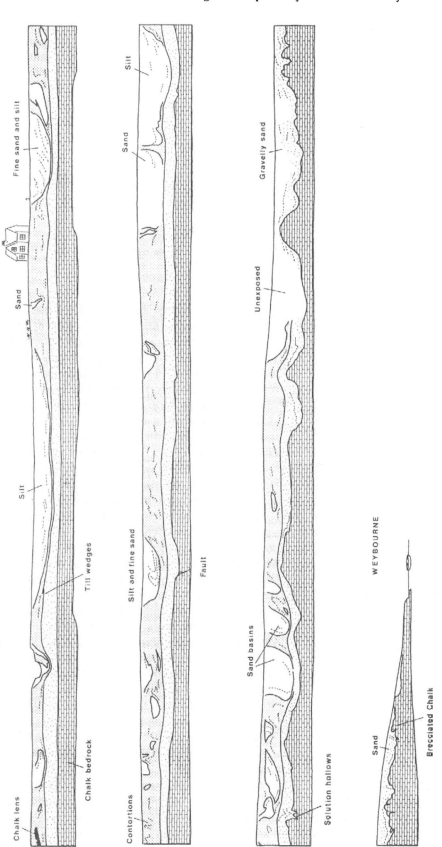

Figure 143 (continued).

Figure 144: The cliff at Weybourne; 4 m of Chalk overlain by 2 m of preglacial sands and a sand-filled basin in the contorted till sequence. The wall on the left is the western fence of the former coast guard station (Photograph: 1986).

Figure 145: Chalk-rich layers within the till sequence west of Sheringham (Photograph: 1986).

Slater was the first to conduct a detailed survey of the cliff exposures between Sheringham and Weybourne. His manuscript maps of 1926, which were never published, were made available to the authors by Professor R.G. West to allow comparison with the

recent situation. The evaluation disclosed that a considerable number of form elements present in the 1926 mapping are still visible in the cliffs today (fig. 144). Because the cliff has retreated since Slater's time by about 50 m, the relative displacement of the forms allows us to ascertain the strike directions of some of the major form elements. The predominant strike direction seems to be NW-SE.

The second survey of the exposures by Dhonau & Dhonau (1964) differs in style from the original mapping by Slater. It is more strongly generalised and less suitable for the re-identification of major form elements than the original survey by Slater.

Reid (1882) had already discovered that the second Cromer Till was considerably more chalk-rich than the over- and underlying tills. From this fact and the occurrence of major Chalk rafts in the cliff sections further to the east, Banham (1970b: 51) and Banham, Perrin & Davies (1975: 253) came to the conclusion that the Cromer Tills, especially the second Cromer Till, were passing laterally into the Marly Drift. In the cliffs between Sheringham and Weybourne the three Cromer Tills cannot be separated. However, it is obvious that till material which contains major quantities of Chalk floes, is to be found approximately in the middle of the sequence (fig. 145). This material, however, is clearly Cromer Till and not Marly Drift.

A detailed survey of the cliffs between Sheringham and Weybourne by the present authors has demonstrated that in some basins remnants of Marly Drift are found to be incorporated in the disturbed sequence. Between Sheringham and Weybourne such remnants are present in four places. Additional localities are situated further to the east between Sheringham and West Runton, east of Cromer (TG 226421) and at Trimingham (TG 296381). Some of these Marly Drift lenses have been previously described as Chalk rafts. However, a closer investigation of the texture clearly demonstrates that they are till. In all cases the Marly Drift was found in the highest position within the till sequence, i.e. above all other till layers of Cromer Till type, but under the sands which have sunk into the basins (Gimingham Sands). There seems to be no connection with the Chalk floes of the Second Cromer Till.

In the cliff section a number of double sand basins occur, consisting of two sand bodies separated from each other only by thin layers of till (fig. 144). We assume that the lower sand lens in these double basins represents sands from between the second and third Cromer Tills, which have been concentrated into the basins by the post-depositional diapirism.

As suggested by Banham (1975), the diapirism therefore seems to have been caused not so much by the pressure of overlying ice, but in the first instance by the loading following deposition of thick meltwater sand sequences associated with the Marly Drift ice advance. The advance formed the Cromer Ridge that reaches over 100 m high just a few kilometres further to the south. The investigations have therefore demonstrated that the Marly Drift does not pass laterally into the Cromer Tills but that it was formed later, i.e. during the Lowestoft advance.

The ice advances of the Anglian

The question remains whether the generally discrete occurrence of the three different tills, the Cromer Tills, the Lowestoft Till and the Marly Drift, is a result of their more or less simultaneous deposition or whether later erosion has removed a portion of the deposits. A third possibility could be that the ice sheets were areally restricted to parts of East Anglia.

That the Cromer Tills may have originally covered a larger area of East Anglia cannot be excluded. Harmer (1928) mentions that in a number of places a sandy till was found below the Chalky Boulder Clay (Lowestoft Till). None of those localities, one of which was near Bury St. Edmunds, was ever investigated in detail. An originally wider distribution of these tills might account for the presence of Norwegian erratics as far south as Hitchin, Cambridge and Peterborough (Rastall & Romanes, 1909; Harmer, 1928). However, there is no unequivocal evidence. Investigations in East Anglia (West & Donner, 1956; Ehlers, Gibbard & Whiteman, 1987) have

demonstrated that in the Marly Drift two different fabric maxima occur, one representing a northwest-southeasterly direction, and a second, older direction representing a southwest-northeasterly ice movement. Since both directions also occur in glaciotectonic features there can be little doubt that they both represent true ice movement directions and that neither is simply a 'B maximum' (transverse fabric) as suggested by Hoare & Connell (1981).

Although the Marly Drift consists almost exclusively of Chalk material in North Norfolk, its petrographical composition changes further to the south, where, especially in the lower parts of the till, major quantities of Jurassic erratics are to be found, for example at Newton (TF 838160), Beetley (TF 987185), Bittering (TF 928176) and Wymondham (TG 153033). This demonstrates that the lower part of the Marly Drift represents a chalk-rich equivalent of the Lowestoft Till.

The younger part of the Marly Drift on the other hand in a glaciodynamic sense represents the 'Gipping Till' of Baden-Powell (1948) and West & Donner (1956). It was deposited by an ice sheet advancing from the northwest to north. Till fabric measurements at Bawsey (Ventris, 1985; 1986) have demonstrated that the local direction of ice movement at this locality was from about 20°. This resulted in the deposition of Marly Drift in an area west of the Chalk escarpment.

Whilst the older Marly Drift-Lowestoft ice advance probably covered all of East Anglia, it seems possible that the younger Marly Drift advance did not enter the eastern parts of the region. The Cromer Ridge may have formed the ice marginal position of this advance; the ice marginal features in the areas around Blakeney and Holt and the ice-thrust strata at Bilsey Hill (TG 023416) (fig. 146) probably also belong to this advance.

No evidence has been discovered that indicated ice melt between deposition of the older and the younger Marly Drift units. The absence of glacial deposits overlying Hoxnian interglacial deposits at numerous sites such as Hoxne (West, 1956), Athelington (Coxon, 1985a), Barford (Phillips, 1976) and in the Nar Valley (Ventris, 1985; 1986) as well as the results of unpublished investigations of the exposures at Beetley (R.G. West, personal communication) indicate that most of East Anglia was not glaciated after the Hoxnian Stage, i.e. during the Wolstonian.

Conclusions

Investigations of a number of exposures in northern

Figure 146: Ice-thrust Marly Drift at Bilsey Hill (Photograph: 1986).

East Anglia have demonstrated that the Marly Drift does not represent an individual stratigraphical unit but that it is simply a chalk-rich facies of the 'normal' Anglian till, i.e. the Lowestoft Till. The investigations have further demonstrated that there is not one Marly Drift of Norfolk, but at least two different chalk-rich till units. The older of these was deposited by ice advancing from the southwest whilst the younger unit was deposited by ice advancing from a northerly direction. Both tills clearly belong to the Anglian; they represent different advance directions of one and the same ice sheet.

The glacial drifts of northeastern Norfolk

JANE K. HART & GEOFFREY S. BOULTON

Norfolk contains some of the most spectacular lowland glacial geology in the British Isles, including good coastal exposures of glacial sediments and glacially constructed landforms. The study of these deposits is important not only to reconstruct the glacial history of the area, but also to increase our understanding of the dynamics of past and present ice sheets.

There is extensive evidence in Norfolk for glacial deposits from two cold periods, the Devensian and the Anglian Stages (see also Straw, this volume and Ehlers *et al.*, this volume). Most of the sediments and landforms were formed during the Anglian. There is also indirect evidence for earlier glaciations. Fig. 147 shows the location of places mentioned in the text.

Early Pleistocene glaciations

There is some evidence for a glaciation during the Baventian Stage from the occurrence of far-travelled, fresh heavy minerals (Solomon, in Funnell & West, 1962) and cherts from Yorkshire (Hey, 1965) in laminated sublittoral marine clays found at Easton Bavents (Funnell & West, 1962; Norton & Beck, 1972) and Covehithe (West *et al.*, 1980). These heavy minerals are thought to have been derived from a 'North Sea ice sheet' (Bowen *et al.*, 1986).

There are also erratic materials and glacially-fractured sand grains found in a series of river gravels in East Anglia associated with the ancient course of the River Thames (Hey, 1965). These are thought to result from glaciation during an interval between the Baventian and the Anglian Stages (West, 1980a; Rose *et al.*, 1976).

SEDIMENT AND LANDFORMS OF THE ANGLIAN STAGE

The lithofacies of Norfolk

Much of the research on Norfolk tills has centred on distinguishing the different lithofacies (Banham, 1968; Straw, 1979d; Perrin *et al.*, 1979). However, it cannot be assumed that lithological similarity implies stratigraphic equivalence, instead lithofacies indicate different sources of supply. It is considered that the study of lithofacies is only one element in the reconstruction of the glacial history of an area and that other evidence such as ice-movement directions, sedimentary and structural relationships are needed to produce a more detailed pattern of landscape change.

The major lithofacies of East Anglia is the 'Chalky Boulder Clay' or Lowestoft Till which is clay-rich and probably mostly derived from the Kimmeridge Clay. The dominant clasts are chalk and flint, with Jurassic limestones and Bunter quartzite as common minor constituents, together with igneous rocks.

The 'North Sea Drift' lithofacies occurs in the northeast corner of Norfolk, its western limit is almost the same as the western limit of the Crag (fig. 148), and its southern limit is along the River Yare. It is a sandy, silty till facies, which is often highly laminated. It usually has fewer clasts compared with the 'Chalky Boulder Clay', which comprise predominantly flint with rare igneous and metamorphic erratics. This deposit is also known as the Cromer Till or the 'Norwich Brickearth'.

A third lithofacies is the 'Marly Drift' (Reid, 1882). This is found in North Norfolk, is a highly laminated diamicton, much like the 'North Sea Drift' facies except that its dominant laminae are composed of crushed and reconstituted chalk. It is found in areas of Chalk bedrock (Ehlers, Gibbard & Whiteman, this volume).

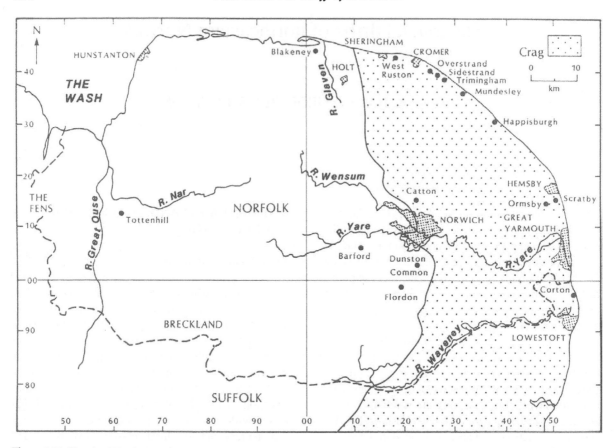

Figure 147: Sketch of Norfolk and limit of the Crag. The border grid of the map corresponds to National Grid 10 km divisions.

The Corton section

There are a few sites where these end members of lithofacies assemblages can be seen in relationship to each other. One site is at Corton (near Lowestoft),

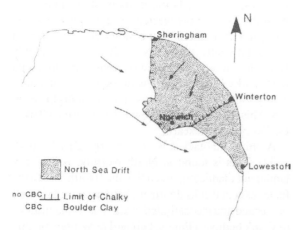

Figure 148: Limit of the 'North Sea Drift' lithofacies, and the 'Chalky Boulder Clay' lithofacies, with postulated ice-flow directions (after Cox & Nickless, 1972).

and for this reason it was designated as the type site of the Anglian Stage (Mitchell *et al.*, 1973). The sediments at Corton are shown in table 16. There are both 'North Sea Drift' facies and 'Chalky Boulder Clay' (Lowestoft Till) in the sequence, separated by sands called the Corton Beds (Pointon, 1978; Bridge, 1988). The Corton Beds are thought to be an outwash sand sequence, possibly a series of distal fans, related to the 'North Sea Drift' glacier (Bridge & Hopson, 1985). The basal 'North Sea Drift' is a laminated deposit with folds and pebble orientations that indicate an ice advance from the northeast (Banham, 1970a). The Lowestoft Till is more homogeneous and therefore lacks folds from which to deduce ice movement direction. A study of the pebble orientation from this till member indicates that it was deposited by an ice advance from the southwest (West & Donner, 1956). It is thought that the Lowestoft Till ice sheet came from the Fenland basin (Perrin *et al.*, 1979) and spread out over East Anglia.

This relationship of 'North Sea Drift' to Lowestoft Till can also be seen near Aylsham and at Scratby (Bristow & Cox, 1973; Hopson & Bridge, 1987) and

Table 16. Type sequence of the Anglian glacial deposits as exposed at the Corton sections (after Boulton *et al.*, 1984).

Age	Name	Deposit
Uncertain	Plateau Gravels	Quartz and quartzite-rich
	Pleasure Gardens Till	Flow till (Banham, 1971)
	Oulton Beds	Varved clays (Banham, 1971)
Anglian	Lowestoft Till	'Chalky Boulder Clay' facies
	Corton Beds	Braided outwash stream deposits from North Sea Drift glacier (Bridge & Hopson, 1985). Shelly fauna, periglacial flora (West & Wilson, 1968), ice wedge casts
	'North Sea Drift'	'North Sea Drift' facies

along the River Waveney (Bridge & Hopson, 1985).

Therefore in southeast Norfolk there is a relatively simple glacial geological record of two ice sheets, an early ice sheet from the northeast and a later from the northwest. However, as investigations continue more complex sequences are found in this area, such as the repeated outwash sediments in the tunnel valleys of south Norfolk (Cox, 1985a). It has been suggested that the two-till relationship is the result of two pene-contemporaneous ice lobes (Bristow & Cox, 1973;

Boulton *et al.*, 1984). However, in northeast Norfolk the pattern is not so simple.

Northeast Norfolk

The sediments of northeast Norfolk contain a number of distinct tills of 'North Sea Drift' lithofacies. There are also a number of glacially-derived landforms, which include the Cromer Ridge end moraine which is thought to represent a long-term glacier still-stand

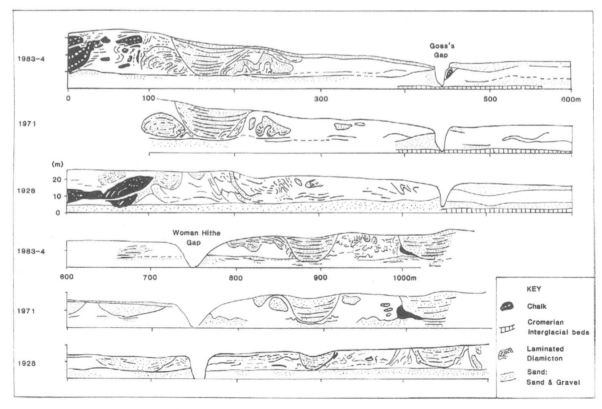

Figure 149: The drift exposure at West Runton. From this the three dimensions of the structures can be seen (1928 – Slater, 1971 – Boulton, 1983–1984 – Hart & Boulton).

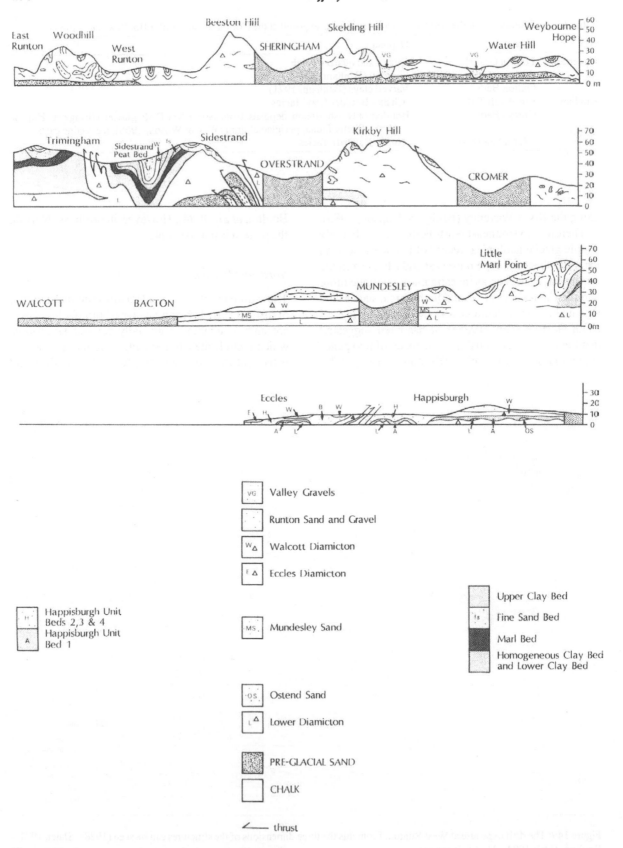

Figure 150: The northeast Norfolk schematic coastal section.

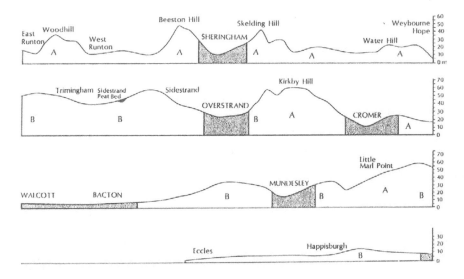

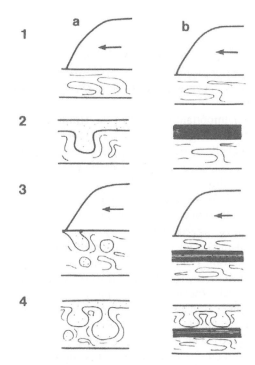

Figure 151: The location of the different tectonic facies.

position. Associated with this feature are kame-like ridges and mounds and an esker south of Blakeney (Sparks & West, 1964; Boulton *et al.*, 1984). There are also other hummocky deposits of tills and sands in north Norfolk, in the Norwich area and along the Waveney valley.

The tills in northeast Norfolk are dramatically deformed and for this reason have been called the 'Contorted Drift'. This is best displayed between Sheringham and Cromer, especially at West Runton (fig. 149). We propose that this sediment was formed by subglacial glaciotectonic deformation as the ice sheet moved over the unconsolidated deposits of the North Sea bed (Boulton & Hart, unpublished; Hart, 1987).

Deformation of subglacial sediments can occur if the strength of these materials is reduced. One way in which this can occur is if meltwater drainage is restricted within the subglacial environment. This can lead to increased porewater pressures and thus a reduction in shear strength. In such a case the sediments would easily deform beneath the glacier. Any initial perturbation in the sediment would induce the development of a fold, and from high shear strain over time these become attenuated and form tectonic laminations. Thus pre-glacial sediments become mixed and incorporated with glacial sediments which are released from the base of the ice sheet.

The features of the resultant diamicton are very similar to those of a purely dynamic metamorphic rock, and the processes occurring beneath the ice sheet are similar to those in a hard-rock shear zone. Thus this deformation can not be used as a stratigraphic indicator.

Figure 152: Formation of the different tectonic facies. The diagram shows two theoretical sites 'a' and 'b'. 1) Subglacial deformation occurs at both sites, producing a laminated diamicton. 2) At site 'a' outwash sand is deposited which sinks into the diamicton, at site 'b' lake clay is deposited. 3) Subglacial deformation occurs. At site 'a' the outwash sand from 'a' is incorporated in the new deformation, and the deformation affects the whole sequence. At site 'b' the lake clay acts as a *décollement* and only the upper part of the sequence is deformed. 4) At both sites an outwash sand is deposited which sinks into the diamicton. Thus at site 'a' there is evidence of high-deformation tectonic facies 'A' and at site 'b' there is evidence of low-deformation tectonic facies 'B'.

Table 17. A new stratigraphy for northeast Norfolk.

Stratigraphic name	Local stratigraphic name			
	Happisburg and Eccles	Mundesley	Trimingham	West Runton
Sidestrand Peat Unit				
Walcott Diamicton Member	Upper sands Walcott Diamicton	Upper sands Upper diamicton	Outwash sands D Upper diamicton Outwash sands C	Runton Sand and Gravel (F4) Laminated diamicton (F3)
*Trimingham Member	Happisburgh Unit	Mundesley Sands	Trimingham Unit	Runton Sand and Gravel (F2)
*Eccles Diamicton Member	Bush Estate Sands Eccles Diamicton/ 'laminated diamicton'			
Happisburgh Diamicton Member	Ostend Sands Happisburgh diamic- ton	Lower diamicton	Outwash sands B? Lower diamicton Outwash diamicton sands A	Laminated diamicton (F1)

*Similar age – Trimingham Substage

It is suggested that the 'Marly Drift' was also formed by subglacial glaciotectonic mixing, but is more chalky because it has passed over a large chalk source.

Another important element of the laminated diamicton are the large blocks of sand and chalk. Some blocks of chalk are over 100 m in length, they can be thought of as large erratics. The blocks were plucked from the bedrock in marginal zones where the subglacial bed is frozen to shallow depth (Boulton, 1972) and then incorporated into the diamicton. Several workers have suggested that the substrate must have been weakened prior to the ice-sheet advance. Banham & Ranson (1965) suggested superficial chalk was weakened by periglacial frost/thaw activity, whereas Peake (personal communication) has suggested rising salt diapirs in the North Sea weakened and fractured the blocks.

Superimposed on the subglacial deformation (F1) at West Runton, there is a second striking deformational style (F2). This consists of large sand synclines which can be demonstrated to be basinal in three dimensions. The sand consists of shelly outwash deposits (Runton Sand and Gravel). These are not depositional basins, but tectonic forms as the basin sides are far too steep to be formed by sedimentary processes. It is suggested that the sand was deposited as the ice sheet retreated in a continuous or near continuous sheet over a dead ice topography. The sand then sank into the underlying still-saturated diamicton to form basins.

It has been suggested (e.g. Banham, 1968) that the

'contortions' in the drift decrease from north to south and cease between Cromer and Overstrand. According to our observations that this is not the case (figs. 150 and 151), and instead there are two distinct

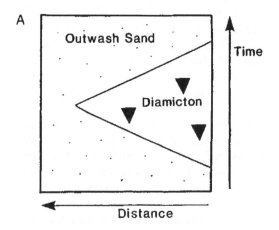

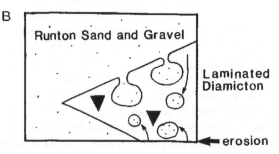

Figure 153: The stratigraphic relationship of the Runton Sand and Gravel to the Laminated Diamicton Unit at West Runton.

tectonic facies. One composed of a highly-deformed tectonic facies 'A', where the whole cliff is deformed, and the other composed of a less-deformed zone, tectonic facies 'A' where there are distinct layers in the cliff. This can be seen in fig. 152. Tectonic facies 'B' occurs where clay-rich or silty sediments predominate in the cliff, these acted as a *décollement* layer, preventing the whole sedimentary succession from being deformed.

Following a reinvestigation of the northeast Norfolk glacial deposits a new stratigraphy has been proposed (Hart, 1987). This is shown in table 17. The deposits of northeast Norfolk are considered to be members of a North Sea Drift Formation. In areas of tectonic facies 'A' distinct deformational episodes can be recognised (F1, F2 etc.) which can be tentatively related to the less-deformed parts of the sequence.

This stratigraphy differs from that proposed by previous authors (Reid, 1882; Solomon, 1932a; Banham, 1968) because of today's clearer understanding of glaciotectonic processes and glacial environments. There is no single type site for the area, because the lateral variability of glacial environments is by definition too great for any 'layer cake' model to be of any value. The new model proposes that there were two main glacial advances over north Norfolk which have been called *1st North Advance* and *2nd North Advance* which deposited the Happisburgh Diamicton Member and the Walcott Diamicton Member respectively (Hart, 1987). These advances were separated by a relatively ice-free period lasting at least 2,000 years during which the sediments of the Trimingham Member were deposited (the *Trimingham Substage*). On table 17 it can be seen how the re-

gional stratigraphy of northeast Norfolk is related to the local stratigraphies of individual sites. The double appearance of the Runton Sand and Gravel at West Runton demonstrates the problem of fitting stratigraphies to glaciated areas. The stratigraphic relationship of the Runton Sand and Gravel to the Laminated Diamicton Unit is shown in fig. 153. Runton Sand and Gravels that predate the first ice advance have been completely eroded. They are no longer *in situ*, but are found as tectonic inclusions within the Laminated Diamicton. Runton Sands and Gravels which postdate the first ice advance both overlie and are locally incorporated into the diamicton.

The lowermost group of deposits comprise the Happisburgh Diamicton Member. This consists of a diamicton (produced by subglacial deformation) and outwash sands and gravels. These sediments were deposited by a glacier advancing from the northwest. At Happisburgh the surface of the diamicton is undulating. Lake deposits (Happisburgh Unit) fill these hollows indicating that the undulations represent an original topography, which has been effected by later diapirism. The antiforms consist of linear ridges parallel to the ice front. It is proposed that these are push moraines.

Overlying this there are deposits that relate to the Trimingham Substage which are grouped together as the Trimingham Member (table 17). During this time the area north of Happisburgh was ice-free, but the Eccles area was covered by an ice sheet from the south (*South Advance*) which deposited the Eccles Diamicton and produced proglacial deformation structures south of Happisburgh. At Trimingham a large proglacial lake was formed (*Lake Trimingham*) which had a lateral extent of at least 5 km. The sediments of this lake (Trimingham Unit) recorded the retreat of the first ice sheet and the advance of the second ice sheet. These consist of:

a) varved clays, which indicate the presence of an ice sheet close by,

b) homogeneous clays, which indicate that the lake has become more distal,

c) lake marls,

d) aeolian sands, which were deposited into a shallow pool when the ice front was furthest from the lake,

e) more varved clays, indicating the return of glacial conditions.

At Happisburgh there are also sediments representing a proglacial environment (Happisburgh Unit). These consist of lake clays which were firstly deposited in small lakes between moraines on the surface of the diamicton and which later were depos-

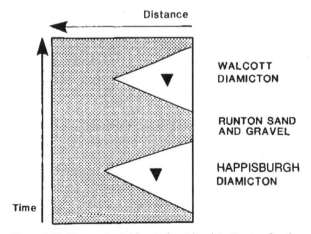

Figure 154: The stratigraphic relationship of the Runton Sand and Gravel, in space and time, to the diamicton beds.

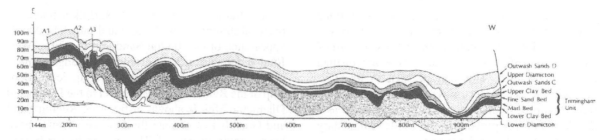

Figure 155: Reconstruction of the section at Trimingham. Listric thrusts in the east (A1, A2, A3), open folding in the east.

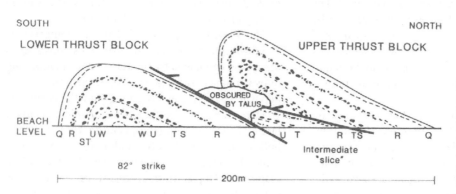

Figure 156: The structure of the chalk rafts at Sidestrand, looking along core of thrust block (reproduced from a photo). This is diagrammatic only – not to scale. Letters refer to actual flint bands (after Peake & Hancock, 1961).

ited in one large lake. The sediments in the lake grade upwards initially into deltaic, and subsequently into channel deposits.

The Trimingham Unit becomes more sandy towards the south, while the Happisburgh Unit becomes more sandy towards the north. At Mundesley, between these two sites, the sediments comprise silty sand. It contains deltaic elements similar to the silty sand units of the Happisburgh and Trimingham units. It also displays sedimentary characteristics of a proglacial alluvial outwash sand. It is suggested therefore, that these Mundesley Sands are the lateral equivalent of the Happisburgh and Trimingham units and represent the distal facies of outwash during the retreat of the 1st N Advance during the Trimingham Substage.

During this time period it may be that the outwash sands at West Runton (Runton Sands and Gravel) were deposited and the associated F2 deformation occurred. These sands represent relatively proximal outwash deposits which relate to both advance and retreat phases of both ice sheets (fig. 154) and therefore cannot be used as a stratigraphic marker horizon except in places where the stratigraphy is undisturbed (tectonic facies 'B').

Overlying the deposits of the Trimingham Member is the Walcott Diamicton Member. This also consists of a diamicton produced by subglacial deformation, with associated outwash sands. These result from an ice sheet that initially came from the northeast but later changed direction and flowed to the southeast.

There is evidence that this ice sheet readvanced during its general retreat. It readvanced as far south as Trimingham. In the west the ice sheet remained stationary and thick outwash sands accumulated, which produced the Cromer Ridge end moraine (at Telegraph Hill), with a steep northern flank representing an ice contact slope and a gentle southern slope representing the sandur deposit (Sparks & West, 1964). In the east where the ridge is less prominent, the ice sheet did not remain stationary, and there was proglacial deformation, consisting of large listric faults and open folding including chevron folding of the graded sediments (Hart, 1987) (figs. 155 and 157). This was probably also responsible for the overturning and thrusting of the large chalk blocks at Sidestrand (fig. 156).

After the final deformation at Trimingham, small lakes or ponds formed in hollows on the surface of the glacial deposits left as the ice retreated. These lake infills contain a pollen assemblage of late glacial and pre-temperate aspect. Thus, at Trimingham there is a nearly continuous sequence of sediments spanning

Figure 157: Photograph of the chevron folds at Trimingham (K. Martinez, 1985).

deglaciation and the start of an interglacial. The pollen evidence indicates that these sediments should be equated in part with the Hoxnian Stage (Hart & Peglar, unpublished; Hart, 1987).

This demonstrates that the 'North Sea Drift' ice was present in north Norfolk during the late Anglian and thus the Lowestoft Till and the 'North Sea Drift' are penecontemporaneous. It also shows that the Cromer Ridge is of Anglian age.

Associated with the Cromer Ridge is the Blakeney esker. Because glaciofluvial landforms have a relatively low chance of survival beneath ice, the eskers that remain must have been formed near the margin of the last ice sheet in the area. Recent work suggests that the palaeoflow within the esker was from northwest to southeast (Gale & Hoare, 1986). It has been suggested that the esker is incised into the landscape. It also has a meandering form which indicates an englacial or supraglacial origin; and flexure cracks infilled with gravel which suggest that the esker was lowered onto the landscape. Gale & Hoare (1986) suggest that the esker was deposited under hydraulic flow in a subglacial or englacial environment. Thus, the most likely suggestion appears to be that of Boulton *et al.* (1984) who suggested that the esker was of englacial origin.

The relationship of the northeast Norfolk deposits to the rest of East Anglia

It has been shown that there were a number of ice sheets covering East Anglia during the Anglian, these included the Lowestoft Till ice sheet from the Fenland basin, and two 'North Sea Drift' ice sheets from the North Sea. However, the origin of the S Advance at Eccles is not yet known, as the diamicton consists of coarse- grained proximal material. It may be part of the 'North Sea Drift' at Corton, or it may be part of the Lowestoft Till. We suggest tentatively that the best correlation is the former, as this correlates with the deposits from Scratby (Banham, 1970a; Hopson & Bridge, 1987), and from this have suggested a provisional North Sea Drift Formation. It is proposed that the term 'North Sea Drift' at Corton be renamed the Corton Till, so that the Name 'North Sea Drift' can be used for the lithofacies and not a particular diamicton.

From the evidence discussed above we propose the following glacial history for the Anglian glaciation in northern East Anglia, shown in fig. 158.

a) An ice sheet (probably but not necessarily of Early Anglian age) advanced from the northeast (Scandinavia) bringing erratics into the North Sea Basin but did not advance onto the present land (fig. 158a).

b) An Anglian ice sheet advanced from the northwest, and reached into northeast Norfolk (1st N Advance). It probably reached Scratby, but not Corton (fig. 158b). The ice sheet retreated and readvanced forming small push moraines at Happisburgh.

c) An ice sheet advanced from the northeast, depositing the 'North Sea Drift' at Corton and the

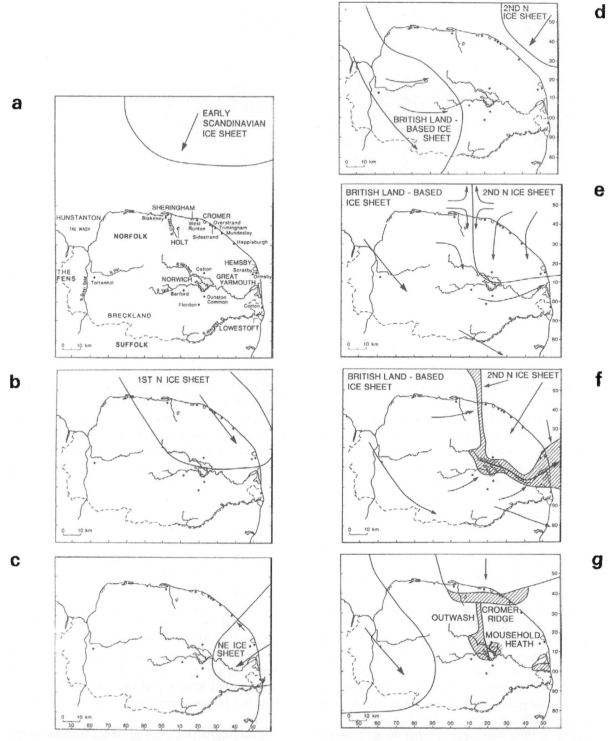

Figure 158: The glacial history of northern East Anglia. a) A possible early Scandinavian ice advance bringing Scandinavian erratics into the area. b) The 1st North Advance, from the NW depositing the Lower Diamicton Group. c) The South Advance, from the NE, depositing the 'North Sea Drift' at Corton and the Eccles Diamicton Member in NE Norfolk. d) Lowestoft Till ice sheet spreading from the Fenland basin. At the same time the 2nd North advance from the NE advances into NE Norfolk depositing the Walcott Diamicton Member. e) The two ice lobes meet, and the flow is changed, the 2nd N Advance in NE Norfolk flows towards the SE. f) The two ice lobes retreat, and outwash deposits are laid along the margins of the ice lobes. g) Further ice retreat, and the Cromer Ridge is formed during a small readvance.

Eccles Diamicton group (S Advance) (fig. 158c). During the retreat of this ice the thick sand and gravel masses were formed which occur in the Waveney Valley and which prograde in an easterly direction. These form kame-like hills along the valley margins and underlie and interdigitate with the 'Chalky Boulder Clay' facies, and overlie (further east) the 'North Sea Drift'.

d) The Lowestoft Till ice sheet advanced from the northwest, which reached into Essex, depositing the Lowestoft Till. At one time it may have flowed outwards from the Fenland basin (fig. 158d).

e) An ice sheet advanced initially from the northeast (2nd North Advance) (fig. 158d). This occurred the same time as the Lowestoft Till ice sheet, and flow lines were diverted towards the southeast (fig. 158e). The 'North Sea Drift' ice sheet would have had a lower slope profile than the land-based Lowestoft Till ice sheet as it was flowing over the soft sediments of the North Sea. Thus in places there may be flow from the Lowestoft Till ice sheet into the 'North Sea Drift' ice sheet, depositing eastern erratics. At this time flow lines within the ice sheet may have come from the southwest, to produce the deformation found by Banham & Ranson (1965) at Weybourne and by Ehlers *et al.* (1987) in northwest Norfolk.

This 2nd North Advance ice sheet spread south of Happisburgh and probably reached Norwich. However, it did not advance very far to the south along the east coast since nowhere is North Sea Drift seen overlying Lowestoft Till.

Shortly after this ice sheet reached its maximum extent, an outwash plain was deposited in a narrow strip running southwards from the north Norfolk coast to Norwich. This was deposited between the 2nd North Advance lobe and the Lowestoft Till ice lobe, described by Cox & Nickless (1972) (fig. 158f). North of Norwich at Mousehold Heath, there is a large mass of outwash sand and gravel, which may provide evidence for a stationary ice front during the ice sheet maximum. The course of the present Rivers Yare and Wensum may have originated as meltwater channels flowing between the two ice lobes.

As the 2nd N Advance ice sheet retreated it left a chaotic hummocky topography over north Norfolk. The Cromer Ridge represents a stationary front of this 2nd North Advance ice sheet as it retreated northwards (fig. 158g).

This schematic interpretation suggests that there were at least four ice lobes which advanced over East Anglia. The ice lobes during the stages shown in fig. 158d and 158e existed at the same time. Three ice lobes (and possibly four - fig. 158a) advanced over the North Sea bed, depositing the 'North Sea Drift' and the 'laminated diamictons' of the north Norfolk coast (including the 'Marly Drift'), whereas the fourth flowed across the Fenland basin depositing a chalk-rich till. It appears that the seabed-based ice lobes reached the area first. This was because ice sheets move faster over the soft sediments (Boulton & Jones, 1979; Alley *et al.*, 1987) such as the bed of the North Sea, possibly in a similar way to the fast Baltic Ice stream that occurred during the Devensian (Boulton *et al.*, 1985). Later, the land-based ice reached the area.

Thus it has been shown that glaciation during the Anglian Stage was much more complex than the proposal shown in table 16. In contrast to the simple sequence of two ice advances separated by an interstadial, which might be deduced from a 'layer cake' interpretation of the Anglian type section at Corton, there are in fact a number of ice sheets of both North Sea and land origin that advanced over East Anglia penecontemporaneously depositing a complex array of interdigitating sediments across the area.

The Devensian glaciation of Norfolk

In contrast to the rich range of Anglian sediments and landforms there is much less evidence of Devensian deposits. At Hunstanton there are a series of ice-contact features that are thought to be of Devensian age, associated with the Hunstanton Till (Boulton, 1977a). They comprise a large esker and a series of kames. The Hunstanton Till is a brown loamy deposit. It has a narrow coastal outcrop (Suggate & West, 1959; Straw, 1960) which extends west to Morston (Gale & Hoare, 1986) and this links with Devensian till found by the British Geological Survey offshore (Balson & Cameron, 1985).

The Devensian deposits were deposited by a lobe of ice that flowed southwards over the soft bed of the North Sea, and extended into the Fenland basin, in a similar manner to, although of course less extensive than, the Anglian Lowestoft Till ice sheet (see also Catt, Straw, this volume).

Acknowledgements

The authors would like to thank Tim Atkinson, Frank Cox and Mary Thornton for their help and encouragement. We would also like to thank Sylvia Peglar for the analysis of pollen samples, and Kirk Martinez for the photography.

The glacial sequence of the southern North Sea

PETER S. BALSON & DENNIS H. JEFFERY

The British Geological Survey (BGS) began the first systematic survey of the geology of the UK Continental Shelf in 1967. Part of this survey has been directed towards the production of 1:250,000 scale geological maps of Quaternary formations (see Balson & Cameron, 1985). In the central part of the southern North Sea, mapping has been in partnership with the Geological Survey of the Netherlands (RGD) which has been jointly responsible for the stratigraphic nomenclature.

This paper summarises our present knowledge of the glacial and interglacial formations of the UK sector of the North Sea south of 55°N. Some of the BGS and BGS/RGD Quaternary map sheets for this area are already published (British Geological Survey, 1986; British Geological Survey et al., 1984; 1986), some are in press and others are presently at various stages of interpretation and compilation. All currently available data has been utilised for this review and to produce the summary maps (fig. 159 and plate 23).

Within the UK sector of the southern North Sea area, the British Geological Survey has collected many thousands of sea bed samples using a Shipek grab, and short cores using gravity corers or vibrocorers. These corers typically penetrate between 1 and 3 m below the sea bed in glacigene diamictons, although penetration of 6 m can occasionally be obtained with the vibrocorer in softer lithologies. The spacing of the sites is variable, but on average they are about 5 km apart. BGS has also obtained thousands of kilometres of shallow seismic records using a variety of acoustic sources. 500J surface-towed boomer and 1kJ sparker have proved the most useful equipment for the examination of the Quaternary succession over most of the area. Spacing of the seismic traverse lines is also variable depending on water depth and sea bed obstructions, but typically lines are 10 km or more apart. An example of part of this seismic network can be seen in Balson & Cameron (1985, fig. 2). A small number of shallow boreholes has also been drilled by the British Geological Survey in order to investigate deeper Quaternary seismo-stratigraphic units. Details of ten BGS boreholes which penetrate glacial or interglacial sediments are given in this review (fig. 163).

Glacial and Interglacial evolution of the Southern North Sea

The Quaternary sediments in the UK sector of the southern North Sea show a dramatic thickening from west to east towards the centre of the basin (fig. 159). The thickest sequence is found within an elongate NNW-SSE trending trough reflecting underlying major structural trends (Caston, 1977). Within this sequence a major hiatus, recognisable in seismic records, separates a 600 m thick sequence of deltaic sediments of Early to early Middle Pleistocene age from a much thinner but more complex succession of glacial, periglacial, marine and brackish-marine deposits of Anglian to Flandrian (Holocene) age.

Sedimentation in the North Sea Basin has thus been dominated by glacial erosional and depositional processes since mid-Quaternary times. Three regional glaciations are thought to have affected Northwest Europe during the Anglian (Elsterian), Saalian (= Wolstonian?), and Devensian (Weichselian) stages. Preserved glacial deposits clearly indicate that contemporary ice sheets covered most of mainland Britain during the Anglian and Devensian glaciations. Ice sheets may (Straw, 1983; Shotton, 1983a) or may not (Sumbler, 1983a) have been equally extensive during the Wolstonian Stage.

Three major episodes of glaciation have been identified within the late Middle and Upper Pleistocene sediments of the southern North Sea. The deposits associated with each episode are broadly similar. They are separated by interglacial marine sediments at a number of critical localities, and by this means have been correlated with the respective glaciations

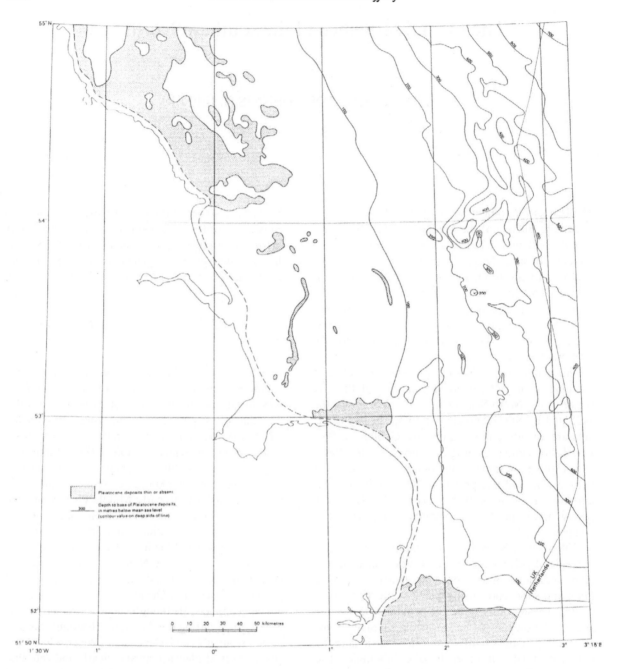

Figure 159: Map showing depth to base Pleistocene in the UK sector of the southern North Sea.

Plates 21–22

21. Stratified till with chalk lenses and flow structures near the base of the Third Cromer Till, West Runton, Norfolk (Photograph: P. Gibbard, 1978).
22. Folding and step faulting of laminated till near the base of the Third Cromer Till, West Runton, Norfolk (Photograph: P. Gibbard, 1978).

PLATE 23

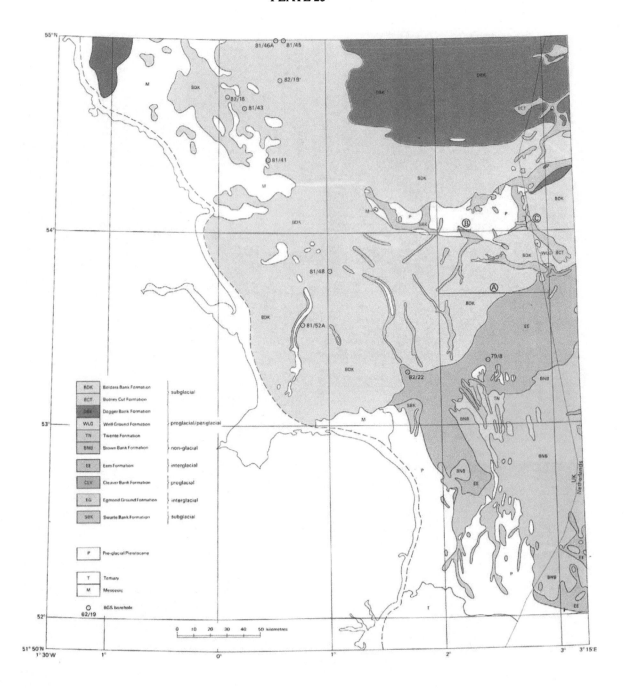

of neighbouring countries (e.g. British Geological Survey *et al.*, 1986).

A characteristic feature of these glaciations offshore was the development of systems of subglacially eroded valleys. All of the valleys have markedly uneven longitudinal profiles. Many appear to be closed at both ends, and may bifurcate or merge laterally with similar valleys, such that each glaciation may be represented by a complex of valleys of common local morphology (fig. 160).

The Anglian (Elsterian) and British Devensian (Weichselian) ice sheets extended much further south in the UK sector of the North Sea than the Saalian glaciation, which appears to have been mainly restricted to areas to the north and east of that considered here (Cameron *et al.*, 1987; Long *et al.*, 1988).

Anglian (Elsterian) Stage

The establishment of full glacial conditions in the southern North Sea during the Anglian (Elsterian) Stage led to the erosion and filling of a major system of subglacial valleys. Deeply incised into Pleistocene and older sediments, they form conspicious features on shallow seismic profiles (fig. 161). The dimensions of the valleys vary greatly; they are up to 12 km wide and locally more than 400 m deep. These valleys have not been identified south of 52°50'N in the British sector, and are most extensively developed between 53° and 54°N and east of 2°E (see fig. 160). Where the valleys are sufficiently well defined to be traced, they have a marked NNW-SSE trend in the centre of the southern North Sea (Balson & Cameron, 1985). They range from long anastomosing valley systems to short isolated oval channels. The valleys are of similar dimensions and morphology to the Elsterian subglacial valley system of continental Europe, described by Ehlers *et al.* (1984). As in continental Europe, it is assumed that they were eroded beneath the ice under very high hydrostatic pressure. All of these Anglian valleys have been completely infilled with sediments, the Swarte Bank Formation. The evidence from seismic profiles is that, in general, much of this fill significantly postdates the cutting of the valleys. Typically the Swarte Bank Formation consists of three units (fig. 162), although not all three are always present in every valley, particularly the smaller ones. The lowest unit, 1, usually has an irregular top, does not necessarily occupy a central position and has 'chaotic' seismic texture with numerous point reflectors. These features indicate a poorly bedded, coarse-grained unit which may be interpreted as glaciofluvial sands or resedimented diamictons derived from the base of the ice sheet and thus penecontemporaneous with glaciation.

Unit 2 is characterised by well defined, parallel to subparallel reflectors, draped over any underlying irregularities. Quiet water sedimentation is therefore inferred, probably under glaciolacustrine conditions. A lacustrine interpretation is supported by site investigation boreholes in the Netherlands sector (Cameron *et al.*, 1987).

Unit 3 typically varies from markedly layered sediments (deposited in relatively still waters) to sediments characterised by inclined reflectors and cut-and-fill structures implying a higher energy environment. The base of unit 3 usually cuts down into unit 2. On the basis of the seismic evidence, it is therefore thought that unit 3 is most probably of shallow marine muds (flats) and sands (channels). Such sediments have been sampled elsewhere in the North Sea (Cameron *et al.*, 1987). Two boreholes (81/45 and 81/46A) at the extreme northern edge of the area here considered and a third borehole to the north of Norfolk (82/22) (see plate 23 for locations) have sampled lower parts of the Swarte Bank Formation at the edges of palaeovalleys. For summary logs of these boreholes, see fig. 163.

By analogy with the valley deposits of northern Germany, the Swarte Bank Formation could have been deposited over a long time span, over several cold and warm periods (cf. Ehlers *et al.*, 1984). The Swarte Bank Formation represents the principal occurrences of Anglian sediment offshore; the sheets of till normally associated with glacial activity on a land surface are conspicuously lacking. This absence is most likely due to post-Anglian erosion, a concept supported by the commonly observed upward truncation of the reflectors in seismic records.

Plate 23

23. Map showing outcrop pattern of glacial and interglacial formations in the southern North Sea. See text for suggested ages of formations. Note: Outcrop annotated 'BDK' at 1°W includes Late Devensian glaciomarine deposits.

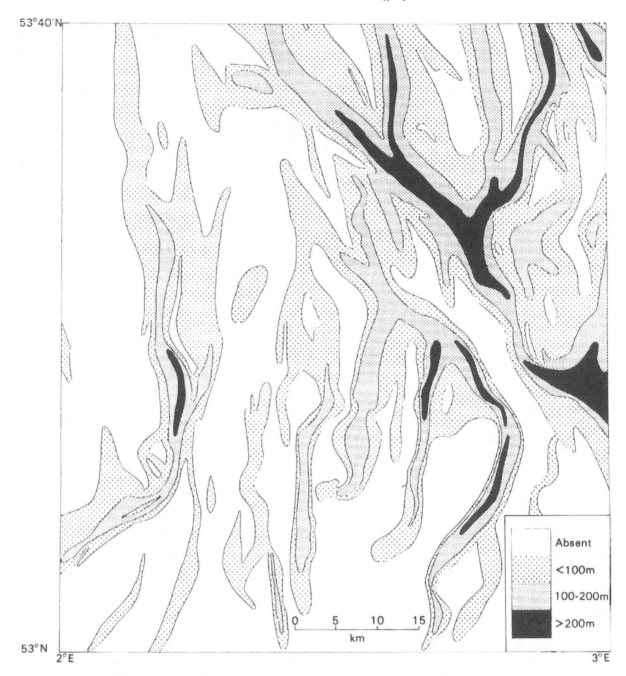

Figure 160: Isopachyte map of the Swarte Bank Formation showing anastomosing valley system.

Hoxnian (Holsteinian) Stage

In the Netherlands' sector of the North Sea, the uppermost component of the Anglian (Elsterian) valley fill commonly comprises sparsely shelly Holsteinian (Hoxnian) marine sands (Egmond Ground Formation) overlying Elsterian glaciolacustrine muds (Laban *et al.*, 1984). Hoxnian sediments are probably

absent in most of the British sector, but marine muds sampled in shallow cores from the margin of an open sea bed depression, the Silver Pit, have been assigned by pollen analysis to the Hoxnian Stage (Fisher *et al.*, 1969). In BGS borehole 81/52A sited nearby (fig. 163), Hoxnian marine beds pass gradationally downwards through glaciomarine muds with dropstones into a thin Anglian (Elsterian) till (Balson & Came-

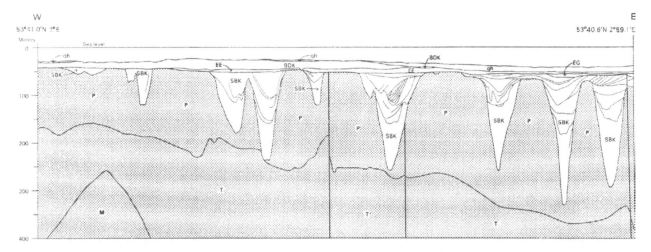

Figure 161: Interpretation of seismic profile along line A showing deeply incised Anglian valleys. Vertical exaggeration: × 50; location and formation abbreviations as plate 23. qh = Holocene. Note also thick sequence of pre-glacial deltaic sediments (P).

ron, 1985), which is similar in appearance and geotechnical properties to the Anglian tills of East Anglia. The marine mud yielded abundant dinoflagellate cysts, most of the species indicating deposition in a shallow-water environment. It also yielded rich and diverse assemblages of Foraminifera of interglacial aspect, similar to those recorded from shallow cores nearby which have been correlated with the Hoxnian on the basis of pollen content (Fisher *et al.*, 1969). None of the shallow cores had sampled Anglian sediments. The succession in the BGS borehole has provided the most complete record yet to be recovered from the offshore area, of sedimentation during the early stages of the Hoxnian climatic amelioration.

Saalian (= Wolstonian?) Stage

During the next major climatic deterioration, the Saalian Stage, extensive ice sheets covered most of NW Europe (Ehlers *et al.*, 1984). The Wolstonian deposits of the West Midlands have been considered by various authors (e.g. West, 1977b: 333) to belong to this period although there is no unequivocal evidence for this assumption. Opposing views have been expressed by Straw (1983) and Sumbler (1983a) as to the extent of this ice sheet in eastern England, but Wolstonian glacial deposits have not yet been unequivocally identified in the UK sector of the southern North Sea. Proglacial silty clays (Cleaver Bank Formation), glaciofluvial fine-grained sands and well-sorted wind-blown sands assigned to the

Saalian (Laban *et al.*, 1984), have been sampled in the Netherlands' sector, and are known to extend into parts of the UK sector. From this, it would seem that periglacial conditions, which allowed the existence of extensive bodies of water, prevailed over most of the British southern North Sea area at that time. There is no evidence of Saalian tills or subglacial valley systems within the area covered in this review.

Ipswichian (Eemian) Stage

Marine sediments of the Ipswichian interglacial stage (the Eem Formation) are widespread in the southern North Sea to the east of 2°E. The sediments are up to 30 m thick, comprise beach, intertidal and shallow marine fine- or medium-grained sands and clays in the west, but may become more fully marine towards the east (Cameron *et al.*, 1987).

Devensian (Weichselian) Stage

During the early stages of climatic deterioration at the onset of the Devensian (Weichselian) Stage, as regional sea level in the southern North Sea fell to about 40 m below its present level (Jelgersma, 1979), brackish-marine silty clays of the Brown Bank Formation were deposited over most of the area between East Anglia and the Netherlands. These silty clays are extensively bioturbated and locally cryoturbated, with thin interbeds of shelly gravelly sand towards their base. In the UK sector the Brown Bank Forma-

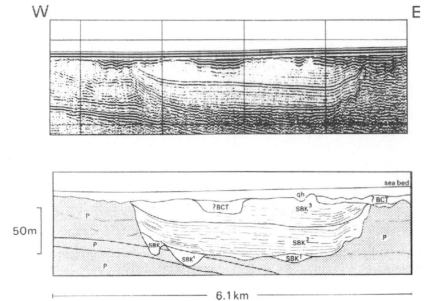

Figure 162: 1kJ sparker profile and interpretation across an Anglian sub-glacial valley along line B (for location and formation abbreviations, see plate 23; qh = Holocene).

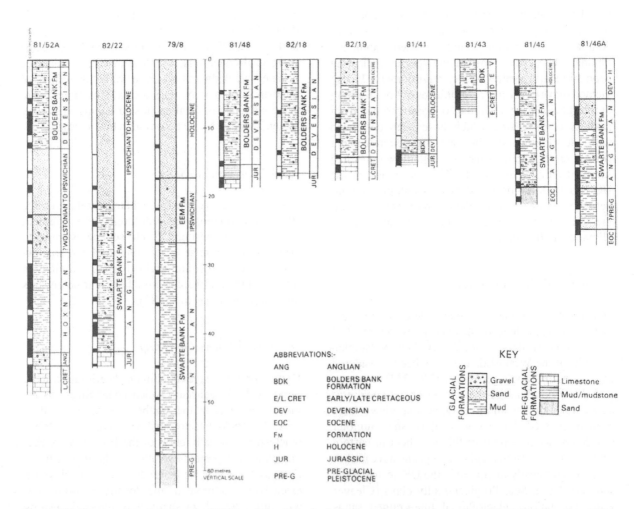

Figure 163: Summary borehole logs for BGS boreholes through glacigene sediments in the southern North Sea. See plate 23 for locations.

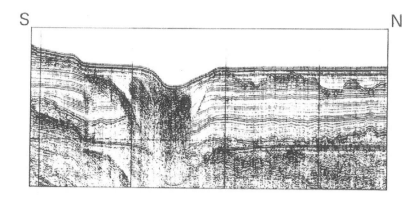

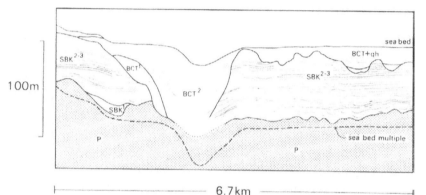

Figure 164: 1kJ sparker profile and interpretation across Devensian subglacial valley along line C (for location and formation abbreviations see plate 23; qh = Holocene). Anglian subglacial valley (filled with SBK) is trending subparallel to line of section.

tion consists of fluviatile current-bedded silts and finely laminated clays which fill late Ipswichian or early Devensian channels up to 20 m deep. The formation has an erosional base. It is mostly between 5 and 10 m thick, but locally in the west may reach more than 25 m. In the east, the brackish-marine beds pass gradationally upwards into lagoonal or lacustrine laminated clays.

By late Devensian times, glacial or periglacial conditions prevailed over most of the offshore area. Devensian clay-rich diamicton, interpreted as lodgement till (Bolders Bank Formation), is at or close to the sea bed over extensive areas to the north of 53°N, resting on Eemian (Ipswichian) marine beds in the east, and passing eastwards into glaciofluvial sands and gravels (British Geological Survey *et al.*, 1986). The pebble component of the Bolders Bank Formation indicates a British source area to the west, so it is likely that this formation is the distal equivalent of the deposits outcropping at Holderness and Hunstanton in eastern England. Within the limits of the till margin was eroded a system of subglacial valleys, radial to the ice front. Although directly analogous both in geometry and in architecture to the palaeovalleys of the Anglian (Elsterian), the Devensian valleys are

generally shallower, narrower and less abundant, and the whole system is less extensive. This probably reflects a thinner ice sheet than during the Anglian Stage. Where these Devensian valleys cut through or along the Anglian valleys, they can be easily distinguished from them (fig. 164). Many of the valleys have been filled by penecontemporaneous glaciofluvial and glaciolacustrine sediments (Botney Cut Formation) and Flandrian marine sediments (Laban *et al.*, 1984), but some, such as the Silver Pit and the Botney Cut itself, have been only partly filled, leaving open depressions up to 70 m deeper than the surrounding sea floor (Donovan, 1973; Eisma *et al.*, 1979). To the east and north east, the Bolders Bank Formation passes laterally into the Dogger Bank Formation, a complex of proglacial diamictons, sands and clays containing a dinoflagellate flora indicating shallow open marine waters (British Geological Survey *et al.*, in press). Typical cores consist of dark grey-brown massive silty clays with dropstones, remnant sand ripples, and minor inter-beds of sand. Late Devensian periglacial wind-blown sands (Twente Formation) have been sampled both south of the till margin and on the surface of the till. Although the Twente Formation is mostly less than 1 m thick it

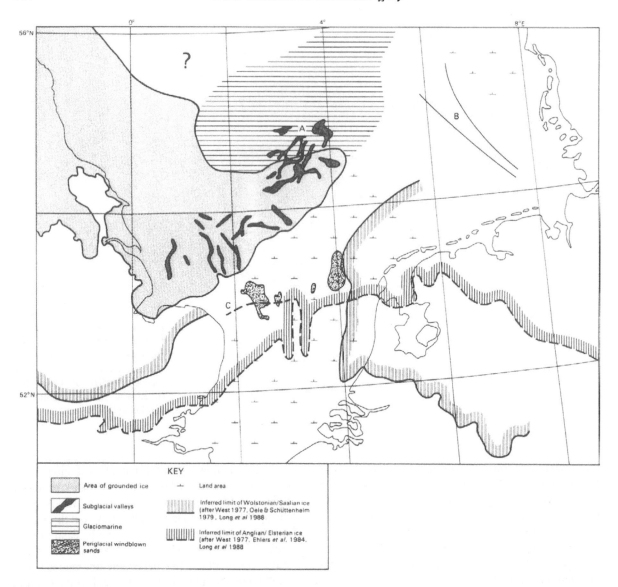

Figure 165: Palaeogeographic reconstruction of the southern North Sea area during the maximum extent of the Devensian glaciation based on known distribution of Devensian sediments. A = Area of fluctuating ice limit or floating ice; B = Helgoland channel (after Eisma *et al.*, 1979); C = Southern limit of Anglian subglacial valleys offshore (after Balson & Cameron, 1985).

may have formed a more extensive deposit prior to the Flandrian transgression.

Flandrian (Holocene) Stage

Peat deposits may have been extensive on the floor o parts of the UK sector of the southern North Sea during the early Flandrian (Balson & Cameron, 1985), but if so, they have been largely removed by erosion, for they have been sampled in only a few

vibrocores north and east of East Anglia and on the Dogger Bank. Most of the present southern North Sea was flooded by the early Flandrian rise in sea level, between 10,000 and 7,000 years BP (Eisma *et al.*, 1981) resulting in deposition of brackish-marine muddy sands and clays during the early stages of the transgression. The modern sea bed sediments are fine- or medium- grained quartzose sands which are gravelly where they overlie Devensian (Weichselian) till and muddy in deeper waters around the Dogger Bank.

These modern sea bed sediments have inherited relict features from the Devensian ice advance. Gravel spreads off the eastern England coast have been interpreted by Robinson (1968) as the remnants of outwash fans formed during temporary halts in the migration of the Devensian ice sheet. The gravels contain pebbles of lithologies found in northern England and southern Scotland, with no evidence of a Scandinavian input. Robinson also thought that the areas of extensive modern sand banks could be related to glaciofluvial sand input. Similarly, Veenstra (1965) believed that gravels on the Dogger Bank were related to the Devensian ice front. The lithological composition here also indicates a British rather than Scandinavian provenance.

Discussion

The presently known distribution of offshore glacial deposits raises some interesting problems. The Devensian glacial deposits are well preserved and have been widely sampled. From fig. 165 the relationship at ice maximum between subglacial, proglacial and periglacial facies is clear, and the palaeogeographic reconstruction is relatively easy. In contrast, no till sheets of the Anglian glaciation have been interpreted from seismic profiles or sampled, yet the Anglian glaciation was more severe than the Devensian if the extent and depth of the palaeovalleys reflect ice thickness. The most likely explanation is that post-Anglian erosion has removed this characteristic glaciation product, leaving glaciolacustrine and glaciofluvial deposits only, to be preserved at lower levels in the palaeovalleys.

What therefore was the situation during the Saalian (Wolstonian?) glaciation? Whereas in the Netherlands the Saalian ice extended over a larger area than did the Elsterian ice (numerous authors, e.g. Oele & Schüttenhelm 1979), in the UK sector of the southern North Sea there is no unambiguous evidence that Saalian ice existed. Indeed, in this southern area only two major groups of glacial deposits can be recognised offshore, an interpretation which accords with the views of Sumbler (1983a) and Cox (1981) for example, in recognising only two glaciations (the Devensian and the Anglian, with the type deposits of Wolston considered as Anglian), but not with the views of other authors such as West (1977b) or Catt (1981).

Acknowledgements

We would like to acknowledge the contribution made to the study of the southern North Sea Quaternary deposits by many of our colleagues in the Marine Geology Research Programme of BGS and Marine Geology division of RGD, and in particular by T.D.J. Cameron (BGS), R. Schüttenhelm and C. Laban (RGD). We would like to thank C.D.R. Evans for critically reading the manuscript. Published with the approval of the Director of the British Geological Survey (NERC).

The tills of southern East Anglia

PETER ALLEN, D. ALLAN CHESHIRE & COLIN A. WHITEMAN

Study of the East Anglian tills and their associated deposits has a long history (table 18). Excluding the 'North Sea Drift' which is confined to northeast East Anglia, schemes of the glacial stratigraphy have progressed from a model involving a single advance and an associated till unit (Trimmer, 1851; Wood, 1870) to one with two units of what would now be recognised (Mitchell *et al.*, 1973) as Anglian and Wolstonian age (Boswell, 1931; Solomon, 1932a; Baden-Powell, 1948; West & Donner, 1956) or, in Essex and Hertfordshire, pre-Anglian and Anglian age (Sherlock & Noble, 1912; Wooldridge, 1938). Subsequently Clayton (1957, 1960) recognised three tills associated with either two or three glacial stages.

With the benefit of recent scientific advances, views have reverted to a monoglacial model, though it is recognised that more than one till may be involved. Most workers consider the tills to be Anglian (Baker, 1971; Turner, 1973; Gibbard, 1977; Perrin *et al.*, 1979), lying in superposition on Cromerian Stage deposits at Corton (West, 1980a), and overlain by Hoxnian Stage deposits at several localities, e.g. Hoxne (West, 1956), Marks Tey (Turner, 1970) and Hatfield (Sparks *et al.*, 1969). Bristow & Cox (1973) assigned the tills to the Wolstonian Stage on the assumption that the Hoxnian and Ipswichian deposits belong to a single temperate stage preceding the Devensian, though Bristow (1985) has subsequently adopted the conventional Anglian age. Straw (1979d, 1983) retains a belief in two advances, one of Anglian age, following the reasons given above, and one of Wolstonian age, based on morphological evidence and by correlation of the tills of central East Anglia with those of Lincolnshire; the latter overlie Hoxnian deposits at Welton-le-Wold and underlie Ipswichian deposits at Tattershall.

In many cases, the differences which often led earlier workers to subdivide the tills in terms of glacial stages are now explained as facies variation within a single advance (Whiteman, 1983, 1987;

Allen, 1984) or fluctuations within one glacial stage (Cheshire, 1983).

ANALYTICAL TECHNIQUES

Harmer (1909) distinguished till provinces on the basis of matrix provenance and Solomon (1932b) used heavy mineral assemblages to distinguish the 'North Sea Drift' from the main till sheet and link them with associated deposits. Baden-Powell (1948) separated tills on the basis of colour and stone counts and West & Donner (1956) confirmed this separation using macrofabric analyses. Clayton (1957, 1960) resorted to altitude, landscape morphology and depth of decalcification to distinguish the three tills in Essex. In the Vale of St. Albans, Gibbard (1977) separated tills on the basis of macrofabric properties, height, lithology and stratigraphic position. Perrin *et al.* (1979), using heavy mineral, carbonate content and textural analyses, concluded that the tills of East Anglia represented a single glacial event.

Many of these approaches made valuable contributions to the understanding of glacial stratigraphy in East Anglia, but it has become clear that only the use of a wider range of techniques can adequately resolve the problems of till genesis and stratigraphy. Consequently the studies summarised here have employed an assemblage of sedimentological techniques, including macrofabric analysis, stone counting, determination of small clast lithology, colour, texture, carbonate content and structure, in conjunction with a consideration of subglacial geology and topography, stratigraphical relationships and till distribution. This has enabled a detailed explanation of till genesis and stratigraphy in southern East Anglia to be developed. However, there are still difficulties in detailed correlation of tills across the region.

THE PHYSICAL SETTING

East Anglia lies on the northern side of the London

Table 18. History of the till stratigraphy of the Lowestoft Formation of East Anglia and the London basin.

Conventional stage names (Mitchell et al. 1973)	Trimmer 1951, 1958	Wood 1970	Wood & Harmer 1972	Whitaker et al. 1978	Harmer 1909, 1910	Solomon 1932a, 1935
——————— MONOGLACIAL ———————						
Wolstonian / Anglian	Upper Till	Upper Glacial	Great Chalky Boulder Clay (Upper Glacial)	Boulder Clay (brick-earth) (gravel and sand) + occasional boulder clay	Chalky Boulder Clay	Chalky Boulder Clay
Pre-Anglian						

Conventional stage names (Mitchell et al. 1973)	Boswell 1931	Wooldridge 1938, 1960 Wooldridge & Linton 1939, 1955	Baden-Powell 1948	West & Donner 1956	Clayton 1957	Clayton & Brown 1958	Clayton 1960	Thomasson 1961	Straw 1979, 1983
——————— MULTIGLACIAL ———————									
Wolstonian	Upper Chalky Boulder Clay		Gipping Boulder Clay	Gipping Boulder Clay	Springfield Till Maldon Till	Upper Till (Springfield Till)	Springfield Till		Wolstonian Till
Anglian	Chalky Jurassic Boulder Clay	Chalky Boulder Clay of the Eastern Drift	Lowestoft Boulder Clay	Lowestoft Boulder Clay	Hanningfield Till	Middle Till Complex / Lower Till	Maldon Till	Chalky Boulder Clay	Lowestoft Till
Pre-Anglian		Older (Chiltern) Drift					Hanningfield Till	Pebbly Clay Drift	

Conventional stage names (Mitchell et al. 1973)	Baker 1971	Turner 1973	Bristow & Cox 1973	Perrin et al. 1973, 1979	Gibbard 1974, 1977	Rose & Allen 1977	Baker & Jones 1980	Allen 1984	Bristow 1985	Whiteman 1987
——————— MONOGLACIAL ———————										
Wolstonian	(Chalky Boulder Clay)*		Springfield Till Maldon Till							
Anglian	(Chalky Boulder Clay)*	Springfield Till Maldon Till		Lowestoft Till	East end Green Till Ware Till	Blakenham Till Barham Till	Springfield Till Ware / Maldon / Quendon Till	Blakenham Till Blood Hill Till Barham Till Creeting Till	Springfield Till (including Hanningfield Till) Maldon Till	Great Waltham Till Newney Green Till
Pre-Anglian										

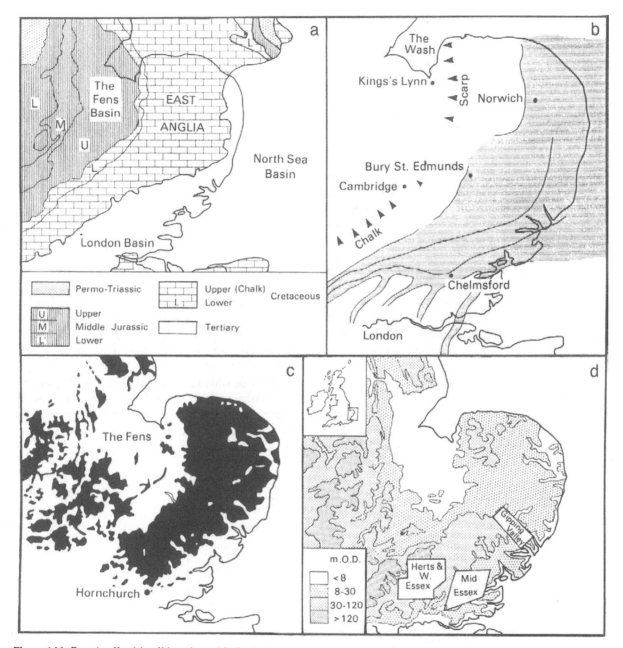

Figure 166: East Anglia: (a) solid geology, (b) distribution of pre-glacial fluvial gravels, (c) till distribution, (d) topography and location of areas discussed in the text.

Basin and extends to the Wash and the North Sea basin. The strike of the bedrock is southwest-northeast within the London Basin, but curves to run south-north and then southeast-northwest in the North Sea. Consequently Jurassic rocks underlie the Wash and Fens and Cretaceous rocks occur in most of the region, overlain by Tertiary sands and clays in the London Basin and by Crags in the North Sea margins (fig. 166a). The pre-Quaternary deposits are overlain

principally by Red Crag, Chillesford (Allen, 1984; Zalasiewicz & Mathers, 1985), Kesgrave (Rose *et al.*, 1976; Hey, 1980) and Ingham (Clarke & Auton, 1982) Formations (fig. 166b). Ice entering East Anglia crossed the Jurassic clays of the Wash and Fens, surmounted the Chalk escarpment and crossed the Crag and the Kesgrave and Ingham Sands and Gravels.

The sub-till lithology and topography had a profound effect on glacier behaviour and the nature of the tills. The tills are finer over the Jurassic clays, sandier over the Kesgrave Sands and Gravels and carbonate-rich over the Chalk (Perrin *et al.*, 1979). However, the till is often absent or thin over the Chalk in west Norfolk and Suffolk, southern Cambridgeshire, east of Thetford and between Bury St. Edmunds and Diss; it is best developed over sand and gravel, including the Cretaceous sands immediately east of the Wash and Fens and the Pleistocene sands and gravels east and south of the Chalk escarpment (fig. 166c). Till structure and fabric will be shown to be influenced by subglacial topography. Glaciofluvial deposits appear to be of lesser importance stratigraphically, being largely restricted to valleys, except in north Norfolk.

Three studies, from Suffolk (Allen), mid-Essex (Whiteman) and Hertfordshire and west Essex (Cheshire) (fig. 166d), will be used to develop the points made above. All these regions are at or near the till margin, but are different in geological or topographical settings and show variation in their tills.

SUFFOLK

Four lithofacies are recognised in Suffolk, forming the Lowestoft Till group (Perrin *et al.*, 1979; Allen,

Table 19. Suffolk: till lithofacies.

	Lithostratigraphic members	Facies
Lowestoft Formation	Haughley Park Gravel	Proximal outwash gravel
	Blakenham Till	Lodgement till
	Bramford Till	Slumped till
	Barham Till	Sheared and deformed till
	Sandy Lane Gravel	Proximal outwash gravel
	Creeting Till	Flow till
	Sandy Lane Gravel	Proximal outwash gravel
	Barham Sand and Gravel	Distal outwash gravel

1984) (fig. 167, table 19). The most extensive and thickest unit is a sheet of lodgement and slumped meltout till (the Blakenham and Bramford Tills respectively) which is considered to cover most of Suffolk. The till sheet is underlain, particularly in valley-side situations, by a banded, sheared basal till, the Barham Till. Flow till (Creeting Till), usually associated with ice-proximal outwash gravels, has been found below the main sheet, but at two locations only.

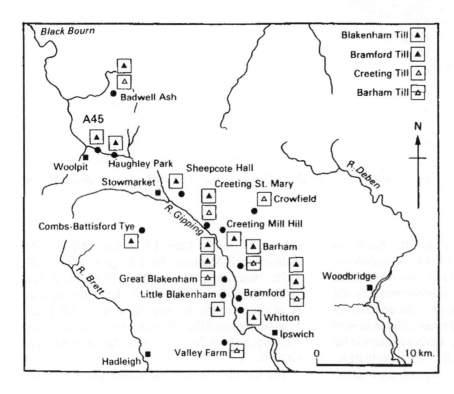

Figure 167: Suffolk: till lithofacies at sites in and near the Gipping Valley.

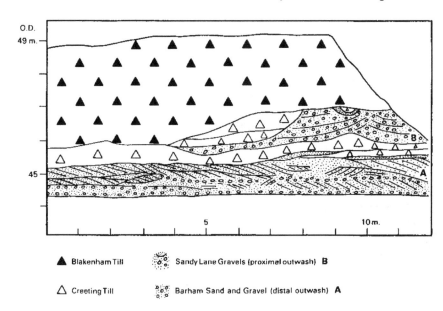

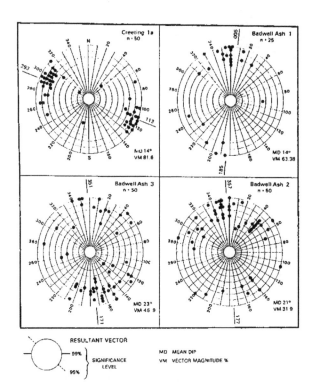

Blakenham Till   **Sandy Lane Gravels** (proximal outwash) **B**

Creeting Till   **Barham Sand and Gravel** (distal outwash) **A**

Figure 168: Badwell Ash: section showing the relationship of the tills to the outwash gravels.

RESULTANT VECTOR

99%
SIGNIFICANCE LEVEL
95%

MD MEAN DIP
VM VECTOR MAGNITUDE %

Figure 169: Creeting Till: macrofabric data.

Creeting Till Member

The Creeting Till, found only at Badwell Ash (TL 9969) and Creeting St. Mary (TM 0955), is crudely bedded, brown and brownish-yellow (7.5YR5/6 –

10YR6/6), occurring in units up to 0.5 m thick and varying from being sandy to silty. At Badwell Ash, two beds of this till were found (fig. 168), with statistically significant north-south vector trends (fig. 169), which differ from the northwest-southeast direction of ice movement established by West & Donner (1956) and Perrin *et al.* (1979). At Creeting, only one bed was found, within glacially deformed sediments and with a northwest-southeast vector trend (fig. 169). The till is identified as a flow till on the basis of its association with outwash (the Sandy Lane and Barham Gravels) which separates it from the Blakenham Till above, its textural variation and its strong macrofabric patterns with variable resultant vectors indicating local controls on the flow direction.

Blakenham Till Member

The Blakenham Till, best exposed at Great Blakenham (TM 102500), is clay-rich ($\bar{x}$ = 33.3%), chalky, predominantly very dark grey (5Y3/1, 5Y3/2, 2.5Y3/2), though the upper part may be weathered to yellowish-brown (10YR5/4, 10YR5/8), and massive. Occasional lenses of sand or gravel occur within the till. It has variable clast preferred orientations, but all relate to the regional direction of ice movement (fig. 170). In detail, east of Bury St. Edmunds, 11 of the 13 macrofabric analyses showed vector strengths significant at the 95% level or better, with a

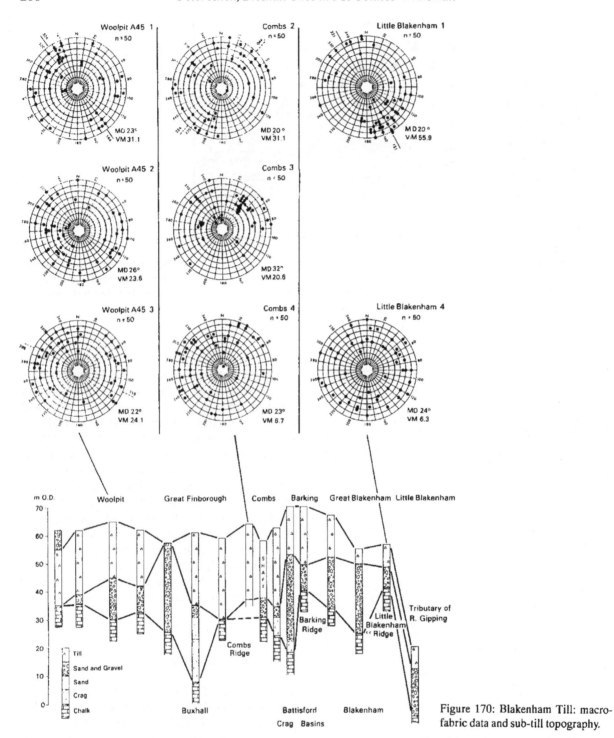

Figure 170: Blakenham Till: macro-fabric data and sub-till topography.

northwest-southeast trend. The remaining two analyses showed the same trend, but the patterns were not statistically significant. In the central area, around Stowmarket, of the 23 macrofabric analyses, only 9 showed statistically significant resultant vectors, five of which were coincident with the direction of ice movement and four transverse. The analyses also showed a reduced number of stones dipping to the southeast compared with the areas to the north and south. In this area, the base of the till rises from

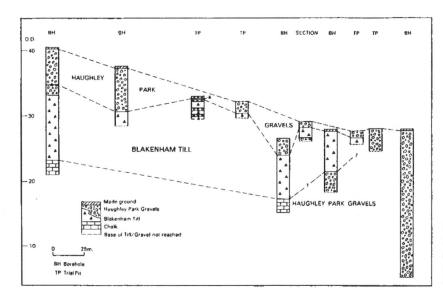

Figure 171: Blood Hill, Little Blakenham: correlation of boreholes and section. Borehole and trial pit data kindly supplied by Suffolk County Council.

c. 25 - 35 m to c. 50 m O.D. within 5 - 10 km. In the southern area, around Great and Little Blakenham, out of 28 macrofabric analyses, only 15 had statistically significant clast preferred orientations, 2 of which had resultant vectors oblique to the direction of ice movement and were adjacent to chalk lenses or smudges (Krüger, 1979). Of the 13 non-significant analyses, 10 had resultant vectors coincident with the regional ice movement.

In the southern area, at Blood Hill, Little Blakenham, part of the till interdigitates with the Haughley Park Gravels (fig. 171). The till is clayey, with a variable light olive to brown matrix (2.5Y5/4 to 10YR5/6). Four macrofabric analyses showed northeast-southwest resultant vector trends signifiicant at the 99% level (fig. 172). The initial impression that the till was a lens within the Haughley Park Gravels, at a higher level than, and with a resultant vector trend at right-angles to, the main Blakenham Till sheet led to the interpretation of it as a flow till (the Blood Hill Till) (Allen, 1984). However, further quarrying has shown the till to be at the same level as the main till sheet. This and the general textural similarity suggest that this till is not a separate unit, but should be regarded as a lateral part of the till sheet, deposited during retarded flow over the Haughley Park Gravels, with the transverse clast preferred orientations occurring as a result.

The till is interpreted as a lodgement or basal meltout till on the basis of its massive nature, its consistent sedimentological and lithological properties, its clay-rich matrix and the general coincidence

between the resultant vectors and the direction of ice movement. In the central area the transverse clast preferred orientations and lack of stones dipping to the southeast suggest compressive flow and shearing as the ice surmounted rising ground.

Bramford Till Member

Till can be traced as a continuous sheet from the plateau areas to the valley sides where it occurs at lower altitudes and its properties become variable, though it is still massive, as at sites such as Great Blakenham (TM 117500), Barham (TM 133515) and Bramford (TM 131483). The till becomes lighter and browner in colour, mostly in the range 10YR5/6 to 10YR7/6, yellowish-brown to yellow, though it can be olive-brown (2.5Y4/4) or, where decalcified, dark yellowish brown (10YR3/4). The till also becomes more variable in composition, being notably richer in sand and deficient in silt and clay. Compared with the plateau till, the valley-side till is deficient in the less durable lithologies such as chalk, limestone and shale. Only 5 of the 14 macrofabric analyses are statistically significant and those have variable resultant vector trends, ranging from 044-224° to 158-338° (fig. 173). The more variable characteristics of this part of the till sheet, particularly the reduced amount of silt and clay, the lack of significant resultant vectors and the variability of those that are significant suggests that this is not a lodgement till but one deposited during a melt phase when fines

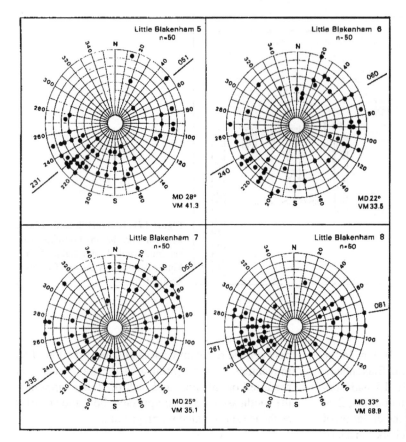

Figure 172: Blood Hill, Little Blakenham: macrofabric data.

were removed and englacial clast preferred orientations were lost by slumping as the debris was deposited, in the manner suggested by Rose (1974) in the Hatfield area.

Barham Till Member

The basal part of the till sheet, where it overlies sands and gravels in a valley-side location, such as Barham, Bramford and Valley Farm (TM 116433), comprises a structureless sandy till followed by sandy, yellowish-brown (10YR5/6) banded till, each band usually being up to a few centimetres thick. At times this till can be so sandy that it is difficult to distinguish from the underlying material. Compared with the Blakenham and Bramford Tills, this till is richer in quartz and deficient in chalk and limestone. Macrofabrics are usually strong and consistently trend northwest-southeast (fig. 174). This till had previously been considered to have been deposited as a flow (Allen, 1984), but its sandy nature, its enriched quartz component, and deficiency of chalk and limestone indicate that the underlying sand and gravel has been actively incorporated into the till. The banding and the very strong clast preferred orientations suggest that the incorporation was achieved by shearing.

Plates 24–27

24. Llanilid. Photomicrograph showing ripple train in fine sands and silts (L3). Note grain-size variation and small-scale faulting near base. Frame height 2 cm (Photomicrograph: R. Donnelly, 1986).
25. Llanilid. Imbricated gravels (L2). Scale 50 cm (Photograph: R. Donnelly, 1986).
26. Llanilid. Undeformed laminated sands, silts and clays (L6) (Photograph: R. Donnelly, 1985).
27. Ffyndaff. Coal clast forming dropstone in silty sand (FD3). Note silt drape (Photograph: C. Harris, 1982).

24	26
25	27

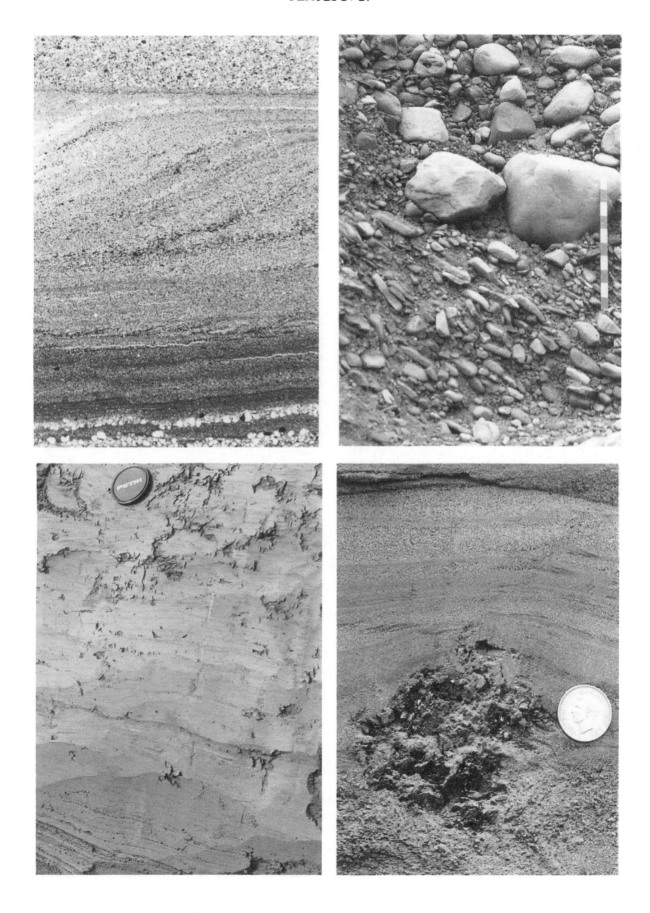

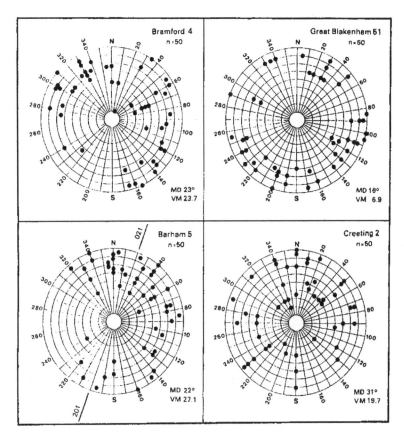

Figure 173: Bramford Till: macrofabric data.

Away from the valley sides, over sand and gravel, the Barham Till is thinner or absent, presenting a much more abrupt junction between the till and the underlying beds. Where the till overlies soft, intertidal sands at Great Blakenham, a sequence of weakly deformed bedrock, penetratively deformed bedrock (deformed till) and then a massive, but thin, sandy till overlies a plane of *décollement*, approximating to the model of Banham (1977).

The 'Lowestoft' and 'Gipping' Boulder Clays

The concept of two tills occurring in Suffolk was formalised by Baden-Powell (1948) who described the older 'Lowestoft Boulder Clay' as being a 'dark type of clay' when fresh (p. 283) and 'dark brown' when weathered (p. 286). The younger, 'Gipping Boulder Clay' was 'yellowish to khaki' when fresh and 'dirty-white' when weathered (p. 286). The stone content of the two tills was also noted to differ, essentially in the proportions of the lithologies present. Although the differentiation of the two tills in East Anglia as a whole was supported by West & Donner (1956) on the basis of macrofabric analyses, in the Gipping valley area both advances were thought to have been from the northwest, so locally the method did not separate the tills.

The dark grey lodgement till, the Blakenham Till, corresponds to Baden-Powell's 'Lowestoft Boulder

Plates 28–29

28. Aber-mawr. Irish Sea till overlying locally derived head. Note jointing and paucicity of clasts in till. Scale 50 cm (Photograph: C. Harris, 1986).

29. Ffos-las. Upper till (FL5). Note strongly developed fabric and high erratic content (Photograph: R. Donnelly, 1985).

28
—
29

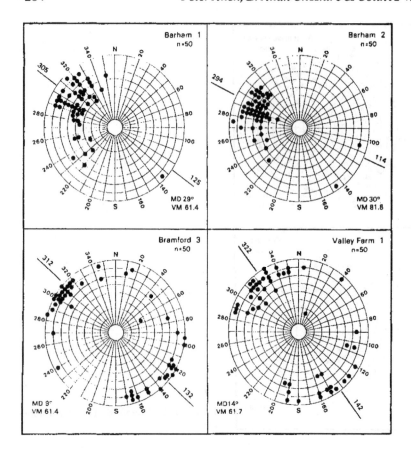

Figure 174: Barham Till: macrofabric data.

Clay'. The valley-side Bramford and Barham Tills meet the 'yellowish to khaki' colour criteria for the 'Gipping Boulder Clay', as also does the upper part of the Blakenham Till, when weathered. Thus the differences between the 'Lowestoft' and 'Gipping Boulder Clays' can be explained in terms of contrasting depositional and weathering histories. It is, however, perfectly understandable why the tills were separated out; the colour differences can be very striking and the lithological differences, at first sight at least, appear to support separation.

Summary

The tills of the Lowestoft Formation of Suffolk comprise various facies. The Blakenham Till, a lodgement or basal meltout till, was deposited as the ice advanced southeastwards across the county. Within the Blakenham Till, an area of compressive flow is identified on the basis of macrofabric analyses in the area around Stowmarket, coinciding with a rise in the sub-till surface. The anomalous sandier till of the Gipping valley described by Perrin *et al.* (1979) is

explained as basal sheared till (Barham Till) and a slumped facies (Bramford Till). The Creeting Till is very localised, representing flow till associated with outwash gravels.

MID ESSEX

Several till units have been recognised in mid-Essex (fig. 175, table 20). It is apparent that they represent facies of the Lowestoft Formation although they are separable into two members on the basis of lithology and macrofabric analyses. No weathering, pedological or interglacial sedimentological evidence for more than one glaciation has been found in spite of careful searching and a second recent survey by the British Geological Survey (Bristow, 1985).

The lower boundary of the till sequence

In Essex tills rest largely on easily deformable Tertiary or Quaternary strata. This has led to some difficulty in the determination of the base of the till

Table 20. Essex: till lithofacies.

Member	Description	Depositional environment	Process	Status	Provenance
Great Waltham Till	Dull yellowish-brown mostly massive very chalk-rich sandy clayey silt	Subglacial	Melt-out? Lodgement	Mature	Exotic
	Dark grey usually massive very chalk-rich sandy clayey silt	Subglacial	Lodgement Shearing	Mature	Exotic
Newney Green Till	Brown banded silty clayey sand becoming moderately chalky upwards	Subglacial	Shearing? Melt-out	Mature	Mixed
	Sandy homogenised bed material	Subglacial	Traction Lodgement	Immature	Local
	Penetratively deformed (cleaved) bed material	Subglacial	Shearing	Immature	Local
Sub-till sediments and palaeosols	Weakly deformed (folded)				
	Undeformed				

sequence which can vary in character over only tens of metres within a quarry. In places the boundary between till and the subjacent sediments is sharp; elsewhere the underlying beds are deformed or interdigitate with till. Banham (1977) has presented a model analogous to metamorphic rock deformation, that closely fits the mid-Essex situation (fig. 176). In this model the principal lithological boundaries are subsidiary to the principal structural boundary (plane of *décollement*) between weakly deformed (folded) bedrock and penetratively deformed (cleaved) bedrock in which clasts possess a strong preferred orientation. The plane of *décollement* thus represents the base of the till sequence where bedrock deformation has occurred.

Newney Green Till Member

Newney Green Till has been recorded at Newney Green (TL 645066), Great Waltham (TL 695135) and Broomfield (TL 723144). Three units of this till conforming to Banham's model have been recognised. Up to 1.87 m of thinly bedded red and grey mottled clayey sand (Valley Farm Soil of Rose & Allen, 1977) and coversand can be traced into suprajacent beds. A plane of *décollement* separates this laminated bed from folded beds beneath and represents the lower boundary of the till sequence. Failure has occurred within clay-rich beds, either the Bt horizon of the Valley Farm Soil or the London Clay, where shear strength is low. Strong clast preferred orientation (fig. 175) trends between WNW-SSE and northwest-southeast. Clast dips are low and clast

lithology reflects that of the subjacent Kesgrave Sand and Gravel. Far-travelled clasts are absent.

At Newney Green this deformation till is overlain by up to 30 cm of structureless, sandy till, generally 7.5YR5/8 in colour and lithologically indistinquishable from the Kesgrave Sand and Gravel. Laterally it was seen to merge into glaciofluvial outwash of the Barham Sand and Gravel. The strong clast preferred orientation trending northwest-southeast and the low mean dip of the clasts (fig. 175) are similar to those of the deformation till. Texturally the till is intermediate between the Kesgrave Sand and Gravel and the overlying till unit (fig. 177). In Banham's (1977) terminology it is local lodgement till. It was produced subglacially by glacial shear stress which initially deformed and then homogenised several local sedimentary units in the traction zone of the deforming bed under conditions of high pore water pressure (Boulton, 1975a).

In turn this unit of the Newney Green Till is overlain by a brown, often prominently banded, unit up to c. 2.0 m thick. It is the lowest calcareous till in the sequence. The content of calcium carbonate equivalent increases upwards but is generally less than 15% of the < 2 mm particle-size fraction. The till is characteristically banded with very slight textural and colour (7.5YR5/6 and 10YR6/6) variations, producing a series of couplets. Sorted laminae only occasionally occur. Bands vary from a few millimetres to about 20 cm but more massive, less obviously banded, areas occur. Clast lithology indicates more local material at the base changing to a greater exotic content, including Chalk and Jurassic rocks, upwards. Strong clast preferred orientation (fig. 175)

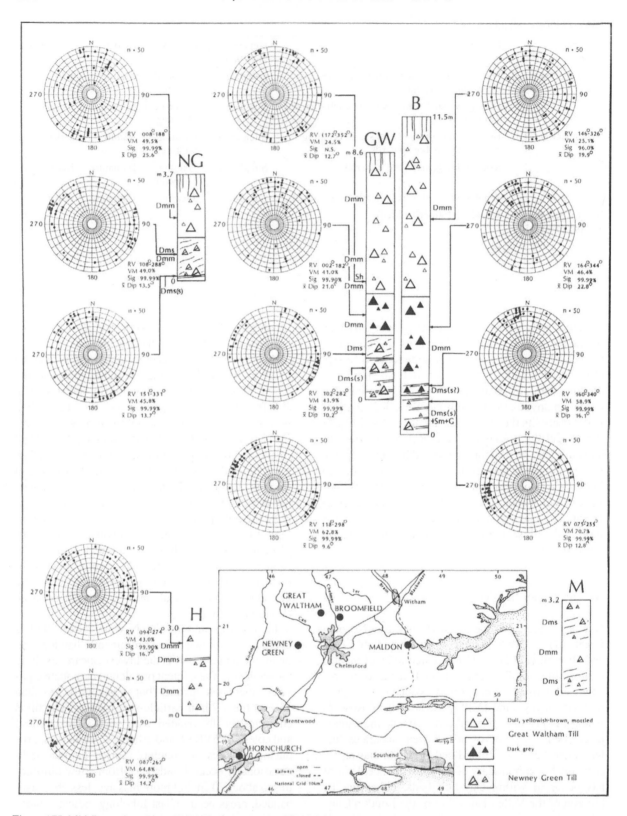

Figure 175: **Mid-Essex: locations, till lithofacies and macrofabric data.**

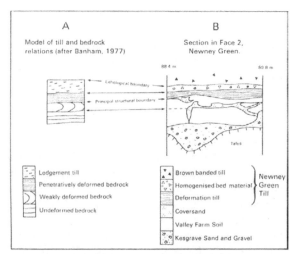

Figure 176: Comparison of the section at Newney Green with Banham's (1977) model of till and soft bedrock relations.

again trends in a direction between WNW-ESE and northwest-southeast.

Banding in tills has variously been recognised by colour (Edwards, 1975), composition (Gibbard, 1980), texture (Menzies, 1979; Lawson, 1981) and structure (Shaw, 1979). Edwards (1975) reported frequent association with erosive contacts and glaciogenic disturbance. Donner & West (1957) found banded tills in Spitsbergen associated with debris-rich ice and transverse clast preferred orientations. Edwards (1975) found mixing of exotic and local material. Many processes have been invoked to explain banding in tills including thrust shearing and meltout (Donner & West, 1957), shearing by glacial plastic flow (Edwards, 1975), folding (Shaw, 1977), meltout (Shaw, 1979), flow (Boulton, 1968) flow with flow-surface washing (Boulton, 1971), meltout and flow (Lawson, 1981), sub-aquatic flow (Evenson *et al.*, 1977), basal meltout into water ('waterlain') (Gibbard, 1980), till washing (Menzies, 1979), basal meltout (Stefan Kozarski, personal communication, 1984) and inverse-grading and 'lodging' (Menzies, 1986).

Possible folding has only been observed at one location, Great Waltham (South), no sedimentary evidence for the presence of the large quantities of water necessary for sub-aquatic flow or meltout into water has been observed and the uniformity of trend of the strong preferred orientation of the clasts is not consistent with flow till deposition. In view of the subjacent deformation, strong clast preferred orientation and low clast dips, freezing-on, or shearing-in by glacial stress followed by confined basal meltout

from stagnant, debris-rich ice is the present preferred explanation of the deposition of this till unit, with the textural couplets possibly being produced by till washing or inverse grading (Menzies, 1979; 1986). However, this member clearly requires further analysis before its mode of origin can be determined with certainty.

Great Waltham Till Member

In mid-Essex, Great Waltham Till directly overlies the Newney Green Till along a sharp or interdigitating boundary or rests on various sub-till sediments. Where Great Waltham Till is thickest (> 3 - 4 m) it can usually be divided into two units, a lower dark grey unit overlain by a dull yellowish-brown, mottled unit. Only the dull yellowish-brown unit is found where the till is less thick.

The dark grey (5Y4/2) unit is usually a uniform, massive, compact, clayey and silty, chalk-rich till typical of the Lowestoft Till throughout East Anglia

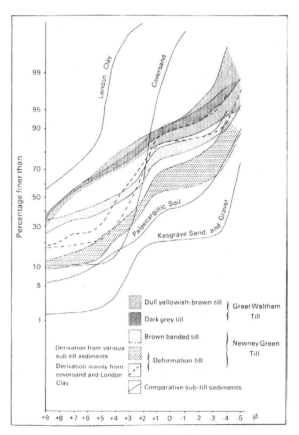

Figure 177: Newney Green and Great Waltham Tills: particle size distributions.

(Perrin *et al.*, 1979). However where it rests directly on a permeable substratum, the basal 20 cm oxidises to 10YR hues. Chalk smudges (Krüger, 1979) are rare, but have been observed at Broomfield; lenses of coarse, chalky gravel have been observed towards the base of this unit. At one location in the north quarry at Great Waltham (TL 686125), the lowest 0.82 m of the dark grey till varied upwards from very dark grey (5Y4/1) brecciated chalky till to texturally banded or strongly sheared till (shears dipping at 40° towards 050°) with well-sorted sand partings near the top. A sharp boundary separated this lower zone from 1.77 m of massive dark grey till. Thin lenses of brown Newney Green Till also occur in the lower levels of the dark grey Great Waltham Till.

Lithologically the dark grey till is dominated by far-travelled chalk, often striated, and fresh flint, with lesser amounts of Jurassic sandstone and shales and rare Permo-Triassic, Carboniferous, igneous and metamorphic rocks. The proportion of chalk clasts in the 8 - 16 mm fraction and the calcium carbonate equivalent in the < 2 mm matrix increases from the base upwards while flint decreases upwards away from the influence of the subjacent tills and sand and gravel.

The preferred orientation of the clasts (fig. 175) indicates an ice movement direction from between NNE and NNW and is most strongly developed in the middle of the unit. Mean clast dips (20 - 25°) are commonly greater than in the underlying Newney Green Till but are still relatively low.

The massive nature of most of the dark grey till, its clast orientation properties, the striated clasts and the chalk smudges are all consistent with subglacial deposition by a lodgement process. Although the large chalky gravel lenses suggest the presence of substantial amounts of meltwater, this does not rule out the lodgement process (Eyles, 1983a). The sharp upper boundary and the shearing in the top of the lower sub-unit of the dark grey till suggests the possibility of ice stagnation followed by renewed forward ice motion later, but it may equally indicate a steep shear stress gradient across this boundary during one depositional episode.

The dark grey till changes gradually into a widespread dull, yellowish-brown, mottled chalky till which always occurs at the top of the till sequence. At Great Waltham (South), Roxwell (TL 655099) and Broomfield, lenses of fine, chalky gravel and sand occur at or near the junction of those tills. The texture of the mottled till is slightly more varied than that of the underlying dark grey unit; its particle-size distribution envelope almost completely encloses that of

the dark grey unit (fig. 177) and its average clast size appears to be larger. The content of chalk clasts and calcium carbonate equivalent is high and increases upwards. Strongly weathered clasts are frequent, the matrix is oxidised, and brown oxidised iron occurs commonly on the surface of the chalk clasts. The uppermost 0.5 m of the till is decalcified and decalcified tongues extend downwards; calcareous nodules occur in the top of the undecalcified till. Small pockets of sand, either weathered clasts or glaciofluvial sand lenses, are common, especially towards the base of this unit. Grey clasts of Jurassic shale are rare. This unit is less compact than the dark grey till and is more jointed. Clast preferred orientation is usually less significant than in the dark grey unit although the general trend is still north-south. The mean dip of clasts is slightly higher. At Great Waltham (North) where it directly overlies the brown Newney Green Till the basal 20 - 30 cm of this till are banded and slightly browner and sandier. At Newney Green thin lenses of Newney Green Till occur within the basal 1 m of the dull yellowish-brown mottled till.

Superficially the two units of the Great Waltham Till are distinct but at Great Waltham (South) lithological properties, calcium carbonate equivalent and the percentage of clasts dipping up-ice show systematic changes upwards through both tills. Similarly the resultant vectors of macrofabric analyses show a consistent change from NNE to NNW upwards through the two units (fig. 175). These characteristics suggest a close association between the tills. Perrin *et al.* (1979) make no great distinction between them other than weathering and they classify both of them as lodgement till. Certainly there is plenty of evidence of weathering in the upper unit, possibly encouraged by enhanced water evacuation through the sand and gravel lenses (Eyles & Sladen, 1981); the question is whether or not post-depositional weathering alone is sufficient to account for the observed differences. The low resultant vector strengths (fig. 175) may indicate some resedimentation as well as weathering, but do not rule out a primary process of lodgement. Alternatively, the bulk of the dull yellowish-brown mottled till may have melted out from ice containing dispersed debris, but the striated clasts do indicate that this till has spent some time at the base of the ice sheet. The paucity of shale and the increase in Chalk suggests an ice flow direction more closely associated with the Chalk outcrop.

The Maldon and 'Hornchurch' Tills

Two isolated occurrences of till, beyond the margin

of the main East Anglian till sheet justify separate discussion on the basis of their historical importance to stratigraphical studies within the region.

Although mid-Essex primarily has one till sheet, divisible into two members, small occurrences of a second 'boulder clay' beneath 'glacial gravels' were recorded by Whitaker *et al.* (1878), especially at Maldon (TL 842067) (Whitaker, 1889), and tentatively attributed to the 'Lower Glacial' of Wood (1870).

Clayton (1957, 1960) incorporated the till at Maldon into his stratigraphic schemes (table 18). Bristow (1985) retained the Maldon Till as a lower till unit separated from the higher Springfield Till by the Chelmsford Gravels (Clayton, 1957; 1960) and Barham Sand and Gravel (Rose & Allen, 1977), both considered to be of glaciofluvial origin and Anglian age.

Recent work (Rose *et al.*, 1978; Whiteman, 1983) has shown, however, that the Chelmsford Gravels were deposited by a former River Thames, prior to the Cromerian Stage. As the Maldon Till is only poorly exposed in valley-side locations or recorded in a few boreholes (Bristow, 1985), its status has been uncertain. However, from preliminary analyses of samples obtained from a recent temporary exposure of the classic section in the Maldon railway cutting, a tentative correlation with the Newney Green Member of the Lowestoft Till Formation is suggested.

The 'Hornchurch Till' has remained one of the most important lithostratigraphic markers in the Thames gravel sequence since its discovery in the Romford to Upminster railway cutting (TQ 547874) by Holmes in 1892. Current investigations, including macrofabric analyses (fig. 175), indicate that the 'Hornchurch Till' can also be tentatively correlated with the Newney Green Member of the Lowestoft Till Formation.

Summary

In mid-Essex, the Lowestoft Till group can be divided into two members each comprising several facies. The Newney Green Till is a complex, relatively sandy, till including (1) a deformation facies, (2) local, homogenised bed material and (3) a banded, possibly sheared, meltout facies deposited at the base of an ice sheet. In this area no convincing evidence of ice retreat has been found but there may have been at least two short hiatuses in glacial deposition as evidenced at Great Waltham (North) by shearing in the top of the brown Newney Green Till and within the

dark grey till. Much of the Great Waltham Till is lodgement till but meltout or secondary resedimentation may also have taken place during the deposition of this till. In view of the virtual absence of evidence for ice retreat in mid-Essex, the apparent change in ice movement direction between the Newney Green Till and the Great Waltham Till may reflect (1) the location of mid-Essex relative to the changing locus of the maximum surface gradient of a single ice sheet, (2) the local influence of topography, (3) resedimentation or (4) the development of transverse clast preferred orientations. The balance of the structural evidence and the orthogonal incidence of the two resultant vectors suggest that a strong transverse clast preferred orientation developed under compressive flow, induced by the rapid incorporation of readily deformable sub-glacial material near the ice margin. The succeeding resultant vectors, parallel to flow, were generated under tension, upstream of the margin, by the same ice sheet moving from a northerly direction.

HERTFORDSHIRE AND WEST ESSEX

In Hertfordshire and west Essex, two groups of deposits have been re-cognised at various times as having a glacial origin. Resting on the lower parts of the Chiltern Hills back-slope, is the so-called 'Chiltern Drift', which contains Bunter quartzite and quartz pebbles. It has been ascribed to deposition by ice moving over or from the Chilterns (Sherlock & Noble, 1912; Sherlock, 1924; Wooldridge, 1938; 1960; Wooldridge & Linton, 1939; 1955; Wooldridge & Cornwall, 1964). Deposits of 'Pebbly Clay Drift' on the South Hertfordshire Plateau were thought to have a similar origin (Thomasson, 1961). Doubts have been cast upon these interpretations (Gibbard, 1974; 1977; Green & McGregor, 1978; Baker & Jones, 1980) and Avery & Catt (1983) showed that at least part of the 'Pebbly Clay Drift' is weathered Anglian till. Moffat & Catt (1986) demonstrated that deposits formerly considered to be 'Chiltern Drift' contained no rocks or minerals that could not be more probably accounted for by fluvial transport before deposition of the Westland Green Gravel.

The remaining group of deposits was termed the 'Eastern Drift' by Wooldridge (1938). On the grounds of overlying Hoxnian deposits near Hatfield, Sparks *et al.* (1969) and Gibbard (1977) recognised these tills as Anglian. They are best known from the Vale of St. Albans, where two or more till units have been recorded (Whitaker, 1889; Salter, 1905; Irving

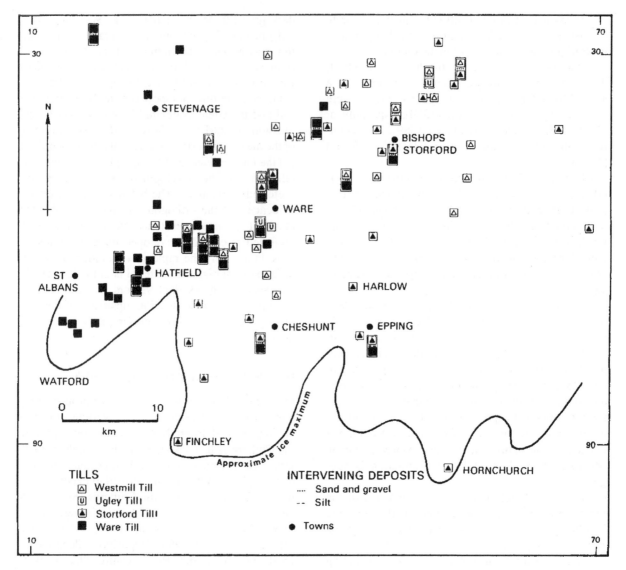

Figure 178: Hertfordshire and west Essex: sample sites and stratigraphic sequences.

& Irving, 1913; Barrow, 1919; Sherlock, 1924; Sherlock & Pocock, 1924; Wooldridge, 1938; Wooldridge & Linton, 1939; 1955; West & Donner, 1956; Clayton, 1957; Clayton & Brown, 1958; Gibbard, 1977).

Where stratigraphic correlation between exposures was discussed or implied, all previous workers believed that upper tills at all sites were equivalent. At sites where only one till was present, it was usually assumed to be the correlative of 'the upper till', so the view became established that the upper till was more extensive than the earlier till(s).

At sites in Hertfordshire and west Essex (fig. 178), up to 21 petrographic variables were determined for

286 till samples and were analysed by multivariate methods in a recent study (Cheshire, 1986). Petrographic properties were found to vary vertically through a sequence of tills more than they varied laterally within an individual till. Using similarity matrices, strong similarity coefficients between neighbouring pairs of samples indicated where equivalence existed between adjacent till sites. Four tills, interpreted as lithostratigraphic units, were recognised (fig. 179).

Till units tend to thin southwestwards, but become thicker and merge northwards and eastwards into a massive till (> 25 m). Although each till unit is generally uniform laterally, it is more variable at its distal limit, and becomes less petrographically distinct

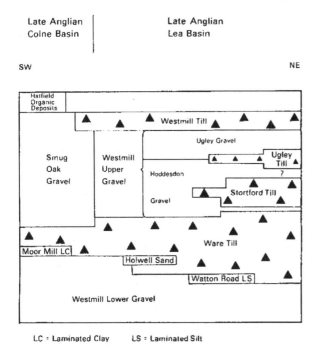

LC = Laminated Clay   LS = Laminated Silt

Figure 179: Schematic representation of the lithostratigraphic units recognised in the Anglian deposits of Hertfordshire and west Essex.

from the tills above or below in the north and east. The organic deposits of Hoxnian age near Hatfield rest above the earliest of the four till units, so that this is the only till for which a pre-Hoxnian age is certain. However, because the petrographic properties of the till units converge northeastwards, and the only deposits found between the tills are gravels of cold aspect (Robinson, 1978), chalky outwash, sand and glaciolacustrine silt, it seems probable that all the tills are of the same Lowestoft Stadial age (West, 1977b).

Ware Till Member

This is the earliest and most extensive till in central Hertfordshire. The stratotype is at Westmill (TL 344158), near Ware, following Gibbard (1977). The till is massive, 6.8 m in the thickest exposure, and very dark grey (10YR3/1) to dark grey (10YR4/1) when fresh, weathering to yellowish brown (10YR5/4) and brown (10YR5/3). Its particle-size characteristics (fig. 180) show a well-defined mode in the +2 phi to +3 phi fraction and a low acid-soluble content throughout. At the stratotype the acid-soluble content in the +1 to -2.5 phi fraction ranges from 21.4 to 40.0%, and the small-clast fraction (+1.5 to -2 phi) has a low quartz:flint ratio (fig. 181), a low proportion

of both *Rhaxella* chert and aggregated grains/crystals (a broad group of rocks which includes all aggregated grains and crystals not consigned to other categories, and includes clasts of igneous, metamorphic and sedimentary origin).

Resultant vectors of macrofabric analyses, where significant at 0.95 level or greater, indicate provenance from the northeast (fig. 182) and strong topographic control in the Vale of St. Albans. They are generally consistent with deposition by lodgement. Exceptions are the low mean dips and low vector magnitudes characteristic of subglacial melt-out (Lawson, 1979) at Westmill, a basal flow till at Holwell Court (TL 277104) and clast preferred orientations reflecting local control of ice-wastage processes near the ice maximum at Bricket Wood (TL 137025) and Harper Lane (TL 1602).

Stortford Till Member

At its stratotype, a borehole (TL 479195) near Bishop's Stortford, the Stortford Till is separated from the Ware Till below by 1.65 m of chalky gravel, by 0.8 m of chalky gravel at Westmill, and by 1.43 m of buff, sandy, clayey silt near the M25/M11 motorway interchange (TL 466001). At the stratotype the till is 12.0 m thick, sufficient to exhibit progressive vertical changes within the unit. The sand content decreases upwards (18.0 to 9.2%) as, conversely, the acid-soluble content in the +1 to -2.5 phi fraction increases upwards from 45.3 to 90.5% (fig. 183). At most other Stortford Till sites only part of this sequence is seen. Small-clast quartz:flint ratios are shown in fig. 184. Colours vary from very dark grey (5Y3/1) at the base to light yellowish brown (2.5Y6/4) where weathered. Clast preferred orientations (fig. 185) indicate lodgement from the NNE in the northwest Essex chalklands (Baker, 1976; 1977) and from the northeast in most of the area. At Harlow (TL 464116), near the Stort-Cam tunnel valley (Woodland, 1970), a significant northwest-southeast resultant vector is found.

Ugley Till Member

The stratotype of the Ugley Till is at Ugley Park Quarry (TL 520280), where it is the lower of two tills. The Ugley Till is massive, over 12.2 m being proved by borings, and dark grey (10YR4/1), but it is the least extensive of the four tills identified. The till is characterised by a very fine sand mode (+4 to +3.5 phi; fig. 186) and an acid-soluble content in the +1 to -2.5 phi fraction ranging from 67.4 to 74.3%. In the

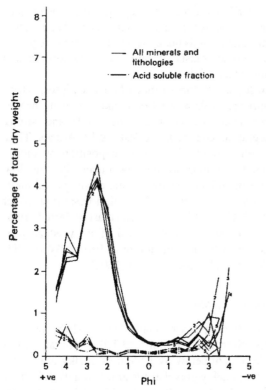

Figure 180: Ware Till: particle-size distributions.

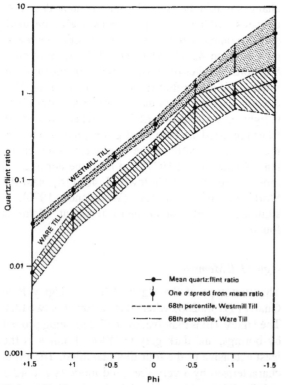

Figure 181: Comparative properties of the quartz:flint ratios of the Ware and Westmill Tills at Westmill.

small- clast fraction (+1.5 to -2 phi) quartz:flint ratios are high (fig. 187), and proportions of aggregated grains/crystals and *Rhaxella* chert are also high. At Ugley Park Quarry the clast preferred orientations are NNW-SSE (fig. 188), possibly influenced by the neighbouring Stort-Cam tunnel valley, but more probably they reflect a general provenance from the NNW in the northwest Essex chalklands. In the Hertford area, lodgement from the northeast or ENE is indicated.

Westmill Till Member

This unit constitutes the uppermost till, the stratotype being at Westmill Quarry (TL 344158). Its particle-size distribution includes a well-defined mode in the +3 to +2.5 phi and +4 to +3.5 phi fractions (fig. 189). The acid-soluble content in the +1 to -2.5 phi fraction ranges from 66.1 to 81.7% at the stratotype. In the +1.5 to -2 phi small-clast fraction, the quartz:flint ratios, the aggregated grains/crystals and the proportion of *Rhaxella* chert are all high.

The till lies invariably at or close to the surface and consequently is weathered light yellowish brown (10YR6/4) to yellowish brown (10YR5/4). Only exceptionally a sufficient thickness is found to preserve relatively unweathered dark greyish brown (2.5Y4/2) till near the base. Macrofabric analyses (fig. 190) and shear planes within the till record a readvance of ice from the north in the northwest Essex and northeast Hertfordshire chalklands, which turned to enter the Vale of St. Albans from the northeast. Fold structures in underlying sediments in the eastern Vale of St. Albans strike northwest-southeast. The till extends southwestwards to Welwyn Garden City and northern Hatfield, and southwards into the lower Lea valley as far as the M25 Motorway.

Discussion

Recognition of the four till lithostratigraphic units, augmented by observed field relations, was effected by subjecting up to 21 petrographic variates from each till sample to principal coordinates and principal components analyses and by analysis of similarity matrices (Cheshire, 1986). The Ugley and Westmill Tills cannot be distinguished on petrographic grounds, but they are separated stratigraphically by up to 9.5 m of the chalky Ugley Gravel Bed at Ugley Park Quarry. Near Hertford (TL 341123 and TL 349129) a till similar to the two tills at Ugley is

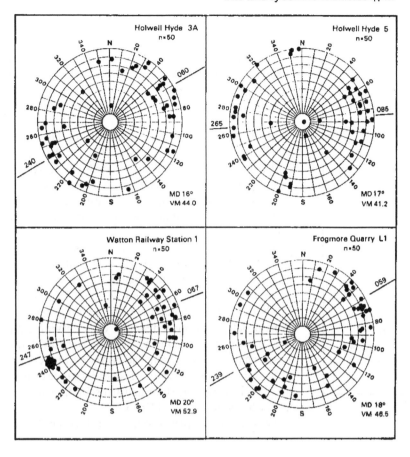

Figure 182: Ware Till: macrofabric data.

overlain by the Ugley Gravel Bed, thus the till is correlated with the Ugley Till. The Ware and Stortford Tills may be distinguished from each other and from the Ugley/Westmill Tills by petrographic criteria.

The sand mode in the Ware Till is higher than in the other three tills, and occurs in the +3 to +2 phi fraction. Its acid-soluble content in the diagnostic +1 to -2.5 phi fraction is lower than the other tills. Compared with the Ugley/Westmill Tills the small-clast fraction (+1.5 to -2 phi) of the Ware Till has a higher quartz content (× 1.08) and a lower content of flint (× 0.39), aggregated grains/crystals (× 0.75) and *Rhaxella* chert (× 0.20).

The Stortford Till contains less medium and coarse sand than the Ware Till and has a sand mode which reduces upwards in magnitude and particle size. Compared with the Ware Till, the Stortford Till has a higher acid-soluble content in the +1 to -2.5 phi fraction, a proportion which increases upwards in the till. While the composition of small clasts (+1.5 to -2 phi) differs only slightly from the Ware Till, the

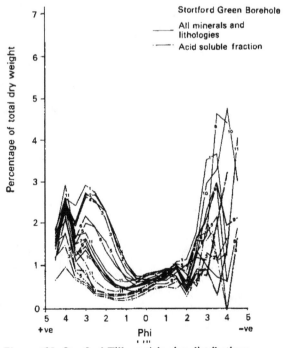

Figure 183: Stortford Till: particle-size distributions.

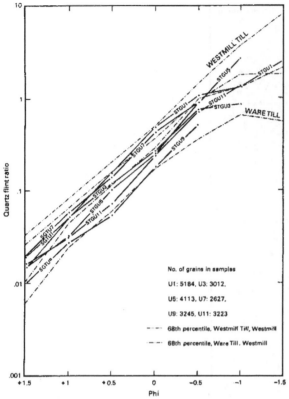

Figure 184: Stortford Till: quartz:flint ratios.

quartz, aggregated grains/crystals and *Rhaxella* chert content is often lower.

The Ugley and Westmill Tills contain larger sand modes than the Stortford Till, occurring in finer fractions (+4 to +3.5 phi and +3 to +2.5 phi) and is deficient in medium sand compared with the Ware Till. The acid-soluble content (+1 to -2.5 phi), while generally comparable with the Stortford Till, averages 2.65 times that of the Ware Till. Of the Ugley and Westmill Till acid-insoluble small-clast fraction, flint, aggregated grains/crystals and *Rhaxella* chert are more abundant than in the Ware and Stortford Tills and the quartz:flint ratios are also consistently higher (fig. 181).

The extent of each of the four tills is shown in fig. 191. The Ware Till extends furthest southwest and the Stortford Till furthest south to Finchley and Hornchurch, but the Ugley and Westmill Tills do not reach the Anglian till margin. Clast preferred orientations usually indicate lodgement on a curving path with provenance from the north in the Hertfordshire-Essex chalklands and from the northeast in the Vale of St. Albans. The presence of four tills suggests that the main Lowestoft ice lobe had an oscillating margin in Hertfordshire and west Essex.

Twenty-six samples taken in Suffolk and east/

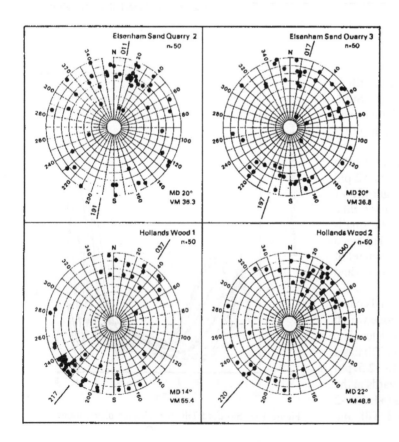

Figure 185: Stortford Till: macrofabric data.

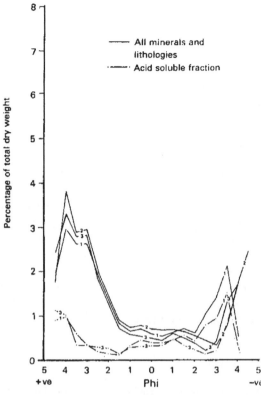

Figure 186: Ugley Till: particle-size distributions.

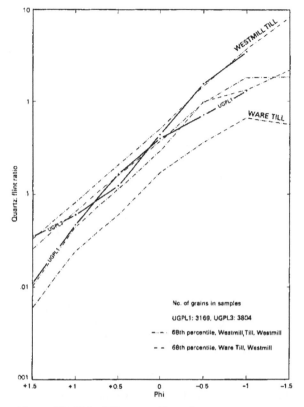

Figure 187: Ugley Till: quartz:flint ratios.

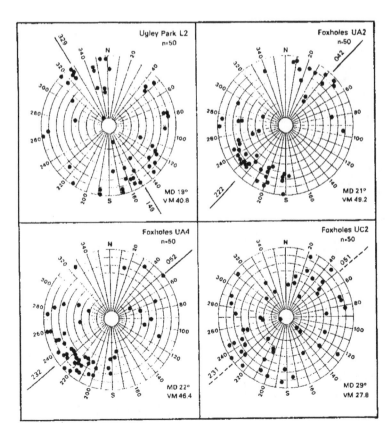

Figure 188: Ugley Till: macrofabric data.

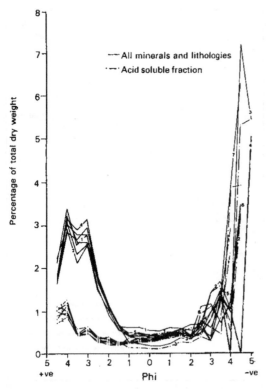

Figure 189: Westmill Till: particle-size distributions.

central Essex, and analysed by the methods used in Hertfordshire and west Essex, show similarities with the Stortford and Ugley/Westmill Tills (Cheshire, 1986). Although the petrographic distinctions by which the tills may be identified become less clear northeastwards, the most frequent correlation of these samples is with the Stortford Till.

Summary

Each till unit exhibits a strong degree of lateral petrographic homogeneity, and is interpreted as a lithostratigraphic unit. Although flow, slumped or geliflucted till lithofacies are known at one or more sites of each till, the great majority of exposures showed macrofabric properties characteristic of lodgement processes. Thus each of the four tills is interpreted as the deposit of successive advances of the ice sheet from the north or northeast into the Hertfordshire and west Essex area.

CORRELATION

All four tills of Hertfordshire, but particularly the

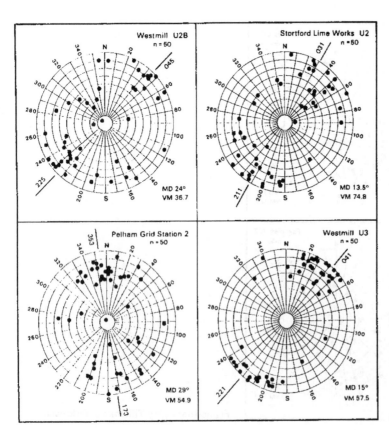

Figure 190: Westmill Till: macrofabric data.

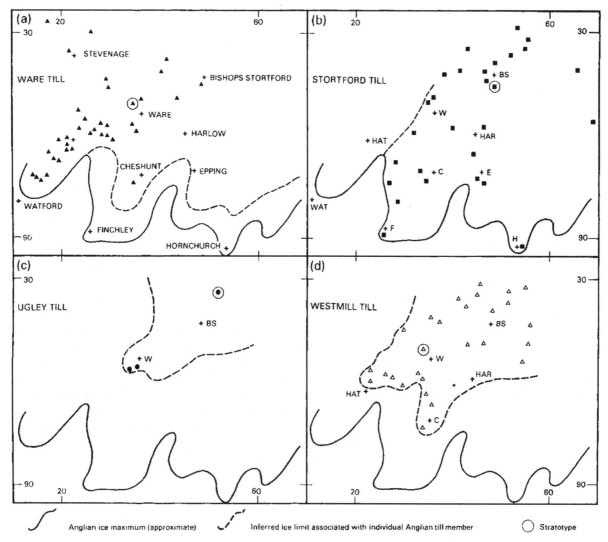

Figure 191: Glacial limits inferred from the Hertfordshire tills.

Stortford Till, correlate well with the Great Waltham and Blakenham Tills (table 21), being dark grey, chalk- and clay-rich and possessing clast orientation patterns fitting the models of radial ice advance over East Anglia (West & Donner, 1956; Perrin *et al.*, 1979; Straw, 1983). These tills were laid down largely by lodgement processes associated with the main southward advance of the Anglian ice.

The Bramford Till of Suffolk is seen as a valley-side facies of the till sheet, representing deposition from decaying ice. This facies has not been recognised in Essex or Hertfordshire.

The Newney Green and Barham Tills are considered equivalents, being sand-rich, banded and having strong clast preferred orientation properties. They may be sub-sole drift, basal melt-out deposits from

sheared, debris-laden ice or deformation tills.

No flow-till units equivalent to the Creeting Till were found in Essex or Hertfordshire.

Table 21. Correlation of the Anglian tills of Hertfordshire, Essex and southern Suffolk.

Hertfordshire and West Essex	Mid-Essex	Suffolk
Westmill Till Ugley Till Stortford Till Ware Till	Great Waltham Till	Blakenham Till
–	–	Bramford Till
–	Newney Green Till	Barham Till
–	–	Creeting Till

These correlations are somewhat tentative in view of the varying criteria and techniques that have been used. Further work may produce evidence that necessitates a revision.

CONCLUSION

The southern margin of the Anglian till sheet is dominated by very dark grey (5Y3/1) lodgement till, having similar stone counts and particle-size properties in each of the areas studied. It forms the Blakenham Till in Suffolk, the Great Waltham Till in Essex and all four tills in Hertfordshire. The clast preferred orientation patterns of these tills fit the models proposed by West & Donner, 1956, Perrin *et al.* (1979) and Straw (1979d, 1983) of ice fanning out over East Anglia from the Wash, approaching Suffolk from the northwest, Essex from the north and Hertfordshire from the northeast.

However, the detail in each area differs. In Suffolk, the Blakenham Till, a lodgement facies, shows a zone of compressive flow with transverse clast preferred orientation or dominantly up-glacier dips of elongate clasts where the ice surmounted higher ground in the Stowmarket area, while the slumped Bramford Till suggests decay of ice in the Gipping valley. The sheared basal meltout Barham Till occurs principally in valley-side locations. Despite the variation in the tills, only one ice advance is recognised.

In mid-Essex only two till members have been differentiated. The Newney Green Till is a series of units associated principally with meltout following basal deformation and shearing responsible for a prominent transverse clast preferred orientation. It is overlain by lodgement and meltout facies, the Great Waltham Till. Although the tills are attributed to a single stage, with ice flowing from only one direction, there may have been at least two hiatuses in deposition.

In Hertfordshire and west Essex, basal meltout and flow tills are of minor importance and, in contrast to mid-Essex and Suffolk, four lodgement tills are recognised, each representing a local advance controlled by the palaeogeography. The contrast between the oscillating ice margin in Hertfordshire and west Essex and the apparently more static margin in mid-Essex and Suffolk may be accounted for by the presence of the higher ground in the former area. The thinner ice and higher ground of the north Hertfordshire chalklands probably impeded forward movement in Hertfordshire more than in the less restricted mid-Essex and Suffolk areas. This probably caused the ice margin in Hertfordshire to retreat markedly in response to relatively minor lowering of the surface of the ice sheet over East Anglia, whereas further east little evidence of fluctuation of the margin has been observed.

The glacial deposits of South Wales

CHARLES HARRIS & ROBERT DONNELLY

The sediments described in this chapter are of Devensian age (Bowen, 1973b; 1974). A raised beach deposit found in the succession at many coastal exposures, underlying the glacial deposits (e.g. fig. 197) is of critical chronostratigraphical significance. Amino acid dating of the contained shells (Davies, 1983; Bowen, 1984, Bowen *et al.*, 1985) together with Uranium-series dating of associated speleothem fragments in the sea cave Minchin Hole (Sutcliffe & Currant, 1984), has confirmed a Last Interglacial (Ipswichian) age of around 125 ka. The beach is correlated with Oxygen Isotope Stage 5e (Bowen, 1984).

An earlier glacial episode affected the whole of South Wales and during this period Irish Sea ice carried erratics into the Bristol Channel that have been reported as far east as Cardiff (for example Bowen *et al.*, 1986). The deposits of this earlier event are however poorly exposed, extensively reworked, and of little sedimentological interest.

The glacial sediments of South Wales are largely known through coastal exposures, although gravel pits provide some sections in fluvioglacial deposits. In this chapter, in addition to coastal sections, temporary exposures at sites of opencast coal extraction are described. These sites are unique in that they provide extensive inland exposures, and hence new information on the complexity of glacial sequences in this area. The locations of all sites described in this chapter are shown in fig. 192.

Southwest Wales

The glacial sediments exposed along the north Pembrokeshire coast overlie periglacial slope deposits ('head') and in places raised beach deposits (John, 1965; 1970). They tend to plug the coastal valleys and are thin on hill tops and headlands. A particularly useful section is at Aber-mawr (fig. 192) where Quaternary sediments up to 13 m thick are exposed on the northern side of the bay. Sedimentary characteristics described by John (1965) are illustrated in fig. 193. The basal head is variable, but largely derived from the Cambrian shale bedrock. The overlying purple-grey Irish Sea till is massive, silt-rich, matrix-supported and calcareous (plate 28). Occasional lenses of sand and irregularly bedded silt in the upper part produce a foliated appearance in places. The till contains fragments of marine molluscs and carbonised wood. Clasts, which include erratics from Scotland, the Lake District and North Wales, are relatively dispersed, weakly orientated and well rounded (fig. 193). The till is overconsolidated, plastic and well jointed. Its lower boundary is erosional and irregular, with some incorporation of the underlying head.

A sequence of highly disturbed sands and gravels up to 5 m thick overlies the till. The sands often occur as ripple laminated units of well sorted material. The gravels sometimes include lenses of poorly sorted diamicts containing relatively large boulders, and the unit as a whole presents a chaotic appearance with sands, gravels and boulders juxtaposed. A variable deposit described by Jehu (1904) as 'Rubble Drift' and composed largely of angular and subangular local material, but including some well rounded erratics, forms the uppermost part of the sequence (fig. 193). Fabrics show strong downslope preferred orientations.

The main till unit may be interpreted as a lodgement till deposited by ice moving onshore from the Irish Sea basin. The sands and gravels represent fluvioglacial sediments deposited in an ice-contact environment, and include a component of melt-out or flow till. The material above the sands and gravel is a head deposit which incorporates some glacial and fluvioglacial material (John, 1965; 1970).

At a number of sites in Southwest Wales, larger bodies of meltwater deposits occur, usually in valley side or valley bottom locations. A kame terrace at Mullock Bridge near Milford Haven (fig. 192) com-

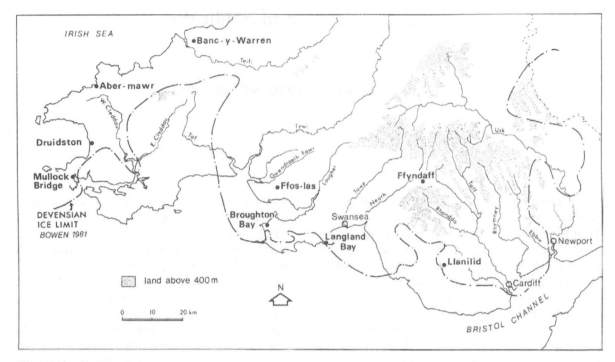

Figure 192: Location map.

prising current bedded sand and gravel was described by John (1970). The most extensive meltwater deposits, however, are at Banc-y-Warren, near Cardigan (fig. 192). Here Helm & Roberts (1975), recognised three cross-bedded gravel units separated by two ripple-laminated sand units. Gravel-filled channels were common within the sands. The cross-bedded units were interpreted as delta foresets. The sands were considered to represent bottomsets produced by settling from density underflows, subsequently channelled by turbidity currents. Helm & Roberts (1975) proposed three incomplete cycles of deltaic sedimentation into a standing water body with three successively higher water levels. Normal faults observed throughout the sequence (fig. 194) were considered to result from gravity-controlled extension.

In contrast, Allen (1982) recognised only two lithostratigraphic units, the lower 30 m thick unit comprising ripple and parallel laminated sands (fig. 195) and the upper 10 m to 12 m thick unit consisting of cross-bedded gravels. In the lower unit climbing ripple sequences were interrupted by gravel-filled channels. These channels contained intact angular clasts of sand identical to the surrounding material. This lower unit was interpreted as a subaerial outwash deposit. Erosion of the frozen sandur in early summer yielded the clasts of sand observed in the

channel-fill gravels. The upper gravels were interpreted as delta foresets, but the lack of bottomsets suggested sedimentation in a body of flowing water rather than a lake. The style of faulting was considered by Allen to be incompatible with Helm & Roberts' interpretation, but indicated disturbance due to melting of an extensive mass of stagnating glacier ice buried during outwash accumulation.

Observations in 1985 and 1986 (Allen, personal communication) have shown a 2 m thick sandy till overlying the upper gravels, and ice wedge casts penetrating the gravels from the contact with the till. The latter indicate permafrost in the uppermost gravels prior to deposition of the till. The till may represent either a slight readvance of the Irish Sea ice, or the emplacement of flow till over the gravels from an exposed stagnant ice mass.

Gower

Two coastal exposures of Devensian glacial sediments occur in Gower, in the east at Langland Bay and in the west at Broughton Bay (fig. 192) (Campbell, 1984). At Langland up to 10 m of very crudely stratified diamict overlies raised beach and head material. In places, the matrix is silty and compact, but

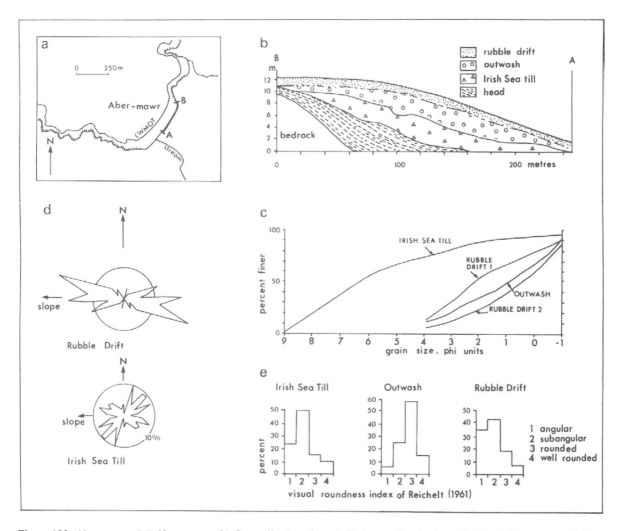

Figure 193: Aber-mawr. a) Drift exposure; b) Generalised section; c) Grain-size distributions (Rubble Drift and outwash from John, 1965; Irish Sea till sampled 1986, tested by wet sieving and Malvern Instruments M3 particle analyser); d) Fabric data (from John, 1965); e) Clast roundness (from John, 1965).

Figure 194: Banc-y-Warren. Section through sands and gravels showing normal faulting (Photograph: J. R. L. Allen, 1982).

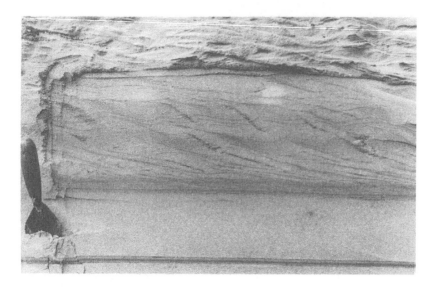

Figure 195: Banc-y-Warren. Climbing-ripple cross-lamination with parallel-laminated sand below (Photograph: J. R. L. Allen, 1978).

Figure 196: Banc-y-Warren. Large scale planar cross beds in upper delta gravels (Photograph: J. R. L. Allen, 1982).

the majority of the section resembles a sandy melt-out till of local South Wales origin.

At Broughton Bay a more complex sequence is described in detail by Campbell (1984) and Campbell *et al.* (1982). Here two till units which overlie raised beach and head deposits are in turn overlain by soliflucted till, colluvium and blown sand (fig. 197). The lower till consists of a 2 m thick matrix-supported diamict containing marine shell fragments and carbonised wood. It is interrupted by impersistent layers of sand and gravel up to 1.3 m thick,

together with pockets of laminated silty clay. Scanning electron microscopy has shown surface textures on many sand grains to be of marine character (Campbell, 1984). Fabrics show variable strengths of preferred orientation, with a dominant direction slightly east of north to slightly west of south (fig. 197). Clast dip directions vary markedly (fig. 197), reflecting deformation of the unit as a whole into gentle basin and dome structures. These structures have a maximum amplitude of 3 m, the elongated axes trending NNE to SSW.

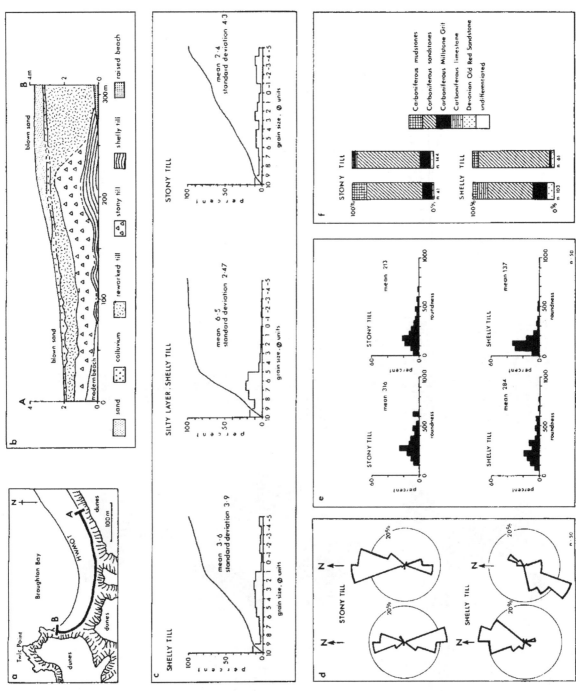

Figure 197: Broughton Bay. Data from Campbell (1984). a) Drift exposure; b) Generalised section; c) Grain-size distributions; d) Fabric data; e) Cailleux roundness; f) Clast lithologies.

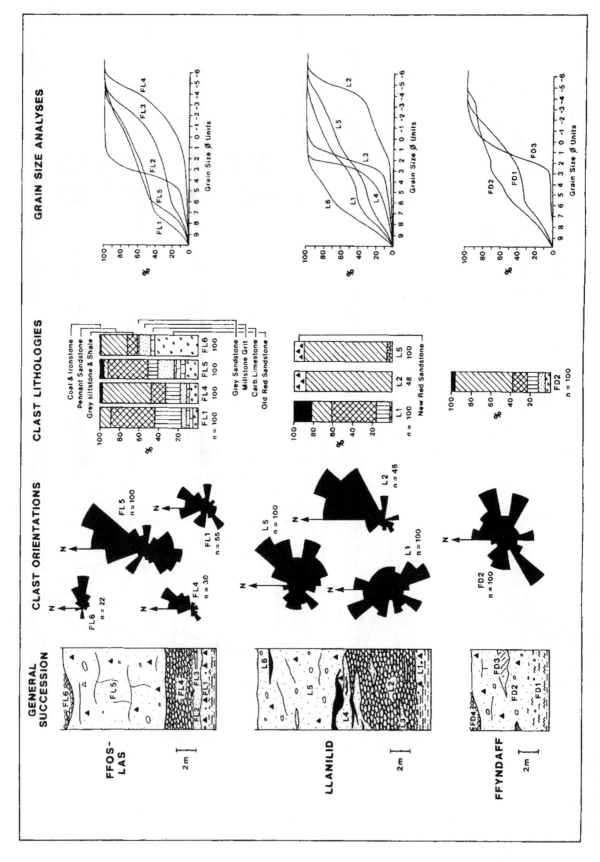

Figure 198: Coalfield Opencast Sites. Sedimentary characteristics.

The overlying till is up to 5 m thick and consists of a massive, yellowish-brown diamict. The unit is stonier than the shelly till, less compact, and in places clast-supported. Lithologically, and in terms of clast roundness, it resembles the lower till (fig. 197). Clasts are strongly orientated, north to south. SEM analysis showed a high proportion of glacially fractured and crushed sand grains (Campbell, 1984).

The Broughton Bay tills were apparently deposited close to the ice margin by a combination of lodgement and melt-out processes. The lower till was enriched with marine sediments incorporated as the glacier crossed Carmarthen Bay. The discontinuous thin beds of sand and gravel suggest release by basal melting. The folding may have been inherited from basal ice structures or have formed during a slight reactivation of the ice front. The stony till and laminated silty clays were subsequently deposited in a melt-out environment.

Along the South Gower coast, between Rhossili and Langland Bay, periglacial head deposits, often incorporating pre-Devensian till, overlie the Ipswichian raised beach (Henry, 1984a; b).

Coalfield

The general succession, stratigraphic relationships, and sedimentological characteristics of the glacial deposits exposed at three sites of opencast coal mining are illustrated in fig. 198. The sediments exposed represent deposition in a range of glacial environments.

Ffos-las. The Ffos-las site is situated just to the north of the village of Trimsaran in the NE-SW trending Gwendreath Fawr valley (fig. 192). Carboniferous Coal Measures bedrock is overlain to the south and east by younger Pennant Measures, and bounded to the north and west by Millstone Grit, Carboniferous Limestone, and Old Red Sandstone.

The siltstone and shale bedrock is often highly shattered up to 2 m below rock head. The overlying unit (FL1) is typically 40 - 60 cm thick, brown-grey, massive, overconsolidated, matrix-supported, and very poorly sorted (fig. 199). Clasts dominantly comprise angular local Coal Measure material, and show a NE to SW preferred orientation (fig. 198).

Overlying the lower diamict (FL1) with an erosional base is a coarsening upwards sequence of sands and gravels (FL2, FL4). The sandy part of the sequence is up to 1 m thick and ranges from fine-grained and well-sorted ripple-laminated sands to coarse-grained poorly sorted sands. The coarse sands grade up into 2 to 3 m of clast-supported, crudely stratified pebble gravels with a sandy matrix.

Often interbedded with the sands near the base of the unit are discontinuous lenses of material (FL3) very similar to the underlying grey diamict. They are, however, not so compact, depleted in fines and are devoid of any large boulders.

Resting on top of the sands and gravels with a

Figure 199: Ffos-las. General succession. Shattered bedrock passes into lower till (FL1) approximately 0.5 m above spade. Overlying the lower till are sands (FL2) grading up into gravels (FL4) which are in turn overlain by the upper till (FL5) (Photograph: R. Donnelly, 1984).

particularly sharp base is the major unit in terms of volume exposed at this site (FL5). It is a massive, jointed and overconsolidated, brown-grey, matrix-supported, very poorly sorted diamict up to 25 m thick. Clasts, although dominantly of local material, include a relatively high proportion of Carboniferous Limestone, Millstone Grit and Old Red Sandstone (fig. 198). A strong fabric (plate 29) indicates ice flow from the NNE (fig. 198).

At the top of the Ffos-las succession, preserved in channels and hollows, are silts, sands and coarse gravels (FL6). The silts and sands are fairly well-sorted, occasionally laminated, and contain rare dropstones. The gravels are generally massive, well-sorted, and often display a well-developed fabric.

This sedimentary sequence is produced by oscillations in the position of an ice front. Initial ice advance from the NE produced the crushed bedrock or comminution till seen at the base of the succession just prior to the subglacial lodgement of the lower till. The ice front then retreated sufficiently to allow the accumulation of the sands and gravels as proglacial outwash sediments. The grey diamict lenses interbedded with the sands are almost certainly flow tills derived from melting ice either at the ice margin, or on the outwash plain.

A major readvance from the NNE then deposited the thick upper diamict as another subglacial lodgement till. Final retreat produced only limited outwash with gravels, sands and silts accumulating in channels and hollows on the till plain.

Llanilid. Llanilid opencast site is located approximately 3 km east of Pencoed (fig. 192). Pennant Sandstone forms the highland to the north, but to the south and east are important exposures of Triassic New Red Sandstone, Devonian Old Red Sandstone, and Carboniferous Limestone. An 'older drift' has been recorded in this area by previous workers (e.g. Bowen, 1974), but there is no reason to believe that any of the deposits described here are older than late Devensian.

Deformed and crushed Coal Measure sediments again form the basal unit. This comminuted material is either interbedded with or grades into an overlying diamict. The diamict (L1) is a grey, compact, matrix-supported sediment with clasts dominantly of local Coal Measure material, and with a fine clayey silt matrix almost certainly derived from the same source. Maximum recorded thickness is approximately 4 m, but the till is discontinuous, tending to be limited to bedrock hollows. Fabric and clast lithologies (fig. 198) suggest local ice moving from the NW.

Above the Lower Till (L1), with a typically erosional base, is a sequence of mainly massive, coarse, clast-supported gravels (L2). Occasionally interbedded with the gravels are layers of fairly well-sorted sands and silts (L3) (plate 24). Sedimentary structures in the gravels include imbrication (plate 25), grading, and rare planar cross-bedding, and in the sands and silts ripple cross-lamination, horizontal lamination, and planar and trough cross-stratification. Fabric and palaeocurrent readings together with the greater abundance of Triassic clasts indicate water flowing from the NE. These sands and gravels reach a maximum thickness of 15 m. Degrees of deformation in the gravels vary from simple clast fracture to almost complete clast disintegration, and in the sands, from small-scale faulting and folding (fig. 200) to almost complete mixing of the beds (fig. 201).

Generally overlying, but in part stratigraphically equivalent to the sands and gravels, are the sediments comprising the Upper Till facies (L4, L5 and L6). This facies is composed of various sediment types that together produce a maximum thickness of over 30 m. It is for the most part made up of a relatively sandy, red, brown or grey-brown matrix-supported diamict. In places it is dense and compact, in others less compact and structureless. Fabric measurements suggest ice movement from the NE (fig. 198). Interbedded or interfingering with the diamict are lenses or bodies of red-brown sands, silts and clays. They are present in considerable thicknesses (up to 8 m), particularly near the base (fig. 202). The finer units often display well-developed laminations. Deformation in these sorted units is restricted to the lower part of the Upper Till facies (fig. 202). Those nearer the surface appear not to have been disturbed in any way (fig. 203 and plate 26).

As at Ffos-las, the environments of deposition at Llanilid resulted from oscillations in the position of an ice front. Initial ice advance from the north produced the communition till, and deposited the lower grey diamict as lodgement till. Any existing 'older drift' would probably have been reworked during this phase. Retreat then allowed the development of a relatively extensive and dominantly proximal sandur system where the coarse gravels and sands accumulated.

Ice from the NE then introduced sediments containing significant amounts of red Triassic material. Red-brown sands, silts and clays were laid down in standing bodies of water that were mainly proglacial, but probably also partly subglacial. Periodic ice advance and overriding was however still important,

Figure 200: Llanilid. Small-scale intensive faulting and folding of silty sands (L3) (Photograph: C. Harris, 1983).

depositing till by lodgement and melt-out, and producing glaciotectonic deformation of the underlying glaciofluvial and glaciolacustrine deposits. Final stagnation allowed the accumulation of increased thicknesses of till, and where meltwater collected in small lakes or ponds, undeformed laminated sediments were deposited.

Ffyndaff. The Ffyndaff opencast site is situated approximately 5 km west of Hirwaun on the north crop of the South Wales Coalfield (fig. 192). To the north is the high land of the Brecon Beacons composed of Old Red Sandstone fringed by Carboniferous Limestone. To the south is the prominent Pennant Sandstone escarpment.

Highly deformed or crushed bedrock again forms the base of the succession (FD1). This unit is sometimes 2 - 3 m thick, with a gradational lower and upper contact. Passing up from the lower unit is a grey-brown, matrix-supported, overconsolidated diamict containing more rounded clasts (FD2). Notably, however, Pennant Sandstone dominates, Old Red Sandstone clasts are rare, and Carboniferous Limestone is apparently totally absent (fig. 198). Fabric measurements show a weakly developed preferred orientation, the strongest component being east-west.

This relatively uncomplicated situation is recorded at the majority of the exposed sections. Occasionally, however, bodies of sand and laminated silt can be seen within the till. One such sand body (FD3) was exposed over a distance of 34 m, with a maximum-thickness of 3 m (fig. 204). Excavation showed the

Figure 201: Llanilid. Intense deformation of sands and gravels (L3 and L2) (Photograph: R. Donnelly, 1985).

Figure 202: Llanilid. Red-brown sands (L4), silts and clays. Note glacio-tectonically produced box-fold (Photograph: R. Donnelly, 1985).

Figure 203: Llanilid. Laminated silts and clays (L6) draped over diamict (L5). Note dropstone upper centre of photograph (Photograph: R. Donnelly, 1985).

unit to be only 8 to 10 m wide. It was underlain by a loose sandy diamict. The lower boundary was irregular with a thin silt band passing upwards into faulted ripple-laminated silty fine sands. Above these, coarse cross-bedded sands containing pebble-sized coal clasts were observed with a strongly sinusoidal erosional lower boundary. The cross beds passed laterally into finer-grained climbing ripples which followed the underlying sinusoidal boundary (fig. 205). The cross-bedded unit was overlain by faulted ripple-laminated sands containing occasional dropstones (plate 27). Palaeocurrent directions were con-

sistently towards the northeast (fig. 205).

At one locality a sandy gravel unit (FD4) was observed in a crude channel form cut into the upper surface of the diamict.

The glacial history represented by the deposits at Ffyndaff is important because unlike the other sites the locality is well within proposed Devension ice limits. As at Ffos-las the lower units of crushed bedrock or comminution till and the overlying matrix-supported diamict represent initial ice advance with its associated subglacial deformation and deposition. Clast lithology indicates a coalfield

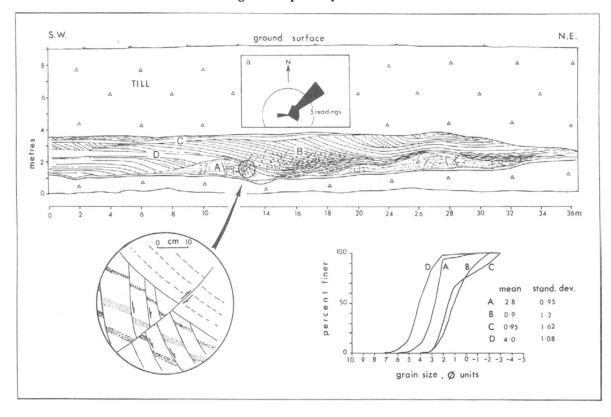

Figure 204: Ffyndaff. Sand body (FD3) within diamict (FD2). Inset (a) palaeocurrent data. Grain-size data, mean and standard deviation in phi units.

Figure 205: Ffyndaff. Cross-bedded and ripple-laminated sands (FD3). Frame height approximately 4 m (Photograph: C. Harris, 1982).

source, but fabric suggest west-east diversion of northward flowing coalfield ice due to confluence with ice flowing south from the Brecon Beacons.

The bodies of sorted material including the sands, silts and clast-supported diamicts probably represent a relatively warm period when ice stagnation and melting allowed accumulation of these deposits in sub- or englacial cavities. The diamict which overlies the sorted material was subsequently deposited following renewed ice accumulation and activity. The sandy gravels described at the top of the sequence represent glaciofluvial outwash after final ice retreat.

A correlation between the three sites described reveals strong similarities between Ffos-las and Llanilid, with basically two till units separated by a sequence of proglacial outwash deposits. At Ffyndaff the subglacial meltwater deposits are found at what may possibly be an equivalent stratigraphic level to the outwash deposits at the other two sites. Ameliorations and deteriorations in climate during the late Devensian are therefore perhaps recognisable at each of the three sites described.

Conclusion

The variety of glacial deposits described in this chapter reflect variations in depositional environment and source area. Lodgement, melt-out and flow tills are described together with meltwater deposits including coarse gravels, well-sorted sands and fine laminated silts and clays. This range of depositional environments makes it very difficult to construct a regional Devensian lithostratigraphy, the only alternative being a simple chronostratigraphy.

Acknowledgements

The authors wish to thank British Coal Opencast Executive for permission to work at the Ffos-las, Llanilid and Ffyndaff opencast sites. The help and cooperation of British Coal officers at these sites is acknowledged.

Glacial deposits of the Isles of Scilly

JAMES SCOURSE

The existence of pebbles of probable glacigenic derivation on the northern Isles of Scilly (fig. 206) has been known for over a century. Smith (1858) first recorded the occurrence of these foreign chalk-flints and greensand at Castle Down, Tresco; "the flints and greenstones varied little in size, ranging from that of a hen's egg to that of a blackbird - how they got to Scilly was a mystery which it was for gentlemen of more scientific knowledge than he professed to explain" (Bishop, 1967). Such a gentleman proved to be Whitley who, in 1882, interpreted these foreign stones as glacial in origin.

This early discovery of erratic material on the Islands was helped by the fact that the solid geology is exclusively granite, Scilly representing the highest parts of an almost completely submerged *cupola* which is the westernmost extension of the South West Variscan batholith.

The stratigraphical context of the glacial erratics was first explored by Barrow (1906), who took Chad Girt on White Island, St. Martin's as his type-site for the Pleistocene of the Scillies (fig. 207). Two years earlier Barrow had exhibited a striated boulder from the Islands to the Geological Society (1904). Barrow observed that in cliff section the erratics were usually set within a fine silty matrix, often cemented by iron oxides, and that this 'glacial deposit' was both underlain and overlain by head deposits, the whole resting on a raised beach (fig. 207).

Mitchell & Orme (1967) undertook a detailed re-examination of the Pleistocene stratigraphy of the Islands, and proposed a somewhat more complicated sequence than Barrow:
1. Upper Head
2. Raised Beach (Porth Seal)
3. Glacial Deposit
4. Raised Beach (Chad Girt)
5. Shore Platform

Mitchell & Orme divided Barrow's 'glacial deposit' into two facies, till and outwash gravel, and identified an ice limit running through the northern Islands based on the distribution of these sediments (fig. 206). By comparing their basic stratigraphy with similar sequences in southwest England and in Ireland, Mitchell & Orme claimed a Gipping (Wolstonian) age for the glacial deposits. This interpretation is largely based on the suggestion that the erratic-free Chad Girt Raised Beach is Hoxnian in age by correlation with other beaches at a similar altitude elsewhere e.g. the Courtmacsherry Raised Beach in southern Ireland (see also Warren and Hoare, this volume). Direct chrono-lithostratigraphical interpretation then dictates that the Porth Seal Raised Beach should be Eemian (Ipswichian) in age.

Bowen (1969, 1973b) argued with this interpretation, suggesting that the lenticular mode of the glacial material and its association with coastal valleys was more consistent with a solifluction origin. He later suggested (1981) that the critical stratigraphy identified by Mitchell & Orme (1967) at the Porth Seal site was "inferred, and superposed, and granite corestones have been interpreted as a marine deposit". Bowen (1973b) regards the single raised beach of Ipswichian age, the soliflucted glacial sediments having been originally emplaced during the Wolstonian.

The dating of the glacial material on the Scillies by both Mitchell & Orme (1967) and Bowen (1973a) is thus heavily dependant on the interpretation of the number and age of the stratigraphically juxtaposed raised beach units, and thereby correlation with neighbouring regions.

Stratigraphical context of the glacial deposits

In order to establish a local lithostratigraphic framework independent of the raised beach-tied stratigraphies erected in southern Ireland and Wales, and to obtain more sedimentological information on the glacial sediments themselves, Scourse (1985, 1986) mapped the extant sections on the sixteen largest Islands. Organic sediments found interbedded in the

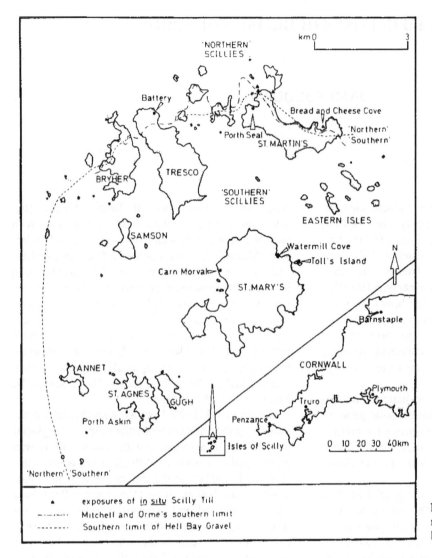

Figure 206: The Isles of Scilly: location map, critical sites and the southern limit of the Hell Bay Gravel.

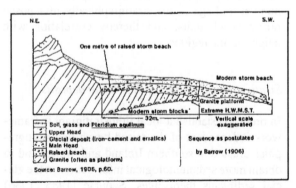

Figure 207: Stratigraphy identified by Barrow (1906) at Chad Girt, White Island. St. Martin's (re-drawn from Mitchell & Orme, 1967, fig. 4, p.67).

sequence have helped to provide a chronology of events through twenty-nine radiocarbon determinations. Two pre-existing thermoluminescence (TL) dates (Wintle, 1981) have also assisted in the establishment of a radiometric chronology. Local inter-site correlations have been strengthened by palynological analyses of the organic deposits, and this has also assisted palaeoenvironmental reconstruction.

The defined units (Scourse, 1985; 1986) have been incorporated into two lithostratigraphic models for the 'southern' Scillies and 'northern' Scillies (fig. 208); these areas can be regarded as 'extra-glacial' and 'glacial' respectively. The southern limit of the

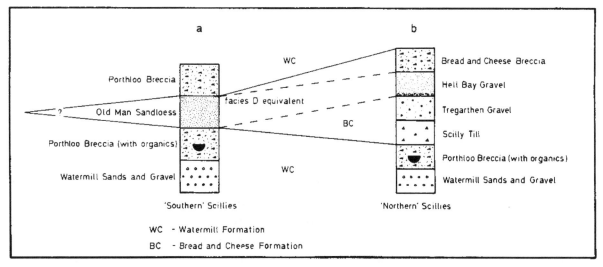

Figure 208: Lithostratigraphic models for the southern and northern Scillies, and their correlation. The Bread and Cheese Formation comprises all members containing abundant foreign material introduced during the Dimlington Stadial, the Watermill Formation being dominated by material of local derivation.

Hell Bay Gravel defines the boundary between these two areas, and corresponds closely to the ice limit identified by Mitchell & Orme (1967).

Overlying the raised beach sediments of the Watermill Sands and Gravel in the southern Scillies is the Porthloo Breccia, a unit of variable soliflucted material derived entirely from the weathering of the granite bedrock (Scourse, 1987). Organic deposits have been found towards the base of this unit at five sites (fig. 206); Carn Morval (SV 905118; plate 31), Watermill Cove (SV 925123), Toll's Island (SV 931119), Porth Askin (SV 882074) and Porth Seal (SV 918166). These are thought to represent the infillings of small ponds or lakes impounded by active solifluction sheets or lobes.

Radiocarbon determinations from these organic sequences (table 22) are critical since the organic sequences are interpreted as pre-dating the units related to the glacial advance, the Scilly Till, the Tregarthen Gravel, the Hell Bay Gravel and the Old Man Sandloess (fig. 208). If accepted as reliable these dates therefore together provide a maximum age for the glaciation.

All the organic sites are permanent open sections thus increasing the potential for contamination of the radiocarbon samples by modern rootlets and groundwater infiltration. After thorough cleaning, samples were taken from two locations within most of the organic units at specific levels, from the open cleaned face ('external samples', table 22), and from a horizontally-bored sample up to 3 m within the unit ('internal samples', table 22). The humic alkali extracts were measured in addition to the solid plant detritus residues permitting a degree of mutual control which assists in the identification of the extent and sources of contamination. As expected, the majority of extract dates are significantly younger than the residue dates suggesting a degree of contamination of the sediment by younger humic materials.

The pollen diagrams from the organic sites are all very similar in recording tundra grassland vegetation, and represent the earliest vegetational record for the Scillies. If the radiocarbon determinations are ac-

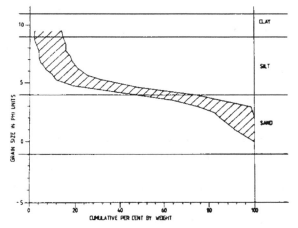

Figure 209: Grain-size envelope for the Old Man Sandloess Member.

Table 22. Radiocarbon determinations on organic sediments within the Porthloo Breccia.

Site	Bed	Sample location	Sample fraction	^{14}C date yr BP	Lab. no. Cambridge (Q)
Carn Morval	2b	extern.	extract	$19,860^{+220}_{-210}$	2357
Carn Morval	2b	extern.	residue	$24,490^{+960}_{-860}$	2356
Carn Morval	2f	extern.	extract	$19,300^{+120}_{-120}$	2359
Carn Morval	2f	extern.	residue	$21,500^{+890}_{-800}$	2358
Carn Morval	2b	intern.	extract	$21,640^{+270}_{-260}$	2446
Carn Morval	'lowest layer'	extern.	bulk sample	$26,550^{+700}_{-650}$	2176
Carn Morval	'middle layer'	extern.	bulk sample	$20,630^{+480}_{-450}$	2177
Watermill Cove	3a	extern.	extract	$23,250^{+1720}_{-1420}$	2361
Watermill Cove	3a	extern.	residue	$27,800^{+1770}_{-1450}$	2360
Watermill Cove	3c	extern.	extract	$23,030^{+1275}_{-1100}$	2363
Watermill Cove	3c	extern.	residue	$24,900^{+430}_{-410}$	2362
Watermill Cove	3a	intern.	extract	$31,770^{+850}_{-770}$	2447
Watermill Cove	3a	intern.	residue	$33,050^{+960}_{-860}$	2408
Watermill Cove	3c	intern.	extract	$28,870^{+590}_{-550}$	2406
Watermill Cove	3c	intern.	residue	$26,680^{+1410}_{-1200}$	2407
Porth Askin	1a	extern.	bulk sample	$25,920^{+590}_{-550}$	2178
Porth Askin	1a	extern.	extract	$20,960^{+180}_{-180}$	2371
Porth Askin	1a	extern.	residue	$23,980^{+1400}_{-1200}$	2370
Porth Askin	1a	intern.	extract	$22,960^{+625}_{-580}$	2413
Porth Askin	1a	intern.	residue	$24,550^{+500}_{-470}$	2412
Porth Seal	4b	extern.	extract	$16,440^{+120}_{-120}$	2365
Porth Seal	4b	extern.	residue	$18,780^{+260}_{-250}$	2364
Porth Seal	4d	extern.	extract	$11,200^{+1550}_{-1300}$	2367
Porth Seal	4d	extern.	residue	$15,450^{+120}_{-120}$	2366
Porth Seal	4b	intern.	extract	$34,500^{+885}_{-800}$	2410
Porth Seal	4d	intern.	residue	$25,670^{+560}_{-530}$	2409
Bread and Cheese Cove	1a	extern.	extract	$7,880^{+180}_{-180}$	2369
Bread and Cheese Cove	1a	extern.	residue	$9,670^{+65}_{-65}$	2368
Bread and Cheese Cove	1a	intern.	extract	$7,830^{+110}_{-110}$	2411

cepted as reliable they appear to indicate deposition of the organic material between $34,500^{+885}_{-800}$ (Q-2410) and $21,500^{+890}_{-800}$ (Q-2358) years B.P.; the pollen diagrams are broadly similar to other spectra of this age from elsewhere in NW Europe (Bell *et al.*, 1972; Morgan, 1973; West, 1977c).

In the southern Scillies the Porthloo Breccia is overlain by the Old Man Sandloess, a coarse aeolian silt (Catt & Staines, 1982) with subdominant fine sand and minor amounts of clay (fig. 209), from which two TL dates, both of $18,600^{+3,700}_{-3,700}$ (QTL 1d and 1f) B.P. have previously been published (Wintle, 1981).

In the northern Scillies the Porthloo Breccia is overlain by three units which are interpreted as all related to one glacial event. At four sites a massive,

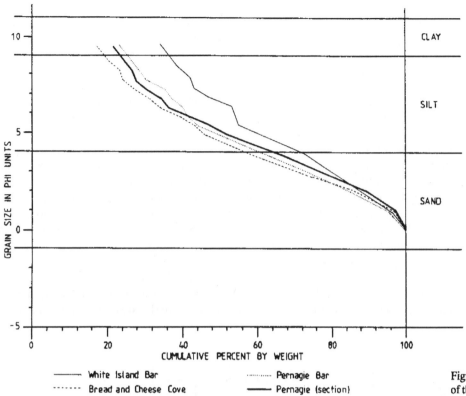

Figure 210: Grain-size analyses of the Scilly Till Member.

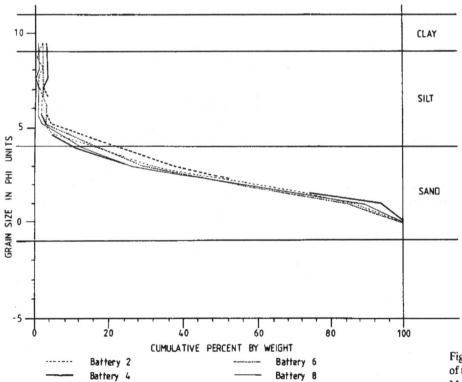

Figure 211: Grain-size analyses of the Tregarthen Gravel Member.

very poorly sorted (fig. 210) clast- and clay-rich pale brown diamicton has been identified as the Scilly Till. At Bread and Cheese Cove (fig. 206) this material occurs in association with a matrix-supported sandy gravel, the Tregarthen Gravel (fig. 211) which has an erratic assemblage consistent with the underlying Scilly Till (fig. 212). This latter material also occurs in isolation elsewhere. Aeolian loessic sedimentation in association with the glacial advance is interpreted to have been responsible for the deposition of the Old Man Sandloess in the southern Scillies. The relative coarseness (fig. 209) of this unit is perhaps a function of proximity to its glacially-derived source material. The mineralogy of the Scilly Till is very similar to the Old Man Sandloess with the exception of the loss of some weatherable minerals in the latter (Catt, 1986b), suggesting a genetic association between the two units.

The Hell Bay Gravel, a coarse silt / fine sand, matrix-supported gravel containing a wide variety of glacially-derived striated erratics (fig. 213) with a considerable proportion of local material, is very widespread in the northern Scillies (fig. 206). This material is interpreted as a solifluction deposit derived from the Scilly Till, Tregarthen Gravel, and, in particular, the Old Man Sandloess; grain-size analyses of the Old Man Sandloess and the matrix of the Hell Bay Gravel are almost identical (fig. 214), as is the mineralogy of the two units (J.A. Catt, personal communication 1983). The Hell Bay Gravel represents Barrow's 'glacial deposit'. In situations where the glacially-derived units were totally stripped from the land surface by solifluction, weathered granite was once again soliflucted, this constituting the Bread and Cheese Breccia. In the southern Scillies this period of solifluction is represented by the upper Porthloo Breccia.

If this evidence is accepted it indicates an ice advance as far as the northern Isles of Scilly during the Dimlington Stadial (Rose, 1985) of the Late

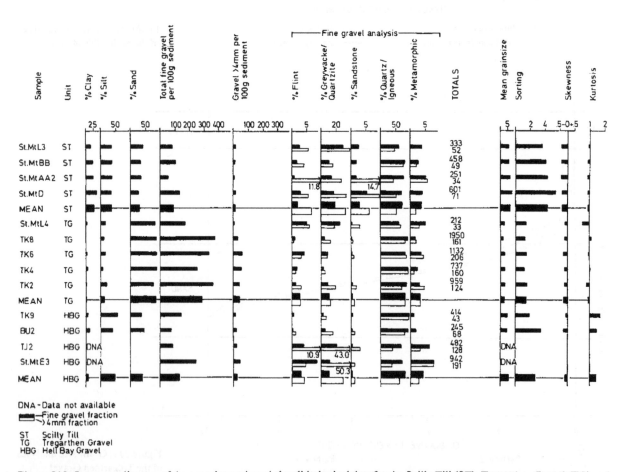

Figure 212: Summary diagram of the granulometric and clast lithological data for the Scilly Till (ST), Tregarthen Gravel (TG) and Hell Bay Gravel (HBG) Members.

Figure 213: Striated erratic of Palaeozoic greywacke from Hell Bay, Bryher (collected by Dr. R.W. Hey). At least three different striae directions can be discriminated on this clast. The 'a' axis of the clast measures 15 cm.

Devensian Substage around 18,600 ±3,700 (QTL 1d and 1f) years B.P. (Wintle, 1981). It is important to recognise, however, that as erratic material does occur within the Watermill Sands and Gravel this glaciation is likely to have been the last and not the only glacial event to have influenced the Islands.

The evidence which supports this model of a Devensian glaciation clearly conflicts in a number of important respects with Mitchell & Orme's sequence of events. The major differences include:

1. Recognition of only one raised beach unit stratified with other sediments.

2. Recognition of widespread loessic sediments in

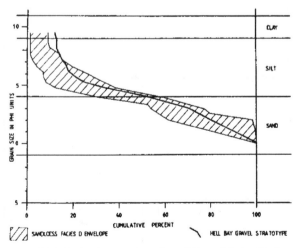

Figure 214: Grain-size envelope for soliflucted facies (facies D) of the Old Man Sandloess Member compared with the grain-size curve for the matrix of the Hell Bay Gravel stratotype.

the Scillies, and an interpretation of a sedimentary suite of till, outwash gravel and loess related to a single glacial event.

3. Independent radiometric dating rather than relative dating based on the inferred ages of raised beach units.

Discussion

As with all interpretations in historical geology, this model of the Devensian glaciation of the Isles of Scilly is but a working hypothesis. The evidence on which it is based is clearly far from unequivocal, but it is impossible to reconcile with a Wolstonian age for the glaciation. The fundamental points in the argument are:

1. The reliability of the radiocarbon and TL dates.

2. The validity of the lithostratigraphic correlations between sites and islands.

3. The *in situ* status of the Scilly Till.

Taken as a group (table 22) it is suggested that the radiocarbon determinations form a fairly consistent group given the problems of analysing pre-Flandrian material. Both the sediments and the pollen spectra from the organic sequences are clearly cold stage, and probably periglacial in their affinities; they are not interglacial or postglacial.

It is possible to question the between site and island lithostratigraphic correlations. There is only one organic deposit which directly underlies the Scilly Till anywhere, at Bread and Cheese Cove (fig. 206). This is also a very unsatisfactory site in that it has provided clearly aberrant radiocarbon determinations (table 22). It could therefore be argued that the Porthloo Breccia in the northern Scillies correlates with the upper Porthloo Breccia in the southern Scillies, thereby overturning the stratigraphic relationship between the radiometric dates and the Scilly Till. This would imply an extremely complicated depositional history, for example, two major phases of loess deposition, extremely steep gradients in depositional environments over short distances, and it would conflict with the mineralogical association between the Scilly Till and Old Man Sandloess (Catt, 1986b).

Characteristics of the Scilly Till

The *in situ* status of the Scilly Till is an important factor in the argument. A detailed examination of its physical and mineralogical characteristics has been undertaken at the type-site, Bread and Cheese Cove (fig. 206). Some of these characteristics, and comparisons with other related units, are given in fig. 212.

At Bread and Cheese Cove the Scilly Till is dark yellowish brown (10YR 4/4) drying to light yellowish brown (10YR 6/4). It is largely non-calcareous, but the mineralogical data (Catt, 1986b) indicates that it is not heavily weathered, containing a number of easily weatherable minerals such as muscovite, glauconite, chlorite, biotite, augite, apatite, olivine and calcite. However, it does not contain any calcareous micro- or macrofossils, though it does contain abundant siliceous sponge spicules (D.A. Jenkins, personal communication, 1986) suggesting a marine derivation with subsequent partial decalcification.

The till is crudely stratified with a number of sub-horizontal iron-stained sand partings. The contained clasts are striated and faceted, and consist of a wide variety of lithologies, including Cretaceous flint, Variscan greywackes and quartzites, red sandstones and schistose metamorphic rocks in addition to local granitic material (fig. 212). The sediment

also contains small specimens of glauconitic micrite of probable Miocene age (A. Morton, personal communication, 1983) derived from the Jones Formation offshore to the north (Pantin & Evans, 1984).

The till is extremely poorly sorted, containg 43% sand, 38% silt and 19% clay, and is strongly coarse-skewed. Three features set the till off sharply from the underlying Porthloo Breccia, the high clay content, the lower granule and clast concentrations and the rich erratic assemblage (fig. 215). The abundance of clay and silt, the occurrence of sponge spicules and the distinctive erratic assemblage are all consistent with a derivation from the offshore area to the north of the Scillies.

A coarse lag of angular granite boulders occurs at the base of the Scilly Till at its contact with the Porthloo Breccia, whilst the upper contact with the Bread and Cheese Breccia is clearly solifluction.

Three clast macrofabric diagrams from the Scilly Till, one from the underlying Porthloo Breccia and

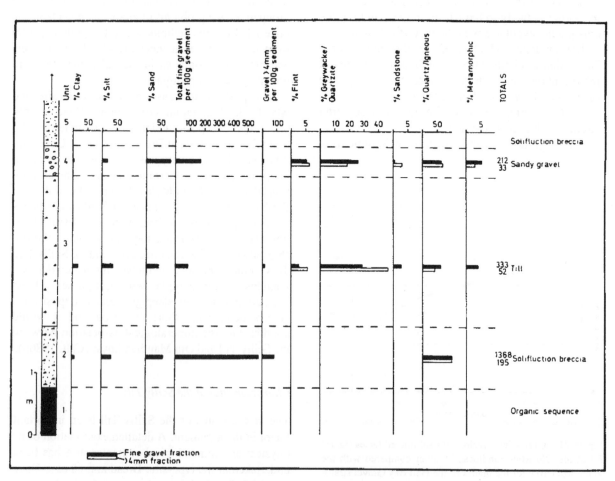

Figure 215: Stratigraphy, granulometry and clast lithological data at Bread and Cheese Cove, St. Martin's.

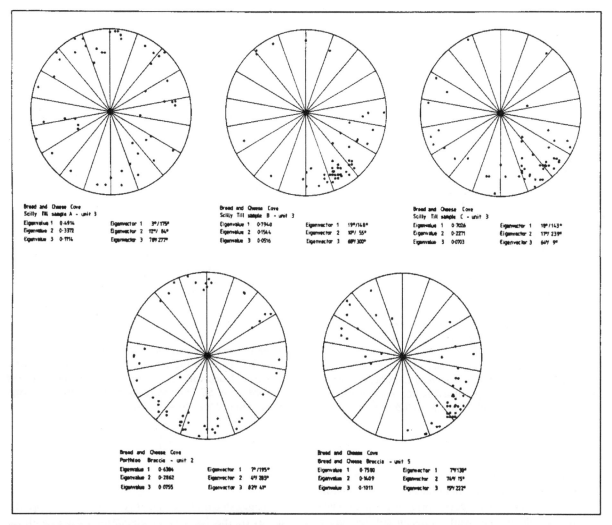

Figure 216: Fabric analyses from the Scilly Till, Porthloo Breccia and Bread and Cheese Breccia Members, at Bread and Cheese Cove, St. Martin's.

one from the overlying Bread and Cheese Breccia are presented in fig. 216. Sample A was from the base of the unit, sample B from the middle and sample C just beneath the soliflucted upper contact. Sample A shows no strong preferred orientation (Eigenvalue 1: 0.4914; Eigenvector 1: 3°/175°) but the weak orientation is in sympathy with the underlying Porthloo Breccia. Sample B, however, has a strong (Eigenvalue 1: 0.794) SW/NE preferred orientation (Eigenvector 1: 19°/148°), as does sample C (Eigenvalue 1: 0.7026; Eigenvector 1: 18°/143°). The overlying Bread and Cheese Breccia has a similar fabric signature, with strong (Eigenvalue 1: 0.758) NW/SE preferred orientation (Eigenvector 1: 7°/130°). The strong fabrics from the Scilly Till samples B and C and the Bread and Cheese Breccia differ, however, in

their vectors from the Porthloo Breccia.

The S1 and S3 eigenvalues (Mark, 1973) for all fabrics from Bread and Cheese Cove have been plotted against fabric data from modern glacigenic sediments (Dowdeswell & Sharp, 1986) in fig. 217. This indicates that while samples B and C plot close to 'meltout till' and 'undeformed lodgement till', sample A is significantly different in resembling 'glacigenic sediment flow', as does the Bread and Cheese Breccia. The Scilly Till fabrics have also been plotted against Rose's (1974) data on 'lodgement' and 'slumped' till from Hertfordshire (Dowdeswell & Sharp, 1986) in fig. 218. Samples B and C clearly resemble lodgement till whilst sample A plots within the 'slumped' till category.

In terms of fabric then, samples B and C are typical

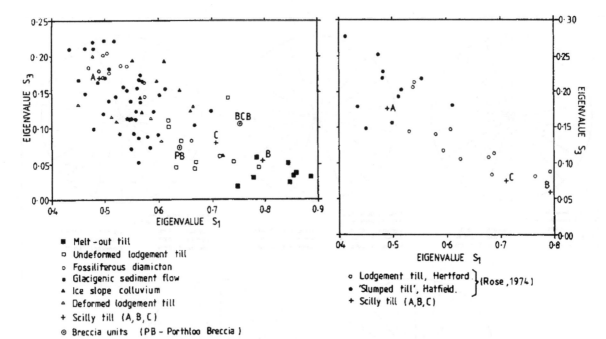

Figure 217: Fabric eigenvalue S1 and S3 parameters from the Scilly Till, Porthloo Breccia and Bread and Cheese Breccia Members plotted with eigenvalue data from modern glacigenic sediments as compiled by Dowdeswell & Sharp (1986).

Figure 218: Fabric eigenvalue parameters S1 and S3 from the Scilly Till, Porthloo Breccia and Bread and Cheese Breccia Members plotted with eigenvalue data from lodgement and slumped till in SE England (data in Dowdeswell & Sharp, 1986, compiled from Rose, 1974).

for lodgement till from both modern and fossil contexts, but the validity of this interpretation is cast in some doubt when the breccia fabric results are considered. Sample A may well represent soliflucted or remobilised till, whereas samples B and C may represent lodgement recording an ice movement from the NW, which accords with additional fabric data from the Scilly Till at the nearby White Island Bar site (Scourse, 1985; 1986). The strong preferred orientation of the Bread and Cheese Breccia compared with the Porthloo Breccia may represent solifluction in response to a glacial modification of the local slope.

Whilst the Scilly Till is undoubtedly of glacigenic origin from an offshore source area, its precise mode of deposition must remain an open question. It is certainly different in a number of fundamental characteristics from undoubted soliflucted till, as represented by the Hell Bay Gravel, which suggests that it may well be *in situ*.

Conclusion

It is suggested that the evidence reviewed to support a model of the Devensian glaciation of the Scillies, though equivocal, is nevertheless more substantial than the evidence presented to support a Wolstonian age for the glaciation. The hypothesis of a Devensian glaciation is not new; John (1971) and Synge (1977a, 1985) both having raised the possibility in speculative discussion. The evidence reviewed here supports such an hypothesis.

The dating of the glacigenic material on the Scillies raises many questions concerning the stratigraphy of the adjoining regions, both onshore and offshore. Some of the offshore questions are addressed by Scourse, Robinson & Evans (this volume).

Glaciation of the central and southwestern Celtic Sea

JAMES SCOURSE, ERIC ROBINSON & CHRISTOPHER EVANS

Over the last two decades a mass of data has been collected on the Quaternary sediments of the Irish and Celtic Seas by the Marine Geology Unit of the British Geological Survey (BGS; formerly Institute of Geological Sciences, IGS). A number of publications arising directly from this work have addressed the problems of the offshore Quaternary sequence, and the correlation of this sequence with the much-studied but controversial onshore stratigraphy (Garrard & Dobson, 1974; Delantey & Whittington, 1977; Garrard, 1977; Pantin & Evans, 1984). The new data which has recently emerged (Scourse, this volume) concerning the character and age of the glacigenic sediments on the Isles of Scilly has stimulated a detailed appraisal of the glacigenic sediments of the Central and Southwestern Celtic Sea (fig. 219).

The general picture revealed in the Irish Sea by the IGS in the early 1970's was of two units of till separated by thick sequences of temperate interglacial marine sediments. Garrard & Dobson (1974) regarded the upper till sheet as being of Devensian age, and proposed southern limits for this upper till sheet in the vicinity of the entrance to St. George's Channel (fig. 220), the underlying marine sediments being interpreted as Ipswichian in age. During the deglaciation of the Late Devensian, the shallower regions of the Irish Sea experienced an intensive period of erosion which produced a number of meltwater channels; these subsequently acted as traps for the deposition of estuarine sediments associated with the early stages of the Flandrian transgression (Garrard & Dobson, 1974). Garrard (1977) reports small and isolated outliers of 'glacial drift' found as far south as the Bristol Channel. These consist of 'lag gravels', small patches of erratic pebbles and cobbles with the fines winnowed out by strong tidal currents. Garrard (1977) also suggested definite limits for the upper of the Irish Sea till sheets (Devensian) in the St. George's Channel area (fig. 220).

In 1976 Doré identified the Neogene/Pleistocene

boundary in the Nymphe Bank area of the northern Celtic Sea (fig. 220), but these 'Neogene' deposits were reinterpreted as Devensian glacigenic sediments by Delantey & Whittington (1977), the boundary itself being reinterpreted as the Devensian glacial limit (fig. 220).

Pantin & Evans (1984) identified two main Quaternary formations in the Central and Southwestern Celtic Sea, the late Pliocene/early Pleistocene Little Sole Formation and the Late Devensian/early Flandrian Melville Formation (fig. 221). The Melville Formation consists mainly of tidal deposits, but at a number of sites it also contains glacigenic sediment. Pantin & Evans also report occasional small 'mounds' of possible glacigenic material and scattered boulders on the sea bed as revealed by side-scan sonar, indicating 'ice-rafting'. The occurrence of these scattered boulders has also been noted by Hamilton et al. (1980).

The cores of 'glacigenic sediment' were samples by the Continental Shelf Division of the IGS between 1974 and 1981. The glacigenic samples were recovered exclusively using vibrocorers and were positioned using Decca Main Chain or Pulse 8 navigation systems (Pantin & Evans, 1984).

These sediments and features occur between 300 and 500 km to the southwest of the suggested limits of Devensian material in the St. George's Channel area (fig. 220). This paper reviews the data that has emerged from the detailed analysis of this glacigenic material. These data come only from the UK sector of the shelf, the Irish and French sectors being largely unknown.

Site locations

The coring sites from which 'glacigenic' sediments have been recovered are shown in fig. 219. The fourteen sites are situated between the submarine Haig Fras granite outcrop and the edge of the conti-

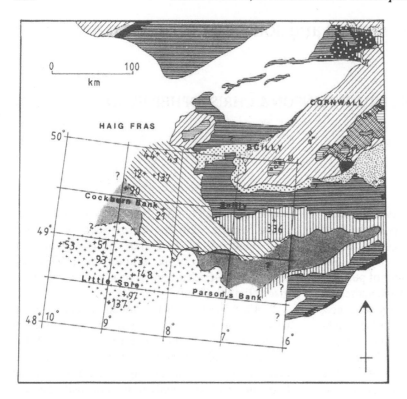

Figure 219: Central and Southwestern Celtic Sea - location map, BGS vibrocoring sites and solid geology.

nental shelf to the southwest, which in this region lies at the relatively low level of −185 to −205 m O.D. (Pantin & Evans, 1984). This is an area twice the size of East Anglia.

The continental shelf in this area dips gently to the southwest and is almost featureless apart from the large linear tidal sand ridges (Stride, 1963; Bouysse *et al.*, 1976; Pantin & Evans, 1984; Belderson *et al.*, 1986) and isolated boulders (Pantin & Evans, 1984) which are considered below. In general the samples recovered from the deepest water were in the southwest, and the shallowest from the northeast; therefore the deepest sample, 48/$\overline{09}$/137 (48° latitude N, 9° longitude W, BGS sample 137), cored at a water depth of −211 m O.D., was also the sample located furthest to the south; *vice versa*, sample 49/$\overline{09}$/44 was cored in the north at a water depth of only −125 m O.D. Sample 48/$\overline{09}$/137 was therefore recovered from beyond the defined 'shelf edge break' (fig. 221).

There is some evidence for glaciomarine sedimentation in the deeper water beyond the defined shelf break. Day (1959) reports possible glaciomarine sediments at 48°39′N 10°35′W in 1419 m of water on the continental slope, and pebbles have been recovered from the top few metres of sediment at DSDP site 548 on the Goban Spur in 1251 m of water

(Graciansky *et al.*, 1985). Ice-rafted material has also been recovered from the open ocean to the west (Ruddiman & McIntyre, 1973; 1981b), and as far south as the continental slopes off Morocco and Portugal (Kudrass, 1973).

Stratigraphy

The glacigenic sediments constitute a member of the Melville Formation, a unit underlain by the Early Pleistocene upper Little Sole Formation and overlain by recent sediments (fig. 221). In no instances did the cores penetrate through the identified glacigenic samples of the Melville Formation into an underlying unit of that or an older formation. Evans & Hughes (1984) indicate that the glacigenic samples mostly overlie Neogene strata (fig. 219).

Most of the glacigenic samples were recovered from bathymetric 'lows' between the linear tidal sand ridges, whilst one sample, 49/$\overline{09}$/44, was from the flank of a sand ridge (figs. 221 and 222). The recent sediments overlying the glacigenic material have been classified into two layers, 'A' and 'B' (Pantin & Evans, 1984). Layer 'A' consists of the superficial mobile sediments, layer 'B' a relatively coarse pavement underlying layer 'A'. Boulders over 1 m in

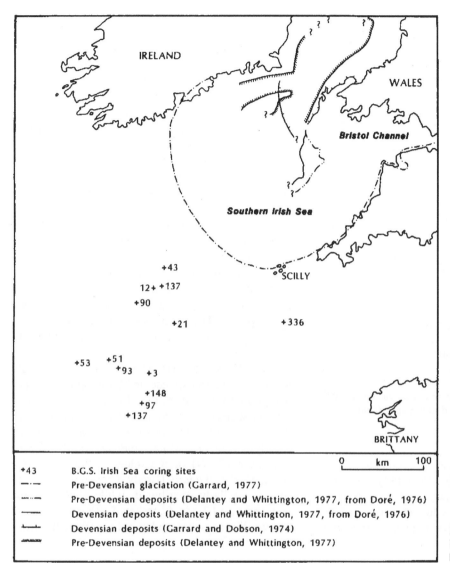

IRELAND

WALES

Bristol Channel

Southern Irish Sea

+43

12+ +137

+90

+21

+336

SCILLY

+53 +51

+93 +3

+148

+97

+137

BRITTANY

0   km   100

+43   B.G.S. Irish Sea coring sites
─·─   Pre-Devensian glaciation (Garrard, 1977)
─··─ Pre-Devensian deposits (Delantey and Whittington, 1977, from Doré, 1976)
───   Devensian deposits (Delantey and Whittington, 1977, from Doré, 1976)
⊥   Devensian deposits (Garrard and Dobson, 1974)
⊥   Pre-Devensian deposits (Delantey and Whittington, 1977)

Figure 220: Postulated glacial limits in the Southern Irish Sea based on offshore evidence.

diameter are widely scattered across the area; these have been retrieved on the anchors of the drilling ships (figs. 223 and 224), are identifiable on the side-scan sonar and have been observed in submarine photographs (Hamilton *et al.*, 1980). They probably represent the larger clasts attributable to layer 'B'.

In most cases the vibrocorer was not able to penetrate more than a metre or two into the sea bed due to the coarse-grained nature of the sediments. This is in itself a measure of the glacigenic character of the material. At a few sites layers 'A' and 'B' were penetrated and a few centimetres of the underlying sediment recovered. At 49/$\overline{09}$/44, however, over a metre of material was obtained.

Evidence from both sampling and acoustic devices

suggests that the glacigenic sediments do not occur as a sheet, but form isolated patches or mounds, separated by the tidal sand ridges and exposures of non-glacigenic Melville Formation units.

Facies classification and distribution

The various characteristics and analyses of the samples described enable their classification into two facies, 'A' and 'B'. Samples 48/$\overline{09}$/137, 48/$\overline{10}$/53, 48/$\overline{09}$/97, 48/$\overline{10}$/57 and 49/$\overline{07}$/336 cannot be ascribed to facies 'A' or 'B' with any certainty.

Facies 'A' is represented by samples 49/$\overline{09}$/43, 49/$\overline{09}$/12, 49/$\overline{09}$/21 and 49/$\overline{09}$/137 and the lower

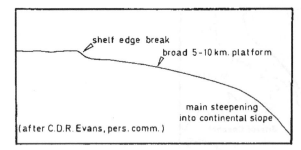

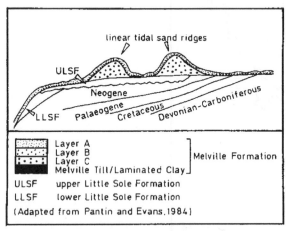

Figure 221: Central and Southwestern Celtic Sea - continental shelf edge morphology (above) and stratigraphy (below).

subunit of 49/$\overline{09}$/44. This facies is overconsolidated (Lambert & Khowaja, 1978), structurally homogeneous, containing abundant fine gravel, with between 8 and 54 granules per 100 g sediment, and abundant pebbles > 4 mm. The matrix is very poorly sorted and consistently coarse skewed. Facies 'A' samples are either devoid of microfauna, or contain very few specimens at least some of which can be attributed to reworking. Two of the samples contain *Hiatella* sp. and three contain barnacle fragments; these are attributable to reworking. The clast lithological assemblage is fairly consistent, with, in general, greywackes/ quartzites > quartz/igneous > metamorphics > sandstones > flint > chalk.

Facies 'A' is interpreted as either proximal glaciomarine sediment or lodgement till. Samples assignable to facies 'A' were recovered from between –127 and –157 m O.D.

Facies 'B' is represented by samples 48/$\overline{09}$/148, 49/$\overline{09}$/90, 48/$\overline{10}$/93 and 48/$\overline{09}$/3. Facies 'B' samples are not overconsolidated but plastic silty clays, contain very small amounts of fine gravel, between 0 and 6 granules per 100 g sediment, and sometimes display well-developed fining-upwards laminae and sand pods. The matrices are moderately to moderately-poorly sorted, and consistently very coarse skewed.

Facies 'B' stands apart from 'A' most distinctively in containing rich ostracod and foraminiferal faunas. Samples 48/$\overline{09}$/3, 48/$\overline{09}$/148, 48/$\overline{10}$/93 and 49/$\overline{09}$/90 contain species which today are not found south of the Arctic Circle. The ostracod *Rabilimis mirabilis* (Brady) occurs no nearer than the fjords of East Greenland (76°N) or the inlets of the White Sea coast. *Krithe glacialis* (Brady, Crosskey & Robertson) is similarly distributed, while *Cytheropteron montrosiense* (Brady, Crosskey & Robertson) is not known alive anywhere today, but appears to be Arctic

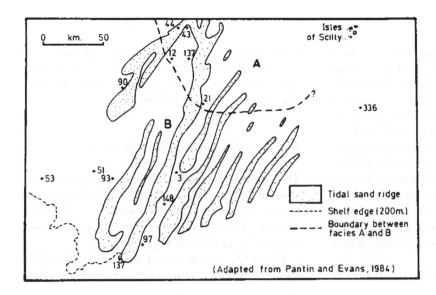

Figure 222: Central and Southwestern Celtic Sea - distribution of linear tidal sand ridges, boundary between facies A and B, and vibrocoring sites.

Figure 223: Boulder recovered on anchor of drilling ship at 49°01.57'N/8°36.80'W. The rope has a diameter of 2.5 cm.

Figure 224: Boulder recovered on anchor of drilling ship at 48° 57'N/8° 47'W. The rope has a diameter of 2.5 cm.

by its association with other organisms (Whatley & Masson, 1979).

Each of these Arctic species is represented by a wide range of valve sizes constituting a growth series of instars making up almost a complete life history; *R. mirabilis* is represented by seven moults in sample 48/$\overline{09}$/3 (fig. 225). This indicates deposition under extremely quiet conditions with practically no current activity, otherwise the wide range of valve sizes would have been dispersed. Additional species of the genera *Cytheropteron, Acanthocythereis, Elofsonella, Heterocyprideis* and *Jonesia* suggest affinities with the pre-Ipswichian open sea marine fauna, if not the precise age, of the Bridlington Crag of Holderness (Catt & Penny, 1966; Neale & Howe, 1975).

The abundance of the foraminifers *Islandiella islandica* (Nørvang) and *I. helenae* (Feyling-Hanssen & Buzas), species which constitute the 'High Arctic' assemblage of Feyling-Hanssen (1982), support the low water temperatures suggested by the ostracods. Sample 49/$\overline{09}$/90 contained five valves of the mollusc *Yoldiella* (=*Portlandia*) *fraterna* (Verrill & Bush),

also an Arctic indicator (Nordsieck, 1969; A. Warén, personal communication, 1983).

Facies 'B' sediments have been recovered from between −135 and −183 m O.D. They are interpreted as probably distal glaciomarine in origin, having formed in very quiet conditions beneath icebergs or an ice-shelf which provided a fairly continuous rain of sediment to the sea floor. The fining-upwards laminae probably represent discrete sediment input events, and the sand pods formerly frozen sand clasts.

In general the facies 'A' samples are grouped in the north with a transition to more distal glaciomarine conditions at about 49°30'N between −127 and −145 m O.D. (fig. 222). This transition is reflected in fig. 226c-g with the facies 'B' samples being characterised by high clay, high silt but low sand values, this also being reflected in the sorting coefficient and mean grain-size plots. These diagrams also reflect the proximal glaciomarine or lodgement till (facies 'A') affinities of some of the samples towards the continental shelf edge, especially 48/$\overline{10}$/53 and 48/$\overline{09}$/137.

By analogy with glaciomarine sedimentation models developed for the George V-Adelie continental shelf in Antarctica these samples may represent residual glaciomarine sediments containing much reworked material typical of the continental shelf edge and slope (Domack, 1982).

Environmental and dating interpretation

The complexities of the glaciomarine sedimentary environment (Drewry & Cooper, 1981; Drewry, 1986) militate against over-precise interpretations of suites of interbedded glacial and glaciomarine sediments. This caution is increased in situations where there is little stratigraphic control, as in this case. There is no independent dating evidence that the separate samples were deposited during the same event, though their geomorphological context would suggest that this is the case.

If the assumption is made that the material was deposited during one event, some very general environmental conclusions can be drawn. Evidence from the lithological analysis of the clasts in facies A, the presence of Neogene lignite in some of the samples and calcareous nannoplankton analysis of chalk erratics (J. Young, personal communication, 1983), suggests that an ice body moved from the northeast towards the southwest across the area between the Haig Fras and Scilly Isles granite outcrops, eroding a variety of rock types including Turonian and Miocene sediments. There is no evidence of very far travelled (i.e. > 100 km) material in any of the samples, thus falsifying the hypothesis that the sediment could be derived from North Atlantic tracking icebergs as suggested by the widespread distribution of ice-rafted sediments in the open ocean to the south and west (Kudrass, 1973; Ruddiman & McIntyre, 1973; 1981b).

At around 49°30'N a change in depositional environment occurred, either a grounding line representing a transition from grounded to floating ice, or simply a change from proximal to more distal glaciomarine conditions. This transition occurred between −127 and −145 m O.D. The evidence from sample 49/$\overline{09}$/44 suggests that marine conditions gradually became more predominant through time, extending further to the north and perhaps causing the floatation and calving of former grounded ice.

Both the biological and lithological evidence from facies 'B' suggests very quiet conditions during deposition; this is surprising given the exceptionally rough conditions, often associated with southwesterly gales, currently experienced on this part of the shelf. One possible explanation for this is deposition beneath an ice shelf rather than from icebergs (Vorren *et al.*, 1983). However, such an hypothesis is difficult to reconcile with glaciological theory. Ice shelves have a physically constrained minimum thickness of around 200–250 m (Sugden & John, 1976; Paterson, 1981). If −135 m O.D. is taken as the fossil grounding line, to float an ice shelf at least 200 m thick sea level must have stood at around +30 m O.D. assuming isostatic rebound of 0.33 x ice thickness and a 4:1 ratio of submergent to emergent ice. There is no evidence in the region as a whole for shorelines higher than present sea level post-dating the Ipswichian Stage. In addition, if the ice were greater than about 100 m the Scillies would have been overridden rather than being only marginally covered by ice (Scourse, this volume). Given these data it is very difficult to envisage deposition beneath an ice shelf in this area.

The most likely hypothesis for these sediments is therefore ice-rafting. The conditions indicated by the ostracods perhaps suggest that basal meltout from icebergs was more important than slumping or dumping (Vorren *et al.*, 1983). The ice-rafting hypothesis is favoured by Pantin & Evans (1984) who refer to the 'mounds' of sediment recorded by the side-scan

Figure 225: Scanning electron micrographs (SEM) of Ostracoda. The broken bar scale is 0.5 mm. (1a) *Cytheropteron montrosiense* Brady, Crosskey & Robertson 1874; right valve: Range - Pleistocene, extinct, arctic cold water. (2) *Cytheropteron simplex* Whatley & Masson 1979, left valve: Range - Pleistocene, extinct, arctic cold water including the Errol Clay of Tayside. (3) *Cytheropteron* cf. *arcuatum* Brady, Crosskey & Robertson 1874, left valve: Range - Pleistocene, extinct. This form has a non-spinose, ornamented ala. (4) *Cytheropteron* cf. *vespertilio* (Reuss) 1850, right valve: Range - Miocene to Recent, deep water near Hebrides. This form has an ornamented ala compared with other Reuss species. The solid bar scale is 1.0 mm. (5a) *Rabilimis mirabilis* (Brady 1868), left valve, instar A- VI. (5b) *Rabilimis mirabilis* (Brady 1868), left valve, instar A-III. (5c) *Rabilimis mirabilis* (Brady 1868), left valve, instar A-II. (5d) *Rabilimis mirabilis* (Brady 1868), left valve, instar A-I. (5e) *Rabilimis mirabilis* (Brady 1868), left valve, adult. (6a) *Acanthocythereis dunelmensis* (Norman 1865), right valve, instar A-III. (6b) *Acanthocythereis dunelmensis* (Norman 1865), right valve, adult male. (6c) *Acanthocythereis dunelmensis* (Norman 1865), right valve, adult female. The presence of extremely small instars and the preservation of small delicate spines testify to the *in situ* character of this Arctic water ostracod fauna. These specimens are lodged with the British Museum (Natural History) Palaeontology Collections.

James Scourse, Eric Robinson & Christopher Evans

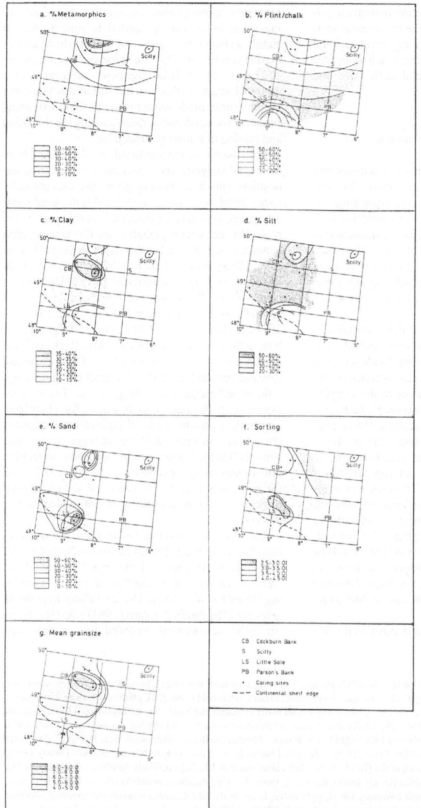

Figure 226: Distribution maps of selected granulometric and lithological characteristics of the glacigenic samples.

sonar as indicating individual iceberg 'dumps'. Some of these features may have such an origin, but at least some of the recovered glacigenic samples suggest quieter conditions and therefore an originally more extensive cover.

The geomorphological context of the glacigenic samples cored in the 'lows' between the tidal sand ridges, and in one case on the side of a sand ridge, suggests that the glacigenic deposition took place *after* the main period of formation of the sand ridges. Pantin & Evans (1984) regard the sand ridges as having formed in around 60 m of water by the Huthnance mechanism (Huthnance, 1982) during the Late Devensian/Flandrian transgression. Recent numerical modelling of the M_2 tidal streams in the Celtic Sea with a sea level lowered by 100 m has generated tidal currents twice as strong as those at present, sufficient to generate and maintain the ridges now found as fossil forms (Belderson *et al.*, 1986). This work therefore implies generation of the features during the early phase of sea level rise from the extreme minima reached during the last cold stage (Bloom *et al.*, 1974). Sand ridges of this size, up to 60 m high and 50 km long, could not have survived subaerial exposure, so the glacigenic material must have been deposited since the formation of the sand ridges and without any intervening low stand of sea level. This would imply a Late Devensian age not only for the sand ridges themselves, but also the overlying glacigenic sediments.

This reconstruction also implies the survival of the sand ridges during the succeeding glacial event. The sand ridges stop fairly abruptly along a northwest/southeast line about 100 km to the southwest of the Scillies in the vicinity of the facies 'A'/'B' transition (fig. 222). This perhaps suggests that any sand ridges that may have existed to the northeast of their present limit were eroded out by the grounded ice sheet, iceberg rafting taking place over, and on the flanks of, the sand ridges to the southwest. Sample 49/09/44 is crucial in this argument; there is no doubt that it is located on the side of a sand ridge. Not only was it the shallowest sample to be recovered, at −125 m O.D., but it was positioned accurately using a Decca Pulse 8. The existence of a sub-unit of facies 'A' material at the base of the sample suggests that it may mark the grounded ice limit on the margin of one of the surviving sand ridges. However, the sand ridges terminate abruptly to the east as far south as 48°N off Ushant in the French sector (Bouysse *et al.*, 1979), and it seems unlikely that this marks the continuation of the ice limit this far south in the absence of any positive sedimentary evidence of glaciation. The northeastern

terminus of the sand ridges could perhaps be of multicausal origin, erosional in the north and depositional in the south.

The acoustic character of the sand ridges suggests that they are composed of sands and gravels, though no samples exist to confirm this. A possible explanation for the source of this massive amount of sediment, given the now known proximity of glacigenic material, must be as subaquatic outwash from an advancing glaciomarine ice terminus. This material may then have been redistributed by shelf processes, such as the Huthnance mechanism or enhanced tidal streams, during the succeeding phase of rising sea level, and the glacigenic sediments deposited during a deglacial phase involving floatation and calving, probably also associated with rising sea level. The overall impression is that the maximum advance of ice in this area was during a sea level stand somewhat higher than the absolute glacio-eustatic minima commonly cited (Bloom *et al.*, 1974), even taking glacio-isostatic depression into account. This advance of Irish Sea ice therefore appears to be out of phase with global glacio-eustatic fluctuations.

It is difficult to provide an explanation for the exceptionally quiet conditions during glaciomarine deposition. During a phase of lowered sea level the wave base would have modified the sea bed more frequently, resulting in an even more erosive environment than at present. However, regional climatic conditions may have been fundamentally different during the Late Devensian with an altered pattern of depression tracks associated with blocking anticyclones over ice masses (French, 1976), resulting in calmer conditions on this part of the shelf. Alternatively or additionally a broad pack ice cover may have produced tranquil conditions at the sea bed. These suggestions stand as a working hypothesis to be tested by future work.

Regional reconstruction

The simplest and most likely regional reconstruction of glacial events in the whole region during the Late Quaternary correlates the offshore glacigenic sediments with the Scilly Till (Scourse, this volume). The Scilly Till resembles the offshore glacigenic samples both lithologically (Scourse, 1985) and mineralogically (Catt, 1986b). The altitudes of the various sedimentary bodies in the region can be reconciled with each other and with glaciological theory. Assuming an offshore grounding line around −135 m O.D. and ice thicknesses around 100 m, global stadial sea level would have stood somewhere

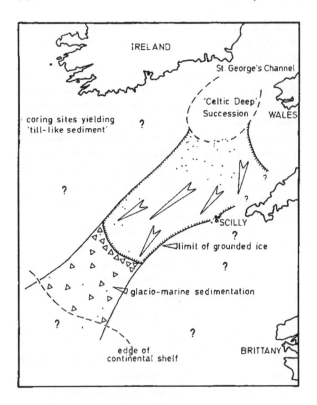

Figure 227: A reconstruction of the Celtic Sea ice lobe and glaciomarine terminus at 19,000 B. P. Dots represent vibrocoring sites yielding glacigenic sediment.

between 100 and 50 m below the present, with a post-rebound shoreline below present sea level. These figures are therefore consistent with the lack of a raised Late Devensian shoreline on the Scilly Isles or in Cornwall. Such a shoreline could only have been generated by ice thicker than 250 m, in which case the Scillies would have been easily overridden, or by a contemporary global sea level close to present sea level which is thought highly unlikely during the Late Devensian.

Recent BGS cores and seismic evidence from the continental shelf in the Celtic Sea to the north of 50°N supports the idea of an ice lobe advancing from the northeast. Though samples of glacigenic sediment from this area have not yet been analysed in detail, the mapping of their occurrence (fig. 227) strongly suggests a link with the ice advance further south. This inferred lobe originates immediately to the south of St. George's Channel from the 'Celtic Deep succession', a tectonically controlled basin up to 250 m thick which includes a number of till units. The southern limit of the Celtic Deep succession is at around 51°10′N (R. Wingfield, personal communication, 1984). Future work will be directed at establishing the stratigraphical, spatial and temporal relationships of this material with the samples to the south which have been analysed in detail.

Glaciomarine deposits of the Irish Sea Basin:
The role of Glacio-Isostatic disequilibrium

NICHOLAS EYLES & A. MARSHALL McCABE

Late Pleistocene sedimentary sequences of glaciomarine origin are well exposed around the margins of the Irish Sea Basin. These are of major interest, not only as a record of glacial and non-glacial marine environments but because they illustrate general principles that can be employed in the form of facies models to other glaciated shallow marine basins.

Late Devensian (broadly equivalent to Late Wisconsinan or Weichselian) sequences in the western and central portions of the basin demonstrate the importance of crustal depression and high relative sea-levels at the margins of large ice sheets (glacio-isostatic disequilibrium; Andrews, 1982). Whereas the existence of vertical crustal movements, resulting from post-glacial rebound, has been identified from raised beaches (e.g. Synge, 1977a; Carter, 1982), it has only recently been established that many beaches are cut into glaciomarine sediments and therefore the extent of late glacial submergence is much greater. The next phase of work, identifying the spatial and temporal relationship between crustal movement, water depths and glaciomarine sedimentation patterns across the basin is only just starting. The broader development of such glacio-isostatic facies modelling, as has been completed for other basins (e.g. Andrews, 1978; Belknap et al., 1987; Scott et al., 1987) is important because most of the Quaternary sedimentary record lies on continental shelves; glacio-isostatic depression and submergence of shelf areas occurs many hundreds of kilometres distant from ice-sheet centres (Peltier, 1982 and references therein) thereby creating deep peripheral 'foreland' basins (c.f. Beaumont, 1978; Quinlan & Beaumont, 1984) in which thick successions of glacial and marine sediments can be preserved. The interpretation of these sequences offers new challenges to Quaternary geologists, because patterns of sedimentation are controlled by crustal behaviour, water depth changes, and sediment supply, rather than climatically driven and regionally synchronous glacier advance/retreat cycles.

Our purpose here is to illustrate typical glaciomarine sequences which may represent depositional conditions recorded elsewhere in Britain. It should be noted that the interpretation of several of these sites remains controversial; many are still viewed as the product of subaerial sedimentation at an ice sheet margin moving across a basin exposed by glacio-eustatic sea-level lowering (Dackombe & Thomas, this volume). Our data is, so far, limited to the western (Irish) and central (Isle of Man) portions of the basin but clearly has major ramifications for future investigations employing modern facies analysis techniques along the Welsh and English coastal margins. A full treatment of Late Devensian glaciomarine environments in the Irish Sea Basin is presented by Eyles & McCabe (1989a).

Patterns of ice flow and deglaciation in the Irish Sea Basin

The general source areas and principal directions of ice flows in the Irish Sea Basin are shown in fig. 228. At the time of maximum glaciation the basin was fed by ice lobes from the Irish lowlands to the west, and by lobes moving south along the axis of the basin from mainland sources. Ice flows were coeval and moved southwards along the centre of the basin toward the Celtic Sea. On the western side of the basin the South of Ireland End Moraine Complex (SIEM) is considered to mark the southern limit of Late Devensian ice (Charlesworth, 1928a; Mitchell, 1976; McCabe, 1985). A lobe probably extended to the south over St. George's Channel with the eastern ice margin reaching the south Welsh coast at Milford Haven and along the Bristol Channel at Camarthen Bay, Swansea Bay and east of Cardiff (fig. 228). The areas immediately south of the limit are covered by 'Older Drifts' which are thought to pre-date the Late-Devensian glaciation but the sequences await proper facies treatments. In particular there is a critical need

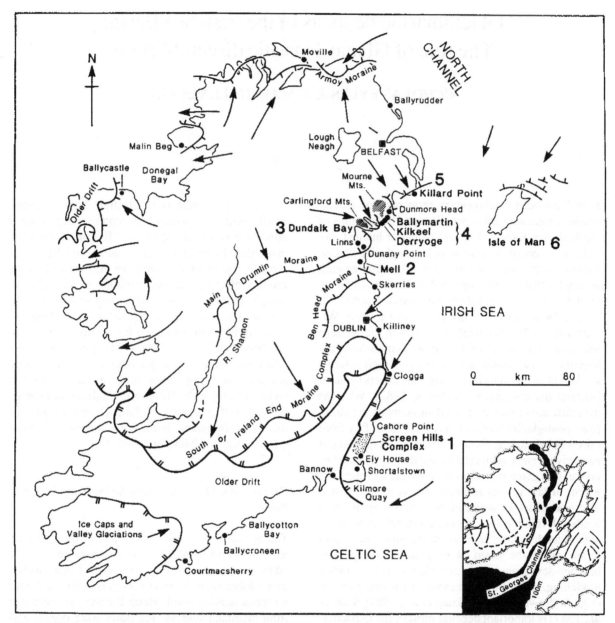

Figure 228: Source areas, flow directions and Late Devensian ice limits in the Irish Sea Basin. The 100 m water depth contour defines a deep channel orientated north-south. Faunal data from the Mell site indicate glacio-isostatic submergence of up to 150 m during glacier retreat. Numbers 1 to 6 refer to location of glaciomarine sequences described in the text.

to evaluate the relationships between large ice-contact depositional systems such as the Screen Hills complex (fig. 228) and fine-grained depositional sequences along the Irish coast to the south and west, outside the SIEM and traditionally regarded as the deposits of separate glaciations. The same remarks apply to deposits along the English coast, in North Devon and the Scilly Islands (see below).

During deglaciation there were high relative sea levels in the basin as a result of glacio-isostatic depression. Glacial sedimentation occurred in tide-water ice-proximal settings as the ice margin retreated northwards (plate 30). Raised beach notches are common geomorphic features in the basin (e.g. Stephens & McCabe, 1977; Synge, 1977a; Carter, 1982) and it now appears that underlying sediments are marine or glaciomarine in origin indicating that the extent of crustal depression and submergence is much greater than considered previously. Thus, the raised beaches are only a partial, and essentially postglacial, record of sea levels (e.g. Carter, 1982). The best record of vertical movements in the basin is stored in the glacial sedimentary stratigraphy into which the beaches are cut. At present, only at one site, Mell (fig. 228), can former water depths be constructed with certainty (see below). It is probable that patterns of crustal response and of submergence varied across the basin as identified in other more well-studied areas affected by glacio-isostatic submergence (Andrews, 1978; Boulton *et al.*, 1982; England, 1983; Belknap *et al.*, 1987; Scott *et al.*, 1987). Consequently a wide range of ice-margin types may have been present.

During deglaciation, separation of ice lobes occurred in the general vicinity of the Irish coastline. The presence of a deep channel in the west central Irish Sea Basin (St. George's Channel - North Channel; fig. 228) may have created a calving bay and the rapid separation of Irish and mainland ice (cf. R. H. Thomas, 1977). The most prominent sequences formed at this time are located around the Screen Hills, Dundalk Bay, South County Down, Killard Point, Mell and on the Isle of Man (fig. 228). All these deposits are complex sedimentologically and consist of thick sediment wedges, commonly preserved within substantial moraine ridges, dominated by lenticular or tabular units of interbedded diamicts, muds, sands and gravels. They are subaqueous in origin may have formed as the retreating tidewater ice margin stabilised in areas of shallow water. The sequence at Dundalk Bay forms part of a much larger ice-marginal system which borders the drumlin belt of Ireland (McCabe *et al.*, 1986, fig. 6). Drumlinisation was characterised by the rapid movement of large volumes of subglacial sediment toward the ice sheet margin (Dardis & McCabe, 1983; Eyles & McCabe, 1989b); such 'surge-type' conditions may explain the source and volume of the sediments deposited subaqueously at the tidewater margin of the Dundalk Bay ice-lobe (fig. 228). The same setting may also have obtained in the area of Killard Point. Surge conditions may have been induced as a result of calving along the marine portions of the ice sheet.

Previous investigations and their limitations

A major, continuing problem with studies in Quaternary and pre-Quaternary glaciated basins is conflicting usage of the term till(ite), a term which we argue to have a specific genetic meaning which consequently prohibits its use as loose descriptive term for poorly-sorted sediments containing clasts (see N. Eyles *et al.*, 1983; 1984 for discussion). In the Irish Sea Basin, the term 'Irish Sea Till' has been traditionally used in 'umbrella' fashion to refer to any fine-grained, often laminated muddy sediment showing varying clast contents, variable content of foraminifera and molluscs and an intimate relationship with marine sediments, all features at odds with their supposed origins as 'basal tills'. Hence, the Irish Sea Drifts have long been recognised as posing especial problems of interpretation to Quaternary stratigraphers, i.e. the 'enigma' of the Irish Sea Drifts.

Previous work in the basin has been characterised by an overwhelming emphasis on 'stratigraphy' rather than facies analysis. The difference is essentially 'splitting', where each lithological change in a section is arbitrarily accorded a distinct stratigraphic entity and any poorly-sorted sediment containing clasts is labelled 'till', and 'lumping' where emphasis is placed on detailed analysis of facies and bed contacts in order to define depositional sequences and to establish the extent of sequence boundaries. Historically, faunal and sedimentological evidence for marine submergence, has been largely disregarded in favour of arguments invoking reworking of older fauna and glaciolacustrine conditions. This was because workers could not envisage anything other than a *eustatic* mechanism whereby high sea levels could co-exist with full glacial conditions (e.g. Thomas & Dackombe, 1985: 446). However, the existence of high relative sea levels resulting from glacio*isostatic* downwarping (Andrews, 1978, 1982, see below) was identified by Colhoun & McCabe (1973) and Synge (1977a) and ongoing work around the basin confirms the regional significance of high relative sea levels created by isostatic submergence.

The Irish Sea Tills can now be recognised as heterogenous assemblages of glacial and glacially-influenced marine sediments deposited in proximal and distal settings around retreating tidewater ice margins in varying water depths (plate 30). Under these conditions, a wide range of lithofacies are produced as a result of mud deposition and ice-rafting, downslope resedimentation, and reworking by floating ice, tidal, storm and meltwater currents. Facies are complexly arranged within larger depositional systems most commonly being large ice-contact subaqueous outwash fans and major push ridge structures marking minor oscillations in the position of the ice margin. A brief review of ice-contact and distal sedimentation at the margins of tidewater glaciers is in order before we illustrate some of the Irish Sea sequences.

a

b

Figure 229: Massive 'rain-out' diamict facies; (a) Dundalk Bay (fig. 228, plate 30), Riverstown section (McCabe, 1986), massive facies with disorganised clast fabric and mud matrix overlain by stratified facies with muddy sand matrix, recording variable but weak bottom current activity; (b) massive facies from the Isle of Man (section 0582 of C.H. Eyles & N. Eyles, 1984). Note silt/sand stringers. Stratified facies are defined where such interbeds comprise more than 10% of unit thickness.

Sedimentation in the ice-contact tidewater environment

Many recent papers have described sedimentation patterns around modern tidewater glaciers, principally in fjord settings in Alaska (Powell, 1981; 1984; Mackiewicz *et al.*, 1984), Arctic Canada (Gilbert, 1982; 1985) and Spitsbergen (Elverhøi *et al.*, 1983). The equivalent data base for tidewater sedi-

mentation around ice sheet margins on extensive shallow marine shelves is by necessity reliant on palaeoenvironmental interpretations from older sequences (Nystuen, 1976; Vorren *et al.*, 1983; Cheel & Rust, 1982; Domack, 1983; 1984; Anderson *et al.*, 1984; Visser, 1983a, b, c; Lea, 1985; Osterman & Andrews, 1983).

In the ice-proximal tidewater setting, marine processes operate on sediments supplied by both the

c

d

Figure 229 (continued): (c) ice-rafted mass of bouldery gravel entombed within massive diamict; Dundalk Bay, Dunay Point section (McCabe, 1986); (d) ice-rafted gravel pocket deformed in muds and massive rain-out facies diamict. Dundalk Bay, Castlecarragh section (McCabe, 1986).

glacier and by large meltstreams that vent subaqueously (plate 30). Around the latter, ice-contact fans develop and may coalesce to form a continuous apron-like fringe along the ice front (Rust, 1977). Where there is sufficient suspended sediment concentrations, meltstreams may generate quasi-continuous turbidity currents (=density underflows) but more commonly meltwaters rise along the ice front and transport fines, up to coarse-sand size, as suspended sediment plumes (Elverhøi *et al.*, 1983; Gilbert, 1983; Powell, 1984). Tidal interaction with these plumes is known to produce laminated muds ('cyclopels' of Mackiewicz *et al.*, 1984); massive mud blankets are also reported (Elverhøi *et al.*, 1983).

Debris contained within the ice margin is dumped sporadically during retreat and, together with other frontal accumulations, may be bulldozed into boula

b

Figure 230: Stratified diamict facies recording interplay of 'rain-out' and current activity. (a) stratified facies at South County Down; (b) stratified facies, west coast of Isle of Man, south of Orrisdale Head (section 2382: C.H. Eyles & N. Eyles, 1984). Note abundant ice-rafted boulders, and 'nest' of gravel (arrowed). Note massive unstructured appearance of uncleaned part of section.

dery morainal banks or large push ridges (Powell, 1981; Lea, 1985). These proximal facies presumably interfinger with basal sediments on their proximal sides but transitions between ice-proximal and sub-glacial facies await documentation. Whether sufficient volumes of supraglacial debris can accumulate to produce significant debris flows is debateable (Powell, 1983). The data of Nemec *et al.* (1984) documenting subaerial to subaqueous debris flow facies on fan deltas may be applicable. Because of very rapid sedimentation and instability caused by ice melt, berg scour, bulldozing, wave processes and basin seismicity, sediment gravity flow is a ubiquitous, sometimes dominant, subaqueous process. Downslope resedimentation can mix different sediment types together producing a very wide range of matrix- and clast-supported diamict facies (C.H. Eyles *et al.*, 1985). Structural highs generated by bull-

a

b

Figure 231: Lithofacies complexes of rain-out diamict facies and pebbly sands and gravels deposited by marine traction currents. (a) Isle of Man; section 0982; C.H. Eyles & N. Eyles, 1984; (b) Isle of Man, section 0782.

dozing may constrain sedimentation to local basins resulting in rapid lateral facies changes. Soft-sediment deformation structures are ubiquitous as a result of rapid and frequently episodic sedimentation. Wave, storm, tidal and biogenic processes around tidewater ice-sheet margins in open shelf settings must be of great importance but their documentation awaits detailed facies studies.

In view of the tendency of early work to label any poorly-sorted sediment containing clasts as 'till', diamict-forming processes are of especial significance since an extremely broad range of poorly-sorted lithofacies is produced in subaqueous ice-contact settings. Diamicts accumulate *in situ* as a result of the interplay between mud sedimentation from plumes, ice rafting and bottom current activity (figs. 229 and 230). This group of diamict deposits has been termed 'rain-out' facies and typically occurs

Figure 231 (continued): (c) Isle of Man, section 1582.

within lithofacies 'continuums' comprising sandy, gravelly and fine-grained marine facies (fig. 231). A disorganised clast fabric, common soft-sediment 'load' deformation on upper surfaces, dewatering structures and interbedded basal contacts are typical; marine microfauna may or may not be present as a result of reduced salinities around ice margins (Miller, 1953; C.H. Eyles *et al.*, 1985 for references). Massive facies may on detailed cleaning show faint stratification in the form of sand or silt stringers (e.g. fig. 229b). Recent work shows that the clast and magnetic fabrics e.g. natural remanent magnetism (NRM) and anisotropy of magnetic susceptibility (AMS), of rain-out facies is distinct from lodgement tills and other subglacial diamictons (Domack & Lawson, 1985; Dowdeswell & Sharp, 1986; Eyles *et al.*, 1987). Stratified 'rain-out' facies (fig. 230) tend to have a coarser matrix texture and are deposited where traction currents operate to remove matrix or prevent deposition, and to transport clasts as bedload. 'Outsized' ice-rafted clasts and debris masses can be

readily identified by associated deformation structures (plate 32a, b, c).

There are many opportunities for slope failure and the deposition of diamicts and other facies by sediment gravity-flows. Resedimented diamict facies, moved as debris flows, typically show a lenticular or channeled cross-sectional geometry and are identified by the presence of flow noses, brecciated sediment masses, the frequent presence of organised clast or grain fabrics, and internal grading and flow banding (fig. 232d, e, f). Basal contacts are usually-sharply conformable but local substrate incorporation and erosion is frequently observed (Jorgenson, 1982; Visser, 1983a, b; Postma *et al.*, 1983; Broster & Hicock, 1985; Piper *et al.*, 1985; Miall, 1985; C.H. Eyles, 1987; N. Eyles, 1987).

Application to the Irish Sea Basin

Having briefly reviewed what is currently known of

a

b

Figure 232: Resedimented facies, deposited by sediment gravity flow. (a) graded pebbly sands showing normal 'coarse-tail' grading; Isle of Man; (b) inverse-graded gravels deformed by loading into underlying sands; Killard Point.

glacially- influenced marine sedimentation at tide-water ice margins, six well-exposed glaciomarine sequences on the western margin of the Irish Sea Basin are now briefly described. These are listed below in order of their relative age.

1) *Screen Hill Complex*. The Screen Hills forms part of a complex moraine system formed as Irish Sea ice decayed north along the coastal zone of County Wexford. Previous work on the Screen Hills by Tho-mas & Summers (1983), identified a depositional evironment similar to that earlier argued for the Isle of Man *viz.* a subaerial glacial environment (see below). The model employed was that developed at Spitsbergen glacier margins transporting large volumes of supraglacial debris by Boulton & Paul (1976) and Paul (1983). Work by Huddart (1981a, b) which identified a well-preserved cold-climate fora-miniferal fauna within the sequence and likely gla-

c

d

Figure 232 (continued): (c) subaqueous debris flow with large boulder protruding from bed top rafted
by flow; Derryoge; (d) interbedded subaqueous debris flow deposits and mud facies; Killard Point.

ciomarine conditions was dismissed by Thomas &
Summers (1983) on the grounds that the fauna were
likely reworked. Thus, marine muds sometimes laminated, showing few erratic clasts have been interpreted as lodgement tills (e.g. Macamore Till;
Thomas & Summers, 1983; Shellag Till, Thomas,
1977; see McCabe *et al.*, 1986 for other examples).
New palaeontological work here and at other sites
along the coast shows mixed foraminiferal populations. As a result of reworking older fauna occurs
together with an *in situ* population dominated by
Elphidium clavatum, common at present day around
tidewater glaciers in Spitsbergen. The same lithofacies ('Irish Sea Tills') are present along the entire
south coast of Ireland as far west as Cork. These
display most of the sedimentary characteristics found
in ice-distal glaciomarine environments (C. H. Eyles
et al., 1985).

e

f

Figure 232 (continued): (e) channeled debris flow deposit; Dundalk Bay; (f) brecciation of resedimented diamict facies as a result of porewater expulsion; Isle of Man.

A recurring theme in the Irish Sea Basin is the close relationship between ice-proximal subaqueous sedimentation and repeated oscillations of the ice margin. At the Screen Hills, short-lived advances (surges?) extensively deformed the surrounding sediment piles, creating a series of imbricated structural highs (see fig. 239). The formation of large ridges promoted gravity-flow activity and constrained subsequent sedimentation resulting in stratigraphic complexity and rapid lateral facies change along section. Sequences at both the Screen Hills and northern Isle of Man show a similar structural pattern of intensely-folded coarsening-up (marine mud → sand → gravel) sequences overlain by thick subaqueous 'type examples of deposition by cold based, polar glaciers' carrying a considerable basal and englacial debris load, with permafrost playing a major role in allowing extensive structural deformation during ice

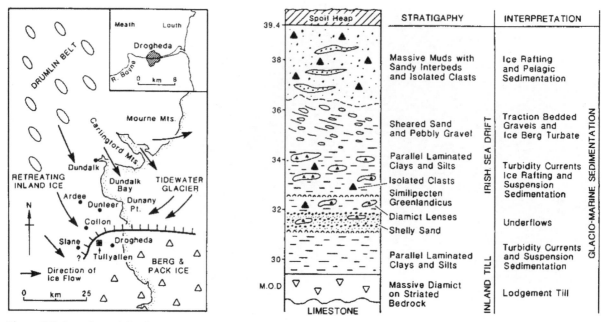

Figure 233: Location and detail of glaciomarine sequence exposed at Tullyallen, Drogheda (based on Colhoun & McCabe, 1973).

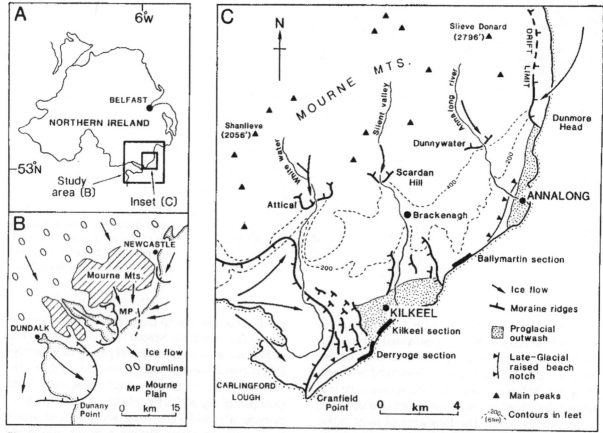

Figure 234: (A) Location of glaciomarine sections on the southern flanks of the Mourne Mountains. (B, C) Late Devensian ice flows and pattern of deglaciation around the Mourne Mountains.

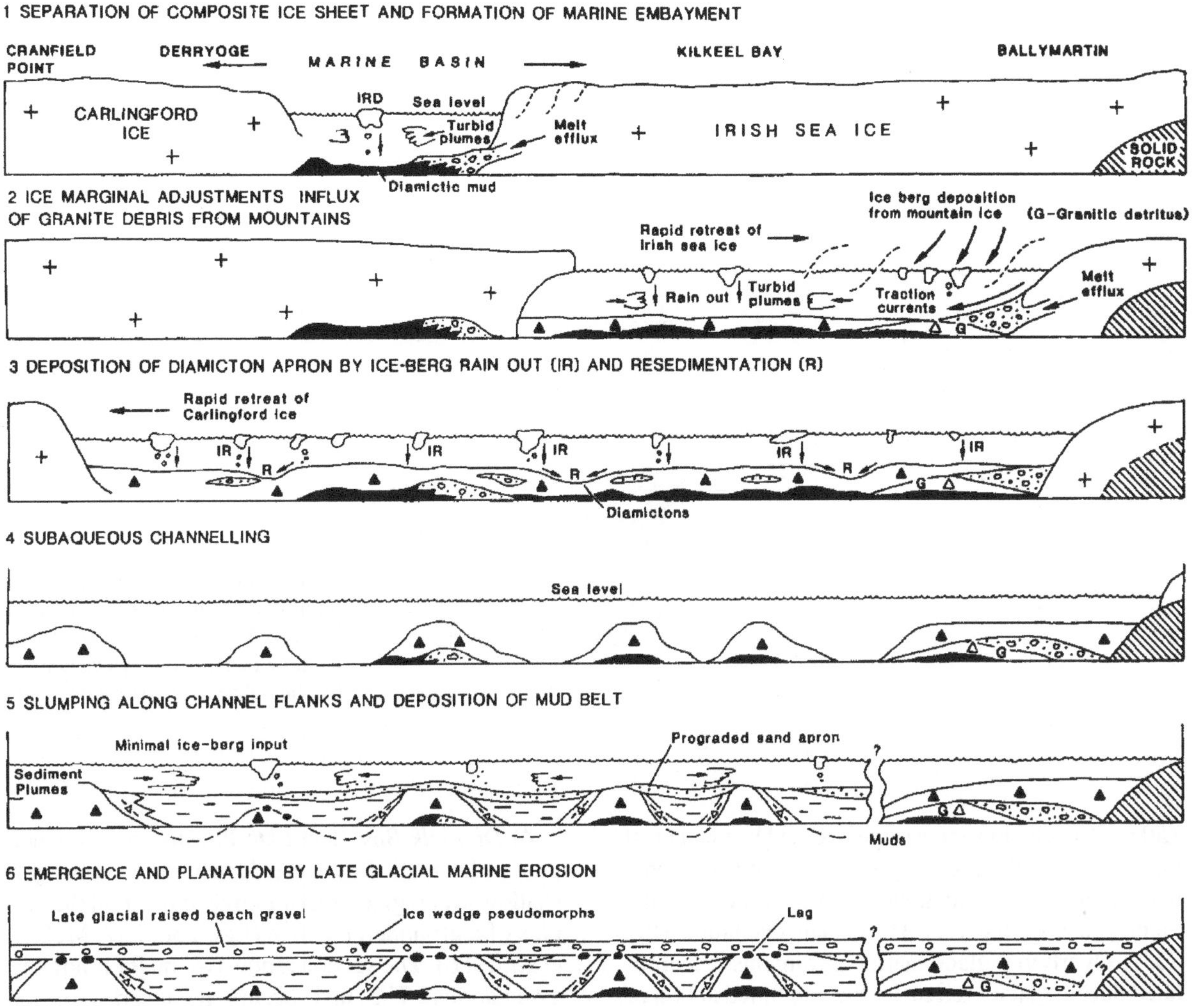

Figure 235: Postulated evolution of glaciomarine sequences south of the Mourne Mountains, between Cranfield Point in the south and Ballymartin to the north (fig. 228).

push. However, the existence of permafrost and glaciers of polar regime in the Irish Sea Basin is not in keeping with evidence elsewhere in Britain. Instead, we would emphasise tidewater glaciers overriding soft marine sediments (C. H. Eyles *et al.*, 1985a) with ice-margin oscillations in response to changes in water depths and substrate sediments during retreat in the manner identified by Andrews (1973) and Hillaire-Marcel *et al.* (1981). Many structures identified as 'glaciotectonic' are also the result of large-scale soft-sediment loading involving sands, gravels and muds. Typically shoreface sands and wave-worked gravels are loaded into marine muds as a consequence of emergence and falling relative sea-level.

2) *Mell.* Colhoun & McCabe (1973) described

extensive exposures of glaciomarine silts, clays and gravels from Drogheda, County Louth (e.g. Tullyallen; fig. 233). These deposits, up to 40 m above modern sea level, contain a diverse *in situ* cold water-fauna of Mollusca, Crustacea, Ostracoda, Foraminifera, Hydrozoa and sparse remains of marine algae. Most significant is the presence of beds of the mollusc *Similipecten* (*Arctinula*) *greenlandicus* which is thought to live in large colonies in cold waters at depths of 20–70 m. Recent amino acid ratios from valves in the sequence indicates a Late Devensian age (Bowen & McCabe, unpublished). Silts are associated with overlying muds containing dropstones and were previously identified as Irish Sea *Till.* Values of isostatic depression of this part of the Irish Sea Basin at this time may have been as much as

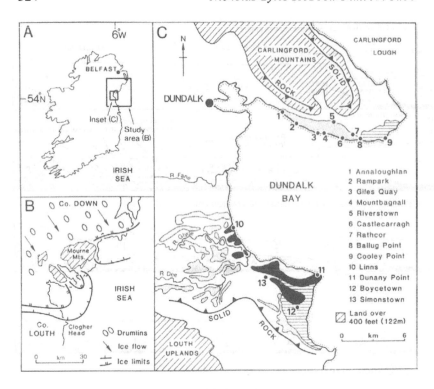

Figure 236: (A) Location of glacio-marine sections on the margins of Dundalk Bay. (B, C) Late Devensian ice flow patterns of deglaciation and location of described sections.

150 m. The glaciomarine sequences at Drogheda rest on lodgement till, overlying striated bedrock, deposited by inland ice which moved offshore during the main phase of the Late Devensian glaciation (fig. 228). Overlying glaciomarine sediments accumulated during glacier retreat by the combined action of ice- rafting and deposition of mud from suspended sediment plumes.

3) *South County Down.* The complex glaciogenic sequences along the coast of south County Down at Ballymartin, Kilkeel and Derryoge have traditionally been interpreted in terms of multiple ice-sheet advance and retreat cycles. Recent work on the sedimentology of these deposits suggests that they form a single depositional complex of glaciomarine origin (McCabe & Hirons, 1986). Sedimentation occurred when the topographic barrier of the Mourne Mountains forced the last composite ice sheet to split in the vicinity of the Mourne Plain (fig. 234) resulting in one ice lobe that retreated into Carlingford Lough and another that retreated northeast along the Irish Sea coastal zone. A depositional scenario is shown in fig. 235. Vertical sequences comprising subaqueous outwash, 'rain-out' diamict facies (e.g. fig. 230a) and emergent beach deposits may be typical of increasingly ice-distal sedimentation in a glacio-isostatically depressed basin (e.g. Boulton *et al.*, 1982).

4) *Dundalk Bay.* The Late Pleistocene sequences around Dundalk Bay were deposited in a variety of shallow-water glaciomarine environments at the margins of a grounded ice lobe (figs. 236, 237; McCabe *et al.*, 1986). The sections show two major lithofacies associations. The first occurs in a series of ice-proximal morainal banks formed at the southern margin of the ice lobe on the flanks of the Lough Uplands (fig. 236). The morainal banks at Dunany Point and Linns (fig. 237a) consist mainly of stony muds deposited from plumes and floating ice, interbedded with rhythmites deposited by turbidity currents, cross-bedded gravels deposited as subaqueous outwash and massive boulder gravels brought in by sediment gravity flows.

The second association accumulated in a narrow arm of the sea adjacent to the northeastern margin of the ice lobe, on the southern margins of the Carlingford Mountains. Deposition took place as a series of coalescing, ice-proximal subaqueous fans; sands and gravels are interbedded with rain-out diamict facies (fig. 229a, c, d) and sediment gravity flow deposits (figs. 232e, 237b). Sequences on both the north (fig. 237a) and south (fig. 237b) side of the Bay are draped by muddy diamict assemblages deposited from suspended sediment plumes and ice-bergs. At several localities the upper parts of the sequence have been

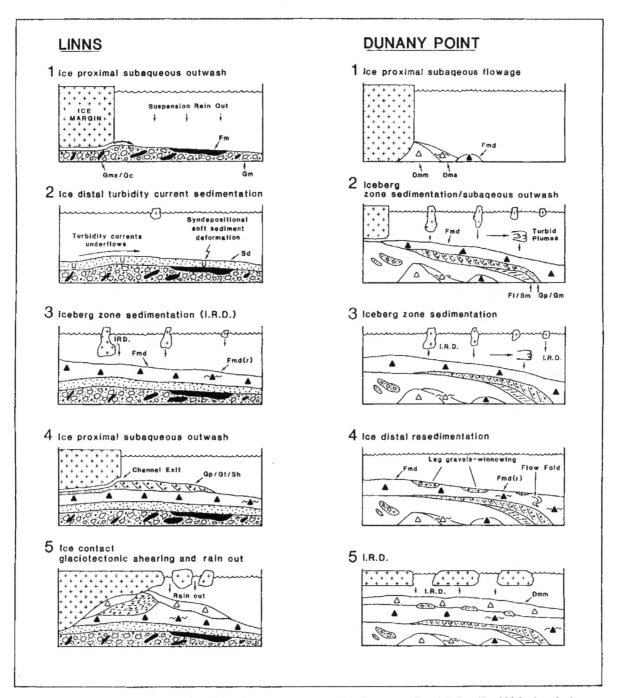

Figure 237: (a) Postulated evolution of glaciomarine sequences exposed at sites around Dundalk Bay (fig. 228 for location).

thrust and tectonised during oscillations of the ice lobe margin (fig. 237b).

5) *The Killard Point Moraine, County Down*. This moraine lies on the margins of the drumlin belt southeast of Lough Neagh and is associated with deposition from an ice-sheet margin which moved-southeastwards from central Ulster (McCabe *et al.*,

1984). The moraine complex records localised fan sedimentation against a retreating tidewater ice margin; sequences are dominated by sediment gravity-flow facies within coarsening-up sequences (fig. 238). Massive and variably stratified diamicts (fig. 232d) deposited as subaqueous debris flows, and interbedded with red muds, define the base of indi-

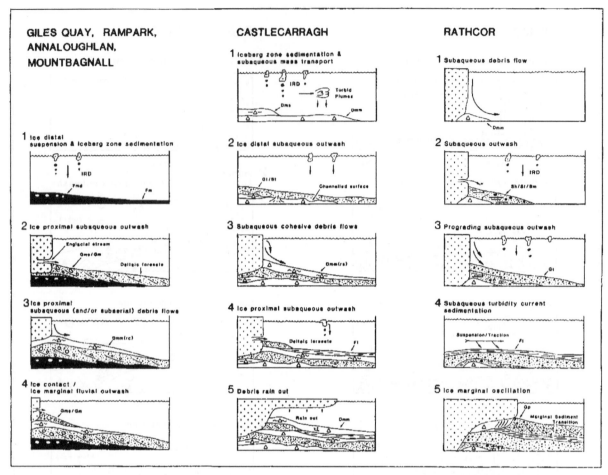

Figure 237 (continued): (b) Postulated evolution of glaciomarine sequences exposed at sites around Dundalk Bay (fig. 228 for location).

vidual sequences. Diamicts are matrix-rich, show common slump folds, and are overlain by parallel-laminated and massive sand facies, deposited by density underflows, containing occasional mud-lenses, 'outsized' ice-rafted clasts and diamict pel-lets. Sands are in turn overlain by normally and inversely-graded gravels preserved within nested channels (fig. 232b). Lenses of red mud occur throughout and were probably deposited from sus-pended sediment during interruptions of sediment

Plates 30–31

30. Sedimentation in the ice-contact glaciomarine environment; a composite model based on the recent literature and field observations in the Irish Sea Basin (see text). Suspended sediment plumes not shown. 1) glacitectonised marine sediments, 2) lensate subglacial diamicts (lodgement till?), 3) coarse-grained 'rain-out' diamict facies deposited by the interplay of mud deposition and ice-rafting with varying influence of current activity (e.g. fig. 230), 4) fine- grained 'rain-out' diamict facies deposited by the interplay of mud deposition and ice-rafting, minor current activity (e.g. fig. 3), 5) proximal bouldery outwash, 6) variable sands and gravels deposited on subaqueous fans, 7) sediment gravity flow; a ubiquitous process affecting all the above, 8) supraglacial debris where present. Note creation of local sub-basins by bulldozed structural highs and occurrence of berg 'turbation'. The effects of storm, wave and tidal processes on sedimentation is as yet poorly understood.   30

31. The Carn Morval organic sequence situated towards the base of the Porthloo Breccia.   31

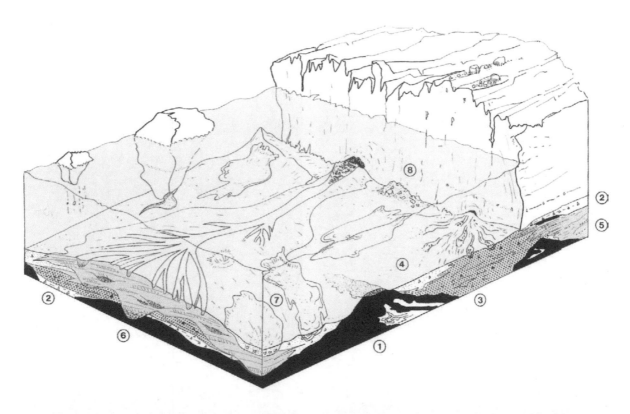

PLATE 32

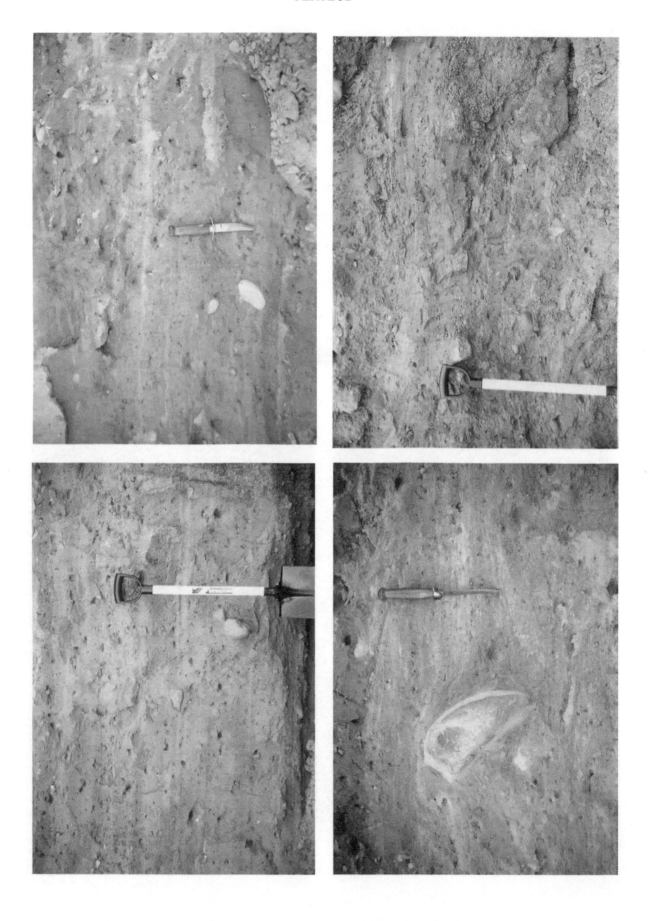

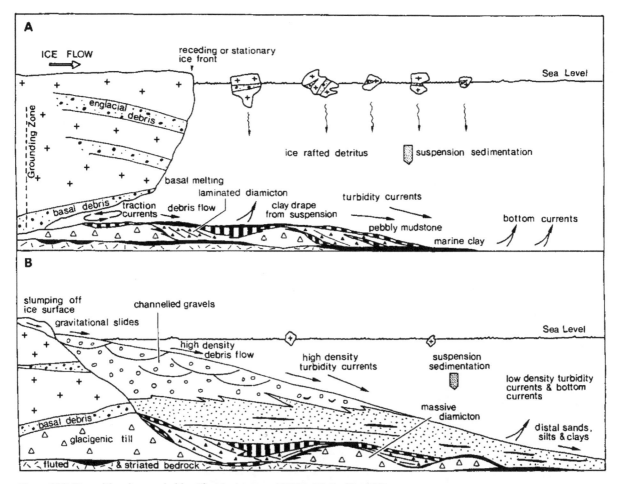

Figure 238: Depositional scenario identified in the area of Killard Point (fig. 219).

gravity-flow activity. The Killard Point section is important because it illustrates the rapid accumulation of sediments in proximal glaciomarine environments, depositional oversteepening and subsequent downslope sediment gravity flow activity (fig. 238).

6) *The Isle of Man.* Exposures along the northern Isle of Man coast provide excellent outcrops through coarse-grained diamict facies deposited as a result of the interplay between ice-rafting, deposition from suspended sediment plumes and bottom currents.

The island occupies a central location in the north-

ern half of the basin about 45 km offshore from Killard Point, and comprises a mountain core (621 m elevation) fringed to the north by an extensive (150 km²) raised marine foreland. This low relief foreland has an upper elevation of about 15 m and is underlain by Pleistocene sediments resting on a marine planated rock surface, at about -45 m O.D., that dips to the north. A substantial push ridge (the Bride Moraine) crosses the northern part of the foreland and is exposed in a section at Shellag Point (fig. 239). To the south of the ridge a coarsening-upward marine se-

Plate 32

b	d
a	c

32. Stratified diamict facies recording interplay of 'rain-out' and current activity. (a, b) East coast of Isle of Man; (c) ice-rafted clast in striated diamict; (d) stratified facies composed of bands with different colour and matrix textures. These units form complexes up to 30 m thick and several kilometres in length on the Isle of Man.

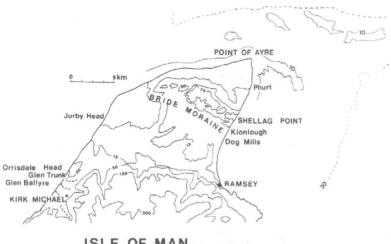

ISLE OF MAN

Figure 239: (a) Location of Isle of Man; contours in metres above and below sea level. The presence of elongate submarine banks parallel to the Bride Moraine suggests repeated push ridge construction (e.g. plate 33) during recession.

quence of pelagic muds with dropstones, shoreface sands and gravels can be identified; the sequence has been glaciotectonised into a series of thrusted folds (plate 33). Resting on the northern ridge flanks is an offlapping complex of gravels, sands and coarse-grained (sandy) diamicts deposited as part of a suba-queous outwash complex adjacent to a marine ice margin (plates 30 and 33).

Diamict facies are poorly consolidated with a muddy sand matrix and are massive, crudely and well stratified (figs. 229 and 230 and plate 32). A distinc-tive orange/brown colour is derived from sandstone source rocks. Massive facies on detailed cleaning commonly show thin (<1 cm) stringers of sand or silt,

reworked shell debris and no preferred clast orienta-tion. Stratified facies show a crude banding on a scale of <1 cm to >20 cm (fig. 230) comprising facies with different colours and/or matrix texture; bedding is generally horizontal but steeper dips are associated with flow noses and brecciated diamict masses; water-escape pipes, together with gravel and sand 'load' structures are common. Large 'outsized' clasts deform underlying stratification and are draped by overlying strata. Interbeds of massive, current-bedded and deformed sand and gravel up to 1 m thick are present; bimodal (tidal) palaeocurrent direction can be identified but many sedimentary structures have been deformed by post-depositional com-

b

c

Figure 239 (continued): (b) The northern raised marine foreland of the Isle of Man. (c) Bride Moraine at Shellag Point.

paction. There are marked lateral and vertical facies changes between diamict, pebbly sands and a wide range of subaqueously-deposited sand-and-gravel facies that include wave-rippled and graded facies (fig. 231). In many cases discrimination between diamict and other pebbly facies is difficult and is arbitrarily based on mud content (<10% for diamicts). Coarse-grained diamicts form a series of offlapping, irregularly thickening and thinning blankets draped on the northern flanks of the Bride Moraine. Diamict units are separated by and inter-bedded with sands and gravels some of which fill large 'sag' basins loaded down into the underlying diamict.

Earlier work identified coarse-grained diamicts as a series of stacked melt-out tills, residual deposits from downwasting ice, and as 'flow tills', material sedimented from the ice margin (G.S.P. Thomas, 1977). In contrast, critical evidence for a subaqueous origin is the nature of the interbeds (sands and gravels deposited by traction currents and normally-graded gravels deposited by turbidity currents), thickness

Figure 239 (continued): (d) Stacked thrusts within the Bride Moraine composed of marine mud → sand → gravel sequences (see plate 33).

Figure 239 (continued): (e) Thrusted sands on the south margins of the Bride Moraine.

(up to 30 m) and lateral extent (up to 5 km) of diamict units, unambiguous evidence of ice-rafting and the occurrence of diamict lithofacies within continuums of marine sediments showing bimodal and unidirectional palaeocurrents (fig. 231).

The primary mechanism for depositing massive diamict facies is the rapid deposition of sand and mud from plumes of suspended sediment, with coarser ice-rafted debris (plate 30 and fig. 229). Stratified facies (plate 32 and fig. 230) represent deposition under conditions of variable currents and/or changing sediment supply. Ongoing work shows that stratified diamicts are bimodal in texture, in the pebble and fine-sand sizes, as a result of deposition of pebbles from bedload and sand from suspension. This appears to be in response to current velocities fluctuating around a critical value for rolling and suspension.

The offshore record

Offshore data collected by the British Geological Survey confirm the importance of glaciomarine sedimentation. Pantin (1977) and Garrard (1977) have described 'Irish Sea Tills' over a wide extent of the Irish and Celtic Seas; stratigraphic units that were successfully cored are characterised by a high matrix and low clast content, frequent stratification and lamination, the presence of common soft-sediment deformation structures in the form of diapirs and load casts and an intimate association with thick marine muds containing foraminiferal faunas typical of cold marine waters of low salinity.

Shallow seismic records across the southern Irish Sea Basin reveal networks of narrow (<2 km) steep-sided channels that are locally overdeepened as much as 130 m and are over 100 km in length (Whittington, 1977). These are identified as tunnel valleys cut subglacially by meltwaters under high hydrostatic heads. Exposures south of Dublin at Bray, provide a window into the fill of one channel and shows ice-proximal glaciomarine sediments. The offshore portions of the channels are filled by distal glaciomarine deposits; channels appear to have been exposed by rapid retreat of the ice margin from the basin in response to rising sea-level (Eyles & McCabe, 1989a, b). Ongoing work demonstrates that much of the offshore record is the distal equivalent of ice-proximal deposits accumulated close to the modern coast as the ice-sheet rapidly retreated out of tide-water.

Offshore sediments having the same 'Irish Sea Till' characteristics are widespread *south* of the accepted limit of Devensian glaciation (fig. 228) and are currently attributed to Pre-Devensian glaciation (Garrard, 1977: 75). However, the deposition of distal glaciomarine sediments, deposited far to the *south* of the accepted limit of Late Devensian glaciation is probable (cf. Scourse *et al.*, this volume) and should be tested by detailed examinations of onshore exposures of 'Older Drifts' in the south of Ireland and in southwest Britain (e.g. North Devon, Fremington; Scilly Isles).

Discussion

The recognition of glaciomarine conditions and high relative 'glacial' sea-levels in the Irish Sea Basin has ramifications for British Quaternary studies which extend beyond the basin itself. Primarily, it is evidence of crustal depression around the British-Scandinavian Ice Sheet and marine submergence. As we have seen, these ideas are not yet widely accepted for the Irish Sea Basin and still less further afield on the British mainland, but the opposite position, that there was no glacio-isostatic depression beyond the ice-sheet margin cannot surely be maintained in the face of overwhelming data that shows crustal depression between 100 and 200 m at the margins of large ice sheets (Andrews, 1982, 1987; Peltier, 1982 for review). Andrews (1982: 3) has stated that 'the primary control on sea-level variation at the margin of an ice sheet is largely a function of glacio-isostatic disequilibrium'. The debate as to high relative sea levels during deglaciation of the Irish Sea Basin is very reminiscent of debate elsewhere, such as in the Canadian Arctic, where marine submergence during full glacial conditions is now well established (e.g. England, 1983).

It can be predicted that many Middle and Late Pleistocene sequences will be seen henceforth to be the product of marine sedimentation around grounded or partially floating ice margins. McCabe *et al.* (1986) have recently identified the deposits of tidewater glaciers in County Mayo, on the west coast of Ireland. 'Shelly drifts' had long been identified in the area but careful field descriptions and analysis of contained micro-fauna show that they are *in situ* deposits. 'Shelly drifts' are described at many coastal localities around the British mainland (e.g. Mitchell *et al.*, 1973). To take a single example, we suggest that a prime candidate for re-examination are the Middle Pleistocene 'North Sea Drift' sequences exposed around the North Norfolk coast (Eyles *et al.*, 1988).

With regard to the Irish Sea Basin, the next step is to identify the history of Devensian isostatic movements across the basin from the sedimentary and geomorphological record and ultimately to derive glacio-isostatic facies models. The message, even if a trifle obscured, is that 'glacial' sedimentary sequences in the basin can no longer be interpreted directly in terms of simple climatically-driven ice advance/retreat cycles. We suggest that the major control on ice-margin behaviour and the style of deposition was marine flooding as a result of ice-sheet loading and glacio-isostatic disequilibrium.

Acknowledgements

This research was funded by the Royal Society, the Research Committee of the University of Ulster and the Natural Sciences and Engineering Research Council of Canada.

Glacial deposits and Quaternary stratigraphy of the Isle of Man

ROGER V. DACKOMBE & GEOFFREY S. P. THOMAS

The glacial deposits of the Isle of Man were regarded by Kendall (1894) to be of such an extraordinary and diverse character as to be of interest unsurpassed in the United Kingdom. Traditionally, they have been divided into two suites. A suite of local insular origin is restricted to the Manx uplands and comprises tills, heads, soliflucted tills, slope gravels and screes. A foreign suite (fig. 240), derived by successive ice advances from source areas in western Scotland, forms an extensive lowland area in the north and comprises a wide range of tills, outwash and lacustrine deposits exceeding 250 m in thickness. The latter suite is exceptionally well exposed in a series of coastal cliffs bounding the northern lowland (fig. 241).

Early work on the glacial deposits was undertaken by Cumming (1846, 1854), Kendall (1894) and Lamplugh (1903). More recent stratigraphic investigations include those of Smith (1930), Slater (1931), Wirtz (1953), Mitchell (1965), Thomas (1976, 1977, 1985b) and Dackombe & Thomas (1985). Sedimentological investigations have been reported by Eyles & Eyles (1984) and Thomas et al. (1985) and analyses of glaciodynamic structures by Slater (1931) and Thomas (1984a). The Late Devensian biogenic deposits have been examined by Erdtman (1925), Mitchell (1965), Dickson et al. (1970), Watson (1971) and Joachim (1977).

Although the lithostratigraphy (fig. 242; table 23) is well established (Thomas, 1976; 1977) the chronology is poorly defined and with the exception of dates from organic sediments overlying the glacial sequence, which indicate that the Devensian Irish Sea ice sheet had cleared the island by c. 15,000 B.P., the island is, so far, devoid of any other dated horizons. The only major datum is the buried cliff and shoreplatform at Ballure which has been assigned to the Ipswichian interglacial on the grounds that nowhere in the complex of glacial deposits that overlie it is there any evidence for a subsequent temperate interlude. The chronology of the thick sequence of glacial deposits beneath sea level in the north of the island, which includes a supposed marine sequence (Lamplugh, 1903), is also unknown but probably pre-Devensian.

Until recently, most workers have accepted the view that the foreign glacial deposits of the island, as elsewhere in the Irish Sea basin, were deposited by land ice. Thomas (1976, 1977 and 1985b) and Pantin (1978) have suggested on palaeontological grounds, however, that when Devensian ice was retreating from the basin the margin may have been partially marine. Using a sedimentary facies approach, Eyles & Eyles (1984) have extended this concept and argue that the whole of the foreign Devensian glacial succession in the Isle of Man is glaciomarine in origin.

The Upland deposits

These deposits thinly clothe the Manx uplands and are solely derived from the Ordovician Manx Slates which underlie them. A wide range of deposit types occur and all show an intimate relationship to the slope forms. Thus, on interfluves and summits the deposits comprise a thin, open-work, angular rubble with little matrix. On lower slopes, and beneath the characteristically asymmetric terraces that occupy many of the valleys, the deposits are thicker, more compact, have a higher proportion of silty matrix and display a crude off-slope stratification (fig. 243). Clasts are sub-angular to edge-rounded, occasionally lightly striated and possess an off-slope fabric throughout the vertical sequence. Beneath more extensive terrace slopes these deposits are intercalated with sequences of massive stony clay and well-sorted, edge-rounded, open-work gravels. The contact between these deposits and the bedrock is commonly marked by a layer of angular rock rubble (Thomas, 1977).

The age and origin of the local deposits is unclear,

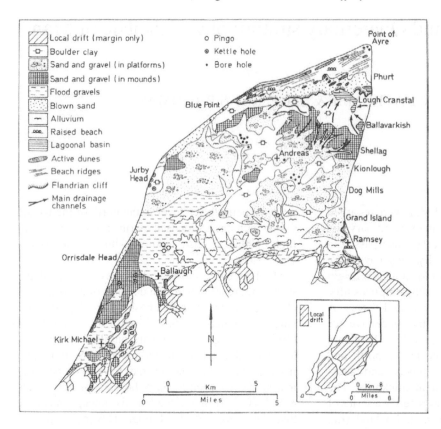

Figure 240: The drift deposits of the northern plain of the Isle of Man.

Figure 241: The east coast of the Isle of Man showing the main ridge of the Bride Moraine separating proglacial outwash and lacustrine deposits in the south (left) from the undulating topography of the readvance tills to the north.

though their off-slope stratification and fabric and the general asymmetric development favour a periglacial origin. If this is so, then a crucial question is whether they originated by *in situ* breakdown of solid rock during severe periglacial conditions when the up-lands were ice-free, or whether they arise, at least in part, from the rapid periglacial reworking of locally derived tills during the retreat of the Irish Sea ice sheet. Given the generally accepted conception of the geometry and thickness of the Devensian Irish Sea

Table 23. Stratigraphic nomenclature for the Quaternary of the Isle of Man (from Thomas, 1977).

Group	Formation	Member	Type site
Ayre	Point of Ayre	Cranstal Silts (CSS) Ayre Beach (AB)	Lough Cranstal, NX455025 BSG Pit, Point of Ayre, NX461041
	Curragh	Curragh Peat (CP)	Curragh, SC365951
	Moorland	Upland Peat (UP) Sulby Gravel (SRG)	Beinn-y-Phott, SC380865 Sulby Glen, SC380915
Surface	Ballaugh	Ballaleigh Debris Fan (BDF) Wyllin Debris Fan (WDF) Ballyre Debris Fan (BYDF) Ballure Debris Fan (BLDF) Ballaugh Debris Fan (BHDF) Crawyn Sand (CS) Jurby Kettles (JK) Wyllin Kettles (WK)	Glen Mooar, SC303894 Glen Wyllin, SC310908 Glen Ballyre, SC314914 Ballure, SC458934 Ballaugh, SC340950 Crawyn, SC340950 Jurby Head, SC346990 Glen Wyllin, SC310908
	Upland	Upper Stratified Head (USH) Upper Gravel (UG) Mid Stratified Head (MSH) Massive Head (MH) Lower Gravel (LG) Lower Stratified Head (LSH) Rock Rubble (RR)	} Upper Sulby, SC384874
		Upper Blue Head (UBH) Lower Blue Head (LBH) Ballure Slope Wash (BSW) Brown Head (BH) Ballure Scree (BS)	} Ballure, SC458934
		Mooar Head (MRH) Mooar Scree (MS)	Glen Mooar, SC303894 Glen Beeg, SC300890
	Jurby	Andreas Platform Gravel (APG)	Dog Mills, SC455985
		Trunk Till (TT) Trunk Gravel (TG)	} Glen Trunk, SC317923
		Cranstal Till (CT)	Cranstal, NX468020
		Ballaquark Sand (BQS) Ballaquark Till (BQT)	} Ballaquark, NX465013
		Jurby Sand (JS) Jurby Till (JT)	} Jurby Head, SC343980
	Orrisdale	Dog Mills Series (DMS) Ballure Clays (BC) Kionlough Till (KT)	Dog Mills, SC455985 Ballure, SC458934 Kionlough, SC455987
		Ballavarkish Sand (BVS) Ballavarkish Till (BVT) Ballavarkish Marginal Series (BVMS)	} Ballavarkish NX462007
		Orrisdale Gravel (OG) Orrisdale Sand (OS) Orrisdale Till (OT)	} Orrisdale Head, SC319930
	Shellag	Kionlough Gravel (KG)	Kionlough, SC457990
		Shellag Gravel (SG) Shellag Sand (SS) Shellag Till (ST)	} Shellag Point, NX460000
		Wyllin Sand (WS) Wyllin Till (WT)	} Glen Wyllin, SC309906
		Ballure Till (BT)	Ballure, SC458934
Sub-surface	Sub-surface	Middle Sands (MSS) Middle Boulder Clay (MBC)	} Borehole 16 (Smith 1930). Kiondroughad, NX315015
Basement	Basement	Ayre Marine Silts (AMS) Lower Sand (LSS) Lower Boulder Clay (LBC)	} Borehole VI (Lamplugh 1903). Point of Ayre, NX465050

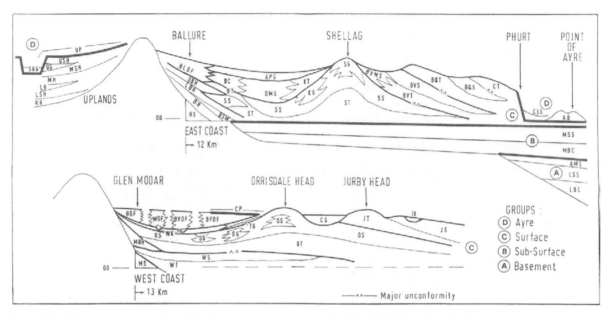

Figure 242: Schematic representation of the Quaternary succession in the Isle of Man. For identification of members see table 23 (from Thomas, 1977).

Figure 243: The upland suite of local deposits at Druidale (SC 355881). The off-slope stratification within the heads which blanket the uplands is clearly shown.

ice-sheet (Boulton *et al.*, 1977), the latter seems more likely (Dackombe & Thomas, 1985).

The foreign diamicts

The interpretation of the diamicts is the subject of some controversy (Eyles & Eyles, 1984; Thomas & Dackombe, 1985, Eyles *et al.*, 1985). Although there

are sound reasons why the 'modern facies approach' (Eyles & Eyles, 1984) should be used to analyse these deposits, their division of the Manx sequence into a coarse-grained and fine-grained diamict associations camouflages the considerable differences in texture and structure which exist within each group. The formal lithostratigraphy of Thomas (1976) is no more satisfactory in this respect in that units are defined by a combination of bounding unconformi-

Figure 244: Boulder and cobble pavements in the Orrisdale Till at Jurby Head.

Figure 245: Bullet-shaped boulder with striated upper surface and 'plucked' down-ice face in the Orrisdale Till at Jurby Head.

ties and a general consistency of lithology. A wide variety of diamict facies occur in the Manx sequence and few of them, in the current state of our knowledge of glacial depositional processes, can be interpreted with assurance.

Perhaps the least controversial type is that represented by the vast bulk of the Orrisdale Till member (Thomas, 1977) which crops out extensively along the west coast of the island (fig. 242; table 23). Exceptionally well-exposed at Jurby Head (fig. 242), this is a massive, sandy, matrix-supported diamict. It

is occasionally bedded and texturally very uniform with only very occasional included waterlain sediments. Where these bands do occur they are always near the top of the unit which at its maximum is in excess of 23 m thick. Lenticular bedding is present in some sections, but is often only visible in fresh cliff falls where the presence of sub-horizontal discontinuities is emphasised. Only very rarely do the bed limits consist of thin laminae of waterlain sediment. Conjugate jointing is present and the material is well compacted forming, in contrast to some of the less

Figure 246: Stone collision in the Orrisdale Till at Jurby Head. The pick points in the direction of ice-flow. The left-hand clast has collided with the right-hand one pushing it along, causing shattering and leaving a streaked-out trail of fine-gravel detritus behind and beneath the left-hand clast.

competent diamict facies, vertical or overhanging cliff facets. Very strong evidence for an origin by subglacial lodgement comes from the consistent orientation of striae on the top surfaces of the clasts which form boulder and cobble pavements within the diamict body (fig. 244). Similarly, the consistent orientation of striae on top of isolated boulders, particularly of limestone, together with their 'classic bullet shape' (fig. 245) and long axis orientation suggests stable, deep subglacial conditions during deposition. The presence of a considerable ice thickness is also exemplified by the occurrence, admittedly only rarely, of stone collisions which have resulted in the crushing and streaking out of one of the clasts (fig. 246). There can be little doubt that this facies is indeed of subglacial origin and not, as suggested by Eyles & Eyles (1984), a product of density under-flows and pelagic rain-out in an ice-proximal position in a marine environment. However, while this facies was being deposited under thick ice back from the ice-margin there is no reason why the margin should not stand in shallow water, either a fresh-water lake or a shallow marine incursion.

More uncertainty surrounds the origin of the diamict facies exemplified in the Isle of Man by the Shellag and Wyllin members. These massive, silt- and clay-rich, clast-poor diamictons contrast with the rest of the Manx sequence but they are, however, very similar to other 'Irish Sea Tills' seen throughout the Irish Sea Basin. As elsewhere in the basin these two units occur only at around sea-level except where

tectonically disturbed as at Shellag Point (fig. 247) and in common with other members of the Manx succession contain shell fragments and occasional complete specimens. One section on the west-coast is a remarkably prolific provider of shell material and when viewed along with presence of a shelly marine fauna in the overlying sands, the '*Turritella* Sands' of Lamplugh (1903), it is difficult to fault his contention (*op cit*, p. 428) that the deposit is "strips of old sea bottom, dragged up and only partly incorporated with the boulder clay". Although usually assumed to be of subglacial lodgement origin (Thomas, 1977), the alternative interpretation of the deposits as of glacio-marine origin (Dackombe, 1978; Eyles & Eyles, 1984) is attractive given the uniformity of this facies throughout the basin, its position close to contemporary sea level and the absence of unambiguous evidence of terrestrial subglacial deposition. Unfortunately, the small area and poor quality of exposures make it difficult to carry out any sound sedimentological research into these deposits, in the Isle of Man at least.

In contrast to the 'Orrisdale' and 'Shellag' types which dominate their respective stratigraphic units, the Ballavarkish Till (Thomas, 1976) comprises a range of facies unique within the island. Five facies, based on subtle variations in particle size distribution and internal structure have been identified (Dackombe & Thomas, 1985) and each generally occupies a characteristic position in the vertical sequence. Facies 'A' forms a transition zone from the underlying Shellag Sands and consists of a sequence of thin bands of matrix-rich diamict which interdigitate with well-sorted fine sands (plate 34). The sands are occasionally rippled or cross-bedded and the whole often shows mild deformation in the form of gentle crenulation or overfolding of cross-bedding. Facies 'B' (plate 35) is a rather poorly laminated and rather heterogeneous matrix-rich sandy diamict often containing deformed inclusions of the underlying sands. These inclusions are often streaked out into thin sand laminations or gently folded into detached recumbent fold closures. In facies 'C' the diamict becomes more homogeneous and develops a laminated structure. This lamination is caused by subtle variations in the amount and colour of fines and the passage from one lamina to the next is usually diffuse and gradational rather than sharp. Only rarely is there any evidence for current reworking as a cause of the lamination. The clasts sometimes penetrate the lamination in the manner of dropstones, but more often they lie with their AB planes in the plane of the lamination and cause the lamination to diverge evenly both above and below (plate 36).

Facies A, B and C, usually in that order from the base of the unit, are succeeded by facies 'D' which forms the bulk of the thickness of the Ballavarkish

Till. Facies 'D' (plate 37) is similar in most respects to the typical 'Orrisdale' facies described above except that clast clusters are rather rare and there is less consistency in either clast fabric or the orientation of striae on embedded clasts. This may be partly accounted for by incomplete re-orientation of fabrics due to override and gentle folding occasioned by the deposition of the succeeding Ballaquark Till. Facies 'E', rarely seen due to the unconformity which truncates the Ballavarkish Till consists of a series of interbedded sands and diamicts similar to facies 'A' but with rather less evidence of deformation.

Eyles & Eyles (1984) ascribe all these facies to the same general mode of deposition, that of pelagic rain-out and deposition from density underflows in an ice-proximal position. Variations in the degree of lamination reflect the variable impact of current winnowing. Dackombe (1978) on the other hand interprets the main body of the unit composed of facies 'D' as being subglacial lodgement till and the lower facies (A, B, and C) as the result, during advance, of overriding, deformation and incorporation of a frontal apron consisting of mixed fluvial sediments and re-sedimented diamicts.

The Trunk Till (fig. 242; table 23) is the most variable in terms of facies assemblage of all the major

Figure 247: Glaciotectonic structures at Shellag Point show the Shellag Till to the lower right in each of the inclined thrust slices.

Figure 248: Interlaminated sands and muddy sands of the Trunk Till. Recumbent folds can be seen in the centre right.

Figure 249: Flow overfold (above tape) in the sandy facies of the Trunk Till.

Figure 250: Sp facies in the Orrisdale Sands.

Figure 251: St facies in the Orrisdale Sands.

diamict units. It consists of a series of contiguous lenses mainly of sandy diamict, which occupy a similar stratigraphic level within the upper part of the Orrisdale Sands. Very rapid vertical and lateral facies changes occur in all of the lenses and although massive, sandy matrix-rich diamicts occur, the main facies consists of thinly interlaminated sands and muddy sands (fig. 248). Clast content is variable and bedding and lamination are common. So too are gravel lags, thin interbeds of rippled or cross-stratified sand and bands of massive or laminated mud. All of these

characteristics, along with the low strength and variable direction of clast fabrics, suggest that these are re-sedimented deposits. Strong evidence for re-sedimentation comes from the common occurrence of flow overfolds (fig. 249), disconnected fold closures, deformed ripple lamination, convoluted bedding and other load and dewatering structures. The presence of large-scale inclined, lenticular beds of diamict, the upper surfaces of which display current winnowing and which pass laterally at their lower ends into laminated muds suggests that at least some

Figure 252: Sr facies in the Orrisdale Sands.

parts of this unit consist of diamict lobes discharged into standing water bodies.

The foreign glaciofluvial deposits

A wide variety of foreign glaciofluvial, glaciolacustrine and related deposits are exposed in the long coastal cliff sections in the north of the island and include ice-front alluvial fan, ice-marginal trough, proximal delta and distal lake-floor environments.

On the west coast, around Orrisdale Head, Thomas *et al.* (1985) have described a sequence of sediments deposited in a series of diachronous marginal sandur troughs formed on unstable, ice-cored supraglacial topography. Facies types are those typical of sand-dominated braided channels and include planar (Sp, fig. 250) and trough cross-stratified sands (St, fig. 251), upper and lower flow regime horizontally laminated sands (Sh) and a wide variety of types of rippled sands (Sr, fig. 252). Sedimentation was con-trolled by dead-ice ridges running parallel to the ice margin, but as buried ice melted out the topographic constraints diminished and individual sandur trough systems widened and coalesced. Sediments within the individual troughs are characterised by very rapid lateral and vertical facies variation caused by channel migration, abandonment and adjustment to a rapidly changing ice and sediment substrate. Facies assem-blages consequently show marked lateral transitions from high-energy, laterally impersistent, gravel-dominated longitudinal bar sequences to laterally extensive, sand-dominated overbank-flood se-quences.

In the initial stages of trough filling, meltwater streams directed parallel to the ice margin by the confining flanks of dead-ice ridges deposited coarse facies assemblages during successive flood events. As the sediment accumulated it abutted the margins of the ice-cored ridges and sheets of flow-till were introduced. Few of these diamicts were preserved but some sequences show alternation of diamict and

Plate 33

33. Depositional scenario interpreted for the glaciomarine sequences of the northern Isle of Man. (A) Glaciotectonism around oscillating ice margin. (B) Deposition of subaqueous outwash fans, sands and gravels together with associated diamicts (as a drape on the rear side of the push ridge (the Bride Moraine)). Height of ridge about 75 m. (1) mud, (2) sand, (3) gravel, (4) 'rain-out' diamicts, sands and gravels as in figs. 2, 3, 4 and 5 and subject to downslope resedimentation. Precise water depths are not known.

PLATE 33

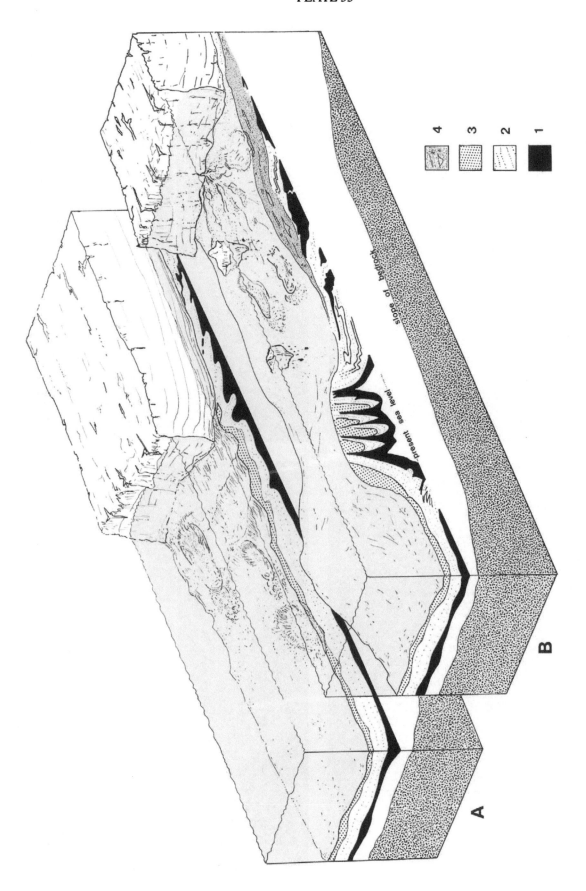

coarse clastics equivalent to Paul's (1983) Type I sediment assemblage. As sedimentation continued and as the dead-ice ridges melted down the sandur system widened. At this stage only a portion of the trough system was active except at peak flow. Consequently peripheral areas accumulated fine-grained facies assemblages which only formed at high flood. With the decline of these floods, sedimentation reverted to the lower parts of the trough where sediment accumulation was slower due to regular and repeated channel incision. Thus, through much of the history of the filling of individual troughs the position of major channel systems remained relatively constant. As the sequence built up however, the importance of dead-ice ridge control declined and in a number of troughs abrupt upward-fining in facies assemblage is indicative of the sudden abandonment of the channel system and its subsequent occupation only during peak floods. In other cases, coarse facies assemblages shift upward and laterally to transgress other basins. This resulted from the elimination of topographic constraints and the consequent coalescence of sandur systems.

Elsewhere on the west coast similar sequences pass into extensive lacustrine sediments. Thus at Killane, the floor of a wide trough separating two major morainic ridges is occupied by extensive varved clay. This clay displays regular couplets of graded silt and clay often separated by thin interbeds of fine sand. Individual couplets vary in thickness but the ratio of coarse to fine is maintained for considerable distance. At two persistent horizons in the sequence, groups of couplets show slumping and flow folding indicative of substrate movement caused by the melting of buried ice. Elsewhere, numerous small imatra stones, or concentrically banded concretions formed around a node of calcite occur. On the northern margin of the basin, the varve sequences are overlain by coarse steeply-dipping gravel foresets.

Estuarine/inter-tidal deposits

On the east coast the situation is rather different and stratified sediments are only seen in any quantity

south of the re-advance moraine of the Bride Hills (fig. 241). At and south of Shellag Point (fig. 247), massive, coarse gravels tectonically disturbed by overthrusts dipping towards the north (Thomas, 1985; Allen, this volume) pass into upward-fining sequences of massive, fine gravels, cross-bedded sands, parallel-laminated sands and rippled sands. These in turn pass at Dog Mills into a series of laterally extensive laminated silts and clays and massive and laminated fine sands that show extensive soft-sediment deformation and dewatering structures. The overall transition demonstrates a passage from proximal ice-front alluvial fan, through sandur to shallow subaquatic deposition. Foraminifera from the latter sequence suggest that the water body was shallow estuarine-intertidal rather than lacustrine (Thomas, 1977; Dackombe & Thomas, 1985).

Discussion

The superb exposures in the drift deposits of the Isle of Man are of considerable interest both in their own right as the products of a variety of glacial environments and in their regional significance as crucial evidence for the reconstruction of the pattern and style of Late Devensian deglaciation in the Irish Sea Basin. By far the most important issue that remains to be resolved is whether deglaciation in the area was accompanied by an active marine margin as suggested by Thomas (1977), Eyles & Eyles (1984) and McCabe *et al.* (1984). If this is the case then there must be a substantial revision of our conception of the palaeo-environmental conditions of the time and in particular the question of water depths must be addressed. Although they quote no figure for sea level the reconstruction of Eyles & Eyles (1984) suggests a relative level at least 200 m above present. While it is acknowledged that the evidence for sea levels during the Lateglacial in the northern basin of the Irish Sea is spread very thinly, the data that is available and soundly dated for Northern Ireland (Carter, 1982) and for the Solway coast (Bishop & Coope, 1977) do not permit such excessive amounts of isostatic relief. Indeed Jardine (1971) concludes

Plates 34–37

34	35
36	37

34. Facies 'A' of the Ballavarkish Till.
35. Facies 'B' of the Ballavarkish Till.
36. Facies 'C' of the Ballavarkish Till.
37. Facies 'D' of the Ballavarkish Till.

that marine waters in the Solway Firth area were at a level *below* the present sea-level while the Late Devensian ice front lay in the vicinity of the present north coast of the Solway.

Even if the deposits south of Dog Mills are accepted as glacio-estuarine the water depths required suggest a relative sea level of the order of only 15 m above present. This level is much more easily accomodated within the present knowledge of sea-level change. For example Orford (in McCabe & Hirons, 1986) suggests that the prominent notch cut into the till around Dundrum at about 14 m O.D. dates from 15,000 B.P. and that the maximum postglacial sea-level was not much above this. If such a sea level is accepted then it seems unlikely that it would have influenced the dynamics or depositional activity of the ice margin except inasmuch as parts of the proglacial environment would be drowned while others where sedimentation was rapid would emerge above the sea. Certainly the deposition of the thicker diamicts would hardly be affected by the buoyancy effects of 20 m of water on an ice mass that must have been at least a couple of hundred metres in thickness some short way back from the margin.

Lamplugh was remarkably perceptive when he wrote in 1903 that the Isle of Man presented "an unrivalled field for the study of the conditions that ruled in the northern part of the basin of the Irish Sea during the Glacial Period" and is "pre-eminently an area wherein the various theories by which the drift phenomena of the Irish Sea basin have been explained, may be put to the test". He would, no doubt, be surprised and perhaps disappointed to find that 85 years later the same issues are still being hotly debated.

Glacial deposits of southwest Ireland

WILLIAM P. WARREN

For the purposes of this paper southwest Ireland is taken as Counties Waterford, Cork, Kerry, Clare and Limerick and south Tipperary (fig. 253). It encompasses most of the area of the so-called 'Munsterian' glacial deposits (see Synge, 1979e), a large part of the Southern Irish End Moraine of Lewis (1894) and Charlesworth (1928a) and some of the area within its limits. It also includes the entire area known to have been influenced by the independent Kerry/Cork ice sheets (Wright & Muff, 1904; Farrington, 1947; 1954; Warren, 1977). Like much of the rest of Ireland the glacial sediments of most of this area have never been mapped in detail. The exceptions are referred to below.

Southwest Ireland is underlain chiefly by rocks of Devonian and Carboniferous age (fig. 253). Devonian sandstones and shales extend from the peninsulas of Corca Dhuibhne and Iveragh in west Kerry, to Waterford Harbour and south Tipperary in the east. In south Cork these pass up into Carboniferous shales and sandstones. South of the co-called Armorican front the rocks form east-west to northeast-southwest aligned anticlinal ridges and synclinal troughs which reflect the strong Hercynian fold pattern that dominates the regional structure. In a general sense Carboniferous limestones and shales floor the troughs while the ridges are formed of the older Devonian rocks, mainly sandstone. In the northern part of the area, in Counties Clare, Limerick and Tipperary the glacial sediments are underlain chiefly by Carboniferous limestone with a number of inliers of Devonian sandstone and Silurian shale. The northwest of the area is underlain largely by Upper Carboniferous shales and sandstones.

As it is impossible to integrate current stratigraphic interpretations with the traditional model as outlined by Mitchell *et al.* (1973) the stratigraphic model and nomenclature proposed by Warren (1985) are used here. The last cold stage is referred to as the Fenitian Stage, the last warm stage (interglacial) is referred to as the Gortian and a previous cold stage, the Bally-

bunnionian is also referred to. Where the terms 'Midlandian' and 'Munsterian' are used they are in quotation marks and are used only in discussing deposits previously attributed to those stages. They carry no chronostratigraphic significance.

Many myths have developed with regard to the glacial sediments, particularly those south of the Southern Irish End Moraine. Perhaps the most important of these is the suggestion that the glacial sediments have been dramatically altered by solifluction processes and that constructional features, such as moraines, eskers, kames and drumlins, have been removed from the landscape by these processes (see Mitchell, 1977). In fact there is no evidence that such features were ever widespread in the area south of the moraine (the 'Munsterian' area) and while there is evidence of large scale solifluction in the higher Namurian hills of north Kerry and west Limerick (see below) there is very little evidence of such activity in Counties Cork, Waterford, south Tipperary and southwest Clare. In addition kame and kettle features and eskers survive in near pristine state in east County Cork, and well-formed kame hills are common further north in County Cork near Watergrasshill and on the high ground extending east from Watergrasshill into County Waterford. Such features also occur between Mitchelstown and Mallow in north County Cork.

The suggestion that the tills in the 'Munsterian' area are more leached of carbonates (Synge, 1970a) is, in a general sense, true but the fact that the 'Midlandian' boundary was extended far south of the End Moraine so as to exclude areas of tills high in carbonate from the 'Munsterian' area, irrespective of the position of the End Moraine begs the question as to why the 'Munsterian' tills contain fewer carbonates. The use of the level of carbonates in tills as an index of age has thus become a circular argument. The dangers inherent in this argument are outlined elsewhere (Warren, 1985; Rose, Hodgson & Warren, this volume).

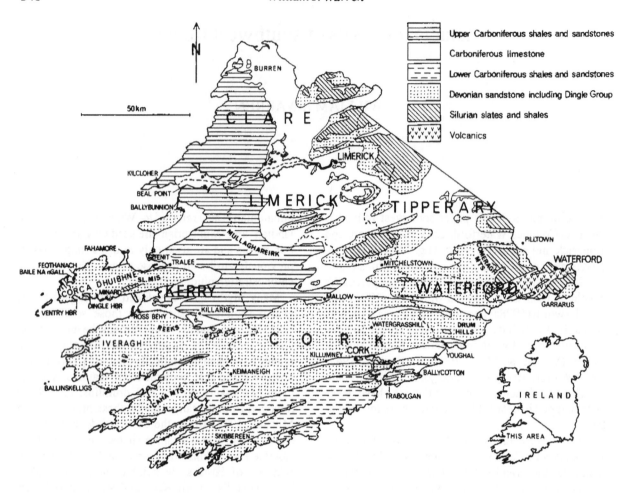

Figure 253: Southwest Ireland: localities referred to and bedrock geology.

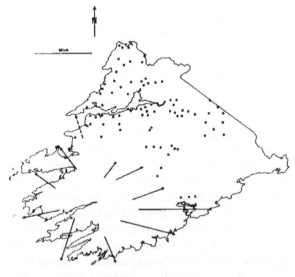

Figure 254: The extent of Kerry/Cork ice and the distribution of Galway Granite erratics.

This area contains glacial deposits of three major provenances: The Irish midlands and Galway (Connemara) to the north, the Irish Sea Basin to the northeast and the Kerry/Cork Mountains. Galway (Connemara) Granite erratics are spread widely through the area, north of a line from Fahamore to Cork City (fig. 254). Granite erratics near Ballinskelligs in south Kerry may also be of Galway provenance. Limestone erratics on Devonian sandstone in Tipperary, Cork and Waterford and volcanic erratics from County Limerick in south Limerick and Tipperary further testify to an ice movement from the midlands (Synge, 1970a). Till containing shell fragments, flint, Leinster Granite and even Ailsa Craig microgranite occurs, particularly in coastal areas as far west as Ballycotton on the south coast of Cork. The northern extent of the Kerry/Cork ice cap can be deduced from the extent of indicator erratics such as Lough Guitane felsite (Warren *et al.*, 1986). Its eastern and northeastern extent is much more difficult to determine owing to the similarities in erratic suites associated with this ice cap and those associated with ice from the north. The pattern of striae has, however, been used to indicate the outline extent of Kerry/Cork ice (Wright & Muff, 1904). Small independent ice caps and valley glaciers developed in many of the mountains of the area. Farrington (1947) outlined extensive areas which he regarded as unglaciated by extraneous ice. More recent information suggests that the unglaciated area should be restricted to the summit areas of the higher mountains (Lewis, 1974; Bryant, 1977; Warren *et al.*, 1986).

Stratigraphy

The well-exposed sections of the south and southwest coasts have provided the main stratotypes for the lithostratigraphic subdivision of the Irish glacial deposits. A classic paper by Wright & Muff (1904) identified the major localities of stratigraphic importance and they outlined the stratigraphic relationships of the major units. Their main focus of attention was the raised beach sediments which rest on a raised marine-cut platform between west County Cork and County Wexford. Regarded by Wright & Muff as preglacial, in the sense that it predates all of the glacial deposits in the area, the beach, the Courtmacsherry Formation, is now regarded as belonging to the last interglacial (Bowen, 1973a; Warren, 1985). Thus the glacigenic deposits overlying it are related to the last glaciation, the Fenitian of Warren (1985). More complex interpretations have, however, been offered (see Synge, 1981).

The common stratigraphic sequence on the south coast as defined by Wright & Muff (1904) is

4. Upper Head
3. Till
2. Main Head
1. Raised Beach

The provenance of the till unit in this sequence varies. West of Cork Harbour it is a local till of western provenance (the Garryvoe Formation of Warren, 1985). East of the harbour it is either the shelly till of Irish Sea Basin provenance (the Ballycroneen Formation), the till of northern provenance (Bannow Formation) or, in east Cork, the Garryvoe Formation. In places two distinct till units are seen in vertical succession, as at Ballycotton Bay where the Garryvoe Formation overlies the Ballycroneen Formation (Wright & Muff, 1904; Farrington, 1954) or at Garrarus, County Waterford where the Ballyvoyle Till

(Bannow Formation) overlies the Ballycroneen Formation (Quinn, 1984). The lateral extent of these units is not always clear but approximations can be made.

Deposits of the Kerry/Cork ice sheet

The Garryvoe Formation (the till associated with ice that extended from the Kerry/Cork Mountains) seems to extend as far east as Youghal Harbour and overlies the Ballycroneen Formation (Wright & Muff, 1904; Farrington, 1954). The Fenit Formation which intervenes stratigraphically between the Garryvoe Formation and the Courtmacsherry Formation has, in places, undergone fabric alteration and compaction under the influence of overriding glacier ice (glaciodiagenesis) and may in these situations be regarded as a glacigenic unit (Warren, 1987a). The Garryvoe Formation has a red/brown matrix and is characterised by phenoclasts of Devonian sandstone and shale, and Carboniferous limestone, shale and chert. It seems clear that it relates specifically to the Fenitian Stage. Similarly the Drum (till) Formation which takes an identical stratigraphic position in relation to the raised beach deposits at Ross Behy, County Kerry (Warren, 1977; 1978), seems to be its precise equivalent. The northern extent of this unit is problematic. Felsite erratics, in a till with Devonian sandstone and Carboniferous chert, associated with ice from the Kerry/Cork Mountains extends to at least 150 m O.D. on the southern flanks of Sliabh Mis and seems to have passed through the lower cols in the mountains of Corca Dhuibhne (Warren *et al.*, 1986) and to have extended as far north as Fahamore. They, along with Devonian sandstone erratics, reach at least 300 m on the southern slopes of the Mullaghereirk Mountains, about 30 km north of their outcrop locality. North of Sliabh Mis, at Carrigaha, till of southern provenance containing Inch conglomerate and chert erratics (Tonakilly Formation), overlies a raised beach deposit. The raised beach is interpreted as equivalent to the Courtmacsherry Formation and the till is therefore regarded as Fenitian (Warren *et al.*, 1986). The stratigraphy at Fenit on the northern shore of Tralee Bay suggests, however, that Kerry/Cork ice of Fenitian age did not extend to there (see below). The situation may have been complicated by earlier ice movement from the same source and subsequent reworking of erratics originally transported by far-travelled ice (for example, Inch Conglomerate on the northern side of Corca Dhuibhne) by later local glaciers. Certainly the stratigraphic sequence at Cuan Lathaí, near Ross

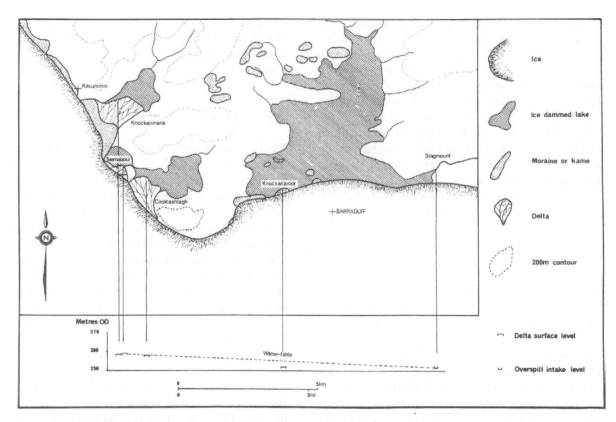

Figure 255: Kilcoolaght kame terrace and ice marginal delta and the related ice margin.

Behy suggests an early and a later ice movement from the Kerry/Cork Mountains (Warren, 1977). The earlier movement may pre-date the raised beach deposits of last interglacial age. Felsite erratics 20 - 30 km from their source at Lough Guitane may relate to either event. The absence of any biostratigraphic evidence or a reliable inland isochronous lithostratigraphic horizon to parallel the Courtmacsherry raised beach makes inland stratigraphy difficult. Erratics of probable southern provenance (Kerry/Cork Mountains) are common in tills as far to the northwest as Minard and Dingle Harbour (Warren *et al.,* 1986) but again the stratigraphy is problematic. There is no evidence of glaciation post-dating the (Courtmacsherry) raised beach deposits at Dingle Harbour, Ventry or Baile na nGall where head deposits alone overlie the beach (Bryant, 1966; Warren *et al.,* 1986).

The maximum known extent of ice centred on the Kerry/Cork Mountains is outlined in fig. 256. A considerable concentration of Devonian conglomerate erratics at Kilcloher in southwest Clare (Finch & Synge, 1966) strongly suggests an ice movement from a southerly direction and may indicate a much

more extensive Kerry/Cork ice sheet then has hitherto been recognised. Various recessional, or readvance ice positions are shown also in fig. 256. The Kilcummin line is the clearest of these. It is marked by large accumulations of ice-marginal deposits. At Coolcashlagh a large kame terrace grades into a large ice-marginal delta (fig. 255). and further north the Kilcummin Moraine abuts the higher hills of Upper Carboniferous sandstone and shale. This moraine zone is marked by large ice-marginal meltwater channels. Outside the Kilcummin Moraine complex, till of southern provenance with Devonian sandstone and felsite erratics is typically overlain by a mantle of geliflucted debris derived from that till. It is possible that this geliflucted deposit can be correlated with the Main Head of Wright & Muff (1904) (the Fenit Formation of Warren, 1985), thus suggesting that this area was not glaciated during the last glaciation (Fenitian Stage) and that the Kilcummin Moraine represents the maximum extent of Fenitian ice (see Warren, 1977). It is however equally possible that the thick slope deposits simply reflect the susceptibility of the clay-rich till, derived from Namurian shales, to gelifluction processes. Thus, stratigraphically, the

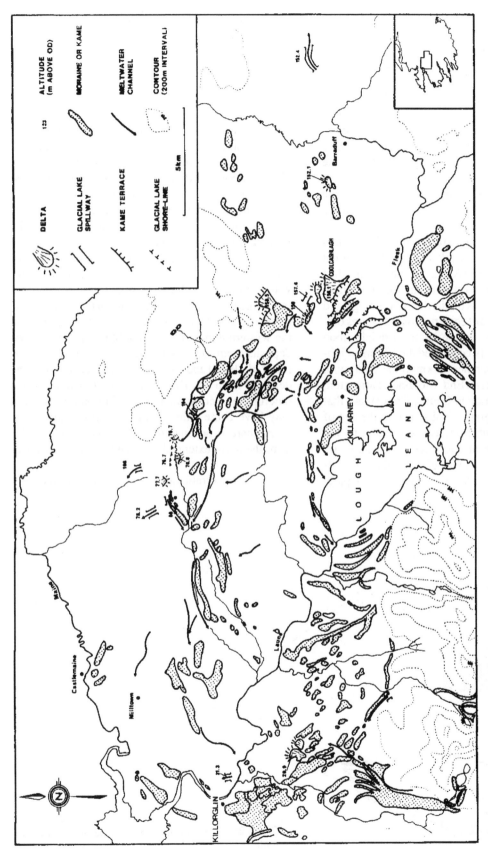

Figure 256: Moraines and glaciofluvial features showing the pattern of ice retreat in the Killarney area.

geliflucted till facies might represent the Upper Head of Wright & Muff (1904).

During its retreat from the Kilcummin line, the margin of the Kerry/Cork ice cap left a remarkable legacy of arcuate moraine ridges, glacial meltwater channels and glaciolacustrine deposits (fig. 256) which confirm what the pattern of striae and ice-moulded surfaces in the highland areas to the south indicate: that the ice dispersed, not from the higher mountains of the MacGillycuddy's Reeks or the mountains of Iveragh, but, as was recognised by Maxwell Close (1867), from an accumulation zone south of the Reeks in the valley of the Kenmare River.

Farrington (1959) regarded the Killumney Moraine 13 km west of Cork City as the terminal eastern moraine of Kerry/Cork ice during the last cold stage and correlated it with the Southern Irish End Moraine on the basis of similar morphology and the absence of significant constructional features outside it. This was exactly the same principle upon which he had previously correlated the 'Keimaneigh-Skibbereen line' (which lies 30 km to the west) with the same Southern Irish End Moraine (Farrington, 1936). Ongoing mapping in County Cork shows that there is neither stratigraphic substance to the age of this moraine nor morphological substance to the claim that no significant constructional features lie outside it. It remains a moot point whether the Killumney Moraine is penecontemporaneous with parts

of the Southern Irish End Moraine, which Synge (1970a) has shown to be very much a diachronous feature.

Deposits of ice from the north

The most dramatic indication of ice from the north is the distribution of Galway Granite erratics (fig. 254). They are not common in the Burren of north Clare (Finch & Synge, 1966) but this is an area of sparse glacial deposits and Galway Granite is a widespread erratic in south Clare, north Kerry and north Limerick and extends into north Cork and is occasionally found as far south as Cork City.

Lewis (1974), citing evidence of chert erratics noted by Bryant (1966), has suggested that the ice which transported the large granite erratic at Faha-more (the furthest west of the known Galway Granite erratics) pushed on-shore as far west as Feothanach at the western end of Corca Dhuibhne. Some of these deposits seem to relate to local valley glaciers (Warren et al., 1986) and some of the chert erratics may be derived from a Devonian conglomerate which outcrops at Feothanach. Any Carboniferous chert may have come either from the north or the south. Thus, there is no unequivocal evidence that the ice from the north extended any further west than Faha-more.

As with the erratics associated with the Kerry/Cork

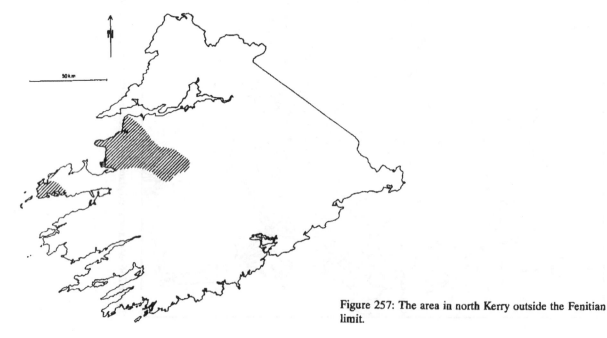

Figure 257: The area in north Kerry outside the Fenitian limit.

icecap, the Galway Granite erratics cannot be identified as relating exclusively to one specific glacial event, although Farrington (1965) thought that where they occurred in Fenitian sediments they had been reworked from earlier deposits. In north Kerry and Limerick they occur both inside and outside the area influenced by Fenitian ice (fig. 257; Warren, 1980).

At Ballybunnion in north Kerry a till composed largely of Carboniferous shale, sandstone and limestone (the Ballybunnion Formation) underlies raised beach deposits thought to be the equivalent of the Courtmacsherry (raised beach) Formation of the south coast (Warren, 1981; 1985). The raised beach sediments are overlain by a geliflucted facies derived from the underlying till. The Ballybunnion Formation is therefore thought to pre-date the Fenitian Stage. Along the north shore of Tralee Bay geliflucted deposits with facies derived from glacial deposits also overlie raised beach deposits (Warren, 1985). As there is no evidence of post-beach glaciation at either site, the area between Ballybunnion and Fenit is thought not to have been glaciated during the Fenitian Stage. The full extent of this area is not known (Warren, 1980) but it is clear that some of the Galway Granite erratics of north Kerry fall within it (fig. 257). It is not clear, however, whether those Galway erratics that occur within the sphere of Fenitian glaciation were re-worked or transported directly from Galway. It is possible too that some of the erratics were transported by floating ice, for example the large boulder at Fahamore and the boulders on the foreshore at Ballinskelligs.

At Camp and Kilshannig, near Fahamore on the northern shore of Corca Dhuibhne a light brown till (Camp Formation) with Carboniferous cherty shale of northeastern to eastern provenance underlies a till of southern provenance with Inch Conglomerate erratics (Tonakilly Formation). Neither the relationship between this deposit and the raised beach nor that between the Tonakilly Formation and the Galway Granite erratics is clear. The Camp Formation is possibly associated with the granite erratics, as some are found in overlying gravels at both sites.

Limits of ice from the north

It is possible that an early movement of ice from the north extended as far southwest as Ballinskelligs. This achievement would have necessitated a strong movement of ice at least 600 m thick over the mountains of west and south Kerry and it would clearly have extended far beyond the present south coast. Given that part of north Kerry remained ice-free during the Fenitian, such an ice movement would have occurred during an earlier cold stage.

Although there is not general agreement on the matter, it is suggested that Fenitian ice extended far south of the Southern Irish End Moraine of Lewis (1894) and Charlesworth (1928a). The stratigraphic reasons for this were outlined by Bowen (1973a) and Warren (1979a; 1985). There is, without doubt, a difference in both the lithology and the morphological expression of the glacial sediments on either side of the moraine. Sediments are generally thinner and there are fewer eskers and drumlins to the south. But this is obviously a reflection of glacial process rather than of age difference. Although it does not negate a difference in age of any degree, it is not evidence of significant age difference any more than the striking change in sediment character and topography that occurs in passing from the gravel-strewn midlands with its eskers to the drumlin belts on the margins (see Synge, 1979e). In southwest Ireland only the area around Ballybunnion and Fenit and part of the western tip of Corca Dhuibhne can be said with confidence not to have been glaciated during the Fenitian. This area, based on Fenit and Ballybunnion, probably extends inland to include much of the uplands of north Kerry (fig. 257).

With the exception of the summit areas of the higher mountain ranges the entire area north and east of a line from Beal Point in north Kerry to Youghal Harbour in east Cork was glaciated by ice from the north during the Fenitian stage (fig. 258). The Ballyvoyle Till of Watts (1959a) regarded as a member of the Bannow Formation by Warren (1985) was deposited by this ice as it moved over County Waterford and off the south coast. The southern limit of this deposit must lie off the south coast. The Southern Irish End Moraine is a composite diachronous feature representing halt or readvance phases as the ice-front retreated northwards (Synge, 1970a). Acid volcanic erratics of southern provenance in till at Pilltown near Carrick-on-Suir in County Waterford (Collins, 1982) await detailed study; they may suggest at least a local complication of the picture in this area. The Mothel Till with limestone erratics (Watts, 1959a), which occurs south of Pilltown and at much higher levels than the Pilltown erratics confirms that Pilltown was overrun by ice from the north. Thus, the southern erratics at Pilltown may have been reworked.

Ballycroneen (Till) Formation

The 'shelly boulder-clay' recorded by Wright & Muff (1904) along the south coast of Ireland (the Ballycro-

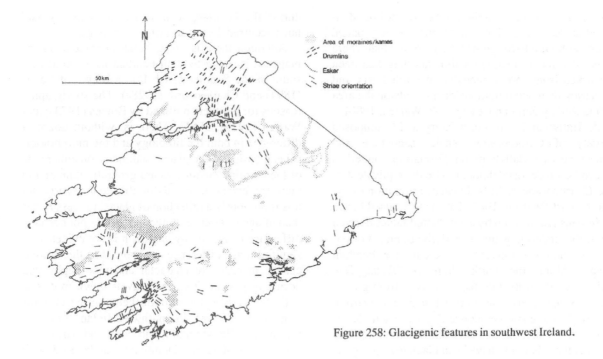

Figure 258: Glacigenic features in southwest Ireland.

neen Formation of Warren, 1985) occurs also along the east coast. Marine shells have also been recorded in association with glacigenic deposits as far inland as County Laois, County Kildare and north County Tipperary. The occurrences of this till however are apparently limited to coastal areas, and this has raised many questions as to its genesis. Both shelf ice (Synge, 1977a) and subaqueous sedimentation (Huddart, 1981b) have been suggested, for different facies, and marine deposition associated with a tidewater glacier was suggested by McCabe *et al.* (1986) for shelly diamictons on the north coast of County Mayo. McCabe (1987) also suggested subaquatic flow deposition in association with a tidewater glacier for the Ballycroneen Formation on the south coast.

The Ballycroneen Till outcrops intermittently in coastal cliff sections along the south coast as far west as Trabolgan Bay, just east of Cork Harbour, and it extends at least 2 km inland from the present coast at Ballycotton Bay. It is typically a massive grey/brown diamicton, rich in clay and silt, with shell fragments and sparse phenoclasts; but at some localities, both on the east coast (Hoare, 1975) and in the lee (west) of headlands on the south coast, it contains stoney facies in which the phenoclasts are dominantly locally derived. Wright & Muff (1904) recorded erratics of Irish Sea Basin origin, and occasional Ailsa

Craig microgranite phenoclasts are found in the Ballycroneen Till.

Its restricted occurrence, in low-lying areas close to the present coast, is difficult to explain but postdepositional factors should not be discounted in considering its distribution. It is extremely mobile in a saturated state and would be susceptible to rapid mass-wasting in hill-slope positions particularly under periglacial conditions. In addition, except in the limestone-floored troughs where it mostly occurs, this deposit is likely to be quickly leached of carbonates and thus lose its distinctive shell fragments, particularly in situations where groundwater has low pH values. Without the aid of surviving noncalcareous erratics any *remanié* deposit would be difficult to identify. Flint erratics in the Drum Hills of west Waterford, although they are not associated with shell fragments, may thus represent an incursion of Irish Sea basin ice (Quinn, 1984). Such an incursion would not be consistent with shelf ice. In addition, if, following Orheim & Elverhøi (1981), we accept that shelf ice carries very little debris more than 70 km from its grounding line, it is difficult to conclude that all the Ballycroneen Till occurrences along a 150 km coastal stretch relate to shelf ice, unless the grounding line paralleled the present southern shoreline. Thus, despite the absence of striae associated with this deposit on the south coast, but bearing in mind its sheared contact with the underlying sediments on the south coast of County Wexford, it is probable that many, though not all, of the shelly facies with flint and other erratics, on the south coast, relate to a ground-based ice sheet emanating, in part at least, from the Irish Sea basin. The suggestion of deposition from a tidewater glacier in association with a high relative sea level (McCabe, 1987) is purely speculative and finds no independent support.

Deposits of mountain glaciers

Almost all of the upland areas above 600 m in southwest Ireland contain cirque basins. In excess of one hundred cirque basins, most of which contain moraine ridges and mounds have been identified. Many of the cirque basins in the mountains of Kerry and west Cork were inundated by the Kerry/Cork ice sheet, yet later retained autochthonous ice which produced small but very distinct moraine ridges within the cirque basin. Such cirque basins are particularly common in the Caha Mountains between Kenmare and Bantry.

On the northern side of the MacGillycuddy's Reeks a large valley glacier extended from a cirque complex. The maximum extent of this glacier is not known but it had receded significantly by the time of the deposition of the Kilcummin Moraine of the Kerry/Cork ice-cap (Warren, 1979c). The Mountains of Corca Dhuibhne contained an ice cap that extended on to the lowlands both to the north and south of the mountains (Lewis, 1974; Warren *et al.*, 1986). The relationship between this ice cap and the larger ice sheet is not clear, but it deposited a large part of the glacigenic deposits on the northern side of the peninsula during the Fenitian Stage (Warren *et al.*, 1986). The Comeraghs seem to be the only other mountainous area large enough to have produced anything more than cirque glaciers. Coumshingaun Conglomerate erratics in local tills on the low ground to the east of the mountains were apparently used to indicate that cirque glaciers coalesced to form a piedmont glacier in the area of Rathgormack (see Hull, 1878).

The ages of the cirque deposits are not known, some may be Lateglacial, but in the absence of clear biostratigraphy this cannot be confirmed (see Warren, 1979c). It is clear, however, that the mountains of Kerry did not harbour a lateglacial icecap on the scale of the Loch Lomond advance of western Scotland (see Watts, 1963) during the equivalent Ballybetagh Substage (Nehanagan Stadial of Mitchell *et al.*, 1973; see Coxon & Gray, this volume).

Glacial deposits and landforms of central and western Ireland

PETER COXON & PHILIP BROWNE

Other than local detailed studies little Quaternary mapping has been carried out (and published) in either the Central Lowlands or the west of Ireland since the work of Synge in the 1960s and early 1970s and the publication of the 1977 INQUA guide to Western Ireland (Finch, 1977). The production of fig. 259 has relied heavily on this work, on available LANDSAT imagery and on the few more recent local studies that are acknowledged in the text. Fig. 260 is designed to accompany fig. 259 and to show locations referred to in the text. The stratigraphy of the glacial deposits for Ireland as a whole is discussed elsewhere (see Hoare, this volume) and only important local sequences are referred to here.

Western Ireland

Ice mass accumulation in western Ireland has followed a complex sequence of events that are particularly hard to unravel because of a lack of biostratigraphic marker horizons, detailed lithostratigraphic work and absolute dates. It is generally accepted (e.g. Mitchell et al., 1973; Mitchell, 1976; Finch, 1977 and McCabe, 1985 and 1987) that two major glaciations occurred; the Munsterian (= Wolstonian/Saalian) and the Midlandian (= Devensian/Weichselian).

However, no undoubted interglacial deposits separating these two glacial episodes (and hence equivalent Ipswichian/Eemian) have been recorded leaving the stratigraphy uncertain. Detailed information regarding the organic deposit known as the Curaun (also referred to in the literature as Corraun) Interglacial (site 17, fig. 260) on a tributary of the Cartron River on the northeastern side of Corraun Hill described by Synge (1968) (table 24) has never been published in detail and the deposit has not been unequivocally shown to be of interglacial status.

On the other hand interglacial deposits of Gortian age are well represented in the area (Jessen et al., 1959; Watts, 1985) at Boleyneendorrish near Gort (site 14, fig. 260). Another Gortian site

(site 15, fig. 260) at Derrynadivva near Castlebar contains an excellent interglacial record (Coxon & Hannon, unpublished). The interglacial sequence ends abruptly, as it does at many Gortian sites, with inwashed inorganic material suggesting a rapid deterioration in climatic conditions during the latter part of the interglacial cycle. Biostratigraphic evidence (Watts, 1985) suggests an Hoxnian (= Holsteinian) correlation for the Gortian, but the relationship of Gortian Stage sediments to overlying tills let alone to a firm lithostratigraphy has not been properly elucidated leaving the stratigraphic position of the Gortian open to question (Warren, 1979a; 1985).

For western Ireland (northern and western Mayo and western Galway) the work of Synge and others enables a working stratigraphy to be proposed (Mitchell et al., 1973). This is presented in table 24.

Glacial sediments

McCabe (1985, 1987) discussed the glacial geomorphology and stratigraphy of Ireland as a whole and noted that although the distribution of major landform units was well known, the landform - sediment associations were far less well understood. This is certainly true for western Ireland where no modern sedimentological studies have been published. For this reason these authors have mapped landform types and then taken individual examples of typical sediment types found at particular locations or over wider areas while realising that complexity and variation in glacial sedimentation prevents generalisation regarding sediment - landform association.

The pattern of glaciation in western Ireland has generally been regarded as follows: Firstly there are older tills exposed predominantly in areas outside of those of the Last Glaciation (for example area 16, fig. 260). The older tills, such as the Erris Till, are weathered and in places cryoturbated but are not represented by 'fresh' glacial landforms (Herries Davies & Stephens, 1978). Their age is uncertain, being

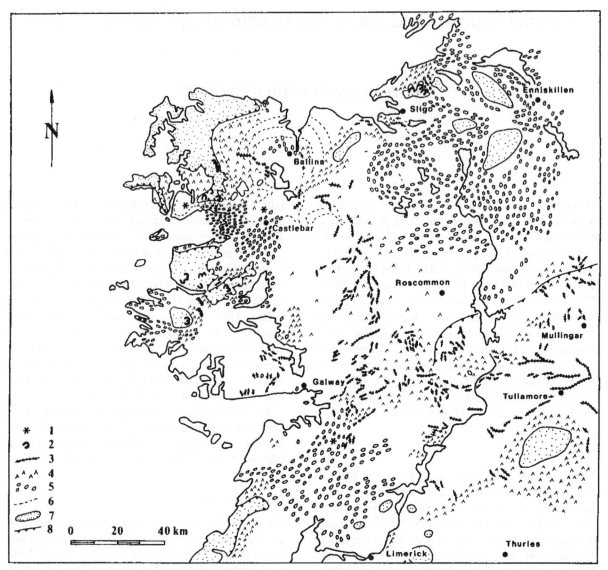

Figure 259: Map of central and western Ireland showing major glacial depositional features. The map was drawn using a wide range of published material, available LANDSAT TM imagery (see detail on plates 43 and 44) and personal observations by the authors. Sources included: Browne (1982), Chapman (1970), Charlesworth (1928), Finch (1977), McCabe (1985), McCabe *et al.* (1986), Orme (1967), Sollas (1896) and Synge (1968, 1978 and 1979). 1 = interglacial deposits; 2 = cirque moraines; 3 = esker; 4 = sand and gravel spreads, including kames; 5 = drumlins; 6 = moraines (after Charlesworth, 1928); 7 = areas with little or no Midlandian deposits; 8 = major ice limits.

dependant on the Curaun deposit (site 17, fig. 260) for stratigraphic position, but Synge (1968) regarded them as Munsterian. There are also obvious problems in dating these tills on surface morphological criteria and the presence or absence of cryoturbation structures.

Secondly, there are the deposits referred to the Last Glaciation (Midlandian) and these provide a complex sequence of events in mountain and lowland areas. The discussion below uses the outline stratigraphy of table 24 but areas such as Sligo, Leitrim, east Galway,

Roscommon and Clare have simply not been the subject of recent research (for example Browne, 1982) and this has provided difficulties in correlation. Landform and sediment types are taken in turn and their age has, of necessity, been considered as a secondary part of the discussion.

Till and moraines

Much of western Ireland is covered by a veneer of

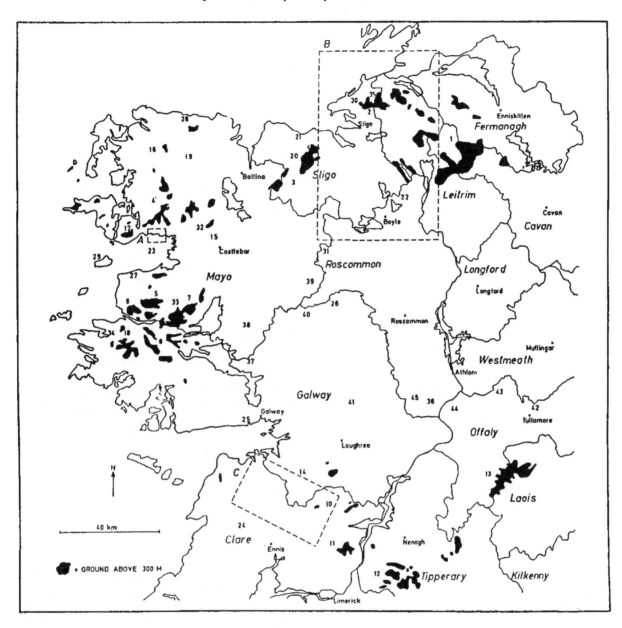

Figure 260: Map of central and western Ireland (to complement fig. 259) showing locations referred to in the text. The insets refer to more detailed figures and to satellite images. Key: 1 = Cuilcagh Plateau; 2 = Benbulbin / Truskmore plateau; 3 = Slieve Gamph or Ox Mountains; 4 = Nephin Beg Mountains; 5 = Sheefry Hills; 6 = Mweelrea Mountains; 7 = Partry Mountains; 8 = Maumturk Mountains; 9 = Twelve Pins; 10 = Slieve Aughty Mountains; 11 = Slieve Bernagh Mountains; 12 = Silvermine Mountains; 13 = Slieve Bloom Mountains; 14 = Boleyneendorrish; 15 = Derrynadivva; 16 = Erris; 17 = Coraun deposit (interglacial/interstadial?); 18 = Kylemore Lough, County Galway; 19 = outwash sands and gravels, County Mayo; 20 = moraine at Lough Easky, County Sligo; 21 = granite erratic north of the Ox Mountains; 22 = area of streamlined moraines and drumlins in Counties Roscommon and Leitrim; 23 = area of drumlins in Clew Bay; 24 = area of drumlins, County Clare; 25 = Seaweed Point, near Galway, County Galway; 26 = area of deglaciation features, County Roscommon; 27 = drumlins near Cleahy rocks (near Louisburgh), County Mayo; 28 = area around Belderg, County Mayo; 29 = Clare Island, County Mayo; 30 = kames near Grange, County Sligo; 31 = kames near Ballyhaunis and Ballaghaderreen; 32 = kames near Beltra Lough, County Mayo; 33 = kame terrace delta south-west of Strahlea Bridge, County Mayo; 34 = delta moraines near Letterfrack, County Galway; 35 = Gleniff cirque and moraines, County Sligo; 36 = till exposures below sand and gravel near Clonmacnoise, County Offaly; 37 = Pollnahallia, County Galway; 38 = area of hummocky terrain near Kilmaine, County Galway, 39 = Dunmore-Ballyhaunis esker system; 40 = Dunmore, County Galway; 41 = Tullamore-Daingean esker system (western end); 42 = Tullamore-Daingean esker system (eastern end); 43 = esker near Clara, County Offaly; 44 = esker near Clonmacnoise, County Offaly; 45 = esker near Ballinasloe, County Galway.

Table insets: A = fig. 264; B = plate 43; C = plate 44.

Table 24. Quaternary stratigraphy in western Ireland.

General glaciation	Cirque glaciation
Midlandian	
1. Roundstone Upper *Salix herbacea* Clay	Acorymore Innermost Moraines – Achill IV (Nahanagan Stadial)
2. Roundstone Mud	
3. Roundstone Lower *Salix herbacea* Clay	
4. Ballycastle-Mulrany Moraine	Achill III
5. Newport Till	
6. Roscahill Till	
7. Oldhead (Louisburgh) Peat	
8. Ballycroy Moraines	Annafrin Outer Moraine – Achill IIb (Athdown)
9. Killadoon Till	
10. Belderg Shelly Till	
11.	Achill IIa (Brittas)
12.	Acorrymore Outer Moraine – Achill I (Aughrim)
13. Curaun (Corraun) Inter- glacial	
Munsterian	
14. Erris Till	
15. Gort Upper Solifluction Till	
Gortian	
16. Gort Mud, Derrynadivva Mud	
17. Gort Fine Sandy Clay	
18. Gort Lower Solifluction Gravel	

diamicton, often filling cavities in the lee-side of *roche moutonnées* or other rock projections (for example site 18, fig. 260; fig. 261). It is not possible to map such a sporadic cover, which is assumed to be till, except to differentiate between surface expres-

sions of the cover, i.e. streamlined drumlins and non-oriented moraine. In very large areas the cover of glacial material is very patchy indeed and the land-scape is ice scoured.

Areas of undulating and hummocky terrain have frequently been recorded in the literature as 'kames'. Such areas are mapped on fig. 259 as 'sand and gravel spreads including kames' after Synge (1979f). Often exposures are not available, but where they are, ob-servations have suggested that many of these areas of 'kames' are morainic, containing tills and other de-posits along with associated stratified outwash sands and gravels (for example area 19, fig. 260).

The deposits in these regions are highly variable spatially. McCabe (1985) referred to these non-oriented forms as disintegration moraine associated with former ice-marginal positions and downwasting of ice. Synge (1968) mapped area 19 (fig. 260) as dead-ice moraine, while Charlesworth's (1928b) plate 5 depicts arcs of recessional moraines that he described as 'kettle moraines' on the lowlands and as 'irregular, tumultuous ridges and mounds' on moun-tain slopes. Fig. 262 shows a rare section in one of Charlesworth's moraines south of Lough Easky (site 20, fig. 260) and fig. 263 shows a granite erratic carried north from the Ox Mountains (to site 21, fig. 260) that indicates the ice advanced out from the Roscommon lowlands that lie to the south. It is clear that the areas mapped as 'spreads of sand and gra-vel...' on fig. 259 contain a variety of ice disintegra-tion landforms and sediments.

Drumlins and other streamlined features

The spectacular drumlin fields of Ireland are well represented in this area. Fig. 259 shows the genera-lised distribution of drumlins and streamlined moraines with concentrated areas around Counties Leitrim and Roscommon (area 22, fig. 260 and plate 43), Clew Bay (area 23, fig. 260 and fig. 264 - Inset A on fig. 260) and County Clare (area 24, fig. 260 and plate 44 - Inset C on fig. 260).

The morphological continuum of streamlined forms including drumlins can be seen on LANDSAT

Plates 38–39

38. View of delta moraines, near Letterfrack, Kylemore Valley, County Galway (34) (Photograph: P. Coxon).

39. A raft of Carboniferous limestone 4 m thick and 24 m long overlying the sand deposit at Pollnahallia (37) (Photograph: P. Coxon).

Figure 261: Stratified poorly-sorted moraine in the lee of a bedrock projection, Kylemore Lough, County Galway (18). The staff is marked in 0.5 m intervals (Photograph: P. Coxon).

Figure 262: A push moraine at Lough Easky including ice-distal outwash to the left of the picture (20) (Photograph: P. Coxon).

Plates 40–42

40. *Roche moutonnée* on lowland near Tully Cross (34) (Photograph: P. Coxon).
41. A rare exposure of limestone-rich lodgement till within a streamlined feature near Clonmacnoise, County Offaly (36) (Photograph: P. Coxon).
42. Seaweed Point (25). Detail of diamicton of possibly glaciomarine origin within a drumlin section. The staff is marked in 0.5 m intervals (Photograph: P. Coxon).

40	
41	42

Figure 263: Granite erratic from the Ox Mountains on the Sligo coast (28) (Photograph: P. Coxon).

images. Their morphological complexity and juxtaposition to each other and to a range of glacial features cannot be discussed here. The internal structures of drumlins are also variable and the recent work of McCabe and others (Dardis & McCabe, 1983; 1984; McCabe, 1985) has shown that detailed sedimentological research is the route to a better understanding of the origin of drumlins. The drumlins on fig. 259 include a wide range of features at many scales including rock drumlins, rock-cored drumlins and those composed entirely of glacigenic sediments. Examples of the latter form include drumlins in Galway Bay, plate 42.

The drumlins in the Clew Bay area are composed of Synge's (1968) Newport Till overlying Roscahill Till. Synge considered these two tills to represent a single glacial episode which probably coincided with the Drumlin Readvance of the Last Glaciation (possibly occurring some 17,000 years B.P.). The ice appears to have originated from the area of County Roscommon (area 26 on fig. 260) streaming out northwestwards, westwards and southwestwards to produce the streamlined features. The drumlins mapped on fig. 259 do not all necessarily date to the same period but they must have been produced by actively moving ice streams late in the Last Glaciation. Fig. 265 is a section in lee-side deposits within a drumlin on the south side of Clew Bay (site 25, fig. 260). Work by McCabe, Haynes & McMillan (1986) near Belderg in north Mayo (site 28, fig. 260) describes a suite of glaciomarine sediments associated with the drumlin ice limits at heights of up to 80 m O.D. Radiocarbon dates indicating an age of ca.

17,000 years for these glacial sediments are reported by McCabe *et al.* This work will undoubtedly lead to a reinterpretation of many glacigenic sequences (previously grouped as lodgement tills) and explain many high level deltaic sequences (see for example Kylemore below).

Other streamlined forms include the extensive areas of ice-scoured bedrock including much of west Galway where for example in Connemara, the lack of glacial deposition and widespread erosion gives rise to 'knock and loughan' scenery.

Glaciofluvial and glaciolacustrine deposits

Deglaciation in Ireland involved retreat of the ice sheet leaving a succession of moraine systems running across the country (McCabe, 1985). The ice decay in western Ireland is poorly understood and complicated by different mountain and lowland decay systems but a general pattern emerges showing retreat of ice back into County Roscommon (area 26, fig. 260) and into mountain areas. Most of the impressive sand and gravel features date from this retreat, but not all, some are associated with glacial lakes ponded between or at the edge of active ice masses. Some examples are outlined below.

Eskers

The esker systems of western Ireland are strongly

Figure 264: Aerial photograph of drumlins in Clew Bay. See inset A on fig. 260 for location. Reproduced from Geological Survey Aerial Photography by permission of the government.

associated with ice retreat patterns and eskers in the Central Lowlands, and the two areas are discussed together under 'Central Ireland' below.

Kames, deltas and undifferentiated sand and gravel

Some isolated mounds of sand and gravel with distinct flat surfaces have been found to be deltaic in origin (delta kames) in a number of areas, for example around Grange (site 30, fig. 260) in County Sligo (Browne, 1982), between Ballyhaunis and Ballaghaderreen (area 31, fig. 260) and near Beltra

Lough (fig. 266; site 32, fig. 260). Another deltaic sequence is found as an elongated kame terrace near Srahlea Bridge (site 33, fig. 260).

A particularly fine sequence of delta moraines at Kylemore (site 34, fig. 260) is related to ice advance along Kylemore valley producing a moraine into standing water at the exit of the valley onto lowland near Letterfrack. Plate 38 is a view of the deltas and fig. 267 of the foresets that can be seen in exposures. The origin of the standing water is uncertain; is it marine, proglacial or a lake between two ice margins? The answer to this problem is complicated by the presence of ice-moulded features in the lowland

Figure 265: Stratified diamicton and laminated muds in a drumlin section on the south side of Clew Bay (27). This section is in the lee-side of the drumlin which is predominantly composed of a stratified diamicton over 25 m in thickness. The staff is marked in 0.5 m intervals (Photograph: P. Coxon).

Figure 266: Foreset deposition in a delta kame, near Beltra Lough, County Mayo (32) (Photograph: P. Coxon).

(e.g. the *roche moutonnée* in plate 40). These suggest an ice movement inland from the sea (Kinney, 1986). Further work in this area may reveal ice movement directions not previously considered. The influence of higher sea level during maximum Midlandian glaciation as proved by McCabe *et al.* (1986) further north is another possible control of the Kylemore delta that would require further research.

Cirque glaciation and associated moraines

The mountain ranges of western Ireland are well endowed with cirques. Plate 43 shows the upland areas of Sligo and Leitrim and the glacial troughs dissecting this region are obvious as are numerous shadowed areas in north-eastern corners that are cirques. Larger moraines from these cirques are

Figure 267: Cross section of fore-sets within delta moraine, near Letterfrack, Kylemore Valley, County Galway (34). The staff is marked in 0.5 m intervals (Photograph: P. Coxon).

marked on fig. 259. They were formed, for example, in the Benbulbin/Truskmore Mountains (area 2, fig. 260), the Nephin Beg Range (area 4, fig. 260) and the Mweelrea Mountains (area 6, fig. 260). The cirque moraines have not been dated anywhere in the region but are considered to represent more than one ice advance or recessional phase (table 24; Synge, 1968). Some detailed mapping of cirque moraines has been carried out (Kenyon, 1982; 1986) but is partly unpublished. Some moraines in the area are considered to be Nahanagan Stadial in age (see Hoare, this volume).

Central Ireland

The area referred to here as central Ireland is that of the western part of the Central Lowlands (Herries Davies & Stephens, 1978) which is underlain for the most part by Carboniferous Limestone. The lowland area of Counties Roscommon and east Galway is important because it has been identified as the southwestern end of an accumulation area for ice in both the Munsterian and the Midlandian glaciation by many authors (McCabe, 1985) and the retreat of the Midlandian ice can be seen to have occurred back towards an ice axis in the same region. The ice advance and retreat that have had the most profound effects are presumably those that occurred late in the Last Glaciation. The separation of the Late Midlandian ice advances into stages delimited by end moraines is discussed by Herries Davies & Stephens

(1978) and Mitchell (1976). There has been little comment on why ice should have accumulated on the lowland to form such an extensive source.

The stratigraphy of this area is relatively simple but this may reflect the lack of recent studies. The glacial sediment cover varies in type and thickness. The authors of the few existing local studies have assumed that the general pattern of patches of till overlain locally or completely by spreads of sand and gravel represents ice advance, deposition and ice disintegration respectively. These sediments have not been dated (there has been no lithostratigraphic or biostratigraphic work carried out) and are assumed to belong to the Last Glacial maximum advance and retreat.

Drumlins and other streamlined features

Localised exposure suggests that ice, producing streamlined till features and scoured bedrock surfaces, radiated out of these western lowlands into surrounding regions. This actively moving ice has been responsible for many of the lowland drumlin swarms of western Ireland referred to above that can be seen on fig. 259. The carriage of erratics northwards, westwards and southwards from the NE-SW trending area of the ice axis supports this radially expanding ice movement (for example Charlesworth, 1928).

In general within Counties Roscommon, Longford, Westmeath, Offaly and eastern Galway patterns

of active ice movement are not well expressed in the landscape. Occasional drumlins and small swarms of drumlins (many of them rock or rock-cored) occur locally and bedrock exposure is usually streamlined, but much of the region is mantled by ice disintegration deposits that are referred to below.

Although exposures are plentiful in the sand and gravel deposits this is not true of the till deposits which are exposed only in occasional shallow pits and roadside cuttings. Borehole evidence shows that the depth to bedrock is highly variable over the region (Mitchell, 1980; 1985). Plate 41 shows a lodgement till within a drumlin (site 36, fig. 260) from west of Clonmacnoise. Such rare till exposures show that much evidence of active ice in central Ireland may be obscured. At Pollnahallia, near Headford in County Galway (site 37, fig. 260) a small quarry has exposed silica sand below a limestone-rich till with a fabric indicating ice movement from the southwest. This site is of particular interest because the sand is a glaciofluvially reworked aeolian deposit filling a 25 m deep gorge in the Carboniferous Limestone. At the base of the gorge is 5 m of organic clay that contains a pollen assemblage characteristic of the Late Pliocene or Early Pleistocene (Coxon & Flegg, 1987). Plate 39 illustrates the ice-rafting of sheets of Carboniferous Limestone.

Figure 268: Well-sorted and faulted sands and gravels in an esker section, Dunmore, County Galway (40). Section is 1.5 m high (Photograph: P. Coxon).

Glaciofluvial, glaciolacustrine and ice disintegration features

Hummocky terrain comprising mostly poorly sorted

Figure 269: Inverse grading in an esker near Ballinasloe, County Galway (45) (Photograph: P. Coxon).

Figure 270: Foreset deposition within an esker (Tullamore-Daingean system) near Clara, County Offaly (43). Water flow was towards the east (Photograph: P. Coxon).

sand and gravel with associated tills and diamictons covers a large area of the west central lowlands. The unclassified area of the western lowlands on fig. 259 is predominantly composed of this landform type. Shallow pits provide the only exposures and consequently only generalisations can be made about the sediments.

During deglaciation ice retreat to an ice dome in east Galway left behind two very large systems of esker ridges and numerous smaller ones. One esker system, McCabe's (1985) Dunmore-Ballyhaunis system, shows water flowed northwards (area 39, fig. 260) and the deposits are indicative of a polygenetic origin for the esker, including deltaic deposition (McCabe, 1985). Fig. 268 shows ice contact deposits from Dunmore, County Galway (site 40, fig. 260). The second major esker system, McCabe's Tullamore-Daingean system, is even larger (areas 41 and 42, fig. 260) and shows water flow towards the east. This system is one of steep-sided ridges in a dendritic pattern that narrows to the east (fig. 259).

The sediments in these eskers are well exposed in numerous quarries and often show massive cores covered by foreset deposits. Frequently well sorted climbing ripple sequences are exposed as are ice collapse features (especially large faults) and coarse sediments (produced by ice-wall collapse) along the flanks of the ridges.

If a general mode of formation could be put forward then it appears that the esker ridges have been produced as delta fronts, possibly controlled by within-ice water tables, have retreated back towards remnant ice domes to the south and west. The deltas often owe their form and sediment associations to closed conduit conditions. The eskers are complex and they are associated with the spreads of sand, gravel and kettled moraine that surround them in these lowland areas. Figs. 269 and 270 show sections in eskers that are typical of these lowland features which are heavily exploited for their sand and gravel content.

Figure 270. Foreset deposition within an esker (Tullamore Drainage system) near Clara, County Offaly (O), water flow was towards the east (Photograph: E. Cox).

The glacial stratigraphy and deposits of Eastern Ireland

PETER G. HOARE

Eastern Ireland is defined here as the counties of Louth, Meath, Dublin and Wicklow (fig. 271a), an area of approximately 6,100 km². The region includes parts of the geomorphic provinces known as The Tertiary Igneous Mountains, The Cavan-Down Hill Country, The Central Lowland and The Leinster Axis (Davies & Stephens, 1978) (fig. 271b); it rises from sea-level to the summit of Lugnaquillia Mountain, County Wicklow (T 032917), at 926 m O.D.* The Leinster Mountains played a most significant role in Eastern Ireland's glacial history: ice-caps developed over them on several occasions during the Quaternary; and, although they may have been overwhelmed by one ice-sheet, they interrupted the progress of at least two others. Further details concerning the geomorphology of Eastern Ireland are given by Whittow (1974) and by Mitchell (1976, 1986). The major topographic subdivisions reflect the underlying solid geology; this is described by Charlesworth (1963a, 1966), Whittow (1974), Naylor *et al.* (1980) and by Holland (1981). Although thick sequences of glacial material mask bedrock in many locations, drift is often remarkably thin elsewhere. Accounts of the geological evolution of the region during the Quaternary have been provided by Mitchell (1972, 1976, 1986), Mitchell *et al.* (1973), Bowen (1973b), Huddart (1977b), Synge (1977a, 1979a), Davies & Stephens (1978) and McCabe (1979, 1986b). Synge also made a significant contribution to the description of the area by Davies & Stephens (1978).

It was in Eastern Ireland that Close (1865, 1867,

1874, 1878) laid the foundations of the study of the country's glacial geology. The relationship between the ice-caps that formed over the Leinster Mountains and the ice-sheets that came from outside the region (fig. 272) was determined by Farrington in a series of publications that began in 1934. Hoare (1975, 1977c and d) and McCabe & Hoare (1978) confirmed much of Farrington's pioneering work, but also added important details. Synge first made a significant contribution to the understanding of the glacial history of Eastern Ireland in 1950; he suggested modifications to some of Farrington's findings. McCabe (1971, 1972, 1973), Colhoun & McCabe (1973) and McCabe & Hoare (1978) unravelled the complex Quaternary record in Counties Louth and Meath. Finally, elevated shoreline features have been employed to decipher some aspects of the glaciation of Eastern Ireland; see Martin (1955), Stephens (1957), Stephens & McCabe (1977), Synge (1977a and f, 1981), and Devoy (1983), for example.

The number and extent of the glacial events to have affected Eastern Ireland are still the subjects of debate. However, studies of deep-sea cores have shown that continental areas invariably contain an incomplete record of these episodes. The absence of marker horizons and of geochronometrically-dated beds within the Quaternary sequence of the region has led to ill-founded speculation regarding the timing of glaciations; views have changed with great rapidity, and formerly-abandoned ideas have been revived. Many workers have been preoccupied with assigning deposits to specific Quaternary cold and warm stages and with making interregional correlations, with the result that the confirmation of local sequences of events has been neglected. Some of the uncertainties and contradictions that emerge in the summary that follows may also be explained by the small amount of detailed analytical work that has been undertaken in the field and in the laboratory.

Sites referred to in the text are located on figs. 273 and 274.

*Quaternary workers in Ireland have measured the heights of features against a variety of datums. This is of little consequence where mountain summits are concerned, but is of considerable importance in connection with raised beaches and related phenomena. Elevations of marine features given in this contribution relate to Ordnance Datum at Belfast (O.D. (B)) and are taken from Devoy (1983) who has discussed the problem in some detail.

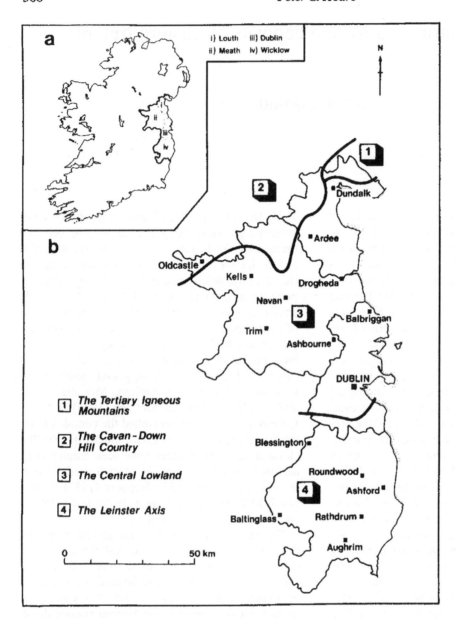

Figure 271: a) The location of Eastern Ireland; b) The region's geomorphic provinces (after Davies & Stephens, 1978).

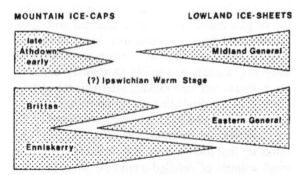

Figure 272: The sequence of glacial events at the northern end of the Leinster Mountains (after Farrington, 1934 *et seq.*). The 'Early Athdown' Glaciation of this scheme occupies the same position as the Aughrim Glaciation of Farrington and Synge.

The start of the Quaternary record

The oldest known Quaternary material in Eastern Ireland may be represented by the cold-loving marine molluscs, similar to those of the Waltonian Crag of East Anglia, that were transported within glacial erratic 'rafts' of Crag sediment (Mitchell, 1972) or were reworked into outwash of (?)Midlandian (=Devensian?) age (Mitchell *et al.*, 1973). Further details are given by Hoare (this volume).

The greater part of the sequence of glacial deposits appears to post-date the formation of a well-developed marine-abrasion platform that occurs at a number of sites in the region (see Martin, 1955;

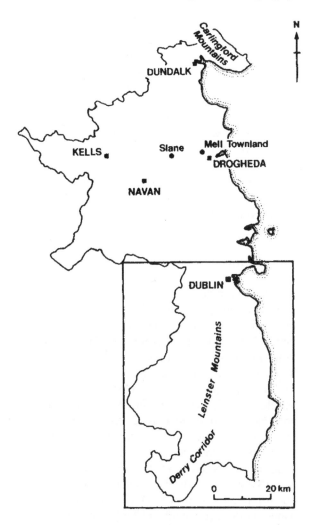

Figure 273: Location of the sites in the northern part of Eastern Ireland that are mentioned in the text; those that occur in the inset area are shown in fig. 274.

Stephens, 1957, and Synge, 1964; 1977a; 1981, for example; see also plate 4 and fig. 16 in Hoare, this volume). It has been described as a '4 m coastal platform' (Davies & Stephens, 1978: 127), and as rising to 11 m O.D. (Synge, 1981: 307) and to 12-20 m O.D. (Synge, 1981: 307). According to Devoy (1983), it has a maximum elevation of 12-14 m O.D. (B) at Wicklow Head (T 3492) and at Arklow Head (T 2070), County Wicklow. Further details of this feature may be found in Hoare (this volume).

The Enniskerry/Clogga Glaciation

Farrington (1934, 1942, 1944) explained the distribu-

tion of certain granite erratics at the northern end of the Leinster Mountains by proposing the development of an Enniskerry ice-cap (fig. 272). A till in the valley of the Cookstown (or Glencullen) River, County Wicklow (O 238182), may also be related to this event (Farrington, 1944). Subsequently, Farrington (1954) described till and glacial gravel from coastal sections in Counties Wicklow and Wexford. Synge (1964) referred to the till as the Clogga Till and gave details of exposures from Wicklow Head, County Wicklow, to Cahore Point, County Wexford (T 2247). This material may be the oldest Quaternary drift in Eastern Ireland. A river gravel at Lucan, County Dublin (O 0335), contains many clasts of Leinster Granite and may have been derived in part from sediment of Enniskerry Glaciation age (Farrington, 1964), although the ice is not generally considered to have advanced this far from the mountains.

The Clogga Till is rich in Leinster Granite erratics. Limestone clasts were recorded by Farrington (1954) but not by Synge (1964). Synge (1964) noted weathered flint (but may have intended to refer to chert); Synge & Huddart (1977) and Huddart (1981a) did

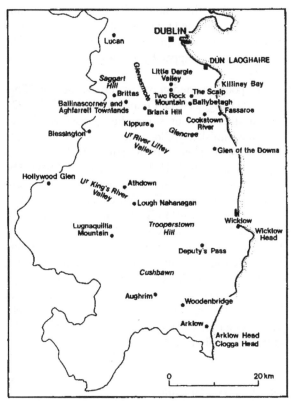

Figure 274: Location of sites in the southern part of Eastern Ireland that are referred to in the text.

not find erratics of flint, shell or limestone, and Synge (1964, 1977a) and Huddart (1981a) referred to the non-calcareous nature of the till. Bedrock striations, *roches moutonnées* and the lithology of the Clogga Till indicate ice-movement from the west or northwest (Farrington, 1954; Synge, 1964; Davies & Stephens, 1978). Farrington (1954) believed the Enniskerry ice-cap was responsible for the drift in the coastal locations, whereas Synge (1973, 1979a, 1981) envisaged a much more substantial Clogga ice-sheet. Davies & Stephens (1978) suggested that these two ice-masses may have co-existed. According to Synge (1979a), the ice-sheet moved southeastwards from a centre of dispersion extending from County Mayo to County Tyrone, passed through the Derry Corridor in southern County Wicklow, and into the Irish Sea Basin. Glacial sediments in the Derry Corridor contain clasts of chert and silicified limestone (Synge, 1973). Synge (1981) believed the ice covered "...even the highest ground..." (p. 307). However, Leinster Granite erratics were transported northwestwards from the parent outcrop and now lie to the north and west of Brittas, County Dublin (O 0321), beyond the generally accepted limit of the Brittas Glaciation (see below). This distribution was linked by Hoare (1975) and by McCabe & Hoare (1978) with a Slievethoul (=Enniskerry) ice-cap and cannot be explained by the southeasterly advance of Synge's Clogga ice-sheet. The surface extent of the Clogga drift as mapped by Synge (1981) is shown on fig. 29.

The poorly-sorted gravel which overlies the Clogga Till at Clogga Head, County Wicklow (T 252692), and at other exposures, may be outwash associated with the Enniskerry/Clogga ice-mass (Martin, 1955; Davies & Stephens, 1978; Huddart, 1981a) or a beach (Synge, 1964; 1970a; 1977a and c; 1981). The gravel rests on a marine-abrasion platform that truncates both the Clogga Till and the adjacent bedrock (Synge, 1977c; 1981) (fig. 16, Hoare, this volume). The weathering of the Clogga Till, the excavation of the platform and the accumulation of the gravel may have occurred during a temperate stage (Synge, 1964) which may have been the Ipswichian (Synge, 1975b; 1977a), although Synge (1970a) stated that "...no weathering horizon has been found..." (p. 38). The Clogga Head exposure is discussed further by Hoare (this volume).

The Drogheda Glaciation

McCabe (1971, 1972, 1973) and Colhoun & McCabe

(1973) identified the Drogheda Glaciation of the northern part of Eastern Ireland from the presence of till at two sites immediately northwest of Drogheda, County Louth (O 0975). Bedrock striae, till fabric analyses and the lithology of the till indicate that an ice-sheet advanced eastwards and southeastwards from a zone of dispersion which extended from the mountains of western Counties Mayo and Galway to Lough Neagh (Davies & Stephens, 1978). It is not possible to link this event with other episodes in the region, although the Drogheda Till occupies the same stratigraphic position as the Clogga drift of Counties Wicklow and Wexford.

A marine transgression?

The Mell Formation consists of laminated clays and silts, sands and poorly-sorted gravels of glaciomarine origin; it overlies the Drogheda Till in Mell Townland (O 074764), near Drogheda. The rich marine fauna includes molluscs, foraminifera, ostracods and hydrozoa, many of which are of arctic aspect. The Mell Formation may be *in situ* (McCabe, 1971; 1972; Mitchell, 1972; Synge, 1977a; 1979a): the considerable extent (approximately 2 km^2) and relatively undisturbed nature of the sediments lend support to this proposal. However, it may have been transported by ice outwards and upwards from the bed of the Irish Sea (Synge, 1977f): the elevation of the single outcrop (29.4 - 40.0 m O.D. (B)) and the depth of water involved (up to 90 - 110 m) favour the 'erratic raft' thesis. Synge (1977a) equated the Mell Formation with the Derryvree 'Interstadial' deposits of County Fermanagh which were radiocarbon-dated to 30,500 B. P. (Colhoun *et al.*, 1972; McCabe, 1986b), a suggestion that does not correspond with the general view of the timing of Quaternary episodes in Eastern Ireland (see below).

The influence of a Scottish ice-sheet

An ice-sheet of Scottish origin is normally held to have been responsible for the next glacial event to influence the region. This is the Eastern General Glaciation of Farrington (1944) (fig. 272), the Ballycroneen (County Cork; W 9362) Glaciation of Mitchell (1960), the Tullyallen (County Louth) Glaciation of McCabe (1972, 1973) and of Colhoun & McCabe (1973), and the Gormanstown (County Meath; O 178679) Glaciation of McCabe & Hoare (1978). The ice advanced westwards and southwestwards across the present coastline of Eastern Ireland (Charlesworth, 1928a; Farrington, 1944; 1954; Mitchell, 1960; McCabe, 1972; 1973; Hoare, 1975; 1977a and b; McCabe & Hoare, 1978). The Tullyallen Till, which was deposited during this event, overlies the Mell Formation and the Drogheda Till near Drogheda (McCabe, 1971; 1972; 1973; Colhoun & McCabe, 1973).

The term 'Irish Sea' has been applied to one or more ice-masses that developed in Scotland and then advanced into the Irish Sea Basin and spread onto the margins of Ireland. It has also been used to refer to drift deposited on the bed of the basin by ice which was centred over west-central Ireland, passed between the Carlingford/Mourne Mountains and the Leinster Mountains, and moved offshore. Further, it has been linked with glacial material which was laid down by this ice when it pushed back onto the southeastern part of the Irish landmass. In view of these confusing and inconsistent uses of the term 'Irish Sea', it is proposed that it should be abandoned.

The most widespread deposit laid down by this Scottish ice-sheet is a tough, 'chocolate' (Farrington, 1954; Synge, 1964; Davies & Stephens, 1978), 'brown' (Synge & Stephens, 1960), 'red-purple-grey' (Synge, 1963b), 'purple-grey' (Synge, 1964), 'purple' (Synge & Stephens, 1960; Synge, 1970a), 'grey' (Huddart, 1981a), 'brownish-grey' and 'grey-brown' (Davies & Stephens, 1978) or brown (7.5YR 4/3-6 on the Standard Soil Color Chart) (Hoare, 1975) calcareous till. The variegated nature of the material not only reflects changes in the underlying bedrock but also the failure by many early workers to employ a colour chart. The till has been described both as containing only a limited amount of gravel-size clasts and as 'stoneless' (Wright & Muff, 1904; Synge & Stephens, 1960; Synge, 1963b; 1964; 1970a; Stephens *et al.*, 1975; Culleton, 1978a and b; Davies & Stephens, 1978; Huddart, 1981a), whereas in reality the texture is highly variable (see McCabe, 1973; Hoare, 1975, and Hanrahan, 1977, for example).

A small proportion of the erratics found in drift associated with this ice-sheet are of chalk, flint, marine shell and Ailsa Craig microgranite. Lamplugh *et al.* (1903), Mitchell *et al.* (1973) and Synge (1977a) documented a shelly upper till of northwestern provenance in Killiney Bay (O 260234 - 267195), County Dublin, although it is clear that the shells must have been reworked from the lower till (Farrington, 1944). Reference to the presence of shells in the upper till has created a misleading impression of the

source of that material (but see Hoare, 1975, 1977a and b, and this volume). The formal description 'shelly' should only be employed where it is meaningful (see also the discussion of the stratigraphy in Shortalstown Townland below).

Drift deposited by the Scottish ice is commonly found as a number of superimposed tills, some of which may be separated by glaciofluvial sand and gravel (McCabe, 1971; 1972; Hoare, 1972; 1975; 1977a and b; McCabe & Hoare, 1978). This phenomenon is perhaps best illustrated by the Killiney Bay exposure, although it has been argued that only the lower part of this succession may have been laid down by this ice-mass (Synge, 1977a; 1981; Davies & Stephens, 1978). Several till facies have been recognised but are probably of little temporal significance as they do not occur in the same sequence at all sites (McCabe, 1973; Hoare, 1975; McCabe & Hoare, 1978). The complex stratigraphy may have resulted from a series of glacial oscillations, from deposition by an ice-mass that contained debris at different levels (McCabe, 1972; Hoare, 1977a and b) or as a consequence of subaqueous sedimentation (Eyles & Eyles, 1984; Eyles *et al.*, 1985b; McCabe, 1985; 1986b). The Ballycroneen Till of the south coast of Ireland (Mitchell, 1960), a possible correlative of the drift of Scottish origin in the eastern part of the country, may also have accumulated as a "...marine till..." (Synge, 1981: 310) beneath a floating ice-shelf (Synge, 1979a; 1981; Devoy, 1983).

The following observations provide some indication of the extent of Scottish ice in the region, although its limit is not known with any degree of certainty. The ice-sheet overwhelmed much of Counties Louth and Meath (McCabe, 1972; 1973), but may have been prevented from advancing beyond Slane, County Meath (N 9674), by the Irish Drogheda ice-mass (McCabe, 1972). The distribution of distinctive erratics (see above) suggests that the Scottish ice covered the greater part of County Dublin (see below), although it may also have come into conflict with inland ice (Hoare, 1975). Glacigenic deposits of northeasterly provenance occur at Blessington, County Wicklow (N 9814), 40 km from the present coastline (Farrington, 1960), at 305 m O.D. on the slopes of Saggart Hill, County Dublin (N 0223), 24 km from the coast (Farrington, 1949), and in southwest County Dublin (Hoare, 1975; 1977c and d). Marine shells in drift in the Irish midlands may have been carried there by the same ice-sheet (Oldham, 1844; Farrington, 1949).

Farrington (1957b) believed the Scottish ice submerged the northern part of the Leinster Mountains to a height of 500 m O.D. Davies & Stephens (1978) suggested that an ice-cap may have prevented it from rising above 550 m O.D. in the same area. Close (1867) noted erratics at 537 m O.D. on the summit of Two Rock Mountain, County Dublin (O 172224), and Hoare (1975) recorded them up to 610 m O.D. on the northern flank of Kippure Mountain, County Dublin (O 1115). Thus the ice-sheet does not appear to have completely overwhelmed this upland region.

Davies & Stephens (1978) suggested that the ice may be linked with a drift-limit of 150 m O.D. near Wicklow (T 3193) and of 61 m O.D. near Arklow (T 2473), County Wicklow. Taken in association with the corresponding figure of 610 m O.D. at the northern end of the Leinster Mountains, and less than 30 km away, this indicates an unlikely glacial gradient of 13 - 14 m km^{-1}. A great deal of work is required on the eastern side of these uplands to determine the true extent of Scottish and of other ice-masses.

No glaciofluvial deposits can be positively linked with the final retreat of the ice-sheet across Eastern Ireland. However, impressive rock-cut meltwater channels such as the Glen of the Downs (O 2610) and Hollywood Glen (N 8903), County Wicklow, and the Scalp (O 2120), on the boundary of Counties Dublin and Wicklow, may have formed at this time (Synge & Stephens, 1960; Hoare, 1975; 1976). A large lake, impounded within the catchment of the River Liffey by Scottish ice situated in the Irish Sea Basin, may have drained via the Glen of the Downs, the Scalp and the Deputy's Pass, County Wicklow (T 2390) (Synge, 1979a).

The Brittas Glaciation

The withdrawal of the Scottish ice was succeeded, after an interval of unknown length and climate, by the expansion of the Brittas ice-cap over the Leinster Mountains (Farrington, 1934; 1942). The majority of workers accept that the granite-rich moraines near Brittas, County Dublin (O 032217), define the maximum extent of this event (Farrington, 1942; Synge, 1970a; Davies & Stephens, 1978), although Mitchell (1972) considered that they mark only the final stages of glacial retreat. Granitic glaciofluvial or glaciolacustrine material rests on till of probable Scottish provenance at the base of a complex delta in Ballinascorney and Aghfarrell Townlands, County Dublin (O 0621) (Hoare, 1975; 1977c and d), a feature first described by Farrington (1942, 1957b). Farrington (1968) recorded gravel deposits of Brittas Glaciation

age that were laid down by a small glacier in the Little Dargle Valley (O 1624), County Dublin.

Part of the extent of the Brittas ice-cap is shown on fig. 29 which is taken from Synge (1981). It is clear from the map that the relationship between this ice and the ice-sheets that advanced from the Irish midlands (see below) and from the Irish Sea Basin remains to be firmly established.

The chronology of the early part of the glacial sequence

Farrington (1944, 1957b) believed that the Enniskerry, Scottish and Brittas glaciations were closely related in time. He explained the "...small assortment of far-travelled rocks ... from the north-east..." (Farrington, 1957b: 24) which occurs in the Clogga Till by suggesting that part of the load of this ice-sheet had been introduced into debris being carried by the Enniskerry ice-cap. It is easier to envisage the ice-cap eroding older drift to obtain this material, but none is known which might have acted as a source. The Clogga Till never overlies till of Scottish provenance, which may be seen as evidence that the ice-sheet was the more powerful (Farrington, 1957b; Davies & Stephens, 1978). On the other hand, Synge (1973) suggested that the break in the succession between the deposition of Clogga drift and of the overlying (Scottish) till "...seems to have been considerable." (p. 567). Farrington (1944) proposed that the delta of Brittas Glaciation age at Fassaroe, County Wicklow (O 237180), accumulated in water ponded in the Cookstown Valley by an ice-sheet as it retreated into the Irish Sea Basin. Rather more substantial evidence is required before the timing of the three ice-masses can be established.

Colhoun & McCabe (1973) believed that the Drogheda Till, the Mell glaciomarine deposits (which contain some far-travelled clasts of Scottish/Irish Sea Basin provenance) and the Tullyallen Till accumulated within a relatively short space of time during the Munsterian (=Wolstonian?) Stage.

More than a century of fieldwork has failed to reveal temperate organic material within the drift succession of Eastern Ireland. Despite this, the majority view is that the glacial events described above took place before the Ipswichian Stage (see Farrington, 1944; 1957b; Mitchell, 1957; 1960; Synge & Stephens, 1960; Davies & Stephens, 1978, and McCabe, 1986b, for example). However, Davies & Stephens (1978) also suggested that the Brittas Glaciation occurred during the Midlandian. Synge

(1977a) considered that Scottish ice advanced to Ballycroneen, County Cork, approximately 22,000 years ago; and Hill & Prior (1968), Synge (1973, 1979a, 1981) and Warren (1985) also placed this glaciation in the Midlandian Stage. A fuller consideration of the timing of these events is given by Hoare (this volume).

THE LATER PART OF THE GLACIAL SEQUENCE

The Aughrim Glaciation

The expansion of an (?)Early Midlandian ice-cap over the northern part of the Leinster Mountains was first suggested by Farrington (1954, 1957b) (fig. 272). The proposal was based on the presence of large, perched, blocks of Leinster Granite such as those on Trooperstown Hill (T 1894) and on the northern flank of Cushbawn (T 1384), County Wicklow, and of moraines outside the limit of the (Early) Athdown Glaciation but well within the maximum extent of the Brittas ice-cap. The 'Early Athdown' and Athdown moraines display similar topographic freshness and are difficult to separate.

Synge (1973, 1975a, 1977a, 1979a, 1981) and Davies & Stephens (1978) identified an (?)Early Midlandian Aughrim or Athdown-I Glaciation. The Aughrim Glaciation was restricted to the development of valley glaciers such as the Avonbeg glacier which advanced to Aughrim, and the Avonmore glacier which extended to Woodenbridge, County Wicklow (T 1877) (Synge, 1973; 1977a; 1979a). Figure 29 (which is based on the work of Synge, 1981) suggests that the Aughrim glaciers advanced beyond the limit of the Brittas ice-cap. Synge (1973, 1975a) and Davies & Stephens (1978) described deltas, moraines and outwash terraces, although Synge (1977a) considered that "...no distinct drift landforms survive..." (p. 208)!

Synge (1973, 1977a) suggested that the concentrations of granite erratics indicated a Midlandian date for the event, of possibly 50,000 years ago(Synge, 1973: 569), although he subsequently proposed a Saalian/Connachtian (=Munsterian?) age (Synge, 1981). The arrangement of Aughrim Glaciation moraines and those associated with the expansion of ice-sheets from the Irish midlands and from the Irish Sea Basin (fig. 274) shows that the local glaciation was the earliest of the three (Synge, 1973; Davies & Stephens, 1978), although it may not have taken place (Davies & Stephens, 1978)!

A further glaciation by ice from the Irish midlands

An ice-sheet advanced from a curved axis of dispersion extending from Lough Rea, County Galway (R 6010), to Lough Neagh (J 0070) (Lamplugh *et al.*, 1903; Farrington, 1939). This 'Midland General' ice (fig. 272) first passed eastwards (McCabe, 1972) and then southeastwards across Eastern Ireland. Evidence of the Late Midlandian age of the event comes from Derryvree Townland, County Fermanagh (H 361390), where the Maguiresbridge Till (McCabe, 1969b) rests on organic material that was radiocarbon-dated at 30,500 B.P. (Colhoun *et al.*, 1972). It is assumed by many that the youngest drift in lowland areas of Eastern Ireland is of approximately the same age as the Maguiresbridge Till.

Although the limestone-rich deposits associated with this Late Midlandian ice-sheet are patchily distributed in Counties Louth, Meath and Dublin, they overlie glacigenic sediment of Scottish/Irish Sea Basin provenance at many sites. Drift from the more recent episode is absent from certain coastal areas where glacial material laid down by the earlier ice-sheet lies at the surface (McCabe, 1972; 1973; Hoare, 1975; 1977a and b; McCabe & Hoare, 1978). This outcrop pattern has led to the suggestion that there was a late readvance of Scottish ice (see below).

A broken line of moraines which may mark the maximum southward extent of the Irish ice-sheet was discovered by Lewis (1894) and mapped in detail by Charlesworth (1928a) who called it the Southern Irish End-Moraine. The equivalent feature at the northwestern end of the Leinster Mountains may be represented by the Hacketstown Moraine or by the Blessington Moraine (Synge, 1973; 1977a; 1981) (fig. 29). McCabe (1971, 1972, 1973), Hoare (1975 and this volume) and McCabe & Hoare (1978) described the numerous moraines that mark still-stands in the northerly and northwesterly retreat of the ice across Eastern Ireland. Further details of this event and of glacial limits at the northern end of the mountains are given by Hoare (this volume).

A further advance by a Scottish/Irish Sea Basin ice-sheet

The upper part of the Quaternary sequence at Killiney Bay and at other coastal sites in Eastern Ireland may date from the 'main' advance of a Scottish ice-sheet (see above), or result from a subsequent advance or readvance of ice onto a coastal strip a few kilometres wide (Davies & Stephens, 1978) or consist of material that was deposited by an ice-sheet that emerged from the Irish Sea Basin but was reworked by ice from the Irish midlands (Synge, 1977a, 1979a, 1981). There are no constructional forms which might record the maximum extent of a late and limited advance by Scottish ice. Further sedimentological and geotechnical work is required to resolve this controversy.

However, two shelly tills occur in Shortalstown Townland, County Wexford (T 0214), and appear to be separated by pollen-bearing estuarine sands of (?)Ipswichian age (Colhoun & Mitchell, 1971). Notwithstanding the following uncertainties, this stratigraphy may indicate an older and a significantly younger glaciation of Scottish provenance:

(i) the units at Shortalstown are greatly disturbed and may not be *in situ*;

(ii) the organic material is not indubitably of Ipswichian age (Synge, 1977a);

(iii) the ice that deposited the upper, shelly, till may not have originated in Scotland. Instead, it may have accumulated over the Irish midlands, advanced into the Irish Sea Basin through the 'corridor' between the Carlingford/Mourne Mountains and the Leinster Mountains, and pushed onto the County Wexford coast. There is no reason why this material should not have been deposited by an essentially Irish ice-sheet.

The interpretation of the stratigraphy in Shortalstown Townland remains uncertain; it is considered further by Hoare (this volume).

Unweathered limestone-rich drift occurs to elevations of 60 - 244 m O.D. on the eastern flank of the Leinster Mountains (Synge, 1964; 1970a; 1973; 1977a; 1981). Synge associated this material with the maximum extent of an advance of 'Irish Sea' ice during the Late Midlandian Glenealy substage (fig. 29). Again, the ice concerned may have advanced from a source-area in the Irish midlands. This episode may be equivalent to the Hacketstown substage on the western side of the mountains (Synge, 1973; 1981; Davies & Stephens, 1978) or may post-date it

(Synge, 1981). Further discussion of the evidence is provided by Hoare (this volume).

The Athdown Glaciation

The last major glacial episode to influence the northern end of the Leinster Mountains was associated with the expansion of an ice-cap and of corrie glaciers (Charlesworth, 1928a; 1937). A more comprehensive account of the event was given by Farrington (1934, 1966a) who called it the Athdown Glaciation (fig. 272). Synge (1981) referred to Athdown-II valley glaciers. An end moraine and delta at Athdown, County Wicklow (O 0602), mark the farthest extent of the ice in this area (Farrington, 1934). The Athdown Glaciation was first thought to have preceded the invasion of the Leinster Mountains by ice from the Irish midlands (Farrington, 1934; Synge & Stephens, 1960), but was then considered to have succeeded it (Farrington, 1934; 1966a; Jessen & Farrington, 1938; Synge, 1970a). Davies & Stephens (1978) proposed that the ice-cap attained its fullest development at the same time as the ice from the Irish midlands stood at the Blessington Line; this latter position was considered by them to mark a readvance of the ice after a withdrawal from the Hacketstown Line. It is clear that further work is necessary to clarify the contradictions on this subject that appear in the literature.

Both Charlesworth and Farrington demonstrated that the main development of the ice-cap was to the east of the major watershed, although important exceptions occurred in the upper reaches of the King's River and River Liffey catchments. The Enniskerry, Brittas and Aughrim ice-caps were generally rather more extensive than the Athdown ice-cap (fig. 29). Farrington (1934) believed that the most northerly extent of Athdown ice was at the head of Glencree, County Wicklow (O 1516) (see below). Charlesworth (1928a) suggested that small corrie glaciers

Plates 43–44

43. LANDSAT TM image of parts of Counties Sligo, Leitrim and Roscommon showing streamlined features, drumlins (on the lower part of the image) and ice erosional features (in the Benbulben / Truskmore Mountain area). For location see inset B on fig. 260. Reproduced from an original with kind permission of Environmental Resources Analysis Ltd., Trinity College, Dublin.

44. LANDSAT TM image of part of County Clare showing streamlined features, drumlins (in two fields coalescing to the south-west of the Slieve Aughty Mountains) and ice erosional features (including the pink area which is ice-eroded bare rock of the Burren). For location see inset C on fig. 260. Reproduced from an original with kind permission of Environmental Resources Analysis Ltd., Trinity College, Dublin.

43
───
44

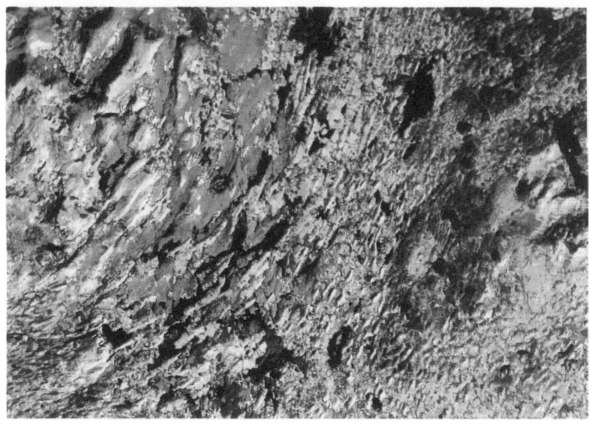

PLATE 45

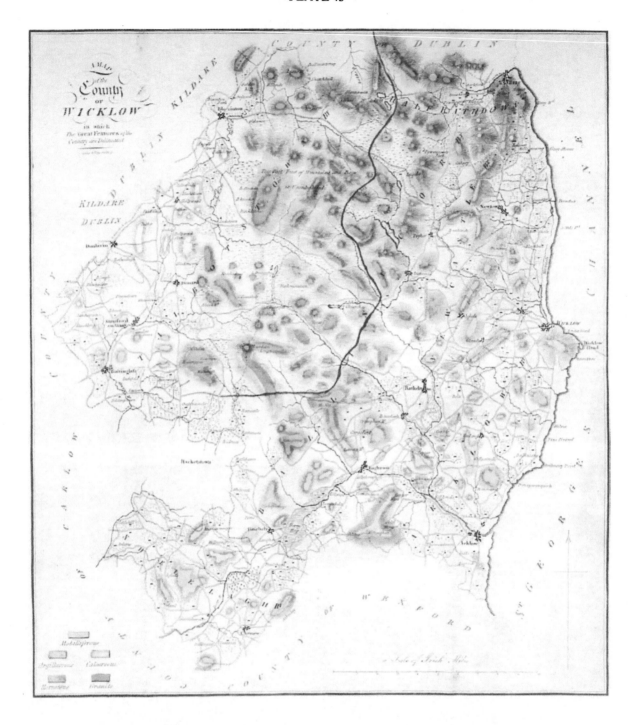

existed in the upper part of Glenasmole, County Dublin (O 1020), but did not show them on a later map (Charlesworth, 1937). However, Synge (1971) and Hoare (1975) demonstrated that a small glacier formed on the northern slopes of Kippure Mountain, County Dublin, and extended a short distance down the Dodder Valley, Glenasmole, to a moraine at Brian's Hill (O 108195). A well-developed granite-rich outwash terrace (O 1021) emerges from the distal face of this moraine and may be traced north-wards for approximately 800 m. Davies & Stephens (1978: 136) mistakenly ascribed a Brittas Glaciation age to this terrace.

Glaciers occupied the corrie basins that are now the sites of Upper and Lower Lough Bray (O 1315) in upper Glencree, and Lough Nahanagan (Turlough Hill) (T 078903) at the head of Glendasan, County Wicklow, during the Athdown Glaciation (fig. 29). Massive block moraines mark the maximum extent of this ice and conceal the bedrock lips of the corries.

The Nahanagan Glaciation

Small block moraines lie on the ice-proximal slopes of Athdown Glaciation moraines at Upper and Lower Lough Bray (Warren, 1970), and also occur on Lugnaquillia Mountain (Farrington, 1966a), and at Lough Nahanagan. They record the latest glacial event to have affected the Leinster Mountains. The series of three moraines which are situated inside the massive Athdown Glaciation moraine at Lough Nahanagan was discovered by Hoare in 1969 and described by Synge (1977d) and Colhoun & Synge (1980). Fragments of organic matter of Woodgrange Interstadial age (see Morrison & Stephens, 1965, and Singh, 1970, for example) within the Nahanagan Stadial morainic gravels have been dated to 11,500 and 11,600 B.P. (Synge, 1977d; Colhoun & Synge, 1980). The *Salix herbacea* clay at Ballybetagh, County Dublin (O 202203), may be equivalent in age to these glacial features (Farrington, 1966a; Watts, 1977; Mitchell, 1976). Warren (1985) suggested the adoption of Ballybetagh as the type-site for the Naha-nagan Stadial, although the term 'stadial' means a period of ice advance (see Sparks & West, 1972: 133;

Andrews, 1975: 182, and Lowe & Walker, 1984: 8, for example) and none took place there at this time. The Nahanagan Stadial is discussed in detail by Gray & Coxon (this volume).

Summary and conclusion

It is impossible to be sure of the number of glacial events that have helped to shape the landscape of Eastern Ireland, and only the Nahanagan Glaciation can be dated with any degree of precision. The ortho-dox view of the drift succession is that the lower part belongs to the Munsterian Stage, the remainder to the Midlandian. These Quaternary stages have tradit-ionally been correlated with the Wolstonian and the Devensian of Britain. However, there is no convinc-ing evidence of the Ipswichian Stage, only rather poor substitutes such as signs of so-called 'inter-glacial weathering'. Even the resolution of the strati-graphic position of the Gortian Stage (see Hoare, this volume) will not clarify the situation unless links can be established between sites where Gortian material occurs and those that reveal the glacial sequence of Eastern Ireland.

Until such time as stratigraphic markers and geochronometric dates become available, attempts must be made to establish a reliable succession of Quaternary events in the region. Eastern Ireland has probably been examined in more detail than any other part of the island, and yet very little progress has been made since Farrington's pioneering work of the 1930s, and there is at present a great deal of confusion. A number of important exposures require re-examination; even the application of relatively unsophisticated laboratory and field analytical meth-ods such as petrological counts, the determination of particle-size distributions and of clast shapes may be of considerable benefit.

Acknowledgements

The author is most grateful to Dr S.J. Gale for sug-gesting a number of improvements to this paper and to Ms R. H. A. Smith for drawing the figures.

Plate 45

45. Fraser's (1801) map of County Wicklow.

Critical topics

Mapping glacial deposits in Britain and Ireland

WILLIAM P. WARREN & ALBERT HORTON

Geological maps have been produced under government auspices in both Britain and Ireland for more than 150 years. The Geological Survey of England and Wales and the Geological Survey of Ireland had separate beginnings. The Geological Survey of England and Wales was founded in 1835 as a branch of the British Ordnance Survey (Board of Ordnance) (Flett, 1937) and the Geological Survey of Ireland traces its separate origins to the geological mapping undertaken by the Ordnance Survey of Ireland which was begun in 1829 (Davies, 1983). In 1845 the two surveys were amalgamated to form the Geological Survey of Great Britain and Ireland. Surveying in Scotland commenced in 1855 and an Edinburgh Office was established in 1867. The Geological Survey of Ireland later separated from the British Survey in 1905, a situation which became irrevocable after independence in 1921. In 1947 the Geological Survey of Northern Ireland was founded as a Northern Ireland Government Office and has been staffed by geologists from the British Geological Survey.

Great Britain

Glacial deposits were not shown on the early 1:63,360 maps, but the acceptance in the 1860s of Agassiz's theory of the glacial origin of many 'drift' deposits provided the first stimulus to map their distribution. The surveyors recognised three lithological units which still remain the main components of today's maps. These were: Boulder Clay, Glacial Sand and Gravel, and a fine-grained glacial deposit, variously described as Brickearth, Brick Clay, Loam (now interpreted as Glacial Lake Deposits or Glacial Laminated Clays and Silts). Subsequently, the broader terms, 'Glacial Drift, undifferentiated' and 'Morainic Drift' were established and all these drift deposits were mapped as lithogenic units, in the same way as the pre-Quaternary (solid) formations. Although first available in about 1854, the general use,

at the turn of the 19th century, of 1:10,560 base maps provided the British surveyors with an opportunity to map in greater detail and the space to show subdivisions of the glacial deposits and also the significant morphological features.

In Great Britain two styles of maps evolved, each related to particular types of landforms. In the upland areas, particularly in Scotland, the glacial deposits belong to the last glaciation and retain many of their original landforms. Drainage channels associated with the last ice sheet can be clearly recognised and directions of ice flow determined from the orientation and form of glacial striae, *roches moutonnées* and crag-and-tail landforms. In some areas drumlin outlines were noted. Thus maps of upland areas included morphological features, indicated by appropriate symbols. In the lowland areas of Britain the glacial deposits were formed by ice sheets on continental scales rather than by coalescing mountain glaciers. The underlying rocks were very much softer and did not give rise to sculptured landforms other than massive channels or depressions, which were subsequently infilled with glacial deposits. In consequence only lithogenic formations could be defined.

Most of the first half of the 20th century was a period of stagnation in development of new ideas on the mapping of glacial deposits. The late 1950s, however, witnessed a massive growth of interest in Quaternary studies within the universities which has extended into the Geological Surveys.

Financial constraints have meant that the main expansion in the mapping of glacial deposits has been in the search for sand and gravel, but this, together with the development of the site investigation industry, has resulted in major advances in knowledge of the associated deposits. As a result much greater diversity of lithological units is now shown on the published maps. Awareness of the needs of the potential users has resulted in the inclusion of descriptions of the glacial deposits and illustrations to show the nature of the sub-Quaternary (rockhead) surface. The objectives of mapping Quaternary deposits are being questioned and new techniques developed which will enhance the traditional lithogenic mapping.

Ireland

Prior to state-financed geological mapping, deposits were recorded and mapped generally as 'Drift' but in at least one case in Ireland limestone-dominated glacial deposits in County Wicklow (plate 45) were mapped as an integral part of the geology under the heading 'Calcareous' (Fraser, 1801). There is no reason to assume that Fraser thought the area was underlain by limestone bedrock (it is not) rather, as Davies (1983) has suggested, it is likely that Fraser wished to represent the soil parent material. Sampson's (1802) map of the geology of County Londonderry included 'Heath and Bog, Clay, Sand, Gravel and Rich Loams', but Tighe (1802) in his map of County Kilkenny added only 'Limestone Gravels and Inundated Wet Ground' as integral parts of the geology. Portlock's (1843) pioneering map and report, centred on County Londonderry, describes what are now known as glacial and glaciofluvial deposits in some detail, but he rejected the notion of land ice as an agent of sediment deposition there.

In its early years the Geological Survey of Ireland collected soil and subsoil samples in parallel with the bedrock mapping programme, but the practice ceased after about eight years and no 'drift' map or soil map was produced. The earliest (county) maps of the Geological Survey of Ireland, at a scale of 1:126,720, include shaded and stippled areas depicting 'drift' which was generally referred to as 'Tertiary Clays and Gravels' on the maps, but which is in fact glacial deposits. As Irish field mapping was done on maps of the 1:10,560 scale from the outset (maps at this scale were available in Ireland as early as 1833 and national coverage was complete by 1846), it was possible to record much information on the 'drift' deposits; both the field sheets and the published 1:63,360 maps depict many significant morphological features. The midland sheets show in great detail the distribution of the eskers and the accompanying memoirs, published in the 1850s, 1860s and 1870s devote much of their texts to the description of these features (Foot & O'Kelly, 1865).

The published 1:63,360 maps generally depict the Quaternary or 'drift' deposits by means of a stipple overlay, which was itself an afterthought prompted by the Survey's retail agents in Dublin, who intimated that the absence of a 'drift' representation would inhibit sales (Davies, 1983). On many of the maps of the midlands the eskers were depicted by hachuring and appear as very prominent features. The glacial and glaciofluvial deposits remained undifferentiated on the maps until the initial survey was completed in 1890, but the memoirs commonly offered good descriptions of both.

Seven years after the completion of the initial survey, J. R. Kilroe (1897) outlined the type of 'drift' map which he envisaged would meet the requirements of the agriculture industry. He later produced 'A description of the Soil Geology of Ireland' and an accompanying map entitled 'Surface Geology of Ireland' (Kilroe, 1907). The map distinguished only five types of 'post-Tertiary' deposits: blown sand, peat, alluvium, raised beaches and drift.

In 1901 the Geological Survey of Ireland first turned its attention to producing a series of special 'drift' maps for the areas around the chief towns of Ireland. Although the memoirs to accompany the maps contain detailed descriptions of the glacial deposits, they generally distinguish only 'Blown Sand, Alluvium, Estuarine Alluvium, River Gravel, Gravelly Delta, Glacial Sand and Gravel, Boulder-clay and Head' (Lamplugh *et al.*, 1903; 1904; 1905; 1907; Wilkinson *et al.*, 1908). The legends in this series of maps show some flexibility; for example, in the Dublin sheet distinction is made between 'Boulder-clay containing much Limestone' and 'Clayey drift of mainly non-calcareous material' while the general term boulder-clay is applied without distinction on the other maps. The agricultural requirements outlined by Kilroe (1897) were ignored in what were referred to as the 'city maps'. W. B. Wright's (1922) map of the Killarney district, based on his work there in 1911-1912, follows the format of the city maps.

In the absence of maps from the Geological Survey, Charlesworth (1928a), beginning with his map of the south of Ireland, produced a series of maps depicting the distribution of deposits of deglaciation in Ireland to accompany his papers on the topic. These were all simple monochrome line maps which, though valuable in the absence of more detailed works, did not adequately illustrate the distribution of the glacial or glaciofluvial deposits in the areas covered.

In the 1950s the Geological Survey of Ireland again turned its attention to mapping Quaternary deposits and the first modern map showing their distribution was prepared by the late F. M. Synge (1966a) of the Geological Survey and published by the Soil Survey. This map was produced to meet the requirements of the Soil Survey and it is therefore not surprising that it resembles more the style of map outlined by Kilroe (1897) than those of the 'city' series. This map illustrates four subdivisions of till on the basis of phenoclast petrography (pebble counts) in addition to 'morainic drift, fluvio-glacial sand and gravel, alluvium and peat'. It also indicates symbolically glacial striae, eskers, drumlins and meltwater channels. In the 1950s the attention of the Geological Survey of Ireland was focused almost entirely on mapping Quaternary deposits (Meenan & Webb, 1957) although it was then a very small body with only seven full-time geologists.

Since the early 1960s the Geological Survey of Northern Ireland has produced a series of 1:63,360 and more recently 1:50,000 scale, maps of the Quaternary geology. They are based on a re-surveying programme and subdivide the glacial deposits into Outwash Sands and Gravels, Glacial Sand and Gravel and Till (Boulder Clay). Striae, *roches moutonnées*, glacial meltwater channels, eskers and, on some maps, drumlins are also shown. These categories of deposits and features are very similar to those shown on the earlier 'city maps' of the Geological Survey of Ireland.

A 1:250,000 scale map of the Quaternary geology of the north of Ireland has been prepared by the Geological Survey of Northern Ireland. This divides the glacial deposits into Till, Sand and Gravel, Glacial Outwash, and Glacial Lake Clays.

In 1969 a specialist Quaternary Section was established in an expanding Geological Survey of Ireland to pursue the Quaternary mapping programme. It was recognised that modern Quaternary maps should provide data that would meet the more diverse requirements of an increasingly specialised 'clientele' in the areas of:

1. Soils
2. Sand and gravel/aggregate resources
3. Groundwater supply
4. Geotechnical assessments (including waste disposal)
5. Mineral exploration
6. Planning
7. Education.

Principles of mapping

As has always been the case, state-sponsored geological mapping projects must be designed to provide a useful service to the public. They must, in order to do this effectively, be prepared by geologists who are aware of both current academic developments in the science and the needs of the users.

The traditional geological map is based on lithostratigraphy, such that the stratigraphic units are characterised by a certain lithological type or a combination of lithological features (Hedberg, 1976). These include lithology, grading characteristics, internal structure and thickness. Fossil assemblages, if present, are generally used to establish a biostratigraphic succession and to determine the relative age of the deposits. Macrofossils may be used also in the definition of lithological units.

Glacial deposits pose problems in the definition of lithological units. For example, should tills be differentiated on maps or simply grouped as a single unit? If differentiated, between tills, on what basis should this be done? Much will depend on the type of mapping being carried out, the purpose for which it is done and the level of detail required. Thus, while an academic research programme may require genetic subdivision of tills (lodgement tills, flow tills, melt-out tills, etc.) Geological Survey mapping programmes may not have the resources to carry out the detailed analyses that are needed for such subdivisions. Descriptive classification on the basis of grading characteristics and the phenoclast petrography (pebble counts), generally provides more useful information to the commercial user and also helps to refine the lithostratigraphy. Ultimately, glacial deposits can only be treated as any other geological unit on a geological map and the same stratigraphic principles must be applied. Thus the basic unit is the formation which must be mappable and have a sufficient distinctive lithology to be readily identified as a separate unit; its basis must be descriptive not interpretative. Thus, a specific till formation cannot be defined as relating to a particular glacial event. Subsequent interpretation may indicate that a till formation may transgress the boundaries of one or more glacial events.

Morphological mapping is an important aspect of any programme aimed at mapping glacial deposits. Sedimentary units may have distinctive morphological expression, but mapping of morphostratigraphic units is avoided where possible as this may lead to serious conflict with orthodox geological mapping and interpretation. Morphological mapping has at times forced mappers to work in a rigid stratigraphic framework which has resulted in geologists attempting to use unsatisfactory criteria in characterising lithological units (Warren, 1985). Nevertheless a morphologically-defined unit may separate two lithologically distinct stratigraphic units (such as two different tills or a till and a glaciofluvial deposit) and this will be illustrated on the map.

The published map, ideally at the scale 1:50,000, will depict the Quaternary geology, provide information (either on the map, on inset maps or diagrams or in the accompanying explanation) on morphology, ice-direction indicators and thickness of lithological units. Such maps are often criticised because they may not be easily understood by non-geologists who may lack an understanding of geological terminology and the geologist's conceptual framework. As planners, engineers and agricultural scientists form an important group of clients their requirements both in terms of content and clarity of presentation must be met as well as those of exploration geologists, the land-search geologist in the aggregate industry, the hydrogeologist, the academic geologist and the leisure user. There is a need to improve communication between the geological community and these clients (Brook & Marker, 1986) and map production should be geared to bridge any such communication gap.

As there is considerable overlap in the requirements of the various client areas (table 25) a priority test of the main map components desired by the clients can be constructed by scoring each component for every client area that requires it (table 26). It is clear, however, that a single map cannot provide all of each client's needs. There is a limit, for example to the number of engineering properties that can be measured or that can be illustrated in a standard mapping programme so as to provide a good degree of predictability. Geotechnical maps of Quaternary deposits, which define units having similar engineering properties (particularly moisture content, index properties, strength, consolidation, compaction and particle-size characteristics), are formulated so that the units are graded to indicate their respective ease of extraction and stability, and their suitability as foundation and building material. These maps are based largely on the Quaternary geology map but they necessitate a specialist engineering input both in the data-gathering exercise and at the map-formulation stage. Similarly, thematic maps (including geological hazard), aquifer maps, bulk mineral-resource maps and geochemical maps, all specialist in character, are based on the geological map, but may be produced at scales different from the geological map. It is important that field measurements meet as many of the requirements of these areas as possible.

Table 25. Data needs set against the chief client areas showing the relative demand for each type of data.

	Resources bulk minerals	Physical planning	Ground-water	Soils / agri-culture	Geotechnical industry	Mineral exploration	Education / leisure
Distribution of Quaternary sediments		+	+	+	+	+	+
Quaternary geomorphology	+	+	+	+	+		+
Classification of sediments	+	+	+	+	+	+	+
Thickness of units	+	+	+		+	+	
Composition of sediments (stone count)	+	+	+	+		+	+
Particle size distribution	+		+	+	+	+	
Stratigraphic arrangement	+		+		+	+	+
Depth to bedrock		+	+		+	+	
Direction and distance of transport						+	+
Structure of sediments	+		+			+	+
Carbonate content	+		+	+		+	
Degree of weathering	+			+		+	+
Geochemistry			+	+		+	+
Geotechnical characters		+	+		+		

The geological surveys in Britain and Ireland have started the production of specialist maps. The Belfast Special Sheet (Bazley & Manning, 1971) provided a geotechnical characterisation of all the outcropping formations as well as a contours on rockhead. In Scotland in particular, and initially at Glenrothes, in Fife (Nickless, 1982) a number of thematic maps have been derived from the traditional geological maps supported by specialist interpretations. Individual maps show the distribution of lithology of unconsolidated deposits, their engineering properties and thickness, the depth to water table in them, the thickness of sand and gravel and the rockhead geology together with surface contours. Other maps include the potential of sand and gravel exploitation and the nature of foundation conditions throughout the area. The Geological Survey of Ireland has not yet published any such maps although a map of potential aquifers and an aquifer-protection map have been prepared using existing Quaternary data for County Galway by the Groundwater Section and copies are on sale to the general public from the open file (Daly, 1985).

The British Geological Survey has no rigid policy on what should be on 'drift maps' but is entering a period of greater flexibility in the definition and classification of surficial deposits. It is left to the surveyor to determine what can be economically achieved within the finances available for each mapping project. The Geological Survey of Ireland is developing a style similar to that used in Norway where a combination of colour coding and symbols allows not only classification of deposits but also some characterisation. The Geological Survey of Ireland places strong emphasis on characterising till units on the basis of the most common contained phenoclast type, with the main colour classification used on the published maps reflecting this (plate 46). The maps, including derivative thematic maps, are a distillation or index of data collected by the geologist, usually at a larger scale, and they indicate that further information may be available in file. In most cases they are regarded as a first reference for the applied user and not a substitute for site investigation.

Field procedure

The general approach to mapping glacial deposits is the same as that for all other rocks. The purpose of all mapping is to define the outcrop distribution and

Table 26. Types of data required to produce the basic Quaternary geological map and the chief derivate thematic or specialist maps.

	Quaternary geological map	Bulk mineral resource map	Potential aquifer map	Geo-technical map	Engineering geology / hazard map	Geo-chemical trace element map	Geomorpho-logical map
Distribution of Quaternary sediments	+		+	+	+	+	+
Quaternary geomorphology		+	+	+	+		+
Classification of sediments	+	+	+	+	+	+	+
Thickness of units	+	+	+	+	+	+	+
Composition of sediments (stone count)	+	+	+		+	+	
Particle size distribution	+	+	+	+	+		
Stratigraphic arrangement	+	+	+	+	+	+	
Depth to bedrock	+		+	+	+	+	
Direction and distance of transport	+					+	
Carbonate content	+	+	+				
Degree of weathering	+	+		+	+	+	
Geochemistry			+			+	
Geotechnical characters				+	+		

form of the unit being surveyed. In the field the most readily recognisable features of a sediment are colour, lithology and grading, and these determine whether the unit is sufficiently distinctive to be differentiated and traced across country. Subsequent laboratory work, particularly pebble counting, particle-size and heavy-mineral analyses may result in further refinement of classification. Field recognition of particular sedimentary characteristics or post-depositional structures (glaciotectonic) may also help as may the morphological expression of a surface unit. In addition, fabric analysis and palaeocurrent determinations will provide evidence of palaeoenvironments and pebble counts will help to determine transport distance and direction.

In mapping large areas where outcrop is scarce and poorly exposed, genetic classification of sediments, particularly diamictons, is difficult. A good description of the sediments is therefore necessary. A field description on the map reading: 'Compact fissile diamicton with polished and striated stone pavement and shear structures' may be interpreted in the field notebook as a lodgement till and 'laminated sand, silt and clay beds with gravel lenses and clusters of dropstones' may be interpreted in the notebooks or on the final map as a waterlain till. But 'diamicton with angular to subangular local shale, sand/silt matrix and rare chert erratics' might be a till, a reworked till or simply a slope deposit incorporating occasional water-transported erratics. It may be possible to categorise such a deposit as mapping progresses, but on many occasions it may not. It is therefore very important to preserve a sedimentological description of the deposit rather than simply map it as 'till' or 'head'.

When beginning a field programme the first task, before starting field work, is to define the objective in terms of area and data required. Maps of appropriate scale (1:10,000 or 1:10,560 scale maps are used in both the British and Irish Surveys) are ordered and aerial-photograph coverage of the area is obtained. The aerial photographs are used at all stages of the project; from preparatory interpretations through field surveys, to the final desk compilation. An archival search is made for all relevant data including published papers and any available unpublished records. This data and, in suitable areas, information from the aerial photographs, may be used to establish a tentative outline geology. A reconnaissance visit also helps in planning the field programme.

In the field items of indispensible equipment include map case, compass, clinometer, abney or similar level, tape, notebook, small pick-end hammer, knife, large pick, spade, hand auger with >20 mm bit,

acid bottle, standard soil colour chart and safety helmet. Only factual data are recorded on the field map. These include the location of all exposures in quarries or pits, streams, ditches and banks and the sites of auger holes, boreholes and trenches. Important sections are numbered on the map and given a National Grid Reference when recorded in the notebook. There may be a need for additional trenches and deeper boreholes through critical parts of the sequences. Because of the current shortage of funds the British Geological Survey undertakes very little shallow drilling at the present time, despite the fact that major gains in knowledge were formerly achieved during the sand and gravel resources assessment programme. In contrast the Dutch and German State geological surveys drill regularly-spaced boreholes during their main investigations. The Geological Survey of Ireland relies heavily on natural and temporary exposures associated with house building and other engineering works. Little drilling is done, but occasionally a mechanical digger is used in trenching.

Boreholes in Quaternary deposits are generally drilled by percussive methods with the collection of disturbed samples and undisturbed lengths of core. The latter can be taken continuously throughout the borehole if desired. With the development of the triple-tube core-barrel, rotary drilling can now provid continuous cores of some unconsolidated deposits.

The first stage of mapping Quaternary deposits is identical to that used in mapping solid rock formations. A rapid reconnaissance, including several cross-traverses of the survey area is made to establish the lithological types of glacial deposits present and determine which types are sufficiently distinctive and mappable. It will also establish whether or not morphological mapping will be viable. Major landforms will be readily recognisable and smaller-scale feature mapping will provide the basis for the definition of contacts or boundaries between deposits, because the slight differences in cohesiveness and induration of individual sediments can produce slight changes in topography. These very small features may show either concave or convex curvature and can generally be traced across country. The relationship of the feature to the geological contact between the deposits can be fixed by augering or digging and hence the boundary can be confirmed.

Geophysical techniques can be used to define boundaries between deposits of differing physical properties. The latter are reflected in the lithology so that the boundary between lithologically contrasting units can be defined where exposure is poor. Resistivity surveys can be particularly useful in delineating sand/gravel bodies; the Wenner array is commonly used in Ireland. Shallow seismic and gravity surveys are useful for defining the form of the rockhead surface and the thickness of Quaternary sediments. Conductivity measurements have been used in East Anglia to define the outcrop and thickness of both non-consolidated and bedrock formations (see Mathers *et al.*, this volume).

Although aerial photography is used as a matter of routine in mapping projects in both Britain and Ireland, remote sensing has not proved to be an effective tool in either country. Improvements in computerised satellite imagery interpretation may in the future allow its use at the level of detail necessary for Quaternary mapping.

Mapping techniques differ depending on the nature of the sediments and their physical relief. In upland areas of Britain and in most of Ireland, the glacial deposits form clear depositional features (drumlins, eskers, moraine ridges), whereas in large areas of lowland Britain, particularly in the English Midlands and East Anglia such features are rare. Whether the relatively subdued topography of these parts of England represents post-depositional degradation is a moot point. In East Anglia the areas covered by glacial deposits give rise to broad plateaux, dissected by relatively steep-sided valleys, on the flanks and floors of which the sedimentary sequence may be exposed. In Ireland the glacial deposits of the midlands form clear geomorphological features and those of peripheral areas, particularly to the south of the midlands, which are poor in such features lack the dissected character of East Anglia. Where deep valleys occur they rarely show a good Quaternary sequence.

There are very few exposures in arable areas in Britain and information has to be gleaned from inspection of soil types, drainage ditches, animal burrows, roots of fallen trees, post-hole debris and hand augering. Here contact mapping is the only method of surveying. Following a reconnaissance of the area, an initial traverse is made from which the surveyor then predicts the alignment of the boundaries between units to a place at least 100 m away. A second controlled traverse is then made, either along a compass bearing, or parallel to the first one but from another reference point, for example along a hedge or road. The position of the boundaries can be confirmed by auger holes at points along the traverse, the locations and data from which are marked on the map. The procedure is repeated until the boundaries

are closed. If there is another similar boundary in the field it will generally prove economical to include both contacts in the traverse. If not, cross traverses may be extended to ensure that only a single type of deposit lies within the closed boundary. Morphogenic mapping can be applied where fresh landforms remain. In hilly or mountainous areas much less augering is undertaken.

In Ireland morphogenic mapping is an important element in mapping glacial sediments, but as in Britain recording of lithological contacts determines the unit boundary. Perhaps due to a preponderance of medium to coarse sediment and the difficulty of penetration very little augering is done. The common occurrence of small fields with a high frequency of open drains which can easily be cleaned with pick and spade, provides valuable and often extensive exposure. Occasionally a mechanical digger is used to trench sediments from which detailed analyses are required. This practice, which is used extensively in Finland, is being increasingly used in Ireland in order to obtain a clear appreciation of the sediment and to carry out any necessary fabric analyses. A systematic sampling programme is operated in Irish mapping projects with a minimum sampling density of 1 per 8 km^2. Samples are obtained from existing exposure or from exposure obtained in the course of the mapping programme.

The process of cross-traversing is still used by both surveys particularly in areas of low relief, but greater relief often results in more abundant exposure and boundaries are more readily established by other means.

In the upland areas landforms generally play a much more important part in elucidating the geological history. In such areas meltwater channels, and glacially-aligned features, such as striae, *roches moutonnées* and crag-and-tail features are also noted. In highland areas sediments are commonly too coarse for adequate sampling, thus a crude morphogenic map with little or no sedimentary control may be the best that can reasonably be produced.

Wherever possible, observations are recorded on the face of the map, but expansive detail and notes on the geological significance are entered in the accompanying field notebook. All field observations on the maps are made in pencil but at the end of each day they are made permanent in ink (plate 47) and at the end of the field season the data are transferred to maps of a scale suitable for compilation or final publication. Compilation of all data including laboratory analyses is done using the colour and symbol scheme decided on for the final published map.

More detailed information on field mapping techniques is included in the Geological Society's handbook 'Basic Geological Mapping' (Barnes, 1981).

Mapping offshore is much less precise. In marine areas boundary delineation is controlled by the number and location of sample points. Where thick sequences occur seismic profiling helps to define lithological units and to establish stratigraphic successions (see Balson & Jeffery, this volume).

Map production

In the Geological Surveys of both Britain and Ireland the traditional production scale has been 1:63,360 and is now 1:50,000 but larger and smaller scales have been used in response to specific demand. The scale of 1:50,000 is the smallest that will produce a map that will be of use in all of the identified client areas (table 25). Smaller scales do not allow the smaller sedimentary units and features to be clearly shown, and at 1:100,000 most esker ridges in central Ireland would suffer substantial distortion and their depiction would suggest a much larger feature than exists, or, alternatively, many ridges would have to be excluded from the map. The amalgamation of units into mixed groups showing, for example, 'till with sand and gravel' would be of very limited use to the client. Both the planner and the extractive industry require clear distinction between till and sand/gravel.

The British Geological Survey has for over a century recognised the major glacial drift divisions; boulder clay, boulder clay and undifferentiated drift, morainic drift, glacial sand and gravel, glacial laminated clay, glacial silt, glacial lake deposits, fluvioglacial sand and gravel and older sand and gravel. These are shown by colour code and symbol. In 1885, Jukes-Browne (p.74) hinted at the need to separate boulder clay types, but only in the last twenty years has detailed mapping led to increasing subdivision of the drift sequence and tills or sand and gravels of different lithological type and/or age. The traditional symbols are retained but used in conjunction with alphabetical suffixes for individual deposits. Thus, on the Warwick Sheet (184, published 1984) the contemporaneous but lithologically distinct Oadby (O) and Thrussington (T) Tills are clearly defined (plate 48). Three types of glacial sand and gravel, the Wolston Sand and Gravel (W), the Baginton Sand and Gravel (B) and the Hillmorton Sand and Gravel (H) are also separated. Similar detail has been published on several East Anglian geological maps, for example, Bury St. Edmunds Sheet (189, 1982)

and Braintree Sheet (223, 1982). Although only one boulder clay is shown on the face of the Kings Lynn and the Wash Sheet (145 and parts of 129, 1978), the distribution of the younger Hunstanton Till is clearly shown on the accompanying figure. On the forthcoming Diss Sheet (175) the Chalky Boulder Clay and Starston Till will be shown on the map face as part of a very detailed drift sequence. However, on the Hereford Sheet (198, in preparation) the sequence shown on the map has been greatly simplified and the complex sequence is only hinted at on an accompanying diagram.

More detailed information is available on special project maps. For example, the maps accompanying the report of the Engineering Geology of the Upper Forth Estuary (Gostelow & Browne, 1986), which was commissioned by the British Department of the Environment, includes a map at 1:50,000 which shows the distribution of a large variety of glacial and late glacial sediments (plate 49). This is accompanied by maps at the same scale showing 'Drift thickness contours' (depth to rockhead) and 'Contours to upper surface of the glacial deposits'. The older maps of the Edinburgh District have been reconstituted and published as a 1:25,000 Special Sheet which shows the distribution of the glacial deposits, of glacial striae, glacial drainage channels and buried drift-filled valleys, and rockhead contours. Although the colour-printed edition shows only the solid geology in colour, the drift information is clearly visible as a red overprint.

The Geological Survey of Northern Ireland's Belfast Special Sheet (Bazley & Manning, 1971) also shows detailed information on rockhead contours.

The Geological Survey of Ireland's forthcoming 1:50,000 Quaternary geology maps (plate 46) are being designed to be useful in all the client areas listed above. The scheme includes colour coding to reflect the dominant petrographic class of phenoclasts. Thus, a till dominated by Lower Palaeozoic phenoclasts is shown purple and a limestone-dominated gravel, green with blue dots (plate 46). Lithologies are distinguished by symbols which, for example, differentiate between till with a fine-grained matrix (clay/silt), till with a sandy matrix, gravelly till and blocky till. Head is also identified by a symbol, and sand and gravel will be distinguished from till using a combination of colour and symbol coding. Alluvium, blown sand, and peat are colour-coded, and ice-direction indicators are symbolised. Depth to bedrock, interpreted direction(s) of ice movement and stratigraphic cross-sections are included in inset panels. Certain geomorphological features such as drumlins will be symbolised on the map but, for the most part, features such as eskers, moraines and kames will be indicated in an inset panel. In some cases matrix characteristics of till units will determine colour coding as for instance in the Ballycroneen Till Formation (Irish Sea Till) where matrix composition characterises the unit.

The first of these maps will cover a large part of County Wicklow (see plate 45) and this will be followed by a map of the Cork Harbour area, the Dublin area, and of part of County Kerry.

Future maps

Glacial deposits cover almost the whole of Ireland and a major portion of Britain. Most public enquiries to the Geological Survey of Ireland relate to these. Furthermore, the sand and gravel industry alone is of probably greater value to both economies than the metallic minerals industry (although it does not carry the same mystique) and the majority of major and minor engineering projects are built on or encounter glacial sediments. Thus, as increasing emphasis is placed on the necessity for applied and specialist maps, increasing attention will be paid to mapping Quaternary deposits which hitherto have not had a high priority in the mapping programmes in either Britain or Ireland.

As demand increases for aggregates, groundwater resources, waste disposal facilities, engineering foundations and geochemical surveys more and more specialised data will be required to identify specific applied characteristics of Quaternary deposits and, in many cases, to define the range of properties of the glacial sediments of both Britain and Ireland. These demands will have to be met initially by the field geologists who now produce the data for standard Quaternary geology maps. Subsequently, in order that specialist maps may be produced, compaction tests, laboratory shear strength tests, permeability tests, particle-size analysis of samples and geochemical tests will be undertaken as a matter of routine.

The traditional lithostratigraphic map will remain the basic reference but it may be published as part of a folio of maps, many of which will be derived from the basic lithostratigraphic map and have a specialist applied orientation. There will be greater subdivision of the major lithologies on the basic map, and it is likely that geomorphological data will be published on a separate sheet from the geological data or on a transparent overlay.

Specialist or thematic maps, some at larger scales than 1:50,000, mostly 1:10,000, directed at a specific client area will become more common, possibly as outline monochrome maps which will be reproduced from open files on demand as is done for example by the Québec Ministère de l'Énergie et des Ressources in relation to its compilation map series. They will reflect the client areas outlined in tables 25 and 26 and they will be used for the most part by the client as a preliminary reference to target zones which need more detailed site investigation.

Acknowledgement

Published by permission of the Director of the Geological Survey of Ireland and the Director of the British Geological Survey.

Geotechnical properties of glacial deposits in lowland Britain

MICHAEL ANTHONY PAUL & JOHN ANTHONY LITTLE

In this chapter we shall review the geotechnical properties of the deposits in selected lowland areas of Great Britain. For reasons of space we have chosen to concentrate on a limited number of properties and on relatively few areas; however, we believe that the principles which underlie our analysis are of general application. The properties we have selected are those which are most commonly reported in the engineering literature; namely, the Atterberg limits, the natural water content and the undrained shear strength. Modern soil mechanics has shown that these simple properties are of fundamental importance in understanding the more complex aspects of soil behaviour (see for example Atkinson & Bransby, 1978). Unfortunately, there are very few areas where such a survey of the full geotechnical properties has been made. Consequently, we feel that it is possible to deal with the deposits of the British Isles as a whole only in simplified terms, and we have not attempted to discuss (for example) the stress, strain and yield behaviour of the materials, since these are not known in sufficient detail for any sensible regional conclusions to be drawn.

GEOTECHNICAL PROPERTIES OF TILL: AN OVERVIEW

Processes

The geotechnical properties of a unit of till in the field are the result of three basic sets of processes. The *erosion and transport phase* endows the material with its basic composition and grading; the *deposition* phase brings about changes in the grading, controls the initial water content of the till and generates a number of macroscopic structures; the *post-deposition* phase may then alter any or all of the above, and may introduce macroscopic features of its own. This process-based framework has been discussed at length elsewhere (Boulton, 1975b; Boulton

& Paul, 1976; Paul, 1977) and so only an outline will be given here.

The gross geotechnical character of a till is determined by the substrate that has been eroded during its formation. This leads to an immediate distinction between *clast dominant* tills that have formed by the erosion of crystalline or coarse sedimentary bedrock and *matrix dominant* tills that result from the erosion of fine-grained lithologies. In lowland Britain the majority of tills are matrix dominant (Derbyshire, 1975). The boundary between clast and matrix sizes may be drawn on the basis of one or more breaks that commonly occur in the complete grading curve of tills (McGown, 1971; McGown & Derbyshire, 1977). In clast dominant tills the larger particles form a continuous structural framework which causes the till to behave as an essentially granular material whereas in a matrix dominant till the clasts 'float' in the matrix which governs the overall properties.

Transport processes govern the detailed grading of the matrix fraction in the early stages of its evolution. These often generate a consistent relationship between the Atterberg limits that causes them to fall on a well defined band on a plasticity chart. Boulton & Paul (1976) referred to this band as the T-line, although it is not certain that there is one unique relationship that is obeyed by all tills.

During deposition the constituent particles of the till are brought together. Those deposited by subglacial lodgement usually have a packing that gives a density close to the maximum obtainable in engineering compaction tests (McGown & Derbyshire, 1977). This causes lodgement till to be incompressible in standard oedometer tests, behaviour which has been interpreted in terms of a substantial preconsolidation pressure, although the mechanism involved is not one of simple loading. Those tills that merely melt-out from interstitial ice have an open structure of low density (McGown & Derbyshire, 1977) which may be disrupted early in its history by shear failure due to excess pore water pressure. In the supraglacial envi-

389

ronment this process leads to the formation of various resedimented deposits including so-called 'flow tills' (Boulton, 1968; Lawson, 1979), a geotechnical model for whose formation has been presented elsewhere (Paul, 1981). The resedimentation process causes the grading to be altered by the partial segregation of particle sizes during flow, and for this reason many such deposits do not fall on a T-line.

Landforms and deposits

The tills in many lowland aras of Britain share several geotechnical similarities of a general nature, although the details of their sequences or properties may well show local variations. For this reason it is convenient to use a model that summarises these similarities. We term this idealised description the *lowland model*, although it is not, however, implied that any one area will necessarily exhibit all its features. The areas of Britain to which this model can be applied are shown in fig. 275. We recognise within the general model a number of component *elements* of engineering significance. It is possible to relate these to inferred depositional processes and we find that in many cases they can be identified with the facies classification proposed by Francis (1983); where appropriate we have adopted his nomenclature. The assemblage of the elements into one model is based on the work of Boulton & Paul (1976), Eyles (1983a), Eyles & Menzies (1983) and Paul (1983).

The simplest case (fig. 276a) arises when deposits with a streamlined upper surface lie directly on rockhead (bedrock surface). In such a case the deposits can be divided between a *rockhead element* and a *basal element*. In detail the rockhead element may consist of (1) smoothed and striated pavements; (2) fractured rock overlain by locally derived, clast dominant diamicton ('deformation till'); (3) channels with mixed infillings of clay, sand, gravel or diamicton, which may exceptionally date from previous glacial episodes. The geotechnical character of the rockhead element is extremely varied, and depends greatly on the underlying local bedrock and the direction of ice movement across it. There are few useful generalisations that can be made about this element: the investigation of its nature and extent is one of the most troublesome and yet important aspects of engineering site investigation in glaciated terrain.

The overlying basal element usually consists of several discrete, superimposed diamictons (4) separated by shear surfaces and channel fillings in the manner described by Eyles *et al.* (1982). These units are almost certainly composed of till that has been emplaced by subglacial lodgement. The upper few metres are usually weathered (5) and the surface often exhibits a characteristic streamlined morphology (6). A common development is the local presence of infillings in the low points of the surface (7): these are often laminated clays of proglacial or submarginal origin.

The geotechnical character of the basal element shows a number of consistent features, despite the variety of possible source materials. The grading curves of the tills are usually of a similar bimodal or multimodal form, and as a result the Atterberg limits obey the broadly consistent relationship discussed earlier, although the exact position of the best fitting line varies regionally. The natural water content of the till lies slightly below the plastic limit in the unweathered material, giving a liquidity index that is fairly consistent (-0.1 to -0.3). In the weathered material the liquidity index increases to a value around 0.3. At the small scale the undrained shear strength follows the variation in water content; at the large scale the strength is usually dominated by the effects of fissures (see below).

Fig. 276b shows a more complicated development at the margin of the former ice sheet. Again the rockhead (1) and basal (2) elements are present, but they are partially concealed beneath a *proglacial element* and a *marginal element*. The proglacial element consists of outwash (3) and lacustrine deposits, and may be superimposed upon or incised into the earlier basal element. The marginal element is typically a complex belt of outwash (5) interbedded with and capped by resedimented deposits (4) of supraglacial origin. The lower part of the element may contain melted out deposits (6) in various states of disturbance and may conceal fragments of the underlying basal element (7). The element as a whole often develops in association with large scale obstructions to ice movement, for example a rising bedrock surface (8).

The geotechnical character of the marginal element is considerably more varied than that of the basal element. The gross behaviour is often dominated by the sequence of deposits, in particular the presence of laminated clays. Since resedimented diamictons may differ from their parent material to a greater or lesser extent due to the sorting efficacy of the flow process, their gradings and index properties can differ also to a greater or lesser degree. Their natural water content is very dependent on their position in the kamiform topography of the marginal belt, and on the local drainage that results from the interbedded outwash or is inhibited by underlying

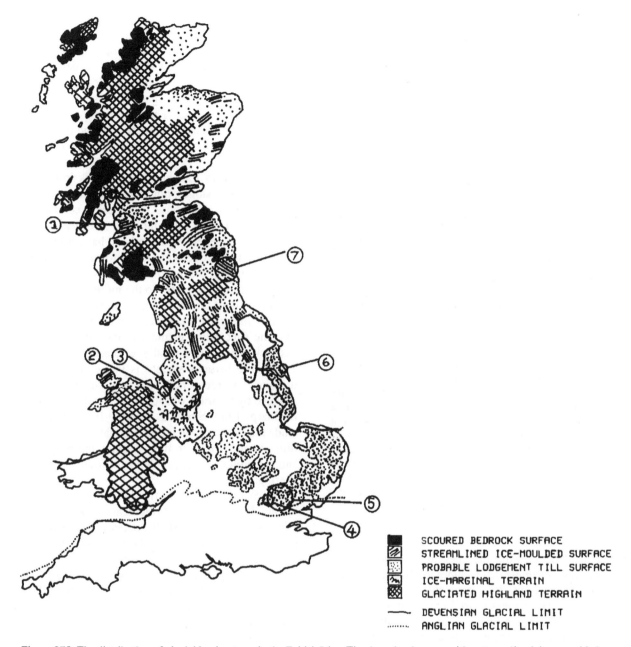

Figure 275: The distribution of glacial landsystems in the British Isles. The deposits shown as either streamlined, ice- moulded or probable lodgement till surfaces are the subject of this chapter (Source: Eyles & Dearman, 1981). Key to sites discussed in the text: (1) West-central Scotland; (2) and (3) Cheshire plain (2 = Chester); (4) and (5) Hertfordshire (4 = Watford); (6) Cowden; (7) Northumbrian plain.

clays; their undrained shear strength broadly follows these variations in water content.

A review of selected areas of Great Britain

Since it is obviously not possible to discuss the totality of British deposits, we have selected a small number of areas that illustrate features which we believe to be of general importance and for which subsurface data are available in the literature. The streamlined till plains of Northumberland and Central Scotland provide excellent and relatively simple examples of the subglacial elements of the

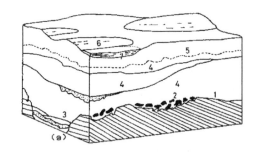

Figure 276: Block diagrams to illustrate the component elements of the lowland model used in the text. (a) Simple succession of subglacial deposits. The numbered features show (1) smoothed and (2) fractured bedrock surface cut by an infilled channel (3). This is overlain by (4) the basal element with (5) a weathered upper zone. At the top the basal element shows a streamlined upper surface (6) which may be infilled (7); (b) the rockhead (1) and basal (2) elements are shown in conjunction with the marginal element. This includes (3) outwash and (4) hummocky kamiform topography that is built on (5) buried outwash and (6) diamicton layers. The basal element may still be present at depth (7). See text for further discussion (Source: based on Eyles, 1983).

lowland model, while the Cheshire basin provides an analogous example that includes the effects of supra-glacial deposition. We also consider two problematic areas: the Holderness coastal area whose genesis is open to discussion, and the geotechnically-distinct chalky tills of the Anglian glaciation.

The Northumbrian Plain

This is the general area that lies north of the River Tyne and is bounded to the west by the rising ground of the northern Pennines and to the north by the Cheviot hills. It is a relatively low lying coastal plain that shares similarities with adjacent areas to the north and south. The subsurface conditions are known principally from coastal exposures and from major opencast coal mines at Butterwell and Ack-lington, and from temporary excavations around the Newcastle conurbation and other urban centres.

The area presents a streamlined, ice-moulded surface that is believed to be of subglacial origin (cf. Eyles *et al.*, 1982). Below this surface is developed a lodgement till unit which, although clearly composite in detail, can be broadly regarded as a single entity for geotechnical purposes. This unit displays several of the features associated with the lowland model that was introduced earlier, and so will be described in this context.

The Rockhead Element. The bedrock in the area consists of sandstones, mudstones and limestones of Carboniferous age whose engineering condition is determined largely by lithology. Arenaceous sediments are in general shattered and have been incorporated into the overlying till to form a clast dominant assemblage that extends upwards for one to two metres before giving way to the more typical matrix dominant till of the succeeding basal element. Argillaceous materials are better comminuted and pass into the succeeding element more rapidly, and the bedrock may often be locally tectonised (Eyles *et al.*, 1982). The bedrock surface is incised by eastward draining, overdeepened channels which are now plugged by a variety of subglacial sediments (Anson & Sharp, 1960; Smith & Francis, 1967) which include till, outwash and laminated clays. These mixed fillings vary rapidly in their engineering properties, and can present major problems for foundation engineering.

The Basal Element. In this area most of the sediments above the rockhead element can be referred to this element, which usually extends to the present ground surface and there displays a variety of stream-lined bedforms. It is possible to recognise an unweathered and a weathered facies whose boundary may lie as deep as eight metres. The properties of these two facies have been reported by Eyles & Sladen (1981), Sladen & Wrigley (1983) and Russell & Eyles (1985).

The unweathered facies is a dark grey matrix dominant diamicton. As shown in figures 277a and b the particle size lies in the range sand 30–50%, silt 30–50%, clay 10–25% and the Atterberg limits lie in the range liquid limit 35–60%, plastic limit 0–20%. These values fall in a broad band on the plasticity chart. The weathered facies is a red-brown matrix dominant diamicton which is depleted in carbonate relative to the underlying unweathered facies. This may be due to the presence in the unweathered till of disseminated sulphides probably derived from the Coal Measures. These may have encouraged the formation of the weathered facies by the formation of sulphuric acid as a by-product of their oxidation to iron oxides and hydroxides, which was then responsible for the removal of the carbonates (Taylor, 1978; Russell & Eyles, 1985). In the weathered facies the particle size distribution is enriched in the clay fraction (30–50%) compared with the unweathered facies, which in turn increases the Atterberg limits to liquid limit 35–60%, plasticity index 15–40%.

The strength of the till is best discussed in terms of typical depth profiles (fig. 277c, based on Eyles & Sladen, 1981, and Eyles *et al.*, 1982). The profile shows that in the unweathered facies the natural water content lies in the range 10–15% which corresponds to a small negative liquidity index (–0.2 to –0.05). These values appear typical of subglacial lodgement till. In the weathered facies the water content rises to around 15–25%, and the corresponding liquidity index to 0 to 0.3. The increased liquidity index corresponds to a reduction in undrained strength to around 100–200 kPa from its value of around 250–300 kPa in the unweathered material. We note, however, that the *in situ* strength of lodgement till is very commonly controlled by the density and orientation of fissures (see below) and thus that the liquidity index may not be necessarily a complete measure of the undrained strength in the field.

West-central Scotland

This area covers the lowlands around and to the southwest of Glasgow, extending into northern Ayrshire. The geotechnical properties of the tills in this region have been presented in a series of publications by McGown, McKinlay and their coworkers (McKinlay *et al.*, 1974; McGown *et al.*, 1975; McKinlay *et al.*, 1975; McGown, 1985) from which this account has mainly been drawn.

Where the glacigenic deposits form the present land surface they usually present a streamlined morphology in which classical drumlin forms can be locally abundant. This morphology suggests strongly that the deposits are of subglacial origin. Beneath this surface lies a till sheet that is geologically divisible on the basis of colour and provenance, but which is indivisible geotechnically, other than into a weathered and an unweathered facies. The boundary between the two lies at a depth of two to three metres in most cases, and is not associated with a consistent change of colour.

The Rockhead Element. The bedrock surface is not usually well exposed except in temporary excavations. Much of the low ground is floored by Carboniferous rocks including limestones, calciferous sandstones and coal measures. The rockhead can possess a local relief of tens of metres, and its condition reflects both lithology and the direction of ice movement relative to the local strike. For example sandstones are often in a fractured condition with substantial incorporation into the overlying till (McGown, 1985) whereas limestones tend to exhibit striated pavements which may extend over several tens of metres. In the Irvine area the rockhead is known to be channelled and excavations for the Clyde tunnel encountered an overdeepened channel that was filled with a variety of glacial sediments (Morgan *et al.*, 1965). It is very likely that such filled channels are widespread beneath the area.

The Basal Element. The work of McGown *et al.* (1975) indicates that there is essentially a single geotechnical unit in the area, albeit differentiated into an unweathered and a weathered facies.

The unweathered facies is a matrix dominant diamicton whose colour may range from grey-black to red-brown depending on the source and may contain up to 25% of clay-size particles (fig. 278a). The liquid limit ranges from 25–35% and the plasticity index from 10–20%. When plotted on the plasticity chart (fig. 278b) the tills do not fall in such a well defined band as those from the other regions considered here. The weathered facies is a matrix dominant diamicton of variable colour. In comparison with the unweathered facies the clay size fraction is more variable in amount, and may exceed 30% in some cases. As might be expected the weathered material has more variable Atterberg limits (liquid limit 20–50%, plasticity index 5–20%) and individual results are scattered widely over the plasticity chart.

Fig. 278c shows a typical depth profile (based on McKinlay *et al.*, 1974). In the lower, unweathered facies the natural water content lies in the range 10–20% with mean 13%: the liquidity index of most samples is around –0.1 to –0.2 in consequence. In the weathered facies the natural water content may increase to over 30%, with a mean value of 20% and as a result the liquidity index of the material is usually in the range 0 to 0.2. This is very similar to the situation in the Northumbrian Plain. There is a general increase in strength with depth as a result of the decreasing liquidity index, however it must also be noted that there is an important scale effect due to the presence of fissures. This effect has been extensively studied in this area (McGown & Radwan, 1975) and it is known that the fissures both reduce the strength of larger samples and impart an anisotropy of

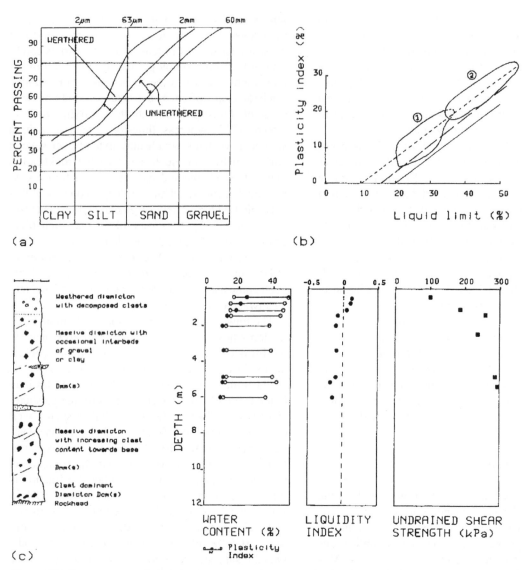

Figure 277: Summary geotechnical data for the Northumbrian plain. (a) Particle size curve, (b) plasticity chart, (c) representative geotechnical profiles. In (b) envelope (1) is for unweathered till and (2) is for weathered till. The lines shown in (b) are the A-line (solid), the marine clay line (long dashes: Skempton, 1970) and the T-line (short dashes: Boulton & Paul, 1976). The lithofacies coding shown in (c) is that of Eyles *et al.*, 1983 (Source: Eyles & Sladen, 1981).

strength. The operational strength of the till is only approached by samples of 230 mm diameter and larger, which may possess only about one half to one quarter of the strength of the intact material.

The Cheshire Lowlands

The Cheshire-Shropshire basin is that lowland area of northwest England that lies between the Pennines and the Welsh borderland and which extends southwards to the Severn gorge at Ironbridge. In this

section we shall consider only a part of this region: the northeastern quadrant that is defined to the east by the Pennine margin and to the west by the mid-Cheshire ridge, and which extends southwards to the line of the Woore-Whitchurch moraine. This area suffered invasion by so-called Irish Sea ice during the Late Devensian but was free of the influence of Welsh ice that was felt further to the west. The present account draws on data from the work of Alderman (1959), Al-Shaikh-Ali (1975; 1978), and Al-Shaikh-Ali *et al.* (1981), and places it in a new engineering geological framework. The geological

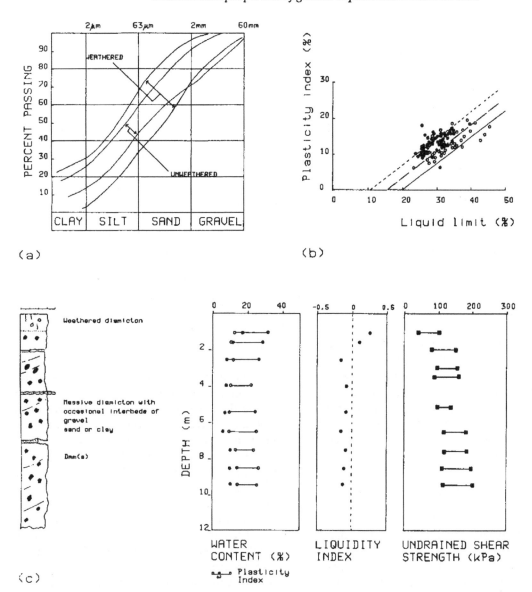

Figure 278: Summary geotechnical data for West-central Scotland. (a) Particle size curve, (b) plasticity chart, (c) representative geotechnical profiles. In (b) the solid symbols are data for unweathered till, the open symbols are for weathered till. The lines shown in (b) are the A-line (solid), the marine clay line (long dashes: Skempton, 1970) and the T-line (short dashes: Boulton & Paul, 1976). The lithofacies coding shown in (c) is that of Eyles *et al.*, 1983 (Sources: McGown *et al.*, 1975; McKinlay *et al.*, 1975).

background has been synthesised from Evans *et al.* (1968), Taylor *et al.* (1963) and Poole & Whiteman (1966), together with the original observations of one of the present authors (MAP).

There are three general surface morphologies present: (i) an ice-moulded, streamlined surface that occurs mainly in the southwest of the area around the flanks of the mid-Cheshire ridge; (ii) a very subdued, undulating surface that is found in the northern and central areas; (iii) a well developed hummocky belt

that lies parallel to the bedrock escarpment of the Staffordshire coalfield and which continues in an arc around the Pennine flank northwards to Stockport and Manchester. We suggest that the major part of the basin is filled by subglacial sediments (principally lodgement till) whose streamlined surface may be obscured by a veneer of proglacial (often lacustrine) deposits. At the southern and eastern margins the obstruction that was presented by the rising ground caused compressive ice flow and the consequent supraglacial release of debris to form a characteristic suite of hummocky landforms that are built largely of outwash interbedded with and capped by resedimented deposits ('flow tills'). In order to accomodate this increased complexity it is neccessary to add a *marginal element* to the lowland model and to discuss its geotechnical character separately.

The Rockhead Element. Over most of the area the bedrock consists of sandstones, marls and evaporites of Triassic age. Subsurface data and a limited number of exposures suggest that the till-bedrock contact is similar to that described in earlier sections. The harder lithologies are generally broken and the fragments incorporated to form a locally clast dominant till of limited thickness. The softer lithologies may show evidence of glaciotectonism and may be intimately sheared with the overlying till. In the northern part of the area Howell (1965) has demonstrated the existence of a system of subglacial channels, and occasional deep bores elsewhere (for example at Sydney, Ettily Heath and Audlem) have penetrated pockets of complexely interbedded tills and outwash that are interpreted as the fillings of similar channels.

The Basal Element. In the western part of the area the rockhead element is succeeded by a till that extends to the ground surface and there exhibits a streamlined morphology. When traced to the east and north this morphology is replaced by an indistinctly undulating surface that is often associated with the outcrop of stoneless laminated clays, although as noted by Evans *et al.* (1968) the distinction between till and laminated clay is often arbitrary. For this reason the geotechnical properties of the two materials are here discussed together.

Investigations for the M56 Motorway (Al-Shaikh-Ali, 1975; 1978; Al-Shaikh-Ali *et al.*, 1981) have revealed subsurface conditions that may be generally typical of this element in the western area. The clay-sized fraction varies between 15 and 45% (fig. 279a) and on a plasticity chart (fig. 279b) both tills and clays plot along a T-line; the clays fall towards the higher end due to their greater liquid limits. The range of liquid limits is typically 20% to 50%, with the division between tills and clays being at about 40%. The natural water content of the till in general lies at or below the plastic limit (around 10–15%), whereas that of the clay is more variable. As would be expected the undrained strength follows variations in the liquidity index (fig. 279c) albeit with scatter that is probably the result of fissuring.

Measurements of the overconsolidation ratio by Al-Shaikh-Ali *et al.* (1981) have shown that below a depth of about six metres the deposits around Chester

are normally consolidated. This is very unusual if these deposits are truly lodgement tills; the possibility is raised that they may be either glaciolacustrine (cf. the common reports of laminated clays in west Cheshire) or have been derived from aqueous deposits without undue disturbance. The proximity of the site to the Irish Sea basin may be significant in this respect (and see the later section on Holderness).

Towards the east the subglacial element is replaced at surface by the kamiform deposits of the marginal element. Exploratory boreholes and temporary exposures in the Manchester and Stockport areas (Alderman, 1959) suggest that it can still be traced at depth, and is here represented by tills which are found to lie in contact with rockhead ('lower boulder clay' in the then terminology). Fig. 279b shows that the plasticity characteristics of these tills are very similar to those of the western area. The undrained shear strength of these tills are typically in the range 150–250 kPa, which is close to but a little higher than those of the western area.

The Marginal Element. To the east and south the edges of the basin are dominated by the broad kamiform topography of the Whitchurch-Woore moraine and its northward extension. Sections through these deposits are relatively infrequent: however, a general synthesis may be made from the work of Alderman (1959), Poole & Whiteman (1966), Taylor *et al.* (1963) and Yates & Moseley (1967) together with the geophysical work of McQuillan (1964). This synthesis shows that the dominant sediment is outwash, usually sand, and that diamictons are present as widespread, discontinuous sheets that lie within or upon the sands as drapes and topographic fillings. Their general thickness is limited to a few metres and is controlled by the morphology of the surface on which they lie. We interpret this evidence to indicate an origin in an ice-marginal environment, and suggest that many of the diamictons are flowed or resedimented deposits ('flow tills'). This view has been discussed at greater length by Paul (1983).

The geotechnical properties of the tills are summarised in figs. 280a to 280c. The plasticity data show that once again the majority of the samples fall along a T-line, with those described as 'glacial clays' at the higher end. The liquid limits of the tills proper mainly lie in the range 20–40%, and the plasticity indices in the range 12–25%. The upper ends of these ranges are a little higher than those quoted for lodgement tills. The natural water contents are usually between 12–20%: this is normally at or slightly below the plastic limit. Since the tills commonly occupy topographic highs and are underdrained by sand, the

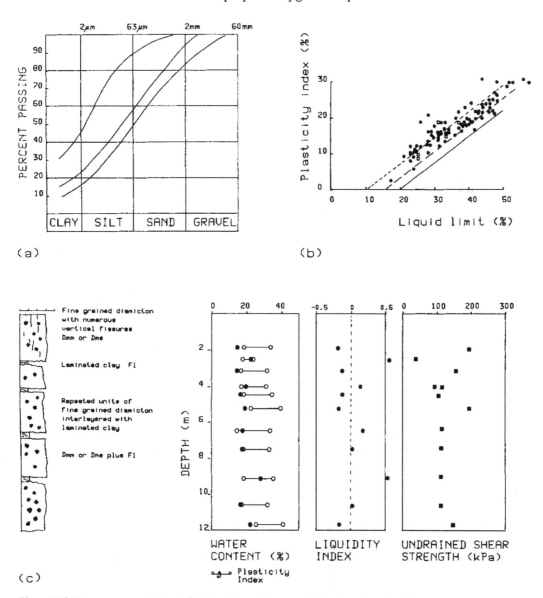

Figure 279: Summary geotechnical data for the basal element of the Cheshire plain. (a) Particle size curve, (b) plasticity chart, (c) representative geotechnical profiles. In (b) the open symbols are results from the eastern area, the remainder are from the western area. The lines shown in (b) are the A-line (solid), the marine clay line (long dashes: Skempton, 1970) and the T-line (short dashes: Boulton & Paul, 1976). The lithofacies coding shown in (c) is that of Eyles *et al.*, 1983 (Sources: Alderman, 1959; Al-Shaikh-Ali, 1975; 1978; Al-Shaikh-Ali *et al.*, 1981).

in situ water content largely reflects the local drainage. It has been noted that the undrained strength of the diamictons in this element (around 100–150 kPa) is usually less than the strength of those in the basal element (around 150–250 kPa).

The Holderness Coastal Plain

The establishment of a test bed in till at Cowden,

Holderness, by the Building Research Establishment in 1966, has enabled a considerable amount of high quality geotechnical data to be obtained from this single site (Marsland, 1980; Marsland & Powell, 1980; Powell *et al.*, 1983; Marsland & Powell, 1985).

Opinion is divided about the probable origin of the Devensian till sequences on this part of the east coast (see Madgett & Catt, 1978, for a summary). Some theoretical reconstructions (for example Boulton *et*

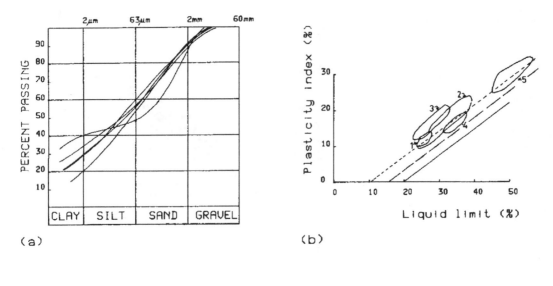

(a)   (b)

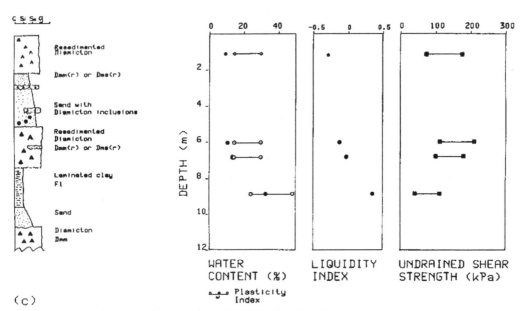

(c)

Figure 280: Summary geotechnical data for the marginal element of the Cheshire plain. (a) Particle size curve, (b) plasticity chart, (c) representative geotechnical profiles. In (b) the numbered outlines are data envelopes for (1) Rudheath; (2) Sandbeach; (3) Congleton; (4) Manchester and (5) Holmes Chapel. The lines shown in (b) are the A-line (solid), the marine clay line (long dashes: Skempton, 1970) and the T-line (short dashes: Boulton & Paul, 1976). The lithofacies coding shown in (c) is that of Eyles *et al.*, 1983 (Sources: Alderman, 1959; Evans *et al.*, 1968; Paul, unpublished data).

al., 1977) postulate a surge lobe to explain the distribution of the Devensian till sheet which extends southwards from the Yorkshire coast to Holderness, Lincolnshire and North Norfolk due to the difficulty of incorporating this configuration of ice front into any plausible steady-state model. If this postulate is correct, the deposits of the Holderness lowland may be a combination of normal subglacial sediments and ice-pushed, possibly overridden, surge deposits.

There is as yet no adequate model to describe such deposits (although cf. Sharp, 1982) and so we shall confine our discussion to a simple account of the soil properties.

Fig. 281a shows the grading of typical material from the Cowden test site, and fig. 281b shows its associated plasticity characteristics. It is noticeable that although the points fall on a reasonable line, this line is closer to the line for silty marine clays (Skemp-

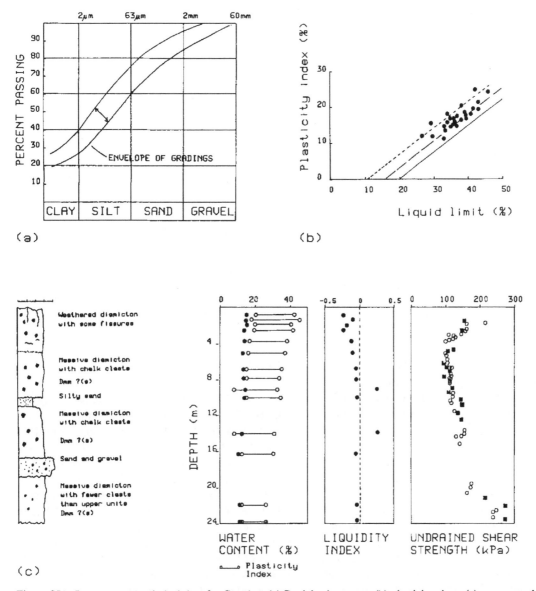

Figure 281: Summary geotechnical data for Cowden. (a) Particle size curve, (b) plasticity chart, (c) representative geotechnical profiles. The open symbols in the profile of undrained strength represent results from pushed samples, the closed symbols the results from plate tests. The lines shown in (b) are the A-line (solid), the marine clay line (long dashes: Skempton, 1970) and the T-line (short dashes: Boulton & Paul, 1976). The lithofacies coding shown in (c) is that of Eyles *et al.*, 1983 (Source: Marsland & Powell, 1985).

ton, 1970) than is usually the case for till. The variation of the natural water content and Atterberg limits is shown in fig. 281c: the data show the till to have a liquidity index between 0.24 and –0.38, with most values in the range –0.1 to 0. In its upper few metres desiccation has reduced this value to around –0.3 to –0.4. The grading shows approximately 35% sand and gravel, 22% silt and 30% clay; the activity (Skempton, 1953) is therefore 0.67. Total carbonate contents vary between 8.5% (top, weathered till) to

over 25% at 23.5 m depth.

Below a depth of about five metres the till contains few fissures. Electron micrographs reveal that although microfissures are present, their length is restricted and the fabric shows evidence of mixing during deposition. It would thus appear that the Cowden diamictons show features that are in some respects not usual for lodgement tills.

Undrained shear strengths determined from tests on good quality pushed samples (and tested within a

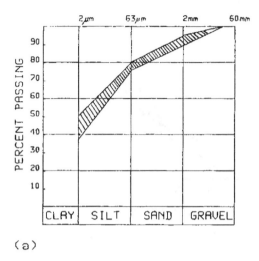
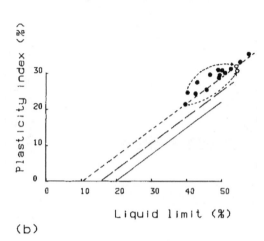

(a)   (b)

Figure 282: Grading and plasticity data for the chalky tills of the Vale of St. Albans. (a) Particle size curve, (b) plasticity chart. (a) shows the envelope of curves from Little (1984). The points in (b) that are connected by arrows show the effect of laboratory decalcification (open circles) on previously tested samples (closed circles). The lines shown in (b) are the A-line (solid), the marine clay line (long dashes: Skempton, 1970) and the T-line (short dashes: Boulton & Paul, 1976) (Source: Little, 1984).

few days of sampling) are shown in fig. 281c, as are the undrained shear strengths determined from the results of 865 mm diameter plate loading tests on the site. There is generally good agreement between the two sets of results, which might result from the absence of fissures; a point of particular note is the relatively low strength of the material. This is corroborated by *in situ* measurements of the horizontal stress in the ground at this site which have, in general, shown the horizontal and vertical effective stresses to be equal, thus indicating a condition of light overconsolidation. Atkinson (1985) has estimated the preconsolidation pressure using a comparison of the intact and remoulded strengths and suggests that between the depths of 3.4 and 7.1 m the pressure was between 387 and 665 kPa.

An interesting comparison may be made between the diamictons of Holderness and those of western Cheshire. In both cases the deposits are located close to a marine basin across which their parent glacier advanced, and both are only lightly overconsolidated. The speculation can be made that the low overconsolidation ratios which characterise both these materials (and which are not usually associated with lodgement till) might have resulted from their derivation from the marine basin, whether emplaced by the postulated surge (in the Cowden case) or by some other process (for a discussion of other geologic and environmental factors which might need to be considered in a full explanation of the degree of overconsolidation of till see Mickelson *et al.*, 1979).

The Vale of St. Albans

The Vale of St. Albans occupies the relatively wide (up to 7.5 km) gently undulating plain between the towns of Watford in the southwest and Ware to the northeast. The ground in the Vale generally lies between 65 m and 85 m O.D., occasionally down to 50 m O.D., rising to over 150 m O.D. to the north and south. The Vale is dissected by a number of rivers which rise to the inclined backslopes of the Chilterns to the north and drain towards the Thames to the south. The northern flank is bounded by the prominent northeast to southwest outcrop of Chalk that forms the Chiltern Hills and the southern boundary by the low northward-facing ridge of the Lower Tertiaries (Reading Beds, London Clay) which form the South Hertfordshire Plateau.

The Vale is largely infilled by an unconformable mantle of Pleistocene sediments which comprise a suite of proto-Thames (pre-Anglian) sands and gravels, tills and proglacial outwash deposits of Anglian age. Although the classic streamlined topography that characterises the Northumbrian Plain and West-Central Scotland is absent, we agree with the view of Perrin *et al.* (1979) and Eyles & Dearman (1981) that the extensive dissected till sheets covering and underlying much of East Anglia, of which those in the Vale are a component part, are probably the remnants of former lodgement till plains. However, since we feel that there is insufficient evidence to give a detailed treatment of the separate elements, we again restrict

Figure 283: A sample of the matrix dominant chalky till from the Vale of St. Albans after triaxial testing. Abundant chalk fragments of all sizes can easily be seen (Photograph: J.A. Little).

ourselves to a simple account of the soil properties at various sites in the area.

The glacigenic deposits of the Vale are derived from Mesozoic sediments, notably those of Jurassic and Cretaceous age. The tills therefore tend to be matrix dominant and additionally contain considerable quantities of chalk. The absence of a chalk content in the other tills which have been described indicates that the chalky nature of these tills is a particular distinguishing characteristic of which account must be taken (but see also the section on Holderness). For this reason we have chosen to adopt a different approach in this section and purpose to describe typical geotechnical properties of these calcareous tills. The descriptions which follow are principally based on the work carried out by one of the authors (JAL) in the Vale of St. Albans (see for example Little, 1984; Little & Atkinson, 1985; Atkinson & Little, 1985) together with the data published by Marsland (1977).

Gibbard (1974, 1977) and Cheshire (1981, 1983; see also Allen, Cheshire & Whiteman, this volume) describe chalky lodgement tills (Anglian) found in the Vale of St. Albans between Watford and Ware in Hertfordshire. Little (1984) has reported the engineering properties of these tills and a summary of their basic properties together with descriptions of these soils is given in table 27 and fig. 282. The grading of the tills indicates 10–15% sand, 40% silt and 40–50% clay. The results of the sieving and sedimentation test designed to quantify the amount and distribution of chalk and acid solubles in the range 16 mm down to 0.063 mm for one of these tills

are given in table 28. The very chalky nature of this till (total acid soluble content approximately 35%) can clearly be observed by an examination of these data (see also fig. 283).

The variation of undrained shear strength (determined *in situ* using a hand shear vane) with liquidity index for selected of these tills is shown in fig. 284. The peak undrained shear strengths varied between approximately 400–500 kPa; remoulded shear strengths between approximately 65–100 kPa. In this figure note how the *in situ* values of remoulded shear strength compare with the (extrapolated) values of undrained shear strength of the reconstituted (< 0.425 mm) matrix determined in the laboratory using a motorised shear vane. Data from site investigations from western Cheshire (Al-Shaikh-Ali, 1975; 1978) are also shown for comparison, and it may be seen that these also continue the laboratory trend.

Although relatively high peak undrained shear strengths are a feature characteristic of low plasticity lodgement tills (Fookes *et al.*, 1975; Eyles & Sladen, 1981; Sladen & Wrigley, 1983) and are consistent with the very low natural water contents of these soils, the peak values for the chalky tills are somewhat higher than reported in the other areas we have considered, and the range of values for the ratios of peak to remoulded shear strength (3.72 to 5.78) are clearly at variance with the commonly held view that tills possess little or no sensitivity (e.g. Sladen & Wrigley, 1983). The evidence from the mineralogical, grading, oedometer and strength tests carried out on these tills strongly suggests the presence of a calcareous cement (probably calcite) in the matrix of these chalky soils. Whilst its precise nature and form have not been examined in detail, its effects have been observed by comparing the engineering properties of these chalky tills with those of a naturally decalcified horizon of the till in the Vale. The presence of such a cement in the *in situ* soil would also satisfactorily explain the apparently high sensitivity of these tills. By contrast the Cheshire results (fig. 284) lie much closer to the trend line for remoulded soils, which suggests that cementation does not influence their strength to the same extent.

A comparison may be made with the deposits at the western end of the Vale near Watford. At the site of the Building Research Station, Marsland (1977) has described 11.5 m of fissured, reddish-brown (with grey mottling) to light to medium brown 'chalky boulder clay' whose grading is 5–25% sand, 25–35% silt, 50–60% clay. The average Atterberg limits are liquid limit 47%, plastic limit 19% and the natural water contents varied from 17–24%, thus

Table 27. The composition of the chalky tills from various sites in the Vale of St. Albans (source: Little, 1984).

Till description	Illite	Ca – Montmoril-lonite	Kaolinite	Amorphous iron and alumina hydrates	Quartz	Calcite	Acid soluble at 0.002 mm	W%	LL	PL	LI	G₂	γ kN/m³	A	% sand	% silt	% clay
Holwell Hyde TL 265116 HH(a): Very dark grey (5Y 3/1) stiff and very stiff, becoming hard CLAY with fine, medium and coarse (up to 100 mm) rounded to subrounded fresh white chalk clasts with clearly marked glacial striations. Occasional angular fragments of medium to large flint and a Jurassic fauna are present. This till is most probably a facies of the Ware Till.	19	9	8	62	18.2	10.0	8.86	18	40	18	0	2.71	21.2	0.571	21	40	39
HH(b): Light grey (5Y 7/1) and light yellowish-brown (2.5Y 6/4) to brownish-yellow (10YR 6/6) stiff and very stiff, becoming hard CLAY with angular fragments (up to 40 mm) of flint and rounded to subrounded fragments (up to 60 mm) of fresh white chalk. Occasional flecks of soft pink (5YR 7/4) powdery silty clay (Red Chalk) and Jurassic bivalves are present. This till is most probably a correlative of Gibbard's (1977) Eastend Green Till.	26	15	9	50	26.3	14.9	10.60	16	43	18	−0.008	2.73	21.4	0.671	22	40	38
Foxholes TL 337125 F(b): Oxidised and almost completely decalcified yellowish-brown (10YR 5/6) becoming dark greyish-brown (10YR 4/2) stiff becoming very stiff sandy CLAY with small to medium-sized (up to 30 mm) fragments of angular and rounded to subrounded flint and with rare small flecks (1 to 2 mm) of light grey (10YR 7/1) to white (10YR 8/1) soft chalk. This till represents a naturally decalcified horizon of the Ware Till.	17	12	8	63	27.2	0	0	20	58	22	−0.056	2.74	20.4	0.726	24	26	50
Westmill TL 344158 W(b): Ware Till. Dark greyish-brown (2.5Y 4/2) very stiff, occasionally hard sandy CLAY with small (<7.5 mm) rounded to subrounded clasts of fresh white chalk with small, medium and large (occasionally 100 mm) angular shattered fragments of flint.	20	15	7	58	23.1	4.3	6.46	17	49	19	−0.067	2.74	21.7	0.653	21	33	46
W(d): Eastend Green Till. Light olive-grey (5Y 6/2) to brown (10YR 5/3) very stiff, occasionally hard CLAY with strong brown (7.5YR 4/6) mottling and streaking and small (2 to 3 mm) patches of pink (5Y 7/4) silty clay (Red Chalk). There are many small, medium and large (occasionally 150 mm) rounded clasts of relatively fresh chalk showing surface striations. Medium to large, irregularly shaped occasionally shattered flints abound and bivalves are present.	17	14	7	62	18.0	12.6	12.57	16	45	16	0	2.73	21.8	0.723	15	45	40
	% by weight of the total clay content				% by weight of the total sample												

Table 28. The chalk and acid soluble content of the chalky till from Holwell Hyde, Vale of St. Albans (source: Little, 1984).

Till HH(a) Particle size (mm)	(phi)	Chalk retained (%)	∑ Chalk (%)	Acid soluble (%)	∑ Acid soluble (%)	Till HH(b) Particle size (mm)	(phi)	Chalk retained (%)	∑ Chalk (%)	Acid soluble (%)	∑ Acid soluble (%)
16.0	(−4)	1.3	1.3	1.3	1.3	16.0	(−4)	10.4	10.4	11.2	11.2
11.2		3.6	4.9	3.6	4.9	11.2		0.9	11.3	0.9	12.1
8.0	(−3)	4.1	9.0	4.1	9.0	8.0	(−3)	4.2	15.5	5.3	17.4
5.6		2.5	11.5	2.5	11.5	5.6		2.8	18.3	4.5	21.9
4.0	(−2)	2.4	13.9	2.4	13.9	4.0	(−2)	1.3	19.6	2.4	24.3
2.8		1.4	15.3	1.4	15.3	2.8		0.9	20.5	1.3	25.6
2.0	(−1)	1.3	16.6	1.4	16.7	2.0	(−1)	0.5	21.0	0.7	26.3
1.4		0.99	17.59	1.1	17.8	1.4		0.9	21.9	2.0	28.3
1.0	(0)	0.41	18.00	0.7	18.5	1.0	(0)	0.5	22.4	0.9	29.2
0.71		0.53	18.53	0.6	19.1	0.71		0.4	22.8	0.7	29.9
0.50	(+1)	0.45	18.98	0.5	19.6	0.50	(+1)	0.3	23.1	0.6	30.5
0.30				0.32	19.92	0.30				0.77	31.27
0.18				0.52	20.44	0.18				1.19	32.46
0.15				0.18	20.62	0.15				0.55	33.01
0.063	(+4)			1.66	22.28	0.063	(+4)			1.53	34.54

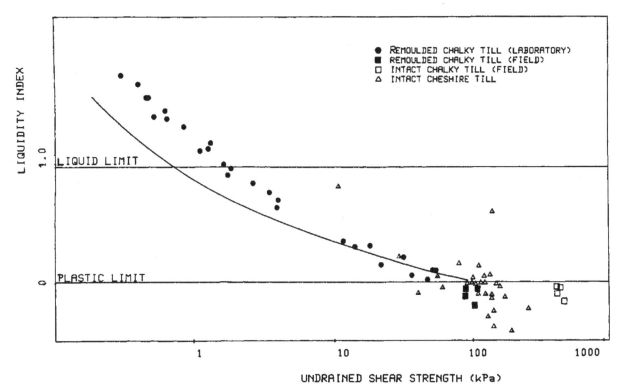

Figure 284: The relationship between liquidity index and undrained shear strength for chalky tills of the Vale of St. Albans and for the Cheshire basal element. The solid curve is that proposed for London Clay by Skempton & Northey (1952) which is shown for comparison (Sources: Alderman, 1959; Al-Shaikh-Ali, 1975; 1978; Little, 1984).

producing a soil with a liquidity index ranging from −0.07 to 0.18. The chalk content varied from 13.4% (at the top) to 30.9% (bottom). Undrained shear strengths determined from 98 mm triaxial and 865 mm diameter plate tests indicate somewhat lower values for shear strength (maximum approximately 300 kPa) compared with those quoted above for the other tills in the Vale. This could well be a reflection of the highly fissured nature of the tills described by Marsland.

Conclusions

This review has shown that in lowland Britain a relatively simple model can be used to generalise certain engineering properties of the tills. On a stan- dard plasticity chart tills of supposed subglacial ori- gin share similarities in their index properties, which are believed to arise primarily as a result of their history of transport. There are also similarities in their profiles of *in situ* water content and undrained strength. This is mainly the result of depositional and post-depositional processes which have determined the water content, and hence the undrained strength. This is controlled by the liquidity index in a manner common to all soils, although when the soil is undis- turbed it may also be increased on the one hand by cementation and reduced on the other by fissuring.

Our review has also emphasised to us the relative lack of sophisticated test results on tills, despite their ubiquitous occurrence. We believe that a detailed understanding of their behaviour in the context of modern soil mechanics to be well overdue.

Styles of Anglian ice-marginal channel sedimentation:
As revealed by a conductivity meter and extendable augers

STEPHEN JOHN MATHERS, JAN ANTONI ZALASIEWICZ & GARY PAUL WEALTHALL

Meltwater drainage at the margins of ice-masses is primarily controlled by the topographic distribution of both the ice and the surrounding terrain.

Where the land surface slopes away from an ice margin, meltwaters are free to drain from the ice-masses often depositing extensive alluvial fans and outwash sandar. In contrast, ice-masses that abut rising ground (as for example along the lateral margin of a valley glacier) tend to focus surface meltwater and run-off into the furrow formed between the hillside and the ice (Russell, 1893). If a longitudinal gradient is present and the ice is 'cold' and unfissured (acting as an impermeable seal) extensive marginal drainage may develop. Important studies of such marginal drainage systems include Tarr (1909), von Engeln (1911), Mannerfelt (1949), and Sissons (1960). The deposits of such drainage may occur in channels incised into the bedrock or be preserved as kame-terraces. Where an ice-mass terminates on gently sloping or level ground the meltwaters are likely to be ponded in depressions leading to proglacial lacustrine deposition. In detail, marginal drainage networks may be influenced by structural features of the ice such as ice-cored ridges (Boulton, 1972). Useful reviews of ice-marginal meltwater drainage are given by Price (1973), Embleton & King (1975) and Sugden & John (1976).

In Suffolk the Anglian ice-sheet advanced southeastwards over a subdued but incised proglacial landscape. The main outwash deposits appear to be largely confined to several major drainage tracts which were the precursors of the major modern river valleys. In the higher ground between these major drainage tracts, systems of smaller-scale meltwater channels are present. These channels are commonly deeply incised into Pleistocene sands and gravels and infilled with substantial amounts of clay-rich glacigenic deposits including tills and fine-grained waterlain sediments.

The marked physical contrast between the channel infills and the neighbouring sediments has enabled the form of the channels to be accurately defined by the systematic measurement of ground conductivity. In addition, the sedimentary sequences infilling the channels have been investigated, largely through the use of extendable augers. This account describes how the data gathered using such techniques enables the construction of models describing the evolution of such channel systems and how different types of infill may be related to channel position *vis-à-vis* the maximum extent of Anglian ice.

Investigative techniques

The two principal techniques used were: the systematic measurement of ground conductivity using a Geonics EM 31 Terrain Conductivity Meter; and the use of extendable augers.

In recent years portable conductivity meters have been developed which, unlike traditional methods, do not involve the use of electrode arrays. These portable meters, which include the Geonics EM 31 used in this study (fig. 285), operate on the electromagnetic inductive principle (for further details see Zalasiewicz *et al.*, 1985 and McNeill, 1980). Their portability and the direct read-out of conductivity values enables data to be gathered rapidly.

The meter (fig. 285) comprises two coils, 3.7 m apart, housed at the ends of a rigid boom, with a control panel and power source located in the centre. Values of ground conductivity are displayed on the control panel, expressed in millimhos per metre (mmhos/m). The 'depth of exploration' of the equipment in normal operating mode is about six metres (McNeill, 1980). The equipment will measure accurately ground conductivity from about 1 mmho/m to about 100 mmhos/m; at higher conductivity values the reading is well below the true ground conductivity. The effective range of values encompasses most sedimentary rocks. The EM 31 is carried by means of a shoulder sling and in operation is 4 m long and weighs 9 kg. In southeast Suffolk conductivity read-

Figure 285: The Geonics EM 31 Terrain Conductivity Meter.

ings were taken at walking speed, every 10–50 m, on a network of traverses between 100–300 m apart depending on accessibility and local geological complexity. In a single day's work at least 500 readings can be obtained, which in most cases are adequate to cover 1–2 km² of ground. Care was taken in using the equipment since buried metal objects and wire fences markedly affect the readings, causing them to fluctuate wildly. Ground beside roads and buildings was also surveyed with caution. The size of the equipment makes its use difficult in areas with numerous hedges, fences or walls.

The meter is most effective in areas where the underlying deposits have markedly contrasting physical properties. In southeast Suffolk clay-rich deposits such as the Lowestoft Till (chalky boulder clay) are characterised by high conductivity values while adjacent sand and gravel dominated deposits give rise to low conductivity values. The results are easy to interpret in such a setting, with the magnitude of the value giving an idea of clay thickness (see below). More difficulty is experienced where lithologies show less contrast or where there are interbedded sequences; here considerable calibration may be

Plates 46–47

46. Part of geologist's draft of 1:50,000 Quaternary geology map of the Wicklow District showing limestone-dominated gravels (green with blue dots), Lower Palaeozoic shale-dominated gravel (green with purple dots), Cambrian/Precambrian sandstone-dominated gravel (green with red dots), chert-dominated gravel (green with yellow dots), Silurian-dominated till (purple), Silurian-dominated till with shell fragments and flint erratics in calcareous matrix (purple and brown stripe), Cambrian/Precambrian-dominated till with shell fragments and flint erratics in calcareous matrix (red and brown stripe), alluvium (orange), littoral and blown sand (yellow) and peat (dark brown). Also shown are striae (black arrow on circle) and glacial meltwaters channel (red arrow). Black triangle with a line under the base indicates silt/clay-dominated till; black triangle with a 'v' under the base indicates stoney till. Areas with rock at or near the surface are shown white. Surveyed by the late F. M. Synge and W. P. Warren and drawn by W. P. Warren.

47. Part of 1:10,560 field sheet 49, County Limerick (type area for the Ballylanders Moraine), showing gravels (green), limestone-dominated till (blue), Devonian sandstone-dominated till (red), Silurian-dominated till/head (purple), peat (brown), alluvium (orange), areas of thin Quaternary cover (white). Heavy colours represent bedrock outcrop. Surveyed and drawn by the late F. M. Synge, Geological Survey of Ireland.

46
———
47

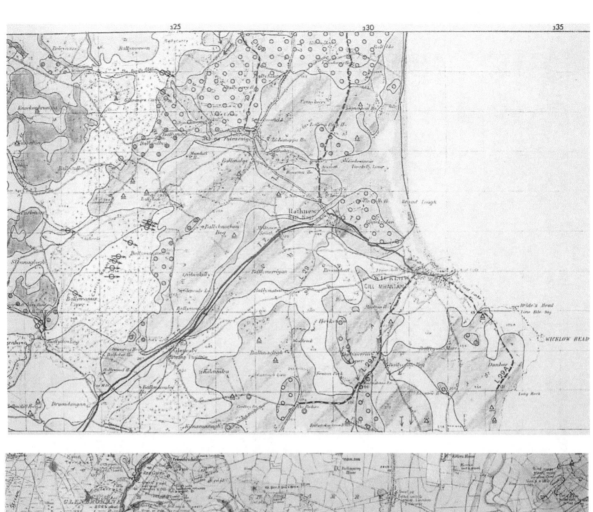

PLATE 48

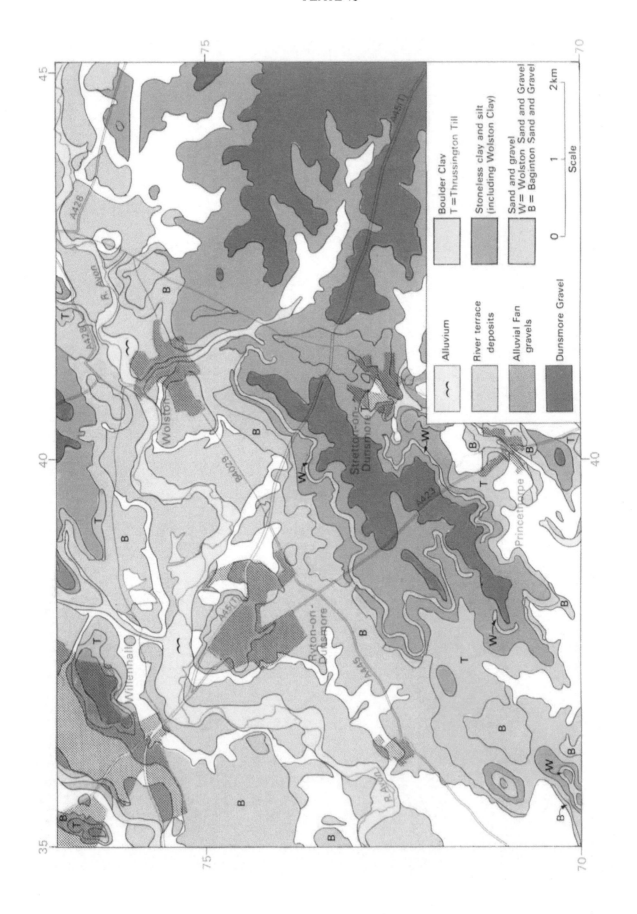

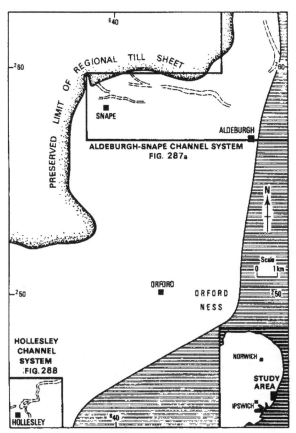

Figure 286: Location of area, and position of the Aldeburgh-Snape and Hollesley channel systems relative to the preserved edge of the regional Anglian till sheet.

needed to resolve patterns and interpret the data.

The greatest constraint on the technique is the pronounced increase in conductivity values associated with near-surface groundwater, especially where it is brackish or saline. In southeast Suffolk low-lying areas within valleys are generally excluded from contouring (see below) as the groundwater enhancement of the readings blankets any changes due to lithological contrast.

Calibration of the conductivity data, and detailed lithological investigations, were carried out using extendable augers. These augers have interchangeable cutting tools for various types of unconsolidated sediment (see Mathers & Zalasiewicz, 1984) and obtain samples that, while disturbed, still allow the recognition of bedding structures. Clearly, the

samples are unsuitable for the investigation of engineering properties, but they can be used for particle size analysis and for mineralogical and micropalaeontological studies.

The depth to which these augers operate depends largely on the ground conditions. Holes up to 10 m deep can be drilled into unconsolidated dry sands, though waterlogged sands cannot generally be drilled effectively. Stiffer materials such as the Lowestoft Till can be drilled to depths of 5–6 m although large pebbles may stop progress. Coarse gravels are very difficult to penetrate. A days work normally allows some five or six 4 m deep augerholes to be drilled, the sequences being recorded graphically as drilling proceeds.

In southeast Suffolk the use of this combination of techniques as a basis enabled effective investigations into the glacigenic sediments. Complementary methods include the examination of natural sections, where present, the excavation of trial pits (to obtain fabric measurements and details of sedimentary structures), and the drilling of deeper (shell and auger) boreholes.

Channel systems

Two main systems of channels have been detected (fig. 286). The channels described from the Aldeburgh-Snape area (Mathers & Zalasiewicz, 1986) are characterised by a consistent pattern of sedimentary infill (see summary below).

A second system of channels occurs in the area around Hollesley (fig. 286) some 16 km southwest of Aldeburgh. This system lies in a more isolated position with regard to the regional distribution of Anglian deposits (fig. 286). Furthermore, the pattern of infill differs from that observed in the channels of the Aldeburgh-Snape area. The description and interpretation of the channel system from the Hollesley area forms the main part of this account.

Aldeburgh-Snape channel system

The channel infill facies model with which the Hollesley channel sediments may be compared was

Plate 48

48. Drift deposits around Stretton on Dunsmore, part of GBS Sheet 184 (Warwick).

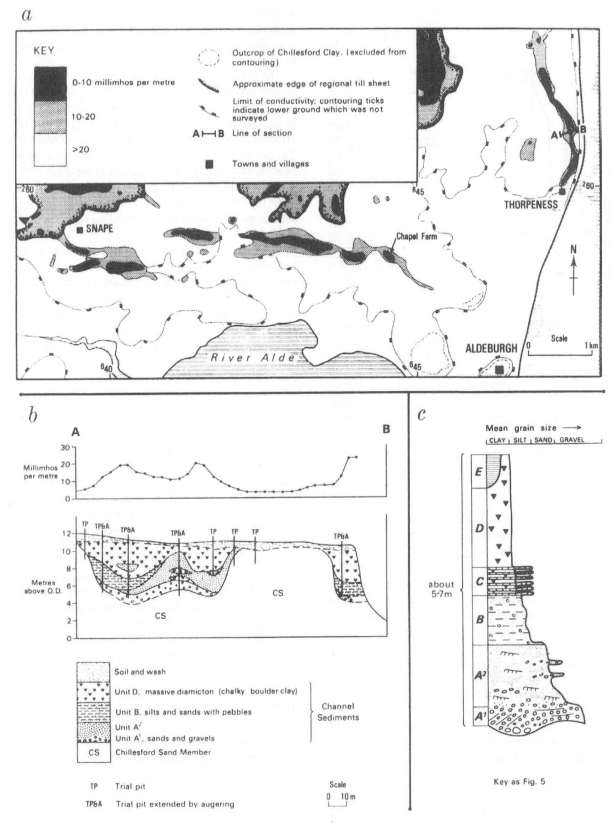

Figure 287: The Aldeburgh-Snape channel system (after Mathers & Zalasiewicz, 1986). a) conductivity contour map; b) representative channel cross-section; c) idealised facies sequence.

established in the Aldeburgh-Snape area (Mathers & Zalasiewicz, 1986; see figs. 286 and 287 herein). The conductivity contour map shown in fig. 287a demonstrates the edge of the regional till sheet in that area, and beyond that the channels, which are picked out as a series of high-conductivity zones some 100–200 m wide and up to 4 km long. These high-conductivity zones are produced by the clay-rich channel infill, which contrasts with the sandy deposits into which the channels are incised. The ground between the edge of the till sheet and the channels contains small high-conductivity patches which represent erosional remnants of an original thin sheet of glacigenic deposits.

The channels are symmetrical or asymmetrical in section (fig. 287b) with a maximum observed depth of 8 m (though locally they are likely to be deeper). A longitudinal (southwards) gradient of 1:500 was established for the easternmost, north-south aligned channel.

Channel infill was investigated along a series of traverses, including that shown in fig. 287b, and these showed a consistent pattern (fig. 287c). This comprises a fining-upwards sequence of waterlain sands and gravels which pass upwards, commonly through interlayering, into a massive diamicton (Lowestoft Till). The channels were interpreted as an ice-marginal drainage system whose linear nature suggests control by the actual ice-front. The infill represents ice-marginal drainage followed by ice advance, with subglacial till deposition plugging the remainder of the channel system (Mathers & Zalasiewicz, 1986).

Hollesley channel system – form and distribution

The ground conductivity map of the area around Hollesley, Suffolk (fig. 288) reveals the presence of several well-defined narrow tracts of high conductivity. The contoured area shown in fig. 288 generally lies between 15 and 20 m above O.D. and slopes are gentle. In the southern and eastern parts of the area the ground is lower lying. At these lower elevations, all the conductivity readings are strongly enhanced by groundwater and near-surface London Clay, and have thus been excluded from the contouring. Anglian glacial deposits have probably been removed from the lower ground by erosion.

Deep augering across the high conductivity ano-

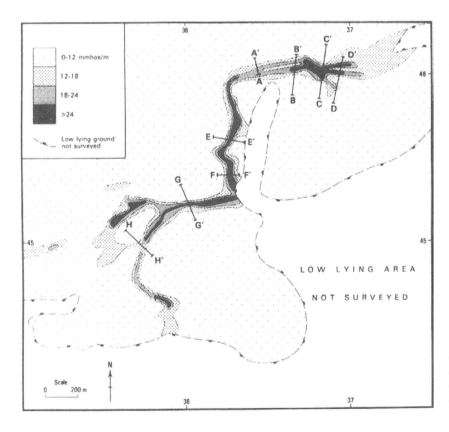

Figure 288: Conductivity contour map of the area around Hollesley, and location of detailed conductivity profiles and extendable auger traverses.

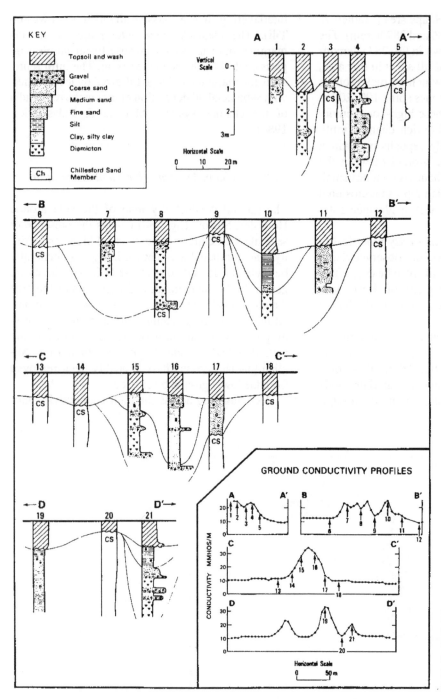

Figure 289: Lithological logs of sediments infilling the northern-most parts of the Hollesley channel system, and detailed conductivity profiles. For location of traverses see fig. 288.

malies indicates that they are produced by clay-rich sediments comprising tills, silts and clays. These deposits constitute a major part of the infilling of a series of channels (figs. 289 and 290). The presence of these had not been recognised before; earlier surveying of the area (Whitaker & Dalton, 1882) had identified the presence of chalky till at several points but had failed to appreciate the linear form of the

outcrops. Most of these channels are incised into the broadly level higher ground; however, short steeper channel segments occur running downslope to the edge of the lower-lying ground. The channel system lies some 5–7 km southeast of the preserved southern edge of the regional sheet of Lowestoft Till (chalky boulder clay) shown in fig. 286. The intervening area comprises a broad plain believed to be largely under-

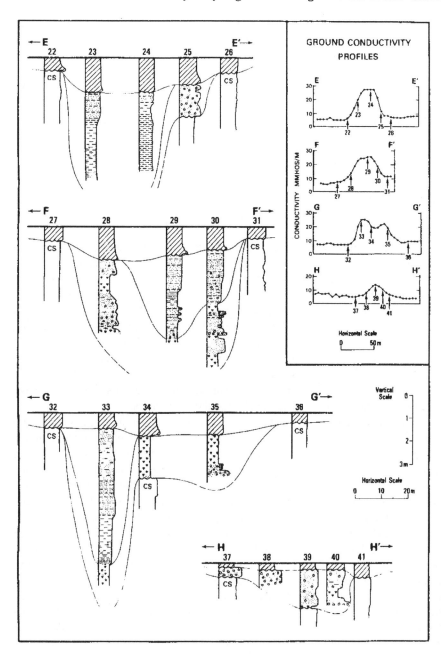

Figure 290: Lithological logs of sediments infilling the central and southern parts of the Hollesley channel system, and detailed conductivity profiles. For location of traverses see fig. 288.

lain by the Middle Pleistocene fluvial Kesgrave Formation (Rose & Allen, 1977). These deposits rest on the marine fine-grained Chillesford Sand Member of the Norwich Crag Formation *sensu* Zalasiewicz & Mathers (1985). The Kesgrave Sands and Gravels of this area may have been subjected to limited reworking by glacial meltwater; compositionally however they remain distinct from Anglian proximal outwash deposits. The systematic measurement of ground conductivity across this area indicates that high conductivity (clay-rich) deposits such as till are absent.

The distribution of the channels broadly coincides with an area where the Kesgrave Sands and Gravels are absent. The channels are incised into the Chillesford Sand Member of the Norwich Crag Formation. In plan the channels form a network of short, variably aligned, moderately straight segments (fig. 288). These segments are 50–100 m wide and up to 800 m in length. The depth of channel incision is variable, and has been proved to range from 2 m (traverse H-H[1], fig. 290) to in excess of 7.7 m (traverse G-G[1], fig. 290). The channel sides are commonly steep with their slopes exceeding 20°. The longitudinal gradient of the channel thalweg between traverse H-H[1] and G-G[1], fig. 290, exceeds 1 in 40. The topography suggests that the majority of the other channel segments have much smaller gradients, of less than 1 in 100.

Locally, the channel form is complex, in particular along the most northerly channel segment (fig. 288) which comprises two closely-spaced parallel channel furrows along much of its length (figs. 288 and 289). In addition to the well-defined channels there are two elongate anomalies of intermediate conductivity in the west of the contoured area (fig. 288). These are likely to represent portions of further channels which may be linked to other channels via segments infilled with low conductivity sands and gravels. Traverses G-G[1] and F-F[1] are located across short channel segments which descend towards the head of an incised valley (fig. 288) from which any Anglian deposits have been eroded.

Sedimentary infill

The sedimentary infill of the Hollesley channels can be subdivided into a diamicton/sand and gravel facies and a silt/clay facies. These two facies have a variable distribution, and at any one point either can constitute the entire channel infill. In augerholes where both sequences were encountered, the silt/clay facies was always found to overlie the diamicton/sand and gravel facies. The thickest developments of the silt/clay facies occur along the channel axis (figs. 289, 290).

(i) *Diamicton/sand and gravel facies.* The diamicton layers comprise a grey-black silty clay matrix in which clasts of chalk, flint, quartzite and other exotic pebbles are set. This deposit is variably consolidated and, where present close to the surface, is extensively oxidised and decalcified. Most of the diamicton appears lithologically identical to the Lowestoft Till that outcrops over much of central East Anglia (Perrin, Rose & Davies, 1979). The diamicton is irregularly intercalated with clastic sediments which are predominantly sands and gravels. These clastic sediments exhibit rapid vertical variations in grain size from coarse chalk/flint gravels to medium and fine sands; rarely, thin silt and clay layers containing sporadic pebbles are developed.

(ii) *Silt/clay facies.* These deposits comprise silty clays which, in places, are finely laminated. Rare pebbles of flint and quartz to 30 mm diameter have been observed in parts of some sequences. An overall coarsening upwards trend from silty clay to sandy silt

has been observed in most of the augerholes in which thick sequences of these deposits were encountered. Sequences near the channel margins tend to be thinner and possess a higher sand content than more axial sequences.

In addition to these two main facies, there is a persistent surficial layer of fine sand which is pebbly and clayey at its base. This is probably a windblown or solifluxed deposit, unrelated to the main phases of infill.

Interpretation and discussion

(i) *Channel incision.* The network of channel forms recognised in the Hollesley area is interpreted as a drainage system. The shape and pattern of this drainage is radically different from that of the extensive pre-Anglian Kesgrave Formation, the deposits of which represent a broad northeastwards-sloping braidplain (Rose & Allen, 1977). By contrast, the Hollesley channels comprise an incised drainage network which is largely infilled by Anglian glacigenic sediments. The channels are thus thought to have been cut by glacial meltwaters.

The reticulate pattern of the drainage network and the closely parallel nature of channel elements in the northernmost segment indicates a controlled rather than free pattern of drainage. The major controlling factor is likely to have been the distribution of ice. The isolated position of the channels, beyond the main body of preserved Anglian deposits, may indicate that they were cut close to the maximum extent of Anglian ice. Channels with a broadly similar morphology have been described from the nearby Aldeburgh-Snape area (Mathers & Zalasiewicz, 1986; figs. 286 and 287 herein) and interpreted as true ice-marginal channels on the basis of their morphology, distribution and pattern of infill. It is likely that most of the channel segments in the Hollesley area have a similar origin though the more reticulate pattern of the Hollesley channels may indicate control by structural features within the marginal ice. The presence of two closely-spaced sub-parallel channel elements along the northern channel segment perhaps represents slight shifts in the position of the ice-margin; alternatively, differential ablation of clean and dirty ice has been observed to lead to the initiation of several closely-spaced marginal meltwater channels (Schytt, 1956; Boulton, 1972).

The short steeply-graded channel segments may have acted as chutes which allowed the escape of meltwater into the adjacent lower-lying ground. Any evidence of the Anglian glacial processes operating within this lower ground has, however, been removed by subsequent erosion.

(ii) *Active phase of channel infill.* Within the diamicton/sand and gravel facies the diamicton layers represent tills and indicate the proximity of the ice during their deposition; their variable thickness and consolidation suggests formation by several processes. Massive diamicton identical to the Lowestoft Till represents direct deposition from ice whilst the thin layers of siltier and sandier till present within thick sequences of sands and gravels may have been deposited as waterlain tills or as debris flows. The rapid variations of grain size present in the associated clastic sediments indicate highly variable flow. These deposits are regarded as glaciofluvial sediments.

The interbedding of diamicton and sand and gravel shows no discernable pattern; this is in contrast to their distribution within the Aldeburgh-Snape channels (Mathers & Zalasiewicz, 1986) where they form part of an ordered sedimentary progression (figs. 287c and 291). It is envisaged that at this stage a complex series of glacial sub-environments were present within the Hollesley channel system. Blocks of ice may have fallen into the channel and stagnated, producing tills, and impeding the drainage resulting

in the variable glacigenic sequences observed.

(iii) *Abandonment phase of channel infill.* Following the first phase of infill the channels ceased to be major routes for meltwater discharge, probably because of the withdrawal of ice from the immediate vicinity. At this stage the channels comprised a series of depressions separated by masses of irregularly distributed till and sand and gravel. Thus, compared with the Aldeburgh-Snape system, no significant overriding by ice seems to have taken place following channel incision; this may indicate that the Hollesley channels were cut nearer to the Anglian ice limit. These depressions were then infilled by finer-grained deposits of the silt/clay facies (fig. 291).

There is little direct evidence of nearby ice during the later phase of infill although the sporadic pebbles may have been dropped in from floating blocks of ice. However, the abundance of fine laminated sediment indicates a periodic input, suggesting derivation from meltwaters as deglaciation proceeded. Impounding of the meltwater may have taken place either through the irregular distribution of sediments in the channels or through the presence of stagnating ice in the lower reaches of the channel system. Locally, infilling of the depressions may have continued after the regional deglaciation.

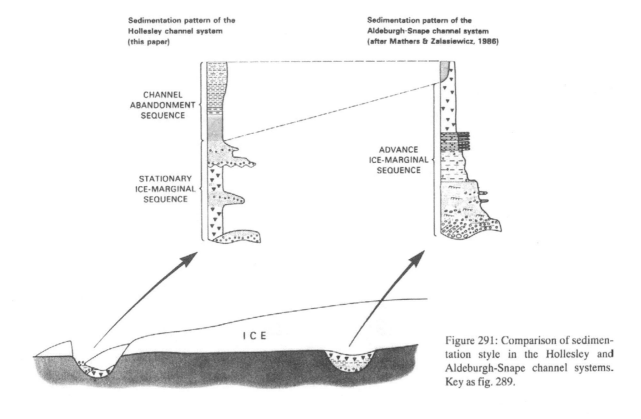

Figure 291: Comparison of sedimentation style in the Hollesley and Aldeburgh-Snape channel systems. Key as fig. 289.

Conclusions

(i) The effectiveness of using a combination of conductivity mapping and deep augering to investigate glacial terrain is demonstrated.

(ii) Small-scale channels near Hollesley, Suffolk, have a similar morphology to channels from the Aldeburgh-Snape area, but possess a different style of sedimentary infill. The two channel systems represent part of the formerly extensive Anglian ice-marginal drainage network.

(iii) The infill of the Hollesley channels is variable but represents two main phases:

(a) an 'active' phase of ice-marginal sedimentation,

(b) an 'abandonment' phase of channel sedimentation.

By comparison with the Aldeburgh-Snape system, channel incision and infill may have taken place closer to the Anglian ice limit (fig. 291).

Till lithology in Ireland

WILLIAM P. WARREN

The petrography of their phenoclasts has long been used to characterise, and distinguish between tills on the basis that it must be reflective of the geology of the terrain over which the glacier or ice sheet has passed. Perceived changes or differences in petrographic character in tills in vertical sequence have been interpreted as indicating differing direction of former ice movement (Jessen *et al.*, 1959). Contained heavy mineral suites have also been used, on the same principle, to characterise glacigenic sediments (Stevens, 1959). Texture (often total particle-size frequency) has also been used as have colour and geochemistry, particularly carbonate content (see below).

In the Geological Survey of Ireland tills are characterised on the basis of the dominant petrographic class of phenoclasts at a given size range, and the basic colour code distinguishes between them on the published maps. The subdivision is purely lithological and has no stratigraphic implications. Other characteristics are noted but often in a qualitative way and are not usually used in the definition of the till units. The characteristic Irish till might be regarded as a Carboniferous limestone-dominated (Synge, 1979d) black/brown till with about 80% matrix (< 2 mm) and up to 40% matrix carbonate (see McCabe, 1973; Hoare, 1975). Such a till can be seen right across the Irish midlands from the Atlantic to the Irish Sea. This type of characterisation is of limited stratigraphic use but is of direct practical relevance in many of the applied areas to which the Survey must address itself (Warren & Horton, this volume) and, being a record of observational data and non-interpretative, is a valuable primary document. It would seem impractical to try to characterise tills over large areas in any other way.

Although the first modern Irish map drawn on the basis of till phenoclast petrography was that of County Limerick (Synge, 1966a), tills have in a general way been characterised on a visual assessment of their phenoclast petrography, (as limestone drift or sandstone drift etc.) on the field maps and Memoirs of the Geological Survey since the middle of the last century. Farrington (1942) was the first to characterise glacigenic deposits in Ireland on a quantitative basis. This was done in an attempt to distinguish between deposits of mountain ice caps and those associated with midland icesheets on the western fringes of the Wicklow Mountains. As the area he studied is underlain by Lower Palaeozoic greywackes and slates with granite hills to the east and limestone lowlands to the west and north, the potential for lithological distinction is clear. Two suites of glaciofluvial sediments were distinguished, one with high granite percentages and low limestone and the other with high limestone, and low granite. These were interpreted as representing deposits of mountain and lowland ice respectively and formed the basis of an outline stratigraphy for the area. A similar exercise was conducted on the east side of the mountains (Farrington, 1944) and this contributed further to the stratigraphy of the area which was outlined:

5. Athdown Mountain Glaciation
4. Midland General Glaciation
3. Brittas Mountain Glaciation
2. Eastern General Glaciation
1. Enniskerry Mountain Glaciation

The observations and analyses of, and distinction between, units in both studies were sound and sufficiently clear not to warrant any complex statistical treatment of the data. However, some of the interpretations placed on them might be questioned (see below).

Since then analysis of phenoclast petrography has been used as the key element in distinguishing between glacigenic stratigraphic units (Synge, 1966a; 1970b; 1979d; Finch & Synge, 1966; McCabe, 1973; Hoare, 1975; Culleton, 1978b; Warren, 1985). It should be clear, however, that problems can arise in stratigraphic interpretation and correlation based on lithological differences between units.

Principles

Lateral and vertical facies change can occur rapidly within till units that have an otherwise obvious sedimentary unity or depositional relationship as facies association models of *inter alios* Boulton (1972) have demonstrated. Some lithological change may simply reflect the variety that might be expected, for example, from deposits of valley glaciers that originated in basins of differing bedrock geology. In ice sheets, changes in flow dynamics, flow patterns and temperature characteristics may be reflected in both vertical and lateral changes in entrained debris which will when deposited, provide similar variety in tills and associated deposits (Broster & Dreimanis, 1981). Mode of deposition and depositional environment will influence facies variation and add to the total lateral and vertical mix in glacigenic deposits. Indeed glacial deformation of subjacent and adjacent glacial and nonglacial sediments may be such that although not actually transported by nor deposited by or from glacial ice they must be regarded as an integral part of the glacigenic sedimentary suite (Warren, 1987a).

The dangers inherent in ascribing discrete vertical lithological units such as those in a till-gravel-till sequence, to separate glacial episodes are now well recognised (Boulton, 1972). But caution must also be applied when dealing with vertical till sequences based on petrographic characteristics of till phenoclasts as at Gort, County Galway (Warren, 1979a). Equally, sediments showing very similar characteristics may be interpreted, following detailed examination, to result from a variety of processes in a variety of environments. For example the shelly diamictons and associated glaciolacustrine and glaciofluvial sediments of the eastern, southern and northwestern coastal areas have at different times and in different areas been interpreted as tills relating to terrestrial ice (McCabe, 1973), tills and/or glaciomarine sediments associated with marine shelf ice (Synge, 1977a; Huddart, 1981b), shelf ice near its grounding line in an ice-marginal lake (Thomas & Summers, 1982) or land-based ice moving offshore into a marine environment (McCabe *et al.*, 1986).

Within the constraints imposed by these restrictions, broad till units can be characterised on the basis of their lithological composition. Indeed it is the only basis upon which major stratigraphic till units can be recognised and readily mapped in the field. They should initially be objectively mapped and their depositional environment interpreted as best it can, but they must finally be interpreted as tills and recorded on published maps as such on the basis of their

lithological homogeneity. The more complete the lithological characterisation the more useful is the data in any applied sense. From an academic viewpoint such characterisation can give considerable insight to the pattern and extent of glacial events, provided it is not used as a basis for attempts to age-date deposits or as a sole means of delimiting specific glacial events.

Synge (1979d) characterised all tills of Ireland on the basis of the dominant petrographic class of phenoclasts. For the most part this simply reflects the immediately subjacent bedrock. However, as might be expected there is a carry-over on to the adjacent bedrock types down-ice from the source. The degree of carry-over is neither regular nor predictable. In southeast Limerick the carry-over distance of Carboniferous limestone (as the dominant petrographic class from the limestone lowlands) on to the Devonian sandstone foothills of the Galty Mountains was found to lie between 2.4 and 4 km (Synge, 1970b). Fig. 292 shows a sudden drop in the percentage limestone count from 80% to 20% over this distance. The degree to which this observation is general for southeast Limerick is not clear, for sampling in the area underlain by Devonian sandstone was not as dense as for the area underlain by Carboniferous limestone, nor was it extended beyond the point at which a dramatic fall in limestone was noted.

Preliminary studies further north, in the Irish midlands, show a precisely similar situation, in the Devonian sandstone foothills of the Slieve Blooms. Here, limestone-dominated till and glaciofluvial gravels ex-

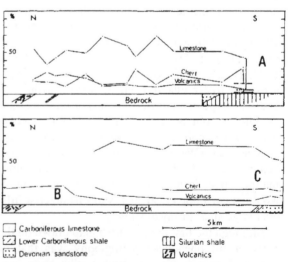

Carboniferous limestone
Lower Carboniferous shale
Devonian sandstone

Silurian shale
Volcanics

5 km

Figure 292: Limestone, chert and volcanic phenoclast proportions in tills of southeast Limerick. See fig. 294 for the lines A and B (from Synge, 1970a).

tend at least 3 km within the Devonian sandstone and Silurian shale foothills and to an altitude of about 180 m. Beyond about 3 km, however, Devonian sandstone becomes increasingly dominant until, at about 7 km within the boundary and above 300 m O.D., although chert is common in the sandstone/shale-dominated tills, fresh limestone is, in the 5.6–11.2 mm fraction, completely absent.

Along the southwest margin of the Leinster Granite in Counties Carlow and Wicklow limestone-dominated calcareous till extends as much as 21 km east of the known limestone/granite contact. Although it is possible that this situation may have been contributed to by hidden limestone outliers, it is likely that topograpic considerations are important. The limestone-dominated till extends across a rolling basin underlain by granite to the foothills of the Wicklow Mountains where the limestone content drops sharply and granite becomes dominant.

Patches of granite-dominated till occur within the limestone-dominated unit, chiefly on the Nurney ridge, a low granite-controlled ridge at the margin of the granite and limestone (fig. 293). These patches are thin and generally occur above 100 m O.D. while the broad spread of limestone-dominated till lies in

the Tullow basin below this level. To the east of this basin the 100 m contour marks approximately the eastern boundary of limestone till.

Carboniferous limestone phenoclasts

Although Carboniferous limestone is the most common petrographic component of the glacigenic deposits of Ireland, it is also the most easily weathered. It is not unusual in particular situations for the particles of all but phenoclasts of boulder size to be leached of all carbonates. While the typical brown, clayey and extremely friable residue of the phenoclasts may retain the shape of the clast (as 'ghosted limestone') in some cases, in others all that is left is a cast with some residue at the bottom. In a till from which limestone and carbonate content have been significantly weathered, depending on texture, there will be some degree of collapse of the non-carbonate till matrix. Thus it can be very difficult to estimate the former limestone content of a weathered or leached till.

In southeast Limerick, Synge (1970b) attempted to distinguish between tills of different stages (Midlandian/Munsterian) on the basis of the percentage of contained limestone phenoclasts. This was based on the assumption that the tills with a low limestone component previously had higher levels but that these had been reduced through weathering. The inferred difference in degree of weathering was interpreted as a function of age; the older tills were more weathered and therefore had a lower limestone component than the younger tills: On the same principle Finch (1971) redrew the supposed limit of last glaciation ice in south Tipperary, some 12–32 km beyond the Southern Irish End Moraine of Lewis (1894) and Charlesworth (1928a), to coincide with the southern limit of limestone-dominated till.

In both cases the proposed ice margins are closely coincident with a dramatic change in the underlying geology and associated topography. In Limerick it is at the base of the anticlinal inlier that protrudes through the Carboniferous limestone to expose Devonian sandstone and Silurian shales. Here the southern limit of fresh limestone-dominated till is broadly coincident with a belt of kame-and-kettle topography that forms part of the Southern Irish End Moraine. This was regarded by Synge (1970a) as the end moraine of last glaciation ice, but where limestone-dominated till occurs south of this feature, particularly in southwest Limerick, Synge (1966a, b; 1970a) extended the glacial limit to the margin of the

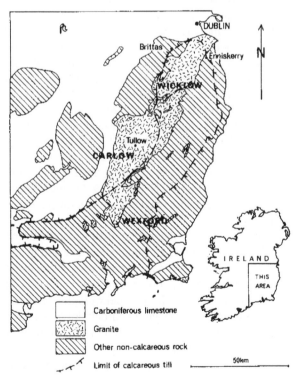

Figure 293: Till limits and bedrock geology in southeast Ireland.

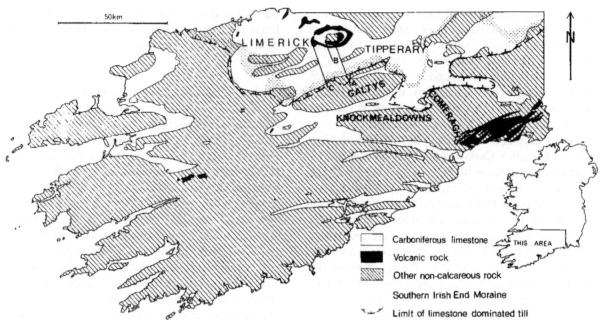

Figure 294: Supposed ice limits and bedrock geology in the South of Ireland.

limestone-dominated till. In south Tipperary the Comeragh and Knockmealdown Mountains also represent anticlinal ridges of Devonian sandstone and Silurian shale bounded on their northern fringes by Carboniferous limestone. Here the limit of limestone-dominated till, as drawn by Finch (1971), coincided with the foot of the steep escarpments that mark the northern slopes of the Comeraghs and Knockmealdowns. Thus there is a very strong coincidence between the limit of last glaciation ice as drawn by Synge (1979d) and the contact between Carboniferous limestone and the Lower Palaeozoic and Upper Carboniferous rocks of the south of Ireland (fig. 294).

Observations in County Kerry, however, raise serious questions as to the basis on which such limits are drawn. Here Warren (1978, 1985) has shown that even in areas within the most conservative and generally accepted limits of last glaciation limestone has been completely leached from till to depths in excess of 7.0 m. A number of factors were recognised as influencing the degree and extent of decalcification of glacial deposits: (a) the availability of a source of acidic or soft groundwater or surface run-off, (b) the size of the non-calcareous petrographic component in the deposit, (c) the permeability of the deposit. At Pallis near Killarney in Kerry, all three factors favoured decalcification. Surface-water drainage, and probably some groundwater movement coming off the steep escarpment of Namurian shale and

sandstone could flush through a permeable till rich in Devonian sandstone. Work carried out in the area of the Leinster Granite/Carboniferous limestone contact confirmed this process (Warren, 1979b).

It is not surprising therefore that in analogous situations in southeast Limerick and in south Tipperary, significant leaching of the limestone element in the tills should have occurred. A precisely similar situation occurs in the foothills of the Slieve Blooms where the limestone component in the till ceases to be dominant above about 180 m O.D. and about 3 km from the nearest limestone outcrop (Warren, 1987b). If the principles applied by Synge (1966a, 1987b) and Finch (1971) hold good in the Slieve Blooms then these mountains must be regarded as having remained largely unglaciated during the last glaciation. This, however, was not Synge's (1970a) view and is hardly consistent with what is known of the pattern of glaciation.

Further work is needed to distinguish between sudden decrease in till carbonate due to leaching and decrease due to the distance decay factor. Work in the Carlow/Wicklow area (Warren, 1979b) suggests an exponential increase in the petrographic element derived from the underlying bedrock down-ice from the point at which it is encountered. This reflects a corresponding decrease in the petrographic classes from outside the boundary of the underlying bedrock-type. In the case of carried-over carbonate phenoclasts the final rapid increase in a relatively acidic

Table 29. Percentage limestone, chert and sandstone content in two till samples in southeast Limerick (data from Synge, 1966b).

	'Older till'	'Younger till'
Limestone	6	44
Chert	14	30
Sandstone	58	13
Limestone/chert ratio	1:2.33	1:0.68
Chert/sandstone ratio	1:4.14	1:0.43

petrographic component will accelerate the leaching potential and provide an even stronger contrast between carbonate-dominant sediments and those almost entirely lacking in carbonate. This process is clearly illustrated by Warren (1987b: fig. 17).

In such cases, where a decrease in limestone is measured in a till which extends from an area underlain by limestone on to a relatively acidic substrate, there will be a corresponding increase in the petrographic element derived from that substrate. There will also be a corresponding decrease in petrographic types associated with the limestone. In the case of the Irish midlands chert is perhaps the most common such. As chert (SiO_2) is much more resistant to acid, chemical weathering than limestone it will remain long after the limestone has been weathered out. It will be at least as durable as Devonian sandstone in any given size range. Thus, it is the proportionate amounts of these petrographic types one to another that is important in determining whether limestone decrease is due to distance decay or weathering. If the limestone/chert ratio remains constant it is likely that the decrease is due to simple carry-over decay, but if the relative amount of chert to limestone increases dramatically it is likely that much of the limestone has been weathered out. The data in table 29, taken from Synge (1966), if consistent over a broad area, suggest both a dramatic carry-over decay *and* a very significant degree of leaching. This is illustrated by both the extreme increase in the limestone/chert ratio (from 1:0.68 to 1:2.33) and an even more extreme increase in the chert/sandstone ratio (from 1:0.43 to 1:4.14). The proportion of chert over limestone has risen by a factor of 3.4 while the proportion of sandstone over chert has risen by a factor of 9.6.

Total petrographic characteristics

Apart from characterising tills by the dominant petrographic unit, attempts have been made to finger-print tills on the basis of their total petrographic characteristics (Farrington, 1944; Finch & Synge, 1966; Synge, 1966; McCabe, 1973; Hoare, 1975; Culleton, 1978b). The assumption that till units of differing ages or glacial advances can be thus distinguished is very often unfounded (see Broster & Dreimanis, 1981). Furthermore, in the Irish literature there are no examples of objective statistical analysis of petrographic data. Indeed the application of tests of variance will in all the cases quoted above illustrate that many of the classes identified are statistically unfounded and that there is no basis for distinguishing one unit from another, much less interpreting glacial events on the basis of this distinction.

Similarly textural data has frequently been used in characterising tills (McCabe, 1973; Hoare, 1975; Culleton, 1978b). Such data has not been grouped or compared in any statistically meaningful way and total particle-size data has rarely been used or compared either graphically or statistically, nor has the probability been taken into account that till from the same sources but resulting from differing depositional mechanisms and environments, may vary texturally.

Culleton (1978b) analysed the relative phenoclast petrography and the texture of the glaciogenic deposits of south Wexford and published all of the lithological data, both textural and petrographic, upon which he distinguished between two glaciogenic formations: the Blackwater Formation of Irish Sea basin provenance and the Bannow Formation of inland, northwestern provenance. An examination of the data presented suggests that on the basis of petrography and texture there is little to substantiate the distinctions drawn between many of the members within each formation and, in many instances, between the formations. A broad distinction can be made between his formations on the basis of carbonate content and pH values. The question arises as to whether the western limit of carbonate-rich till is an adequate basis on which to distinguish between tills of two separate ice streams each of which would have flowed over calcareous substrata and in relation to which a boundary can be drawn to its calcareous element (fig. 293). The area of non-calcareous till in southwest Wexford cannot, on the basis of its relatively low calcareous content be said to relate to either of the presumed ice streams (fig. 293). Given that shell fragments and flints occur in some of the sediments ascribed to the Bannow Formation it is likely that ice from the Irish Sea basin extended further west than the line drawn by Culleton (1978b).

It would appear that in south Wexford lithological analyses were used to attempt to substantiate the existing framework of glacial events. It is a pity that the data gathered was not assessed in order to test this framework.

A critical examination of Farrington's (1942, 1944) work in north Wicklow referred to above, supports in broad terms the lithological units identified but must question the interpretation placed upon them. At Enniskerry Farrington (1944) identified seven separate lithological units. These were based primarily on relative phenoclast petrography and secondarily on observed (but not measured) texture. Thus two tills of identical petrographical composition were distinguished on degree of stoniness, and a clean delta gravel was distinguished from a morainic gravel of similar composition. Four glacial events were interpreted at this locality: two ice advances from the mountains, one from the midlands and one from the Irish Sea basin. A reassessment of this data suggests that all of the deposits, save one, might relate to a single event. The one local deposit, a till composed entirely of local petrographic types, probably relates to a local, mountain ice sheet while all of the others with varying amounts of granite and limestone and varying limestone/chert ratios may relate to ice from the Irish Sea basin. The varying amounts of specific petrographic types can be explained in terms of changing depositional environments (Synge & Warren, unpublished). Limestone/chert ratios can be explained in terms of postglacial weathering, applying the principles outlined above with reference to southeast Limerick.

Conclusions

Apart from the applied usefulness of a knowledge of the total lithological characteristics of glacial deposits, statistically valid patterns in statistically valid data can provide useful distinction between sedimentary units. Facies models cannot be constructed except on the basis of lithology and all too frequently visual field assessments of dominant grain-size or petrographic class simply reflect the most striking element in a given sedimentary unit. Thus a small number of large clasts may lead to a sedimentary unit being described as coarse or a small number of visually striking clasts may be overestimated as a percentage of the whole.

Changes in lithology, however, no matter how striking cannot be used as a basis for chronostratigraphic interpretation. Where the stratotype clearly indicates a chronostratigraphic sequence, for example an interglacial deposit separating two tills, which have been characterised lithologically on a statistically valid basis, these can then be mapped out and may be distinguished elsewhere in vertical sequence without the intervening presence of the interglacial sediments. This presumes that some form of lithological homogeneity has been identified and validated.

The distribution and stratigraphy of drumlins in Ireland

A. MARSHALL McCABE

The term drumlin (Gaelic *druim*) was introduced into glacial literature by the Rev. Maxwell Close in 1867. In Ireland this term has been used to describe a variety of subglacial bedforms which range from well-developed streamlined forms with blunt, stoss ends and tapering lee ends (tadpole forms) to smooth oval mounds (equiaxial forms) and mammillary hills. In the north and west of Ireland there is almost a continuous presence of drumlins on low ground within the so-called 'drumlin readvance moraine' (Synge, 1969) (fig. 295).

In Ireland drumlin studies have concentrated on four main themes:

1. Drumlins are used as directional indicators in the reconstruction of regional ice flow patterns (Chapman, 1970; Farrington, 1965; Finch, 1971; Finch & Synge, 1966; Finch & Walsh, 1973; Hill, 1970; 1971; 1973; Hill & Prior, 1968; McCabe, 1969b; 1985; Synge, 1968; 1969; 1970; Vernon, 1966; Wright, 1912). For example, Vernon (1966) considered that the drumlins of east County Down were formed by a dominant ice flow from a centre in the Lough Neagh basin and a lesser flow of ice from a Scottish source. The idea of competing ice masses is based on long axes variability within one continuous drumlin field and is not supported by lithofacies modelling of drumlin sediments. Vernon concluded that the County Down drumlins represent wave phenomena in the ice and are concentrated in belts perpendicular to ice pressure.

2. Measurements and statistical analysis of drumlin parameters from maps, air photographs and direct field observation have been made in an attempt to explain drumlin formation (Hill, 1970; 1971; 1973; Vernon, 1966). Hill (1973) has shown that drumlin density rises from the inner margin of the field (County Down) to a maximum at its centre and then declines towards the terminal zone. Within this swarm alternating zones of high and low drumlin density perpendicular to ice movement appear to underlie the regional density trend. Variability has

been attributed to variations in stress and basal debris load. However, in a general sense, it is important to note that morphological measurements and their statistical evaluation are a poor surrogate for the range of subglacial processes and depositional environments which operate at the base of an ice sheet.

3. Contrasts in the texture of drumlin-forming tills and the presence or absence of bedrock cores have been observed throughout the drumlin belt. For example, in east County Down drumlin sediments include a substantial component of fine-grained Irish Sea drift (Hill & Prior, 1968). In contrast the drumlins of northern and western County Down are mainly rock-cored and veneered by drift.

4. In recent years various workers have concentrated on the internal geometry and sedimentary sequences within drumlins. One of the main aims of this type of research is to examine the range of sequential subglacial processes and environments which contribute to drumlin formation (see section on lithofacies associations below).

Dating

The drumlin morphology of Ireland was created towards the end of the last cold stage (26,000–12,000 years B.P.) (Mitchell *et al.*, 1973). Estimates of the age of the maximum extent of drumlin-forming ice vary from 19–13,000 years B.P. (Stephens & McCabe, 1977) to 16–15,000 years B.P. (Synge, 1977). At present only two sites can be used to place drumlin formation in a stratigraphic and ^{14}C context.

1. The drumlin limit in north County Mayo is marked by a raised glaciomarine delta complex. Shelly muds from this complex have been ^{14}C dated at 17,000 B.P. (McCabe *et al.*, 1986). It is almost certain that the moraines bordering the drumlin belt are approximately synchronous throughout the northern part of the island (McCabe, 1985). This interpretation is based on the marked lateral continuity of

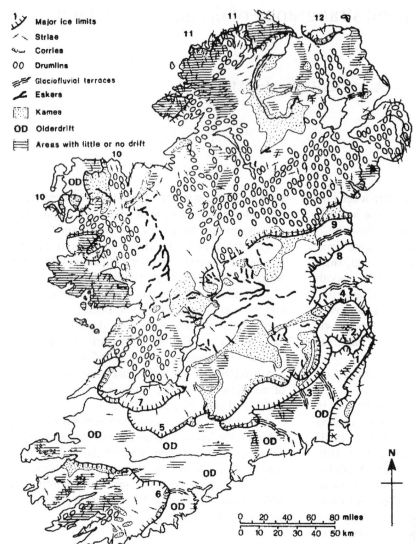

Legend:
- Major ice limits
- Striae
- Corries
- OO Drumlins
- Glaciofluvial terraces
- Eskers
- Kames
- OD Olderdrift
- Areas with little or no drift

Figure 295: Quaternary geology of Ireland (redrawn from Synge, 1979).

ice-marginal moraine/outwash systems and the fact that the drumlin ice limits acted as major barriers to the penetration inland of the late-glacial marine transgression on an isostatically depressed land surface (Stephens & McCabe, 1977; Synge, 1977; McCabe *et al.*, 1986).

2. Several drumlins in County Fermanagh are composed of two till sheets separated by an *in*

situ interstadial surface which has been dated at $30,500^{+1170}_{-1130}$ years B.P. (fig. 296) (Colhoun *et al.*, 1973; McCabe *et al.*, 1978). At all these sites only the upper till is drumlin-forming whereas the lower till is an erosional remnant.

It is therefore important to note that we can date only the maximum extent of drumlin-forming ice to c. 17,000 B.P. and not all of the remnants of earlier till

Plate 49

49. Drift geology of the upper Forth Estuary, part of BGS report 'Engineering Geology of the upper Forth Estuary'.

PLATE 49

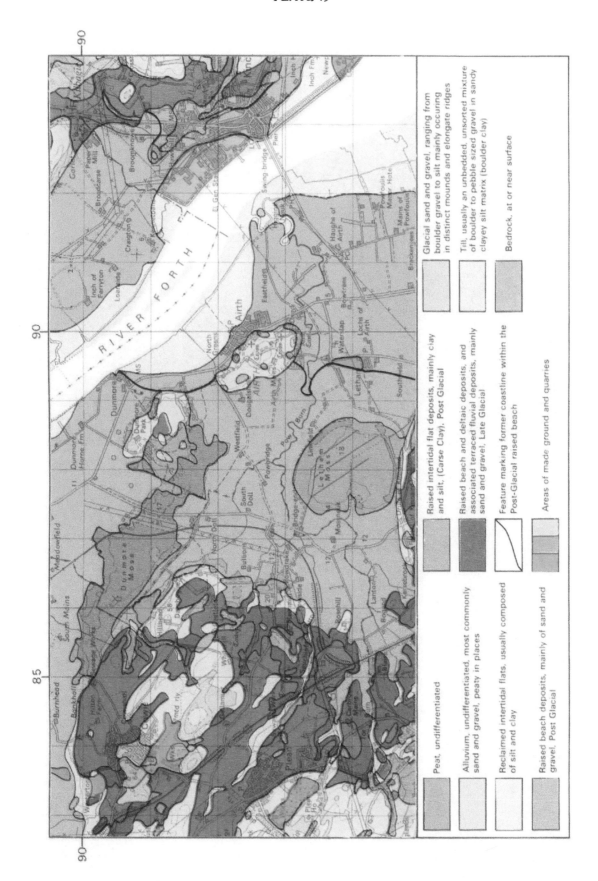

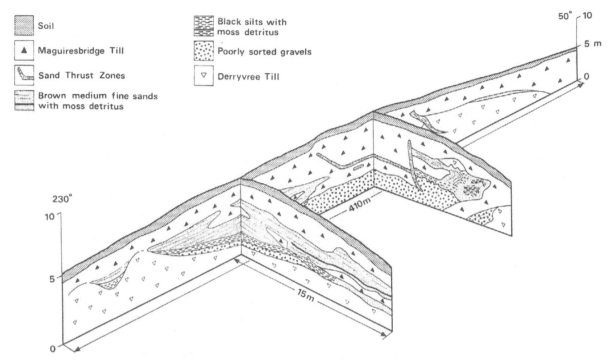

Soil

▲ Maguiresbridge Till

Sand Thrust Zones

Brown medium fine sands with moss detritus

Black silts with moss detritus

Poorly sorted gravels

▽ Derryvree Till

Figure 296: Internal geometry of the Derryvree drumlin, County Fermanagh (from McCabe, 1985).

sheets which sometimes occur in drumlin cores (fig. 296).

The drumlin ice sheet

The almost continuous swarms of drumlins on the northern and western lowlands (fig. 295) (c. 150 m O.D.) are associated with general westward and upland to lowland shifts in the main zones of ice dispersion toward the end of the last cold stage (Midlandian = Devensian or Weichselian) (Synge, 1969; McCabe, 1985). By 17,000 B.P. the main ice sheet had contracted from the South of Ireland End Moraine northwards and northwestwards to the moraine complexes

fronting the major drumlin swarms. This moraine has been termed the 'Drumlin Readvance' (Synge, 1968; 1969; 1970). Drumlins are generally absent in areas south of this morainic line which are dominated by till plains and glaciofluvial systems. There is no unequivocal stratigraphic or organic evidence to support the idea of a major readvance which is underpinned only by streamlining of the substrate and shift in axes of ice dispersion.

The limits, areas of ice dispersion and flow lines associated with this ice sheet are well-documented from extensive field mapping, notably by F. M. Synge (Chapman, 1969; Farrington, 1936; 1965; Finch, 1971; Finch & Walsh, 1973; Finch & Synge, 1966; McCabe, 1969b; Stephens & McCabe, 1977; Steph-

Plates 50–51

50. Lithofacies types present in a drumlin at Spiddal, Galway Bay. Note the cake-like layering of the sequence and the absence of glaciotectonic structures. General view of drumlin. Ice flow was towards the reader. The figure shows a basal unit of stratified diamicton at the base of the drumlin which has been channelled and is overlain by a thick unit of interbedded diamictons and silts (debris flow and sheet flow). Thick diamicton units which grade laterally into stratified units form a carapace over the form. The entire sequence is subaquatic in origin and may represent ice-proximal glaciomarine deposits overridden by ice at the head of Galway Bay.

51. Laterally continuous bed of massive mud (suspension deposit) between massive diamicton beds (rain-out deposits) in a drumlin at Spiddal, Galway Bay.

50
——
51

ens & Synge, 1966; Stephens *et al.*, 1975; Synge, 1968, 1969, 1970; Synge & Stephens, 1960; Wright, 1912). Lines of ice flows are associated with at least five major domes of ice dispersion in Donegal, Lough Neagh basin, Omagh basin, south Mayo and east Galway (fig. 295). Ice flows from these centres coalesced over much of the northern lowlands and built up fairly continuous lines of moraine at their margins. In many cases the drumlin swarms occur within 1 km of the limiting moraines but distances of 2–6 km are more common. In some cases the drumlin-forming ice has advanced marginally over the frontal outwash complexes. This situation is best seen along the margins of the large bays of the western seaboard. In typical situations rapid ice loss by calving and possibly relative fluctuations in sea

level resulted in a forward motion of the ice and drumlinisation of the subaqueous outwash. Calving bay situations like this may lead to lowering of ice sheet profiles and eventually trigger ice sheet disintegration (McCabe *et al.*, 1984).

Deglacial conditions following the 'drumlin event' were varied and depended largely on distance from centres of ice dispersion, topographic controls, ice supply and the magnitude of ice loss from a wide variety (terrestrial-subaquatic) of ice marginal environments. Three main patterns seem to emerge from these controls:

1. In south-eastern Ulster the ice wasted rapidly following drumlinisation and left little evidence of glaciofluvial activity. The absence of sand and gravel complexes and meltwater scouring may point to the former presence of major englacial drainage systems in this area. This type of regional stagnation possibly of an ice sheet with a very low profile may be linked to rapid ice-marginal wastage by calving into the Irish Sea basin (McCabe *et al.*, 1984).

2. In many areas of western Ireland areas of hummocky gravel moraine, linear gravel moraine and large esker systems occur along linear tracts between major drumlin swarms. They are rarely superimposed. The Dunmore esker system (fig. 6b in Mc-Cabe, 1985) occurs in such a situation and is, in essence, an interlobate deposit between two major drumlin swarms. Clearly it developed in a fairly thin ice mass with open segments, ice-contact delta environments and short subglacial sections.

3. Ice-contact landforms superimposed on drumlins are rare but do occur in topographic depressions near centres of ice dispersion. In these areas thick ice could be maintained, and during stagnation superimposition occurred locally. Examples from the Clogher Valley and central Ulster typify this type of landform assemblage.

In general the patterns of associated glaciofluvial landforms and lateral moraines suggest that, during drumlinisation, the ice sheet profiles were relatively flat and much of the ice had already wasted away. This contrasts markedly with the thick ice masses which existed during the main (earlier) phase of glaciation and probably implies that a marked change in glaciological conditions occurred at the ice-substrate interface at c. 17,000 B.P. If drumlinisation is effected beneath an ice sheet with a relatively flat profile, ice sheet movement and extension must be facilitated by a soft-sediment bed condition and large amounts of meltwater storage at or near the ice/substrate interface. In effect the ice sheet could be largely floating on saturated sediment and in a surg-

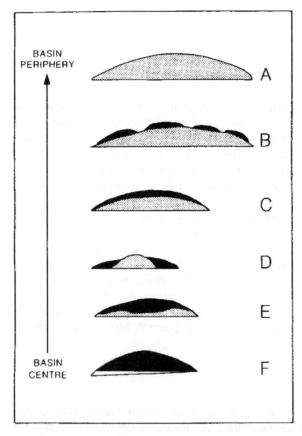

Figure 297: Spatial variation in glacial bedform geometry in central Ulster (from Dardis, 1985). (A) streamlined hill, (B) streamlined hill with superimposed drumlins, (C) rock drumlin with thin till veneer, (D) till drumlin with central *roche moutonnée* form, (E) till drumlin superimposed on minor *roche moutonnée* form, (F) till drumlin superimposed on unconsolidated tills and pre-Quaternary sediments.

ing condition. Extreme glaciological conditions of this type probably have no modern analogue and may, in part, explain the rather restricted occurrence of drumlins in glaciated terrains.

Drumlin morphology

It is difficult in many cases to relate drumlin morphology to processes of drumlinisation for a wide variety of reasons. One major reason is that drumlins *in statu nascendi* are often small and their internal structure is largely unknown. Field evidence from Ireland indicates that a range of forms exist which can in some instances be linked with sediment geometry, inferred ice-sheet history or basal conditions at the ice-substrate interface. Some general conclusions are apparent:

1. Drumlin bedforms are an element within a complex continuum of streamlined forms composed of either rock or drift in any combination.

2. Drumlins vary greatly in length, breadth and height but tend to attain similar dimensions in any given area.

3. Spatial variations in drumlin/bedform geometries vary from centres of ice dispersion towards the margins of depositional basins. For example, Dardis (1985) has shown that bedform geometry varies systematically in central Ulster, with a decrease in magnitude and an increase in the amount of lodgement/meltout till components within drumlins from the periphery to the centre of the drumlin field in the Lough Neagh basin (fig. 297).

4. The classical notion that drumlins have a steep stoss side and gently elongated lee slopes is not the general case. These forms are best developed in areas of low drumlin density, on low ground and near ice-marginal positions which ended in water bodies. They tend not to be well-developed near centres of ice dispersion and are traditionally believed to be associated with active ice streaming. Other common drumlin forms are torpedo-, barkhanoid-, spindle- and parabolic-shaped. In the majority of cases these forms contain significant stratified sequences on their lee-side flanks. They tend to occur in linear zones transverse to ice flow (Dardis *et al.*, 1984).

5. Certain types of spindle- and parabolic-shaped drumlins are large-scale migratory bedforms as evidenced from proximal to distal sediment transformations (Dardis *et al.*, 1984).

6. Drumlin bedforms occur in close juxtaposition with a wide range of subglacial forms. These include hummocky till deposits and large till ridge complexes

transverse to ice flow. The latter are similar to rogen moraines though, in zones of high drumlin density, where drumlin bedforms coalesce and are superimposed a similar effect occurs. This situation will occur where linear groups of drumlins coalesce giving rise to a landform complex similar to drumlinised rogen moraine.

Morphological diversity in any bedform assemblage is to be expected. Often it is difficult to account for variability in form because of the complexity of subglacial systems and the lack of good exposure. The former involves drift availability and type, basal velocity variations, pore-water pressures and time-dependent thermal contrasts. In any case the final drumlin construct is an end product of time-dependent systems which do not fully reflect the conditions responsible for till agglomeration, drumlin genesis or geotechnical change. For example, the origin of large *en échelon* rows of drumlins with well-defined proximal hooks near Clew Bay remains unexplained (Hanvey, personal communication).

Lithofacies associations

The diversity of drumlin structure is not in doubt but their genesis is. This problem prompted detailed facies analysis of drumlin sequences in an attempt to provide more accurate information on the processes and subglacial environments which contribute to drumlin formation (Dardis & McCabe, 1983; Dardis *et al.*, 1984; Dardis, 1985; McCabe, 1985). Stratigraphic complexity in drumlins indicates that they cannot be viewed from a simple geotechnical viewpoint alone or from classical modes of erosion/deposition but in terms of dynamic subglacial systems and a range of sub-environments. At least six main facies associations are now known from the drumlins of Ireland. Each association reflects a distinct event or depositional environment identified from drumlin stratigraphy. It is recognised that not all facies associations occur throughout the drumlin belt.

Facies Association 1 (Forms with a core of older-dated till). The drumlins around Maguiresbridge, Co. Fermanagh contain cores (fig. 296) of till (pre-dates 30,500 years B.P.) which are thought to be early Midlandian in age (McCabe & Hirons, 1986). It is possible that in these cases elements of the older drift landscape acted as obstacles which influenced local patterns of till lodgement and local streamlining (fig. 296).

Facies Association 2 (Overridden ice-marginal

Figure 298: Lithofacies types present in a drumlin at Spiddal, Galway Bay. Mud (suspension sedimentation) interbedded with thick diamicton units (rain-out). Note presence of boulder (dropstone?) in the muds.

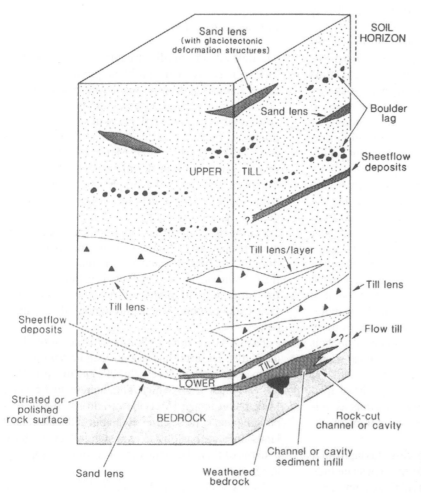

Figure 299: Major components of the subglacial lodgement/meltout till facies association in drumlins in central Ulster (from Dardis, 1985).

Figure 300: Glacial morphology of the Poyntz Pass channel. The location of the Jerretts Pass. Glebe Hill and Tandragee drumlins are indicated by X, Y and Z respectively (from Dardis & McCabe, 1983).

subaqueous facies). Many drumlins which occur along the margins and heads of major coastal embayments of Western Ireland are composed of cake-like, tabular and lensate units of interbedded diamictons, sands, gravels, muds and diamictic muds (plates 50, 51 and fig. 298). Lithofacies continua are a common element of the stratigraphic architecture and indicate that the sequences formed by a combination of subaqueous processes which include rain-out, suspen-

sion sedimentation, debris-flow activity and traction-current activity.

The sequences are not disturbed by glaciotectonics and are similar sedimentologically to the ice-proximal subaquatic outwash sequences which border the drumlin belt of eastern Ulster (McCabe *et al.*, 1984; 1987; McCabe, 1986a). The evidence at present suggests that the ice-marginal outwash has been streamlined by a westerly movement of drumlin ice into Clew Bay and Galway Bay. These advances are probably local in context and may be associated with calving-bay situations with high rates of ice wastage which triggered off a marginal response in an attempt to restore equilibrium. Alternatively, a slight lowering of sea level would initiate a similar response. This type of ice-marginal activity does not occur in eastern parts of the main drumlin belt and may also be a function of distance from centres of ice dispersion.

Facies Association 3 (Subglacial lodgement/ meltout). The majority of drumlins examined show subtle vertical changes in till texture and lithology but little variation in texture between stratigraphically similar till units (Dardis, 1982; 1985). In most cases the lower till units are rich in fines and overlie a variety of glacially eroded strata. The upper tills are generally sandy in texture. Typical sedimentary (fig. 299) sequences bear no obvious relationship to drumlin form. Some sequences look like banded or foliated tills which reflect processes of lodgement, meltout, debris flowage and sheet flowage forming successive waves of buried till micro- and macro-surfaces. Dardis (1985) has also shown that the thickness of lodgement/meltout till units increases from the periphery to the centre of the depositional basin in Ulster (fig. 297).

This association commonly contains ancillary inter-till sand layers/lenses which are interbedded with stratified or massive diamicton units (Dardis & McCabe, 1987). Where stratification is present, units dip down-ice at very low angles and do not show evidence of glacially induced deformation structures. Sand beds are thought to be a result of sheet flood events under conditions of ephemeral subglacial meltwater flow. Interbedded, poorly-sorted or graded diamictons formed by resedimentation of subglacially-deposited material (cohesive debris flows). Interbedded sequences of this type probably occurred during oscillating phases of high and low meltwater discharge. In many cases it is extremely difficult to identify cohesive debris-flow units which have undergone minimal sorting or transportation from *in situ* basal-till units in restricted exposures and where

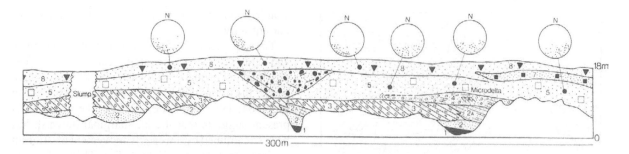

Figure 301: Sedimentary facies of the Jerretts Pass drumlin. (1) bottomset beds, (2) topset beds, (3) foreset beds, (4) topset beds, (5) lower till facies, (6) sediment gravity flow facies, (7) massive diamicton facies, (8) upper diamicton facies (from Dardis & McCabe, 1983).

A

facies transitions are absent.

Facies Association 4 (Subglacial channel sedimentation). Three main styles of subglacial channel sequences have been identified from the cores of drumlins which formed in or near to large-scale, subglacial meltwater systems (fig. 300) (Dardis & McCabe, 1983). The different sedimentary facies represent deposition in an hierarchial flow system related to the size of the channel and magnitude of subglacial meltwater activity.

1. Drumlins with the largest stratified cores occur along the axes (fig. 300) of large tunnel valleys (e.g. Jerrets Pass, fig. 301). The 'sand core' consists of a bottomset (rhythmytes), foreset (cross-stratified, avalanche front gravels) and topsets (horizontally-bedded gravels) which is typical of prograded Gilbert-type deltas (Gilbert, 1885). However, the

B

Figure 302: Drumlin sediments. (A) Derryvree drumlin, County Fermanagh. The dark unit of laminated silt and sand contains organic detritus which has been dated as $30,500^{+1170}_{-1130}$ years B.P. This interstadial surface separates upper (Maguiresbridge) and lower (Derryvree) regional till sheets. (B) Jerretts Pass drumlin, County Armagh. A stratified core is composed of steeply-dipping (away from view) planar gravel foresets overlain by a thick till unit which forms a continuous carapace over the drumlin.

C

D

Figure 302 (continued): (C) Stratified diamicton from a drumlin on Clare Island, Clew Bay. Note the presence of continuous beds of parallel-laminated sand interbedded with diamicton units. (D) Lee-side stratified gravels from Mullantur drumlin, County Armagh.

poorly-developed distribution of topsets suggests that the sand core originated as a single bedform or megadune which migrated down-channel (cf. McDonald & Vincent, 1972). The overlying streamlined till carapace seems to be associated with basal meltout with no major hiatus or amalgamated surface separating it from the sand core.

2. The cores of drumlins on the flanks of large tunnel valleys are sedimentologically complex (Dardis & McCabe, 1983). Basically they consist of basal till on ice-moulded bedrock overlain by a wide range of stratified sediments (debris-flow deposits, grain-flow deposits, current-bedded deposits, rhythmytes, turbidites). The overlying till carapace is a probably meltout deposit and lithostratigraphically similar to the upper tills at Jerrets Pass.

3. Drumlins located along minor tributaries of the meltwater system generally have stratified cores of fine-grained rhythmytes (turbidites/suspension deposits) interbedded with tabular units of stratified diamictons (debris flow) (fig. 302).

The stratigraphic position (fig. 303) of the sand

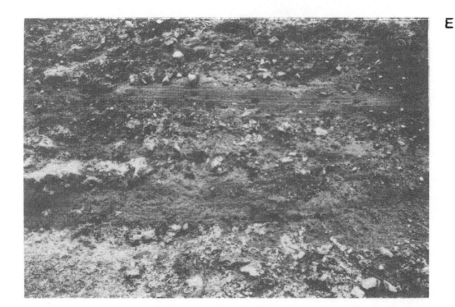

E

F

Figure 302 (continued): (E) Inter-bedded rhythmytes (parallel-laminated silts and sands) and debris-flow units (diamictons), Glebe Hill drumlin, County Armagh. This complex forms the drumlin core and was deposited in a minor channel, tributary to the main Poyntz Pass system. (F) Thinly-bedded silts and sands which commonly form the end member of lee-side stratification sequences, Derrylard drumlin, County Armagh. Sequences of this type reflect ephemeral sheet-flow conditions.

cores indicates that they formed prior to regional meltout of basal debris and drumlin-streamlining events. It is likely that regional meltout was initiated by the blocking of subglacial channel networks causing a change from arterial to sheetflow at the base of the ice mass (Dardis, 1985).

Facies Association 5 (Lee-side sedimentation). Lee-side stratification sequences are now recognised to be a common feature from widely-spaced areas of the Irish Drumlin belt (fig. 304). In Ulster, detailed sedimentological analyses of lee-side sequences show well-defined proximal to distal sediment transformations from massive till to stratified diamictons interbedded with a wide range of sand and gravel lithofacies (Hanvey, 1987; see figs. 304, 305, 306, 307 and 308). A depositional model summarises the range of mass-flow processes which are thought to occur during lee-side cavity sedimentation (fig. 309).

The presence of lee-side sequences of this type require pre-existing till nuclei to facilitate the formation of lee-side cavities (in ice or till, see figs. 304 and

G

H

Figure 302 (continued): (G) Bouldery diamicton formed by debris flow on the ice-distal side (between horns) of a barkhanoid drumlin, Buncrana, County Donegal. (H) Stratified diamicton (debris flows) interbedded with parallel-laminated sands (sheet flow), Buncrana drumlin, County Donegal. This is a transitional facies between ice-proximal massive till and ice-distal stratified units.

309). The fact that matrix-rich diamictons are inter-bedded with the lee-side sequences from the leading point of drumlins indicates that cavity infilling was probably contemporaneous with streamlining. This condition is well illustrated by the Buncrana drumlin-which is barkhanoid in style with interbedded dia-mictons and sands infilling the lee zone between the horns (figs. 302 G, H and 307). The glaciological implications of lee-side cavity development have been discussed fully by Dardis *et al.* (1984), who consider that they increase towards a point of super-

cavitation (Lliboutry, 1979) with ensuing catastro-phic glacier slip and drumlin streamlining.

Facies Association 6 (Remobilised/Superimposed till facies). Dardis (1982, 1985) has shown that drum-lins in certain areas display a slightly hummocky topography as a result of superimposition of supra-glacial sediments. The latter are generally coarse-grained and local in origin. They occur only in zones subject to wholesale stagnation (west-central Ulster) and are absent near active-ice regions (east-central Ulster). Post-drumlinisation slope processes are also

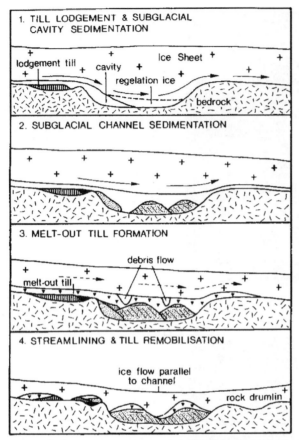

Figure 303: Simplified depositional model of facies arrangement in the sand-cored drumlins within and adjacent to the Poyntz Pass glacial and drainage channel (from Dardis & McCabe, 1983).

common with major resedimentation of tills down drumlin flanks especially where the drumlins were submerged in proglacial lakes.

Discussion and conclusions

Although the examples of facies associations and inferred palaeoenvironments are mainly from Ulster, field observations indicate that similar features and events can be identified from widely-spaced sites within the drumlin belt of Ireland. Facies analysis indicates that most drift drumlins are largely subglacial erosional bedforms created by active ice during the last deglacial phase. Dardis (1985) has clearly demonstrated that neither subglacial deformation, squeezing of detritus into basal ice cavities or basal deposition was directly responsible for drumlinisation.

Sedimentary sequences within drumlins suggest a sequential series of events:

1. Erosion of pre-existing till sheets and bedrock.

2. Localised lodgement and subglacial channel sedimentation in Nye-type channels.

3. Widespread subglacial deposition (lodgement/ meltout) which is associated with a reduction in hydraulic transmissibility at the ice-substrate interface when arterial drainage systems were blocked by sediment.

4. Sheetflow and lee-side sequences occurred when localised ponding of water occurred immedi-

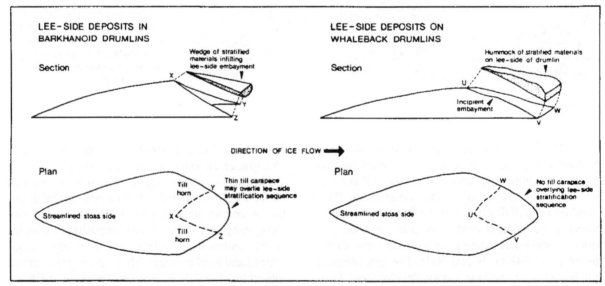

Figure 304: Stratigraphic position of lee-side stratification sequences in barkhanoid and whaleback drumlin forms (from Dardis *et al.*, 1984).

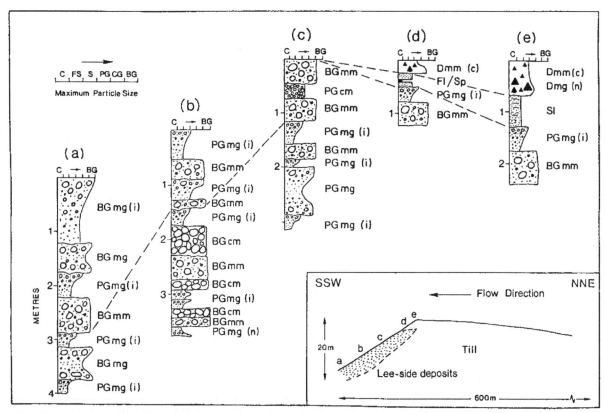

Figure 305: Proximal-distal lithofacies relationships in the Mullantur drumlin. (F) mud, (FS) fine sand, (S) medium and coarse sand, (PG) pebbly gravel, (CG) cobble gravel, (BG) boulder gravel.

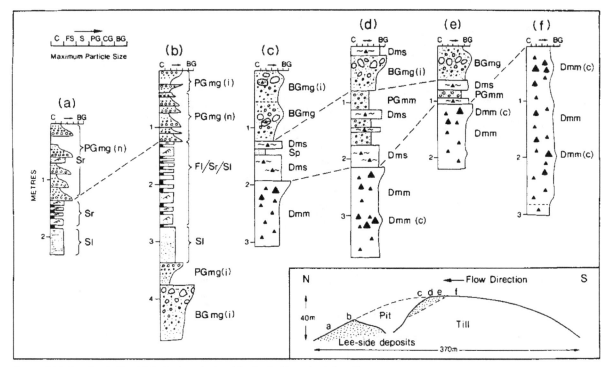

Figure 306: Proximal-distal lithofacies relationships in the Derrylard drumlin.

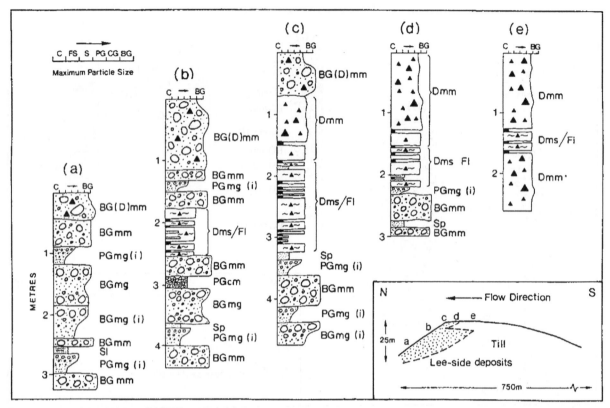

Figure 307: Proximal-distal lithofacies relationships in the Buncrana drumlin. Facies 'BG(D)mm' is intermediate in composition between 'BGmm' and 'Dmm' (from Dardis *et al.*, 1984).

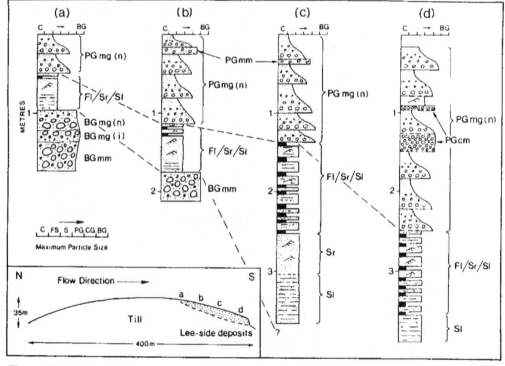

Figure 308: Proximal-distal relationships in the Derrinraw drumlin (from Dardis *et al.*, 1984).

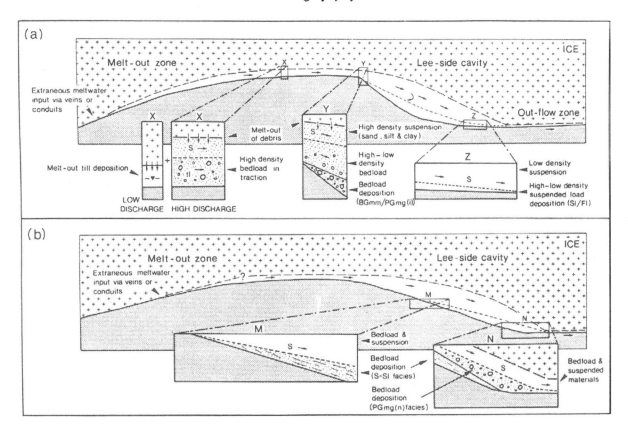

Figure 309: Depositional palaeoenvironments of lee-side stratification sequences associated with (a) barkhanoid drumlins, showing inferred hydrodynamic conditions, (b) whaleback drumlins.

ately prior to and during drumlinisation.

5. High basal meltwater pressures may have effected a total uncoupling leading to uplift of the ice sheet (cf. Röthlisberger & Iken, 1981). Uplift would probably initiate a surge-type flow during which streamlining occurred. Dardis (1985) has pointed out that under these conditions (ponded water, high meltwater pressures, high ice velocity) of very low basal shear stresses, drumlins have a high potential for preservation. In a surging glacier these conditions occur near the 'trigger zone' (Robin & Weertman, 1973) which separates a cold downglacier zone from a temperate upstream zone. If streamlining occurs near a trigger zone during uncoupling similar drumlin forms will be created in linear zones transverse to ice flow. In the field similar drumlin forms and densities tend to occur in zones transverse to ice flow and may be associated with this general mechanism. It could be argued that inward migration of successive trigger zones would create a large drumlin field.

Facies analysis of drumlin sequences help to explain the internal geometry and multiple 'till' units found within the drumlins of Ireland. In most cases their sequence context indicates a complex suite of depositional events rather than a series of ice oscillations. Where this is the case multiple units within drumlins should not be used for stratigraphic subdivision/correlation or changes in flow patterns (cf. Chapman, 1970; Colhoun, 1971a; Fowler & Robbie, 1961).

The synopsis indicates that drumlins are best examined using a facies-analysis approach. The results to date imply that rather special and sequential depositional and glaciological events occurred prior to and during drumlin streamlining. It is highly likely that these conditions are not fully recognised in modern analogues and seem to have only occurred over very restricted areas covered by Late Pleistocene ice.

Sedimentology of glaciofluvial deposits

IAN D. BRYANT

An oft neglected topic in the description of glacial sequences has been the detailed subdivision and interpretation of glaciofluvial outwash bodies. Until relatively recently the sedimentology of such sediments was poorly understood, however, in the past fifteen years or so considerable advances have been made in the interpretation of the complex sedimentary associations which may be deposited in such environments. The purpose of this article is to:

a) briefly review key sedimentological studies of modern and ancient glaciofluvial systems, b) review the application of these studies to Quaternary sequences in Britain and Ireland and, c) indicate the potential for further applications of such an approach.

Deposition in glaciofluvial environments

In recent years considerable progress has been made in the development of models of glaciofluvial sedimentation (compare reviews of Church & Gilbert, 1975; Miall, 1983 and Smith, 1985). The formulation of such models necessarily involves study of both modern environments (table 30) and ancient sequences (table 31).

Glacial river systems are characterised by fluctuating discharge at a variety of time scales as a consequence of:

a) long term changes in ice volume or proximity of the ice front, b) seasonal changes in rates of glacial ablation and, c) daily/hourly changes in the amount of precipitation and ice melt.

In addition many systems are subject to floods associated with the drainage of ice-dammed lakes (*jökulhlaups*) recurring on an annual or longer time-scale. Most glacial rivers are also characterised by high but variable rates of sediment supply due to release of material from the glacier itself (Hammer & Smith, 1983; Mosley, 1988) or by erosion of weakly consolidated glacial deposits in the catchment (Church & Ryder, 1972). As a consequence of this variation in discharge and sediment supply the deposits of glacial river systems are extremely variable both spatially and in vertical sequences. Generally, most glaciofluvial systems show a proximal to distal gradation from coarse to fine-grained sediments. This trend is usually accompanied by a proximal to distal transition from deposition in gravelly, multi-channel braided streams, through sandy braided streams to fine-grained single-channel systems (table 32). This simplified model should, however, be regarded with some caution, for whilst it is representative of many of the better-documented modern glaciofluvial systems these are located principally in alpine or sub-polar regions. In these areas downstream increases in bank stability associated with the establishment of a vegetation cover may play an important role in controlling channel morphology. In high latitudes (and possibly during maximum stadial conditions) such an effect is not apparent. Furthermore simple downstream grain-size trends may be complicated by deposition of both coarse and fine materials in proximal settings at the ice margin.

Proximal glaciofluvial sediments are strongly influenced by glacial ice and may occur as:

a) 'eskers' deposited by supra-, sub- or englacial river systems and subsequently left as linear ridges following ice-melt, b) less regular 'kames' deposited as irregular mounds following ice-melt, c) ice-front deltas, d) ice-front outwash fans.

Fluvial sediments in these settings may be complexly interbedded with till, mass-flow deposits, lacustrine and/or aeolian sediments (Fraser & Cobb, 1982). Sediments deposited in glaciolacustrine deltas may closely resemble those deposited by sub-aerial channel systems (Rust & Romanelli, 1975). These relationships may be further complicated by the melt-out of buried glacier ice (Shaw, 1972a and b; Allen, 1982) or naled ice (Cegla & Kozarski, 1977) leading to extensive deformation of the overlying sediments.

Ice-front outwash fans are characterised by rapidly

Table 30. Key studies of modern glaciofluvial environments.

Type of fluvial system	Area	Reference
Outwash fans / sandar	Scott fan, Alaska	Boothroyd 1972
	Scott and Yana fans, Alaska	Boothroyd & Ashley 1975
	Alaska and Iceland	Boothroyd & Nummedal 1978
	Baffin Island, Canada	Church 1972
	Malaspina foreland, Alaska	Gustavson 1974
	Iceland	Krigstrom 1962
	Peyto fan, Canada	McDonald & Banerjee 1971
Outwash rivers / valley sandar	Markarfljot and Landmannalauger sandar, Iceland	Bluck 1974, 1979, 1982
	Knik and Matanuska rivers, Alaska	Bradley *et al.* 1972
	Adventelva, Svalbard	Bryant 1983a
	Okstindelva, Norway	Cheetham 1979
	Baffin Island, Canada	Church & Gilbert 1975
	Tanaelva, Norway	Collinson 1970
	White River, USA	Fahnestock 1963
	Slims River, Alaska	Fahnestock & Bradley 1973
	Hilda Creek, Canada	Hammer & Smith 1983
	Kicking Horse River, Canada	Hein & Walker 1977
	Bossons River, France	Maizels 1979
	West Greenland	Maizels 1983
	Donjek, Canada	Rust 1972, 1975
	Kicking Horse River, Canada	Smith 1974
	Hilda foreland, Canada	Southard *et al.* 1984
	Donjek, Canada	Williams & Rust 1969

Table 31. Key studies of ancient glaciofluvial sequences, excluding Britain and Ireland.

Type of fluvial system	Area	Reference
Ice contact / delta top	Soesterberg, Netherlands	Augustinus & Riezebos 1971
	N. Sjælland, Denmark	Clemmensen & Houmark-Nielsen 1981
	Chicago, USA	Fraser & Cobb 1982
	E. Jylland, Denmark	Houmark-Nielsen 1983
	Netherlands	Ruegg 1977
	Funen, Denmark	Schwan & van Loon 1979
Glacio-lacustrine deltas	Haapajarri, Finland	Aario 1972
	Massachusetts, USA	Ashley 1975
	Massachusetts, USA	Gustavson *et al.* 1975
	Massachusetts, USA	Jopling & Walker 1975
	Newfoundland, Canada	Leckie & McCann 1982
	Veluwe, Netherlands	Postma *et al.* 1983
	Netherlands	Ruegg 1983
	Alberta, Canada	Shaw 1975
Glaciofluvial	Bihar, India	Casshyap & Tewan 1982
	British Columbia, Canada	Clague 1975
	Ontario, Canada	Costello & Walker 1972
	Ontario, Canada	Eynon & Walker 1974
Eskers	Ontario and Quebec, Canada	Banerjee & McDonald 1975
	Manitoba, Canada	Ringrose 1982
	Ontario, Canada	Saunderson 1975, 1977

Table 32. Summary vertical facies models for glaciofluvial systems (modified from Miall 1984). For illustration of facies types, see figure 311.

Vertical sequence type (and main reference source)	Description	Main facies	Minor facies
Scott type (Boothroyd 1972, Boothroyd & Ashley 1975)	Proximal, gravel-dominated rivers	Massive gravels	Planar and trough cross-bedded gravels; planar and trough bedded, ripple laminated sands, laminated and massive muds
Donjek type (Williams & Rust 1969, Rust 1972)	Medial, gravel-sand rivers	Massive and trough bedded gravels; trough bedded sands	Planar cross-bedded gravel; horizontal, ripple and planar cross-bedded sands; laminated and massive muds
Platte type (Smith 1970, 1971, 1972, Blodgett & Stanley 1980)	Distal, sand-dominated rivers	Trough and planar cross-bedded sands	Horizontal, ripple laminated and scour-fill sands; massive gravel; laminated and massive muds
Slims type (Fahnestock 1969, Rust 1978)	Distal	Laminated and massive muds	

shifting, braided, low-sinuousity channels which deposit massive to crudely horizontally stratified gravel sheets as longitudinal bars (Boothroyd, 1972; Gustavson, 1974). Further downstream some of these braid bars may become stabilised giving rise to 'tiered' outwash plains. These 'tiers' or 'levels' have been described from the Donjek (Williams & Rust, 1969) and other glacial rivers (Bryant, 1983b) as comprising:

a) a lower level occupied by the main channels and bars that is exposed only at low water, b) an intermediate level of bars active only at flood stage and, c) a higher level which receives only fine-grained sediment at maximum flood stage.

The sediments comprising the bars in this part of the river system are more structured than their upstream equivalents due to differential sedimentation on different parts of the bar (fig. 310). In their lower part they comprise trough-bedded or massive gravels, overlain by sheets of imbricated gravels, planar cross-bedded gravels and trough or planar cross-bedded and ripple-laminated sands (Bluck, 1979).

Yet further downstream these flow-elongate bars may be replaced by flow-transverse bars deposited by shallow sandy channel systems which merge to form sheets of flowing water at high discharges. Sediments deposited in these systems are dominated by flat and ripple-laminated sands together with planar cross-bedded sands.

Miall (1983) presented a summary model whereby vertical facies models (Miall, 1977; 1978; Rust, 1978) may be used to represent proximal to distal changes in glaciofluvial sedimentary bodies (table 32). In practice these ideal vertical sequence types (based largely on extrapolation from studies of modern rivers) are rarely found in ancient successions (Fraser, 1982; Bryant, 1983a and b; Thomas *et al.*, 1985; Dawson & Bryant, 1987). Several processes act to ensure that these ideal cycles are rarely preserved:

1. Often large areas of glaciofluvial systems are inundated at only the highest of river discharges. During the intervening long periods of exposure these areas are subject to deflation and/or alteration by pedogenic or thermal contraction processes. Consequently stratigraphic breaks in glaciofluvial successions are marked by deflation lags, soil horizons or thermal contraction features such as ice-wedge casts.

2. Commonly aggradational cycles are truncated by fluvial scour as a consequence of fluctuations in discharge and sediment load. Such alternation of aggradation and incision is typical of glaciofluvial channel systems (e.g. Maizels, 1983a).

3. Confinement of channel systems by glacier ice (Shaw, 1972a; Cheel & Rust, 1982) or till ridges (Thomas *et al.*, 1985) may restrict channel movement such that adjacent facies belts are unable to migrate laterally to form conformable vertical sequences.

To summarise, glaciofluvial systems comprise a

Figure 310: Intermediate glaciofluvial sedimentation in the Adventelva, Svalbard. A lateral bar viewed from upstream showing the spatial organisation of sedimentation in the channel. Coarse to medium gravels are deposited in the main channel (MC) and as imbricated sheets on the bar head (BH); medium gravel to fine sand and silt are deposited as planar cross-bedded and ripple-laminated units in the slough channel (SC) and at the bar tail (BT). The area to the left of the photograph represents a higher topographic level which receives only fine-grained sediment deposited at highest river stage (Width of slough channel (SC) 1 - 2 m).

complex of depositional subenvironments giving rise to rapid and often abrupt lateral variation. Studies of modern depositional environments have yielded facies models which attempt to describe this variation. These models have been used to improve our understanding of ancient successions with varying degrees of success (see below). However, the models require critical application due to the frequent disruption of idealised relationships in such a complex depositional setting.

Sedimentology of glaciofluvial deposits in Britain and Ireland

Early descriptions of Quaternary glacial sequences in Britain tended to make the broad distinction between glacial deposits *sensu stricto* and 'outwash', thereby masking the extreme variability in the character of these sediments.

The first detailed analyses of sedimentary facies within glaciofluvial sequences in Britain date from the early 1970's. On the basis of detailed logging of quarry exposures and surface resistivity surveys Helm (1971) demonstrated that deposits on Anglesey, Wales, formerly described as eskers on morpho-

logical evidence, were in fact the result of deposition in lacustrine, fluvial and deltaic sub-environments. Shaw (1972a and b) described Devensian ice-contact sediments in Shropshire. Through detailed analyses of quarry exposures and boreholes he was able to recognise vertically-stacked fluvial successions resulting from stream training by ice walls. Similar stream training but associated with till ridges rather than ice is indicated by a study of ice-contact sediments from the Isle of Man (Thomas *et al.*, 1985).

The frequent difficulty of distinguishing glaciofluvial and glaciolacustrine sediments is illustrated by two contrasting interpretations of a Devensian sequence exposed at Banc-y-Warren, Wales. This sequence was interpreted to result from cyclic progradation of a delta lobe into a deep glacial lake (Helm & Roberts, 1975). However, re-examination of the evidence suggests that the sequence was initially deposited as supraglacial outwash and was subsequently deformed by meltout of the underlying glacial ice. Gravel foresets suggest that any associated lake was neither large nor deep (Allen, 1982). Gravelly top-set deposits of glaciofluvial origin are also described in association with a glacial lake from County Wicklow, Ireland (Cohen, 1979) and from

Figure 311: Lithofacies from coarse-grained glaciofluvial sediments. (a) Trough cross-bedded gravels: Palaeoflow toward the viewer. Ranging pole graduated at 0.2 m intervals. (b) Planar cross-bedded gravels overlain by massive gravels. Palaeoflow from right to left. Trowel 0.25 m long. (c) Massive gravels with some well-imbricated horizons. Palaeoflow from left to right. Lens cap 0.06 m diameter. (d) Trough cross-bedded sands overlying massive gravels and overlain by trough cross-bedded gravels. Palaeoflow toward the viewer. Spade handle 0.1 m wide. (e) Planar cross- bedded gravels (right) passing into planar cross-bedded sands (left) and overlying massive gravels. Palaeoflow in planar cross- bedded unit from right to left. Ranging pole graduated at 0.2 m intervals. (f) Massive gravels with interbeds of flat-laminated sands separated from overlying planar cross-bedded gravels (top of photograph) by a planar erosion surface. Note that the fabric in the gravel immediately beneath the erosion surface has been disturbed by frost processes. Palaeoflow from left to right. Spade handle 0.1 m wide.

subglacial channel deposits (Dardis & McCabe, 1983). Many other Devensian sequences also show evidence of deposition in a complex lateral association of glaciolacustrine, glaciofluvial and glaciodeltaic sub-environments similar to that seen in many modern proglacial areas. Thomas *et al.* (1982) and Thomas (1984b, 1985a) report evidence for the co-existence of these sub-environments from studies of sections in Wales and the Welsh borderland. Likewise, the composite nature of fluvial sediments underlying the Main Terrace of the River Severn, Shropshire has been demonstrated by detailed analyses of quarry sections (Dawson, 1985; Dawson & Bryant, 1987).

Detailed sedimentological investigations of pre-Devensian glaciofluvial sequences are less common. This may be partly explained by the less frequent preservation of these older deposits and is partly a consequence of the difficulty of distinguishing sedimentary structures in often more intensely weathered sediments. A notable exception are fining-upward, coarse-grained channel-fills which have been described from Suffolk. They are interpreted as being in part infilled subglacially on the basis of the interdigitation of diamicton in the upper part of the fills (Mathers & Zalasiewicz, 1986).

These studies highlight several important points:

1. The successions all show complex lithological variation which may often be only adequately assessed by repeated examination of retreating quarry faces or combined analysis of exposures with boreholes and/or surface electrical surveys (see Mathers *et al.*, this volume).

2. Contrary to interpretations based purely on morphological criteria, sedimentological analyses often indicate deposition in a number of complexly interrelated sub-environments.

3. Directional properties measured within the different lithological units often have the potential to indicate the orientation of the sediment bodies themselves.

Further application of sedimentological studies

The above examples serve to illustrate the way in which detailed sedimentological analyses may pro-vide information on the various environments of deposition represented by sedimentary facies present in glaciofluvial sequences. In addition to the intrinsic value of these studies to sedimentologists and palaeo-geographers these studies also have important implications for stratigraphers and to the efficient exploitation of glaciofluvial sediments as an economic resource.

The common assumption that individual morphological elements of the landscape are underlain by sediments laid down in a single depositional environment has often been demonstrated to be incorrect (e.g. Thomas, 1985a). Careful analysis of the sequence of sedimentary structures and delineation of the major surfaces which separate conformable sequences provides a means to subdivide composite sequences into their component parts (e.g. Dawson & Bryant, 1987). Similarly recognition of soil horizons or separate generations of ice-wedge casts (cf. Bryant, 1983a; Seddon & Holyoak, 1985) may also be used to subdivide complex depositional sequences. In this way datable organic remains may be ascribed to their correct stratigraphic order, which may not always be strictly depth-related.

Glaciofluvial sediments represent a major source of land-won aggregate and yet sedimentological knowledge is very rarely used in prospecting for, or developing, such reserves. Simple interpretations of palaeocurrent data measured from quarry sections or application of basic facies models have great potential for predicting likely sediment geometries and could greatly reduce the need for costly trial-hole drilling programmes (Martin & Lovell, 1981).

Conclusions

Glaciofluvial sediments accumulate in a complex variety of sub-environments, however, detailed facies analyses of outcrops, supplemented by borehole and other subsurface data have the potential to interpret these sedimentary sequences in a meaningful way. Several recent studies demonstrate the viability of this approach but considerable scope still exists for the application of these techniques, particularly in the fields of detailed stratigraphy and resource development.

Glaciofluvial landforms

J. MURRAY GRAY

This paper describes the types of, and relationships between, depositional landforms produced by former meltwater streams in Britain and Ireland. Little is said about glaciofluvial sediments (which are considered by Bryant, this volume), and erosional landforms are considered only where relevant to descriptions of depositional landform assemblages. The paper is necessarily a personal view in which examples are drawn mainly from the author's limited knowledge of the vast array of such landforms in Britain and Ireland. Thus many fine examples known to readers will be missing from this paper.

The traditional classification of glaciofluvial landforms, and the one followed here, is into two main groups according to environment of deposition. First there are the ice-contact landforms produced sub-, en-, or supraglacially or ice-marginally, and secondly there are the proglacial features deposited once the meltwater streams have left the ice. In this paper individual landform types are described first, following which spatial relationships between landforms (landform assemblages or 'landsystems') are considered. Finally, the glaciofluvial landscapes that evolve through time are discussed.

Ice-contact landforms

First of all it is worth stressing that, as in many fields of geomorphology, there is often an understandable but misleading over-emphasis on textbook examples of named landforms rather than on the typical landscapes that exist in the field. Thus just as it is hardly surprising that students have the impression that all periglacial tundra landscapes are dominated by pingos, ice-wedge polygons and sorted stripes when these features dominate the textbooks, so the general nature of ice-contact landscapes is rarely considered. The author suspects that the large majority of ice-contact glaciofluvial sediments form rather indeterminate, undulating areas which are difficult to map,

describe or sub-classify. This is hardly surprising given the complexity of ice-contact glaciofluvial environments (e.g. Shaw, 1972a). However, in certain localities well-defined landforms are found:

1. *Kettle holes*. These are not confined to glaciofluvial landscapes but they are common features of such terrain. By definition they are ice-contact landforms being the hollows where ice blocks melted out. Many are the sites of present-day or infilled lakes (fig. 312), though the well-drained nature of glaciofluvial sediments means that they are frequently dry. Sizes vary from a few metres to a kilometre or more in diameter, and some are over 20 m deep, for example the 'unusually large' hollows in the kame terraces at Nannerch in the Wheeler Valley, North Wales (Embleton, 1970: 74). However, often the most impressive are the small, steep-sided ones. A good example occurs by the roadside at Inverlair in Glen Spean, western Scotland which Sissons (1977d: 8) described as 'one of the most pronounced kettle holes in Scotland', while many impressive hollows occur along the front edges of kame terraces at Loch Etive, western Scotland (see below).

2. *Kames*. These are the end product of sedimentation in traps along meltwater drainage routes through glaciers. As Sugden & John (1976: 318-319) pointed out there are various situations in which fluvial sediments will be deposited on, in or under the ice. For example, where meltwater plunges into a moulin a sediment trap will occur at the base of the moulin. Similarly, lakes and pools, uphill sections of tunnels, or debris-rich ice bands crossed by a stream, are all potential sites of glaciofluvial sedimentation. When the ice melts these deposits are lowered onto the ground surface to become mounds of glaciofluvial sediment, termed kames. They may be conical, flat-topped, elongated or irregular and can occur singly or in small or large groups. In Britain and Ireland most are under 20 m high. A fine example of a conical kame occurs by the A 702 road at Dolphinton, south of Edinburgh (fig. 313), while many flat-topped

443

Figure 312: Kettle hole at Ford near Lochgilphead, Argyll, partially infilled by lake sediments that are over 7 m deep. Kame terraces are also illustrated.

Figure 313: The Dolphinton Kame, near Edinburgh, with summit memorial.

Figure 314: Section in the Heacham kame, Norfolk.

kames occur in the Eddleston Valley, also south of Edinburgh (Sissons, 1958b) and in the Kelvin Valley, north of Glasgow (J. Rose, personal communication). An elongated example, 400 m long, 60 m wide, 4 m high, occurs at Heacham in Norfolk, and an exposed section (fig. 314) in it has been interpreted as indicating that the sediments were deposited in a supraglacial channel in which melting and subsidence were active during sedimentation (Boulton, 1977a).

3. *Eskers*. Eskers are in effect a special type of elongated kame formed by rivers flowing along their lengths. They are typically sinuous in detail and have undulating crests. In Britain few are over 20 m high and it is rare to be able to trace an individual esker ridge for more than a few kilometres. Even the longer examples are often dissected into a series of shorter sections (fig. 315). Many of the most impressive eskers in the British Isles are in Ireland. For example, at Trim, northwest of Dublin an esker extends for 14.5 km and for the last 8 km it is over 15 m high (Synge, 1950). Interestingly, some of the more famous Scottish esker groups are locally known as 'kames', for example the Carstairs Kame near Lanark and the Kildrummie Kame near Nairn. Boulton (1972) reinterpreted the former as an 'inverted supraglacial mould', but the site is part of a meltwater drainage system that extends across central Scotland over undulating terrain, and a subglacial interpretation is much more likely (Sutherland, 1984a). Elsewhere in the British Isles well-known examples include the swarm of eskers in lower Glen Feshie (Young, 1975), the systems near Tullamore west of Dublin (Farrington & Synge, 1970) and the Hunstan-ton and Blakeney eskers in Norfolk. Recent work on the last points to deposition under full-pipe flow conditions and confirms its englacial or subglacial origin (Gale & Hoare, 1986; Gray, 1988). A special type of esker is the so-called 'subglacially engorged esker'. This is related to rivers flowing into and under a glacier, for example down a valley side, rather than by the free flow of meltwater rivers towards the ice margin which is the normal situation.

4. *Kame terraces*. These are deposited in the linear depression that often develops between the lateral margin of an ice mass and the adjacent hill slope (fig. 312). Most examples are glaciofluvial, but some are glaciolacustrine as the result of, for example, ice-damming at a valley constriction. They are found along the margins of ice-sheets and ice-caps, but in the case of valley glaciers, the terraces, where they occur, are normally found along both valley sides. Textbook examples of valley kame terrace systems occur at Lochs Etive and Etteridge in the Scottish Highlands, while kame terraces marginal to ice sheets occur in East Lothian (Sissons, 1958a), and in south Lincolnshire (Straw, 1979a).

Proglacial landforms

Since typical kame terraces are formed between ice and ice-free slopes they are in effect transitional between ice-contact and proglacial. However, because of their ice-marginal location they are normally classified as ice-contact. Thus the word 'proglacial' is normally used in the sense of 'beyond the glacier *snout*'.

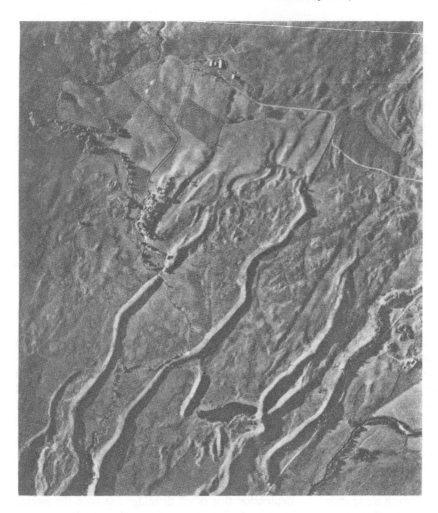

Figure 315: Vertical air photograph of eskers at Dell Farm, near
Nairn (Crown Copyright - RAF
photograph, 1946).

1. *Outwash fans.* These are deposited immediately beyond the snout by rivers escaping from the ice. Thus they are commonly bounded up-ice by an ice-contact slope and down-ice they grade away steeply

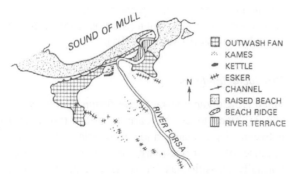

Figure 316: Map of the Glen Forsa outwash fans and related features.

but with decreasing gradient down-valley. Good examples occur at Glen Forsa on the Isle of Mull (fig. 316), and at Salthouse in North Norfolk (Sparks & West, 1964).

2. *Outwash terraces.* In presently glacierised areas outwash fans grade down-valley into outwash trains or *sandar* with their characteristic braided meltwater systems. Outwash terraces are in effect the remains of dissected outwash trains, a good example being the terrace at Homersfield in the Waveney Valley, North Suffolk (Coxon, 1984). Some of the so-called 'river terraces' found along major river valleys in England (e.g. Severn and Trent) are probably outwash rather than river terraces though further work is needed to distinguish these.

Spatial sequences of landforms

As has been hinted above, frequently different types

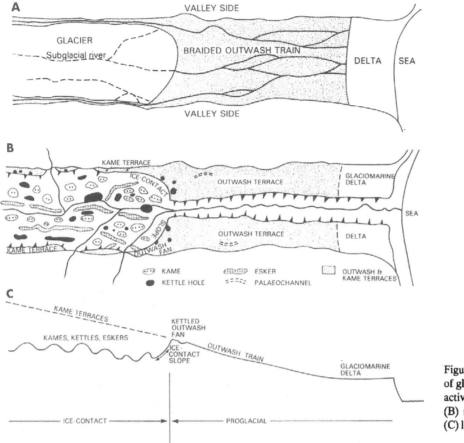

Figure 317: Full spatial sequence of glaciofluvial landforms. (A) an active glaciofluvial environment, (B) sometime after deglaciation, (C) long profile of B.

of glaciofluvial landforms are closely related and it is often possible to identify coherent sequences of landforms extending from ice-contact to proglacial positions and beyond. In modern terminology we can identify a glaciofluvial landsystem whose morphological components have systematic relationships to underlying sediments and structures. Fig. 317 is a map showing the full range of landform associations with the related long profile below. It will be appreciated that it is rare to find the complete assemblage in a single locality, but one area where such examples do occur is the east coast of Scotland between the Firth of Forth and the Moray Firth. Here Sissons & Smith (1965), and Cullingford & Smith (1980), for example, have mapped and levelled landform sequences (see fig. 318) that extend from ice-contact topography giving way down-valley to steeply-sloping outwash terraces which, upon reaching the coast, level off at glaciomarine deltas.

Though the full sequence is relatively rare, parts of it demonstrating important landform relationships, are much more common. For example, the rather

inappropriately named 'delta-kame' or 'delta-moraine' is in effect a drastically foreshortened version of the above full sequence in which the outwash component is missing. They form where a former glacier snout terminated in relatively shallow water with the result that the ice-contact slope is succeeded immediately by a glaciomarine or glaciolacustrine delta. Good examples occur southwest of Dublin where meltwater rivers were discharging from the ice directly into glacial Lake Blessington (Synge, 1977e; Cohen, 1979), while in Britain such deltas occur, for example, at Ebchester in Northumberland (Allen & Rose, 1986) and at Achnasheen in central Ross-shire (Sissons, 1982a).

The term 'kame-and-kettle topography' describes one of the most widespread types of glaciofluvial landscape. Famous examples occur in the Nith valley, north of Dumfries (Stone, 1959), around Blackford between Stirling and Perth and at various other localities along the northern and southern margins of the Midland Valley of Scotland. Eskers and ice-marginal kame terraces are frequently also found in kame-and-

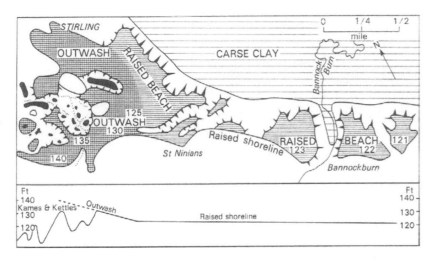

Figure 318: Example of the full sequence near Stirling (from Sissons,
1967).

kettle areas and together constitute an ice-contact
glaciofluvial landform assemblage, for example
around the Glaven valley in North Norfolk (Sparks &
West, 1964). Meltwater channels too are commonly
associated with these landscapes. At Tinto near Lanark, for example, Sissons (1961a) mapped esker
fragments separated by stretches of meltwater channels, thus demonstrating that the glaciofluvial river
reponsible was erosional along part of its course and
depositional in others.

Kettle holes are very common on kame terraces,
particularly along their front edges which were built
out onto the glacier margins. Good examples occur
on the Loch Etive kame terraces (fig. 323) where
some are so heavily kettled that they are almost
unrecognisable as terraces except for the altitudinal
accordance of the inter-kettle areas (Gray, 1975).
Impressive examples of kettled kame terraces also
occur in the Wheeler Valley, North Wales (Embleton,
1970; Brown & Cooke, 1977). A very important
point in the recognition of kame terraces is that
kames, kettles and eskers, including subglacially engorged examples, are often found downslope of the
kame terraces (see fig. 325 below).

The transition from ice-contact to proglacial deposition is illustrated at Glen Forsa, Isle of Mull (fig.
316) where a short esker winds northwards over the
valley floor, climbs an ice-contact slope and suddenly
terminates at the point where the outwash fan begins.
Similarly the eskers at Trim are clearly related to an
ice-sheet margin represented by the Galtrim Moraine
(Synge, 1950).

The most common associations of proglacial landforms include the frequent presence of kettle holes on
outwash fans and the proximal parts of outwash trains

(e.g. Blairgowrie; Sissons, 1967a), and the common
occurrence of palaeochannels on outwash terrace
surfaces. Good examples of the latter occur on the
terraces of the North Esk near Edzell (Maizels,
1983b). Thus here, as in all the above cases, there is a
close association between types of glaciofluvial landform that has a logical explanation in terms of the
former glacial/proglacial depositional environment.

Temporal sequences of landforms

The relationships described so far have attempted to
avoid reference to the fact that changes in the depositional environment occur with time, yet the glaciofluvial landscapes of Britain and Ireland inevitably have been affected by such temporal factors.
This section attempts to introduce this vital complication. As in the previous section we shall begin with
the complete sequence and then deal with examples
of individual parts of this sequence.

In the case of temporal changes, during deglaciation areas that are originally ice-covered eventually
become ice-free. The significance of this for glaciofluvial landforms is that ice-contact deposition
changes to proglacial deposition over time, and
clearly there will be an intermediate stage in which
proglacial rivers will flow between blocks of dead ice
with their associated ice-contact processes. An excellent example of this temporal sequence was
described by Young (1974) in the Glenmore basin
between Aviemore and the Cairngorm Mountains
(fig. 319). He divided the sequence of ice wastage
into three phases:

a) gradual thinning, b) widespread wastage and

Figure 319: Glaciofluvial landforms in the Glenmore basin, near Aviemore (from Young, 1974).

stagnation, c) total stagnation and final wastage.

During the *gradual thinning* phase, the ice sheet was hundreds of metres thick and the drainage system at this stage was essentially ice-directed (Sugden, 1970), i.e. the meltwater channels bear no relationship to the contemporary drainage, but instead slope gently from west to east as ice-marginal or submarginal channels (e.g. 1, fig. 319). However, "as the physical condition of the ice deteriorated the drainage became increasingly through and under the ice, controlled more in its direction by the location of the lowest escape routes available for meltwaters out of Glenmore" (Young, 1974: 154). The phase of *widespread wastage and stagnation* had now been reached and the drainage pattern had switched from ice-directed to topography-directed. There are distinct changes in the trends of eskers and meltwater channels reflecting the opening of successively lower

cols as drainage routes out of the Glenmore basin. For example around 2, the channels were clearly carrying meltwaters towards and through the Pass of Ryvoan, whereas the esker and channel patterns around 3 indicate subsequent northward drainage towards An Slugan, a col c. 75 m lower than the Pass of Ryvoan. Eventually the lowest col, that carrying the present River Spey, became penetrable by meltwater and the eskers and channels switched direction towards it (e.g. 4, fig. 319). At 5 a kame terrace demonstrates the restricted nature of the ice by this stage. The phase of *total stagnation and final wastage* was now reached. Kames, kettles and eskers were formed on the lower hill slopes and valley floors as the stagnating ice became restricted to these areas. Eventually, parts of the valley floors became ice-free and much previously deposited sediment was redistributed by the now proglacial rivers to form extensive outwash spreads along the valley floors. Areas of ice-contact features were preserved above the valley floors and in the lee of hill masses, but the valley floors themselves are dominated by outwash. Two exceptions occur where 'islands' of kame-and-kettle are surrounded by outwash, the meltwater rivers having cut steep bluffs around them (6 and 7, fig. 319). Subsequent changes in meltwater flow led to dissection of the main outwash train and the production of successively lower outwash terraces that have been further dissected by the Holocene River Spey. This is perhaps the best example of a full deglaciation history, but other examples have been studied in Glen Feshie (Young, 1975), and in north Northumberland (Clapperton, 1971a), the latter illustrating the common development of ice-dammed lakes during deglaciation.

In some areas the part of the deglacial sequence involving final ice stagnation has been studied in detail. A good example is the Eddleston Valley, south of Edinburgh where Sissons (1958b) reconstructed the stages of final ice decay. He explained the flat-topped kames in terms of an englacial water table extending through the decaying ice and controlling the upper level of glaciofluvial deposition. Worsley (1970: 97), in referring to the deglaciation of the area south of Shrewsbury, provides an interpretation that includes the "detachment of masses of ice from the active glacier as it retreated, followed by slow wast-

age *in situ* and the aggradation of fluvial materials around them."

The dissection of outwash spreads is very common and in valleys results in fragmentary remains of many different outwash train levels stretching down-valley. In some cases, levelling of the terrace fragments has enabled reconstruction of the changing long profiles of meltwater rivers. For example, between Kilninver and Loch Scammadale south of Oban a sequence of dissected outwash terraces (fig. 321) is related to a rapidly falling relative sea-level due to strong glacio-isostatic uplift at the time (Gray & Sutherland, 1977). The upper series of terrace fragments originates at c. 65 m O.D. at Lagganbeg and slopes down-valley at a gradient of c. 6 m/km to Kilninver where the terraces level off at c. 41 m O.D. for a distance of over 500 m (fig. 320). This is interpreted as indicating outwash deposition related to a relative sea-level at 41 m at which level a glaciomarine delta was formed (fig. 317 above). The ice-front subsequently retreated 1.5 km to Loch Scammadale where a small group of kames, kettles, eskers and kame terraces is succeeded down-valley by outwash terraces that are clearly related to a level well below the upper terrace. It is suggested that while the ice-front retreated by 1.5 km, relative sea-level fell by at least 12 m (S4 to T53, fig. 320) and probably by over 20 m. A similar sequence is identifiable a few kilometres farther

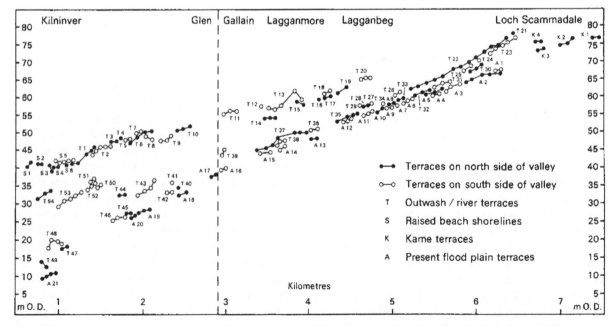

Figure 320: Terrace long profiles between Loch Scammadale and Kilninver, near Oban (from Gray & Sutherland, 1977).

Figure 321: Outwash terraces at Kilninver.

south at Kilmartin, and again provides a good example of the divergent down-valley long profiles that result from meltwater flow during a period of falling relative sea-level (Gray & Sutherland, 1977).

Outwash terraces related in the main to falling lake rather than sea-levels were described from the North Esk Valley south of Edinburgh by Kirby (1969b). During deglaciation the ice sheet in the area split, one section retreating southwards to the Southern Uplands, the other retreating northwards towards the Firth of Forth. Meltwater from the former filled the floor of the North Esk valley with outwash but only as far as the latter ice margin which impounded the meltwater in an ice-dammed lake. The lake level was controlled by an outlet col through the ridge to the east, and as the damming ice retreated to the north, successively lower cols were revealed. The end product is a series of dissected outwash terraces that extend progressively farther down valley (fig. 322).

Successively lower kame terraces are also formed during deglaciation. For example, at Loch Etive levelling of terrace fragments (fig. 323) along both north and south sides of the loch revealed that two main terrace series occur on both sides with traces of lower series (fig. 324). These were formed as the stagnating glacier gradually shrank towards the centre of the valley and were probably produced during dissection of the Achnacree outwash plain along the line now occupied by the narrow loch entrance channel. Interestingly, the terrace gradients are steeper on the north side of the loch (5 - 6 m/km) than on the

south side (3.5 - 4.5 m/km) (Gray, 1975). In East Lothian, Sissons (1958a) described a series of four kame terrace levels between 320 and 244 m formed during downwasting of the ice margin.

Some points of interpretation

Finally, two important points concerning glaciofluvial landform interpretation may be made in the light of some of the above descriptions. First, since kame terraces can resemble outwash terraces (or indeed river terraces) when occurring in dissected form along the sides of a valley, it is worth briefly discussing possible ways in which they may be distinguished. The crucial difference is that outwash terraces (and river terraces) have been carved out of a complete valley floor infill, whereas kame terraces were deposited on the valley sides when the central strip of the valley contained a glacier. Thus:

a) as explained above, kame terraces often have other ice-contact glaciofluvial landforms situated downslope of them. This is not possible with dissected outwash (or river) terraces (fig. 325),

b) kame terraces are often kettled along their front margins and are higher here than at the valley side slopes. Outwash terraces (unlike river terraces) may be kettled but generally not preferentially at their front edges, and they tend to be highest at their lateral margins (fig. 325),

c) outwash terraces (and river terraces) are frequently altitudinally closely matched across the val-

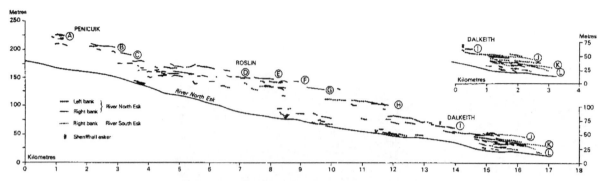

Figure 322: Long profiles of outwash terraces of the Midlothian Esk (from Kirby, 1969).

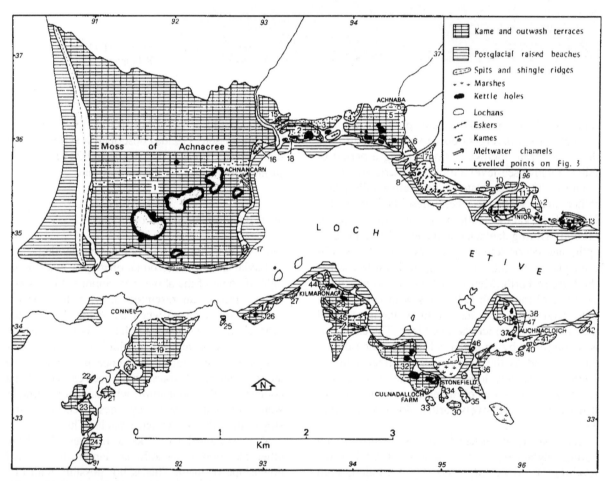

Figure 323: Map of the Loch Etive kame terrace system, near Oban (from Gray, 1975).

ley, whereas kame terraces are often not so matched since each valley side operates as a largely independent hydrological system (see the Loch Etive example described above), and

    d) kame terraces may merge down-stream into

delta terraces formed when the ice-marginal rivers entered temporary ice-dammed lakes.

    There is a growing tendency towards the idea that kame terraces are commonly found in glaciated valleys (for example Fookes *et al.*, 1978; Eyles, 1983a),

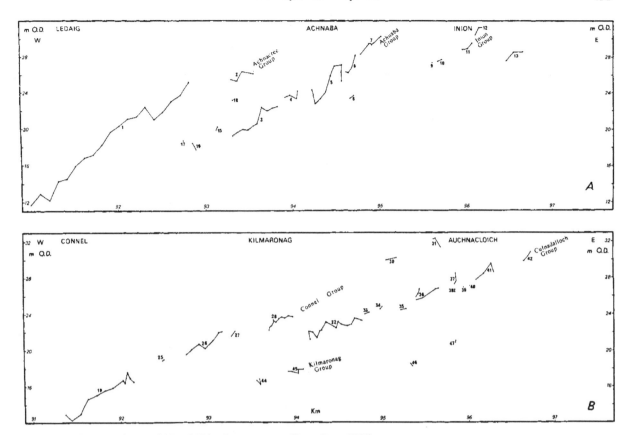

Figure 324: Long profiles of the Loch Etive kame terraces (from Gray, 1975).

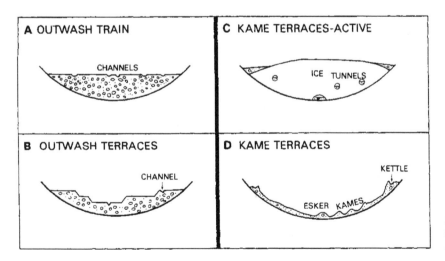

A OUTWASH TRAIN	C KAME TERRACES-ACTIVE
CHANNELS	ICE TUNNELS
B OUTWASH TERRACES	D KAME TERRACES
CHANNEL	KETTLE / ESKER KAMES

Figure 325: Diagram to illustrate some differences between kame and outwash/river terraces.

whereas in many cases if terraces are present they are more probably dissected outwash or river terraces.

The second point on the interpretation of glacio-fluvial landforms arises from the Glenmore example of Young (1974) described above. In terms of spatial sequences/landform associations it will be appreciated that it is generally not possible to find kames, eskers etc. *on* outwash or kame terrace surfaces.

However, the appearance of this can occur where outwash or kame terrace sediments are deposited around ice-contact features. As well as the Glenmore 'islands' described above, other examples include kame terrace/esker associations at Gart, southeast of Callander.

It should be clear from the above discussion that glaciofluvial landform development is very complex, but that with careful mapping and levelling it is possible to make detailed reconstructions about the nature of deglaciation.

Acknowledgements

I am grateful to Jim Rose for commenting on the first draft of this paper and to Leslie Milne for drawing most of the figures.

Deformation structures in British Pleistocene sediments

PETER ALLEN

Pleistocene deposits are found disturbed in a variety of ways, but as yet little work has been done in Britain to assess the relative importance of the various processes. Disturbance by glaciotectonic activity is important, but may not be the dominant process; loading and gravity are likely to be equally important. Periglacial processes (other than wedging, cryoturbation and cambering) are also relevant.

Local descriptions of disturbed Pleistocene deposits have been reported periodically and some examples are indicated below (fig. 326). In the Gipping Valley, Suffolk, Boswell (1913, 1927) and Slater (1927) described glaciotectonic disturbances of the sediments in the spurs and interpreted them as indicating the passage of ice down the valley. Banham (1973) explained the glaciotectonic origin of interbedded Chalk, Lower London Tertiaries and Pleistocene sands and till at Claydon, while Boulton (1973) similarly explained asymmetric folding at Creeting St. Mary. Allen (1983, 1984) noted that not all the disturbances at these sites were of glaciotectonic origin and that those that were did not indicate that the passage of ice was down-valley. Precise measurement of the disturbances, combined with till macrofabric work, indicated that ice movement was from the northwest, crossing the valley obliquely. Sketch sections illustrating the Contorted Drift in north Norfolk have been made by Slater (in an unpublished notebook; see Boulton *et al.*, 1984), Dhonau & Dhonau (1963) and Hart (Boulton *et al.*, 1984). Detailed descriptions and interpretations, showing the disturbances to be partly of glaciotectonic origin and partly the result of loading have been provided by Banham (1966, 1970a), Banham & Ranson (1965) and Banham *et al.* (1975). In the Midlands, Bishop (1958) described a graben at Stretton-on-Fosse within the Paxford Gravels, Shotton (1963) observed faulting of the sediments along the M45 motorway and Rice (1981b) describes a sequence of tills, clays, sands and gravels dislocated by a series of glaciotectonic folds, faults and slide planes at Dunton Bassett which are part of a more extensive zone of deformation in southern Leicestershire. McMillan & Browne (1983) interpreted a repeating and partly reversed sequence of beds at Bellshill, east of Glasgow, as glaciotectonically overturned. Thomas (1984a) and Thomas & Summers (1984) ascribed the structures of the Blackwater Formation, Co. Wexford, Ireland and of the Bride Moraine in the Isle of Man to a combination of glacially-induced folding, thrusting and diapiric action.

Overviews putting glaciotectonic and other disturbances into perspective are few. Slater (1926, 1943) formulated the view that in the earlier stages of the development of glaciotectonic disturbances, drift-laden ice pressed against a local outcrop, forming folds in that rock. Subsequently thrusting and overfolding occurred and the material was incorporated into the ice to be redeposited elsewhere. These ideas were applied to sections in Suffolk, Norfolk, Cheshire, and, in 1931, to the Bride Moraine. A brief review of various structures throughout Europe was made by Charlesworth (1957). Shotton (1965), examining sections at Stretton-on-Fosse and east of Rugby, was early in recognising that disturbances in Pleistocene sediments were not necessarily of glaciotectonic origin. He related those at Stretton-on-Fosse to periglacial activity and those east of Rugby to reactivation of geologically old faults during or after the melt phase of the Wolstonian ice. In north Norfolk, Banham (1975) noted that some of the tectonics, in the Cromer Tills, were glacially induced, but others, in the Contorted Drift, were the result of loading. In 1977, he presented a glaciotectonic model for the origin of the Marly Drift.

Glaciotectonic disturbances

Banham (1975, 1977) recognised both exodiamict and endiamict deformation associated with the deposition of lodgement tills. In sequence, there is a

Figure 326: Locations of sites mentioned in text.

till is considered to be at the base of (D), but the structural boundary is put at the lower limit of the penetrative deformation, at the base of (C) in the full sequence. The latter boundary is considered the more important. Thus (C), structurally within the till but not part of the diamict, is described as an exodiamict glaciotectonite, while (B) is comprised only of exo-diamict glaciotectonic structures.

Examples from Essex and Suffolk, associated with the base of the Lowestoft Till, occurring on a small scale of less than 3 m, are described in this volume (contribution by Allen *et al*). In Suffolk, on a scale of 5 - 10 m, the structures at Claydon (fig. 328) (Banham, 1973) and at Great Blakenham, Face 3 (fig. 329) (Allen, 1983; 1984), can be interpreted as exodiamict glaciotectonites, while those at Creeting St. Mary (fig. 330) (Boulton, 1973) appear to be exodiamict glaciotectonic structures. In Norfolk, on a scale of several kilometres, Banham (1977) considers the base of the intermediate-type Marly Drift to be the lithological base of the Lowestoft Till and the base of the Cromer-type Marly Drift to be its structural base (fig. 331). The latter, therefore, is an exodiamict glaciotectonite.

The deformation sequence may depart from that of Banham's model as the exodiamict glaciotectonic structures can, in certain cases, be separated from the glaciotectonite. In part of Face 2 at Great Blakenham, below lodgement till (fig. 332), beds IV to X are affected by cryoturbation and have been deflected to the southeast by ice drag. The sand of bed III is

progression (fig. 327) from undeformed bedrock (A), through folded bedrock (B) and penetratively cleaved bedrock (C) to lodgement till (D), though (B) and (C) may be missing. The lithological base of the

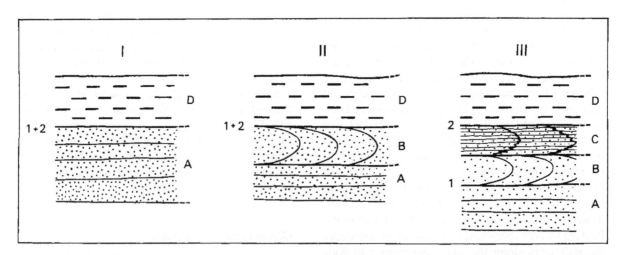

Figure 327: Till - soft bedrock relations. (A) undeformed bedrock; (B) weakly deformed bedrock (non-penetrative exodiamict glacitectonic structures); (C) penetratively deformed bedrock (exodiamict glacitectonite); (D) lodgement till (often an endiamict glacitectonite); (1) principal structural boundary; (2) principal lithological boundary. Reprinted from Banham (1977) by permission of Norwegian University Press (Universitetsforlaget AS), Oslo.

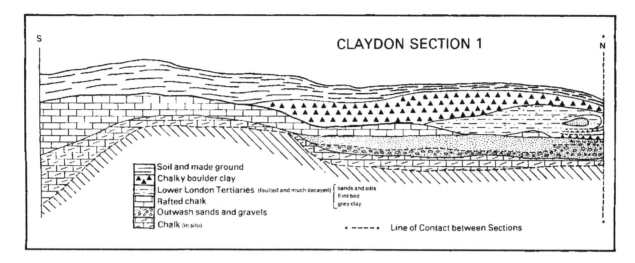

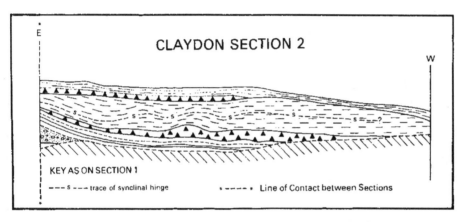

Figure 328: Sections at Claydon, Gipping Valley, Suffolk (after Banham, 1973).

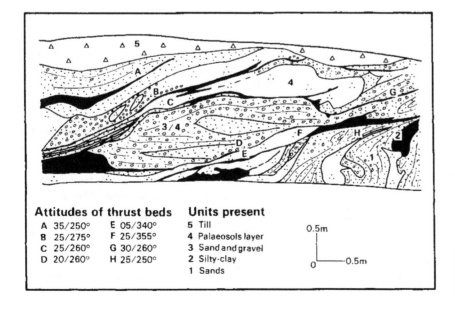

Attitudes of thrust beds

A 35/250°	E 05/340°
B 25/275°	F 25/355°
C 25/260°	G 30/260°
D 20/260°	H 25/250°

Units present

5 Till
4 Palaeosols layer
3 Sand and gravel
2 Silty-clay
1 Sands

0.5m

0 — 0.5m

Figure 329: Section at Great Blakenham, Gipping Valley, Suffolk (after Boulton, 1973).

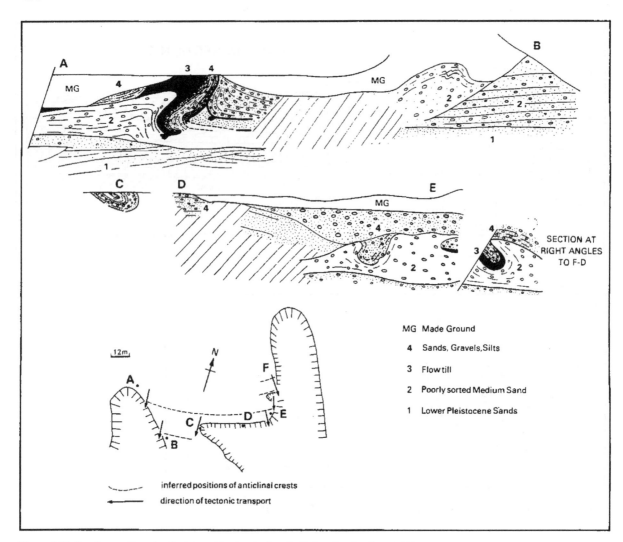

Figure 330: Sections at Creeting St. Mary, Gipping Valley, Suffolk (after Boulton, 1973).

undisturbed. Beds 1 and 11 involve silty-clay and show faulting and shearing reflecting the northwest to southeast passage of ice over this part of East Anglia. The last beds are, therefore, deformed by exodiamict glaciotectonic structures, separated by undisturbed beds from glaciotectonically deformed beds above.

Banham's (1977) model shows a range of conditions from the full sequence described above to a simple situation in which lodgement till directly overlies undeformed bedrock. In the Gipping Valley, the transition from the complex to the simple situation can occur over distances as short as ca. 100 m and the presence of a clay substrate appears to be a prerequisite for the complex situation to occur. At Great Blakenham and Creeting, the glaciotectonic defor-

mation occurs above the College Farm Silty-clay (Allen, 1983; 1984) and at Claydon above the clay of the Lower London Tertiaries; in each case the deformation is greatly reduced or absent within a short distance where the clay is absent. In clays, high pore water pressures can build up to the extent that the overlying sediments can be thought of as floating on the clay and readily subject to sliding and deformation by gravity or slight lateral pressures (Rubey & Hubbert, 1959; Mathews & Mackay, 1960; Banham, 1975). Boulton (1974) indicates that such a condition could also occur as ice moves over substrates of high (e.g. sand) and low (e.g. clay) transmissibility, the latter acting like dams to create high pore water pressures on the up-glacier side. Deformation is particularly associated with ice movement by compress-

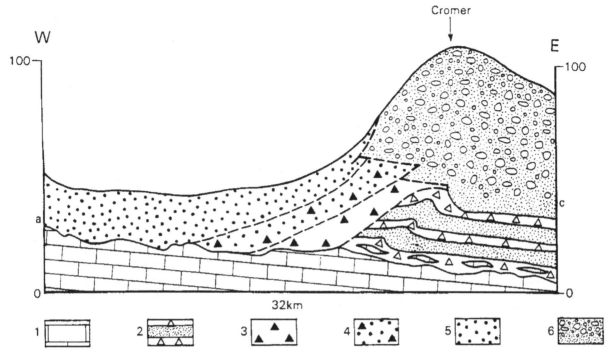

Figure 331: Hypothetical east-west section across north Norfolk immediately after retreat of the Lowestoft (Anglian) ice. Vertical scale in metres (exaggerated). (1) chalk; (2) Cromer Tills; (3, 4 and 5) Marly Drift of Cromer, Intermediate and Lowestoft lithological types, respectively; (6) gravels and sands. Reproduced from Banham (1975) with permission of the Liverpool Geological Society.

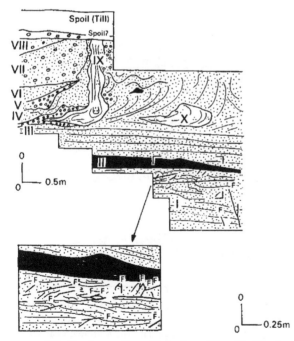

Figure 332: Section at Great Blakenham (Face 2), Gipping Valley, Suffolk.

ive flow, which occurs in association with scarps and ice margins (Kupsch, 1962; Moran, 1971), and in constricted valleys (de Jong, 1967). Banham (1975) combines these ideas into a generalised compressive flow model (fig. 333).

A more complex version of this model is given by Thomas (1984a) in the Isle of Man. The structures of the Bride Moraine are interpreted in terms of an ice readvance over the Shellag Formation, a sequence of till, sand and gravel, each of which was affected to a different degree by permafrost. Subsequently, high pore water pressures were created during the ice readvance. The readvance initially caused folding, but with increasing pressure, low-angle overthrusting occurred in the marginal areas and, as the pore water pressure increased, diapiric action occurred attenuating and shearing the folds, creating a series of mushroom-shaped, sheared isoclines (fig. 334).

Endiamict deformation is described from within the three Cromer Tills (North Sea Drift) by Banham (1975) in the form of flat, sheared-out folds (fig. 335, a-b). Within the First Cromer Till, the folds trend northwest-southeast, indicating ice movement from

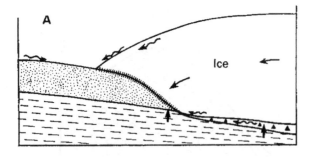

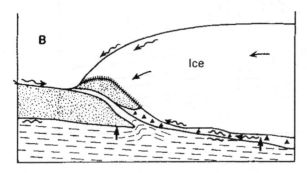

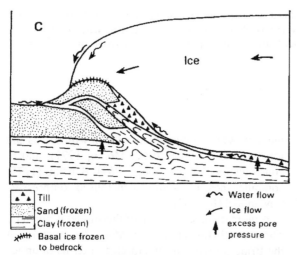

Figure 333: Banham's idealised compressive flow model of deformation. (A) Before significant deformation, (B) Ice and bedrock loading on wet clay and the lateral push of the ice have caused the movement of a raft of frozen sand, (C) Ice, raft and bedrock loading have caused the detachment of a second raft. Reproduced from Banham (1975) with permission of the Liverpool Geological Society.

the northeast. In the Second Cromer Till, the trend is northeast-southwest, indicating movement from the northwest and in the Third, the trend is north-south/ northeast-southwest, with ice movement from the west/northwest. Some of these endiamict structures are further deformed into steep, diapiric folds, pro-

bably induced by the overlying Gimingham Sands and Britons Lane Gravels (fig. 335c-d).

Loading

The best British examples of deformation of Pleistocene sediments resulting from loading are described from north Norfolk by Banham (1975) and Banham *et al.* (1975). Broad synclinal basins of Gimingham Sands have sunk into the underlying Cromer Tills, part of which have risen to compensate (fig. 336). In the east, in the Bacton - Mundesley area, the deformation affects only the uppermost, Third, Cromer Till. Between Overstrand and Cromer, the base of the Gimingham Sands has suffered a relative fall of up to 40 m and the Second Cromer Till is involved in the deformation. West of Cromer, all three Cromer Tills are deformed and the base of the Gimingham Sands

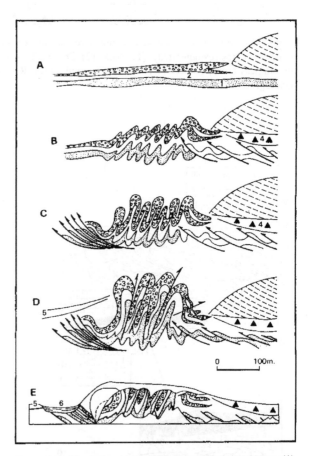

Figure 334: Evolution of the Bride Moraine, Isle of Man. (1) Shellag Till, (2) Shellag Sand, (3) Shellag Gravel, (4) Ballavarkish Till, (5) Kionlough Till, (6) Marginal channel. Reprinted from Thomas (1977) by permission of Norwegian University Press (Universitetsforlaget AS), Oslo.

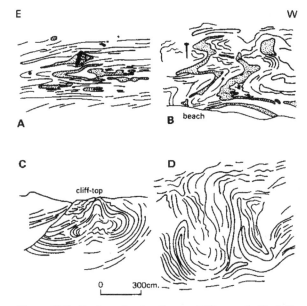

Figure 335: Sections in the Cromer Tills, north Norfolk. Ornament: black, flint; stippled, reconstituted chalk; solid lines, Contorted Drift (with Weybourne Crag at base). (A) and (B) primary folds in Contorted Drift, (C) and (D) primary folds with (originally) sub-horizontal axial planes, folded by secondary folds about steep axial planes. Reproduced from Banham & Ranson (1965).

descends to within a few centimetres of the underlying Cromer Forest Bed Series. The basins in which the Gimingham Sands lie alternate with domes of till; dip measurements on the base of sands show the folding to be periclinal in style. Within the basins, the Gimingham Sands show relatively little disturbance, but the tills below the basins show compensating thinning and outward flow into the adjacent diapiric domes, sometimes involving large Chalk rafts from within the till. The limbs of the folds are normally steep and often overturned, indicating the mushroom form of the domes. The beds below the plane of decollement at the base of the deformed sediments are relatively undisturbed.

This style of deformation has been ascribed to the loading effect of overlying ice (Kazi & Knill, 1969) but the weight of the more dense Gimingham Sands and Britons Lane Gravels is now thought more likely to have been the cause of the diapirism (Banham, 1975; Hart & Boulton, 1984). Other descriptions of this style of loading on such a scale are lacking, except for a limited example from Great Blakenham, Face 2 (fig. 337) (Allen, 1983; 1984).

Infrequent, large scale faulting (> 10 m) of Pleistocene deposits, exposed during excavations for the M45 motorway near Kilsby in Northamptonshire and the M1 motorway near Narborough in Leicestershire, has been described by Shotton (1965) (fig. 338). In the latter case, the faults could be traced down into the Trias beneath and the pattern of the faulting broadly reflected the Charnian and Caledonian trends of structures found within the Trias. The faults are ascribed to reactivation of the pre-Pleistocene structures, caused by updoming associated with isostatic recovery following the melting of the Wolstonian ice sheet.

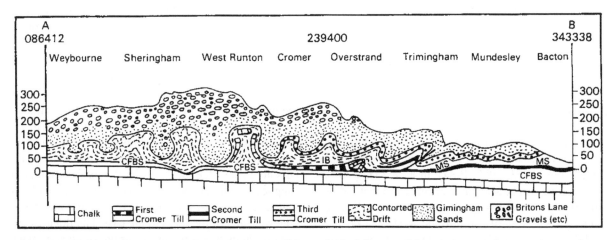

Figure 336: Diagrammatic section to show the structures of the north Norfolk coast. Selected representative structures are drawn true-scale; vertical scale of profile exaggerated and given in feet as on existing maps. The degree of diapiric deformation of the Contorted Drift (etc.) increases as the thickness of the overburden of sands and gravels increases. (CFBS-Cromer Forest Bed Series; IB-Intermediate Beds; MS-Mundesley Sands). Reproduced from Banham (1975) with the permission of the Liverpool Geological Society.

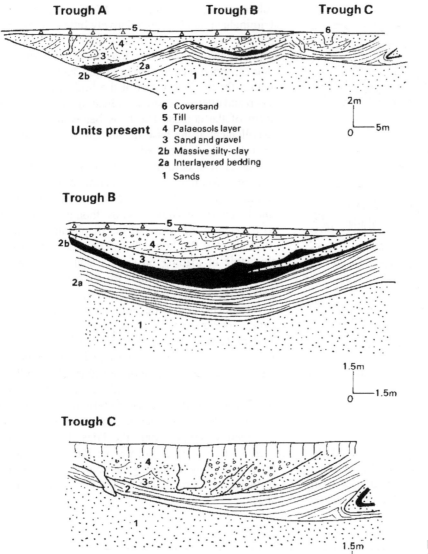

Units present

6 Coversand
5 Till
4 Palaeosols layer
3 Sand and gravel
2b Massive silty-clay
2a Interlayered bedding
1 Sands

Figure 337: Section at Great Blakenham (Face 2), Gipping Valley, Suffolk.

Gravity

Gravity-induced deformation is described from Barham, on the eastern side of the Gipping Valley, Suffolk (Allen, 1973; 1983; 1984). In Face 3 (fig. 339), the bedding can be seen to be disturbed so that it dips apparently westwards at 10° and involves a flexure with a downthrow of 2.4 m to the west. The flexure and the dip of the beds are associated with extension and collapse of the sediments to the west, into the valley. At the south end of Face 4 (fig. 340), the independent overfolds of the loess and of the sand, with closures to the north, are thought to be apparent.

Separate folding of the loess and of the sand is difficult to explain as the loess would have to be overfolded before the sand was deposited, which in turn would have to be overfolded, without further disturbance of the loess. A simpler explanation is to suggest a style of folding as shown schematically in fig. 340. Thus, the true closure of the sand may be to the west and of the loess to the east. The flexures are again thought to relate to gravity-induced mass movement to the west, down the valley slope. As till is involved, the movements are considered to post-date the glacial episode.

At Blood Hill, Little Blakenham, Suffolk, study of

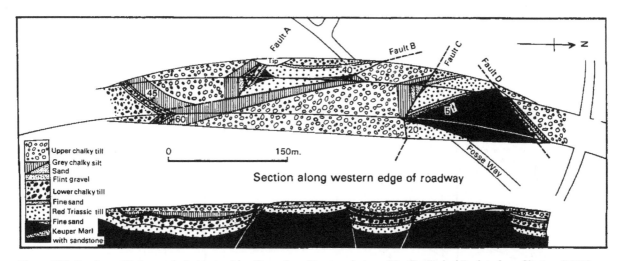

Figure 338: Section at Narborough, Leicestershire. Reproduced by permission of the Geological Society from Shotton (1965).

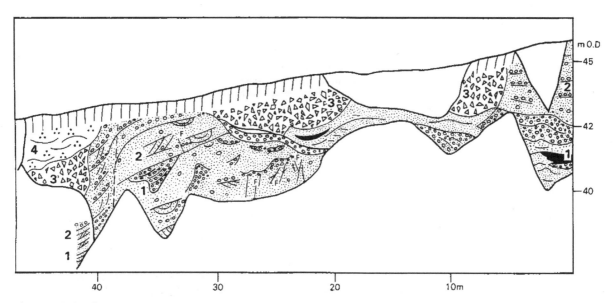

Figure 339: Section at Barham (Face 3), Gipping Valley, Suffolk. (4) coversand, (3) head, (2) Barham Sand and Gravel (1) Kesgrave Sand and Gravel, (F) fault.

large- and small-scale structures in glacigenic sediments arranged into two anticlines showed interesting contrasts. These structures are shown in some detail in fig. 341 and in summary in fig. 342. The structures on limbs A and B of anticline I correspond to those described by Kehle (1970) for gravity gliding (fig. 343). The small normal fault (A5) is attributed to pulling away from the crest area. Gravity folding (A3), associated with upturning of beds (A4) immediately adjacent, indicates movement down the dip. Further downdip is a sequence of imbricate reverse

faults (A1) as expected in the lower area, analogous to the shears in the toe of a mudflow. The reverse fold (steeper limb updip) and fault (A2) are considered to be due to impeded movement of the leading material along the underlying thrust, causing the following material to under-ride. Thus the upper part of limb A shows a structure associated with extension and the lower part structures associated with shortening, indicating that there was a net movement of the sediment body downdip.

Limb B similarly shows extension in its upper part

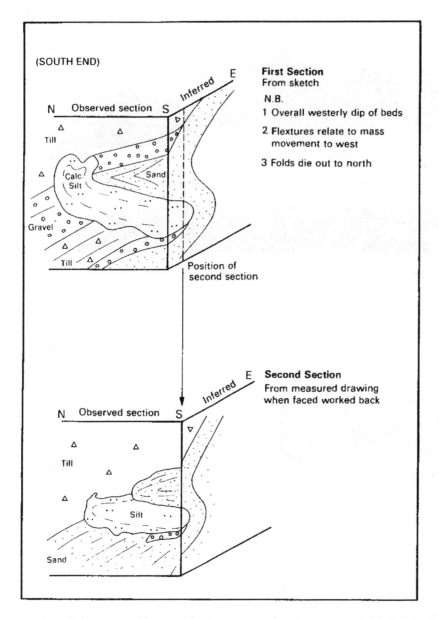

(SOUTH END)

E

Inferred

First Section
From sketch
N.B.
1 Overall westerly dip of beds

2 Flextures relate to mass
movement to west

3 Folds die out to north

N   Observed section   S

Till

Calc.
Silt

Sand

Gravel

Till

Position of
second section

E

Inferred

Second Section
From measured drawing
when faced worked back

N   Observed section   S

Till

Silt

Sand

Figure 340: Section at Barham (Face 4), Gipping Valley, Suffolk.

with a 'pull-apart' fault graben (B1). Immediately downdip are faults with no measurable movement along them (B2). A reverse fold in the same area also indicates impedance of movement. Further downdip, imbricate reverse faults occur, initially small-scale (B3), then large-scale (B4). Adjacent to the thrusts is an unusual structure, most readily described as having a zig-zag pattern (B5), probably due to micro-faulting. In the lowermost area is a series of at least six folds (B6). All the structures from B2 to B6 indicate shortening. Thus, as in limb A, it is argued that there has been a net movement of the sediment body downdip.

The structures in limbs A and B, which are gravity induced, imply the pre-existence or contemporary formation of the anticlinal structure (I). When and how anticline I was formed is not known, but as the structures are interpreted as being due to gravity movements, both to the north and to the south, adjustment to an unstable base is indicated.

The structures on limbs C and D correspond to those described by Sanford (1959) associated with vertical movement of basement material (fig. 344). Limb C is affected by a series of four normal faults (C1, C2), three of which are irregular in that they are very steep, curved and multiple. Associated with this

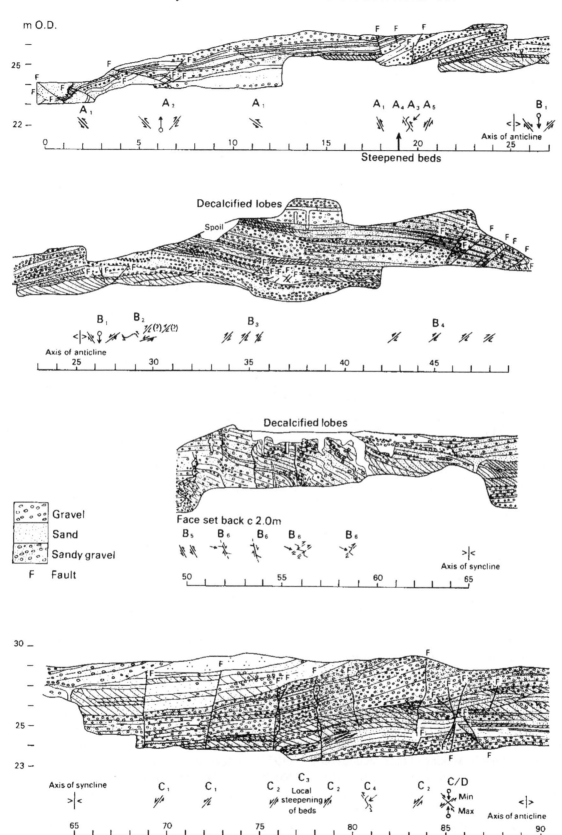

Figure 341: Section at Little Blakenham, Gipping Valley, Suffolk.

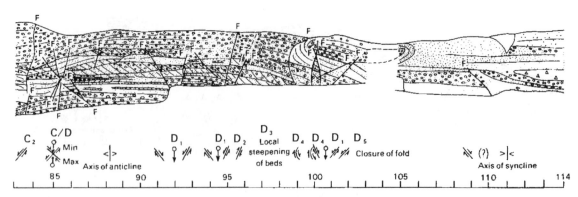

Figure 341 (continued).

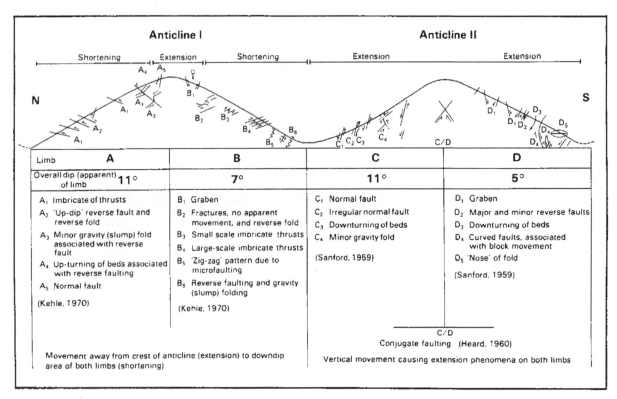

Figure 342: Summary of structures at Little Blakenham.

faulting is downturning of beds (C3). The steepness of the faults, their curvature and the downturning of beds suggest accommodation to vertical movement (Sanford, 1959), either by sinking of the synclines or uplift of the anticline. The stellate fault pattern at the apex of the anticline (C/D) is interpreted as being conjugate and related to vertical pressures (Heard, 1960) (fig. 345), suggesting syn-depositional faulting.

On the south side of anticline II, limb D, are three mini-graben (D1), reverse faulting (D2), downturning of beds (D3) and two curved faults (D4). This assemblage corresponds to that noted by Sanford (1959) associated with vertical uplift of a central block (fig. 344). This, combined with the conjugate faulting, indicates that the anticline was created, in part at least, by relative uplift due to lateral sinking of the synclines. Thus, again, the structures must be related to an unstable base.

The changes in thickness and number of beds

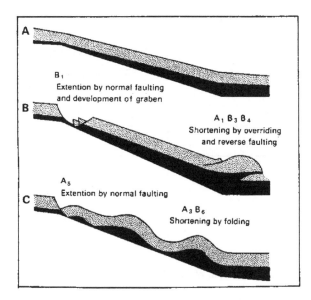

Figure 343: Kehle's model for gravity gliding. (A) Rock sequence before slide (lower unit is assumed to be mobile shale or salt), (B) Formation of pull-apart grabens updip and imbricate thrusts downdip, (C) Single pull-away with reverse drag updip and folding. Reproduced with permission from Kehle (1970).

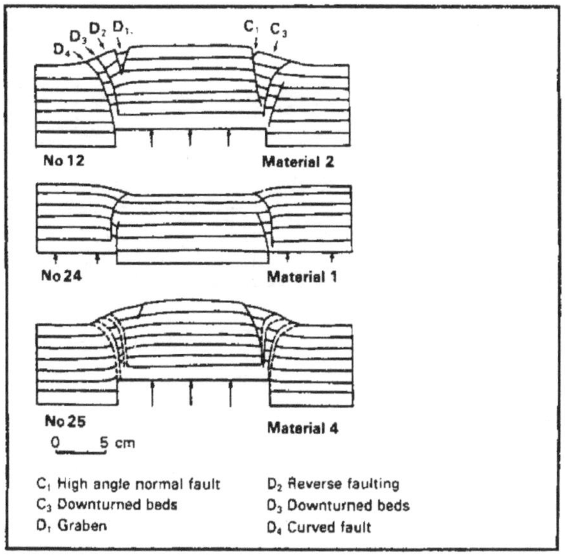

Figure 344: Sanford's models of faulting associated with uplift. Reproduced with permission of the Geological Society of America, from Sanford (1959).

either side of faults A1, A2, B4, C2, D1 indicate either transcurrent movement along the faults or that fault movement was contemporary with deposition. The elliptical structure, D5, probably results from the formation of a fold, possibly associated with gravitational movement into or out of the exposed face.

Many of the structures, e.g. A1, A3, B1-3, are of very limited extent and affect only one or two beds and even the larger structures rarely transgress all the beds. This suggests that the movements were occurring during deposition of the sediments.

In accounting for the structures as a whole, one

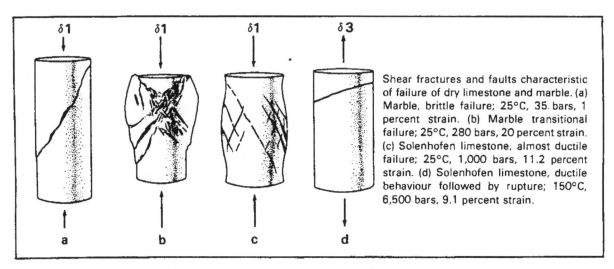

Figure 345: Heard's model of faulting associated with vertical compression. Reproduced with permission of the Geological Society of America from Heard (1970).

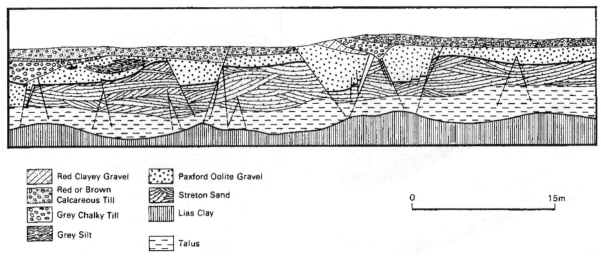

	Red Clayey Gravel		Paxford Oolite Gravel
	Red or Brown Calcareous Till		Streton Sand
	Grey Chalky Till		Lias Clay
	Grey Silt		
			Talus

0   15m

Figure 346: Section at Stretton-on-Fosse, Warwickshire. Reproduced by permission of the Geological Society from Shotton (1965).

possibility is that the synclines were created by melting of buried ice or compaction of the lower sequence, but alternatively the anticlines could have been created by upward movement of the underlying till complementing adjacent loading, either by the weight of the sand and gravel or by ice.

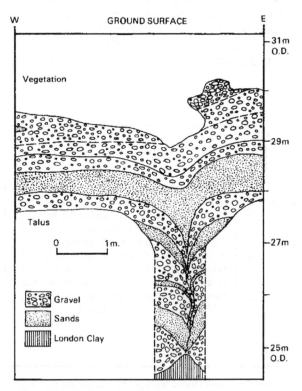

Figure 347: Section at Aveley, Essex.

Periglacial disturbances

A study of small-scale faulting in Pleistocene deposits at Stretton-on-Fosse, Warwickshire (Shotton, 1965) revealed a sequence of normal faults defining a series of graben and horst structures which showed a relationship to undulations in the underlying Lias (fig. 346). Where the Pleistocene deposits were let down in troughs, the Lias rises directly below. This situation is unusual as lowering of a body of sediment is usually compensated by lateral upwelling of the underlying sediments by diapiric action, as in north Norfolk. Similar situations, but with collapse structures above and diapirism immediately below, have been noted in association with the London Clay. At Stebbing in Essex, the structure affects glacigenic and Pleistocene fluvial sediments. Quite frequently terrace deposits are affected, as at Aveley, Essex (Wiseman, 1978) (fig. 347) and Shakespeare Farm pit, St. Mary's Hoo, north Kent (Bridgland, 1983). Where exposures allow, the structures can usually be traced for several metres or tens of metres as linear features.

The sense of movement from both above and below is vertical along one plane. Shotton (1965) relates the structures to tension faulting associated with shrinkage of frozen ground during the Pleistocene. Normally shrinkage of gravel in the periglacial environment would be limited, but above clay or silt a high water table could occur to account for the contraction. In such a situation, vertical fissures would allow contemporaneous movement from below and above, most probably during melt phases when the clay below would be plastic, though Muller (1947) notes that the shear strength of clays initially increases upon freezing but soon decreases and approaches those of ice as the pores are progressively filled with frozen water.

In the case of Stebbing, similar structures have been recorded (Whiteman *et al.*, 1983) and their surface expression is similar to that of gulls associated with cambering in the Midlands (Kellaway, 1972). At Aveley, the structures occur towards the outer margin of terrace gravels, approximately parallel to the terrace front. Either periglacial contraction or cambering could be invoked in this case. At Shakespeare Farm pit and in other pits in the area, in plan the structures occur at various angles, suggesting they are associated with periglacial patterned ground, and so with thermal contraction.

Comment

The late 1980s is an appropriate time to review the ways in which Pleistocene deposits can be disturbed. Prior to the early 1970s, very little work had been done on the subject in Britain apart from that of Slater (1926, 1927, 1943), Charlesworth (1957) and Shotton (1965) and the related work of Kellaway (1972) on valley bulges. Since then work in this field has increased to the extent that there is now recognition that a variety of processes can affect Pleistocene deposits, but the amount of relevant research being done in Britain is still relatively small. It is hoped that this review will stimulate a greater interest in the examination and interpretation of disturbances in Pleistocene sediments.

Geochemical properties of glacial deposits in the British Isles

CYNTHIA V. BUREK & JOHN M. CUBITT

Sedimentary rocks are redeposited as a result of erosion and chemical alteration of pre-existing rocks. Glacial deposits, classified as sedimentary rocks, are for the most part mechanical in origin and normally contain rock fragments rather than individual mineral grains. Thus all glacial deposits are geochemically immature. In general only the finer fraction of tills will contain a rich assemblage of both major and trace elements, while the sands and gravels – outwash sediments will be dominated by SiO_2 and its associated trace elements Ti, Zr and Be and minor amounts of K, Na, Ca and Al, (not associated with the aluminosilicates). The finer-graded glacial sediments, for example the outwash sands, loess and finer fraction of till are sorted as regard to the various minerals present due to the difference of density and resistance to abrasion. Table 33 lists the geochemistry of the more common minerals of glacial deposits and table 34 shows some typical sedimentary and igneous rock geochemistry, which are often associated with them (Aubert & Pinta, 1977; Freedman, 1975; Gough et al., 1979; Kabata-Pendias, 1968; Krauskopf, 1967; Kubota, 1967; Swaine & Mitchell, 1960). However it is the presence of the minerals within the glacial deposits which influences the distribution of elements available. Table 35 lists some of the major and trace elements which are associated with these minerals and are available to be released on weathering (Mitchell, 1964). The geochemistry of glacial deposits is therefore directly related to the geology of the bedrock available for incorporation and their weathering capabilities.

Only limited work has been attempted on the geochemistry of individual glacial deposits within the British Isles (Burek, 1978). The major elements have been studied far more extensively than the trace elements, possibly because of the sophistication of the techniques needed for the latter. Thus the use of trace element geochemistry in glacial geomorphology has been limited to exploration in glaciated terrain, most notably in Canada (e.g. Shilts, 1971; 1973a and b; 1977; 1984) and Scandinavia (e.g. Eriksson, 1983), to provenance and petrographic correlation studies of the tills of Derbyshire (Bayliss et al, 1979; Burek, 1985a; 1985b; Burek & Cubitt, 1979; Ineson, 1969) and Quaternary sediments in general in Suffolk (Lloyd et al., 1981; Peachey et al., 1984; 1985). The lack of geochemical evidence led to analyses being performed on selected tills from the Lake District, Isle of Man, Derbyshire, West Wales, the Scilly Isles, East Anglia, Midlands and the Home Counties (table 36). The data generated was subjected to multivariate statistical analysis, i.e. cluster analyses and principal component analysis.

The results clearly demonstrate a division of British tills based on their location and calcareous nature (fig. 348). Tills from the western side of the Pennines group together be they from Lake District or the Isles of Scilly and show the extent and influence of the Irish Sea ice. Included in this group are the sandy till from Great Waltham, the Ware Till and the decalcified Orford Till from southeast Suffolk. In each case the CaO values are less than 10% and Sr values less than 150 ppm. The tills of eastern and southern England form a group based on the Lowestoft Till distribution. These tills are rich in CaO and Sr. So the major division, amplified by multivariate statistics, arises from CaO in the composition of the tills – a calcareous versus non-calcareous division. Separate from these are the tills of Derbyshire which bear little resemblence in their geochemistry to either of the other two groups. These tills contain too much calcareous material to be closely correlated with the non-calcareous group, but do not contain the same suite of other major and trace elements to be included with the chalky tills. This supports the hypothesis proposed elsewhere (Burek, 1978) that the influence of the North Sea and Irish Sea ice played little part in the development of the tills in the North Derbyshire area and that these deposits are primarily of local origin.

Table 33. Chemical composition of some silicate and clay minerals normally present in glacial deposits (in percent).

	SiO₂	Al₂O₃	MgO +FeO	CaO	Na₂O +K₂O	H₂O
Quartz	100	–	–	–	–	–
Muscovite	47	37	–	–	12	4
Biotite	36	20	30	–	10	4
Orthoclase feldspar	65	18	–	–	17	–
Albite feldspar	68	20	–	–	12	–
Plagioclase feldspar	54	29	–	12	5	–
Amphibole	40	10	30	15	3	2
Hornblende	40	10	30	12	1–2	1
Olivine	40	–	60	–	–	–
Pyroxene	50	–	35	15	–	–
Apatite	50	3	23	20	–	–
Allophene	34	31	–	2	–	33
Kaolinite	45	38	1	1	0.7	14
Halloysite	44	39	0.1	–	0.2	16
Montmorillonite	51	20	4	2	0.1	23
Vermiculite	36	11	34	0.5	–	20
Illite	49	29	4	1	8	9
Glauconite	53	6	26	0.1	8	9
Chlorite	27	25	36	0.2	–	12

Source: Degens (1965), Open University (1981).

Geochemistry

Since so little work has been attempted on the geochemistry of Quaternary sediments it is worth considering the role major elements and more important trace elements might play. Special emphasis will be placed on the North Derbyshire tills.

Aluminium is the third most abundant element in the earth's crust after oxygen and silicon. It is a prime constituent of aluminosilicate minerals such as feldspar and pyroxene in the form of a tetrahedral combination. However on weathering aluminium progressively acquires more octahedral coordinations with hydroxyls which produce clay minerals.

During weathering Al ions are released from feldspars for example, and form a hydrated stage almost immediately. Silicon is released as H_4SiO_4 which joins with the Al to form a silica-aluminium copolymer becoming a clay. The two-layered clays kaolinite, halloysite and dickite are favoured by weathering conditions where water is abundant and circulation rapid. The solution concentration is low. Three-layered clays illite and montmorillonite form when drainage is slower and water less abundant. Therefore the pH of the weathering solution is alkaline and the solution concentration higher. The solubility of Al is low, which promotes the retention in weathering products of low solubility. It is thus contained in the most stable minerals in the weathering zone – the two-layered clays, the usual constituents of sandy tills.

Calcium is the fifth most abundant element in the earth's crust. It is the major constituent of limestones. In very high purity limestones such as the Bee Low Limestone of Derbyshire which have 98.5% $CaCo_3$, the equivalent CaO is 55.18% (Cox & Bridge, 1977). CaO and $CaCO_3$ are soluble at acidic pH values and this leads to leaching in a wet temperate climate.

Fresh tills from over limestone or chalk will consistently have high CaO values due to the erratic content (limestone) and the subsequent breakdown due primarily to physical weathering which would supply the CaO to the fine matrix of the deposit. Subsequent chemical weathering would redistribute this through a profile.

Within the chalky tills of East Anglia, the CaO levels are high due to the nature of the fine fraction and the weathering capabilities of the chalk itself as opposed to the lower CaO content (18%) of the Derbyshire tills. The latter may be influenced by aeolian deposition, the material having been derived from the chalky tills of Lincolnshire and the east or the chalklands themselves.

Iron, the fourth most abundant element in the earth's crust after silicon, oxygen and aluminium, is a primary collector of trace elements during weathering, e.g. Cu and Co substitute for Fe^{2+} in the silicate structure. Many ferric oxides are transported as coatings on clays.

Potassium is normally associated with clay minerals especially in shales and sandstones, which is a function of K-feldspar (orthoclase) and K-micas e.g. illite and glauconite. The weathering of K-feldspar is a function of the pH, Al and Si concentration of the soil solution and the rate of formation of hydrated-aluminosilicates. It is used in the formation of illite or adsorbed onto other clay minerals. In this way K is retained in the breakdown fraction of rocks in the soil.

Magnesium is the eighth most abundant element in the crust. It is usually contained in the orthosilicates such as fosterite or in dolomite, chlorite or glauconite. It can replace Ca in $CaCO_3$ as in dolomitic limestone or be adsorbed onto clay mineral particles.

Silicon. Silica, is the most common of all the major oxides in the crust, is contained in almost all main rock types and is stable under most environmental conditions (less than 9 pH).

There are three types of silicates, based on weathering conditions. Firstly, those that dissolve congruently, i.e. olivines, pyroxenes; secondly, those

Table 34. Comparison of bedrock and till geochemistry with superficial sediment data.

	CaO	SiO_2	Al_2O_3	MgO	K_2O	Fe_2O_3	TiO_2	MnO	Ba	Cd	Co	Cr	Cu	Ga	Li	Mo	Ni	Pb	Sn	Sr	V	Zn	Zr
Limestone Eyam (12)	54.3	1.58	0.15	0.32	0.05	1623		282					8					39		200		53	10
Monsal Dale (105)																							
Upper Pale	54.3	1.1	0.09	0.38	0.02	240		114					12					6				19	
Dark (96)	53.3	2.45	0.26	0.64	0.05	581		114	$				15				8	7		550		27	
Lower pale (105)	54.4	0.62	0.08	0.27	0.01	333		137					12					4		300		20	
Bee Low (79)	56.0	0.15	0.01	0.21	0.0	199		136					6					2		200		17	
Woo Dale (14)	55.4	0.16	0.03	0.26	0.01	146		86					9					15		200		12	
						Total Fe																	
Olivine dolerite sill (9)	8.6	49.0	14.1	8.3	0.44	10.6	1.74	1500	80		35	310	66	18	25		230			220	170		100
Millstone Grit (11)	0.01	97.9	0.57	0.06	0.16	0.45	0.36	600	10		20	60	20	10		2	45	75–180	5	10	85		500
Dolomitic limestone (15)	31.6	1.34	0.06	20.17	0.03	0.34		856			24		15					15		100		209	
Coal Measures (4)							0.3–0.6	600–8500			97	97	55			2	130–200	50	9		85–200		
Triassic NRS (4) (1)							0.3–0.6	2100			25	50	10			2	25	95	8		40	2200	
Superficial deposits (17)	2.4	72.7	14.2	0.9	1.5	4.5		1766	1186	83	22	136	59	17	50	18	100	479		144	195	557	528
Till (98)	18.5	44.7	11.0	1.5	1.3	4.5	0.2	1343	1925		37	152	59	17	78	83	132	399		256	146	365	255
Shale	11.0	58.7	9.8	1.3	1.5	2.6	0.26	832	1473	244	75	162	162	14	45	127	304	39	0	362	945	823	76
Weathered tuff (2)	1.4	55.0	26.6	2.6	3.2	10.8	0.8	1942	166	112	94	512	96	26	98	0	566	64	0	26	322	570	406
Basalt (1)	4.0	40.7	21.15	3.67	4.7	12.3**	0.9	450					55			0		0				260	
Carboniferous arenaceous rocks (58)	0.2	95.15	1.91	0.14	0.2	0.7	0.2	600	10–1500		10	35	10–100	10	31	2	15			10	10–270	30	600

All major oxides except Fe_2O_3 are in % unless otherwise marked. All trace elements in ppm.
( ) is the total number of samples analysed; **Fe_2O_3, not total.
$ combined D_2 limestone; after Burek & Cubitt (1979), Table 1, with additional data from Hanson & Adlam (1985).

Table 35. Release of elements with weathering of certain minerals.

Mineral	Major elements										Trace elements																	
	Mg	Fe	Si	Ca	Al	K	P	F	Na	Ti	Ni	Co	Mn	Li	Zn	Zr	Cu	Mo	Ga	V	Sc	Sr	Sn	Pb	Ba	Rb	Cr	Hf
Olivine	Mg	Fe	Si								Ni	Co	Mn	Li	Zn		Cu	Mo										
Hornblende	Mg	Fe	Si	Ca	Al						Ni	Co	Mn	Li	Zn		Cu		Ga	V	Sc							
Augite	Mg		Si	Ca	Al						Ni	Co	Mn	Li	Zn		Cu		Ga	V	Sc			Pb				
Biotite	Mg	Fe	Si		Al	K					Ni	Co	Mn	Li	Zn		Cu		Ga	V					Ba	Rb		
Apatite				Ca			P	F														Sr		Pb				
Anorthite			Si	Ca	Al								Mn				Cu		Ga			Sr						
Andesine			Si	Ca	Al				Na				Mn				Cu		Ga			Sr						
Oligoclase			Si	Ca	Al				Na								Cu		Ga									
Albite			Si		Al				Na								Cu		Ga									
Garnet	Mg	Fe	Si	Ca	Al								Mn						Ga								Cr	
Orthoclase			Si		Al	K											Cu		Ga			Sr			Ba	Rb		
Muscovite			Si		Al	K		F											Ga	V		Sr			Ba	Rb		
Titanite			Si	Ca						Ti										V			Sn					
Ilmenite		Fe								Ti	Ni	Co								V							Cr	
Magnetite		Fe									Ni	Co			Zn					V							Cr	
Tourmaline	Mg	Fe	Si	Ca	Al		P	F						Li					Ga									
Zircon			Si													Zr												Hf
Quartz			Si																									

Mg Magnesium; Fe Iron; Si Silica; Ca Calcium; Al Aluminium; K Potassium; P Phosphorus; F Fluorine; Na Sodium; Ti Titanium; Ni Nickel; Co Cobalt; Mn Manganese; Li Lithium; Zn Zinc; Zr Zirconium; Cu Copper; Mo Molybdenum; Ga Gallium; V Vanadium; Sr Strontium; Sn Tin; Pb Lead; Ba Barium; Rb Rubidium; Cr Chromium; Hf Hafnium; Sc Scandium. Taken from Mitchell (1964).

which dissolve incongruently, i.e. feldspars, and micas giving rise to clay minerals, and thirdly, those which are essentially resistant, i.e. quartz (Wedepohl, 1970). Obviously, the first will disappear from a weathered profile quickly, the second less quickly and the third will form the detrital constituent. In silicates substitution for silica may take place by Al, Ti, Zr or Be.

Manganese is the least common major element (consequently measured in ppm) and is usually associated with iron. However, it is more mobile than iron under reducing conditions and is more soluble at all pH and Eh values (Yaalon *et al.*, 1974). It will accumulate in the A horizon of podsols, brown forest and acid soils because of biological accumulation. It becomes immobilised as MnO_2 by oxidation in clays or organic-rich horizons. Mn is prone to leaching in a poorly-drained soil (Mitchell, 1964) and a typical tundra soil will contain:

Soil horizon	Manganese content
A_1	0.099%
B	0.09%
C	0.14%

(Source: Vinogradov, 1959)

Under permafrost conditions when soils are water-logged and vegetation is sparse, manganese is easily leached.

Manganese has a positive association with Ca, Ba, Zn and Co and a negative relationship with Ag and Cd.

Table 34 shows some of the average major element geochemical analyses for some typical sedimentary and igneous rocks. However rocks are merely physical aggregates of minerals and when analysed each mineral contributes a proportion of elements to the bulk composition of the rock involved. In general glacial deposits derived from shales will have a diverse geochemistry because most trace elements are concentrated and enriched in them, while those of a sandstone origin will be depleted in all but SiO_2 and its associated trace elements. In Derbyshire tills, the fine fraction is dominated by ground-up shale and this is reflected in the amount of clay material present. However the glaciers also traversed numerous mineral veins present in the area and so the levels of elements associated with these is high in the matrix (i.e. lead, barium etc.). Derbyshire tills contain little far-travelled material, and so the fine fraction is directly correlated with the distance the incorporated material has travelled and the resistance to weathering of the rock types involved. The high percentage of iron can be attributed to the weathering of the basalts and the calcium to the weathering of the

Table 36. Representative geochemical analyses for British glacial tills.

Location:	CBC Huncote	Irish Sea Abermawr	Lea Basin Westmill Quarry	Lake District Thornsgill Till	Isle of Man Wyllin	Derbyshire Shining Bank Quarry	Essex Great Waltham Lowe-stoft Till	Isles of Scilly Bread and Cheese Cove Till	Suffolk Orford Till
Age:	Wolstonian	Devensian	Anglian	Devensian	Devensian	Wolstonian	Anglian	Devensian	Anglian
Major elements %									
SiO_2	42.60	61.40	58.00	60.30	55.40	40.30	46.40	73.10	ND
TiO_2	0.62	0.68	0.66	0.87	0.87	0.27	0.40	0.95	0.56
Al_2O_3	12.20	11.20	13.20	18.70	14.60	10.30			
Fe_2O_3	9.50	4.50	5.60	7.20	6.50	4.00	3.50	5.00	6.36
MnO	0.12	0.08	0.04	0.20	0.11	0.09	0.04	0.10	0.03
MgO	1.60	1.90	1.20	0.80	2.60	1.90	1.00	1.10	1.00
CaO	13.70	7.20	7.60	0.30	5.20	25.50	19.50	0.10	0.67
Na_2O	0.40	0.80	0.20	0.80	1.20	ND	0.30	0.80	ND
K_2O	1.94	2.88	2.24	2.89	2.85	1.25	1.31	2.33	ND
P_2O_5	0.29	0.14	0.15	0.32	0.32	ND	0.13	0.30	ND
Trace elements ppm									
Zr	165	218	187	186	196	236	115	203	350
Sr	222	181	137	69	164	307	247	51	83
Ga	14	15	18	24	16	10	10	11	19
Zn	84	66	90	88	78	344	47	48	117
Ni	46	35	50	41	48	127	31	21	66
Cr							88	80	110
V	134	102	155	105	130	156	103	93	170
Ba	284	356	342	420	468	2353	207	293	350
Co	ND	ND	ND	ND	ND	32	ND	ND	26
Pb	ND	ND	ND	ND	ND	300	ND	ND	27
Mo	ND	ND	ND	ND	ND	54	ND	ND	2
Cd	ND	ND	ND	ND	ND	84	ND	ND	ND

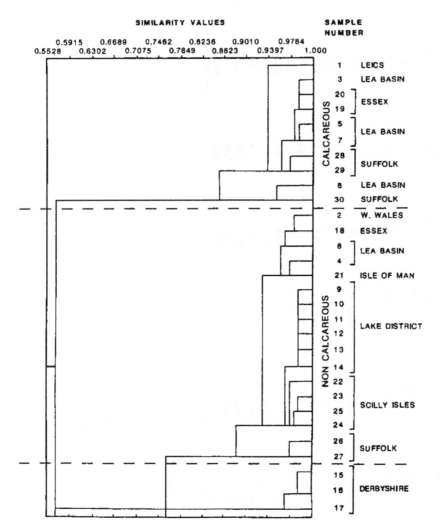

Figure 348: Dendrogram of British till geochemistry.

limestone erratics (Burek, 1978).

Each major element has its own associated traces and we will deal with some of the more important here.

Titanium is the second most abundant minor element in the lithosphere, and readily substitutes for silicon because of the similar ionic size and same valency state. SiO_2 is stable in the minerals muscovite, ilmenite and magnetite but can also form from weathered olivine which leads to a Ti-rich residue. This can then be adsorbed onto montmorillonite clay minerals.

With respect to soils there is an accumulation of Ti in the B horizon of podsols but loessic and calcic soils are poor in Ti.

Barium is a common trace element in the crust. It usually occurs as the mineral baryte but other hosts are the alkali feldspars, albite and orthoclase. It is

rapidly lost during the initial weathering of feldspars. It does not readily enter into magnesium minerals (Day, 1963) but because of a similar ionic radius, it can replace Sr, K and Pb, in particular the K in illite. The leaching of K is greater than Ba because of a preferential adsorption of Ba onto the clay minerals (Short, 1961) and this gives rise to a weathering ratio Ba:K which will increase with weathering (Burek, 1979; 1985a).

In carbonate minerals Ba can replace Ca in the aragonite series. In pedogenesis, podsolisation causes an increase in Sr and a decrease in Ba based solely on climatic conditions. This is illustrated by Vinogradov (1959) in tundra soils with a ratio of Ba:Sr = 1.5 and podsols 8.8.

In Derbyshire the barium content of the superficial sediments and bedrock is high, due to the extensive mineralisation of the limestones. Thus the availabil-

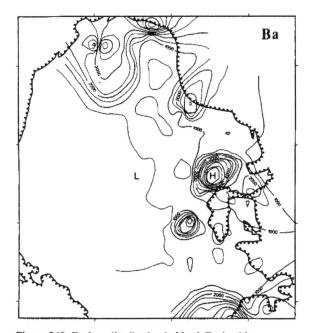

Figure 349: Barium distribution in North Derbyshire.

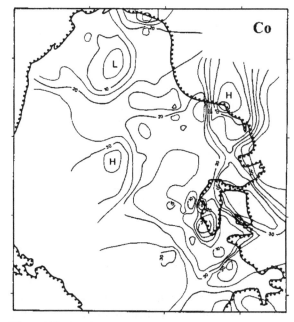

Figure 350: Cobalt distribution in North Derbyshire.

ity of barium for the tills and the 'silty drift' (Pigott, 1962) is great. Both have values over 3,000 ppm. Fig. 349 details the distribution of barium in ppm within 48 cm of the surface. No sample is less than 18 cm from turf level. Much of the area is over 500 ppm, however three areas of high values occur. To the south of Castleton (SK 150829) and on Bradwell Moor (SK 140800), high concentrations indicate baryte present in the mineralisation. This is also true of the area south of Northing 650. The highest value is over 6,000 ppm and is related to the superficial deposits over Longstone Edge (SK 200730) where baryte and fluorite are common, even in the A horizon of the soil. South-southwest of this area is another high, which corresponds to the old lead mining area near Sheldon. In areas such as this, mining and reworking has mixed the previously dumped material within the superficial sections, contaminating the whole profile. However, it is still true that above the mineralised veins, physical weathering will break down the vein into its constituent parts and chemical weathering will release those soluble elements into the weathering cycle. It is worth noting here that $BaSiO_2$ is more soluble than $BaSO_4$ and that under Mediterranean, hot humid temperate climates, Ba is the second most mobile element in the sequence Sr > Ba > Co, Ni, Cr, Cu, V > Ti (Yaalon *et al.*, 1974). It is by looking at trace elements in this detail that we can understand the geochemical distribution of

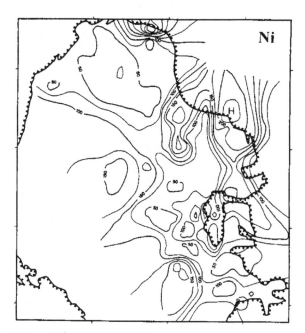

Figure 351: Nickel distribution in North Derbyshire.

superficial deposits in a small area. We will now examine cobalt, chromium, strontium and nickel in the same detail. The distribution of Zn, Pb, V, Mo and Cd in Derbyshire glacial deposits have been dealt with in detail elsewhere (Burek & Cubitt, 1979; Briggs & Burek, 1985).

Cobalt is normally found in basalts and basic soils. It has a similar ionic radius to Fe and can substitute for it. Co resembles Ni and there is a strong association of Mn and Co in minerals. Augite and olivine are the two chief hosts for Co. It has a very small adsorption capacity of quartz, feldspar and kaolinite and consequently it is deficient in sandstones and glacial deposits derived from sandstone.

Under strong weathering conditions cobalt is lost to a soil profile and migrates downwards. Impeded drainage also mobilises Co (Mitchell, 1964). Thus cobalt is used in the weathering ratio Ni:Co which decreases with weathering in acid soils (Burek, 1985a).

In Derbyshire the average quantity of Co is 28 ppm; this is consistent with the shales, dolerites and sandstones. The tills are on average slightly higher in Co (37 ppm), due to the higher input of ground shale in the fine matrix compared with loess and 'silty drift'. This point is amplified by fig. 350. The only higher value which occurs on the computer contour map is 50 ppm near Eyam (SK 217765) which is associated with a high Ni value (fig. 351). One reason for this could be the recent uncovering of the limestone in this area and the cobalt has not yet adapted to the conditions by moving down the profile. The low over Bradwell Moor (SK 140800) which is also present for Ni, Cr and V is probably due to strong weathering conditions which leach out these particular elements. Table 37 shows this relationship.

Chromium is associated with Fe^{3+} and Al in sedimentary rocks. Under weathering conditions it acts like vanadium (Yaalon *et al.*, 1974). It is associated with basaltic rocks and with ferric iron in the minerals magnetite and ilmenite. However, it is also concentrated in micas and clays, especially illite (Nicholls & Loring, 1962; Le Riche, 1968). Under oxidising condition, with an increase in Fe^{3+}, chromium becomes more soluble, but considerable leaching is still necessary to concentrate chromium in the lower horizons of a profile.

In Derbyshire, stream sediment sampling (Nichol

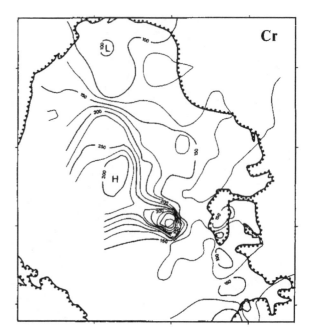

Figure 352: Chromium distribution in North Derbyshire.

et al., 1970) gave values for the limestone of less than 70 ppm. When compared with bedrock values this is reasonable. However, the superficial sediment average is 146 ppm and the tills slightly higher (152 ppm) reflecting the shale input which tends to elevate the values.

Fig. 352 summarises the elemental distribution. The high value (420 ppm) which occurs at Lees Bottom, in a temporary road section, is considered to be associated with the high muscovite and montmorillonite peaks obtained by X-ray diffraction (Burek, 1978). It is close to a weathered basalt and solifluction has occurred on the slope behind the section.

Nickel has a siderphile affinity and can substitute for iron and magnesium in minerals (Day, 1963; Yaalon *et al.*, 1974). It reacts in a similar manner to cobalt but does not oxidise like Fe and Co. Nickel is concentrated in shales and the organic fraction of marine shells. It is abundant in soils and weathering products from ultrabasic rocks (up to 5,000 ppm; Swaine, 1955), because of a concentration in olivines, pyroxenes, biotites and augites. It is also strongly bonded in the lattice of certain clay minerals. Under strong weathering conditions Ni moves down the soil profile although only Ni^{2+} ions are mobile.

In Derbyshire, stream sediment sampling recorded reasonably high levels (70–150 ppm) especially over areas draining shales. Table 34 shows the deficiency

Table 37. Mobility relationship of trace elements (with increasing weathering and subsequent movements in a profile).

Initial weathering	Moderate weathering
Li> Ca> Sr> Ba> Mn, Zn> Co, Ni, Cr, Pb, Cu, V > Ti	
Strong weathering Li, Sr, Ba, Sn, Pb, Zr	
Cr, V, Co, Ni	

Adapted from McLaughlin (1955), Yaalon *et al.* (1974a).

of nickel in the limestones and its relative abundance in all other rock types except sandstones. Nickel occurs widely as bravoite (Fe, Ni, Sn) inclusions in fluorite in mineral veins. Superficial deposits gave a high mean reading as did the tills. These were similar to the shale, dolerite and Coal Measure values. The lowest value recorded (6 ppm) is in sample 26 in a quarry crevice near Pin Dale, the highest value (1,295 ppm) occurred immediately above the Longstone Mudstone. Both these values are in keeping with the above discussion.

Fig. 351 details Ni distribution. Notable features are the high over Eyam, the bulging south of high values to incorporate Wardlow Mires, and the low centred over Bradwell Moor. The relatively high values throughout the area are due to the basically derived soils.

Nickel is used in the weathering ratios Ni:Co and Mg:Ni. The former increases with weathering, the latter decreases (Burek, 1985a).

Strontium has a high affinity with calcium (Parekh *et al.*, 1977) and in pure limestones can reach 3,000 ppm (Harrison & Adlam, 1985). It is rare in sandstones but high in shales and the weathering products of igneous rocks because relative to calcium, it is adsorbed onto clay minerals (Turekian & Kulp, 1956). Strontium is rapidly released during the initial weathering stages of feldspar, especially the plagioclases, and the clay minerals are the weathering product. Thus the background level of Sr on the Derbyshire limestone plateau is high.

The ionic radius (1.12 Å) falls between Ca^{2+} (0.99 Å) and Ba^{2+} (1.34 Å) but has greater affinity for Ca because it is more abundant.

In Derbyshire values for all deposits are reasonably high and for tills specifically, very high, showing the influence of shales and insoluble residue of limestone on the deposits. However, when examined closely, the figures fall short of the average shale or limestone values. The influence of sandstones on the deposits is great, especially on the loess and this causes the values of strontium to be low. In the case of the tills however the value is much closer to the shale data. The highest concentration noted is that of 1,593 ppm which occurs in a 'silty drift' sample above a lead vein over the very pure Bee Low Limestone (1,000 ppm; Harrison & Adlam, 1985). The presence of high peaks of calcite, baryte and fluorite together with a lack of feldspar accounts for the high value.

Fig. 353 shows the dispersal of the element and highlights the above discussion.

An erratic sand and gravel block within the till at Raper Pit (SK 218652; see Burek, 1977b, and this volume) illustrates the high and low potential of strontium. The current-bedded sand with a relatively low value (190 ppm) and the other sands and gravels (171 and 203 ppm respectively) compared with the till (greater than 311 ppm) show the affinity Sr has with clay minerals and calcite and not with quartz and sands (although it must be admitted that the values in this specific case are slightly higher than would normally be expected because the sands are slightly calcareous. It serves to illustrate the point however).

Sr is used in the weathering ratio Ba:Sr (Burek, 1985a).

In conclusion it can be seen that in rock weathering, isomorphic replacement can take place if the following four conditions are fulfilled:

(i) Similar ionic radius; tolerance of 10% e.g. Co^{2+} (0.82 Å) can replace Fe^{2+} (0.83 Å),

(ii) Similar ionic charges.

(iii) Similar coordination number (Ba^{2+}, coordination number 8 cannot replace Fe^{2+}, coordination number 6),

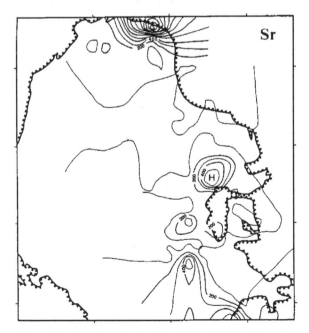

Figure 353: Strontium distribution in North Derbyshire.

Table 38. Major absorbents and their associated trace elements.

Organic matter: Ni, Co
Iron monosulphide: Pb, V, Ba, very little with Co, Ni
Clay minerals (especially illite): Ba, Sr, Pb, V, Li, Ga
Within lattice: V, Li
Interlayered position: Ba, Sr, Pb

(iv) Electrical neutrality must be preserved in resulting structure (Carroll, 1970b).

It is important to note that adsorption is more important than precipitation. The three major adsorbents are outlined in table 38.

Cold climatic conditions

Under cold climates the chief breakdown of the rock surface occurs as physical weathering but chemical weathering is not necessarily brought to a standstill (McLaughlin, 1955; Tedrow & Ugolini, 1966; Carroll, 1970a). In tundra areas for example, periodic leaching leads to an accumulation of more resistant grains. This occurs in the warmest part of the year above the permafrost, as this acts as a barrier isolating soils from the underlying weathering rocks.

Tedrow & Ugolini (1966) in a study of Antarctic soils and pedogenic processes, emphasise the lack of the biotic element. All soils are ahumic. This inhibits the concentration of Pb and Sn (which are dependent on organics to move up the soil profile) and is one major difference between glacial or periglacial and temperate climates. Ahumic soils in general release Ca, Mg, K, Na, Cl and S from the parent rock and make them available for redistribution under glacial or periglacial conditions. Table 39 gives a typical ahumic profile. Algal growth is recorded at $< 0°C$ (32°F), and bacteria down to 3°C (39°F). Thus in the northern hemisphere during the cold periods of the Pleistocene, pedogenic processes would be influenced by four factors; climate, topography, lithology and time. Limited chemical weathering can take place in snow-free areas during the warm summer months when air and rock temperatures rise above freezing.

To summarise, glacial physical weathering may initiate breakdown of the parent material and allow a greater surface area to be available for limited chemical weathering during the summer. The prime factor in the above discussion is time. Chemical weathering of rock at low temperatures is slow.

Weathering ratios

A weathering ratio is the relative weathering relationship of one element plotted against another element. Thus by analysing different pairs of elements which are chemically related to each other, down a section, a ratio is produced. This can be used to plot a graph of relative weathering *versus* time represented by the vertical sequence of a previously undifferentiated section.

Weathering ratios used in the study of tills have been found in the literature (Burek, 1979; 1985a). Seven different ratios have proved useful but more have been outlined in table 40.

The weathering ratios for 18–48 cm data from Derbyshire have been contoured in order to give a visual representation of the ratios. These maps can be looked upon either as a distribution of sediment age, if time is the most important factor in weathering or as sediment susceptibility to climatic influences. The highs and lows in all the subsequent ratio maps bear no relationship to bedrock geology, to geomorphology or topography (see fig. 119 in Burek, this volume). Because of the nature of the sediments and the sampling depth, vegetation should have only limited influence. The chief influential factor on these weathering ratios is time.

$Ga:Al_2O_3$

Fig. 354 shows the distribution of this ratio, which decreases with increased weathering. In the superficial sediments (117 samples) the values range from 0.85–1.85. The most intense weathering (the lower values) occurs in the northwest near Buxton (SK 060735) and the area southwest of Longstone (SK 182715). As a general trend the north and northwest is more intensely weathered than the centre or the east.

Table 39. Typical ahumic soil profile. Colour 10 YR 4/3; location: Antarctica.

Brown medium sandy loam	Ca	Mg	Na	K	P	SO$_4$	Cl
6.4 – 7.6 cm	–	240	1910	160	9	1.2	1100
12.7 – 22.9 cm	–	270	1240	31	11	2.4	900

All values in ppm except SO$_4$, which is given in percent (after Tedrow & Ugolini, 1966).

Table 40. Effect of weathering on weathering ratio.

Ga:Al	decreases with weathering
Ni:Co	increases with weathering
Mg:Ni	decreases with weathering
Fe:Co	decreases with weathering
Ba:K	increases with weathering
Ti:Al	increases with weathering
Ba:Sr	increases with weathering
V:TiO$_2$	decreases with weathering
Fe:Mn	decreases with weathering
Cr:TiO$_2$	decreases with weathering
Sr:Ca	increases with weathering
Ba:Ca	increases with weathering

Data compiled from various sources.

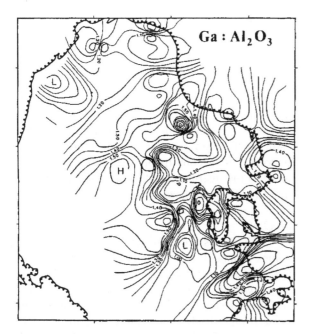

Figure 354: Ga:Al$_2$O$_3$ distribution in North Derbyshire.

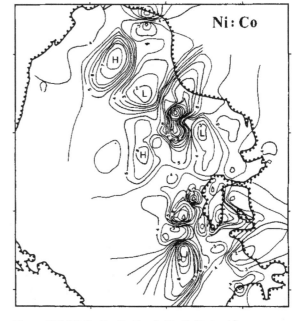

Figure 355: Ni:Co distribution in North Derbyshire.

Older sediments lie to the northwest of the area in question and in a broad band from Eyam to Cales Dale (east of Monyash). This broad band is broken by 'islands' over Monsal Dale and Longstone. Areas off the limestone to the east are less weathered. The relationship between these ratios and the uncovering of the limestone from beneath a shale cover is unclear. If the relationship is straightforward, all weathering ratios should decrease (in this specific ratio) from the shale/limestone boundary to a large sink over the oldest limestone revealed. This is not the case. Thus it must be the age of the sediments themselves which is depicted. We are looking at an areal distribution of loess and 'silty drifts' of different ages.

Ni:Co

Fig. 355 shows the distribution of the elemental ratio for Ni:Co. The values range from 1–13. This map emphasises the increased weathering of Bradwell Moor (also shown in Ba:Sr, fig. 356) and a high to the west of Wardlow Mires – the Tideswell and Litton area. To the south is a slight dome at Sheldon. A general trend indicates a central arcuate ridge with values falling to the east and west. The ridge is cut through in three places. These areas do not correspond to topographical lows and are not influenced by the bedrock geology. It is inferred that they are places of 'recent' i.e. Postglacial deposition which have had

less chance to weather than surrounding areas.

Ba:Sr

Ba:Sr (fig. 356) gives values more consistent with podsolic soils (8.8) than with tundra soils (1.5) and shows that chemical weathering has moved elements

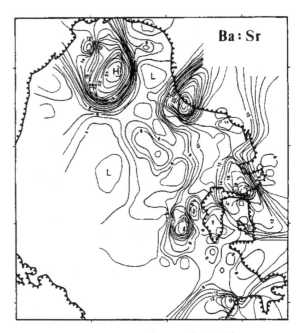

Figure 356: Ba:Sr distribution in North Derbyshire.

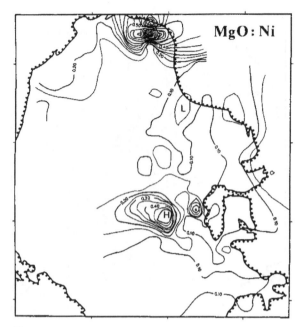

Figure 357: MgO:Ni distribution in North Derbyshire.

does not indicate increased weathering because of the falling ratio values along Longstone Edge. The general lowering of values west with a tongue extending due north and a possible tongue northwest would tend to indicate either recent deposition and therefore less elemental movement or disturbance of formerly stable material.

The low values south of Bradwell Moor are interesting in that they cover an area of hilltops and dale bottoms. Thus geomorphology in this area is immaterial to the distribution. A general trend surface is indicated from northeast to southwest.

MgO:Ni

Fig. 357 shows the distribution of MgO:Ni. High values around Castleton must be attributed to the high nickel content in the weathered profiles associated with lead mining contamination. This map correlates well with high Mo and Pb. Ratio values range from 0.05–1.20 but most of the area is below 0.30 except Castleton and Sheldon. Both can be attributed to human influence. Sheldon and Pin Dale are old lead mining areas and contamination is probably the cause of these peaks. Thus this ratio, which should decrease with increased weathering is of little value in the North Derbyshire area.

Fe:Co

Fig. 358 shows the distribution of the weathering ratio Fe:Co in the 18–48 cm sediments. This ratio

within the top 36 cm since the end of the Devensian. Values range from 4–27. The highs over Bradwell Moor, around Hucklow and the eastern end of Longstone Moor show that weathering has either continued longer in those areas or that it was more intense. The fact that they all occur on high elevations

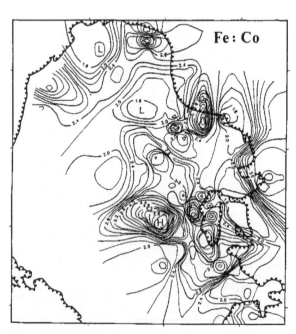

Figure 358: Fe:Co distribution in North Derbyshire.

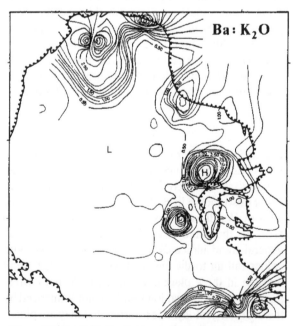

Figure 359: Ba:K$_2$O distribution in North Derbyshire.

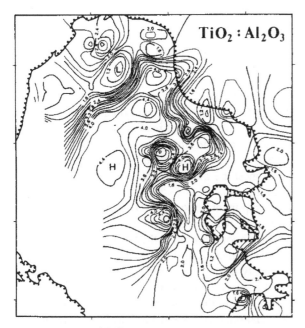

Figure 360: TiO_2:Al_2O_3 distribution in North Derbyshire.

TiO_2:Al_2O_3

This ratio increases with increased weathering as Al_2O_3 is lost relative to TiO_2. The values on fig. 360 range from 0.8–4.8. The highs occur in the centre and lows around the edge. This ratio may be reflecting the unveiling of the limestone and recent weathering. It is the only ratio which displays this pattern and bears no similarity to any of the other relationships.

However, it must be remembered that many of the TiO_2 values were considered high (beginning of the chapter) which would tend to elevate the ratios. The second difficulty is the presence of anatase (TiO_2) in the 'clay wayboards' (weathered tuffs and dolerites; see 'Geological background' in Burek, this volume, and Walkden, 1972) which would account for the high availability of Titanium as this element tends to be residual.

Till geochemistry

The tills of Britain fall into three groups based on the geochemistry, which in turn reflects the sum of weathered erratics and the type of matrix i.e. clay or sand, calcareous or siliceous. This, in turn, is related to the geology over which the ice has travelled, and the weathering capabilities of the rock types involved, i.e. basalt weathers three times faster than granite (Carroll, 1970).

It is important to remember that initial weathering is geochemical and inorganic and that secondary organic weathering will only occur once pedogenesis has started. Chemical weathering occurs in frost-free areas primarily but such weathering has been recorded from Antarctica (Tedrow, 1966). Arctic conditions are ahumic. Therefore here only four factors are influencing pedogenesis: lithology, climate, topography, and time. The importance of rock type and summer temperatures is shown by the fact that Fe^{2+} and Mn^{2+} move across temperature boundaries and oxidise to Fe^{3+} and Mn^{4+} stabilising ahumic soils. The soil-forming processes release Ca, Mg, K, Na, Cl and S from the parent material and redistribute them. These conditions could be represented in the British Quaternary. However, the age of the glacial deposits is of paramount importance because weathering takes place only very slowly under these conditions. Interglacial temperate periods are the main intervals during which weathering occurs and older tills have been subjected to one or more of these events and are therefore geochemically more mature. Time is an important factor. $CaCO_3$ decreases with time in loesses of all ages. CaO decreases with an increase in

should decrease with increased weathering. The values obtained range from 1.2–6.6. Areas of intensified weathering occur over Bradwell Moor excluding the northeast edge as far south as Chee Dale, southeast of Wormhill and secondly over Eyam.

This pattern resembles the Ga:Al_2O_3 distribution (fig. 354). Areas of younger and less well developed weathering occur to the west of Eyam and in a broad band from Taddington to Longstone. Values rise sharply off the limestone to the east.

Ba:K_2O

Fig. 359 shows the distribution of the weathering ratio Ba:K_2O. Values range from 0.25–5.00. Highs (more intense weathering) occur over five main areas, all correlating very closely with the barium distribution in fig. 348, although now the highs are amplified. These five areas are Bradwell Moor, a small area near Hucklow, Longstone Edge, the area around Sheldon and the southern edge around Youlgreave. All are lead mining localities.

It is interesting to note that the Ba:K_2O (fig. 359) and the Ba:Sr (fig. 356) ratios do not resemble each other except in a very general way. As the Ba:K_2O so strongly resembles the Ba distribution, it is not considered that this ratio is a good weathering indicator in this area, because it merely duplicates results.

precipitation (leaching) which would also occur once the climate had ameliorated from a full glacial condition.

The complexity of the data and difficulty of interpretation has neccessitated the use of multivariate statistical analysis which was first applied to the data obtained from the Derbyshire deposits (Burek, 1978).

North Derbyshire tills

Four datasets were used to test the different groups of samples. Dataset one contained 117 loess and 'silty drift' samples, dataset two 98 till samples and dataset three the points (116) used to draw out the contoured maps of the different elemental distribution. Dataset four acted as a reference guide. A factor analysis was performed on the four datasets in order to simplify the geochemical interpretation and reduce the number of variables from 26 to a meaningful and interpretable number. To highlight the axes obtained, a varimax rotation was performed. This was followed by a Q-mode cluster analysis which groups the samples on the basis of similar geochemistry. Only 21 variables were used: TiO_2, Ag, Be, Bi, Ge and Sn were omitted from the statistical data either because of the small variation in the samples (which would bias the results) or lack of reliability in the original geochemical data.

Results

In dataset one there is a strong positive correlation between Al_2O_3 and Ga (reflecting the clay mineralogy) with subsidiary Li, V and Co, between CaO, Sr, Mo, Cd, Ba and Cu, with secondary Pb and Zn, between Fe, Cu and Cr, between MnO and Ni and finally between SiO_2 and Zr.

This represents the carbonate association with mineralisation, a heavy mineral assemblage and a detrital association respectively. The strong negative association between CaO and SiO_2, the carbonate and non-carbonate fractions, and Zr with Cr and Mo are reflecting the high associations between these two commonly grouped elements. The mineralogy of the loess and 'silty drift' shows this strong silica content with no calcite present. However the background grouping of mineralised vein deposits with a little CaO shows the incomplete insoluble residue mixing i.e. the 'silty drift'. These groupings emphasise the cryoturbation of the loess with the insoluble residue to form 'silty drift' (cf. Burek, this volume).

To help the understanding of the geological significance of the component axes and to emphasise the influential relationships, a varimax rotation of the seven axes was performed. Fig. 361 illustrates the varimax axis loadings and fig. 362 shows the maximum variance accounted for on the first two axes. In fig. 361 the main component on axis one represents the high association of insoluble residue from mineralised veins and limestones. The negative loadings of SiO_2, K_2O represent the lack of these elements in the limestone and insoluble residue, but not in the clay minerals.

Component 2 seems to represent the high association of Fe, Cu and Cr in basalts and lavas of the area and their contribution to the 'silty drift' and loess.

Component 3 represents the shale component in the 'silty drift'. The high MnO which can substitute for Ca in carbonates or as an oxide with a pyrite structure and the negative SiO_2 value precludes this association. Component 3 is therefore related to component 4.

Component 4 represents the carbonate / non-carbonate or detrital relationship. This could be interpreted as limestone against sandstone control on the loess and 'silty drift'.

Component 5 is difficult to understand in this context. The association of MgO to Ga and K_2O could relate to feldspar or dolomite input.

Component 6 shows the clay mineral association and Component 7 highlights the vanadium and molybdenium emphasising the shale component.

Therefore the geochemical associations developed in the loess and 'silty drift' are found to include seven meaningful relationships. Firstly the geochemistry of the 'silty drift' and loess is influenced by an insoluble residue component, then successively in decreasing importance, a basic igneous component, a manganese component, a carbonate component, a magnesium-potassium component, a clay mineral component and lastly a vanadium-molybdenium component.

In the till dataset two, there is a strong positive correlation (table 41) between CaO, Sr and Cd, between Al_2O_3, Fe, K_2O and SiO_2, between Co, Cr, Cu and Ga and finally between Mo and Ni. R-mode principal components analysis produced six significant components (eigenvalues > 1.0) which accounted for 82.7% of the total variance. A varimax rotation was completed on each of the axes to emphasise the relationships between the variables (fig. 363).

The main component on axis one represents the influence of shale and dolerite sills on the till samples. In order to test this primary association a subsi-

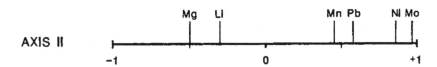

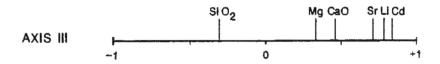

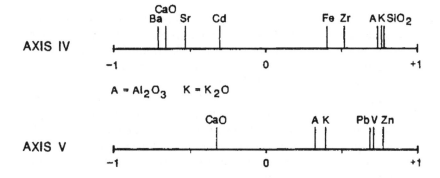

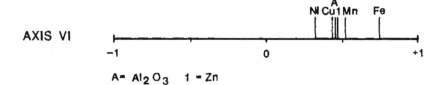

Figure 361: Varimax axis loadings of geochemical variables for dataset 1.

diary dataset was instigated which contained all the till samples and 13 different rock types. Because of the nature of the data it was only possible to correlate eight variables: the major oxides, Cu and MnO. However, the resulting Q-mode cluster analysis plotted the shale samples firmly with the tills and not with the

dolerite (Burek, 1985b). Thus the tills themselves are composed of comminuted shale, but in classifying the tills, the dolerite content is significant.

The second component represents a negative relationship between Mg and Li and Mn, Pb, Ni and Mo. Lithium commonly replaces magnesium in silicates

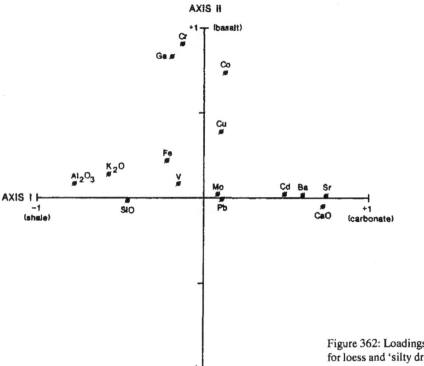

Figure 362: Loadings of variables on the 1st and 2nd axes for loess and 'silty drift'.

especially illite, kaolinite and chlorite (all of which occur in the tills). This association represents clays *versus* non-clays.

Component 3 is a simple carbonate/non-carbonate or detrital association and this is also portrayed in component 4 where a carbonate assemblage has a negative correlation with a clastic group.

Component 5 represents carbonate *versus* oxidised sulphide deposits while component 6 shows high Fe values and associated elements, indicating the proximity of basalt deposits and thus the high content of this rock type in tills. These will weather to release Fe into a till profile.

The scores of till samples were subjected to a Q-mode cluster analysis program to examine the effects of these components on the till samples themselves (fig. 364). Three test samples were included and these have grouped out together as G. All the samples show high correlation, but it is possible to form some meaningful groups. Seven clusters can be recognised. Group A contains all samples from Shining Bank Quarry and Raper Pit except sample 220 (a till sample from southwest of Bakewell village in the Wye valley). Group B includes 3 Stoney Middleton tills and one from Shining Bank Quarry. These possibly represent weathered samples. Group C contains many different localities and probably repre-

sents weathered tills. Group D contains only Shining Bank Quarry samples from 3.5 - 5.3 m and is highly influenced by factor 3. Group E contains only the samples from Pocket Hole Quarry and shows the dissimilarity of older tills in their geochemistry. Group F appear to be related to highly oxidised sulphide deposits (factor 5). Group G contains 'silty drift' and solifluction deposits, the test samples. The dendrogram, while not completely comprehensive, does group the tills mainly by locality and shows the importance local geology has on the geochemistry of the tills. The information gained from the statistical analysis supports the hypothesis that till from Pocket Hole Quarry is substantially different from elsewhere on the plateau and confirms that the tills are primarily derived from the Carboniferous shale of the area.

The reference set shows a high negative correlation between CaO and SiO_2 and a strong positive correlation between Sr, Cd, Ba and CaO and between Co, Cr and Ga. Fig. 365 shows the results of the varimax axis loadings. All are relationships which have been emphasised in the other analyses.

Component 1 is the shale/carbonate relationship

Component 2 is the basalt association

Component 3 is the clastic *versus* carbonate association

Component 4 is the mineralisation element

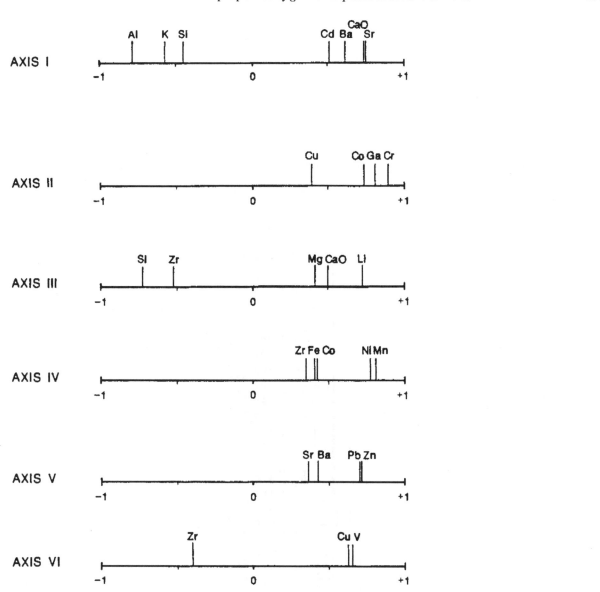

Figure 363: Varimax axis loadings of geochemical variables for Derbyshire till.

Component 5 is the insoluble residue and subsidiary Pb/Zn association

Component 6 is the Cu and V *versus* Zr relationship

Component 7 is the Mo/Ni association.

Three main groups are discerned: till, loess and 'silty drift', and mineralised sediments. The association of elements for each group is slightly different. In the loess and 'silty drift' data the main distinguish-

ing feature is the carbonate and mineralised residue in the insoluble residue and secondly the basaltic component. In the till the chief distinction is the influence of shale and lavas on the samples. In the reference dataset the marked difference is similar to that of the loess dataset, the clays *versus* carbonate and mineralisation association. These are presented in figs. 366 and 367 showing the axis and its associated major and trace elements.

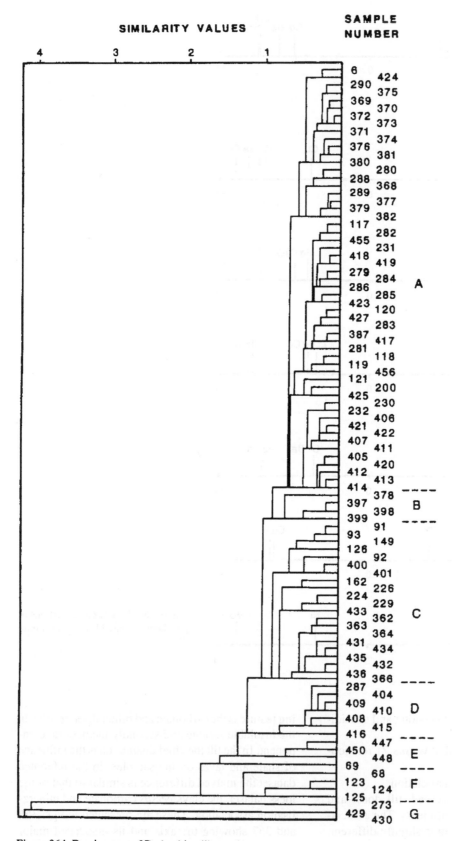

Figure 364: Dendrogram of Derbyshire till samples.

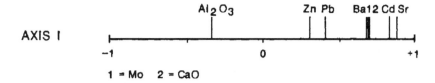

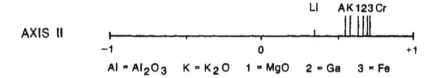

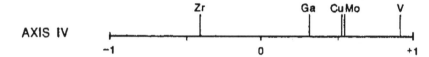

Figure 365: Varimax axis loadings of the geochemical variables for the reference set.

Summary of Derbyshire data

In conclusion the geochemical data falls into distinct patterns, which show an association of elements. The loess and 'silty drift' samples show that an external influence has been instrumental in their deposition. The decrease in many element quantities away from the eastern shale - limestone boundary points to an eastern source for the external influence. The superficial sediments contain many elements not present in the limestone, or in its insoluble residue and therefore they could have only been brought in from outside. The presence of lavas, basalts and a shale outlier complicate the picture. However it is possible to see their contribution to the distribution as amplified by the statistical analysis.

The associations of elements appear to be a grouping of Ba, Sr, Cd and Mo around CaO, a clustering of

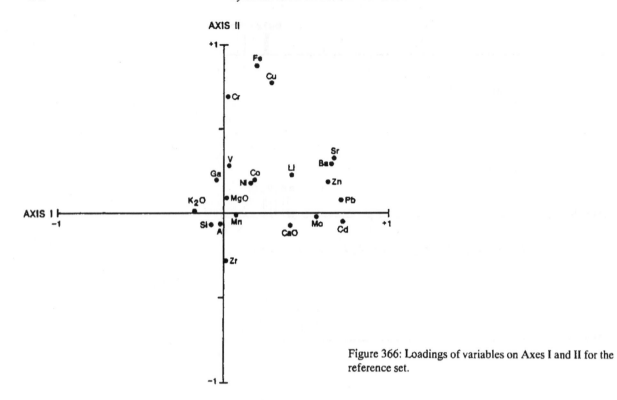

Figure 366: Loadings of variables on Axes I and II for the reference set.

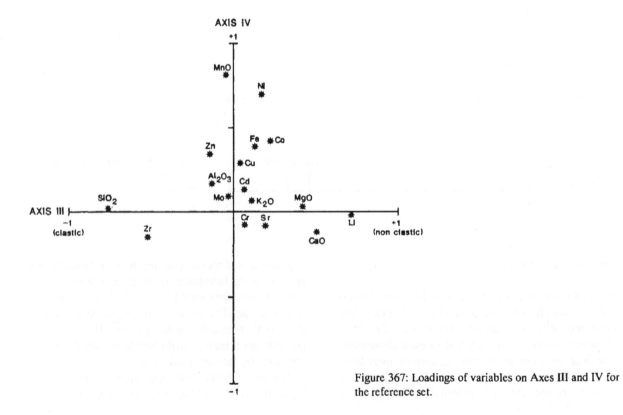

Figure 367: Loadings of variables on Axes III and IV for the reference set.

Fe, Ba, Cd, Cu, Li, Mo, Ni, Pb, Sr, Zn over the Castleton old lead mines, a grouping of MnO, Cu, V over the tills of Long Rake and finally a clustering of SiO_2, MnO, Li, Mo and V over the shale enclave on Northing 70.

The statistical analysis distinguishes three main associations a Cu, Cr, Ga and Co group, a Ba, CaO, Cd, Sr and Mo association and a Pb, Zn relationship. SiO_2 is always grouped with Zr but normally has a negative association with the axes and it is the absence, not the presence, of these two elements which is a distinctive feature. It always has an antipathetic relationship to CaO and carbonate association.

The statistical analysis shows the influence that shale has on the clay content of the till and the importance of the lavas in the loess and 'silty drift'.

Thus general trends can be discerned which point to older sediments present over Bradwell Moor and the western part of the area while younger sediments occur towards the eastern margin. This probably reflects the older residual gravels, cherts and the soils of undisturbed areas as opposed to the younger loess-derived sediments.

Acknowledgements

We would like to acknowledge the following for their help: Drs. J. Rice, A. Cheshire, J. Boardman, J. Scourse, H. Davies and C. Whiteman for providing samples, S. Mathers, F. Cox and J. Arum for literature, N. Marsh of Leicester University, Geology Department for running the analyses, Poroperm-Geochem Ltd., Chester for providing computer, secretarial and graphical assistence and J. Rose for providing the incentive for putting pen to paper.

Glacial deposits of Britain and Europe: General overview

JÜRGEN EHLERS, PHILIP GIBBARD & JIM ROSE

There have been several publications recently that have attempted to synthesize the Pleistocene stratigraphy of Britain and Ireland (e.g. Shotton, 1985; Bowen et al., 1985). Our task here is not to review these previous overviews, but to ask, how do the glacial deposits fit into the overall stratigraphic framework?

The locations of the critical areas dealt with in the regional part of this volume are shown in fig. 368. The slightly higher concentration of articles dealing with the southeast reflects the fact that these areas contain the most complete sequence of Quaternary strata. In almost all other parts of Britain and Ireland, Quaternary deposits are either restricted to the Devensian, or the age and stratigraphic position of older deposits remains controversial.

It is essential to see the British Isles in the context of their setting adjacent to NW Europe and the North Atlantic Ocean. This is particularly important in view of recent discoveries in the offshore regions, especially in the North Sea basin, where the relationship of the British and Continental sequences will ultimately be established.

Since new dating techniques are only just beginning to develop, it may still be too early to attempt a reliable correlation between the glacial deposits of Britain, Ireland and those of continental Europe. Nevertheless an attempt has been made here (table 41). Despite the many uncertainties, there are a few fixed points within the stratigraphic framework which can be used for this purpose.

One, the comparison of the isotope stages of the deep sea record with the palaeomagnetic stratigraphy demonstrates that the Brunhes/Matuyama magnetic epoch boundary, dated at 788,000 years B.P., corresponds with oxygen isotope stage 19 or 20, depending on the reference core used. Since this magnetic boundary can be identified in Cromerian 'Glacial A' of the Dutch sequence (Zagwijn et al., 1971), the Dutch prefer to place the magnetic epoch boundary in stage 20 (Zagwijn, 1985; de Jong, 1988). The IGCP 24 Committee on the other hand put it in stage 19 (Šibrava et al., 1986). By contrast it is now accepted that the Ipswichian/Eemian Stage is represented in the deep sea record by isotope stage 5e. This leaves 6 temperate stages and 7 cold stages between the fixed points, the correlation of which are yet to be determined.

Here the authors have interpreted the Hoxnian/Holsteinian as representing isotope stage 11 for the following reasons: 1. The Wacken/Dömnitzian event is probably a true interglacial comparable with the Hoxnian or the Ipswichian and should logically therefore be represented by a comparable peak in the isotope curve. 2. There have without doubt been two large-scale glaciations of continental Northwest Europe separated by an ice-free interval during the Saalian Stage. No interglacial deposits have been distinguished so far from the interval, but if the isotope curve represents changes in ice volume rather than climate, a period in which northern Germany and Poland were ice-free should be represented by a peak in the isotope curve. 3. Isotope stages 6, 8 and particularly 12 represent major peaks, while stage 10 (the equivalent of the continental Fuhne Stadial in our scheme) seems to have been of shorter duration and minor importance. 4. The Cromerian 'Interglacial IV' is characterised by a strong influence of certain volcanic minerals in River Rhine sediments, associated with the onset of the Eifel volcanism (de Jong, 1988). The Selbergit tuff of the Eifel has yielded a maximum K-Ar radiometric date of 570,000 years B.P. (Frechen & Lippolt, 1965), which indicates that it should be present in stage 14 and younger Rhine sediments.

The part of the stratigraphic record covering the time span between about 750,000 and 1,000,000 years B.P. is far better understood than major parts of more recent history of the earth's. This is because several reversals of the earth's magnetic field occurred during this period. These events are well-dated and assist in providing a sound stratigraphic

Table 41. Proposed correlation of Continental, British and Irish Pleistocene sequences showing suggested equivalence with deep sea and palaeomagnetic stratigraphies. DS = deep sea stages; Int. = Interstadial; Igl. = Interglacial; – – – = magnetic reversal. Correlations of the stages between Tiglian C–3 and C–4c and Eburonian and Cromerian IV with stages in Britain should be considered very tentative.

Britain	Ireland	Continent	Palaeomag.	DS
Flandrian	Littletonian	Holocene		1
Dimlington Stadial				2
Upton Warren Int.				3
Stadial	Midlandian	Stadial		4
Brimpton Int.	(Fenitian)	Odderade Int.		5a
Stadial		Stadial		5b
Chelford Int.		Brørup Int.		5c
Stadial		Stadial		5d
Ipswichian		Eemian		5e
Wolstonian	Munsterian	Warthe Substage		6
		Interstadial		7
		Drenthe Substage		8
		Wacken/Hoogeveen		9
		Fuhne Stadial		10
Hoxnian	Gortian	Holsteinian		11
Anglian		Elsterian II		12
		Miltitzer Int.		
		Elsterian I		
Cromerian s.s.		Cromerian IV		
Beestonian (part)		Glacial C		
		Cromerian III		
		Glacial B		
		Cromerian II		
		Glacial A		
– – – – – – – –		Cromerian I	788 ka	19
		Dorst Glacial		
		Leerdam Igl.		
		Linge Glacial		
– – – – – – – –		Bavel Igl.	– 900 ka	
		Menapian		
		Waalian		
– – – – – – – –		Eburonian	– 1750 ka	
Beestonian (?)				
Pastonian		Tiglian C-5		
Pre-Pastonian/Baventian		Tiglian C-4c		
Bramertonian/Antian (?)		Tiglian C-3 to 4b		
Thurnian		Tiglian B		
Ludhamian		Tiglian A		
– – – – – – – –			– 2200 ka –	
Pre-Ludhamian		Praetiglian/Reuverian		

framework. The Early Pleistocene deposits with normal polarity, which were originally thought to be of Waalian age, have now been reassigned to the newly recognised Bavel Interglacial of the Bavelian Stage. Their deposition must have occurred during the Jaramillo Event, about 900,000-970,000 years B.P. This implies that the Bavel Interglacial must be

either isotope stage 23 or 25. If the Bavelian is stage 23 (the preferred interpretation of the Dutch workers; cf. de Jong, 1988), then the subsequent Leerdam Interglacial should be Stage 21.

Whilst the upper Early Pleistocene is relatively well known, the stratigraphic record of the older part is more fragmentary, and vague correlations based on magnetostratigraphy can only be offered. The oxygen isotope stages for this period have only recently been assigned (N.J. Shackleton, personal communication) but it remains uncertain how these relate to the climatic cycles recognised in shallow marine and land sequences.

However, it should also be said that the correlation in the younger part of table 41 is largely tentative and may have to be revised as better dates become available.

The 'preglacial' record

The 'preglacial' Pleistocene deposits of Britain, particularly in East Anglia, have been discussed in the contributions by Hey and Hart & Boulton (this volume). The British sequence is, of course, far from complete and a satisfactory correlation with the Dutch sequence has not been possible until very recently, mainly because of the lack of precise dates and the identification of significant biostratigraphical events. On the continent there have been major discoveries recently such as the recognision of the Bavel and Leerdam Interglacials (Bavelian Stage; Zagwijn, 1985) and the detection of a complete record of Menapian to Elsterian sediments above the Gorleben salt dome in North Germany (Müller, 1985). In Britain much of the evidence remains to be deciphered. Nevertheless, new results from BGS works in Suffolk strongly suggest that the Red Crag Formation is mostly of Pre-Ludhamian age and that this, contrary to earlier ideas (e.g. West, 1980a), may be assigned to the Netherlands' Praetiglian or Late Reuverian Stages (Zalasiewicz et al., 1988). Later, true cold stage sediments of Baventian age which are thought by some (e.g. Bowen, this volume; Hart & Boulton, this volume; Solomon in Funnell & West, 1962) to include the first evidence of possible ice-rafted detritus in the British North Sea sequence may simply record longshore drift of minerals derived from the North German river system (A. Burger, personal communication). These sediments represent a very early position in the Dutch succession (cf. Funnell, 1987; Zalasiewicz et al., unpublished; cf. Bowen, this volume). Recent investigations of the small mammal faunas

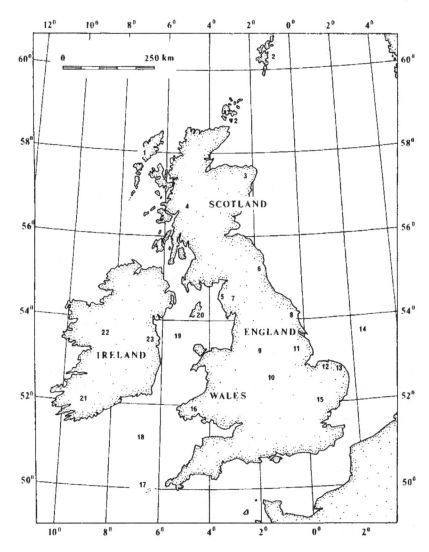

Figure 368: Location map for the regional part of this volume. 1 = Hebrides, 2 = Shetlands and Orkneys, 3 = Buchan, 4 = western Grampians, 5 = Cumbrian lowlands, 6 = Northumbria, 7 = Lake District, 8 = East Yorkshire, 9 = Peak District, 10 = Midlands, 11 = Lincolnshire, 12 = northwest Norfolk, 13 = northeast Norfolk, 14 = southern North Sea, 15 = southern East Anglia, 16 = south Wales, 17 = Isles of Scilly, 18 = Celtic Sea, 19 = Irish Sea Basin, 20 = Isle of Man, 21 = southwest Ireland, 22 = central and western Ireland, 23 = eastern Ireland, 24 = Cheshire and Shropshire lowlands.

have further indicated that a large hiatus is present between the British Pastonian and Cromerian Stages (Mayhew & Stuart, 1985). The implications of this are that there is a considerable intervening time period that may be only partially represented in eastern England and that the Pastonian to Pre-Ludhamian Stages should be equated with the continental Reuverian (Late Pliocene) to Tiglian Stages (Mayhew & Stuart, 1980). Recently a Dutch-British Workgroup has been established to examine the correlation across the southern North Sea during the Early and early Middle Pleistocene (Zagwijn *et al.*, in press). The results of these discussions are included in table 41.

Regarding direct evidence of glaciation there are two horizons in which important evidence is preserved in the Netherlands. The Hattem Beds of the Enschede Formation are the older of these. They comprise gravelly sands deposited by a westward flowing river system during the Menapian (Lüttig & Maarleveld, 1961; 1962; Zandstra, 1971). At this stage the east German and Polish rivers were diverted westwards for the first time, bringing Thüringerwald and Erzgebirge erratics to the Netherlands. The Hattem Beds also contain Scandinavian pebbles such as Rapakivi Granite, Stockholm Granite, Dalarna Porphyries and Småland Granites, possibly transported by river ice (Zandstra, 1983). According to Bijlsma (1981) the Baltic River which had been flowing into the southern North Sea since the Pliocene, ceased to exist after the Menapian Stage. This implies that the Baltic basin may have been initiated by glacial scour at this time.

The second indicator of glacial conditions is found

in the Weerdinge Member of the Urk Formation (Cromerian 'Glacial C'). The sediments, which apart from eastern and some southern (Rhine) material, contain Scandinavian erratics, have been interpreted as glaciofluvial deposits, derived from ice in the vicinity (Ruegg & Zandstra, 1977; Ruegg, 1983). No tills of this age have been reported from anywhere in North Germany. However, at Harreskov, central Jylland (Denmark) a till has been found beneath early Middle Pleistocene interglacial deposits (Sjørring, 1983). Several Early and Middle Pleistocene pre-Elsterian glaciations have been also reported from Poland (Rzechowski, 1985), the oldest being TL-dated to greater than 700,000 years B.P. and thought to have taken place during the Jaramillo Event. This corresponds closely with the Early Pleistocene till discovered in the northern North Sea Basin by Sejrup *et al.* (1987).

Drillings in the North Atlantic have shown that ice-rafting had already begun by the Pliocene/Pleistocene boundary, about 2.4 million years B.P. (Shackleton *et al.*, 1984). Furthermore between Norway and Greenland drift ice occurred - at least periodically - as early as 3-5 million years B.P. (Thiede *et al*, 1986). On Iceland, a marked climatic deterioration occurred with the onset of the first glaciation at 3.1 million years B.P. (Einarsson & Albertsson, 1988). Clear evidence of a glacial climatic influence such as sand blocks, ice wedge casts and far-travelled (Scandinavian) erratics, some of them clearly striated, have been found in the Kaolin Sand sequence on the Isle of Sylt in northwest Germany (von Hacht, 1987; Ehlers, 1987). According to heavy mineral analyses by Burger (1986) the Kaolin Sand should be correlated with the Brunssumian Stage (Lower Pliocene).

Meanwhile in southern Britain a series of fluvial aggradations comprising predominantly quartz-rich gravels of the Kesgrave Formation were deposited by the ancestral Thames (Hey, this volume). These deposits range from the 'Pre-Pastonian a' Substage to the early Anglian and contain ample evidence of cold climate deposition. Finds of volcanic rocks, especially rhyolithic tuffs derived from North Wales (Whiteman, 1983) within these gravels have been interpreted as indicating early glaciation in the Welsh Mountains, the so-called 'Berwyn Glaciation'. Since these erratics have been recovered from several members within the Kesgrave Formation, it has been thought that this glaciation may have been multiphased (Bowen *et al.*, 1986). There is, however, relatively little chronological control on these fluvial aggradations and therefore any detailed interpreta-

tion must await further discoveries. However, recent finds of organic temperate stage sediments at Ardleigh, Broomfield and Little Oakley in Essex (Gibbard *et al.*, unpublished; Bridgland *et al.*, in press) indicate that the younger Kesgrave members span the late 'Cromerian Complex' (early Middle Pleistocene). They provide evidence for two additional stages to the British sequence. The Kesgrave Formation can be traced into the offshore region where it forms part of the fluvial-deltaic Yarmouth Roads Formation (Balson & Cameron, 1985). The latter also includes material derived from the Rhine, Meuse and the North German river system (Zagwijn, 1974; 1985).

Evidence for correlation of the British Cromerian Stage (*sensu stricto*) with the Dutch 'Cromerian Complex' is unclear as a result of the limited knowledge of events and palaeontology during this period. However, at present it appears that the Cromerian *sensu stricto* either may not be represented in the Netherlands and could occur chronologically between Cromerian Interglacials II and IV or could possibly be equivalent to Cromerian II (Westerhoven) (Dutch-British Workgroup: Zagwijn *et al.*, in press).

The Anglian Glaciation

Both in Britain and on the Continent Anglian/Elsterian tills are the first widespread unequivocal evidence of the presence of Pleistocene ice sheets. On the Continent, the Elsterian seems to consist of two separate major ice advances, separated by an ice-free interval (Eissmann, 1975). Although up to five different Anglian tills can be distinguished in parts of East Anglia, no evidence of a major intervening ice-free period between has so far been discovered (cf. Hart & Boulton and Ehlers *et al.*, this volume). The previously identified Corton event is no longer interpreted as an interstadial since it contains no evidence of marked climatic amelioration (West & Wilson, 1968), but appears to represent a change in ice dynamics manifest as local ice front retreat. Recent studies indicate deposition of these sands as a product of glaciofluvial activity (Pointon, 1978; Hopson & Bridge, 1985; Ehlers & Gibbard, this volume).

It has been generally assumed that the end of the Anglian/Elsterian was a rather abrupt change to the temperate conditions of the succeeding Hoxnian/Holsteinian Stage. However, recent investigations in the Schöningen opencast mine in Lower Saxony, West Germany, have suggested three Late Elsterian interstadials (Offleben I, II and Esbeck Interstadials) in which *Betula, Pinus* and *Picea* forests were established (Urban *et al.*, 1988).

Wolstonian – the missing glaciation

The Hoxnian temperate Stage in East Anglia is followed abruptly at many localities by washing of fine sediments into depositional basins. This event is interpreted as indicating the establishment of periglacial conditions. At Hoxne, Wymer (1983) has reported the occurrence of interstadial conditions in beds 4 and 5 of his excavations i.e. overlying the interglacial lacustrine and fluvioperiglacial beds. It would seem possible that this could be the equivalent of the continental Wacken/Hoogeveen/Bantega/ Dömnitz event (Gibbard & Turner, 1988).

Although palynological investigations are not always well suited to the dating of sediments, as pointed out by Bowen (this volume), many temperate events show a very characteristic vegetational sequence, which allows them to be correlated. Among these are the classic Hoxnian sequences at Hoxne, Marks Tey, and elsewhere which can be clearly distinguished from the classic Ipswichian deposits at Bobbitshole (Ipswich), Trafalgar Square, etc. (West, 1980b). In contrast to these there are other Middle or Upper Pleistocene sites which do not fit so readily into the scheme and which may either represent additional stages (like the continental Wacken/ Dömnitzian) or are atypical equivalents of the Hoxnian and Ipswichian Stages. Newly-developed dating techniques may help to elucidate some of the correlation problems.

Because of the lack of absolute dates, the Pleistocene glacial chronology is based largely on 'counting from the top'. Since this method is extremely vulnerable when applied to incomplete sequences, there has been much controversy about the ages assumed for many glacial deposits. One of the fiercest disputes is centred on the Wolstonian glaciation (Ehlers & Gibbard, this volume; Rice & Douglas, this volume).

In contrast to most areas in continental Europe (Ehlers, *et al.*, 1984) in Britain the Wolstonian glaciation seems to have had a more limited extent than the preceding Anglian and the subsequent Devensian glaciations except in the English Midlands. The exact extent of the Wolstonian ice sheet is still not known. Some argue that it reached as far south as Norfolk (cf. Straw, this volume), while others think that its outermost limit was much further north, somewhere in Lincolnshire (cf. Catt, this volume).

In either case there is a striking discrepancy between the situation in the Netherlands with extensive Saalian deposits and where the Elsterian is virtually missing (e.g. ter Wee, 1983a, b) and in Britain, where the Wolstonian is of minor importance but the Anglian ice reached the Thames Valley. Balson & Jeffery's map (this volume) shows 'crossing boundaries' for the Saalian and Elsterian ice sheets in the North Sea. This anomaly stimulated Cox and others (Bristow & Cox, 1973; Cox, 1981) to argue that what were assumed to be Anglian tills were in fact Wolstonian instead. However, this simple solution could not be sustained because in many places, for example at Hoxne and Marks Tey, Hoxnian Stage deposits rest in kettle hole depressions directly upon Anglian glacial deposits. Moreover, palynological studies clearly demonstrate that the Hoxnian was not simply a slightly different representation of the Ipswichian, but a separate interglacial which can be equated with the continental Holsteinian.

Because Wolstonian glacial deposits are seldom well exposed, very little is known about the intra-Wolstonian stratigraphy. In Yorkshire, in the coastal cliff sections, there seems to be just one Wolstonian till (Catt, this volume). With which of the three major Saalian ice advances on the Continent this might be correlated, is still unknown.

Although the general stratigraphical scheme as outlined on table 41 seems to be fairly well established, there may still be room for additional stages. One thing, however, can be said with certainty: in the Dutch, North German and Danish records no indication of more than one Eemian type interglacial event has been found, in contrast to the repeated suggestions of Bowen (e.g. Bowen *et al.*, 1986), unless the Brørup or another of the Early Weichselian Interstadials may fulfil this role. At all the well-investigated localities (such as Oerel, Sonnenberg, Schalkholz, Neu Wulmstorf, Hof Keller in Northwest Germany) the Eemian Stage sediments always directly overlie those of the Saalian.

The Devensian Glaciation

There is general agreement between the British and Scandinavian Weichselian stratigraphy insofar as there seems to have been no major Early Devensian glaciation either in the British Isles (Worsley, this volume) or on the western European Continent. In Sweden, however, Early and Middle Weichselian ice reached south beyond Gothenburg (Hillefors, 1983). In Finland, on the other hand, it is postulated that the ice did not reach beyond the northern central part of the country at this time (Nenonen, 1986; Hirvas & Nenonen, 1987). The Early Weichselian in Siberia was the main glaciation of the stage, and there were only minor glacier advances in what would be the equivalent of the Dimlington Stadial of Britain (Velichko, 1984).

Not only are there problems of connecting the British and Irish stratigraphic sequences with those of continental Europe, but also with one another. The recent introduction of a new Irish stratigraphical terminology (cf. Warren, this volume) does not necessarily make things easier. This is the reason why it was decided to present two different views of the Irish Quaternary in the introductory chapters (cf. Warren and Hoare, this volume).

In detail there is much disagreement between the British and continental climatic records of the last glaciation. Whilst on the Continent there seem to have been at least six interstadials within the Early and Middle Weichselian, of which the earliest (Brørup and Odderade) were the warmest (cf. Behre & Lade, 1986), the situation in Britain appears to be different. The Chelford Interstadial is most probably the equivalent of the Brørup. The Brimpton Interstadial of Bryant *et al.* (1983) is thought to be an equivalent of the Odderade, as well as possibly the Wretton Interstadial of West (1979). The Upton Warren Interstadial (dated at 42,000 B. P.) was too short for trees to invade Britian, but, according to the insect fauna, summers were very warm (Coope & Angus, 1975).

General agreement exists, however, that at the end of the Devensian there was a major glacier readvance, following the Lateglacial or Windermere Interstadial: the Loch Lomond Stadial oscillation (Gray & Coxon, this volume), which is the equivalent of the Younger Dryas Stadial on the Continent.

The extra-glacial record: loess, terraces, river diversions

The advance of continental ice sheets into lowland NW Europe for the first time in the Elsterian Stage had widespread effects. The area overridden by the ice was subjected to total landscape remodelling with old river courses destroyed or buried. An entirely new landscape was formed beneath the ice by glacial and glaciofluvial deposition and erosion, the sculpturing of deep glacial valleys was to have a striking palaeogeographic impact following ice retreat. At the ice margin major river valleys were dammed all

across the region. The Thames and its tributaries were diverted southwards, the Elbe was dammed and the north German rivers were deflected westwards. However, the most striking feature was the development of a massive ice-dammed lake in the southern North Sea basin into which the Thames, Rhine, Meuse, Scheldt and possibly the Ems all discharged (Gibbard, 1988). Overspill of this lake almost certainly initiated the Dover Straits and greatly enlarged the former Channel River system (Reid, 1913; Smith, 1985; Gibbard, 1988; Bowen *et al.*, 1986).

Arrival of the Anglian ice sheet caused burial of the Kesgrave Formation in all but the southernmost part of East Anglia. The Anglian ice sheet advanced into and overrode the contemporary valley of the River Thames in Hertfordshire and also advanced into southern tributary valleys (Ehlers & Gibbard, this volume).

Following the Hoxnian/Holsteinian temperate Stage three major aggradations of gravel and sand were laid down in the Thames system. The Boyn Hill, Lynch Hill and Taplow members and their lateral equivalents can be traced throughout the system and indeed offshore where they are aligned southwards towards the Dover Strait (Gibbard, 1985; 1988). The relationship of these units to post-Hoxnian Stage glaciation is difficult to establish, with the exception of the Upper Thames Taplow equivalent, the Wolvercote Gravel. The latter contains non-local flint thought to be derived from till at Moreton-in-Marsh (Bishop, 1958; Briggs & Gilbertson, 1980). Questions raised about the existence of a Wolstonian glaciation clearly have a bearing on the age of this Moreton Till. In all other respects the Wolstonian Thames terrace aggradations are comparable to those from the Devensian/Weichselian Stage and are predominantly of periglacial fluvial origin.

In comparison to parts of the neighbouring Continent, aeolian deposits, such as loess and wind-blown sand are relatively rare in the British Isles. Pre-Devensian sediments are very fragmentary. The oldest currently recognised are the Barham Coversand and equivalents of Essex and Suffolk, beneath Anglian till (Rose *et al.*, 1985). Wolstonian age loess is currently known from Northfleet (Kemey & Sieveking, 1977), Bobbitshole (West, 1957) and Hitchin (Gibbard, 1974; Gibbard, Catt & Wintle, unpublished). More continuous spreads have been recognised from the Devensian. These occur over much of southern and eastern England, where they form part of a spread of so-called 'Brickearth' sediments (Catt, 1977b). Recent TL dating of these spreads indicate deposition during the Late Devensian (e.g.

Wintle, 1981). Wind-blown sand is present in Lincolnshire and central East Anglia, and here too is of Late Devensian age (Straw, this volume; Catt, this volume; Bennett, 1984). Similar deposits of loess and sandloess are recorded from southwestern England (Catt & Staines, 1982; Harrod *et al.*, 1973; Scourse, 1986 and this volume) as well as further north (Lee, 1979).

Correlation with the deep sea records

The fundamental problem of the land-based Pleistocene stratigraphy is the incompleteness of the record. This shortcoming has to be overcome by improved dating techniques which would provide better understanding of exactly when the preserved strata were formed and by correlation with the deep sea records (Bowen *et al.*, 1989).

In much of the deep sea sedimentation takes place continuously, and the $^{16}O/^{18}O$ isotope composition ratios from foraminiferal tests combined with other indicators allow a reconstruction of the complete climatic history of the Pleistocene and beyond. This record so far has proved extremely valuable in all research directed towards an explanation of the ice-age climatic cycles. However, direct correlation with the land-based stratigraphy is still at a very early stage. Only the correlation of the Eemian/Ipswichian Stage with isotope stage 5e seems to be generally accepted as already mentioned. The position of the Hoxnian/Holsteinian is still uncertain, and anything from stage 7 to 11 still seems possible, depending on which dating method is preferred. However, most workers favour stage 11 as the probable correlative of the Holsteinian, particularly on the basis of magnitude of the transition from the preceding stage 12 which, as Shackleton (1987) shows, was an exceptionally large glaciation. This corresponds with dates obtained by Sarnthein *et al.* (1986). According to Shackleton (1987) oxygen isotoge stages 6 and 16 were also more extreme glaciations of similar or slightly greater magnitude than stage 2, the Late Weichselian/Devensian glaciation. Of these stage 6 may conceivably be a good candidate for a Saalian/Wolstonian glaciation and stage 16 may be that recorded in the early Middle Pleistocene. Nevertheless conformation of these results must await independent dates.

The glacial history of the North Sea

Recent investigations by the British Geological Sur-vey, partly in cooperation with the Dutch Rijks Geologische Dienst, have shed new light on the Quaternary development of the North Sea basin (Balson *et al.*, this volume). One of the surprising results was that the sea floor had been dissected by several generations of deeply incised channels, the origins of which are still under discussion. The seismic profiles have revealed surprisingly little evidence of the occurrence of direct glacial deposits. It seems that the tills are restricted largely to the marginal parts of the basin, and that they are more extensive in the northern North Sea than in the centre or in the south (Cameron *et al.*, 1987). Further, it now seems to be established that the Devensian/Weichselian ice sheets of Britain and Norway never met (Sutherland, this volume).

With the dominance of marine sediments in much of the North Sea it might be tempting to explain the occurrence of Norwegian erratics in Scotland, Yorkshire and East Anglia by drift ice transport. However, regarding the relative abundance of these far-travelled erratics, such as in Holderness (Catt, this volume), this seems rather unlikely. The concentration of Oslo erratics in Holderness is actually greater than in North Germany, where no drift ice transport is envisaged, and such transport would require a high sea level, which under glacial conditions would not occur.

Different styles of glaciation

Although the land-based stratigraphy is very much a relative chronology, some basic patterns of glaciation can be discerned. In continental Europe it has been particularly recognised that the Pleistocene glaciations differed in style. The Elsterian/Anglian seems to have been a period of enormous incision, during which glaciers and their meltwaters cut channels hundreds of metres deep into the underlying substrate. During the Saalian, incision was very limited, but large proglacial sandur plains accumulated in many areas in North Germany and Poland. During the Weichselian/Devensian, channel erosion recurred, but was far more limited in depth and extent than during the previous stage. Little is so far known about the exact reasons for these differences. The most likely explanation would be different climatic conditions, which resulted in different glaciodynamics. On the floor of the North Sea, however, the differences are not so obvious. From the BGS mapping it appears that channels in the Central North Sea were formed during the Elsterian Saalian and Weichselian.

Despite the differences in style and extent, there are a number of features all the major glaciations had in common. On the Continent, where the development of the ice shield was always initiated in the Scandinavian mountains, this gave rise to initial N-S ice movements in the southern marginal areas, i.e. in Denmark and North Germany. Later, as the ice sheet grew thicker, the ice divide migrated eastwards, and ice flow followed the course of the Baltic Sea depression, bringing northeastern and finally eastern ice into North Germany. In Denmark the ice even flowed northwards during the late stages (Sjørring, 1983; Houmark- Nielsen, 1987).

This sequence of glacier movements seems to have been repeated during all three glaciations. The extent of the individual phases differed from time to time, but, as can be demonstrated from provenance studies of the erratics and fabric orientation measurements, they occurred each time. Whether or not similar patterns developed in the British glaciation area is difficult to assess because of the relatively limited exposure of pre-Devensian deposits.

Speed of ice advance

One of the puzzling questions which remains unresolved concerns the speed of Pleistocene ice sheet advance. According to Lundqvist (1983) and Hillefors (1983), southern Sweden was ice-free as late as 24,000 years B. P. Assuming that the ice sheet margin had reached the outskirts of Hamburg by not later than 15,000 years B. P., the ice must have travelled the intervening 450 km at a rate of about 50 m/year. If the Weichselian maximum was at about 20,000 years B. P., as in eastern North America (Fullerton & Richmond, 1985), the advance rate would have been about 75 m/year or more (Ehlers, 1981). If it is assumed that ice from the glacial centre moved to the outermost ice margin, movements of 100-150 m/year have to be considered. That these high progression rates are not just speculative can be shown from dates available from North America. Using radiocarbon dates from trees overridden by the advancing Laurentide ice sheet, Goldthwait (1958, 1959) demonstrated that in Ohio the Wisconsinan ice advanced at rates of between 17-119 m/year. Today, these ice advance velocities are in a range only encountered in the largest ice streams of Greenland and Antarctica, but are unlikely for a major ice sheet as a whole. If a drop in summer temperatures below zero over most of Britain and Northern Europe during Late Weichselian time triggered large-scale 'instantaneous glaciation', this would still not explain the transport of

erratics from Scotland to the southernmost ice margin within one ice advance.

To account for such rapid advances, Boulton has introduced the concept of ice movement by subglacial deformation (Boulton & Jones, 1979). Traces of this kind of movement have been found in many exposures (Stephan & Ehlers, 1983), however, the degree to which this may have contributed to the ice advance seems to have been limited. In most cases the deformed sediments represent material which seems to have been reworked just a few metres to hundreds of metres up-glacier. Some of the deformed sediments, of course, may have been ultimately incorporated in the till, but the limited till thickness militates against excessive reworking of this kind.

Isostatic influence

A map of the Devensian glaciation in Britain and Ireland shows that the ice sheet margin had an outline which, under normal conditions, would not be expected to occur. Ice, like water, always flows downhill. Therefore it is hard to understand why the Devensian ice sheet should have developed a 400 km long lobe down along the east coast towards East Anglia, while it could have flowed much more easily downslope into the central North Sea. A similar lobe has been proposed for the southern Irish Sea region by Scourse *et al.* (this volume). The way to explain this phenomenon seems to be that there was no downward gradient into the North Sea, but that the coastal areas were deeper than the adjoining North Sea floor. This could have occurred as a result of isostatic downwarping.

Little is known about the isostatic behaviour of the areas in question. In Scotland, Postglacial uplift has been reconstructed as in Scandinavia by evaluation of raised beaches (Smith & Dawson, 1983). However, this method can only cover the period from the onset of the Postglacial transgression to the present, not the truly glacial periods. Recent investigations in Scandinavia have revealed that at about 10,300 B. P. central Scandinavia was depressed by about 450 m (Svendsen & Mangerud, 1987). When this value is extrapolated to the phase of maximum downwarping at about 13,000 B. P., values of about 800 m must be envisaged (Mörner, 1980). No comparable evaluation exists, to our knowledge for the British glaciated area. Interpretation of dates from raised shorelines in Scotland, published by Sissons (1977f), indicate maximum isostatic depression of the Western Grampians in the order of 150-200 m during the Devensian.

Not only were the glaciated areas depressed under the ice load, but the adjoining unglaciated areas were also slightly uplifted. The generation of a forebulge has to be seen as a result of horizontal mass transfer in the low viscosity asthenosphere, according to Mörner (1987). He envisages a maximum uplift in the Central North Sea of about 170 m. That uplift occurred in the marginal areas has been shown by Svendsen & Mangerud (1987), who demonstrated a resultant Postglacial downward movement of westernmost Norway (Sunnmøre) in the order of 20 m.

Such deformations of the earth's crust must have influenced flow patterns of the Pleistocene ice sheets. If the comparatively minor Devensian/Weichselian glaciation led to large-scale downwarping of vast areas, then how much larger would the influence of a major glaciation like the Anglian/Elsterian have been? The time available for isostatic equilibrium to develop after the build-up of the ice dome would have been of considerable importance. For the Devensian/Weichselian glaciation of Scandinavia it is assumed that equilibrium was not reached. A similar situation, however, cannot be assumed for earlier glaciations.

Climatic reconstruction

One of the most important tasks of Quaternary research is to provide a better understanding of contemporary global climatic changes. Although modern dating techniques have demonstrated that the Pleistocene deep sea isotope stages were relatively long events, in the order of 40,000 years prior to 735,000 years B.P., and 100,000 years after 735,000 years B.P. (Ruddiman & Raymo, 1988), sudden and dramatic alterations of climatic conditions can occur. This is not solely true for climatic ameliorations, some of which took place so rapidly that the vegetation had no time to adjust (see above), but also for climatic deterioration. As far as can be judged, the vegetational record seems to indicate that we are living in the latter half of the present (Holocene/Flandrian) interglacial and therefore further investigations of the mechanism driving and controlling of the changes are of primary importance. As improved research and dating techniques become available, major results can be anticipated, but it is vital that those who govern us become aware of the need to support this area of research while there is still time. Otherwise it might well be 'that while astronomers are peering into black holes, while physiologists are investigating the molecular basis of neurobiology and geneticists are busy manipulating genes, a very large surging ice sheet will come and prod them in the backside" (West, 1984).

References

Aario, R., 1972. Association of bed forms and palaeocurrent patterns in an esker delta, Haapajärvi, Finland. *Annales Academiae Scientiarum Fennicae* A III 111: 55 p.

Agassiz, L., 1840. On the evidence of the former existence of glaciers in Scotland, Ireland and England. *Proceedings of the Geological Society of London* 3: 327-332.

Aitken, A.M., J.H. Lovell, A.J. Shaw & C.W. Thomas, 1984. The sand and gravel resources around Dalkeith and Temple, Lothian Region. Description of 1:25,000 sheets NT25 and 35 and NT26 and 36. *Mineral Assessment Report of the British Geological Survey* 140.

Aitken, A.M., J.W. Merritt & A.J. Shaw, 1979. The sand and gravel resources of the country around Garmouth, Grampian Region. Description of 1:25,000 sheet NJ36. *Mineral Assessment Report of the Institute of Geological Sciences* 41.

Aitken, A.M. & D.L. Ross, 1982. The sand and gravel resources of the country around Glenrothes, Fife Region. Description of 1:25,000 sheet NO20 and parts of NO21, 30 and 31. *Mineral Assessment Report of the Institute of Geological Sciences* 101.

Aitkenhead, N., J.I. Chisholm & I.P. Stevenson, 1985. Geology of the country around Buxton, Leek and Bakewell. *British Geological Survey Memoir* 111: 168 pp. London: Her Majesty's Stationery Office.

Alabaster, C. & A. Straw, 1976. The Pleistocene context of faunal remains and artefacts discovered at Welton-le-Wold, Lincolnshire. *Proceedings of the Yorkshire Geological Society* 41: 75-94.

Alderman, J.K., 1959. The geotechnical properties of the glacial deposits of northwest England. Unpublished Ph.D. Thesis, University of Manchester.

Allen, J.R.L., 1982. Late Pleistocene (Devensian) glaciofluvial outwash at Banc-y-Warren, near Cardigan (west Wales). *Geological Journal* 17: 31-47.

Allen, P., 1973. Barham, Sandy Lane. In J. Rose & C. Turner (eds.), *Easter Field Meeting, 1973; Clacton*, Cambridge: Quaternary Research Association.

Allen, P., 1983. *Middle Pleistocene Stratigraphy and Landform Development in South-east Suffolk*. Unpublished Ph.D. Thesis, University of London.

Allen, P. (ed.), 1984. *Field Guide (Revised Edition, October 1984) to the Gipping and Waveney Valleys, Suffolk, May 1982*. Cambridge: Quaternary Research Association. 116 pp.

Allen, P. & J. Rose, 1986. A glacial meltwater drainage system between Whittonstall and Ebchester, Northumberland. In M.G. Macklin & J. Rose (eds.), *Quaternary river landforms and sediments in the northern Pennines*: 69-88. Cambridge: British Geomorphological Research Group / Quaternary Research Association.

Alley, R.B., D.D. Blakenship, C.R. Bentley & S.T. Rooney, 1986. Deformation of till beneath ice stream B, West Antarctica. *Nature* 322: 57-59.

Al-Shaikh-Ali, M.M.H., 1975. Full scale pile testing to failure to determine the effect of fissuring in stiff boulder clay in the county of Cheshire. In R. Hoole (ed.), *The Engineering Properties of Glacial Materials*. Proceedings of the Birmingham Symposium: 256-262. (republished by GeoAbstracts 1978).

Al-Shaikh-Ali, M.M.H., 1978. The behaviour of Cheshire basin lodgement till in motorway construction. In *Clay Fills*: 15-23. London: Institution of Civil Engineers.

Al-Shaikh-Ali, M.M.H., A.G. Davies & M.J. Lloyd, 1981. *In situ* measurement of Ko in a stiff, fissured glacial till by hydraulic fracturing. *Ground Engineering*, January, 1981: 19-25.

Ambrose, K. & J. Brewster, 1982. A reinterpretation of parts of the 400 foot bench of South East Warwickshire. *Quaternary Newsletter* 36: 21-24.

Andersen, B.J., 1981. Late Weichselian ice sheets in Eurasia and Greenland. In G.H. Denton & T.J. Hughes (eds.), *The Last Great Ice Sheets*: 1-65. New York: John Wiley & Sons.

Andrew, R. & R.G. West, 1977. Appendix. Pollen Analyses from Four Ashes, Worcs. *Philosophical Transactions of the Royal Society of London* B 280: 242-246.

Andrews, J.T., 1973. The Wisconsin Laurentide Ice Sheet: Dispersal centers, problems of rates of retreat and climatic implications. *Arctic and Alpine Research* 5: 185-199.

Andrews, J.T., 1975. *Glacial systems: an approach to glaciers and their environments*: 191 pp. North Scituate: Duxbury.

Andrews, J.T., 1978. Sea level history of arctic coasts during the Upper Quaternary. *Progress in Physical Geography* 2: 375-407.

Andrews, J.T., 1982. On the reconstruction of Pleistocene ice sheets: A review. *Quaternary Science Reviews* 1: 1-30.

Andrews, J.T., 1987. The Late Wisconsin glaciation and deglaciation of the Laurentide Ice Sheet. In: W.F. Ruddiman & H.E. Wright (eds.), *North America and adjacent oceans during the last deglaciation*. Geological Society of America K-3: 13-38.

Andrews, J.T. & R.E. Dugdale, 1970. Age prediction of glacio-isostatic strandlines based on their gradients. *Bulletin of the Geological Society of America* 81: 3769-3771.

Andrews, J.T., D.D. Gilbertson & A.B. Hawkins, 1984. The Pleistocene succession of the Severn Estuary: a revised model based upon amino acid racemization studies. *Journal of the Geological Society of London* 141: 967-974.

Anson, W.W. & J.I. Sharp, 1960. Surface and rockhead relief features in the northern part of the Northumberland coalfield. *Department of Geography, University of Newcastle Research Series* 2: 23 pp.

Arkell, W.J., 1947. *The geology of Oxford*. Oxford: Clarendon Press.

Ashley, G.M., 1975. Rhythmic sedimentation in glacial lake Hitchcock, Massachusetts, Connecticut. In A.V. Jopling & B.C. McDonald (eds.), *Glaciofluvial and glaciolacustrine sedimentation*. Society of Economic Paleontologists and Mineralogists, Special Publication 23: 304-320.

Atkinson, J.H., 1985. Undrained strength and overconsolidation of a clay till. In M.C. Forde (ed.), *Glacial Tills 85*. Proceedings of the International Conference: 49-56. Edinburgh: Engineering Technics Press.

Atkinson, J.H. & P.L. Bransby, 1978. *The Mechanics of Soils*. 375 pp. Maidenhead: McGraw-Hill.

Atkinson, J.H. & J.A. Little, 1985. Undrained triaxial strength and stress-strain characteristics of reconstituted and undisturbed samples of a lodgement till. *The City University Geotechnical Engineering Research Centre Research Report* GE/85/14.

Atkinson, T.C., T.J. Lawson, P.L. Smart, R.S. Harmon & J.W. Hess, 1986. New data on speleothem deposition and palaeoclimate in Britain over the last forty thousand years. *Journal of Quaternary Science* 1: 67-72.

Aubert, H. & M. Pinta, 1977. Trace elements in soils. *Developments in Soil Science* 7: 395 pp. Amsterdam: Elsevier.

Augustinus, P.G.E.F. & H.Th. Riezebos, 1971. Some sedimentological aspects of the fluvioglacial outwash plain near Soesterberg (The Netherlands). *Geologie en Mijnbouw* 50: 341-348.

Auton, C.A. & R.G. Crofts, 1986. The sand and gravel resources of the country around Aberdeen, Grampian Region. Description of 1:25,000 resource sheets NJ71, 80, 81 and 91 with parts of NJ61, 90 and 92 and with parts of NO89 and 99. *Mineral Assessment Report of the British Geological Survey* 146.

Avery, B.W. & J.A. Catt, 1983. Northaw Great Wood. In J. Rose (ed.), *The Diversion of the Thames*: 96-101. Cambridge: Quaternary Research Association.

Baden-Powell, D.F.W., 1938. On the glacial and interglacial marine beds of Northern Lewis. *Geological Magazine* 75: 395-409.

Baden-Powell, D.F.W., 1948. The chalky boulder clays of Norfolk and Suffolk. *Geological Magazine* 85: 279-296.

Baden-Powell, D.F.W., 1956. The correlation of the Pliocene and Pleistocene marine beds of Britain and the Mediterranean. *Proceedings of the Geologists' Association* 66: 271-292.

Bailey, E.B., C.T. Clough, W.B. Wright, J.E. Richey & G.V.

Wilson, 1924. Tertiary and post-Tertiary geology of Mull, Loch Aline and Oban. *Memoirs of the Geological Survey, Scotland*. Edinburgh: Her Majesty's Stationery Office.

Baker, C.A., 1971. A contribution to the glacial stratigraphy of west Essex. *Essex Naturalist* 32: 313-330.

Baker, C.A., 1976. Late Devensian periglacial phenomena in the upper Cam Valley, north Essex. *Proceedings of the Geologists' Association* 87: 285-306.

Baker, C.A., 1977. Quaternary Stratigraphy and Environments in the Upper Cam Valley: 309 pp. Unpublished Ph.D. thesis, University of London.

Baker, C.A. & D.K.C. Jones, 1980. Glaciation of the London Basin and its influence on the drainage pattern: a review and appraisal. In D.K.C. Jones (ed.), *The Shaping of Southern England*: 131-175. London: Academic Press.

Ballantyne, C.K., 1979. A sequence of Lateglacial ice-dammed lakes in East Argyll. *Scottish Journal of Geology* 15: 153-160.

Ballantyne, C.K., 1984. The Late Devensian periglaciation of upland Scotland. *Quaternary Science Reviews* 3: 311-343.

Ballantyne, C.K., 1986. Protalus rampart development and the limits of former glaciers in the vicinity of Baosbheinn, *Scottish Journal of Geology* 22: 13-25.

Ballantyne, C.K. & J.M. Gray, 1984. The Quaternary geomorphology of Scotland: The research contribution by J.B. Sissons. *Quaternary Science Reviews* 3: 259-289.

Ballantyne, C.K. & T. Wain-Hobson, 1980. The Loch Lomond Advance on the island of Rhum. *Scottish Journal of Geology* 16: 1-10.

Balson, P.S. & T.D.J. Cameron, 1985. Quaternary mapping offshore East Anglia. *Modern Geology* 9: 221-239.

Banerjee, I. & B.C. McDonald, 1975. Nature of esker sedimentation. In A.V. Jopling & B.C. McDonald (eds.), *Glaciofluvial and glaciolacustrine sedimentation*. Society of Economic Paleontologists and Mineralogists, Special Publication 23: 132-154.

Banham, P.H., 1966. The significance of till pebble lineations and their relation to folds in two Pleistocene tills at Mundesley, Norfolk. *Proceedings of the Geologists' Association* 77: 469-474.

Banham, P.H., 1968. A preliminary note on the Pleistocene stratigraphy of North-East Norfolk. *Proceedings of the Geologists' Association* 79: 507-512.

Banham, P.H., 1970a. Notes on Norfolk coastal sections. In G.S. Boulton (ed.), *Field Guide, Norwich*. Cambridge: Quaternary Research Association.

Banham, P.H., 1970b. Appendix. In G.S. Boulton (ed.), *Field Guide Norwich - Easter 1970*: 39-42. Cambridge: Quaternary Research Association.

Banham, P.H., 1971. Pleistocene beds at Corton, Suffolk. *Geological Magazine* 108: 281-285.

Banham, P.H., 1973. Claydon chalk pit. In J. Rose & C. Turner (eds.), *Easter Field Meeting, 1973; Clacton*. Cambridge: Quaternary Research Association.

Banham, P.H., 1975. Glacitectonic structures: a general discussion with particular reference to the contorted drift of Norfolk. In A.E. Wright & F. Moseley (eds.), *Ice Ages: Ancient and Modern*. Geological Journal Special Issue Number 6: 69-94. Liverpool: Seel House Press.

Banham, P.H., 1977. Glacitectonites in till stratigraphy. *Boreas* 6: 101-105.

Banham, P.H., H. Davies & R.M.S. Perrin, 1975. Short field meeting in north Norfolk, 19-21 October 1973. *Proceedings of the Geologists' Association* 86: 251-258.

Banham, P.H. & C.E. Ranson, 1965. Structural study of the Contorted Drift and disturbed Chalk at Weybourne, north Norfolk. *Geological Magazine* 102: 164-174.

Barnes, J.W., 1982. *Basic geological mapping*: 128 pp. London: Geological Society.

Barrow, G., 1904. On a striated boulder from the Scilly Isles. *Quarterly Journal of the Geological Society of London* 60: 106.

Barrow, G., 1906. *The Geology of the Isles of Scilly*. Memoir of the Geological Survey U.K. (England and Wales). London: Her Majesty's Stationery Office.

Barrow, G., 1919. Some future work for the Geologists' Association. *Proceedings of the Geologists' Association* 30: 1-48.

Barrow, G., J.S.G. Wilson & E.H.C. Craig, 1905. The geology of the country round Blair Atholl, Pitlochry and Aberfield. *Memoir of the Geological Survey of Great Britain*.

Battiau-Quenney, Y., 1981. Contribution à l'étude Géomorphologique du Massif Gallois. Unpublished thesis, University of Brittany.

Bayliss, J.M., P.R. Ineson & I.H. Rison, 1979. Extent of Belland ground adjacent to Tideslow and Maiden Rakes - Little Hucklow. *Bulletin of the Peak District Mines Historical Society* 7 (3): 153-157.

Bazley, R.A.B., 1978. Interglacial and interstadial deposits in Northern Ireland. *Report of the Institute of Geological Sciences* No 77/16: 6 pp.

Bazley, R.A.B. & P.I. Manning, 1971. *Geology of Belfast and district. Special engineering geology sheet, solid and drift.* Geological Survey of Northern Ireland.

Beaumont, C., 1978. The evolution of sedimentary basins on a viscoelastic lithosphere: theory and examples. *Geophysical Journal of the Royal Astronomical Society* 55: 471-497.

Beaumont, P., 1971. Stone orientation and stone count data from the Lower Till Sheet, eastern Durham. *Proceedings of the Yorkshire Geological Society* 38: 343-360.

Beckett, S.C., 1981. Pollen diagrams from Holderness, north Humberside. *Journal of Biogeography* 8: 177-198.

Behre, K.-E. & U. Lade, 1986. Eine Folge von Eem und 4 Weichsel-Interstadialen in Oerel/Niedersachsen und ihr Vegetationsablauf. *Eiszeitalter und Gegenwart* 36: 11-36.

Belderson, R.H., N.H. Kenyon & J.B. Wilson, 1973. Iceberg plough marks in the North East Atlantic. *Palaeogeography, Palaeoclimatology, Palaeoecology* 13: 215-224.

Belderson, R.H., R.D. Pingree & D.K. Griffiths, 1986. Low sea-level tidal origin of Celtic Sea sand banks - evidence from numerical modelling of M_2 tidal streams. *Marine Geology* 73: 99-108.

Belknap, D.F., 1987. Late Quaternary sea-level changes in Maine. In S. Nummedal, O.H. Pilkey & J.D. Howard (eds.), *Sea-level fluctuation and Coastal Evolution*. Society of Economic Paleontologists and Mineralogists, Special Publication 41: 71-86.

Bell, A., 1888. On the 'manure' gravels of Wexford. *Report of the British Association for the Advancement of Science*: 133-141.

Bell, A., 1889. On the manure gravels of Wexford. *Report of the British Association for the Advancement of Science*: 92-93.

Bell, A., 1890. On the manure gravels of Wexford. *Report of the British Association for the Advancement of Science*: 410-424.

Bell, A., 1915. The fossiliferous Molluscan deposits of Wexford and North Manxland. *Geological Magazine* 2: 164-169.

Bell, A., 1917. The shells of the Holderness Basement Clays. *The Naturalist*: 95-98, 135-138.

Bell, A., 1919a. Fossils of the Holderness Basement Clays. *The Naturalist*: 57-59.

Bell, A., 1919b. Fossil shells from Wexford and Manxland. *Irish Naturalist* 28: 109-114.

Bell, F.G., G.R. Coope, R.J. Rice & T.H. Riley, 1972. Mid-Weichselian fossil-bearing deposits at Syston, Leicestershire. *Proceedings of the Geologists' Association* 83: 197-211.

Bijlsma, S., 1981. Fluvial sedimentation from the Fennoscandian area into the north-west European basin during the Late Cenozoic. *Geologie en Mijnbouw* 60: 337-345.

Birks, H.J.B., J. Deacon & S. Peglar, 1975. Pollen maps for the British Isles 5000 years ago. *Proceedings of the Royal Society of London* B 189: 87-105.

Birks, H.J.B. & R.W. Mathewes, 1978. Studies in the vegetational history of Scotland. V. Late Devensian and early Flandrian pollen and macrofossil stratigraphy at Abernethy Forest, Inverness-shire. *New Phytologist* 80: 455-484.

Birks, H.J.B. & S.M. Peglar, 1979. Interglacial pollen spectra from Sel Ayre, Shetland. *New Phytologist* 83: 559-575.

Birks, H.J.B. & M.E. Ransom, 1969. An interglacial peat at Fugla Ness, Shetland. *New Phytologist* 68: 777-796.

Birnie, J., 1983. Tolsta Head: further investigations of the interstadial deposit. *Quaternary Newsletter* 41: 18-25.

Bisat, W.S., 1939. The relationship of the 'Basement Clays' of Dimlington, Bridlington and Filey bays. *The Naturalist*: 133-135, 161-168.

Bisat, W.S., 1940. Older and Newer Drift in East Yorkshire. *Proceedings of the Yorkshire Geological Society* 24: 137-151.

Bishop, W.W., 1958. The Pleistocene geology and geomorphology of three gaps in the Midland Jurassic escarpment. *Philosophical Transactions of the Royal Society of London* B 241: 255-306.

Bishop, W.W., 1967. *Discussion* in Mitchell & Orme (1967), The Pleistocene deposits in the Isles of Scilly. *Quarterly Journal of the Geological Society of London* 123: 91.

Bishop, W.W. & G.R. Coope, 1977. Stratigraphical and faunal evidence for Lateglacial and early Flandrian environments in south-west Scotland. In J.M. Gray & J.J. Lowe (eds.), *Studies in the Scottish Lateglacial Environment*: 61-88. Oxford: Pergamon Press.

Blodgett, R.H. & K.O. Stanley, 1980. Stratification, bedforms and discharge relations of the Platte braided river system, Nebraska. *Journal of Sedimentary Petrology* 50: 139-148.

Bloom, A.L., W.S. Broecker, J.S. Chappell, R.K. Matthews & K.J. Mesolella, 1974. Quaternary sea-level fluctuations on a tectonic coast: New Th^{230}/U^{234} dates from the Huon Peninsula, New Guinea. *Quaternary Research* 4: 185-205.

Bluck, B.J., 1974. Structure and directional properties of some valley sandur deposits in southern Iceland. *Sedimentology* 21: 533-554.

Bluck, B.J., 1979. Structure of coarse grained braided stream alluvium. *Transactions of the Royal Society of Edinburgh* 70: 181-221.

Bluck, B.J., 1982. Texture of gravel bars in braided streams. In R.D. Hey, J.C. Bathurst & C.R. Thorne (eds.), *Gravel-Bed Rivers*: 339-355. New York, London, Sydney: Wiley Interscience.

Boardman, J., 1978. Grèze litées near Keswick, Cumbria. *Biuletyn Peryglacjalny* 27: 23-34.

Boardman, J., 1980. Evidence for pre-Devensian glaciation in the northeastern Lake District. *Nature* 286: 599-600.

Boardman, J., 1981. Quaternary Geomorphology of the Northeastern Lake District. Unpublished Ph.D. thesis, University of London: 410 pp.

Boardman, J., 1982. Glacial geomorphology of the Keswick area, northern Cumbria. *Proceedings of the Cumberland Geological Society* 4: 115-134.

Boardman, J., 1983. The role of micromorphological analysis in an investigation of the Troutbeck Paleosol, Cumbria, England. In P. Bullock & C.P. Murphy (eds.), *Soil Micromorphology* 1: 281-288.

Boardman, J., 1984. Red soils and glacial erosion in Britain. *Quaternary Newsletter* 42: 21-24.

Boardman, J., 1985. The Troutbeck Paleosol, Cumbria, England. In J. Boardman (ed.), *Soils and Quaternary Landscape Evolution*: 231-260. Chichester: Wiley.

Boardman, J., 1988. *Classic Landforms of the Lake District*. Sheffield: Geographical Association.

Boardman, J., J. Andrews, D. Gordon, D.D. Harkness, D. Holyoak & J.J. Lowe (unpublished). A peat bed in Mosedale, Cumbria.

Bonny, A.P., S.J. Mathers & E.Y. Hawirth, 1986. Interstadial deposits with Chelford affinities from Burland, Cheshire. *Mercian Geologist* 10: 151-160.

Booth, S.J., 1983. The sand and gravel resources of the country between Bourne and Crowland, Lincolnshire. *Mineral Assessment Report of the Institute of Geological Sciences* 130.

Boothroyd, J.C., 1972. *Coarse grained sedimentation on a braided outwash fan, northeast Gulf of Alaska*. Coastal Research Division, University of South Carolina, Columbia, Technical Report 6: 127 pp.

Boothroyd, J.C. & G.M. Ashley, 1975. Processes, bar morphology and sedimentary structures on braided outwash fans, northeastern Gulf of Alaska. In A.V. Jopling & B.C. McDonald (eds.), *Glaciofluvial and glaciolacustrine sedimentation*. Society of Economic Paleontologists and Mineralogists, Special Publication 23: 193-222.

Boothroyd, J.C., R.A. Levey & M.S. Cable, 1977. Depositional history of the West Malaspina Foreland, Alaska: an aid to interpreting glacial sediments of southern New England. *Geological Society of America Abstracts with programs* 9: 243-244.

Boothroyd, J.C. & D. Nummedal, 1978. Proglacial braided outwash: a model for humid alluvial-fan deposits. In A.D. Miall (ed.), *Fluvial sedimentology*. Canadian Society of Petroleum Geologists Memoir 5: 641-668.

Boswell, P.G.H., 1913. The age of the Suffolk valleys. *Quarterly Journal of the Geological Society of London* 69: 581-620.

Boswell, P.G.H., 1914. On the occurrence of the North Sea Drift (Lower Glacial), and certain other brick-earths in East Anglia. *Proceedings of the Geologists' Association* 25: 121-153.

Boswell, P.G.H., 1916. The petrology of the North Sea Drift and Upper Glacial brick-earths in East Anglia. *Proceedings of the Geologists' Association* 27 (2): 79-98.

Boswell, P.G.H., 1927. The geology of the country around Ipswich. *Memoir of the Geological Survey of the United Kingdom*. London: Her Majesty's Stationery Office.

Boswell, P.G.H., 1931. The stratigraphy of the glacial deposits of East Anglia in relation to early man. *Proceedings of the Geologists' Association* 42: 87-111.

Boulter, M.C., T.D. Ford, M. Ijtaba & P.T. Walsh, 1971. The Brassington formation - a newly recognised Tertiary formation in the Southern Pennines. *Nature*, Physical Science 231: 134-136.

Boulter, M. & W.I. Mitchell, 1977. Middle Pleistocene (Gortian) deposits from Benburb, Northern Ireland. *Irish Naturalists' Journal* 19: 2-3.

Boulton, G.S., 1968. Flow tills and related deposits on some Vestspitsbergen glaciers. *Journal of Glaciology* 7: 391-412.

Boulton, G.S., 1970. On the origin and transport of englacial debris in Svalbard glaciers. *Journal of Glaciology* 9: 213-229.

Boulton, G.S., 1971. Till genesis and fabric in Svalbard, Spitsbergen. In R.P. Goldthwait (ed.), *Till, a symposium*: 41-72. Columbus: Ohio State University Press.

Boulton, G.S., 1972. Modern arctic glaciers as depositional models for former ice sheets. *Journal of the Geological Society of London* 128: 361-393.

Boulton, G.S., 1973. Creeting St. Mary. In J. Rose & C. Turner (eds.), *Easter Field Meeting, 1973; Clacton*. Cambridge: Quaternary Research Association.

Boulton, G.S., 1974. Processes and patterns of glacial erosion. In D.R. Coates (ed.), *Glacial Geomorphology*. Binghampton: Publications in Geomorphology, State University of New York.

Boulton, G.S., 1975a. Processes and patterns of subglacial sedimentation: a theoretical approach. In A.E. Wright & F. Moseley (eds.), *Ice Ages: Ancient and Modern*: Geological Journal Special Issue Number 6: 7-42. Liverpool: Seel House Press.

Boulton, G.S., 1975b. The genesis of glacial tills - a framework for geotechnical interpretation. *The Engineering Properties of Glacial Materials*. Proceedings of the Birmingham Symposium: 52-59 (republished by Geo Abstracts 1978).

Boulton, G.S., 1977a. Heacham kame deposits. In R.G. West (ed.), *East Anglia*: 53-57. X INQUA Congress Field Guide. Norwich: Geo Abstracts.

Boulton, G.S., 1977b. A multiple till sequence formed by a Late Devensian Welsh ice-cap, Gllanllynau, Gwynedd. *Cambria* 1 (4): 10-31.

Boulton, G.S., 1984. Development of a theoretical model of sediment dispersal by ice sheets. *Prospecting in Areas of Glaciated Terrain*: 213-223. London: The Institution of Mining and Metallurgy.

Boulton, G.S., C.T. Baldwin, J.D. Peacock, A.M. McCabe, G. Miller, J. Jarvis, B. Horsfield, P. Worsley, N. Eyles, P.N. Chroston, T.E. Day, P. Gibbard, P.E. Hare & V. von Brunn, 1982. A glacio-isostatic facies model and amino-acid stratigraphy for late Quaternary events in Spitsbergen and the Arctic. *Nature* 298: 437-441.

Boulton, G.S., P.N. Chroston & J. Jarvis, 1981. A marine seismic study of late Quaternary sedimentation and inferred glacier fluctuations along western Inverness-shire, Scotland. *Boreas* 10: 39-51.

Boulton, G.S., F. Cox, J. Hart & M. Thornton, 1984. The glacial geology of Norfolk. *Bulletin of the Geological Society of Norfolk* 34: 103-122.

Boulton, G.S. & A.S. Jones, 1979. Stability of temperate ice caps and ice sheets resting on beds of deformable sediments. *Journal of Glaciology* 24: 29-44.

Boulton, G.S., A.S. Jones, K.M. Clayton & M.J. Kenning, 1977. A British ice-sheet model and patterns of glacial erosion and deposition in Britain. In F.W. Shotton (ed.), *British Quaternary Studies, Recent Advances*: 231-246. Oxford: Clarendon Press.

Boulton, G.S. & M.A. Paul, 1976. The influence of genetic processes on some geotechnical properties of glacial tills. *Quarterly Journal of Engineering Geology* 9: 159-192.

Boulton, G.S., G.D. Smith, A.S. Jones & J. Newsome, 1985. Glacial geology and glaciology of the last mid-latitude ice sheets. *Journal of the Geological Society of London* 142: 447-474.

Boulton, G.S. & P. Worsley, 1965. Late Weichselian glaciation in the Cheshire-Shropshire Basin. *Nature* 207: 704-706.

Bouysse, P., R. Horn, F. Lapierre & F. Le Lann, 1976. Étude des grands bancs de sable du sud-est de la mer Celtique. *Marine Geology* 20: 251-275.

Bowen, D.Q., 1969. A new interpretation of the Pleistocene succession in Bristol Channel area. *Proceedings of the Ussher Society* 2: 86.

Bowen, D.Q., 1970. Southeast and Central South Wales. In C.A. Lewis (ed.), *The Glaciations of Wales and Adjoining Regions*: 197-227. London: Longman.

Bowen, D.Q., 1973a. The Pleistocene history of Wales and the borderland. *Geological Journal* 8: 207-224.

Bowen, D.Q., 1973b. The Pleistocene succession of the Irish Sea. *Proceedings of the Geologists' Association* 84: 249-272.

Bowen, D.Q., 1974. The Quaternary of Wales. In T.R. Owen (ed.), *The Upper Palaeozoic and post Palaeozoic rocks of Wales*: 373-426. Cardiff: University of Wales Press.

Bowen, D.Q., 1978. *Quaternary Geology: A Stratigraphic Framework for Multidisciplinary Work*: 221 pp. Oxford: Pergamon Press.

Bowen, D.Q., 1981. The 'South Wales End Moraine': Fifty years after. In J. Neale & J. Flenley (eds.), *The Quaternary in Britain*: 60-67. Oxford: Pergamon Press.

Bowen, D.Q., 1981. Sheet 1.3. In H. Carter & H.M. Griffiths (eds.), *National Atlas of Wales*. Cardiff: University of Wales Press.

Bowen, D.Q., 1984. Introduction. In D.Q. Bowen & A. Henry (eds.), *Wales: Gower, Preseli, Fforest Fawr. Field Guide*: 1-17. Cambridge: Quaternary Research Association.

Bowen, D.Q., S. Hughes, G.A. Sykes & G.H. Miller, 1989. Land-sea correlations in the Pleistocene based on isoleucine epimerization in non-marine molluscs. *Nature* 340: 49-51.

Bowen, D.Q., J. Rose, A.M. McCabe & D.G. Sutherland, 1986. Correlation of Quaternary Glaciations in England, Ireland, Scotland and Wales. *Quaternary Science Reviews* 5: 299-340.

Bowen, D.Q., J. Rose, A.M. McCabe & D.G. Sutherland, 1988 (in press). Reply to the comments of P. Worsley. *Quaternary Science Reviews* 7.

Bowen, D.Q. & G.A. Sykes, 1988. Correlation of marine events and glaciations on the northeast Atlantic margin. In N.J. Shackleton, R.G. West & D.Q. Bowen (eds.), *The Past Three Million Years: evolution of climatic variability in the north Atlantic region*. The Royal Society, London. (first published in *Philosophical Transactions of the Royal Society of London* B 318: 619-635).

Bowen, D.Q., G.A. Sykes, S. Hughes & G.H. Miller, unpublished. Revised classification of the British Quaternary based on the epimerization of isoleucine in non-marine mollusca.

Bowen, D.Q., G.A. Sykes, A. Reeves, G.H. Miller, J.T. Andrews, J.S. Brew & P.E. Hare, 1985. Amino acid Geochronology of raised beaches in south west Britain. *Quaternary Science Review* 4: 279-318.

Boylan, P.J., 1981. The role of William Buckland (1794-1856) in the recognition of glaciation in Great Britain. In J. Neale & J. Flenley (eds.), *The Quaternary in Britain*: 1-8. Oxford: Pergamon.

Bradley W.C., R.K. Fahnestock & E.T. Rowekamp, 1972. Coarse sediment transport by flood flows on Knik River, Alaska. *Bulletin of the Geological Society of America* 83: 1261-1284.

Bremner, A., 1915. Problems in the glacial geology of Northeast Scotland and some fresh facts bearing on them. *Transactions of the Edinburgh Geological Society* 10: 334-345.

Bremner, A., 1934. The glaciation of Moray and ice movements in the north of Scotland. *Transactions of the Edinburgh Geological Society* 13: 17-56.

Bremner, A., 1939. Notes on the glacial geology of east Aberdeenshire. *Transactions of the Edinburgh Geological Society* 13: 474-475.

Bremner, A., 1943. The glacial epoch in the North-East. In J.F. Tocher (ed.), *The Book of Buchan (Jubilee Volume)*: 10-30. Aberdeen: Aberdeen University Press.

Bridge, D.McC., 1988. Corton Cliffs. In P.L. Gibbard & J.A. Zalasiewicz (eds.), *Pliocene - Middle Pleistocene of East Anglia*, Field Guide: 119-125. Cambridge: Quaternary Research Association.

Bridge, D.McC. & P.M. Hopson, 1985. Fine gravel, heavy mineral and grain-size analysis of Mid-Pleistocene glacial deposits in the Lower Waveney Valley, East Anglia. *Modern Geology* 9: 129-144.

Bridger, J.F.D., 1981. The glaciation of Charnwood Forest, Leicestershire and its geomorphological significance. In J. Neale & J. Flenley (eds.), *The Quaternary in Britain*: 68-81. Oxford: Pergamon.

Bridgland, D.R., 1980. A reappraisal of Pleistocene stratigraphy in north Kent and east Essex and new evidence concerning the former course of the Thames and Medway. *Quaternary Newsletter* 32: 15-24.

Bridgland, D.R., 1983. The Quaternary Fluvial Deposits of North Kent and Eastern Essex: 660 pp. Unpublished Ph.D. Thesis, Council for National Academic Awards, City of London Polytechnic.

Bridgland, D.R., P.L. Gibbard & R.C. Preece (in print). Early Middle Pleistocene interglacial deposits at Little Oakley, Essex, England. *Philosophical Transactions of the Royal Society of London* B.

Briggs, D.J. & C.V. Burek, 1985. Quaternary deposits in the Peak District. In D.J. Briggs, D.D. Gilbertson & R.D.S. Jenkinson (eds.), *The Peak District and North Dukeries, Field Guide*: 17-32. Cambridge: Quaternary Research Association.

Briggs, D.J. & D.D. Gilbertson, 1980. Quaternary processes and environments in the upper Thames valley. *Transactions of the Institute of British Geographers* 5: 53-65.

Bristow, C.R., 1985. Geology of the country around Chelmsford. *Memoir of the Geological Survey of the United Kingdom*: 108 pp. London: Her Majesty's Stationery Office.

Bristow, C.R. & F.C. Cox, 1973. The Gipping Till: a reappraisal of East Anglian glacial stratigraphy. *Journal of the Geological Society of London* 129: 1-37.

British Geological Survey, 1986. 1:250,000 Series. Quaternary Geology. California: Sheet 54°N 00°E.

British Geological Survey & Rijks Geologische Dienst, 1984. 1:250,000 Series. Quaternary Geology. Flemish Bight: Sheet 52°N 02°E.

British Geological Survey & Rijks Geologische Dienst, 1986. 1:250,000 Series. Quaternary Geology. Indefatigable: Sheet 53°N 02°E.

British Geological Survey & Rijks Geologische Dienst (in press). 1:250,000 Series. Quaternary Geology. Silver Well: Sheet 54°N 02°E.

Bromehead, C.E.N., W. Edwards, D.A. Wray & J.V. Stephens, 1933. The geology of the country around Holmfirth and Glossop. *Memoir of the Geological Survey*. London: Her Majesty's Stationery Office.

Brook, D. & B.R. Marker, 1986. Thematic geological mapping as an essential tool in land-use planning. In M.G. Culshaw (ed.), *Planning and engineering geology. Preprints of papers for the 22nd Annual Conference of the Engineering Group of the Geological Society*: 217-224.

Broster, B.E. & A. Dreimanis, 1981. Deposition of multiple lodgement tills by competing glacial flows in a common ice sheet. Cranbrook, British Columbia. *Arctic and Alpine Research* 13: 197-204.

Broster, G.E., A. Dreimanis & J.C. White, 1979. A sequence of glacial deformation, erosion and deposition at the ice-rock interface during the last glaciation; Cranbrook, British Columbia, Canada. *Journal of Glaciology* 23: 283-294.

Broster, B.E. & S.R. Hicock, 1985. Multiple flow and support mechanisms and the development of inverse grading in a subaquatic glacigenic debris flow. *Sedimentology* 32: 645-657.

Brown, C.S., M.F. Meier & A. Post, 1982. Calving speed of Alaska tidewater glaciers with application to Columbia Glacier. *U.S. Geological Survey Professional Paper* 1258 C.

Brown, E.H. & R.U. Cooke, 1977. Landforms and related glacial deposits in the Wheeler Valley area, Clwyd. *Cambria* 4: 32-45.

Browne, M.A.E. & D.K. Graham, 1981. Glaciomarine deposits of the Loch Lomond Stade glacier in the Vale of Leven between Dumbarton and Balloch, west-central Scotland. *Quaternary Newsletter* 34: 1-7.

Browne, M.A.E., D.D. Harkness, J.D. Peacock & R.G. Ward, 1977. The date of deglaciation of the Paisley-Renfrew area. *Scottish Journal of Geology* 13: 301-303.

Browne, M.A.E., A.A. McMillan & D.K. Graham, 1983. A late-Devensian marine and non-marine sequence near Dumbarton, Strathclyde. *Scottish Journal of Geology* 19: 229-234.

Browne, P., 1982. Aspects of the glaciation of the north Sligo-Leitrim border. Unpublished undergraduate dissertation (Geography), University of Dublin, Trinity College.

Browne, P., 1986. Vegetational history of the Nephin Beg Mountains, County Mayo. Unpublished Ph.D Thesis, University of Dublin, Trinity College.

Bryant, I.D., 1983a. Facies sequences associated with some braided river deposits of late Pleistocene age from southern Britain. In J.D. Collinson & J. Lewin (eds.), *Modern and ancient fluvial systems*. International Association of Sedimentologists, Special Publication 6: 267-275.

Bryant, I.D., 1983b. The utilization of Arctic river analogue studies in the interpretation of periglacial river sediments from southern Britain. In K.J. Gregory (ed.), *Background to Palaeohydrology*: 413-431. New York, London, Sydney: Wiley Interscience.

Bryant, I.D., D.T. Holyoak & K.A. Moseley, 1983. Late Pleistocene deposits at Brimpton, Berkshire, England. *Proceedings of the Geologists' Association* 94: 321-343.

Bryant, R.H., 1966. The 'pre-glacial' raised beach in south-west Ireland. *Irish Geography* 5: 188-203.

Bryant, R.H., 1977. Bantry to Waterville. In D.Q. Bowen (ed.), *South and South-west Ireland*. Guidebook for excursion A 15, INQUA X Congress: 23-29. Norwich: Geobooks.

Budd, W.F. & I.N. Smith, 1979. The growth and retreat of ice sheets in response to orbital radiation changes. In *Sea Level, Ice, and Climatic Change (Proceedings of the Canberra Symposium, 1979)*. IAHS Publication 131: 369-409.

Burek, C.V., 1977a. The Pleistocene ice age and after. In T.D. Ford (ed.), *Limestones and caves of the Peak District*: 87-128. Norwich: Geobooks.

Burek, C.V., 1977b. An unusual occurrence of sands and gravels in Derbyshire. *Mercian Geologist* 6: 123-130.

Burek, C.V., 1978. Quaternary deposits on the Carboniferous

Limestone of Derbyshire: 509 pp. Unpublished Ph.D. Thesis, University of Leicester.

Burek, C.V., 1979. Weathering ratios as evidence of cessation of till deposition. *Geological Society of America, Abstracts with Programs* 11 (1): 5.

Burek, C.V., 1985a. The Bakewell Till. In D.J. Briggs, D.D. Gilbertson & R.D.S. Jenkinson (eds.), *The Peak District and North Dukeries, Field Guide*: 43-70. Cambridge: Quaternary Research Association.

Burek, C.V., 1985b. The use of trace element weathering ratios in Pleistocene Geology. *Quaternary Newsletter* 47: 4-18.

Burek, C.V. & J.M. Cubitt, 1979. Trace element distribution in the superficial deposits of Northern Derbyshire, England. *Minerals and the Environment* 1: 90-100.

Burger, A.W., 1986. Sedimentpetrographie am Morsum Kliff, Sylt (Norddeutschland). *Mededelingen van het Werkgroep voor Tertiaire en Kwartaire Geologie* 23: 99-109.

Cameron, T.D.J., M.S. Stoker & D. Long, 1987. The history of Quaternary sedimentation in the UK sector of the North Sea Basin. *Journal of the Geological Society of London* 144: 43-58.

Campbell, R., 1934. On the occurrence of shelly boulder clay and interglacial deposits in Kincardineshire. *Transactions of the Edinburgh Geological Society* 13: 176-182.

Campbell, S., J.T. Andrews & R.A. Shakesby, 1982. Amino acid evidence for Devensian ice, west Gower, South Wales. *Nature* 300: 249-250.

Campbell, S., 1984. The Nature and Origin of the Pleistocene Deposits around Cross Hands and on West Gower, South Wales. Unpublished Ph.D. Thesis, University of Wales.

Cannell, B. & R.G. Crofts, 1984. The sand and gravel resources of the country around Henley-in-Arden, Warwickshire. *Mineral Assessment Report of the Institute of Geological Sciences* 142.

Carroll, D., 1970a. *Rock weathering*: 203 pp. New York: Plenum Press.

Carroll, D., 1970b. Clay minerals: A guide to their X-ray identification. *Geological Society of America. Special Publication* 126: 80 pp.

Carruthers, R.G., 1939. On Northern glacial drifts: some peculiarities and their significance. *Quarterly Journal of the Geological Society of London* 95: 299-310.

Carruthers, R.G., 1947. The secret of the glacial drifts. Part II. Applications to Yorkshire. *Proceedings of the Yorkshire Geological Society* 27: 129-171.

Carruthers, R.G., 1953. *Glacial Drifts and the Undermelt Theory*: 42 pp. Newcastle: Harold Hill.

Carruthers, R.G., G.A. Burnett & W. Anderson, 1930. The Geology of the Alnwick District. *Memoir of the Geological Survey of England and Wales*. London: Her Majesty's Stationery Office.

Carruthers, R.G., G.A. Burnett & W. Anderson, 1932. The Geology of the Cheviot Hills. *Memoir of the Geological Survey of England and Wales*. London: Her Majesty's Stationery Office.

Carter, R.W.G., 1982. Sea-level changes in Northern Ireland. *Proceedings of the Geologists' Association* 83: 7-23.

Carter, R.W.G. & J.D. Orford, 1981. *Field Guide No 4*. Dublin: Irish Quaternary Association.

Casshyap, S.M. & R.C. Tewari, 1982. Facies analysis and paleogeographic implications of a late Paleozoic glacial outwash deposit, Bihar, India. *Journal of Sedimentary Petrology* 52: 1243-1256.

Caston, V.N.D., 1977. A new isopachyte map of the Quaternary of the North Sea. *Report of the Institute of Geological Sciences* 77/11: 1-8.

Catt, J.A., 1977a. The contribution of loess to soils in lowland Britain. *Council for British Archaeology. Research Report* 21.

Catt, J.A., 1977b. Loess and coversands. In F.W. Shotton (ed.), *British Quaternary Studies, Recent Advances*: 221-229. Oxford: Clarendon.

Catt, J.A., 1981. British Pre-Devensian glaciations. In J. Neale & J. Flenley (eds.), *The Quaternary in Britain*: 9-19. Oxford: Pergamon Press.

Catt, J.A., 1982. The Quaternary deposits of the Yorkshire Wolds. *Proceedings of the North of England Soils Discussion Group* 18: 61-67.

Catt, J.A., 1986a. The nature, origin and geomorphological significance of the clay-with-flints. In G. de G. Sieveking & M.B. Hart (eds.), *The scientific study of flint and chert*: 151-159. London: Cambridge University Press.

Catt, J.A., 1986b. Silt mineralogy of loess and 'till' on the Isles of Scilly. In J.D. Scourse (ed.), *The Isles of Scilly, Field Guide*: 134-136. Coventry: Quaternary Research Association.

Catt, J.A., & P.G.N. Digby, 1988. Boreholes in the Wolstonian Basement Till at Easington, Holderness, July 1985. *Proceedings of the Yorkshire Geological Society* 47: 21-27.

Catt, J.A. & P.A. Madgett, 1981. The work of W.S. Bisat F.R.S. on the Yorkshire coast. In J. Neale & J. Flenley (eds.), *The Quaternary in Britain*: 119-136. Oxford: Pergamon Press.

Catt, J.A. & L.F. Penny, 1966. The Pleistocene deposits of Holderness, East Yorkshire. *Proceedings of the Yorkshire Geological Society* 35: 375-420.

Catt, J.A. & S.J. Staines, 1982. Loess in Cornwall. *Proceedings of the Ussher Society* 5: 368-376.

Catt, J.A., A.H. Weir & P.A. Madgett, 1974. The loess of Eastern Yorkshire and Lincolnshire. *Proceedings of the Yorkshire Geological Society* 40: 23-39.

Cazalet, P.C.D., 1968. Investigation of profile character development and distribution of the soils of parts of the southern Pennines. Unpublished Ph.D. Thesis, University of London (Bedford College).

Cazalet, P.C.D., 1969. Correlation of Cheshire Plain and Derbyshire Dome glacial deposits. *Mercian Geologist* 3: 71-84.

Cegla, J. & S. Kozarski, 1977. Sedimentary and geomorphological consequences of the occurrence of naled sheets on the outwash plain of the Gas Glacier, Sorkappland, Spitsbergen. *Acta Universitatis Wratislaviensis* 387: 63-84.

Chapelhowe, R., 1965. On the glaciation in North Roe, Shetland. *Geographical Journal* 131: 60-70.

Chapman, R.J., 1970. The Late-Weichselian glaciations of the Erne basin. *Irish Geography* 6: 153-161.

Charlesworth, J.K., 1924. The glacial geology of the north-west of Ireland. *Proceedings of the Royal Irish Academy* 36 B: 174-314.

Charlesworth, J.K., 1926. The readvance, marginal kame-moraine of the South of Scotland, and some later stages of retreat. *Transactions of the Royal Society of Edinburgh* 55: 25-50.

Charlesworth, J.K., 1928a. The glacial retreat from central and southern Ireland. *Quarterly Journal of the Geological Society of London* 84: 293-342.

Charlesworth, J.K., 1928b. The glacial geology of north Mayo and west Sligo. *Proceedings of the Royal Irish Academy* 38 B: 100-115.

Charlesworth, J.K., 1937. A map of the glacier-lakes and the local glaciers of the Wicklow Hills. *Proceedings of the Royal Irish Academy* 44 B: 29-36.

Charlesworth, J.K., 1939. Some observations on the glaciation of north-east Ireland. *Proceedings of the Royal Irish Academy* 45 B: 255-295.

Charlesworth, J.K., 1955. The Late-glacial history of the Highlands and Islands of Scotland. *Transactions of the Royal Society of Edinburgh* 62: 769-928.

Charlesworth, J.K., 1957. *The Quaternary Era*: 1700 pp. London: Edward Arnold.

Charlesworth, J.K., 1963a. *Historical geology of Ireland*: 565 pp. Edinburgh: Oliver and Boyd.

Charlesworth, J.K., 1963b. Some observations on the Irish Pleistocene. *Proceedings of the Royal Irish Academy* 62 B: 295-322.

Charlesworth, J.K., 1966. *The geology of Ireland: an introduction*: 276 pp. Edinburgh: Oliver and Boyd.

Charlesworth, J.K., 1973. Stages in the dissolution of the last ice-sheet in Ireland and the Irish Sea region. *Proceedings of the Royal Irish Academy* 73 B: 79-86.

Cheel, R.J. & B.R. Rust, 1982. Coarse-grained facies of glaciomarine deposits near Ottawa, Canada. In R. Davidson-Arnott, W. Nickling & B.D. Fahey (eds.), *Research in Glacial, Glacio-Fluvial and Glacio-Lacustrine Systems*: 279-295. Norwich: Geo Books.

Cheetham, G.H., 1979. Flow competence in relation to braiding. *Bulletin of the Geological Society of America* 90: 877-886.

Cheshire, D.A., 1981. A contribution towards a glacial stratigraphy of the Lower Lea valley, and implications for the Anglian Thames. *Quaternary Studies* 1: 27-69.

Cheshire, D.A., 1983. Till lithology in Hertfordshire and West Essex. In J. Rose (ed.), *Diversion of the Thames*: 50-59. Cambridge: Quaternary Research Association.

Cheshire, D.A., 1986. The Lithology and Stratigraphy of the Anglian Deposits of the Lea Basin: 515 pp. Unpublished Ph.D. thesis, Hatfield Polytechnic, Council National Academic Awards.

Chinn, T.J.H., 1979. Moraine forms and their recognition on steep mountain slopes. In Ch. Schlüchter (ed.), *Moraines and Varves*: 51-57. Rotterdam: Balkema.

Church, M., 1972. Baffin Island sandurs; a study of Arctic fluvial processes. *Geological Survey of Canada Bulletin* 216: 208 pp.

Church, M. & R. Gilbert, 1975. Proglacial fluvial and lacustrine environments. In A.V. Jopling & B.C. McDonald (eds.), *Glaciofluvial and glaciolacustrine sedimentation*. Society of Economic Paleontologists and Mineralogists, Special Publication 23: 22-100.

Church, M. & J.M. Ryder, 1972. Paraglacial sedimentation: a consideration of fluvial processes conditioned by glaciation. *Bulletin of the Geological Society of America* 83: 3059-3071.

Clague, J.J., 1975. Sedimentology and paleohydrology of Late Pleistocene outwash, Rocky Mountain trench, south-eastern British Columbia. In A.V. Jopling & B.C. McDonald (eds.), *Glaciofluvial and glaciolacustrine sedimentation*. Society of Economic Paleontologists and Mineralogists, Special Publication 23: 223-237.

Clapperton, C.M., 1970. On the evidence for a Cheviot ice cap. *Transactions of the Institute of British Geographers* 50: 115-126.

Clapperton, C.M., 1971a. The pattern of deglaciation in part of north Northumberland. *Transactions of the Institute of British Geographers* 53: 67-78.

Clapperton, C.M., 1971b. The location and origin of glacial meltwater phenomena in the eastern Cheviot Hills. *Proceedings of the Yorkshire Geological Society* 38: 361-380.

Clapperton, C.M., A.R. Gunson, & D.E. Sugden, 1975. Loch Lomond Readvance in the Eastern Cairngorms. *Nature* 253: 710-712.

Clapperton, C.M. & D.E. Sugden, 1977. The Late-Devensian glaciation of North-East Scotland. In J.M. Gray & J.J. Lowe (eds.), *Studies in the Scottish Lateglacial Environment*: 1-13. Oxford: Pergamon Press.

Clark, R., 1971. Periglacial landforms and landscapes in Northumberland. *Proceedings of the Cumberland Geological Society* 3: 5-20.

Clarke, M.R. & C.A. Auton, 1982. The Pleistocene depositional history of the Norfolk-Suffolk borderlands. *Report, Institute of Geological Sciences* 82/1: 23-29. London: Her Majesty's Stationery Office.

Clayton, K.M., 1957. Some aspects of the glacial deposits of Essex. *Proceedings of the Geologists' Association* 68: 1-21.

Clayton, K.M., 1960. The landforms of part of southern Essex. *Transactions of the Institute of British Geographers* 28: 55-74.

Clayton, K.M. & J.C. Brown, 1958. The glacial deposits around Hertford. *Proceedings of the Geologists' Association* 69: 103-119.

Clemmensen, L.B. & M. Houmark-Nielsen, 1981. Sedimentary features of a Weichselian glaciolacustrine delta. *Boreas* 10: 229-245.

Close, M.H., 1865. Notes on the general glaciation of the rocks in the neighbourhood of Dublin. *Journal of the Royal Geological Society of Ireland* 1: 3-13.

Close, M.H., 1867. Notes on the general glaciation of Ireland. *Journal of the Royal Geological Society of Ireland* 1: 207-242.

Close, M.H., 1874. The elevated shell-bearing gravels near Dublin. *Journal of the Royal Geological Society of Ireland* 4: 36-40.

Close, M.H., 1878. The physical geology of the neighbourhood of Dublin. *Journal of the Royal Geological Society of Ireland* 5: 49-77.

Clough, C.T., G. Barrow, C.B. Crampton, H.B. Maufe, E.B. Bailey & E.M. Anderson, 1910. The geology of East

Lothian. *Memoir of the Geological Survey of Scotland.*

Cohen, J.M., 1976. The sedimentology of varved clays from Glacial Lake Blessington: 227 pp. Unpublished B.A. Dissertation, Dublin, Trinity College.

Cohen, J.M., 1979. Deltaic sedimentation in glacial Lake Blessington, County Wicklow, Ireland. In Ch. Schlüchter (ed.), *Moraines and Varves*: 357-367. Rotterdam: Balkema.

Cole, G.A.J. & T. Hallissy, 1914. The Wexford gravels and their bearing on interglacial geology. *Geological Magazine* 1: 498-509.

Colhoun, E.A., 1970. On the nature of the glaciations and final deglaciation of the Sperrin Mountains and adjacent areas in the north of Ireland. *Irish Geography* 6: 162-185.

Colhoun, E.A., 1971a. The glacial stratigraphy of the Sperrin Mountains and its relation to the glacial stratigraphy of north-west Ireland. *Proceedings of the Royal Irish Academy* 71 B: 37-52.

Colhoun, E.A., 1971b. Late Weichselian periglacial phenomena of the Sperrin Mountains, Northern Ireland. *Proceedings of the Royal Irish Academy* 71 B: 53-71.

Colhoun, E.A., 1981. A protalus rampart from the western Mourne Mountains, Northern Ireland. *Irish Geography* 14: 85-90.

Colhoun, E.A., J.H. Dickson, A.M. McCabe & F.W. Shotton, 1972. A Middle Midlandian freshwater series at Derryvree, Maguiresbridge, County Fermanagh, Northern Ireland. *Proceedings of the Royal Society of London* B 180: 273-292.

Colhoun, E.A. & A.M. McCabe, 1973. Pleistocene glacial, glaciomarine and associated deposits of Mell and Tullyallen townlands, near Drogheda, eastern Ireland. *Proceedings of the Royal Irish Academy* 73 B: 165-206.

Colhoun, E.A. & G.F. Mitchell, 1971. Interglacial marine formation and lateglacial freshwater formation in Shortalstown Townland, Co. Wexford. *Proceedings of the Royal Irish Academy* 71 B: 211-245.

Colhoun, E.A. & F.M. Synge, 1980. The cirque moraines at Lough Nahanagan, County Wicklow, Ireland. *Proceedings of the Royal Irish Academy* 80 B: 25-45.

Collins, J.F., 1982. Soils on Munsterian- and Midlandian-age drifts at Piltown, County Kilkenny. *Journal of Earth Sciences Royal Dublin Society* 4: 101-108.

Collinson, J.D., 1970. Bedforms of the Tana River, Norway. *Geografiska Annaler* 52 A: 31-56.

Connell, E.R., K.J. Edwards & A.M. Hall, 1982. Evidence for two pre-Flandrian palaeosols in Buchan, Scotland. *Nature* 297: 570-572.

Connell, E.R. & A.M. Hall, 1987. The periglacial history of Buchan, Scotland. In J. Boardman (ed.), *Periglacial Processes and Landforms in Britain and Ireland*: 277-285. London: Cambridge University Press.

Connell, E.R., A.M. Hall, D. Shaw & L.A. Riley, 1985. Palynology and significance of radio-carbon dated organic materials from Cruden Bay brick pit, Grampian Region, Scotland. *Quaternary Newsletter* 47: 19-25.

Coope, G.R., 1959. A Late Pleistocene insect fauna from Chelford, Cheshire. *Proceedings of the Royal Society of London* B 151: 70-86.

Coope, G.R., 1977. Fossil coleopteran assemblages as sensit-

ive indicators of climatic changes during the Devensian (Last) cold stage. *Philosophical Transactions of the Royal Society of London* 280 B: 313-340.

Coope, G.R. & R.B. Angus, 1975. An ecological study of a temperate interlude in the middle of the last glaciation, based on fossil Coleoptera from Isleworth, Middlesex. *Journal of Animal Ecology* 44: 365-391.

Coope, G.R., J.H. Dickson, J.A. McCutcheon, & G.F. Mitchell, 1979. The Late Glacial and Early Postglacial deposit at Drumurcher, Co. Monaghan. *Proceedings of the Royal Irish Academy* 79 B: 63-85.

Coque-Delhuille, B. & Y. Veyret, 1984. La limite d'englacement Quaternaire dans le Sud-ouest anglais (Grande Bretagne). *Revue Geomorphologique Dynamique* 33: 1-24.

Cornish, R., 1981. Glaciers of the Loch Lomond Stadial in the western Southern Uplands of Scotland. *Proceedings of the Geologists' Association* 92, 105-114.

Cornish, R., 1982. Glacier flow at a former ice-divide in SW Scotland. *Transactions of the Royal Society of Edinburgh: Earth Sciences* 73: 31-41.

Cornish, R., 1983. Glacial erosion in an ice-divide zone. *Nature* 301: 413-415.

Cornwell, J.D. & R.M. Carruthers, 1986. Geophysical studies of a buried valley system near Ixworth, Suffolk. *Proceedings of the Geologists' Association* 97: 357-364.

Costello, W.R. & R.G. Walker, 1972. Pleistocene sedimentology, Credit River, southern Ontario: a new component of the braided river model. *Journal of Sedimentary Petrology* 42: 389-400.

Coudé, A., 1983. Les cirques glaciares du Iar Connacht (Irelande Occidentale), témoins géomorphologiques des influences climatiques Atlantiques au Pléistocène. *105-Congrés national des Societes savantes, Caen 1980, Publications de la Coéitée des Travaux Historiques et Géographiques*: 131-154.

Coudé, A., 1985. Nature et distribution des dépots glaciares et periglaciaires en Irelande Occidentale: discussion sur l'extension de la Dernière Grande Glaciation. *Bulletin de l'Association Française pour l'étude du Quaternaire* 2-3: 97-104.

Coward, M.P., 1977. Anomalous glacial erratics in the southern part of the Outer Hebrides. *Scottish Journal of Geology* 13: 185-188.

Cox, F.C., 1981. The 'Gipping Till' revisited. In J. Neale & J. Flenley (eds.), *The Quaternary in Britain*: 32-42. Oxford: Pergamon.

Cox, F.C., 1985a. The East Anglia Regional Geological Survey: an overview. *Modern Geology* 9: 103-126.

Cox, F.C., 1985b. The tunnel valleys of Norfolk, East Anglia. *Proceedings of the Geologists' Association* 96: 357-369.

Cox, F.C. & D.M. Bridge, 1977. The limestones and dolomite resources of the country around Monyash, Derbyshire. Description of 1:25,000 resource sheet SK 16: 138 pp. *Institute of Geological Sciences mineral assessment report* 26.

Cox, F.C. & E.F.P. Nickless, 1972. Some aspects of the glacial history of central Norfolk. *Bulletin of the Geological Survey of Great Britain* 42: 79-98.

Coxon, P., 1984. The terraces of the River Waveney. In P. Allen (ed.), *Field Guide (revised edition, Oct. 1984) to the*

Gipping and Waveney Valleys, Suffolk, May 1982: 80-94. Cambridge: Quaternary Research Association.

Coxon, P., 1985a. A Hoxnian Interglacial site at Athelington, Suffolk. *New Phytologist* 99: 611-621.

Coxon, P., 1985b. Quaternary Geology and Gleniff - Protalus rampart. In R.H. Thorn (ed.), *Sligo and West Leitrim. Irish Association for Quaternary Studies Field Guide* 8: 1-12. IQUA. Dublin.

Coxon, P., 1986a. A radiocarbondated Early Post Glacial pollen diagram from a pingoremnant near Millstreet, County Cork. *Irish Journal of Earth Sciences* 8: 9-20.

Coxon, P., 1986b. Lower and Middle Pleistocene deposits in Ireland. *Quaternary Newsletter* 50: 27-28.

Coxon, P. & A.M. Flegg, 1985. A Middle Pleistocene interglacial deposit from Ballyline, Co. Kilkenny. *Proceedings of the Royal Irish Academy* 85 B: 107-120.

Coxon, P. & A.M. Flegg, 1987. A Late Pliocene / Early Pleistocene deposit at Pollnahallia, near Headford, County Galway. *Proceedings of the Royal Irish Academy* 87 B: 15-42.

Craig, A.J., 1978. Pollen percentage and influx analyses in South-east Ireland: A contribution to the ecological history of the Late Glacial period. *Journal of Ecology* 66: 297-324.

Creighton, J.R., 1974. A study of the late Pleistocene geomorphology of north-central Ulster. Unpublished Ph.D. Thesis, Belfast: The Queen's University.

Crofts, R.G., 1982. The sand and gravel resources of the country between Coventry and Rugby, Warwickshire. *Mineral Assessment Report of the Institute of Geological Sciences* 125.

Croll, J., 1870. The boulder clay of Caithness a product of land ice. *Geological Magazine* 7: 209-214; 271-278.

Croll, J., 1875. *Climate and Time in their Geological Relations*. London: Dalby, Isbister & Co.

Culleton, E.B., 1978a: Limits and directions of ice movements in south County Wexford. *Journal of Earth Sciences Royal Dublin Society* 1: 33-39.

Culleton, E.B., 1978b. Characterisation of glacial deposits in south Wexford. *Proceedings of the Royal Irish Academy* 78 B: 293-308.

Cullingford, R.A. & D.E. Smith, 1980. Late Devensian raised shorelines in Angus and Kincardinshire, Scotland. *Boreas* 9: 21-38.

Cumming, J.G., 1846. On the geology of the Isle of Man. Part 2: The Tertiary Formations. *Quarterly Journal of the Geological Society of London* 2: 335-348.

Cumming, J.G., 1854. On the superior limits of the glacial deposits of the Isle of Man. *Quarterly Journal of the Geological Society of London* 10: 211-232.

Dackombe, R.V., 1978. Aspects of the tills of the Isle of Man: 522 pp. Unpublished Ph.D. Thesis, University of Liverpool.

Dackombe, R.V. & G.S.P. Thomas (eds.), 1985. *Isle of Man. Field Guide*. Cambridge: Quaternary Research Association.

Dackombe, R.V. & G.S.P. Thomas, 1985. *Field Guide to the Quaternary of the Isle of Man*: 122 pp. Cambridge: Quaternary Research Association.

Dale, E., 1900. *The scenery and geology of the peak of Derbyshire*. Buxton: Simpson, Marston & Co Ltd.: VI: 106-127.

Dalton, A.C., 1958. The distribution of dolerite boulders in the glaciation of N.E. Derbyshire. *Proceedings of the Geologists' Association* 68: 278-285.

Daly, D., 1985. Groundwater in County Galway with particular reference to its protection from pollution: 48 pp. Geological Survey of Ireland, Internal Report.

Dardis, G.F., 1982. Sedimentological aspects of the Quaternary geology of south-central Ulster, Northern Ireland: 422 pp. Unpublished Ph.D. thesis, Ulster Polytechnic.

Dardis, G.F., 1985. Till facies associations in drumlins and some implications for their mode of formation. *Geografiska Annaler* 67 A: 13-22.

Dardis, G.F. & A.M. McCabe, 1983. Facies of subglacial channel sedimentation in late-Pleistocene drumlins, Northern Ireland. *Boreas* 12: 263-278.

Dardis, G.F. & A.M. McCabe, 1984. Characteristics and origins of lee-side stratification sequences in late Pleistocene drumlins, Northern Ireland. *Earth Surface Processes and Landforms* 9: 409-424.

Dardis, G.F. & A.M. McCabe, 1987. Subglacial sheetwash and debris flow deposits in Late-Pleistocene drumlins, Northern Ireland. In J. Rose & J. Menzies (eds.), *Drumlins*, Proceedings of the 1st International Conference on Geomorphology, Manchester 1985: 225-240. Rotterdam: Balkema.

Dardis, G.F., A.M. McCabe & W.I. Mitchell, 1984. Characteristics and origins of lee-side stratification sequences in Late Pleistocene drumlins, Northern Ireland. *Earth Surface Processes and Landforms* 9: 409-424.

Dardis, G.F., W.I. Mitchell & K.R. Hirons, 1985. Middle Midlandian interstadial deposits at Greenagho, near Belcoo, County Fermanagh, Northern Ireland. *Irish Journal of Earth Sciences* 7: 1-6.

Davies, G.L.H., 1983. *Sheets of Many Colours. The Mapping of Ireland's Rocks 1750-1890*: xiv + 242 pp. Dublin: Royal Dublin Society.

Davies, G.L.H. & N. Stephens, 1978. *Ireland*: 250 pp. London: Methuen.

Davies, H.C., M.R. Dobson & R.J. Whittington, 1984. A revised seismic stratigraphy for Quaternary deposits on the inner continental shelf west of Scotland between 55°30' and 57°30' north. *Boreas* 13: 49-66.

Davies, K.H., 1983. Amino acid analysis of Pleistocene marine molluscs from the Gower Peninsula. *Nature* 302: 137-139.

Dawson, A.G., 1977. A fossil lobate rock glacier in Jura. *Scottish Journal of Geology* 13: 37-42.

Dawson, A.G., 1979. A Devensian medial moraine in Jura. *Scottish Journal of Geology* 15: 43-48.

Dawson, A.G., 1982. Lateglacial sea-level changes and ice-limits in Islay, Jura and Scarba, Scottish Inner Hebrides. *Scottish Journal of Geology* 18: 253-265.

Dawson, A.G., 1983. *Islay and Jura, Scottish Hebrides,* Field Guide: 31 pp. Cambridge: Quaternary Research Association.

Dawson, M.R., 1985. Environmental reconstructions of a late Devensian terrace sequence. Some preliminary findings. *Earth Surface Processes and Landforms* 10: 237-246.

Dawson, M.R. & I.D. Bryant, 1987. Three dimensional facies geometry in Pleistocene outwash sediments, Worcestershire, U.K. In F.G. Etheridge (ed.), *Recent developments in fluvial sedimentology*. Society of Economic Paleontologists and Mineralogists, Special Publication 39: 191-196.

Day, A.A., 1959. The continental margin between Brittany and Ireland. *Deep-Sea Research* 5: 249-265.

Day, F.H., 1963. *The chemical elements in nature*: 372 pp. New York: Reinhold Publishing Company.

De Boer, G., 1945. A system of glacier lakes in the Yorkshire Wolds. *Proceedings of the Yorkshire Geological Society* 25: 223-233.

De Boer, G., J.W. Neale & L.F. Penny, 1958. A guide to the geology of the area between Market Weighton and the Humber. *Proceedings of the Yorkshire Geological Society* 31: 157-209.

De Boer, G., L.F. Penny & J.A. Catt, 1965. Holderness and Spurn Head, 18th to 20th September 1964. *Proceedings of the Yorkshire Geological Society* 35: 294-298.

Degens, E.T., 1965. *Geochemistry of sediments*: 342 pp. New Jersey: Prentice-Hall Inc.

De Jong, J., 1967. The Quaternary of the Netherlands. In K. Rankama (ed.), *The Quaternary*, Volume 2: 301-426. New York: Interscience.

De Jong, J., 1988. Climatic variability during the past three million years, as indicated by vegetational evolution in northwest Europe and with emphasis on data from The Netherlands. *Philosophical Transactions of the Royal Society of London* B 318: 603-617.

Delantey, L.J. & R.J. Whittington, 1977. A re-assessment of the 'Neogene' deposits of the South Irish Sea and Nymphe Bank. *Marine Geology* 24: 23-30.

Derbyshire, E., 1975. The distribution of glacial soils in Great Britain. *The Engineering Properties of Glacial Materials*. Proceedings of the Birmingham Symposium: 6-17. (republished by Geo Abstracts 1978).

Derbyshire, E., C. Foster, M.A. Love & N.J. Edge, 1984. Pleistocene lithostratigraphy of northeast England: a sedimentological approach to the Holderness sequence. In W.C. Mahaney (ed.), *Correlation of Quaternary Chronologies*: 371-384. Norwich: Geobooks.

Devoy, R.J., 1983. Late Quaternary shorelines in Ireland: an assessment of their implications for isostatic land movement and relative sea-level changes. In D.E. Smith & A.G. Dawson (eds.), *Shorelines and Isostasy*: 227-254. London: Academic Press.

Dhonau, T.J. & N.B. Dhonau, 1963. Glacial structures on the north Norfolk coast. *Proceedings of the Geologists' Association* 74: 433-439.

Dickson, C.A., J.H. Dickson & G.F. Mitchell, 1970. The Late Weichselian Flora of the Isle of Man. *Philosophical Transactions of the Royal Society of London* B 258: 31-79.

Dines, H.G., 1946. Geology of the country around Witney. *Memoirs of the Geological Survey of Great Britain*.

Dingle, R.V., 1970. Quaternary sediments and erosional features off the north Yorkshire coast, western North Sea. *Marine Geology* 9: M17-M22.

Dingle, R.V., 1971a. A marine geological survey off the north-east coast of England (western North Sea). *Journal of the Geological Society of London* 127: 303-338.

Dingle, R.V., 1971b. Buried tunnel valleys off the Northumberland coast, western North Sea. *Geologie en Mijnbouw* 50 (5): 679-686.

Dobson, M.R., W.E. Evans & K.H. James, 1971. The sediment on the floor of the southern Irish Sea. *Marine Geology* 11: 27-69.

Dobson, M.R., W.E. Evans & R.J. Whittington, 1973. *The geology of the south Irish Sea*: 35 pp. Report of the Institute of Geological Sciences, No 73/11.

Domack, E.W., 1982. Sedimentology of glacial and glacial marine deposits on the George V - Adelie continental shelf, East Antarctica. *Boreas* 11: 79-97.

Domack, E.W., 1983. Facies of Late Pleistocene glacial marine sediments on Whidbey Island, Washington: an isostatic glacial marine sequence. In B.F. Molnia (ed.), *Glacial-marine Sedimentation*: 535-570. Ney York: Plenum Press.

Domack, E.W., 1984. Rhythmically bedded glaciomarine sediments to Whidbey Island, Washington. *Journal of Sedimentary Petrology* 54: 589-602.

Domack, E.W. & D.E. Lawson, 1985. Pebble fabric in an ice-rafted diamicton. *Journal of Geology* 93: 577-591.

Donner, J.J., 1979. The Early or Middle Devensian peat at Burn of Benholm, Kincardineshire, Scotland. *Scottish Journal of Geology* 15: 247-250.

Donner, J.J. & R.G. West, 1955. Ett drumlinfält på øn Skye, Skottland. *Terra* 2: 45-48.

Donner, J.J. & R.G. West, 1957. The Quaternary geology of Braganset, Nordaustlandet, Spitsbergen. *Skrifter Norsk Polarinstitutt* 109: 1-29.

Donovan, D.T., 1973. The geology and origin of the Silver Pit and other closed basins in the North Sea. *Proceedings of the Yorkshire Geological Society* 39: 267-293.

Donovan, D.T. & R.V. Dingle, 1965. Geology of part of the southern North Sea. *Nature* 207: 1186-1187.

Doré, A.G., 1976. Preliminary geological interpretation of the Bristol Channel approaches. *Journal of the Geological Society of London* 132: 453-459.

Double, I.S., 1924. The petrography of the later Tertiary deposits of the east of England. *Proceedings of the Geologists' Association* 35: 332-358.

Douglas, T.D., 1974. The Pleistocene beds exposed at Cadeby, Leicestershire. *Transactions of the Leicester Literary & Philosophical Society* 68: 57-73.

Douglas, T.D., 1976. The Pleistocene Geology and Geomorphology of Western Leicestershire: 175 pp. Unpublished Ph.D. thesis, University of Leicester.

Douglas, T.D., 1980. The Quaternary deposits of western Leicestershire. *Philosophical Transactions of the Royal Society of London* B 288: 259-286.

Douglas, T.D. & S. Harrison, 1985. Periglacial landforms and sediments in the Cheviots. In J. Boardman (ed.), *Field Guide to the periglacial landforms of Northern England*: 68-75.

Douglas, T.D. & S. Harrison, 1987. Late Devensian periglacial slope deposits in the Cheviot Hills. In J. Boardman (ed.), *Periglacial processes and landforms in Britain and Ireland*: 237-244. Cambridge: Cambridge University Press.

Dowdeswell, J.A. & M.J. Sharp, 1986. Characterization of pebble fabrics in modern terrestrial glacigenic sediments. *Sedimentology* 33: 699-710.

Drewry, D.J., 1986. *Glacial geologic processes*: 276 pp. London: Edward Arnold.

Drewry, D.J. & A.P.R. Cooper, 1981. Processes and models of Antarctic glaciomarine sedimentation. *Annals of Glaciology* 2: 117-122.

Dury, G.H., 1951. A 400 foot bench in south-eastern Warwickshire. *Proceedings of the Geologists' Association* 62: 167-173.

Dwerryhouse, A.R., 1923. The glaciation of north-eastern Ireland. *Quarterly Journal of the Geological Society of London* 79: 352-422.

Earp, J.R. & B.J. Taylor, 1986. Geology of the country around Chester and Winsford. *Memoir of the British Geological Survey*, Her Majesty's Stationery Office.

Eastwood, T., S.E. Hollingworth, W.C.C. Rose & F.M. Trotter, 1968. The geology of the country around Cockermouth and Caldbeck. *Memoir of the Geological Survey of the United Kingdom*. London: Her Majesty's Stationery Office.

Eden, R.A., I.P. Stevenson & W. Edwards, 1957. Geology of the country around Sheffield. *Memoir of the Geological Survey of Great Britain*: 238 pp. London: Her Majesty's Stationery Office.

Edwards, C.A., 1981. The tills of Filey Bay. In J. Neale & J. Flenley (eds.), *The Quaternary in Britain*: 108-118. Oxford: Pergamon Press.

Edwards, K.J. & E.R. Connell, 1981. Interglacial and interstadial sites in north-east Scotland. *Quaternary Newsletter* 33: 22-28.

Edwards, K.J. & W.P. Warren, 1985. Quaternary studies in Ireland: in K.J. Edwards & W.P. Warren (eds.), *The Quaternary History of Ireland*: 1-16. London: Academic Press.

Edwards, M.B., 1975. Late Precambrian subglacial tillites, north Norway. *IXme Congrès International de Sedimentologie, Nice*, Theme 1: 61-66.

Edwards, W., 1937. A Pleistocene strand line in the Vale of York. *Proceedings of the Yorkshire Geological Society* 23: 103-118.

Edwards, W., G.H. Mitchell & T.H. Whitehead, 1950. Geology of the district north and east of Leeds. *Memoir of the Geological Survey*. London: Her Majesty's Stationery Office.

Edwards, W., D.A. Wray & G.H. Mitchell, 1940. Geology of the country around Wakefield. *Memoir of the Geological Survey*. London: Her Majesty's Stationery Office.

Ehlers, J., 1981. (Discussion remark). *Annals of Glaciology* 2: 191.

Ehlers, J., 1983. The glacial history of north-west Germany. In J. Ehlers (ed.), *Glacial Deposits in North-West Europe*: 229-238. Rotterdam: Balkema.

Ehlers, J., 1987. Die Entstehung des Kaolinsandes von Sylt. In U. von Hacht (ed.), *Fossilien von Sylt II*: 249-267. Hamburg: von Hacht.

Ehlers, J., 1988. Skandinavische Geschiebe in Großbritannien. *Der Geschiebesammler* 22: 49-64.

Ehlers, J., P.L. Gibbard & C.A. Whiteman, 1987. Recent investigations of the Marly Drift of northwest Norfolk, England. In J. van der Meer (ed.), *Tills and Glaciotectonics*: 39-54. Rotterdam/Brookfield: Balkema.

Ehlers, J., K.-D. Meyer & H.-J. Stephan, 1984. The pre-Weichselian glaciations of North-West Europe. *Quaternary Science Reviews* 3: 1-40.

Einarsson, T. & K.J. Albertsson, 1988. The glacial history of Iceland during the past three million years. *Philosophical Transactions of the Royal Society of London* B 318: 637-644.

Eisma, D., J.H.F. Jansen & Tj.C.E. van Weering, 1979. Sea-floor morphology and recent sediment movement in the North Sea. In E. Oele, R.T.E. Schüttenhelm & A.J. Wiggers (eds.), *The Quaternary History of the North Sea*, Acta Universitatis Upsaliensis Symposium Universitatis Upsaliensis Annum Quingentesimum Celebrantis 2: 217-231.

Eisma, D., W.G. Mook & C. Laban, 1981. An early Holocene tidal flat in the Southern Bight. In S.-D. Nio, R.T.E. Schüttenhelm & Tj.C.E. van Weering (eds.), *Holocene Marine Sedimentation in the North Sea Basin*. Special Publication of the International Association of Sedimentologists 5: 229-237.

Eissmann, L., 1975. Das Quartär der Leipziger Tieflandsbucht und angrenzender Gebiete um Saale und Elbe. Modell einer Landschaftsentwicklung am Rand der europäischen Kontinentalvereisung. *Schriftenreihe für Geologische Wissenschaften* 2: 228 pp.

Ellis-Gruffydd, I.D., 1977. Late Devensian glaciation in the Upper Usk basin. *Cambria* 4: 46-55.

Elverhøi, A., O. Lonne & R. Seland, 1983. Glaciomarine sedimentation in the modern fjord environment, Spitsbergen. *Polar Research* 1: 127-149.

Embleton, C., 1970. North-eastern Wales. In: C.A. Lewis (ed.), *The glaciations of Wales*: 59-82. London: Longman.

Embleton, C. & C.A.M. King, 1975. *Glacial Geomorphology*, 2nd. edition: 583 pp. London: E. Arnold; New York: Halstead.

England, J., 1983. Isostatic adjustments in a full glacial sea. *Canadian Journal of Earth Sciences* 20: 895-917.

Erdtman, G., 1925. Pollen statistics from the Curragh and Ballaugh, Isle of Man. *Proceedings of the Liverpool Geological Society* 14: 158-163.

Eriksson, K, 1983. Till investigations and mineral prospecting. In J. Ehlers (ed.), *Glacial deposits in north-west Europe*: 107-113. Rotterdam: Balkema.

Evans, C.D.R. & M.J. Hughes, 1984. The Neogene succession of the South Western approaches, Great Britain. *Journal of the Geological Society of London* 141: 315-326.

Evans, H., 1975. The two-till problem in west Norfolk. *Bulletin of the Geological Society of Norfolk* 27: 61-75.

Evans, H., 1976. Aspects of the glaciation of west Norfolk: 246 pp. Unpublished M.Phil. Thesis, University of East Anglia.

Evans, W.B. & R.S. Arthurton, 1973. Northeast England. In G.F. Mitchell et al. (eds.), *A correlation of Quaternary deposits in the British Isles*. Geological Society of London, Special Report No.4: 99 pp.

Evans, W.B., A.A. Wilson, B.J. Taylor & D. Price, 1968. The geology of the country around Macclesfield, Congleton, Crewe and Middlewich. *Memoir of the Geological Survey of Great Britain*. London: Her Majesty's Stationery Office.

Evenson, E.B., A. Dreimanis & W. Newsome, 1977. Subaquatic flow tills: a new interpretation for the genesis of some laminated till deposits. *Boreas* 6: 115-133.

Eyles, C.H., 1987. Glacially influenced submarine-channel sedimentation in the Yakataga Formation, Middleton Island, Alaska. *Journal of Sedimentary Petrology* 57: 1004-1017.

Eyles, C.H. & N. Eyles, 1984. Glaciomarine sediments of the Isle of Man as a key to late Pleistocene stratigraphic investigations in the Irish Sea Basin. *Geology* 12: 359-364.

Eyles, C.H., N. Eyles & A.D. Miall, 1985. Models of a glaciomarine sedimentation and their application to the interpretation of ancient glacial sequences. *Palaeogeography, Palaeoclimatology, Palaeoecology* 51: 15-84.

Eyles, N., 1979. Facies of supraglacial sedimentation on Icelandic and Alpine temperate glaciers. *Canadian Journal of Earth Science* 16: 1341-1361.

Eyles, N., 1983a. *Glacial Geology: an Introduction for Engineers and Earth Scientists*: 409 pp. Oxford: Pergamon.

Eyles, N., 1983b. Glacial geology: a landsystems approach. In N. Eyles (ed.), *Glacial Geology: an introduction for engineers and Earth scientists*: 1-18. Oxford: Pergamon Press.

Eyles, N., 1983c. Modern Icelandic glaciers as depositional models for 'hummocky moraine' in the Scottish Highlands. In E.B. Evenson, Ch. Schlüchter & J. Rabassa (eds.), *Tills and related deposits*: 47-59. Rotterdam: Balkema.

Eyles, N., 1987. Late Pleistocene debris flow deposits in large ice-contact lakes in British Columbia and Alaska. *Sedimentary Geology* 53: 33-71.

Eyles, N., T.E. Day & T. Gavican, 1987. Depositional influences upon the magnetic characteristics of lodgement tills and other glacial diamict facies. *Canadian Journal of Earth Sciences* 24: 2436-2458.

Eyles, N. & W. Dearman, 1981. A glacial terrain map of Britain for engineering purposes. *Bulletin of the International Association of Engineering Geology* 24: 173-184.

Eyles, N., C.H. Eyles & A.M. McCabe, 1985. Reply to Thomas & Dackombe (1985). *Geology* 13: 446-447.

Eyles, N., C.H. Eyles & A.M. McCabe, 1988. Sedimentation in an ice contact sub-aqueous setting: the Mid Pleistocene North Sea drifts of Norfolk, U.K. *Quaternary Science Reviews* 8: 57-74.

Eyles, N., C.H. Eyles & A.D. Miall, 1983. Lithofacies types and vertical profile model; an alternative approach to the description and environmental interpretation of glacial diamict and diamictite sequences. *Sedimentology* 30: 393-410.

Eyles, N. & J. Menzies, 1983. The subglacial landsystem. In N. Eyles (ed.), *Glacial Geology: an introduction for engineers and Earth scientists*: 19-70. Oxford: Pergamon Press.

Eyles, N. & A.D. Miall, 1985. Glacial facies. In R.G. Walker (ed.), *Facies Models*. Geoscience Canada Reprint Series 1: 15-38.

Eyles, N., A.D. Miall & C.H. Eyles, 1984. Lithofacies types and vertical profile models; Reply to Comments. *Sedimentology* 31: 891-898.

Eyles, N. & A.M. McCabe, 1989a. The Late Devensian (22,000 BP) Irish Sea Basin: The sedimentary record of a collapsed ice sheet margin. *Quaternary Science Reviews* 8: 304-351.

Eyles, N. & A.M. McCabe, 1989b. Glaciomarine facies subglacial tunnel valleys: the sedimentary record of glacio-isostatic downwarping in the Irish Sea Basin. *Sedimentology* 36: 431-448.

Eyles, N. & R.J. Rogerson, 1978. A framework for the investigation of medial moraine formation: Austerdalsbreen, Norway and Berendon Glacier, British Columbia. *Journal of Glaciology* 20: 99-114.

Eyles, N. & J.A. Sladen, 1981. Stratigraphy and geotechnical properties of weathered lodgement till in Northumberland, England. *Quarterly Journal of Engineering Geology* 14: 129-141.

Eyles, N., J.A. Sladen & S. Gilroy, 1982. A depositional model for stratigraphic complexes and facies superimposition in lodgement tills. *Boreas* 11: 317-333.

Eynon, G. & R.G. Walker, 1974. Facies relationships in Pleistocene outwash gravels, southern Ontario: a model for bar growth in braided rivers. *Sedimentology* 21: 43-70.

Fahnestock, R.K., 1963. Morphology and hydrology of a glacial stream - White River, Mount Rainier, Washington. *U.S. Geological Survey Professional Paper* 422-A: 70 pp.

Fahnestock, R.K., 1969. Morphology of the Slims River. *Icefield Ranges Research Project Scientific Results* 1: 161-172.

Fahnestock, R.K. & W.C. Bradley, 1973. Knik and Matanuska Rivers, Alaska: a contrast in braiding. In M. Morrisawa (ed.), *Fluvial Geomorphology*: 220-250. New York: State University.

Farrington, A., 1934. The glaciation of the Wicklow Mountains. *Proceedings of the Royal Irish Academy* 42 B: 173-209.

Farrington, A., 1936. The glaciation of the Bantry Bay district. *Scientific Proceedings of the Royal Dublin Society* 21: 345-361.

Farrington, A., 1939. Glacial geology of south-eastern Ireland. *Proceedings of the Geologists' Association* 50: 337-344.

Farrington, A., 1942. The Glacial drift near Brittas, on the border between County Dublin and County Wicklow. *Proceedings of the Royal Irish Academy* 47 B: 279-291.

Farrington, A., 1944. The glacial drifts of the district around Enniskerry, Co. Wicklow. *Proceedings of the Royal Irish Academy* 50 B: 133-157.

Farrington, A., 1945. Notes on the glacial geology of the Glen of Aherlow. *Irish Geography* 1: 42-45.

Farrington, A., 1947. Unglaciated areas in Southern Ireland. *Irish Geography* 1: 89-97.

Farrington, A., 1949. The glacial drifts of the Leinster Mountains. *Journal of Glaciology* 1: 220-225.

Farrington, A., 1954. A note on the correlation of the Kerry-

Cork glaciations with those of the rest of Ireland. *Irish Geography* 3: 47-53.

Farrington, A., 1957a. Glacial Lake Blessington. *Irish Geography* 3: 216-222.

Farrington, A., 1957b. The Ice Age in the Dublin district. *Journal of the Institute of Chemistry of Ireland* 5: 23-27.

Farrington, A., 1959. The Lee Basin. Part One: Glaciation. *Proceedings of the Royal Irish Academy* 60 B: 135-166.

Farrington, A., 1960. In F.M. Synge & N. Stephens, The Quaternary period in Ireland - an assessment, 1960. *Irish Geography* 4: 122.

Farrington, A., 1964. Granite gravel at Lucan, Co. Dublin. *Irish Naturalists' Journal* 14: 212-213.

Farrington, A., 1965. The last glaciation in the Burren, Co. Clare. *Proceedings of the Royal Irish Academy* 64 B: 33-39.

Farrington, A., 1966a. The last glacial episode in the Wicklow Mountains. *Irish Naturalists' Journal* 15: 226-229.

Farrington, A., 1966b. The early-glacial raised beach in county Cork. *Scientific Proceedings of the Royal Dublin Society* A 2: 197-219.

Farrington, A., 1968. A buried moraine in County Dublin. *Irish Naturalists' Journal* 16: 52-53.

Farrington, A. & F.M. Synge, 1970. The eskers of the Tullamore district. In N. Stephens & R.E. Glasscock (eds.), *Irish geographical studies*: 49-52. Belfast: The Queen's University.

Fettes, D.J., J.R. Mendum, D.I. Smith & J.V. Watson (in press). The Geology of the Outer Hebrides. *Memoirs of the British Geological Survey*.

Feyling-Hanssen, R., 1982. Molluscs and other megafossils. In E. Olausson (ed.), *The Pleistocene-Holocene boundary in South West Sweden. Sveriges Geologiska Undersökning* 794: 120-136.

Finch, T.F., 1971. Limits of Weichsel glacial deposits in the south Tipperary area. *Scientific Proceedings of the Royal Dublin Society* B 3: 35-41.

Finch, T.F. (ed.), 1977. *Guidebook for INQUA excursion C16: Western Ireland*: 40 pp. Norwich: Geo Abstracts.

Finch, T.F. & F.M. Synge, 1966. The drifts and soils of west Clare and the adjoining parts of counties Kerry and Limerick. *Irish Geography* 5: 161-172.

Finch, T.F. & M. Walsh, 1973. Drumlins of County Clare. *Proceedings of the Royal Irish Academy* 73 B: 405-413.

Finlay, T.M., 1926. A Tönsbergite boulder from the boulder-clay of Shetland. *Transactions of the Edinburgh Geological Society* 12: 180.

Firth, C.M., 1986. Isostatic depression during the Loch Lomond Stadial: preliminary evidence from the Great Glen, Northern Scotland. *Quaternary Newsletter* 48: 1-9.

Fisher, M.J., B.M. Funnell & R.G. West, 1969. Foraminifera and pollen from a marine interglacial deposit in the western North Sea. *Proceedings of the Yorkshire Geological Society* 37: 311-320.

FitzPatrick, E.A., 1965. An interglacial soil at Teindland, Morayshire. *Nature* 207: 621-622.

Flett, J.S., 1937. *The first hundred years of the Geological Survey of Great Britain*: 280 pp. London: Her Majesty's Stationery Office.

Flinn, D., 1977. The erosion history of Shetland: a review. *Proceedings of the Geologists' Association* 88: 129-146.

Flinn, D., 1978a. The most recent glaciation of the Orkney-Shetland Channel and adjacent areas. *Scottish Journal of Geology* 14: 109-123.

Flinn, D., 1978b. The glaciation of the Outer Hebrides. *Geological Journal* 13: 195-199.

Flinn, D., 1983. Glacial meltwater channels in the northern isles of Shetland. *Scottish Journal of Geology* 19: 311-320.

Flint, R.F., 1930. The origin of the Irish 'eskers'. *Geographical Review* 20: 615-630.

Flint, R.F., 1971. *Glacial and Quaternary Geology*: 892 pp. New York: John Wiley.

Fookes, P.G., D.L. Gordon & I.E. Gigginbottom, 1975. Glacial landforms: their deposits and engineering characteristics. *The Engineering Properties of Glacial Materials*. Proceedings of the Birmingham Symposium: 18-51. (republished by Geo Abstracts 1978).

Fookes, P.G., L.W. Hinch, M.A. Huxley & N.E. Simons, 1978. Some soil properties in glacial terrain - the Taff valley, South Wales. In *The engineering behaviour of glacial materials*: 93-116. Norwich: Geo Abstracts.

Foot, M.J. & J. O'Kelly, 1865. Explanation to accompany sheets 98, 99, 108 and 109 of the one-inch map of the Geological Survey of Ireland. *Memoir of the Geological Survey*: 39 pp. Dublin.

Ford, T.D., 1972. Evidence of early stages in the evolution of the Derbyshire Karst. *Transactions of the Cave Research Group of Great Britain* 14: 73-77.

Ford, T.D., 1985. The Castleton Caves: results of speleothem dating. In D.J. Briggs, D.D. Gilbertson & R.D.S. Jenkinson (eds.), *Peak District and North Dukeries*. Field Guide: 77-83. Cambridge: Quaternary Research Association.

Foster, S.W., 1986. The Late Glacial and Post Glacial history of the Vale of Pickering and northern Yorkshire Wolds. Unpublished Ph.D. Thesis, University of Hull.

Fowler, A. & J.A. Robbie, 1961. Geology of the country around Dungannon. *Memoir of the Geological Survey of Northern Ireland*. Her Majesty's Stationery Office: 274 pp.

Francis, E.A., 1970. Quaternary. In G.A.L. Johnston & G. Hickling (eds.), *Geology of County Durham*: 134-152. Transactions of the Natural History Society of Northumberland 41.

Francis, E.A., 1975. Glacial sediments. A selective review. In A.E. Wright & F. Moseley (eds.), *Ice ages ancient and modern*: 43-68. Liverpool: Seel House Press.

Francis, E.A., 1983. On the classification of glacial sediments. In J. Neale & J. Flenley (eds.), *The Quaternary in Britain*: 237-247. Oxford: Pergamon.

Fraser, G.S. & J.C. Cobb, 1982. Late Wisconsinan proglacial sedimentation along the West Chicago Moraine in northwestern Illinois. *Journal of Sedimentary Petrology* 52: 473-491.

Fraser, R., 1801. *General view of the agriculture and mineralogy, present state and circumstances of the County Wicklow, with observations on the means of their improvement*: xvi + 289 pp. Dublin: Dublin Society.

Fraser, J.Z., 1982. Derivation of a summary facies sequence based on Markov chain analysis of the Caledon outwash: a

Pleistocene braided glacial fluvial deposit. In R. Davidson-Arnott, W. Nickling & B.D. Fahey (eds.), *Research in glacial, glacio- fluvial and glacio-lacustrine systems*: 175-202. Norwich: Geobooks.

Frechen, J. & H.J. Lippolt, 1965. Kalium-Argon-Daten zum Alter des Laacher Vulkanismus, der Rheinterrassen und der Eiszeiten. *Eiszeitalter und Gegenwart* 16: 5-30.

Freedman, J., 1975. Trace element geochemistry in health and disease. *Geological Society of America special paper* 155: 118 pp.

French, H.M., 1976. *The Periglacial Environment*: 309 pp. London: Longmans.

Frost, D.V. & D.W. Holliday, 1980. *The geology of the country around Bellingham*: 112 pp. Her Majesty's Stationery Office.

Fullerton, D.S. & G.M. Richmond, 1986. Comparison of the marine oxygen isotope record, the eustatic sea level record, and the chronology of glaciation in the United States of America. *Quaternary Science Reviews* 5: 197-200.

Funnell, B.M., 1977. Plio-Pleistocene foraminifera of the North Sea basin. In K. Clayton (ed.), *Abstracts, X INQUA Congress, Birmingham*: 151.

Funnell, B.M., 1987. Late Pliocene and Early Pleistocene stages of East Anglia and the adjacent North Sea. *Quaternary Newsletter* 52: 1-11.

Funnell, B.M. & R.G. West, 1962. The early Pleistocene of Easton Bavents, Suffolk. *Quarterly Journal of the Geological Society of London* 118: 125-141.

Gale, S.J., 1985. The Late- and Post-glacial history of the southern Cumbrian massif and its surrounding lowlands. In R.H. Johnson (ed.), *The geomorphology of North-west England*: 282-298. Manchester: Manchester University Press.

Gale, S.J. & P.G. Hoare, 1986. Blakeney ridge sands and gravels. In R.G. West & C.A. Whiteman (eds.), *The Nar Valley and North Norfolk*, Field Guide: 94-95. Coventry: Quaternary Research Association.

Gallois, R.W., 1978. The Pleistocene history of west Norfolk. *Bulletin of the Geological Society of Norfolk* 30: 3-38.

Gardiner, M.J. & P. Ryan, 1964. *Soils of Co. Wexford*. Dublin: An Foras Talúntais.

Garnes, K., 1979. Weichselian till stratigraphy in central South Norway. In Ch. Schlüchter (ed.), *Moraines and Varves*: 207-222. Rotterdam: Balkema.

Garrard, R.A., 1977. The sediments of the South Irish Sea and Nymphe Bank area of the Celtic Sea. In C. Kidson & M.J. Tooley (eds.), *The Quaternary History of the Irish Sea*: 69-92. Liverpool: Seel House Press.

Garrard, R.A. & M.R. Dobson, 1974. The nature and maximum extent of glacial sediments off the west coast of Wales. *Marine Geology* 16: 31-44.

Gaunt, G.D., 1974. A radiocarbon date relating to Lake Humber. *Proceedings of the Yorkshire Geological Society* 40: 195-197.

Gaunt, G.D., 1976. The Devensian maximum ice limit in the Vale of York. *Proceedings of the Yorkshire Geological Society* 40: 631-637.

Gaunt, G.D., 1980. Quaternary history of the southern part of the Vale of York. In J. Neale & J. Flenley (eds.), *The Quaternary in Britain*: 82-97. Oxford: Pergamon.

Gaunt, G.D., 1981. Quaternary history of the southern part of the Vale of York. In J. Neale & J. Flenley (eds.), *The Quaternary in Britain*: 82-97. Oxford: Pergamon Press.

Gaunt, G.D., G.R. Coope, P.J. Osborne & J.W. Franks, 1972. An interglacial deposit near Austerfield, southern Yorkshire. *Report of the Institute of Geological Sciences* 72/4: 13 pp.

Geikie, A., 1863. On the phenomena of the glacial drift of Scotland. *Transactions of the Geological Society of Glasgow* 1: 1-190.

Geikie, J., 1873. On the glacial phenomena of the Long Island or Outer Hebrides. *Journal of the Geological Society of London* 29: 532-545.

Geikie, J., 1878. On the glacial phenomena of the Long Island or Outer Hebrides. *Journal of the Geological Society of London* 34: 819-870.

Gennard, D.E., 1984. A palaeoecological study of the interglacial deposit at Benburb, Co. Tyrone. *Proceedings of the Royal Irish Academy* 84 B: 43-55.

George, T.N., 1970. Discussion. *Proceedings of the Geological Society of London* 1660: 378-380.

Gibbard, P.L., 1974. Pleistocene Stratigraphy and Vegetational History of Hertfordshire: 2 volumes, 286 + 129 pp. Unpublished Ph.D. thesis, University of Cambridge.

Gibbard, P.L., 1977. Pleistocene history of the Vale of St. Albans. *Philosophical Transactions of the Royal Society of London* B 280: 445-483.

Gibbard, P.L., 1979. Middle Pleistocene drainage in the Thames valley. *Geological Magazine* 116: 35-44.

Gibbard, P.L., 1980. The origin of stratified Catfish Creek Till by basal melting. *Boreas* 9: 71-85.

Gibbard, P.L., 1983. The diversion of the Thames - a review. In J. Rose (ed.), *Diversion of the Thames*, Field Guide. Cambridge: 8-23. Quaternary Research Association.

Gibbard, P.L., 1985. *The Pleistocene history of the middle Thames valley*. 155 pp. Cambridge: Cambridge University Press.

Gibbard, P.L., 1988. The history of the great northwest European rivers during the past three million years. *Philosophical Transactions of the Royal Society of London* B 318: 559-602.

Gibbard, P.L. & C. Turner, 1988. In defence of the Wolstonian stage. *Quaternary Newsletter* 54: 9-14.

Gibbard, P.L. & J.A. Zalasiewicz (eds.), 1988. *Pliocene - Middle Pleistocene of East Anglia*, Field Guide: 195 pp. Cambridge: Quaternary Research Association.

Gilbert, G.K., 1885. The topographic features of lake shores. *U.S. Geological Survey 5th Annual Report*: 69-123.

Gilbert, R., 1982. Contemporary sedimentary environments on Baffin Island, N.W.T., Canada: glaciomarine processes in fiords of eastern Cumberland Peninsula. *Arctic and Alpine Research* 14: 1-12.

Gilbert, R., 1983. Sedimentary processes of Canadian Arctic fiords. *Sedimentary Geology* 36: 147-175.

Gilbert, R., 1985. Quaternary glaciomarine sedimentation interpreted from seismic surveys of fiords on Baffin Island, N.W.T. *Arctic* 38: 271-280.

Gilligan, A., 1920. The petrography of the Millstone Grit of Yorkshire. *Quarterly Journal of the Geological Society of London* 75: 251-294.

Girling, M.A., 1974. Evidence from Lincolnshire of the age and intensity of the mid-Devensian temperate episode. *Nature* 250: 270.

Gladfelter, B.G., 1975. Middle Pleistocene sedimentary sequences in East Anglia. In K.W. Butzer & G. Isaac (eds.), *After the Australopithecines - Stratigraphy, Ecology and Culture Change in the Middle Pleistocene*: 225-258. The Hague, Paris: Mouton.

Godard, A., 1965. *Recherches de Géomorphologie en Ecosse du Nord-Ouest*. Paris: Les Belles Lettres.

Good, T.R. & I.D. Bryant, 1986. Fluvio-aeolian sedimentation - an example from Banks Island, N.W.T., Canada. *Geografiska Annaler* 67 A: 33-46.

Godwin, H., 1955. Vegetational history at Cwm Idwal: a Welsh plant refuge. *Svensk botanisk Tidsskrift* 49: 35-43.

Godwin, H. & E.H. Willis, 1959. Radiocarbon dating of the Late- glacial period in Britain. *Proceedings of the Royal Society of London* B 150: 199-215.

Goldthwait, R.P., 1958. Wisconsin age forests in western Ohio, I. Age and glacial events. *The Ohio Journal of Science* 58: 209-219.

Goldthwait, R.P., 1959. Scenes in Ohio during the last ice age. *The Ohio Journal of Science* 59: 193-216.

Goodchild, J.G., 1875. The glacial phenomena of the Eden valley and the west part of the Yorkshire Dales District. *Quarterly Journal of the Geological Society of London* 31: 55-99.

Goodchild, J.G., 1887. Ice work in Edenside and some of the adjoining parts of North Western England. *Transactions of the Cumberland and Westmorland Advancement of Literature and Science* 12: 111-167.

Goodlet, G.A., 1964. The kamiform deposits near Carstairs, Larnakshire. *Bulletin of the Geological Survey of Great Britain* 21: 175-196.

Goodlet, G.A., 1970. Sands and gravels of the southern counties of Scotland. *Report of the Institute of Geological Sciences* 70/4.

Gough, L.P., H.T. Shacklette & A.A. Case, 1979. Element concentrations toxic to plants, animals and man. *U.S. Geological Survey Bulletin* 1466: 80 pp.

Graciansky, P.C. de, C.W. Poag *et al.*, 1985. *Initial Reports of the Deep Sea Drilling Project* 80. Washington: US Government Printing Office. 679 pp.

Gray, J.M., 1972. The Inter-, Late-, and Postglacial shorelines and ice-limits of Lorn and eastern Mull. Unpublished Ph.D. Thesis, University of Edinburgh.

Gray, J.M., 1974. The Main Rock Platform of the Firth of Lorn, Western Scotland. *Transactions of the Institute of British Geographers* 61: 81-99.

Gray, J.M., 1975. The Loch Lomond Readvance and contemporaneous sea-levels in Loch Etive and neighbouring areas of western Scotland. *Proceedings of the Geologists' Association* 86: 227-238.

Gray, J.M., 1978a. Low-level shore platforms in the south-west Scottish Highlands: altitude, age and correlation. *Transactions of the Institute of British Geographers* N.S. 3: 151-164.

Gray, J.M., 1978b. Report of a short field meeting at Oban. *Quaternary Newsletter* 26: 14-16.

Gray, J.M., 1982. The last glaciers (Loch Lomond Advance) in Snowdonia, N. Wales. *Geological Journal* 17: 111-133.

Gray, J.M., 1988. Glaciofluvial channels below the Blakeney esker, Norfolk. *Quaternary Newsletter* 55: 8-12.

Gray, J.M. & C.L. Brooks, 1972. The Loch Lomond Readvance moraines of Mull and Menteith. *Scottish Journal of Geology* 8: 95-103.

Gray, J.M. & J.J. Lowe, 1977a (eds.), *Studies in the Scottish Late Glacial environment*: 197 pp. Oxford: Pergamon.

Gray, J.M. & J.J. Lowe, 1977b. The Scottish Lateglacial environment: a synthesis. In J.M. Gray & J.J. Lowe (eds.), *Studies in the Scottish Lateglacial environment*. Oxford: Pergamon.

Gray, J.M. & D.G. Sutherland, 1977. The Oban-Ford Moraine: a reappraisal. In J.M. Gray & J.J. Lowe (eds.), *Studies in the Scottish Lateglacial environment*. Oxford: Pergamon.

Green, C.P., G.R. Coope, A.P. Currant, D.T. Holyoak, M. Ivanovich, R.L. Jones, D.H. Keen, D.F.M. McGregor & J.E. Robinson, 1984. Evidence of two temperate episodes in Late Pleistocene deposits at Marsworth, UK. *Nature* 309: 778-781.

Green, C.P. & D.F.M. McGregor, 1978. Pleistocene gravel trains of the River Thames. *Proceedings of the Geologists' Association* 89: 143-156.

Green, C.P., R.W. Hey & D.F.M. McGregor, 1980. Volcanic pebbles in Pleistocene gravels of the Thames in Buckinghamshire and Hertfordshire. *Geological Magazine* 117: 59-64.

Green, C.P. & D.F.M. McGregor, 1983. Lithology of the Thames gravels. In J. Rose (ed.), *Diversion of the Thames*. Field Guide: 24-38. Cambridge: Quaternary Research Association.

Green, C.P., D.F.M. McGregor & A.H. Evans, 1982. Development of the Thames drainage system in Early and Middle Pleistocene times. *Geological Magazine* 119: 281-290.

Gregory, K.J., 1962. The deglaciation of eastern Eskdale. *Proceedings of the Yorkshire Geological Society* 33: 363-380.

Gregory, K.J., 1965. Proglacial Lake Eskdale after sixty years. *Transactions of the Institute of British Geographers* 36: 149-162.

Gresswell, R.K., 1967. *Physical Geography*: 504 pp. London: Longmans.

Griffith, R.J., 1835. On the geological map of Ireland. *Report of the British Association for the Advancement of Science*: 56-58.

Gustavson, T.C., 1974. Sedimentation on gravel outwash fans, Malaspina Glacier Foreland, Alaska. *Journal of Sedimentary Petrology* 44: 374-389.

Gustavson, T.C., G.M. Ashley & J.C. Boothroyd, 1975. Depositional sequences in glaciolacustrine deltas. In A.V. Jopling & B.C. McDonald (eds.), *Glaciofluvial and glaciolacustrine sedimentation*. Society of Economic Paleontologists and Mineralogists, Special Publication 23: 264-280.

Gustavson, T.C. & J.C. Boothroyd, 1982. Subglacial fluvial erosion: a major source of stratified drift, Malaspina Glacier, Alaska. In R. Davidson-Arnott, W. Nickling & B.D.

Fahey (eds.), *Research in Glacial, Glaciofluvial & Glaciolacustrine systems*: 318 pp. Norwich: Geo Books.

Hacht, U. von, 1987. Spuren früher Kaltzeiten im Kaolinsand von Braderup/Sylt. In U. von Hacht (ed.), *Fossilien von Sylt II*: 269-301. Hamburg: von Hacht.

Hall, A.R., 1980. Late Pleistocene deposits at Wing, Rutland. *Philosophical Transactions of the Royal Society of London* B 289: 135-164.

Hall, A.M., 1984. *Buchan Field Guide*. Cambridge: Quaternary Research Association.

Hall, A.M., 1985. Cenozoic weathering covers in Buchan, Scotland, and their significance. *Nature* 315: 392-395.

Hall, A.M. & E.R. Connell, 1982. Recent excavations at the Greensand locality of Moreseat, Grampian Region. *Scottish Journal of Geology* 18: 291-296.

Hall, A.M. & E.R. Connell, 1986. A preliminary report on the Quaternary sediments at Leys gravel pit, Buchan, Scotland. *Quaternary Newsletter* 48: 17-28.

Hamblin, R.J.O., 1986. The Pleistocene sequence of the Telford district. *Proceedings of the Geologists' Association* 97: 365-377.

Hamilton, D., J.H. Somerville & P.H. Stanford, 1980. Bottom currents and shelf sediments, southwest of Britain. *Sedimentary Geology* 26: 115-138.

Hammer, K.M. & N.D. Smith, 1983. Sediment production and transport in a proglacial stream. *Boreas* 12: 91-106.

Hanrahan, E.T., 1977. *Irish glacial till: origin and characteristics*: 81 pp. Dublin: An Foras Forbartha.

Hanvey, P.M., 1987. Sedimentology of lee-side stratification sequences in Late-Pleistocene drumlins, north-west Ireland. In J. Menzies & J. Rose (eds.) *Drumlins*, First International Conference on Geomorphology: 241-253. Rotterdam: Balkema.

Harkness, D.D., 1981. Scottish Universities Research and Reactor Centre Radiocarbon Measurements IV. *Radiocarbon* 23: 252-304.

Harkness, D.D. & H.W. Wilson, 1974. Scottish Universities Research and Reactor Centre Radiocarbon measurements II. *Radiocarbon* 16: 238-251.

Harkness, R., 1869. On the Middle Pleistocene deposits. *Geological Magazine* 6: 542-550.

Harmer, F.W., 1902. A sketch of the later Tertiary history of East Anglia. *Proceedings of the Geologists' Association* 17: 416-479.

Harmer, F.W., 1904. The Great Eastern Glacier. *Geological Magazine* V (1): 509-510.

Harmer, F.W., 1909. The Pleistocene period in the eastern counties of England. In H.W. Monckton & R.S. Herries (eds.), *Geology in the Field*: 103-123. London: Geologists' Association.

Harmer, F.W., 1910. The Pleistocene Period in the eastern counties of England. *Jubilee Volume of the Geologists' Association*: 103-123.

Harmer, F.W., 1922. *The Pliocene Mollusca of Great Britain* (Volume II): 900 pp. London: The Palaeontographical Society.

Harmer, F.W., 1928. The distribution of erratics and drift. *Proceedings of the Yorkshire Geological Society* 21: 83-150.

Harrison, D.J. & K.A. McL. Adlam, 1985. The limestones and dolomite resources of the Peak District of Derbyshire and Staffordshire. Description of parts of the 1:50,000 geological sheets 99, 111, 112, 124 and 125. *Mineral Assessment Report* 144: 45 pp. British Geological Survey.

Harrison, W.J., 1898. The ancient glaciers of the Midland counties of England. *Proceedings of the Geologists' Association* 15: 400-408.

Hart, J.K., 1987. The Genesis of the North East Norfolk Drift. Unpublished Ph.D. Thesis, University of East Anglia.

Hart, J., 1988. The glacial sequence. In P.L. Gibbard & J.A. Zalasiewicz (eds.), *Pliocene - Middle Pleistocene of East Anglia*, Field Guide: 158-171. Cambridge: Quaternary Research Association.

Hart, J. & G.S. Boulton, 1984. Glacitectonic deformation at West Runton, north Norfolk coast. *Quaternary Newsletter* 44: 47-48.

Heard, H.C., 1960. Transition from brittle fracture to ductile flow in Solenhofen limestones as a function of temperature, confining pressure and interstitial fluid pressure. In D.T. Briggs & J. Hardin (eds.), *Rock Deformation - a Symposium*. Geological Society of America Memoir 79.

Hedberg, H.D. (ed.), 1976. *International Stratigraphic Guide*: 200 pp. New York: Wiley.

Hein, F.J. & R.G. Walker, 1977. Bar evolution and development of stratification in the gravelly, braided, Kicking Horse River, British Columbia. *Canadian Journal of Earth Sciences* 14: 562-570.

Helm, D.G., 1971. Succession and sedimentation of glaciogenic deposits at Hendre, Anglesey. *Geological Journal* 7: 271-298.

Helm, D.G. & B. Roberts, 1975. A re-interpretation of the origin of sands and gravels around Banc-y-Warren, near Cardigan, west Wales. *Geological Journal* 10: 131-146.

Henry, A., 1984a. The Lithostratigraphy, Biostratigraphy and Chronostratigraphy of Coastal Pleistocene Deposits in Gower, South Wales. Unpublished Ph.D. Thesis, University of Wales.

Henry, A., 1984b. Gower. In D.Q. Bowen & A. Henry (eds.), *Wales: Gower, Preseli, Fforest Fawr. Field Guide*: 18-32. Cambridge: Quaternary Research Association.

Hey, R.W., 1965. Highly quartzose Pebble Gravels in the London Basin. *Proceedings of the Geologists' Association* 76: 403-420.

Hey, R.W., 1976. Provenance of far-travelled pebbles in the pre- Anglian Pleistocene of East Anglia. *Proceedings of the Geologists' Association* 87: 69-82.

Hey, R.W., 1980. Equivalents of the Westland Green Gravels in Essex and East Anglia. *Proceedings of the Geologists' Association* 91: 279-290.

Hey, R.W., 1986. A re-examination of the Northern Drift of Oxfordshire. *Proceedings of the Geologists' Association* 97: 291-301.

Hey, R.W. & P.J. Brenchley, 1977. Volcanic pebbles from Pleistocene gravels in Norfolk and Essex. *Geological Magazine* 114: 219-225.

Hill, A.R., 1968. An analysis of the spatial distribution and origin of drumlins in north Down and south Antrim, Northern Ireland: 329 pp. Unpublished Ph.D. Thesis, Belfast, The Queen's University.

Hill, A.R., 1970. The relationship of drumlins to the direction of ice movement in north Co. Down. In N. Stephens & R. E. Glasscock (eds.), *Irish Geographical Studies*: 53-59. Belfast: The Queen's University.

Hill, A. R., 1971. The internal composition and structure of drumlins in north Down and south Antrim, Northern Ireland. *Geografiska Annaler* 53 A: 14-31.

Hill, A. R., 1973. The distribution of drumlins in County Down. *Annals of the Association of American Geographers* 63: 226-240.

Hill, A. R. & D. B. Prior, 1968. Directions of ice movement in north-east Ireland. *Proceedings of the Royal Irish Academy* 66 B: 71-84.

Hillaire-Marcel, C., S. Occhietti & J.S. Vincent, 1981. Sakami moraine, Quebec: A 500 km long moraine without climatic control. *Geology* 9: 210-214.

Hillefors, A., 1983. The Dösebacka and Ellesbo drumlins - Morphology and stratigraphy. In J. Ehlers (ed.), *Glacial Deposits in North-West Europe*: 141-150. Rotterdam: Balkema.

Hinxman, L.W., R.G. Carruthers & M. Macgregor, 1923. The geology of Corrour and the Moor of Rannoch. *Memoir of the Geological Survey of Great Britain*.

Hirvas, H. & K. Nenonen, 1987. The till stratigraphy of Finland. *Geological Survey of Finland, Special Paper* 3: 49-63.

Hoare, P.G., 1972. The glacial stratigraphy of County Dublin: 287 pp. Unpublished Ph.D. Thesis, Dublin, Trinity College.

Hoare, P.G., 1975. The pattern of glaciation of County Dublin. *Proceedings of the Royal Irish Academy* 75 B: 207-224.

Hoare, P.G., 1976. Glacial meltwater channels in County Dublin. *Proceedings of the Royal Irish Academy* 76 B: 173-185.

Hoare, P.G., 1977a. The glacial stratigraphy in Shanganagh and adjoining townlands, south-east County Dublin. *Proceedings of the Royal Irish Academy* 77 B: 295-305.

Hoare, P.G., 1977b. Killiney Bay. In D. Huddart (ed.), *South East Ireland*, Guidebook for Excursion C 14, X INQUA Congress: 17-21. Norwich: Geo Abstracts.

Hoare, P.G., 1977c. The glacial record in southern County Dublin, Eire. *Journal of Glaciology* 20: 223-225.

Hoare, P.G., 1977d. Aghfarrell. In D. Huddart (ed.), *South East Ireland*, Guidebook for excursion A 14, X INQUA Congress: 26-28. Norwich: Geo Abstracts.

Hoare, P.G. & E.R. Connell, 1981. The chalky till at Barrington, near Cambridge, and its connection with other Quaternary deposits in southern Cambridgeshire and adjoining areas. *Geological Magazine* 118: 463-476.

Hodgson, D.M., 1982. Hummocky and fluted moraines in part of north-west Scotland: 278 pp. Unpublished Ph.D. Thesis, University of Edinburgh.

Hodgson, D.M., 1986. A study of fluted moraines in the Torridon area, N.W. Scotland. *Journal of Quaternary Science* 1: 109-118.

Hodgson, J.M., 1974. Soil Survey Handbook. *Technical Monograph of the Soil Survey* 5: 99 pp.

Holden, W.G., 1977. The glaciation of Central Ayrshire: 486 pp. Unpublished Thesis, University of Glasgow.

Holland, C.H. (ed.), 1981. *A geology of Ireland*: 335 pp. Edinburgh: Scottish Academic Press.

Hollingworth, S.E., 1931. Glaciation of western Edenside and adjoining areas and the drumlins of the Edenside and Solway Basin. *Quarterly Journal of the Geological Society of London* 87: 281-359.

Hollis, J.M. & A.H. Reed, 1981. The Pleistocene deposits of the southern Worfe catchment. *Proceedings of the Geologists' Association* 92: 59-74.

Holmes, R., 1977. Quaternary deposits of the central North Sea, 5. The Quaternary geology of the UK sector of the North Sea between 56 and 58 N. *Report of the Institute of Geological Sciences* 77/14.

Holmes, T.V., 1892. The new railway from Grays Thurrock to Romford. *Quarterly Journal of the Geological Society of London* 48: 365-372.

Hoppe, G., 1974. The glacial history of the Shetland Islands. *Institute of British Geographers Special Publication* 7: 197-210.

Hopson, P.M. & D.McM. Bridge, 1987. Middle Pleistocene stratigraphy in the lower Waveney valley, East Anglia. *Proceedings of the Geologists' Association* 98: 171-186.

Horne, J., D. Robertson, T.F. Jamieson, J. Fraser, P.F. Kendall & D. Bell, 1893. The character of the high-level shell-bearing deposits at Clava, Chapelhall and other localities. *Report of the British Association for the Advancement of Science*: 483-514.

Houmark-Nielsen, M., 1983. Depositional features of a late Weichselian outwash fan; central East Jutland, Denmark. *Sedimentary Geology* 36: 51-63.

Houmark-Nielsen, M., 1987. Pleistocene stratigraphy and glacial history of the central part of Denmark. *Bulletin of the Geological Society of Denmark* 36: 1-189.

Howell, F.T., 1965. Some aspects of the sub-drift surface of some parts of northwest England. Unpublished Ph.D. Thesis, University of Manchester.

Huddart, D., 1967. Deglaciation in the Ennerdale area: a reinterpretation. *Proceedings of the Cumberland Geological Society* 2: 63-75.

Huddart, D., 1970. Aspects of glacial sedimentation in the Cumberland Lowland: 340 pp. Unpublished Ph.D. Thesis, University of Reading.

Huddart, D., 1971a. Textural distinction between Main Glaciation and Scottish Readvance tills in the Cumberland Lowland. *Geological Magazine* 108: 317-324.

Huddart, D., 1971b. A relative glacial chronology from the tills of the Cumberland lowland. *Proceedings of the Cumberland Geological Society* 3: 21-32.

Huddart, D., 1973. The origin of esker sediments, Thursby, Cumberland. *Proceedings of the Cumberland Geological Society* 4: 59-69.

Huddart, D., 1976. *The Screen Hills moraine and the glaciation of Wexford*. Abstract, Irish Conference of University Geographers, University College, Cork.

Huddart, D., 1977a. Gutterby Spa - Annaside Banks Moraine and St. Bees Moraine. In M.J. Tooley (ed.), *The Isle of Man, Lancashire coast and Lake District*. Field guide for excursion A4, X INQUA Congress: 38-40. Norwich: Geo Abstracts.

Huddart, D. (ed.), 1977b. *South East Ireland*, Guidebook for Excursion A 14, X INQUA Congress: 56 pp. Norwich: Geo Abstracts.

Huddart, D., 1981a. Pleistocene foraminifera from south-east

Ireland - some problems of interpretation. *Quaternary Newsletter* 33: 28-41.

Huddart, D., 1981b. Knocknasilloge Member of Wexford: glacio-marine, marine or glacio-lacustrine? *Quaternary Newsletter* 35: 6-11.

Huddart, D., 1981c. Fluvioglacial systems in Edenside. In J. Boardman (ed.), *East Cumbria field guide*: 81-103. Cambridge: Quaternary Research Association.

Huddart, D., 1983. Flow tills and ice-walled lacustrine sediments, the Petteril valley, Cumbria. In E. B. Evenson, Ch. Schlüchter & J. Rabassa (eds.), *Tills and related deposits: genesis, petrography, application, stratigraphy*: 81-94. Rotterdam: Balkema.

Huddart, D. & M.J. Tooley, 1972. *The Cumberland Lowland handbook*: 96 pp. London: Quaternary Research Association.

Huddart, D., M.J. Tooley & P. Carter, 1977. The coasts of north-west England. In C. Kidson & M.J. Tooley (eds.), *The Quaternary History of the Irish Sea*. Geological Journal Special Issue 7: 119-154. Chichester: Wiley.

Hughes, S.A., 1987. The amino stratigraphy of British Quaternary non-marine deposits. Unpublished Ph.D. thesis, University of Wales.

Hull, E., 1855. On the physical geography and Pleistocene phaenomena of the Cotteswold Hills. *Quarterly Journal of the Geological Society of London* 11: 477-496.

Hull, E., 1871. Observations on the general relations of the drift deposits of Ireland to those of Great Britain. *Geological Magazine* 8: 294-299.

Hull, E., 1878. *The Physical Geology and Geography of Ireland*: 291 pp. London: Edward Stanford.

Hunt, C.O., A.R. Hall, D.D. Gilbertson, A. Blackham, C. Williams & H.K. Kenward, 1984. The Palaeobotany of the Late-Devensian sequence at Skipsea Whitow Mere. In D.D. Gilbertson (ed.), *Late Quaternary Environments and Man in Holderness*: 81-108. British Archaeological Reports British Series 134.

Huthnance, J.M., 1982. On the mechanism forming linear tidal sand banks. *Estuarine, Coastal and Shelf Science* 14: 79-99.

Imbrie, J., 1985. A theoretical framework for the Pleistocene ice ages. *Journal of the Geological Society of London* 142: 417-432.

Imbrie, J. & N.G. Kipp, 1971. A new micropalaeontological method for quantitative palaeoclimatology. In K.K. Turekian (ed.), *The Late Cenozoic Glacial Ages*: 71-182. Newhaven: Yale University Press.

Ineson, P.R., 1969. Trace element aureoles in limestone wallrocks adjacent to lead-zinc-barite-fluorite mineralisation in the Northern Pennine and Derbyshire ore field. *Transactions of the Institute of Mining Metallurgy* 78: B29-40.

International Union for Quaternary Research, Commission 2 on Genesis and Lithology of Quaternary Deposits (eds.), 1977. *Key to Glacial Landforms*: 5 pp.

Irving, A. & P.A. Irving, 1913. The Harlow Boulder Clay and its place in the glacial sequence of eastern England. *Report of the British Association*, Section C: 480-481.

Jackson, I., A.N. Lowe, N. Morigi & S.J. Mathers, 1983. The sand and gravel resources of the country around Whitchurch and Malpas, Clwyd, Cheshire and Shrop- shire. *Mineral Assessment Report of the Institute of Geological Sciences* 136: 102 pp. London, Her Majesty's Stationery Office.

James, J.W.C., 1982. The sand and gravel resources of the country north and west of Billingham, Cleveland. *Institute of Geological Sciences, Mineral Assessment Report* 99. London: Her Majesty's Stationery Office.

Jamieson, T.F., 1858. On the Pleistocene deposits of Aberdeenshire. *Quarterly Journal of the Geological Society of London* 14: 509-532.

Jamieson, T.F., 1863. On the parallel roads of Glen Roy and their place in the history of the glacial period. *Quarterly Journal of the Geological Society of London* 19: 235-259.

Jamieson, T.F., 1865. On the history of the last geological changes in Scotland. *Quarterly Journal of the Geological Society of London* 21: 161-203.

Jamieson, T.F., 1882. On the red clay of the Aberdeenshire coast and the direction of ice-movement in that quarter. *Quarterly Journal of the Geological Society of London* 38: 160-177.

Jamieson, T.F., 1906. The glacial period in Aberdeenshire and the southern border of the Moray Firth. *Quarterly Journal of the Geological Society of London* 62: 13-39.

Jansen, J.H.F., T.C.E. van Weering & D. Eisma, 1979. Late Quaternary sedimentation in the North Sea. In E. Oele, R.T.E. Schüttenhelm & A.J. Wiggers (eds.), *The Quaternary History of the North Sea, Acta Universitatis Upsaliensis, Symposium Universitatis Upsaliensis Annum Quingentesimum Celebrantis* 2: 175-187.

Jardine, W.G., 1971. Form and age of Late Quaternary shorelines and coastal deposits of south-west Scotland: critical data. *Quaternaria* 14: 103-114.

Jehu, T.J., 1904. The glacial deposits of northern Pembrokeshire. *Transactions of the Royal Society of Edinburgh* 41: 53-87.

Jelgersma, S., 1979. Sea-level changes in the North Sea basin. In E. Oele, R.T.E. Schüttenhelm & A.J. Wiggers (eds.), *The Quaternary History of the North Sea*. Acta Universitatis Upsaliensis, Symposium Universitatis Upsaliensis Annum Quingentesimum Celebrantis 2: 233-248.

Jessen, K., S.T. Andersen & A. Farrington, 1959. The interglacial deposit near Gort, Co. Galway, Ireland. *Proceedings of the Royal Irish Academy* 60 B: 1-77.

Jessen, K. & A. Farrington, 1938. The bogs at Ballybetagh, near Dublin, with remarks on Late-Glacial conditions in Ireland. *Proceedings of the Royal Irish Academy* 44 B: 205-260.

Joachim, M., 1977. Glen Ballyre. In M.J. Tooley (ed.), *The Isle of Man, Lancashire coast and Lake District*, Field guide for excursion A 4, X INQUA Congress. 33-36. Norwich: Geo Abstracts.

John, B.S., 1965. Aspects of the Glaciation and Superficial Deposits of Pembrokeshire. Unpublished D.Phil. Thesis, University of Oxford. 2 vols.

John, B.S., 1970. Pembrokeshire. In C.A. Lewis (ed.), *The Glaciations of Wales*: 229-265. London: Longman.

John, B.S., 1971. Pembrokeshire. In C.A. Lewis (ed.), *The Glaciations of Wales and Adjoining Regions*: 229-265. London: Longmans.

Johnstone, G.S., 1966. *The Grampian Highlands*, 3rd Edition. British Regional Geology, Institute of Geological Sciences. Edinburgh: Her Majesty's Stationery Office.

Jones, D.K.C., 1981. *Southeast and southern England* (The Geomorphology of the British Isles). London: Methuen & Co.

Jones, R.L., 1977. Late Devensian deposits from Kildale, north- east Yorkshire. *Proceedings of the Yorkshire Geological Society* 41: 185-188.

Jopling, A.V. & R.G. Walker, 1968. Morphology and origin of ripple drift cross-lamination, with examples from the Pleistocene of Massachusetts. *Journal of Sedimentary Petrology* 38: 971-984.

Jorgensen, N.B., 1982. Turbidites and associated resedimented deposits from a tilted glaciodeltaic sequence, Denmark. *Danmarks Geologiske Undersøgelse, Årbog* 1981: 47-72.

Jowett, A. & J.K. Charlesworth, 1929. The glacial geology of the Derbyshire Dome and the S.W. Pennines. *Quarterly Journal of the Geological Society of London* 85: 307-334.

Jukes-Browne, A.J., 1885. The Geology of the south-west Part of Lincolnshire. *Memoir of the Geological Survey*: 180 pp. London: Her Majesty's Stationery Office.

Kabata-Pendias, A., 1968. The sorbtion of trace elements by soil forming minerals. *Roczniki Gleboznawcze* 19: 55-72.

Kazi, A., 1972. Clay mineralogy of North Sea drift. *Nature, Physical Science* 240: 61-62.

Kazi, A. & J.L. Knill, 1969. The sedimentation and geotechnical properties of the Cromer Till between Happisburgh and Cromer, Norfolk. *Quarterly Journal of Engineering Geology* 2: 63-86.

Keach, B.N. & J. Horne, 1881. The glaciation of Caithness. *Proceedings of the Royal Physical Society of Edinburgh* 6: 316-352.

Kehle, R.O., 1970. Analysis of gravity sliding and orogenic translation. *Bulletin of the Geological Society of America* 81: 1641-1664.

Kellaway, G.A., 1972. Development of non-diastrophic Pleistocene structures in relation to climate and physical relief in Britain. *Proceedings of the International Geological Congress, 24th Session, Montreal* 12: 136-146.

Kellaway, G.A., J.H. Redding, E.R. Shephard-Thorn & J.-P. Destombes, 1975. The Quaternary history of the English Channel. *Philosophical Transactions of the Royal Society of London* A 279: 189-218.

Kelly, M.R., 1964. The Middle Pleistocene of North Birmingham. *Philosophical Transactions of the Royal Society of London* 247 B: 553-592.

Kelly, M.R., 1968. Floras of middle and upper Pleistocene age from Brandon, Warwickshire. *Philosophical Transactions of the Royal Society of London* B 254: 401-415.

Kelly, S.R.A. & P.F. Rawson, 1983. Some late Jurassic - mid Cretaceous sections of the East Midlands shelf, as demonstrated on a field meeting, 18-20 May 1979. Appendix: The distribution of Spilsby Sandstone erratics. *Proceedings of the Geologists' Association* 94: 65-73.

Kemp, R.A., 1985. The Valley Farm Soil in southern East Anglia. In J. Boardman (ed.), *Soils and Quaternary Landscape Evolution*: 179-196. Chichester: Wiley.

Kendall, P.F., 1894. On the glacial geology of the Isle of Man. *Yn Lioar Manninagh* 1: 397-437.

Kendall, P.F., 1902. A system of glacier-lakes in the Cleveland Hills. *Quarterly Journal of the Geological Society of London* 58: 471-571.

Kendall, P.F. & H.B. Muff, 1903. On the evidence for glacier-dammed lakes in the Cheviot Hills. *Transactions of the Edinburgh Geological Society* 8: 226-230.

Kenyon, R., 1982. The glaciation of the Nephin Beg Range, Co. Mayo, Eire. Unpublished M.Sc. Thesis, City of London Polytechnic. Council for National Academic Awards.

Kenyon, R., 1986. The glaciation of the Nephin Beg Range, Co. Mayo, Eire. *Quaternary Studies* 2: 14-21.

Kerney, M.P. & G. de G. Sieveking, 1977. Northfleet. In I.R. Shephard-Thorn & J.J. Wymer (eds.), *Southeast England and the Thames Valley*. X INQUA Congress, Guidebook for Excursion A 5: 44-47. Norwich: Geo Abstracts.

Kerr, R.J., 1978. The nature and derivation of glacial till in part of the Tweed basin. Unpublished Thesis, University of Edinburgh.

Kerr, W.B., 1982. Pleistocene ice movements in the Rhins of Galloway. *Transactions of the Dumfries and Galloway Natural History and Antiquarian Society, 3rd Series* 57: 1-10.

Kilroe, J.R., 1897. The distribution of Drift in Ireland in its relation to Agriculture. *Scientific Proceedings of the Royal Dublin Society* 8: 421-429.

Kilroe, J.R., 1907. *A description of the Soil Geology of Ireland*: 300 pp. Dublin: Department of Agriculture and Technical Instruction for Ireland.

Kinahan, G.H., 1865. *Explanation to accompany sheets 115 and 116 of the maps, and sheets 17 and 18 of the longitudinal sections*. Memoir of the Geological Survey of the United Kingdom: 43 pp. Dublin.

Kinahan, G.H., 1874. Glacialoid or re-arranged glacial drift. *Geological Magazine* 1: 111-117.

Kinahan, G.H., 1878. *Manual of the geology of Ireland*: 444 pp. Dublin: Kegan Paul.

Kinahan, G.H., 1894. The recent Irish glaciers. *Irish Naturalist* 3: 236-240.

King, C.A.M., 1976. *The Geomorphology of the British Isles: Northern England*: 213 pp. London: Methuen.

Kinney, P., 1986. Aspects of the glaciation of Kylemore Valley, County Galway. Unpublished undergraduate dissertation (Geography), University of Dublin, Trinity College.

Kirby, R.P., 1968. The ground moraines of Midlothian and East Lothian. *Scottish Journal of Geology* 4: 209-220.

Kirby, R.P., 1969a. Till fabric analyses from the Lothians, Central Scotland. *Geografiska Annaler* 51 A: 48-60.

Kirby, R.P., 1969b. Morphometric analysis of glaciofluvial terraces in the Esk basin, Midlothian. *Transactions of the Institute of British Geographers* 48: 1-18.

Kirk, W. & H. Godwin, 1963. A late-glacial site at Loch Droma, Ross and Cromarty. *Transactions of the Royal Society of Edinburgh* 65: 225-249.

Krauskopf, K.B., 1967. *Introduction to geochemistry*: 721 pp. McGraw-Hill Inc.

Krigstrom, A., 1962. Geomorphological studies of sandur plains and their braided rivers in Iceland. *Geografiska Annaler* 44: 328-346.

Krinsley, D.H. & B.M. Funnell, 1965. Environmental history of quartz sand grains from the Lower and Middle Pleistocene of Norfolk, England. *Quarterly Journal of the Geological Society of London* 121: 435-461.

Krüger, J., 1979. Structures and textures in till indicating subglacial deposition. *Boreas* 8: 323-340.

Kubota, J., 1967. Distribution of cobalt deficiency in grazing animals in relation to soils and forage plants of the United States. *Soil Science* 106 (1): 122-130.

Kudrass, H.-R., 1973. Sedimentation am Kontinentalhang vor Portugal und Marokko im Spätpleistozän und Holozän. *'Meteor' Forschungsergebnisse* C 13: 63 pp.

Kupsch, W.O., 1962. Ice-thrust ridges in western Canada. *Journal of Geology* 70: 582-594.

Laban, C., T.D.J. Cameron & R.T.E. Schüttenhelm, 1984. Geologie van het Kwartair in de Zuidelijke Bocht van de Noordzee. *Mededelingen van de Werkgroep voor Tertiaire en Kwartaire Geologie* 21: 139-154.

Lambert, J.T. & Z.M. Khowaja, 1978. Geotechnical analysis of till-like material from the S.W. Approaches. *Institute of Geological Sciences Engineering Geology Unit Internal Report* No.78/23: 6 pp.

Lamplugh, G.W., 1879. On the divisions of the glacial beds in Filey Bay. *Proceedings of the Yorkshire Geological Society* 7: 167-177.

Lamplugh, G.W., 1882. Glacial sections near Bridlington. *Proceedings of the Yorkshire Geological Society* 7: 383-397.

Lamplugh, G.W., 1883. Glacial sections near Bridlington. Part III. Cliff section extending 900 yards south of the harbour. *Proceedings of the Yorkshire Geological Society* 8: 27-38.

Lamplugh, G.W., 1890a. Glacial sections near Bridlington. Part IV. *Proceedings of the Yorkshire Geological Society* 11: 275-297.

Lamplugh, G.W., 1890b. On a new locality for the arctic fauna of the 'Basement' boulder clay in Holderness. *Geological Magazine* Decade 3, 7: 61-70.

Lamplugh, G.W., 1891. On the drifts of Flamborough Head. *Quarterly Journal of the Geological Society of London* 47: 384-431.

Lamplugh, G.W., 1892. The Flamborough drainage sections. *Proceedings of the Yorkshire Geological Society* 12: 145-148.

Lamplugh, G.W., 1903. *The Geology of the Isle of Man*. Memoir of the Geological Survey of Great Britain: 620 pp. London: Her Majesty's Stationery Office.

Lamplugh, G.W., J.R. Kilroe, A. M'Henry, H.J. Seymour & W.B. Wright, 1903. *The geology of the country around Dublin*. Memoir of the Geological Survey of Ireland: 160 pp. Dublin.

Lamplugh, G.W., J.R. Kilroe, A. M'Henry, H.J. Seymour, W.B. Wright & H.B. Muff, 1904. *The geology of the country around Belfast*. Memoir of the Geological Survey of Ireland: 166 pp.

Lamplugh, G.W., J.R. Kilroe, A. M'Henry, H.J. Seymour, W.B. Wright & H.B. Muff, 1905. *The geology of the country around Cork and Cork Harbour*. Memoir of the Geological Survey of Ireland: 135 pp.

Lamplugh, G.W., J.R. Wilkenson, J.R. Kilroe, A. McHenry,

H.J. Seymour & W.B. Wright, 1907. *The geology of the country around Limerick*. Memoir of the Geological Survey of Ireland: 119 pp.

Land, D.H., 1974. Geology of the Tynemouth District. *Memoir of the Geological Survey of Great Britain*. London: Her Majesty's Stationery Office.

Lawson, D.E., 1979. Sedimentological analysis of the western terminus region of the Matanuska Glacier, Alaska. *Cold Regions Research and Engineering Laboratory Report* 79-9: 112 pp. Hanover, New Hampshire: CRREL.

Lawson, D.E., 1981. Distinguishing characteristics of diamictons at the margin of the Matanuska Glacier, Alaska. *Annals of Glaciology* 2: 78-84.

Lawson, T.J., 1984. Reindeer in the Scottish Quaternary. *Quaternary Newsletter* 42: 1-7.

Laxton, J.L. & E.F.P. Nickless, 1980. The sand and gravel resources of the country around Lanark, Strathclyde Region. Description of 1:25,000 sheet NS94 and part of NS84. *Mineral Assessment Report of the Institute of Geological Sciences* 49.

Lea, P.D., 1985. Late Pleistocene glacitectonism and sedimentation on a macrotidal piedmont coast, Ekuk Bluffs, SW Alaska. *Geological Society of America, Abstracts with Programs* 17: 298.

Leckie, D.A. & S.B. McCann, 1982. Glacio-lacustrine sedimentation on low slope prograding delta. In R. Davidson-Arnott, W. Nickling & B.D. Fahey (eds.), *Research in glacial, glacio-fluvial and glacio-lacustrine systems*: 261-278. Norwich: Geobooks.

Lee, M.P., 1979. Loess from the Pleistocene of the Wirral Peninsula, Merseyside. *Proceedings of the Geologists' Association* 90: 21-26.

Le Riche, H.H., 1968. The location of trace elements in sedimentary rocks and in soils derived from them. *Welsh Soils Discussion Group Report*: 17-29.

Letzer, J.M., 1978. The glacial geomorphology of the region bounded by Shap Fell, Stainmore and the Howgill Fells in East Cumbria. Unpublished M.Phil. Thesis, University of London.

Letzer, J.M., 1981. The Upper Eden valley. In J. Boardman (ed.), *Eastern Cumbria*, Field Guide: 43-60. London: Quaternary Research Association.

Lewis, C.A., 1974. The Glaciations of the Dingle Peninsula, County Kerry. *Scientific Proceedings of the Royal Dublin Society* 5 A: 207-235.

Lewis, C.A., 1985. Periglacial features. In K.J. Edwards & W.P. Warren (eds.), *The Quaternary History of Ireland*: 95-113. London: Academic Press.

Lewis, H.C., 1894. *Papers and notes on the glacial geology of Great Britain and Ireland*: 469 pp. London: Longman, Green & Co.

Little, J.A., 1984. Engineering properties of glacial tills in the Vale of St. Albans: 435 PP. Unpublished Ph.D. Thesis, City University.

Little, J.A. & J.H. Atkinson, 1985. Some engineering properties of Anglian tills in the Vale of St. Albans. In M.C. Forde (ed.), *Glacial Tills 85*. Proceedings of the International Conference: 213-218. Edinburgh: Engineering Technics Press.

Lliboutry, L., 1979. Local friction laws for glaciers: a critical review and new openings. *Journal of Glaciology* 23 (89): 67-96.

Lloyd, J.W., D. Harker & R.A. Baxendale, 1981. Recharge mechanisms and groundwater flow in the chalk and drift deposits of southern East Anglia. *Quarterly Journal of Engineering Geology* 14: 87-96.

Long, D., C. Laban, H. Streif, T.D.J. Cameron & R.T.E. Schüttenhelm, 1988. The sedimentary record of climatic variation in the southern North Sea. *Philosophical Transactions of the Royal Society of London* B 318: 523-537.

Long, D. & A.C. Skinner, 1985. Glacial meltwater channels in the northern isles of Shetland: comment. *Scottish Journal of Geology* 21: 222-224.

Longworth, D., 1985. The Quaternary history of the Lancashire Plain. In R.H. Robinson (ed.), *The geomorphology of North-West England*: 178-200. Manchester: Manchester University Press.

Lovell, J.H., 1982. The sand and gravel resources of the country around Catterick, North Yorkshire. *Institute of Geological Sciences, Mineral Assessment Report* 120. London: Her Majesty's Stationery Office.

Lowe, J.J., 1984. A critical evaluation of pollen-stratigraphic investigations of pre-Late Devensian sites in Scotland. *Quaternary Science Reviews* 3: 405-432.

Lowe, J.J. & M.J.C. Walker, 1980. Problems associated with radiocarbon dating the close of the Lateglacial period in the Rannoch Moor area, Scotland. In J.J. Lowe, J.M. Gray & J.E. Robinson (eds.), *Studies in the Lateglacial of North-west Europe*. Oxford: Pergamon.

Lowe, J.J. & M.J.C. Walker, 1981. The early Postglacial environment of Scotland: evidence from a site near Tyndrum, Perthshire. *Boreas* 3: 281-294.

Lowe, J.J. & M.J.C. Walker, 1984. *Reconstructing Quaternary environments*: 389 pp. London: Longman.

Lowe, J.J. & M.J.C. Walker, 1986. Lateglacial and early Flandrian environmental history of the Isle of Mull, Inner Hebrides, Scotland. *Transactions of the Royal Society of Edinburgh: Earth Sciences* 77: 1-20.

Lüttig, G. & G.C. Maarleveld, 1961. Nordische Geschiebe in Ablagerungen prä-Holstein in den Niederlanden (Komplex von Hattem). *Geologie en Mijnbouw* 40: 163-174.

Lüttig, G. & G.C. Maarleveld, 1962. Über altpleistozäne Kiese in der Veluwe. *Eiszeitalter und Gegenwart* 13: 231-237.

Lundqvist, J., 1977. Till in Sweden. *Boreas* 6: 73-85.

Lundqvist, J., 1983. The glacial history of Sweden. In J. Ehlers (ed.), *Glacial Deposits in North-West Europe*: 77-82. Rotterdam: Balkema.

Lunkka, J.P., 1988. Sedimentation and deformation of the North Sea Drift Formation in the Happisburgh area, North Norfolk. In D.G. Croot (ed.), *Glaciotectonics*:109-122. Rotterdam: Balkema.

Lunn, A.G., 1980. Quaternary. In A. Robson (ed.), *The Geology of Northeast England*: 48-60. The Natural History Society of Northumberland.

Lyell, C., 1840. On the Boulder Formation, or drift and associated Freshwater Deposits composing the Mud-cliffs of Eastern Norfolk. *The London and Edinburgh Philosophical Magazine and Journal of Science*, Third Series, 16 (104): 345-380.

Macgregor, M, 1927. The Carstairs District. *Proceedings of the Geologists' Association* 59: 151-171.

Mackiewicz, N.E., R.D. Powell, P.R. Carlson & B.F. Molnia, 1984. Interlaminated ice-proximal glaciomarine sediments in Muir Inlet, Alaska. *Marine Geology* 57: 113-147.

MacPherson, J.B., 1980. Environmental change during the Loch Lomond Stadial: evidence from a site in the Upper Spey Valley, Scotland. In J.J. Lowe, J.M. Gray & J.E. Robinson (eds.), *Studies in the Lateglacial of North-west Europe*: 89-102. Oxford: Pergamon.

Madgett, P.A., 1975. Re-interpretation of Devensian till stratigraphy of eastern England. *Nature* 253: 105-107.

Madgett, P.A. & J.A. Catt, 1978. Petrography, stratigraphy and weathering of Late Pleistocene tills in East Yorkshire, Lincolnshire and north Norfolk. *Proceedings of the Yorkshire Geological Society* 42: 55-108.

Maizels, J.K., 1979. Proglacial aggradation and changes in braided channel patterns during a period of glacier advance: an alpine example. *Geografiska Annaler* 61 A: 87-101.

Maizels, J.K., 1983a. Proglacial channel systems: change and thresholds for change over long, intermediate and short timescales. In J.D. Collinson & J. Lewin (eds.), *Modern and ancient fluvial systems*. International Association of Sedimentologists Special Publication 6: 251-266.

Maizels, J.K., 1983b. Channel changes, paleohydrology and deglaciation: evidence from some Lateglacial sandur deposits of northeast Scotland. *Quaternary Studies in Poland* 4: 171-187.

Manley, G., 1949. The snowline in Britain. *Geografiska Annaler* 31: 179-193.

Manley, G., 1959. The Late-glacial climate of North-West England. *Liverpool and Manchester Geological Journal* 2: 188-215.

Mannerfelt, C.M., 1949. Marginal drainage channels as indicators of the gradients of Quaternary ice-caps. *Geografiska Annaler* 31: 194-199.

Mark, D.M., 1973. Analysis of axial orientation data, including till fabrics. *Bulletin of the Geological Society of America* 84: 1369-1374.

Marsland, A., 1977. The evaluation of the engineering design parameters for glacial clays. *Quarterly Journal of Engineering Geology* 10: 1-26.

Marsland, A., 1980. The interpretation of *in situ* tests on glacial clays. In D. Ardus (ed.), *Proceedings of the International Conference on Offshore Site Investigation*: 218-288. London: Society for Underwater Technology.

Marsland, A. & J.J.M. Powell, 1980. Cyclic loading tests on 865 mm diameter plates in a stiff clay till. In G.N. Pande & O.C. Zienkiewicz (eds.), *Proceedings of the International Symposium on Soils under Cyclic and Transient Loading* 2: 837-847. Chichester: J. Wiley & Son.

Marsland, A. & J.J.M. Powell, 1985. Field and laboratory investigations of the clay tills at the Building Research Establishment test site at Cowden, Holderness. In M.C. Forde (ed.), *Glacial Tills 85*. Proceedings of the International Conference: 147-168. Edinburgh: Engineering Technics Press.

Martin, J.H., 1980. Loundoun Hill Sand Pit. Snabe Sand Pit.

In W.G. Jardine (ed.), *Glasgow Region*, Field Guide: 57-62. Cambridge: Quaternary Research Association.

Martin, J.H., 1981. *Quaternary glaciofluvial deposits in central Scotland: sedimentology and economic geology*. Unpublished Thesis, University of Edinburgh.

Martin, J.H. & J.P.B. Lovell, 1981. Location and assessment of sand and gravel deposits: a sedimentological approach. *Institute of Mining and Metallurgy Transactions, Section B Applied Earth Science* 90: 52.

Martin, S., 1955. Raised beaches and their relation to glacial drifts on the east coast of Ireland. *Irish Geography* 3: 87-93.

Mathers, S.J. & J.A. Zalasiewicz, 1984. A new approach to drift mapping in East Anglia. *Quaternary Newsletter* 43: 13-17.

Mathers, S.J. & J.A, Zalasiewicz, 1986. A sedimentation pattern in Anglian marginal meltwater channels from Suffolk, England. *Sedimentology* 33: 559-573.

Mathers, S.J., J.A. Zalasiewicz, A.J. Bloodworth & A.C. Morton, 1987. The Banham Beds: a petrologically distinct suite of Anglian glacigenic deposits from central East Anglia. *Proceedings of the Geologists' Association* 98: 229-240.

Mathews, W.H. & J.R. Mackay, 1960. Deformation of soils by glacier ice and the influence of pore pressure and permafrost. *Transactions of the Royal Society of Canada* 54: 27-36.

May, J., 1981. The glaciation and deglaciation of upper Nithsdale and Annandale: 600 pp. Unpublished Ph.D. Thesis, University of Glasgow.

Mayhew, D.F., 1985. Preliminary report of a research project on small mammal remains from British Lower Pleistocene sediments. *Quaternary Newsletter* 47: 1-4.

Mayhew, D.F. & A.J. Stuart, 1986. Stratigraphic and taxonomic revision of the fossil vole remains (*Rodentia: Microtinae*) from the Lower Pleistocene deposits of eastern England. *Philosophical Transactions of the Royal Society of London* B 312: 431-485.

McAdam, A.D. & W. Tulloch, 1985. Geology of the Haddington district. *Memoir of the British Geological Survey, Sheet 33W and part of 41*.

McAulay, I.R. & W.A. Watts, 1961. Dublin radiocarbon dates I. *Radiocarbon* 3: 26-38.

McCabe, A.M., 1969a. A buried head deposit near Lisnaskea, Co. Fermanagh, Northern Ireland. *Irish Naturalists' Journal* 16: 232-233.

McCabe, A.M., 1969b. The glacial deposits of the Maguiresbridge area, Co. Fermanagh, Northern Ireland. *Irish Geography* 6: 63-77.

McCabe, A.M., 1971. The glacial geomorphology of eastern Counties Meath and Louth, eastern Ireland: 382 pp. Unpublished Ph.D. Thesis, Dublin, Trinity College.

McCabe, A.M., 1972. Directions of Late-Pleistocene ice-flows in eastern Counties Meath and Louth, Ireland. *Irish Geography* 6: 443-461.

McCabe, A.M., 1973. The glacial stratigraphy of eastern Counties Meath and Louth. *Proceedings of the Royal Irish Academy* 73 B: 355-382.

McCabe, A.M. (ed.), 1979. *East Central Ireland*, Field Guide: 63 pp. Dublin: Quaternary Research Association.

McCabe, A.M., 1985. Glacial Geomorphology. In K.J. Edwards & W.P. Warren (eds.), *The Quaternary History of Ireland*: 67-93. London: Academic Press.

McCabe, A.M., 1986a. Glaciomarine facies deposited by retreating tidewater glaciers: An example from the Late-Pleistocene of Northern Ireland. *Journal of Sedimentary Petrology* 56: 880-894.

McCabe, A.M., 1986b. Ireland. In D.Q. Bowen, J. Rose, A.M. McCabe & D.G. Sutherland, Correlation of Quaternary glaciations in England, Ireland, Scotland and Wales. *Quaternary Science Reviews* 5: 312-318.

McCabe, A.M., 1987. Quaternary deposits and glacial stratigraphy in Ireland. *Quaternary Science Reviews* 6: 259-299.

McCabe, A.M., G.R. Coope, D.E. Gennard & P. Doughty, 1987. Freshwater organic deposits and stratified sediments between Early and Late Midlandian (Devensian) till sheets, at Aghnadarragh, County Antrim, Northern Ireland. *Journal of Quaternary Science* 2: 11-33.

McCabe, A.M., G.F. Dardis & P.M. Hanvey, 1984. Sedimentology of a Late-Pleistocene submarine-moraine complex, County Down, Northern Ireland. *Journal of Sedimentary Petrology* 54: 716-730.

McCabe, A.M., G.F. Dardis & P.M. Hanvey, 1987. Sedimentation at the margins of a Late Pleistocene ice lobe terminating in shallow marine environments, Dundalk Bay, eastern Ireland. *Sedimentology* 34: 473-493.

McCabe, A.M. & N. Eyles, 1988. Sedimentology of an ice-contact glaciomarine delta, Carey Valley, Northern Ireland. *Sedimentary Geology* 59: 1-14.

McCabe, A.M., D. Gennard, R. Coope & P. Doughty, 1986. Aghnadarragh. In A.M. McCabe & K.R. Hirons (eds.), *South-East Ulster Field Guide*: 142-168. Cambridge: Quaternary Research Association.

McCabe, A.M., J.R. Haynes & N.F. McMillan, 1986. Late-Pleistocene tidewater glaciers and glaciomarine sequences from north County Mayo, Republic of Ireland. *Journal of Quaternary Science* 1: 73-84.

McCabe, A.M. & K.R. Hirons (eds.), 1986. *Field guide to the Quaternary Deposits of South-east Ulster*: 180 pp. Cambridge: Quaternary Research Association.

McCabe, A.M. & P.G. Hoare, 1978. The Late Quaternary history of east-central Ireland. *Geological Magazine* 115: 397-413.

McCabe, A.M., G.F. Mitchell & F.W. Shotton, 1978. An inter-till freshwater deposit at Hollymount, Maguiresbridge, Co. Fermanagh. *Proceedings of the Royal Irish Academy* 78 B: 77-89.

McCann, S.B., 1966. The limits of the Lateglacial Highland or Loch Lomond Readvance along the West Highland seaboard from Oban to Mallaig. *Scottish Journal of Geology* 2: 84-95.

McDonald, B.C. & I. Banerjee, 1971. Sediments and bedforms on a braided outwash plain. *Canadian Journal of Earth Sciences* 8: 1282-1301.

McDonald, B.C. & J.S. Vincent, 1972. Fluvial sedimentary structures formed experimentally in a pipe and their implications for interpretation of subglacial sedimentary environments. *Geological Survey of Canada Paper* 72/27: 31 pp.

McGown, A., 1971. The classification for engineering purposes of tills from moraines and associated landforms. *Quarterly Journal of Engineering Geology* 4: 115-130.

McGown, A., 1985. Construction problems associated with the tills of Strathclyde. *Glacial Tills 85*. Proceedings of the International Conference: 177-186. Edinburgh: Engineering Technics Press.

McGown, A., W.F. Anderson & A.M. Radwan, 1975. Geotechnical properties of the tills in West Central Scotland. *The Engineering Properties of Glacial Materials*. Proceedings of the Birmingham Symposium: 89-99 (republished by Geo Abstracts 1978).

McGown, A. & E. Derbyshire, 1977. Genetic influences on the properties of tills. *Quarterly Journal of Engineering Geology* 10: 389-410.

McGown, A. & A.M. Radwan, 1975. The presence and influence of fissures in the boulder clays of West Central Scotland. *Canadian Geotechnical Journal* 12: 84-97.

McKinlay, D.G., M.J. Tomlinson & W.F. Anderson, 1974. Observations on the undrained strength of a glacial till. *Geotechnique* 24: 503-516.

McKinlay, D.G., A. McGown, A.M. Radwan & D. Hossain, 1975. Representative sampling and testing in fissured lodgement tills. *The Engineering Properties of Glacial Materials*. Proceedings of the Birmingham Symposium: 143-155. (republished by Geo Abstracts 1978).

McLaughlin, R.J.W., 1955. Geochemical change due to weathering under varying climatic conditions. *Geochimica et Cosmochimica Acta* 8 (3): 109-130.

McLean, F., 1977. The glacial sediments of a part of East Aberdeenshire: 225 pp. Unpublished Ph.D. thesis, University of Aberdeen.

McLellan, A.G., 1969. The last glaciation and deglaciation of central Lanarkshire. *Scottish Journal of Geology* 5: 248-268.

McMillan, A.A. & A.M. Aitken, 1981. The sand and gravel resources of the country west of Peterhead, Grampian region. Description of 1:25,000 Sheet NK04 and parts of NJ94, 95 and NK05, 14 and 15. *Mineral Assessment Report of the Institute of Geological Sciences* 58.

McMillan, A.A. & M.A.E. Browne, 1983. Glaciotectonic structures at Bellshill, east of Glasgow. *Quaternary Newsletter* 40: 1-6.

McMillan, N.F., 1938. On an occurrence of Pliocene shells in Co. Wicklow. *Proceedings of the Liverpool Geological Society* 17: 255-266.

McMillan, N.F., 1964. The Mollusca of the Wexford Gravels (Pleistocene), south east Ireland. *Proceedings of the Royal Irish Academy* 63 B: 265-290.

McNeill, J.D., 1980. Electromagnetic terrain conductivity measurement at low induction numbers. *Technical note TN-6*: 15 pp. Geonics Limited, Ontario, Canada.

McQuillan, R., 1964. Geophysical investigations of seismic shot holes in the Cheshire basin. *Bulletin of the Geological Survey of Great Britain* 21: 197-203.

Meenan, J. & D.A. Webb, 1957. The principal scientific institutions of Dublin. In J. Meenan & D.A. Webb (eds.), *A view of Ireland*: 243-254. Dublin: British Association for the Advancement of Science.

Menzies, J., 1979. The mechanics of drumlin formation with particular reference to the change in porewater content of the till. *Journal of Glaciology* 22: 373-384.

Menzies, J., 1981. Investigations into the Quaternary deposits and bedrock topography of central Glasgow. *Scottish Journal of Geology* 17: 155-168.

Menzies, J., 1986. Inverse-graded units within till in drumlins near Caledonia, southern Ontario. *Canadian Journal of Earth Sciences* 23: 774-786.

Mercer, J.H., 1961. The response of fiord glaciers to changes in the firn limit. *Journal of Glaciology* 3: 850-858.

Merritt, J.W., 1981. The sand and gravel resources of the country around Ellon, Grampian Region. Description of 1:25,000 resources sheets NJ93 with parts of NJ82, 83 and 92, and NK03 and parts of NK02 and 13. *Minerals Assessment Report of the Institute of Geological Sciences* 76.

Miall, A.D., 1977. A review of the braided-river depositional environment. *Earth Science Reviews* 13: 1-62.

Miall, A.D., 1978. Lithofacies types and vertical profile models in braided river deposits: a summary. In A.D. Miall (ed.), *Fluvial sedimentology*. Canadian Society of Petroleum Geologists Memoir 5: 597-604.

Miall, A.D., 1983. Glaciofluvial transport and deposition. In N. Eyles (ed.), *Glacial geology*: 168-183. Oxford: Pergamon.

Miall, A.D., 1985. Sedimentation on an early Proterozoic continental margin; the Gowganda Formation (Huronian), Elliot Lake area, Ontario, Canada. *Sedimentology* 32: 763-788.

Mickelson, D.M., L.J. Acomb & T.B. Edil, 1979. The origin of preconsolidated and normally consolidated tills in eastern Wisconsin, U.S.A. In Ch. Schlüchter (ed.), *Moraines and Varves*: 179-187. Rotterdam: Balkema.

Miller, D.J., 1953. Late Cenozoic marine glacial sediments and marine terraces on Middleton Island, Alaska. *Journal of Geology* 61: 17-40.

Miller, G.H., J.T. Hollin & J.T. Andrews, 1979. Amino stratigraphy of UK Pleistocene deposits. *Nature* 281: 539-543.

Mills, D.A.C. & J.H. Hull, 1976. Geology of the country around Barnard Castle. *Memoir of the Geological Survey of Great Britain*. London: Her Majesty's Stationery Office.

Mitchell, G.F., 1948. Two inter-glacial deposits in south-east Ireland. *Proceedings of the Royal Irish Academy* 52 B: 1-14.

Mitchell, G.F., 1951. The Pleistocene period in Ireland. *Dansk Geologisk Forening* 12: 111-114.

Mitchell, G.F., 1957. The Pleistocene epoch. In J. Meanan & D.A. Webb (eds.), *A view of Ireland*: 32-38. Dublin: The British Association for the Advancement of Science.

Mitchell, G.F., 1960. The Pleistocene history of the Irish Sea. *The Advancement of Science* 17: 313-325.

Mitchell, G.F., 1962. Summer field meeting in Wales and Ireland. *Proceedings of the Geologists' Association* 73: 197-213.

Mitchell, G.F., 1963. Morainic ridges on the floor of the Irish Sea. *Irish Geography* 4: 335-344.

Mitchell, G.F., 1965. The Quaternary deposits of the Ballaugh and Kirk Michael districts. *Quarterly Journal of the Geological Society of London* 121: 358-381.

Mitchell, G.F., 1968. Glaciogravel on Lundy Island. *Transactions of the Royal Geological Society of Cornwall* 20: 65-68.

Mitchell, G.F., 1970. The Quaternary deposits between Fenit and Spa on the north shore of Tralee Bay, Co. Kerry. *Proceedings of the Royal Irish Academy* 70 B: 141-162.

Mitchell, G.F., 1972. The Pleistocene history of the Irish Sea: second approximation. *Scientific Proceedings of the Royal Dublin Society* A 4: 181-199.

Mitchell, G.F., 1973. Fossil pingos in Camaross Townland, Co. Wexford. *Proceedings of the Royal Irish Academy* 73 B: 269-282.

Mitchell, G.F., 1976. *The Irish Landscape*: 240 pp. London: Collins.

Mitchell, G.F., 1977. Periglacial Ireland. In G.F. Mitchell & R.G. West (eds.), The changing environmental conditions in Great Britain and Ireland during the Devensian (Last) Cold Stage. *Philosophical Transactions of the Royal Society of London* B 280: 199-209.

Mitchell, G.F., 1980. The search for Tertiary Ireland. *Journal of Earth Sciences Royal Dublin Society* 3: 13-33.

Mitchell, G.F., 1981. The Quaternary - until 10,000 BP. In C.H. Holland (ed.), *A geology of Ireland*: 235-258. Edinburgh: Scottish Academic Press.

Mitchell, G.F., 1985. The Preglacial landscape. In K.J. Edwards & W.P. Warren (eds.), *The Quaternary History of Ireland*: 17-37. London: Academic Press.

Mitchell, F., 1986. *The Shell guide to reading the Irish landscape*: 228 pp. London: Michael Joseph/Country House.

Mitchell, G.F., E.A. Colhoun, N. Stephens & F.M. Synge, 1973. Ireland. In G.F. Mitchell, L.F. Penny, F.W. Shotton & R.G. West (eds.), *A correlation of Quaternary deposits in the British Isles*: 67-80. Geological Society of London, Special Report No 4.

Mitchell, G.F. & A.R. Orme, 1967. The Pleistocene deposits of the Isles of Scilly. *Quarterly Journal of the Geological Society of London* 123: 59-92.

Mitchell, G.F., L.F. Penny, F.W. Shotton & R.G. West, 1973. A correlation of Quaternary deposits in the British Isles. *Geological Society of London, Special Report* 4: 99 pp.

Mitchell, G.H., J.V. Stephens, C.E.N. Bromehead & D.A. Wray, 1947. Geology of the country around Barnsley. *Memoir of the Geological Survey*. London: Her Majesty's Stationery Office.

Mitchell, R.L., 1964. Trace elements in soils. In F.E. Bear (ed.), *Chemistry of soil*: 320-368. Reinhold Publ. Co.

Mörner, N.-A., 1980. The Fennoscandian Uplift: Geological Data and their Geodynamical Implication. In N.-A. Mörner (ed.), *Earth Rheology, Isostasy and Eustasy*: 251-284. Chichester, New York, Bisbane, Toronto: John Wiley & Sons.

Moffat, A.J. & J.A. Catt, 1986. A re-examination of the evidence for a Plio-Pleistocene marine transgression on the Chiltern Hills. III. Deposits. *Earth Surface Processes and Landforms* 11: 233-247.

Moran, S.R., 1971. Glacitectonic studies in drift. In R.P. Goldthwait (ed.), *Till, a symposium*: 127-148. Columbus: Ohio State University Press.

Morgan, A., 1973. Late Pleistocene environmental changes indicated by insect faunas of the English Midlands. *Boreas* 2: 173-212.

Morgan, A.V., 1973. The Pleistocene geology of the area north and west of Wolverhampton, Staffordshire, England. *Philosophical Transactions of the Royal Society of London* B 265: 233-297.

Morrison, M.E.S. & N. Stephens, 1965. A submerged late-Quaternary deposit at Roddans Port on the north-east coast of Ireland. *Philosophical Transactions of the Royal Society of London* B 249: 221-255.

Mortimer, J.R., 1905. *Forty Years' Researches in British and Saxon Burial Mounds of East Yorkshire* Hull: A. Brown & Sons.

Mosley, M.P., 1988. Bedload transport and sediment yield in the Onyx River, Antarctica. *Earth Surface Processes and Landforms* 13: 51-67.

Müller, H., 1985. Altquartäre Sedimente im Deckgebirge des Salzstocks Gorleben. *Zeitschrift der Deutschen Geologischen Gesellschaft* 137: 85-95.

Muller, S.W., 1947. *Permafrost or permanently frozen ground and related engineering problems*: 231 pp. Ann Arbor: Edwards.

Munro, M., 1986. Geology of the country around Aberdeen. *Memoir of the British Geological Survey, Sheet 77 (Scotland)*. London: Her Majesty's Stationery Office.

Murdoch, W.M., 1977. *The glaciation and deglaciation of south east Aberdeenshire*. Unpublished Thesis, University of Aberdeen.

Mykura, W., 1976. *British Regional Geology. Orkney and Shetland*. Her Majesty's Stationery Office.

Mykura, W. & J. Phemister, 1976. The geology of western Shetland. *Memoir of the Geological Survey of Great Britain*.

Naylor, D., 1965. Pleistocene and post-Pleistocene sediments in Dublin Bay. *Scientific Proceedings of the Royal Dublin Society* A 2: 175-188.

Naylor, D., W.E.A. Philips, G.D. Sevastopulo & F.M. Synge, 1980. *An introduction to the geology of Ireland*: 49 pp. Dublin: Royal Irish Academy.

Neale, J. & H.V. Howe, 1975. The marine Ostracoda of Russian Harbour, Novaya Zemlya and other high latitude faunas. *Bulletin of American Paleontology* 65: 381-431.

Nemec, W., R.J. Steel, S.J. Porebski & A. Spinnangr, 1984. Domba Conglomerate, Devonian, Norway: process and lateral variability in a mass flow-dominated lacustrine fan-delta. In E.H. Koster & R.J. Steel (eds.), *Sedimentology of Gravels and Conglomerates. Canadian Society Petroleum Geology Memoirs* 10: 295-320.

Nenonen, K., 1986. Orgaanisen aineksen merkitys moreenistratigrafiassa. *Geologi* 38: 41-44.

Nichol, I., I. Thorton, J.S. Webb, W.K. Fletcher, R.F. Horsnail, J. Khaleelee & D. Taylor, 1970. Regional geochemical reconnaissance of the Derbyshire area. *Institute of Geological Sciences report* No 70/2: 37 pp.

Nicholls, G.D. & D.H. Loring, 1962. The geochemistry of some British Carboniferous sediments. *Geochimica et Cosmochimica Acta* 26: 181-223.

Nickless, E., 1982. *Environmental geology of the Glenrothes district, Fife Region: Description of 1:25,000 sheet No. 20:* 53 pp. Institute of Geological Sciences Report 82/15.

Nobles, L. H. & J. Weertman, 1971. Influence of irregularities of the bed of an ice sheet on deposition rates of till. In R. P. Goldthwait (ed.), *Till, a symposium*: 117-126. Columbus: Ohio State University Press.

Nordsieck, F., 1969. *Die europäischen Meeresmuscheln (Bivalvia) vom Eismeer bis Kapverden, Mittelmeer und Schwarzes Meer*. Stuttgart: Gustav Fischer Verlag.

Norton, P. E. P. & R. B. Beck, 1972. Lower Pleistocene molluscan assemblages and pollen from the Crag of Aldeby (Norfolk) and Easton Bavents (Suffolk). *Bulletin of the Geological Society of Norfolk* 22: 4-31.

Nye, J. F., 1952. The mechanics of glacier flow. *Journal of Glaciology* 2 (112): 82-93.

Nystuen, J. P., 1976. Facies and sedimentation of the Late Precambrian Moelv Tillite in the eastern part of the Sparagmite Region, southern Norway. *Norges Geologiske Undersøgelse* 329: 1-70.

Oele, E. & R. T. E. Schüttenhelm, 1979. Development of the North Sea after the Saalian glaciation. In E. Oele, R. T. E. Schüttenhelm & A. J. Wiggers (eds.), *The Quaternary History of the North Sea*. Acta Universitatis Upsaliensis, Symposium Universitatis Upsaliensis Annum Quingentesimum Celebrantis 2: 191-215.

Oldham, T., 1844. On the more recent geological deposits in Ireland. *Journal of the Geological Society of Dublin* 3: 61-71.

Omand, D., 1973. The glaciation of Caithness. Unpublished M. Sc thesis, University of Strathclyde.

Open University (eds.), 1981. The Earth: Structure, composition and evolution. Block I: 5237, 60.

Orheim, O. & A. Elverhøi, 1981. Model for submarine glacial deposition. *Annals of Glaciology* 2: 123-128.

Orme, A. R., 1966. Quaternary Changes of Sea Level in Ireland. *Transactions of the Institute of British Geographers* 39: 127-140.

Orme, A. R., 1967. Drumlins and the Weichsel Glaciation of Connemara. *Irish Geography* 5: 262-274.

Osterman, L. E. & J. T. Andrews, 1983. Changes in glacial-marine sedimentation in core HY77-159, Frobisher Bay, Baffin Island, N. W. T.: A record of proximal, distal and ice-rafting glacial-marine environments. In B. F. Molnia (ed.), *Glacial-Marine Sedimentation*: 451-494. New York: Plenum Press.

Page, N. R., 1972. On the age of the Hoxnian interglacial. *Geological Journal* 8: 129-142.

Palmer, J., 1966. Landforms, drainage and settlement in the Vale of York. In S. R. Eyre & G. R. J. Jones (eds.), *Geography as human ecology*: 91-121, London: Edward Arnold.

Palmer, L. S., 1931. On the Pleistocene succession of the Bristol district. *Proceedings of the Geologists' Association* 42: 345-361.

Pantin, H. M., 1975. Quaternary sediments of the northeastern Irish Sea. *Quaternary Newsletter* 17: 7-9.

Pantin, H. M., 1977. Quaternary sediments of the northern Irish Sea. In C. Kidson & M. J. Tooley (eds.), *The Quaternary History of the Irish Sea*: 27-54. Liverpool: Seel House Press.

Pantin, H. M., 1978. The Quaternary sediments from the north-east Irish Sea: Isle of Man to Cumbria. *Bulletin of the Geological Survey, United Kingdom* 64: 1-50.

Pantin, H. M. & C. D. R. Evans, 1984. The Quaternary history of the Central and Southwestern Celtic Sea. *Marine Geology* 57: 259-293.

Parekh, P. P., M. P. Dulsri & W. M. Bausch, 1977. Distribution of trace elements between carbonate and non carbonate phases of limestone. *Earth and Planetary Science Letters* 34: 39-50.

Parsons, A. R., 1966. Some aspects of the glacial geomorphology of Northeast Northumberland. Unpublished M.Sc. Thesis, University of Leicester.

Paterson, I. B., 1977. Sand and gravel resources of the Tayside Region. *Report of the Institute of Geological Sciences* 77/6.

Paterson, I. B., M. Armstrong & M. A. E. Browne, 1981. Quaternary estuarine deposits in the Tay - Earn area, Scotland. *Report of the Institute of Geological Sciences* 8 1/7.

Paterson, W. S. B., 1972. Laurentide ice sheet: estimated volumes during Late Wisconsin. *Reviews of Geophysics and Space Physics* 10: 885-917.

Paterson, W. S. B., 1981. *The Physics of Glaciers*, 2nd edition: 380 pp. Oxford: Pergamon Press.

Paul, M. A., 1977. Studies of the influence of depositional processes on certain geotechnical properties of modern tills in Spitsbergen: 317 pp. Unpublished Ph. D. Thesis, University of East Anglia (Norwich).

Paul, M. A., 1981. A geotechnical model for the process of supraglacial deposition. In M. A. Paul (ed.), *Soil Mechanics in Quaternary Science*: 73-86. Cambridge: Quaternary Research Association.

Paul, M. A., 1983. The supraglacial landsystem. In N. Eyles (ed.), *Glacial Geology: An Introduction for Engineers and Earth Scientists*: 71-90. Oxford: Pergamon Press.

Peach, B. N., & J. Horne, 1879. The glaciation of the Shetland Isles. *Quarterly Journal of the Geological Society* 35: 778-811.

Peach, B. N. & J. Horne, 1880. The glaciation of the Orkney Islands. *Quarterly Journal of the Geological Society* 36: 648-663.

Peach, B. N. & J. Horne, 1893. On the occurrence of shelly boulder clay in North Ronaldshay, Orkney. *Transactions of the Geological Society of Edinburgh* 6: 309-313.

Peach, B. N., J. Horne, A. B. Woodward, C. T. Clough, A. Harker & C. B. Wedd, 1910. The Geology of Glenelg, Lochalsh and Southeast Part of Skye. *Memoirs of the Geological Survey of Scotland*. Edinburgh: Her Majesty's Stationery Office.

Peachey, D., J. L. Roberts & B. P. Vickers, 1984, Resistate geochemistry applied to crops and Quaternary sediments from the Woodbridge area of Suffolk. *British Geological Survey, Analytical Chemistry Research Group, Report* 84/7.

Peachey, D., J. L. Roberts, B. P. Vickers, J. A. Zalasiewicz & S. J. Mathers, 1985. Resistate geochemistry of sediments - a promising tool for provenance studies. *Modern Geology* 9: 145-157.

Peacock, J. D., 1966. Note on the drift sequence near Portsoy, Banffshire. *Scottish Journal of Geology* 2: 35-37.

Peacock, J. D., 1970. Some aspects of the glacial geology of west Inverness-shire. *Bulletin of the Geological Survey of Great Britain* 33: 43-56.

Peacock, J. D., 1971a. Terminal features of the Creran glacier of Loch Lomond Readvance age in western Benderloch, Argyll and their significance in the Late-glacial history of the Loch Linnhe area. *Scottish Journal of Geology* 7: 349-356.

Peacock, J.D., 1971b. A re-interpretation of the Coastal Deposits of Banffshire and their place in the Late-glacial history of N.E. Scotland. *Geological Survey of Great Britain Bulletin* 37: 81-89.

Peacock, J.D., 1971c. Marine shell radiocarbon dates and the chronology of deglaciation in western Scotland. *Nature, Physical Science* 230: 43-45.

Peacock, J.D., 1974. Islay shelly till with *Palliolum groenlandicum*. *Scottish Journal of Geology* 10: 159-160.

Peacock, J.D., 1975a. Landslip associated with glacier ice. *Scottish Journal of Geology* 11: 363-364.

Peacock, J.D., 1975b. Scottish late and post-glacial marine deposits. In A.M.D. Gemmell (ed.), *Quaternary Studies in North East Scotland*: 45-48. Aberdeen: Aberdeen University, Department of Geography.

Peacock, J.D., 1977. S. Shian to Fort William. In R.J. Price (ed.), *Western Scotland*. INQUA excursion guide A 12: 38-40. Norwich: Geo Abstracts.

Peacock, J.D., 1980. An overlooked record of interglacial or interstadial sites in north-east Scotland. *Quaternary Newsletter* 32: 14-15.

Peacock, J.D., 1981. Report and Excursion Guide, Lewis and Harris. *Quaternary Newsletter* 35: 45-54.

Peacock, J.D., 1983. Quaternary geology of the Inner Hebrides. *Proceedings of the Royal Society of Edinburgh* 83 B: 83-89.

Peacock, J.D., 1984. Quaternary geology of the Outer Hebrides. *Report of the British Geological Survey* 16: 1-26.

Peacock, J.D., N.G. Berridge, A.L. Harris & F. May, 1968. The geology of the Elgin district. *Memoir of the Geological Survey of Scotland*.

Peacock, J.D. & D.L. Ross, 1978. Anomalous glacial erratics in the southern part of the Outer Hebrides. *Scottish Journal of Geology* 14: 262.

Peake, N.B. & J.M. Hancock, 1961. The Upper Cretaceous of Norfolk. In G.P. Larwood & B.M. Funnell (eds.), *The Geology of Norfolk* (2nd edition): 293-339. Norwich: Geological Society of Norfolk.

Peel, R.F., 1949. A study of two Northumbrian spillways. *Transactions of the Institute of British Geographers* 15: 73-89.

Peltier, R., 1982. Dynamics of the Ice Age Earth. *Advances in Geophysics* 24: 1-146.

Pennington, W., 1970. Vegetation history in the north-west of England: a regional synthesis. In D. Walker & R.G. West (eds.), *Studies in the vegetational history of the British Isles*: 97-116. Cambridge: Cambridge University Press.

Pennington, W., 1975. A chronostratigraphic comparison of Late- Weichselian and Late-Devensian subdivisions, illustrated by two radiocarbon-dated profiles from western Britain. *Boreas* 4: 157-171.

Pennington, W., 1977. The Late Devensian flora and vegetation of Britain. *Philosophical Transactions of the Royal Society of London* B 280: 247-271.

Pennington, W., 1978. Quaternary Geology. In F. Moseley (ed.), *The Geology of the Lake District. Yorkshire Geolog-*

ical Society Occasional Publication 3: 207-225.

Penny, L.F., 1964. A review of the last glaciation in Great Britain. *Proceedings of the Yorkshire Geological Society* 34: 387-411.

Penny, L.F. & J.A. Catt, 1967. Stone orientation and other structural features of tills in East Yorkshire. *Geological Magazine* 104: 344-360.

Penny, L.F., G.R. Coope & J.A. Catt, 1969. Age and insect fauna of the Dimlington Silts, East Yorkshire. *Nature* 224: 65-67.

Perrin, R.M.S., J. Rose & H. Davies, 1979. The distribution, variation and origins of pre-Devensian tills in eastern England. *Philosophical Transactions of the Royal Society of London* B 287: 535-570.

Phillips, L., 1976. Pleistocene vegetational history and geology in Norfolk. *Philosophical Transactions of the Royal Society of London* B 275: 215-286.

Pigott, C.D., 1962. Soil formation and development on the Carboniferous limestone of Derbyshire 1) Parent material. *Journal of Ecology* 50: 145-156.

Piper, D.J.W., J.A. Farre & A. Shore, 1985. Late Quaternary slumps and debris flow on the Scotian Slope. *Geological Society of America Bulletin* 96: 1508-1517.

Pointon, W.K., 1978. The Pleistocene succession at Corton, Suffolk. *Bulletin of the Geological Society of Norfolk* 30: 55-76.

Poole, E.G. & A.J. Whiteman, 1966. The geology of the country around Nantwich and Whitchurch. *Memoir of the Geological Survey of Great Britain*. London: Her Majesty's Stationery Office.

Poole, E.G., B.J. Williams & B.A. Hains, 1968. Geology of the country around Market Harborough. *Memoir of the Geological Survey of Great Britain*.

Portlock, J.E., 1843. *Report on the geology of the County of Londonderry and parts of Tyrone and Fermanagh*: xxxi + 784 pp. Dublin: Andrew Milliken, and Hodges and Smith; and London: Longman, Brown, Green and Longmans.

Postma, G., Th.B. Roep & G.H.J. Ruegg, 1983. Sandygravelly mass- flow deposits in an ice-marginal lake (Saalian, Leuvenumsche Beek Valley, Veluwe, The Netherlands), with emphasis on plug-flow deposits. *Sedimentary Geology* 34: 59-82.

Powell, J.J.M., A. Marsland & A.N. Al-Khafagi, 1983. Presuremeter testing of glacial clay tills. *Proceedings International Symposium on* in-situ *testing of Rocks* 2: 373-378. Paris.

Powell, R.D., 1981. A model for sedimentation by tidewater glaciers. *Annals of Glaciology* 2: 129-134.

Powell, R.D., 1983. Submarine flow tills at Victoria, British Columbia: Discussion. *Canadian Journal of Earth Sciences* 20: 509-510.

Powell, R.D., 1984. Glaciomarine Processes and inductive lithofacies modelling of ice shelf and tidewater glacier sediments based on Quaternary examples. *Marine Geology* 57: 1- 52.

Power, G. & J.B.L. Wild, 1982. The sand and gravel resources of the country south of Horncastle, Lincolnshire. *Institute of Geological Sciences, Mineral Assessment Report* 108: 44 pp.

Preece, R.C., P. Coxon & J.E. Robinson, 1986. New biostratigraphic evidence of the Post-glacial colonization of Ireland and for Mesolithic forest disturbance. *Journal of Biogeography* 13: 487-509.

Price, R.J., 1973. *Glacial and fluvioglacial landforms*: 242 pp. London: Longman.

Price, R.J., 1983. *Scotland's environment during the last 30,000 years*. Edinburgh: Scottish Academic Press.

Pringle, A.W., 1985. Holderness coast erosion and the significance of ords. *Earth Surface Processes and Landforms* 10: 107-124.

Prior, D.B., 1968. The late-Pleistocene geomorphology of north-east Antrim: 316 pp. Unpublished Ph.D. Thesis, Belfast, The Queen's University.

Quinlan, G.M. & C. Beaumont, 1984. Appalachian thrusting, lithospheric flexure and the Paleozoic stratigraphy of the Eastern Interior of North America. *Canadian Journal of Earth Sciences* 21: 973-996.

Quinn, I.M., 1984. *The Glaciation of County Waterford*. Field guide for the Geographical Society of Ireland weekend field trip, 19th and 20th May, 1984. Maynooth: 29 pp.

Rae, D.A., 1976. Aspects of glaciation in Orkney: 466 pp. Unpublished Ph.D. Thesis, Liverpool.

Raistrick, A., 1926. The glaciation of Wensleydale, Swaledale, and adjoining parts of the Pennines. *Proceedings of the Yorkshire Geological Society* 20: 366-410.

Raistrick, A., 1931a. The glaciation of Wharfedale, Yorkshire. *Proceedings of the Yorkshire Geological Society* 22: 9-30.

Raistrick, A., 1931b. The glaciation of Northumberland and Durham. *Proceedings of the Geologists' Association* 42: 281-291.

Raistrick, A., 1934. The correlation of glacial retreat stages across the Pennines. *Proceedings of the Yorkshire Geological Society* 22: 199-214.

Rastall, R.H. & J. Romanes, 1909. On the Boulders of the Cambridge Drift: their Distribution and Origin. *The Quarterly Journal of the Geological Society of London* 65: 246-264.

Read, H.H., 1923. The geology of the country round Banff, Huntly and Turriff. *Memoir of the Geological Survey of Scotland*.

Read, H.H., A. Bremner, R. Campbell & A.W. Gibb, 1925. Records of the occurrence of boulders of Norwegian rocks in Aberdeenshire and Banffshire. *Transactions of the Edinburgh Geological Society* 11: 230-231.

Reichelt, G., 1961. On the field study of the shape and degree of roundness of gravel particles. *Petermanns Geographische Mitteilungen* 105: 15.

Reid, C., 1882. *The Geology of the Country around Cromer*. Memoirs of the Geological Survey of England and Wales: 143 pp. London: Her Majesty's Stationery Office.

Reid, C., 1885. *The geology of Holderness and the adjoining parts of Yorkshire and Lincolnshire*. Memoir of the Geological Survey. London: Her Majesty's Stationery Office.

Reid, C., 1913. *Submerged forests*: 129 pp. Cambridge: Cambridge University Press.

Reid, P.C. & C. Downie, 1973. The age of the Bridlington Crag. *Proceedings of the Yorkshire Geological Society* 39: 315-318.

Rice, R.J., 1965. The early Pleistocene evolution of north-

eastern Leicestershire and parts of adjacent counties. *Transactions of the Institute of British Geographers* 37: 101-110.

Rice, R.J., 1968. The Quaternary deposits of central Leicestershire. *Philosophical Transactions of the Royal Society of London* A 262: 459-509.

Rice, R.J., 1977. Huncote sand and gravel pit. In F.W. Shotton (ed.), *The English Midlands*. INQUA X Congress 1977, Guidebook for Excursion A2: 25.

Rice, R.J., 1981a. The Pleistocene deposits of the area around Croft in south Leicestershire. *Philosophical Transactions of the Royal Society of London* B 293: 385-418.

Rice, R.J., 1981b. Dunton Bassett sand and gravel pit. In T.D. Douglas (ed.), *Field guide to the East Midlands region*. Cambridge: Quaternary Research Association.

Rice, R.J., 1981c. Croft quarry. In T.D. Douglas (ed.), *Field guide to the East Midlands region*. Cambridge: Quaternary Research Association.

Richey, J., 1961. *Scotland: The Tertiary Volcanic Districts*, 3rd Edition. British Regional Geology, Institute of Geological Sciences. Edinburgh: Her Majesty's Stationery Office.

Ringrose, S., 1982. Depositional processes in the development of eskers in Manitoba. In R. Davidson-Arnott, W. Nickling & B.D. Fahey (eds.), *Research in glacial, glaciofluvial and glacio-lacustrine systems*: 117-137. Norwich: Geobooks.

Rise, L., K. Rokoengen, A.C. Skinner & D. Long, 1984. *Northern North Sea. Quaternary geology map between 60°30' and 62°N and east of 1°E*. Institutt for kontinentalsokkelund ersøkelser, Norway.

Robin, G. de Q. & J. Weertman, 1973. Cyclic surging of glaciers. *Journal of Glaciology* 12: 3-18.

Robinson, A.H.W., 1968. The submerged glacial landscape off the Lincolnshire coast. *Transactions of the Institute of British Geographers* 44: 119-132.

Robinson, J.E., 1978. Ostracods from deposits in the Vale of St Albans. *Quaternary Newsletter* 25: 8-9.

Robinson, M., 1977. Glacial limits, sea-level changes and vegetational development in part of Wester Ross. Unpublished Ph.D. Thesis, University of Edinburgh.

Robinson, M. & C.K. Ballantyne, 1979. Evidence for a glacial readvance pre-dating the Loch Lomond Advance in Wester Ross. *Scottish Journal of Geology* 15: 271-277.

Röthlisberger, H. & A. Iken, 1981. Plucking as an effect of water pressures at the glacier bed. *Annals of Glaciology* 2: 57-62.

Rolfe, W.D.I., 1966. Woolly rhinocerous from the Scottish Pleistocene. *Scottish Journal of Geology* 2: 253-258.

Rose, C.B., 1865. On the Brick-earth of the Nar. *Geological Magazine* II: 8-12.

Rose, J., 1974. Small-scale spatial variability of some sedimentary properties of lodgement till and slumped till. *Proceedings of the Geologists' Association* 85: 239-258.

Rose, J., 1975. Raised beach gravels and ice-wedge casts at Old Kilpatrick, near Glasgow. *Scottish Journal of Geology* 11: 15-21.

Rose, J., 1980a. Geilston. In W.G. Jardine (ed.), *Glasgow Region*, Field Guide: 24-29. London: Quaternary Research Association.

Rose, J., 1980b. Day 2. In W.G. Jardine (ed.), *Glasgow Region*, Field Guide. London: Quaternary Research Association.

Rose, J., 1980c. Landform development around Kisdon, upper Swaledale, Yorkshire. *Proceedings of the Yorkshire Geological Society* 43: 201-219.

Rose, J., 1981. Field guide to the Quaternary geology of the south eastern part of the Loch Lomond basin. *Proceedings of the Geological Society of Glasgow* 123: 3-19.

Rose, J., 1985. The Dimlington Stadial / Dimlington Chronozone: a proposal for naming the main glacial episode of the Late Devensian in Britain. *Boreas* 14: 225-230.

Rose, J., 1987. Status of the Wolstonian Glaciation in the British Quaternary. *Quaternary Newsletter* 53: 1-9.

Rose, J. & P. Allen, 1977. Middle Pleistocene stratigraphy in south-east Suffolk. *Journal of the Geological Society of London* 133: 85-102.

Rose, J., P. Allen & R.W. Hey, 1976. Middle Pleistocene stratigraphy in southern East Anglia. *Nature* 263: 492-494.

Rose, J., P. Allen, R.A. Kemp, C.A. Whiteman & N. Owen, 1985. The early Anglian Barham Soil in southern East Anglia. In J. Boardman (ed.), *Soils and Quaternary Landscape Evolution*: 197-229. Chichester: Wiley.

Rose, J. & J. Boardman, 1983. River activity in relation to short-term climate deterioration. *Quaternary Studies in Poland* 4: 189-198.

Rose, J. & J.M. Letzer, 1977. Superimposed drumlins. *Journal of Glaciology* 18: 471-480.

Rose, J., R.G. Sturdy, P. Allen & C.A. Whiteman, 1978. Middle Pleistocene sediments and palaeosols near Chelmsford, Essex. *Proceedings of the Geologists' Association* 89: 91-96.

Rowell, A.J. & J.S. Turner, 1952. Corrie-glaciation in the Upper Eden Valley, Westmorland. *Liverpool and Manchester Geological Journal* 1: 200-209.

Rubey, W.W. & M.K. Hubbert, 1959. Role of fluid pressure in mechanics of overthrust faulting II. Overthrust belt in geosynclinal area of western Wyoming in light of fluid-pressure hypothesis. *Bulletin of the Geological Society of America* 70: 167-205.

Ruddiman, W.F., 1977. Late Quaternary deposition of ice rafted sand in the sub-polar North Atlantic (latitude 40° to 65°). *Bulletin of the Geological Society of America* 88: 1813-1827.

Ruddiman, W.F., 1987. Northern oceans. In W.F. Ruddiman & H.E. Wright jr. (eds.), *The Geology of North America*, vol. K-3: 137-154. North America and Adjacent Oceans during the last deglaciation. Boulder, Colorado: The Geological Society of America.

Ruddiman, W.F. & A. McIntyre, 1973. Time-transgressive deglacial retreat of polar waters from the North Atlantic. *Quaternary Research* 3: 117-130.

Ruddiman, W.F. & A. McIntyre, 1976. Northeast Atlantic palaeoclimatic changes over the past 600,000 years. *Geological Society of America Memoir* 145: 111-146.

Ruddiman, W.F. & A. McIntyre, 1979. Warmth of the subpolar North Atlantic Ocean during Northern Hemisphere ice-sheet growth. *Science* 204: 173-175.

Ruddiman, W.F. & A. McIntyre, 1981a. Oceanic mechanisms for amplification of the 23,000-year ice-volume cycle. *Science* 212: 617-627.

Ruddiman, W.F. & A. McIntyre, 1981b. The North Atlantic Ocean during the last deglaciation. *Palaeogeography, Palaeoclimatology, Palaeoecology* 35: 145-214.

Ruddiman, W.F., A. McIntyre, V. Niebler-Hunt & J.T. Durazzi, 1980. Oceanic evidence for the mechanism of rapid northern hemisphere glaciation. *Quaternary Research* 13: 33-64.

Ruddiman, W.F., M.E. Raymo & A. McIntyre, 1986. Matuyama 41,000- year cycles: north Atlantic ocean and northern hemisphere ice sheets. *Earth and Planetary Science Letters* 80: 117-129.

Ruddiman, W.F. & M.E. Raymo, 1988. Northern hemisphere climate regimes during the past three Ma: possible tectonic connections. In N.J. Shackleton, R.G. West & D.Q. Bowen (eds.), *The Past Three Million Years; evolution of climatic variability in the North Atlantic region*: 1-20. London: The Royal Society (first published in *Philosophical Transactions of the Royal Society of London* B 318: 411-430).

Ruddiman, W.F., C.D. Sancetta & A. McIntyre, 1977. Glacial/ interglacial response rate of subpolar North Atlantic water to climatic change: the record of ocean sediments. *Philosophical Transactions of the Royal Society of London* 280 B: 119-142.

Ruegg, G.H.J., 1977. Features of Middle Pleistocene sandur deposits in the Netherlands. *Geologie en Mijnbouw* 56: 5-24.

Ruegg, G.H.J., 1983. Glaciofluvial and glaciolacustrine deposits in the Netherlands. In J. Ehlers (ed.), *Glacial deposits in North-West Europe*: 379-392. Rotterdam: Balkema.

Ruegg, G.H.J. & J.G. Zandstra, 1977. Pliozäne und pleistozäne gestauchte Ablagerungen bei Emmerschans (Drenthe, Niederlande). *Mededelingen Rijks Geologische Dienst* (NS) 28: 65-99.

Russel, D.J. & N. Eyles, 1985. Geotechnical characteristics of weathering profiles in British overconsolidated clays (Carboniferous to Pleistocene). In K.S. Richards, R.R. Arnett & S. Ellis (eds.), *Geomorphology and Soils*: 417-436. London: George Allen and Unwin.

Russell, I.C., 1893. Malaspina Glacier. *Journal of Geology* 1: 217-245.

Rust, B.R., 1972. Structure and process in a braided river. *Sedimentology* 18: 221-245.

Rust, B.R., 1975. Fabric and structure in glaciofluvial gravels. In A.V. Jopling & B.C. McDonald (eds.), *Glaciofluvial and glaciolacustrine sedimentation*. Society of Economic Paleontologists and Mineralogists, Special Publication 23: 238-248.

Rust, B.R., 1977. Mass flow deposits in a Quaternary succession near Ottawa, Canada: diagnostic criteria for subaqueous outwash. *Canadian Journal of Earth Sciences* 14: 175-184.

Rust, B.R., 1978. Depositional models for braided alluvium. In A.D. Miall (ed.), *Fluvial sedimentology*. Canadian Society of Petroleum Geologists Memoir 5: 605-625.

Rust, B.R. & R. Romanelli, 1975. Late Quaternary subaqueous outwash deposits near Ottawa, Canada. In A.V. Jopling & B.C. McDonald (eds.), *Glaciofluvial and gla-*

ciolacustrine sedimentation. Society of Economic Paleontologists and Mineralogists, Special Publication 23: 177-192.

Rzechowski, J., 1986. Pleistocene till stratigraphy in Poland. *Quaternary Science Reviews* 5: 365-372.

Sabine, P.A., 1949. The source of some erratics from northeastern Northamptonshire and adjacent parts of Huntingdonshire. *Geological Magazine* 113: 241-250.

Salter, A.E., 1905. On the superficial deposits of central and parts of southern England. *Proceedings of the Geologists' Association* 19: 1-56.

Sampson, G.V., 1802. *Statistical Survey of the County of Londonderry, with Observations on the Means of Improvement*: xxvi + 510 pp. Dublin: Dublin Society.

Sandford, K.S., 1926. Pleistocene deposits. In J. Pringle (ed.), The geology of the country around Oxford. *Memoirs of the Geological Survey of Great Britain*.

Sanford, A.R., 1959. Analytical and experimental study of simple geologic structures. *Bulletin of the Geological Society of America* 70: 19-52.

Sarnthein, M., H.E. Stremme & A. Mangini, 1986. The Holstein Interglaciation: Time-Stratigraphic Position and Correlation to Stable-Isotope Stratigraphy of Deep-Sea Sediments. *Quaternary Research* 26: 283-298.

Saunderson, H.C., 1975. Sedimentology of the Brampton esker and its associated deposits: an empirical test of theory. In A.V. Jopling & B.C. McDonald (eds.), *Glaciofluvial and glaciolacustrine sedimentation*. Society of Economic Paleontologists and Mineralogists, Special Publication 23: 155-176.

Saunderson, H.C., 1977. The sliding bed facies in esker sands and gravels: a criterion for full-pipe (tunnel) flow? *Sedimentology* 24: 623-638.

Saunders, G.E., 1968. Glaciation of possible Scottish Readvance age in northwest Wales. *Nature* 218: 76-78.

Schwan, J. & A.J. van Loon, 1979. Structural and sedimentological characteristics of a Weichselian kame terrace at Sonderby Klint, Funen, Denmark. *Geologie en Mijnbouw* 58: 305-319.

Schytt, V., 1956. Lateral drainage channels along the northern side of the Moltne Glacier, north-west Greenland. *Geografiska Annaler* 38: 65-77.

Scott, D.B., R. Boyd & F.S. Medidli, 1987. Relative sea-level changes in Atlantic Canada: Observed level and sedimentological changes vs theoretical models. In D. Nummedal, O.H. Pilkey & J.D. Howard (eds.), *Sea-level Fluctuation And Coastal Evolution*. Society of Economic Paleontologists and Mineralogists, Special Publication 41: 87-96.

Scourse, J.D., 1985. Late Pleistocene Stratigraphy of the Isles of Scilly and Adjoining Regions: 466 pp. Unpublished Ph.D. thesis, University of Cambridge.

Scourse, J.D. (ed.), 1986. *The Isles of Scilly*. Field Guide. Coventry: Quaternary Research Association: 151 pp.

Scourse, J.D., 1987. Periglacial sediments and landforms in the Isles of Scilly and West Cornwall. In J. Boardman (ed.), *Periglacial Processes and Landforms in Britain and Ireland*: 225-236. London: Cambridge University Press.

Seddon, B., 1957. Late-glacial Cwm glaciers in Wales. *Journal of Glaciology* 3: 94-99.

Seddon, M.B. & D.T. Holyoak, 1985. Evidence of sustained regional permafrost during deposition of Late Pleistocene sediments at Stanton Harcourt (Oxfordshire, England). *Proceedings of the Geologists' Association* 96: 53-71.

Sejrup, H.P., I. Aarseth, K. Bjørklund, J. Brigham-Grette, K.L. Ellingsen, E. Jansen, E. Larsen & M.S. Stoker, 1984. Quaternary stratigraphy of the Bosies Bank - Sleipner area, northern North Sea. In I. Aarseth & H.P. Sejrup (eds.), *Abstract Volume. Symposium on the Quaternary Stratigraphy of the North Sea, University of Bergen*: 56-58.

Sejrup, H.P., I. Aarseth, K.L. Ellingsen, E. Reither, E. Jansen, R. Løvlie, A. Bent, J. Brigham-Grette, E. Larsen & M. Stoker, 1987. Quaternary stratigraphy of the Fladen area, central North Sea: a multidisciplinary study. *Journal of Quaternary Science* 2: 35-58.

Shackleton, N.J., 1977. Oxygen isotope stratigraphy of the Middle Pleistocene. In F.W. Shotton (ed.), *British Quaternary studies: recent advances*: 1-16. Oxford: Clarendon Press.

Shackleton, N.J., 1987. Oxygen Isotopes, Ice Volume and Sea Level. *Quaternary Science Reviews* 6: 183-190.

Shackleton, N.J., J. Backman, H. Zimmerman, D.V. Kent, M.A. Hall, D.G. Roberts, D. Schnitker, J.G. Baldauf, A. Desprairies, R. Haomrighausen, P. Huddlestun, J.B. Keene, A.J. Kaltenback, K.A.O. Krumsiek, A.C. Morton, J.W. Murray & J. Westenberg-Smith, 1984. Oxygen isotope calibration of the onset of icerafting and history of glaciation in the North Atlantic region. *Nature* 307 (5952): 620-623.

Shackleton, N.J. & N.D. Opdyke, 1973. Oxygen isotope and palaeomagnetic stratigraphy of Equatorial Pacific core V28-238: oxygen isotope temperatures and ice volumes on a 10^5 year and 10^6 year scale. *Quaternary Research* 3: 39-55.

Shakesby, R.A., 1978. Dispersal of glacial erratics from Lennoxtown, Stirlingshire. *Scottish Journal of Geology* 14: 81-86.

Sharp, M.J., 1982. A comparison of the landforms and sedimentary sequences produced by surging and non-surging glaciers in Iceland. Unpublished Ph.D. Thesis, University of Aberdeen.

Shaw, J., 1971. Mechanism of till deposition related to thermal conditions in a Pleistocene glacier. *Journal of Glaciology* 10: 363-373.

Shaw, J., 1972a. Sedimentation in the ice-contact environment, with examples from Shropshire (England). *Sedimentology* 18: 23-62.

Shaw, J., 1972b. The Irish Sea Glaciation of north Shropshire - some environmental reconstructions. *Field Studies* 4: 603-631.

Shaw, J., 1975. Sedimentary successions in Pleistocene ice-marginal lakes. In A.V. Jopling & B.C. McDonald (eds.), *Glaciofluvial and glaciolacustrine sedimentation*. Society of Economic Paleontologists and Mineralogists, Special Publication 23: 281-303.

Shaw, J., 1977, Tills deposited in arid polar environments. *Canadian Journal of Earth Sciences* 14: 1239-1245.

Shaw, J., 1979. Genesis of the Sveg tills and rogen moraines of central Sweden: a model of basal melt-out. *Boreas* 8: 409-426.

Shaw, R., 1976. Periglacial features in part of the south-east

Grampian Highlands of Scotland. Unpublished Ph.D. Thesis, University of Edinburgh.

Sheppard, T., 1904. Quartzite pebbles on the Yorkshire Wolds. *The Naturalist*: 54-56.

Sherlock, R.L., 1924. The superficial deposits of south Buckinghamshire and south Hertfordshire, and the old course of the Thames. *Proceedings of the Geologists' Association* 35: 1-28.

Sherlock, R.L. & A.H. Noble, 1912. On the origin of the Clay- with-Flints of Buckinghamshire and on a former course of the Thames. *Quarterly Journal of the Geological Society of London* 68: 199-212.

Sherlock, R.L. & R.W. Pocock, 1924. The geology of the country around Hertford. *Memoir of the Geological Survey of the United Kingdom*. London: Her Majesty's Stationery Office. 66 pp.

Shilts, W.W., 1971. Till studies and their application to regional drift prospecting. *Canadian Mineral Journal* 92 (4): 45-50.

Shilts, W.W., 1973a. Glacial dispersal of rocks, minerals and trace elements in Wisconsin till, southeastern Quebec, Canada. *Geological Society of America Memoir* 136: 189-219.

Shilts, W.W., 1973b. Drift prospecting; geochemistry of eskers and till in permanently frozen terrain: District of Keewatin, Northwest Territories. *Geological Survey of Canada Paper* 72-45: 34 pp.

Shilts, W.W., 1977. Geochemistry of till in perennially frozen terrain of the Canadian Shield - applications to prospections. *Boreas* 6: 203-212.

Shilts, W.W., 1984. Till geochemistry in Finland and Canada. *Journal of Geochemical Exploration* 21: 95-117.

Short, N., 1961. Geochemical variations in four residual soils. *Journal of Geology* 69 (5): 534-571.

Shotton, F.W., 1953. Pleistocene deposits of the area between Coventry, Rugby and Leamington and their bearing on the topographic development of the Midlands. *Philosophical Transactions of the Royal Society of London* B 237: 209-260.

Shotton, F.W., 1963. Middle Pleistocene sections in the M 45 motor road. *Proceedings of the Coventry Natural History and Scientific Society* 3: 183-189.

Shotton, F.W., 1965. Normal faulting in British Pleistocene deposits. *Quarterly Journal of the Geological Society of London* 121: 419-434.

Shotton, F.W., 1966. The problems and contributions of methods of absolute dating within the Pleistocene period. *Quarterly Journal of the Geological Society of London* 122: 357-383.

Shotton, F.W., 1967. Age of the Irish Sea Glaciation of the Midlands. *Nature* 215: 1366.

Shotton, F.W., 1976. Amplification of the Wolstonian Stage of the British Pleistocene. *Geological Magazine* 113: 241-250.

Shotton, F.W., 1983a. The Wolstonian Stage of the British Pleistocene in and around its type area of the English Midlands. *Quaternary Science Reviews* 2: 261-280.

Shotton, F.W., 1983b. Observations on the type Wolstonian glacial sequence. *Quaternary Newsletter* 40: 28-36.

Shotton, F.W., 1985. Glaciations in the United Kingdom.

Quaternary Science Reviews 5: 293-297.

Shotton, F.W., A.S. Goudie, D.J. Briggs & H.A. Osmaston, 1980. Cromerian interglacial deposits at Sugworth, near Oxford, England, and their relation to the Plateau Drift of the Cotswolds and the terrace sequence of the upper and middle Thames. *Philosophical Transactions of the Royal Society of London* B 289: 55-86.

Šibrava, V., D.Q. Bowen & G.M. Richmond (eds.), 1986. *Quaternary Glaciations in the Northern Hemisphere. Report of the International Geological Correlation Programme*. Quaternary Science Reviews 5, 514 pp.

Simpson, I.M. & R.G. West, 1958. On the stratigraphy and palaeobotany of a Late Pleistocene organic deposit at Chelford, Cheshire. *New Phytologist* 57: 239-250.

Simpson, J.B., 1933. The Lateglacial readvance moraines of the Highland border west of the River Tay. *Transactions of the Royal Society of Edinburgh* 57: 633-645.

Singh, G., 1970. Late-glacial vegetational history of Lecale, Co. Down. *Proceedings of the Royal Irish Academy* 69 B: 189-216.

Sissons, J.B., 1958a. The deglaciation of part of East Lothian. *Transactions of the Institute of British Geographers* 25: 59-77.

Sissons, J.B., 1958b. Supposed ice-dammed lakes in Britain with particular reference to the Eddleston valley, southern Scotland. *Geografiska Annaler* 40: 159-187.

Sissons, J.B., 1958c. Sub-glacial stream erosion in southern Northumberland. *Scottish Geographical Magazine* 74: 163-174.

Sissons, J.B., 1960. Subglacial, marginal and other glacial drainage in the Syracuse-Oneida area, New York. *Bulletin of the Geological Society of America* 71: 1575-1588.

Sissons, J.B., 1961a. A subglacial drainage system by the Tinto Hills, Lanarkshire. *Transactions of the Edinburgh Geological Society* 18: 175-192.

Sissons, J.B., 1961b. The central and eastern parts of the Lammermuir - Stranraer moraine. *Geological Magazine* 98: 380-392.

Sissons, J.B., 1964a. The Perth Readvance in central Scotland. Part II. *Scottish Geographical Magazine* 80: 28-36.

Sissons, J.B., 1964b. The glacial period. In J. Wreford Watson & J.B. Sissons (eds.), *The British Isles, a systematic geography*: 131-152. Edinburgh and London: Nelson.

Sissons, J.B., 1967a. *The evolution of Scotland's Scenery*: 259 pp. Edinburgh: Oliver & Boyd.

Sissons, J.B., 1967b. Glacial stages and radiocarbon dates in Scotland. *Scottish Journal of Geology* 3: 175-181.

Sissons, J.B., 1969. Drift stratigraphy and buried morphological features in the Grangemouth-Falkirk-Airth area, central Scotland. *Transactions of the Institute of British Geographers* 48, 19-50.

Sissons, J.B., 1972. The last glaciers in part of the South-East Grampians. *Scottish Geographical Magazine* 88: 168-181.

Sissons, J.B., 1974a. A lateglacial ice cap in the central Grampians, Scotland. *Transactions of the Institute of British Geographers* 62: 95-114.

Sissons, J.B., 1974b. Lateglacial marine erosion in Scotland. *Boreas* 3: 41-48.

Sissons, J.B., 1974c. The Quaternary in Scotland: a review.

Scottish Journal of Geology 10: 311-337.

Sissons, J.B., 1976a. A remarkable protalus rampart complex in Wester Ross. *Scottish Geographical Magazine* 92: 182-190.

Sissons, J.B., 1976b. Lateglacial marine erosion in South-East Scotland. *Scottish Geographical Magazine* 92: 17-29.

Sissons, J.B., 1976c. *The Geomorphology of the British Isles: Scotland*: 150 pp. London: Methuen.

Sissons, J.B., 1977a. The Loch Lomond Readvance in the northern mainland of Scotland. In J.M. Gray & J.J. Lowe (eds.), *Studies in the Scottish Lateglacial environment*: 45-59. Oxford: Pergamon.

Sissons, J.B., 1977b. The Loch Lomond Advance in southern Skye and some palaeoclimatic implications. *Scottish Journal of Geology* 13: 23-36.

Sissons, J.B., 1977c. Former ice-dammed lakes in Glen Moriston, Inverness-shire, and their significance in upland Britain. *Transactions of the Institute of British Geographers* N.S. 2: 224-242.

Sissons, J.B., 1977d. *Glen Roy National Nature Reserve: the Parallel Roads of Glen Roy*: 8 pp. Nature Conservancy Council.

Sissons, J.B., 1977e. Palaeoclimatic inferences from Loch Lomond Advance glaciers. In J.J. Lowe, J.R. Gray & J.E. Robinson (eds.), *Studies in the Lateglacial of North-West Europe*: 31-43. Oxford: Pergamon.

Sissons, J.B., 1977f. *The Scottish Highlands*. INQUA X Congress, Guidebook for Excursions A11 and C11: 51 pp. Norwich: Geo Abstracts.

Sissons, J.B., 1978. The parallel roads of Glen Roy and adjacent glens, Scotland. *Boreas* 7: 229-244.

Sissons, J.B., 1979a. The limit of the Loch Lomond Advance in Glen Roy and vicinity. *Scottish Journal of Geology* 15: 31-42.

Sissons, J.B., 1979b. Palaeoclimatic inferences from former glaciers in Scotland and the Lake District. *Nature* 278: 518-521.

Sissons, J.B., 1979c. Catastrophic lake drainage in Glen Spean and the Great Glen, Scotland. *Journal of the Geological Society of London*: 136: 215-224.

Sissons, J.B., 1979d. The Loch Lomond Advance in the Cairngorm Mountains. *Scottish Geographical Magazine* 95: 66-82.

Sissons, J.B., 1979e. The Loch Lomond Stadial in the British Isles. *Nature* 280: 199-203.

Sissons, J.B., 1979f. The later lakes and associated fluvial terraces of Glen Roy, Glen Spean and vicinity. *Transactions of the Institute of British Geographers* N.S.4: 12-29.

Sissons, J.B., 1980. The Loch Lomond Advance in the Lake District, Northern England. *Transactions of the Royal Society of Edinburgh: Earth Sciences* 71: 13-27.

Sissons, J.B., 1981a. Lateglacial marine erosion and a jokulhlaup deposit in the Beauly Firth. *Scottish Journal of Geology* 17: 7-19.

Sissons, J.B., 1981b. The last Scottish ice-sheet: facts and speculative discussion. *Boreas* 10: 1-17.

Sissons, J.B., 1982a. A former ice-dammed lake and associated glacier limits in the Achnasheen area, central Ross-shire. *Transactions of the Institute of British Geographers* N.S. 7: 98-116.

Sissons, J.B., 1982b. The so-called high 'interglacial' rock shoreline of western Scotland. *Transactions of the Institute of British Geographers* N.S. 7: 205-216.

Sissons, J.B., 1983. Quaternary. In G.Y. Craig (ed.), *Geology of Scotland*, 2nd Edition: 399-424. Edinburgh: Scottish Academic Press.

Sissons, J.B. & R. Cornish, 1982. Fluvial landforms associated with ice-dammed lake drainage in upper Glen Roy, Scotland. *Proceedings of the Geologists' Association* 93: 45-52.

Sissons, J.B. & A.G. Dawson, 1981. Former sea-levels and ice limits in part of Wester Ross, northwest Scotland. *Proceedings of the Geologists' Association* 92: 115-124.

Sissons, J.B. & A.H.J. Grant, 1972. The last glaciers in the Lochnagar area. *Scottish Journal of Geology* 8: 85-93.

Sissons, J.B., J.J. Lowe, K.S.R. Thompson & M.J.C. Walker, 1973. Loch Lomond Readvance in the Grampian Highlands of Scotland. *Nature, Physical Science* 244: 75-77.

Sissons, J.B. & D.E. Smith, 1965. Raised shorelines associated with the Perth Readvance in the Forth Valley and their relationship to glacial isostasy. *Transactions of the Royal Society of Edinburgh* 66: 143-168.

Sissons, J.B., D.E. Smith & R.A. Cullingford, 1966. Lateglacial and post-glacial shorelines in South-East Scotland. *Transactions of the Institute of British Geographers* 39: 9-18.

Sissons, J.B. & D.G. Sutherland, 1976. Climatic inferences from former glaciers in the South-East Grampian Highlands, Scotland. *Journal of Glaciology* 17: 325-346.

Sissons, J.B. & M.J.C. Walker, 1974. Lateglacial site in the central Grampian Highlands. *Nature* 249: 822-824.

Sjørring, S., 1983. The glacial history of Denmark. In J. Ehlers (ed.), *Glacial Deposits in North-West Europe*: 163-179. Rotterdam: Balkema.

Skempton, A.W., 1953. The colloidal activity of clays. *Proceedings of the Third International Conference on Soil Mechanics and Foundation Engineering* 1: 57-61. Zürich: Icosomef.

Skempton, A.W., 1970. The consolidation of clays by gravitational compaction. *Quarterly Journal of the Geological Society* 125: 373-412.

Skempton, A.W. & R.D. Northey, 1952: The sensitivity of clays. *Geotechnique* 3: 30-53.

Sladen, J.A. & W. Wrigley, 1983. Geotechnical properties of lodgement till. In N. Eyles (ed.), *Glacial Geology: an introduction for engineers and Earth scientists*: 184-212. Oxford: Pergamon Press.

Slater, G., 1926. Glacial tectonics as reflected in disturbed drift deposits. *Proceedings of the Geologists' Association* 37: 392-400.

Slater, G., 1927. Studies in the drift deposits of the south-western part of Suffolk. *Proceedings of the Geologists' Association* 38: 157-216.

Slater, G., 1931. The structure of the Bride Moraine, Isle of Man. *Proceedings of the Liverpool Geological Society* 68: 402-448.

Slater, G., 1943. Sections and models illustrating the drifts of East Anglia. *Proceedings of the Geologists' Association* 54: 124.

Smalley, I.J., 1966. The properties of glacial loess and the formation of loess deposits. *Journal of Sedimentary Petrology* 36: 669-676.

Smalley, I.J., 1971. *In-situ* theories of loess formation and the significance of the calcium-carbonate content of loess. *Earth Science Reviews* 7: 67-85.

Smith, A., 1858. On the chalk flint and greensand fragments found on the Castle Down of Tresco, one of the Islands of Scilly. *Transactions of the Royal Geological Society of Cornwall* 7: 343-344.

Smith, A.J., 1985. A catastrophic origin for the palaeovalley system of the eastern English Channel. *Marine Geology* 64: 65-75.

Smith, B., 1912. The glaciation of the Black Combe district. *Quarterly Journal of the Geological Society of London* 68: 402-448.

Smith, B., 1930. Borings through the glacial drifts of the northern plain of the Isle of Man. *Summary of Progress Geological Survey* 3: 14-23.

Smith, B., 1931. The glacial lakes of Eskdale, Miterdale and Wasdale, Cumberland and the retreat of the ice during the Main Glaciation. *Quarterly Journal of the Geological Society of London* 88: 57-83.

Smith, D.B., 1981. The Quaternary geology of the Sunderland district, north-east England. In J. Neale & J. Flenley (eds.), *The Quaternary in Britain*: 146-167. Oxford: Pergamon Press.

Smith, D.B. & E.A. Francis, 1967. The geology of the country around Durham and West Hartlepool. *Memoir of the Geological Survey of Great Britain*. London: Her Majesty's Stationery Office.

Smith, D.B. *et al.*, 1973. North-east England. In F. Mitchell *et al.* (eds.), *A correlation of Quaternary deposits in the British Isles*. Geological Society of London, Special report No.4: 22-28.

Smith, N.D., 1970. The braided stream depositional environment: comparison of the Platte River with some Silurian clastic rocks, north-central Appalachians. *Bulletin of the Geological Society of America* 81: 2993-3014.

Smith, N.D., 1971. Transverse bars and braiding in the lower Platte River, Nebraska. *Bulletin of the Geological Society of America* 82: 3407-3420.

Smith, N.D., 1972. Some sedimentological aspects of planar cross-stratification in a sandy braided river. *Journal of Sedimentary Petrology* 42: 624-634.

Smith, N.D., 1974. Sedimentology and bar formation in the Upper Kicking Horse River, a braided outwash stream. *Journal of Geology* 82: 205-223.

Smith, N.D., 1985. Proglacial fluvial environment. In G.M. Ashley, J. Shaw & N.D. Smith (eds.), *Glacial sedimentary environments*: 85-134. Society of Economic Paleontologists and Mineralogists, Short Course Notes 16.

Smith, R.A., 1967. The deglaciation of south-west Cumberland: a reappraisal of some features in the Eskdale and Bootle areas. *Proceedings of the Cumberland Geological Society* 2: 76-83.

Smith, R.F. & J. Boardman (in press). The use of soil information in the assessment of the incidence and magnitude of historic flood events in upland Britain. *Proceedings of the British Geomorphological Research Group Workshop on Floods*. John Wiley.

Smythe, J.A., 1912. The glacial geology of Northumberland. *Transactions of the Natural History Society of Northumberland*, New Series 4: 86-116.

Soil Survey of England and Wales, 1983. *Soil Map of England and Wales Scale 1:250,000. Sheet 1, Northern England*. Harpenden: Soil Survey of England and Wales.

Sollas, W.J., 1896. A map to show the distribution of eskers in Ireland. *The Scientific Transactions of the Royal Dublin Society* 5: 785-822.

Sollas, W.J. & R. Ll. Praeger, 1895. Notes on glacial deposits in Ireland. II Kill-o'-the-Grange. *Irish Naturalist* 4: 321-329.

Solomon, J.D., 1932a. The glacial succession on the north Norfolk coast. *Proceedings of the Geologists' Association* 43: 241-270.

Solomon, J.D., 1932b. On the heavy mineral assemblages of the Great Chalky Boulder-clay and Cannon-shot Gravels of East Anglia and their significance. *Geological Magazine* 69: 314-321.

Solomon, J.D., 1962. In B.M. Funnell & R.G. West, The early Pleistocene of Easton Bavents, Suffolk. *Quarterly Journal of the Geological Society of London* 118: 125-141.

Southard, J.B., N.D. Smith & R.A. Kuhnle, 1984. Chutes and lobes: newly identified elements of braiding in shallow gravelly streams. In E.H. Koster & R.J. Steel (eds.), *Sedimentology of gravels and conglomerates*. Canadian Society of Petroleum Geologists Memoir 10: 51-59.

Sparks, B.W. & R.G. West, 1964. The drift landforms around Holt, Norfolk. *Transactions of the Institute of British Geographers* 35: 27-35.

Sparks, B.W. & R.G. West, 1972. *The Ice Age in Britain*: 302 pp. London: Methuen.

Sparks, B.W., R.G. West, R.B.G. Williams & M. Ransom, 1969. Hoxnian interglacial deposits near Hatfield, Herts. *Proceedings of the Geologists' Association* 80: 243-267.

Stather, J.W., 1922. On a peculiar displacement in the Millepore Oolite near South Cave. *Proceedings of the Yorkshire Geological Society* 19: 395-400.

Stephan, H.-J. & J. Ehlers, 1983. North German till types. In J. Ehlers (ed.), *Glacial Deposits in North-West Europe*: 239-247. Rotterdam: Balkema.

Stephens, J.V., G.H. Mitchell & W. Edwards, 1953. Geology of the country between Bradford and Skipton. *Memoir of the Geological Survey*. London: Her Majesty's Stationery Office.

Stephens, N., 1957. Some observations on the 'interglacial' platform and the early Post-Glacial raised beach on the east coast of Ireland. *Proceedings of the Royal Irish Academy* 58 B: 129-149.

Stephens, N., 1970. The coastline of Ireland. In N. Stephens & R.E. Glasscock (eds.), *Irish geographical studies*: 125-145. Belfast: The Queen's University.

Stephens, N., J.R. Creighton & M.A. Hannon, 1975. The Late-Pleistocene period in north-eastern Ireland: an assessment 1975. *Irish Geography* 8: 1-23.

Stephens, N. & A.M. McCabe, 1977. Late Pleistocene ice movements and patterns of Late- and Post-Glacial shorelines on the coast of Ulster, Ireland. In C. Kidson & M.J. Tooley (eds.), *The Quaternary history of the Irish*

Sea: 179-198. Liverpool: Seel House Press.

Stephens, N. & F.M. Synge, 1965. Late Pleistocene shorelines and drift limits in north Donegal. *Proceedings of the Royal Irish Academy* 64 B: 131-153.

Stephens, N. & F.M. Synge, 1966. Late- and post-glacial shorelines and ice limits in Argyll and north-east Ulster. *Transactions of the Institute of British Geographers* 39: 101-125.

Stevens, L.A., 1959. Studies in the Pleistocene deposits of the British Isles. Unpublished Ph.D. thesis, University of Cambridge.

Stevens, L.A., 1960. The interglacial of the Nar valley, Norfolk. *Quarterly Journal of the Geological Society of London* 115: 291-315.

Stevenson, I.P. & G.D. Gaunt, 1971. Geology of the country around Chapel en le Frith. *Memoir of the Geological Survey of Great Britain*: 444 p. London: Her Majesty's Stationery Office.

Stoker, M.S. & A. Bent, 1985. Middle Pleistocene glacial and glaciomarine sedimentation in the west central North Sea. *Boreas* 14: 325-332.

Stoker, M.S. & D. Long, 1985. Comments on the Quaternary deposits and landforms of Scotland and the neighbouring shelves: a review. *Quaternary Science Reviews* 4: iii-viii.

Stoker, M.S., D. Long & J.A. Fyfe, 1985a. The Quaternary succession in the central North Sea. *Newsletters in Stratigraphy* 14: 119-128.

Stoker, M.S., D. Long & J.A. Fyfe, 1985b. A revised Quaternary stratigraphy for the central North Sea. *British Geological Survey Report* 17/2.

Stone, J.C., 1959. A description of glacial retreat features in mid-Nithsdale. *Scottish Geographical Magazine* 75: 164-168.

Straw, A., 1960. The limit of the 'Last' Glaciation in north Norfolk. *Proceedings of the Geologists' Association* 71: 378-390.

Straw, A., 1961a. Drifts, meltwater channels and ice-margins in the Lincolnshire Wolds. *Transactions of the Institute of British Geographers* 29: 115-128.

Straw, A., 1961b. The erosion surfaces of east Lincolnshire. *Proceedings of the Yorkshire Geological Society* 33: 149-172.

Straw, A., 1963a. The Quaternary evolution of the lower and middle Trent. *East Midland Geographer* 3: 171-189.

Straw, A., 1963b. Some observations on the coversand of north Lincolnshire. *Transactions of the Lincolnshire Naturalists' Union* 15: 260-269.

Straw, A., 1965. A reassessment of the Chalky boulder-clay or Marly Drift of north Norfolk. *Zeitschrift für Geomorphologie* N.F. 9: 209-221.

Straw, A., 1966. The development of the middle and lower Bain valley, east Lincolnshire. *Transactions of the Institute of British Geographers* 40: 145-154.

Straw, A., 1968. A Pleistocene diversion of drainage in North Derbyshire. *East Midlands Geographer* 4: 275-280.

Straw, A., 1969. Pleistocene events in Lincolnshire: a survey and revised nomenclature. *Transactions of the Lincolnshire Naturalists' Union* 18: 85-98.

Straw, A., 1979a. Eastern England. In A. Straw & K.M. Clayton, *The geomorphology of the British Isles: east-ern and central England*: 1-139. London: Methuen.

Straw, A., 1979b. An Early Devensian glaciation in eastern England? *Quaternary Newsletter* 28: 18-24.

Straw, A., 1979c. Age and correlation of Pleistocene deposits in west Norfolk. *Bulletin of the Geological Society of Norfolk* 31: 17-30.

Straw, A., 1979d. The geomorphological significance of the Wolstonian glaciation of eastern England. *Transactions of the Institute of British Geographers* (N.S.) 4: 540-549.

Straw, A., 1980a. An Early Devensian glaciation in eastern England reiterated. *Quaternary Newsletter* 31: 18-23.

Straw, A., 1980b. The age and geomorphological context of a Norfolk palaeosol. In R.A. Cullingford, D.A. Davidson & J. Lewin (eds.), *Timescales in Geomorphology*: 305-315, Chichester: J. Wiley & Sons.

Straw, A., 1982. Sediments, fossils and geomorphology - a Lincolnshire situation. In D.A. Davidson & M.L. Shackley (eds.), *Geoarchaeology*: 317-326, London: Duckworth.

Straw, A., 1983. Pre-Devensian glaciation of Lincolnshire (eastern England) and adjacent areas. *Quaternary Science Reviews* 2: 239-260.

Stride, A.H., 1963. North-east trending ridges of the Celtic Sea. *Proceedings of the Ussher Society* 1: 62-63.

Stuart, A.J. & L.H. van Wijngaarden-Bakker, 1985. Quaternary vertebrates. In K.J. Edwards & W.P. Warren (eds.), *The Quaternary history of Ireland*: 221-249. London: Academic Press.

Sugden, D.E., 1970. Landforms of deglaciation in the Cairngorm Mountains of Scotland. *Transactions of the Institute of British Geographers* 51: 201-219.

Sugden, D.E., 1977. Reconstruction of the morphology, dynamics and thermal characteristics of the Laurentide ice sheet at its maximum. *Arctic and Alpine Research* 9: 21-47.

Sugden, D.E. & B.S. John, 1976. *Glaciers and Landscape: A Geomorphological Approach*: 376 pp. London: Edward Arnold.

Suggate, R.P. & R.G. West, 1959. On the extent of the last glaciation in eastern England. *Proceedings of the Royal Society of London* B 150: 263-283.

Sumbler, M.G., 1983a. A new look at the type Wolstonian glacial deposits of Central England. *Proceedings of the Geologists' Association* 94: 23-31.

Sumbler, M.G., 1983b. The type Wolstonian sequence - some further observations. *Quaternary Newsletter* 40: 36-39.

Sutcliffe, A.J. & A. Currant, 1984. Minchin Hole Cave. In D.Q. Bowen & A. Henry (eds.), *Wales: Gower, Preseli, Fforest Fawr. Field Guide*: 33-37. Cambridge: Quaternary Research Association.

Sutherland, D.G., 1980. Problems of radiocarbon dating deposits from newly deglaciated terrain: examples from the Scottish Lateglacial. In J.J. Lowe, J.M. Gray & J.E. Robinson (eds.), *Studies in the Lateglacial of North-West Europe*: 139-149. Oxford: Pergamon Press.

Sutherland, D.G., 1981a. The raised shorelines and deglaciation of the Loch Long/Loch Fyne area, western Scotland: 460 pp. Unpublished Ph.D. Thesis, University of Edinburgh.

Sutherland, D.G., 1981b. The high-level marine shell beds of

Scotland and the build-up of the last Scottish ice sheet. *Boreas* 10: 247-254.

Sutherland, D.G., 1984a. The Quaternary deposits and landforms of Scotland and the neighbouring shelves: a review. *Quaternary Science Reviews* 3: 157-254.

Sutherland, D.G., 1984b. Modern glacier characteristics as a basis for inferring former climates with particular reference to the Loch Lomond Stadial. *Quaternary Science Reviews* 3: 291-309.

Sutherland, D.G., 1985. Reply to comments of J.D. Peacock and M.S. Stoker & D. Long. *Quaternary Science Reviews* 4: ix-xiii.

Sutherland, D.G., 1986. A review of Scottish marine shell radiocarbon dates, their standardization and interpretation. *Scottish Journal of Geology* 22: 145-164.

Sutherland, D.G., C.K. Ballantyne & M.J.C. Walker, 1984. Late Quaternary glaciation and environmental change on St. Kilda, Scotland, and their palaeoclimatic significance. *Boreas* 13: 261-272.

Sutherland, D.C. & M.J.C. Walker, 1984. A late Devensian ice-free area and possible interglacial site on the Isle of Lewis, Scotland. *Nature* 309: 701-703.

Svendsen, J.I. & J. Mangerud, 1987. Late Weichselian and Holocene sea-level history for a cross-section of western Norway. *Journal of Quaternary Science* 2: 113-132.

Swaine, D.J., 1955. The trace element content of soils. *Commonwealth Bureau of Soil Science Technical Communication* 8: 157 pp.

Swaine, D.J. & R.L. Mitchell, 1960. Trace element distribution in soil profiles. *Journal of Soil Science* 11 (2): 347-368.

Synge, F.M., 1950. The glacial deposits around Trim, Co. Meath. *Proceedings of the Royal Irish Academy 53 B*: 99-110.

Synge, F.M., 1952. Retreat stages of the last ice-sheet in the British Isles. *Irish Geography* 2: 168-171.

Synge, F.M., 1956. The glaciation of North-East Scotland. *Scottish Geographical Magazine* 72: 129-143.

Synge, F.M., 1963a. The Quaternary succession round Aberdeen, North-East Scotland. *Report of the VIth International Congress on Quaternary, Volume III: Geomorphological Section*: 353-361.

Synge, F.M., 1963b. A correlation between the drifts of south-east Ireland and those of west Wales. *Irish Geography* 4: 360-366.

Synge, F.M., 1963c. The glaciation of the Nephin Beg Range, County Mayo. *Irish Geography* 4: 397-403.

Synge, F.M., 1964. Some problems concerned with the glacial succession in south-east Ireland. *Irish Geography* 5: 73-82.

Synge, F.M., 1966a. *Glacial Drift Map of Co. Limerick*. Dublin: An Foras Talúntais.

Synge, F.M., 1966b. Glacial Geology. In T.F. Finch & P. Ryan (eds.), *Soils of County Limerick*: 12-20. Soil Survey Bulletin 16. Dublin: An Foras Talúntais.

Synge, F.M., 1968. The glaciation of west Mayo. *Irish Geography* 5: 372-386.

Synge, F.M., 1969. The Würm ice limit in the west of Ireland. In *Quaternary Geology and Climate*, Publication 1701: 89-92. Washington: National Academy of Sciences.

Synge, F.M., 1970a. The Irish Quaternary: Current views 1969. In N. Stephens & R.E. Glasscock (eds.), *Irish Geographical Studies*: 34-48. Belfast: The Queen's University.

Synge, F.M., 1970b. An analysis of the glacial drifts of southeast Limerick. *Geological Survey of Ireland Bulletin* 1: 65-71.

Synge, F.M., 1971. The glacial deposits of Glenasmole, County Dublin, and the neighbouring uplands. *Geological Survey of Ireland Bulletin* 1: 87-97.

Synge, F.M., 1973. The glaciation of south Wicklow and the adjoining parts of the neighbouring counties. *Irish Geography* 6: 561-569.

Synge, F.M. (ed.), 1975a. *The Quaternary of the Wicklow district*, Field Guide: 21 pp. Dublin: Quaternary Research Association.

Synge, F.M., 1975b. Clogga. In F.M. Synge (ed.), *The Quaternary of the Wicklow district*: 5. Dublin: Quaternary Research Association.

Synge, F.M., 1977a. The coasts of Leinster (Ireland). In C. Kidson & M.J. Tooley (eds.), *The Quaternary History of the Irish Sea*: 199-222. Liverpool: Seel House Press.

Synge, F.M., 1977b. Introduction. In D.Q. Bowen (ed.), *South and southwest Ireland*: 4-8. Guidebook for excursion A 15, INQUA X Congress. Norwich: Geobooks.

Synge, F.M., 1977c. Clogga. In D. Huddart (ed.), *South East Ireland*, Guidebook for Excursion A 14, X INQUA Congress: 21-23. Norwich: GeoAbstracts.

Synge, F.M., 1977d. Lough Nahanagan. In D. Huddart (ed.), *South East Ireland*, Guidebook for Excursion A 14, X INQUA Congress: 24-26. Norwich: GeoAbstracts.

Synge, F.M., 1977e. Blessington. In D. Huddart (ed.), *South East Ireland*: 28-30. X INQUA Congress Field Guide. Norwich: GeoAbstracts.

Synge, F.M., 1977f. Records of sea levels during the Late Devensian. In G.F. Mitchell & R.G. West eds.), The changing environmental conditions in Great Britain and Ireland during the Devensian (Last) Cold Stage. *Philosophical Transactions of the Royal Society of London* B 280: 211-228.

Synge, F.M., 1978a. Pleistocene events. In G.L.H. Davies & N. Stephens, *Ireland*: 115-180. London: Methuen.

Synge, F.M., 1978b. Contributions on the Pleistocene history. In G.L.H. Davies & N. Stephens, *Ireland*. London: Methuen.

Synge, F.M., 1979a. Quaternary glaciation in Ireland. *Quaternary Newsletter* 28: 1-18.

Synge, F.M., 1979b. Killakee. In A.M. McCabe (ed.), *East Central Ireland*, Field Guide: 60. Dublin: Quaternary Research Association.

Synge, F.M., 1979c. In A.M. McCabe (ed.), *Field Guide to East Central Ireland*: 40-48. Dublin: Quaternary Research Association.

Synge, F.M., 1979d. Glacial Drift. In *Atlas of Ireland*: plates 18 and 19. Dublin: Royal Irish Academy.

Synge, F.M., 1979e. Glacial Landforms. In *Atlas of Ireland*: Plate 21. Dublin: Royal Irish Academy.

Synge, F.M., 1979f. Figure 12. In A.M. McCabe (ed.), *Field Guide to East Central Ireland*: 63 pp. Dublin: Quaternary Research Association.

Synge, F.M., 1980. Quaternary period. In D. Naylor, W.E.A. Phillips, G.D. Sevastopulo & F.M. Synge, *An introduction to the geology of Ireland*: 39-42. Dublin: Royal Irish Academy.

Synge, F.M., 1981. Quaternary glaciation and changes of sea level in the south of Ireland. *Geologie en Mijnbouw* 60: 305-315.

Synge, F.M., 1985. Coastal evolution. In K.J. Edwards & W.P. Warren (eds.), *The Quaternary history of Ireland*: 115-131. London: Academic Press.

Synge, F.M. & T.F. Finch, 1966. The drifts and soils of west Clare and the adjoining parts of Counties Kerry and Limerick. *Irish Geography* 5: 161-172.

Synge, F.M. & D. Huddart, 1977. Clogga. In D. Huddart (ed.), *South East Ireland*, Guidebook for Excursion A 14, X INQUA Congress: 21-23. Norwich: GeoAbstracts.

Synge, F.M. & N. Stephens, 1960. The Quaternary period in Ireland - an assessment, 1960. *Irish Geography* 4: 121-130.

Synge, F.M. & W.P. Warren (unpublished). Wicklow District: Quaternary Geology. Dublin: Geological Survey of Ireland.

Tarr, R.S., 1909. Some phenomena of the glacier margins in the Yakutat Bay region, Alaska. *Zeitschrift für Gletscherkunde* 3: 81-110.

Taylor, B.J., I.C. Burgess, D.H. Land, D.A.C. Mills, D.B. Smith & P.T. Warren, 1971. Northern England, 4th Edition. *British Regional Geology*. London: Her Majesty's Stationery Office.

Taylor, B.J., R.H. Price & F.M. Trotter, 1963. The geology of the country around Stockport and Knutsford. *Memoir of the Geological Survey of Great Britain*. London: Her Majesty's Stationery Office.

Taylor, R.K., 1978. Properties of mining wastes with respect to foundations. In F.G. Bell (ed.), *Foundation Engineering in Difficult Ground*: 175-203. London: Newnes-Butterworth.

Tedrow, J.C.F. & F.C. Ugolini, 1966. Antarctic soils. In *Antarctic soils and soil forming processes* 8: 161-177. Antarctic Research series, American Geophysical Union.

Terwindt, J.H.J. & P.G.E.F. Augustinus, 1985. Lateral and longitudinal successions in sedimentary structures in the Middle Mause Esker, Scotland. *Sedimentary Geology* 45: 161-188.

Thiede, J., G.W. Diesen, B.-E. Knudsen & T. Snare, 1968. Patterns of Cenozoic sedimentation in the Norwegian-Greenland Sea. *Marine Geology* 69: 323-352.

Thomas, G.S.P., 1976. The Quaternary stratigraphy of the Isle of Man. *Proceedings of the Geologists' Association* 87: 307-323.

Thomas, G.S.P., 1977. The Quaternary of the Isle of Man. In C. Kidson & M.J. Tooley (eds.), *The Quaternary History of the Irish Sea*: 155-178. Liverpool: Seel House Press.

Thomas, G.S.P., 1984a. The origin of the glacio-dynamic structure of the Bride Moraine, Isle of Man. *Boreas* 13: 355-364.

Thomas, G.S.P., 1984b. A late Devensian glaciolacustrine fan- delta at Rhoseomor, Clywd, North Wales. *Geological Journal* 19: 125-142.

Thomas, G.S.P., 1984c. Sedimentation of a sub-aqueous esker-delta at Strabathie, Aberdeenshire. *Scottish Journal of Geology* 20: 9-20.

Thomas, G.S.P., 1985a. The Late Devensian glaciation along the border of northeast Wales. *Geological Journal* 20: 319-340.

Thomas, G.S.P., 1985b. The Quaternary of the Northern Irish Sea basin. In R.H. Johnson (ed.), *The Geomorphology of North-west England*: 143-158. Manchester: Manchester University Press.

Thomas, G.S.P., 1989. The Late Devensian glaciation along the western margin of the Cheshire-Shropshire Lowland. *Journal of Quaternary Science* 4: 167-181.

Thomas, G.S.P., M. Connaughton & R.V. Dackombe, 1985. Facies variation in a Late Pleistocene supraglacial outwash sandur from the Isle of Man. *Geological Journal* 20: 193-213.

Thomas, G.S.P. & R.C. Connell, 1985. Iceberg drop, dump, and grounding structures from Pleistocene glaciolacustrine sediments, Scotland. *Journal of Sedimentary Petrology* 55: 243-249.

Thomas, G.S.P. & R.V. Dackombe, 1985. Comment on 'glaciomarine sediments of the Isle of Man as a key to Late Pleistocene stratigraphic investigations in the Irish Sea basin' (Eyles & Eyles, 1984). *Geology* 13: 445-446.

Thomas, G.S.P. & A.J. Summers, 1981. Pleistocene foraminifera from south-east Ireland - a reply. *Quaternary Newsletter* 34: 15-18.

Thomas, G.S.P. & A.J. Summers, 1982. Drop-stone and allied structures from Pleistocene waterlain till at Ely House, County Wexford. *Journal of Earth Sciences Royal Dublin Society* 4: 109-119.

Thomas, G.S.P. & A.J. Summers, 1983. The Quaternary stratigraphy between Blackwater Harbour and Tinnaberna, County Wexford. *Journal of Earth Sciences Royal Dublin Society* 5: 121-134.

Thomas, G.S.P. & A.J. Summers, 1984. Glacio-dynamic structures from the Blackwater Formation, Co. Wexford, Ireland. *Boreas* 13: 5-12.

Thomas, G.S.P., A.J. Summers & R.V. Dackombe, 1982. The Late- Quaternary deposits of the middle Dyfi Valley, Wales. *Geological Journal* 17: 297-309.

Thomas, R.H., 1977. Calving bay dynamics and ice sheet retreat up the St. Lawrence Valley system. *Géographie Physique et Quaternaire* 31: 347-356.

Thomasson, A.J., 1961. Some aspects of the drift and geomorphology of south-east Hertfordshire. *Proceedings of the Geologists' Association* 72: 287-302.

Thompson, D.B. & P. Worsley, 1966. A Late Pleistocene molluscan fauna from the drifts of the Cheshire Plain. *Geological Journal* 5: 197-207.

Thompson, K.S.R., 1972. The last glaciers in western Perthshire. Unpublished Thesis, University of Edinburgh.

Thomson, M.E. & R.A. Eden, 1977. Quaternary deposits of the central North Sea, 3. The Quaternary sequence in the west-central North Sea. *Report of the Institute of Geological Sciences* 77/12.

Thorp, P.W., 1981a. An analysis of the spatial variability of glacial striae and friction cracks in part of the western Grampians of Scotland. Quaternary Studies. *City of Lon-*

don *Polytechnic and Polytechnic of North London Occasional Papers* 1: 71-94.

Thorp, P.W., 1981b. A trimline method for defining the upper limit of the Loch Lomond Advance glaciers: examples from the Loch Leven and Glen Coe areas. *Scottish Journal of Geology* 17: 49-64.

Thorp, P.W., 1984. The glacial geomorphology of part of the western Grampians with especial reference to the limits of the Loch Lomond Advance: 427 pp. Unpublished Ph.D. Thesis, City of London Polytechnic, Council for National Academic Awards.

Thorp, P.W., 1985. The glacial geology of the Entiat valley in the eastern North Cascade Range, Washington, U.S.A. *City of London Polytechnic report to the Goldsmiths' Company, London*: 59 pp.

Thorp, P.W., 1986. A mountain icefield of Loch Lomond Stadial age, western Grampians, Scotland. *Boreas* 15: 83-97.

Tighe, W., 1802. *Statistical Observations relative to the County of Kilkenny, made in the years 1800, 1801*: xvi + 644 + 199 pp. Dublin: Dublin Society.

Tillotson, E., 1934. The glacial geology of Nidderdale. *Proceedings of the Yorkshire Geological Society* 22: 215-228.

Tomlinson, M.E., 1925. The river terraces of the lower valley of the Warwickshire Avon. *Quarterly Journal of the Geological Society of London* 81: 137-169.

Tooley, M.J. (ed.), 1977. *The Isle of Man, Lancashire coast and Lake District*. Field guide for excursion A4, X INQUA Congress: 61 pp. Norwich: GeoAbstracts.

Trechmann, C.T., 1915. The Scandinavian Drift of the Durham coast and the general glaciology of Southeast Durham. *Quarterly Journal of the Geological Society of London* 71: 53-82.

Trimmer, J., 1851. Generalisations respecting the erratic Tertiaries or Northern Drift. *Quarterly Journal of the Geological Society of London* 7: 19-31.

Trotter, F.M., 1922. In: Report from the Cumberland district. *Summary of Progress of the Geological Survey of Great Britain* 1922: 62.

Trotter, F.M., 1924. In: Report from the Cumberland district. *Summary of Progress of the Geological Survey of Great Britain* 1924: 80.

Trotter, F.M., 1929. The glaciation of the Eastern Edenside, Alston Block and the Carlisle Plain. *Quarterly Journal of the Geological Society of London* 88: 549-607.

Trotter, F.M. & S.E. Hollingworth, 1932. The geology of the Brampton District. *Memoir of the Geological Survey of the United Kingdom*. London: Her Majesty's Stationery Office.

Trotter, F.M., S.E. Hollingworth, T. Eastwood & W.C.C. Rose, 1937. The geology of the Gosforth district. *Memoir of the Geological Survey of the United Kingdom*. London: Her Majesty's Stationery Office.

Tufnell, L., 1969. The range of periglacial phenomena in Northern England. *Biuletyn Peryglacjalny* 19: 291-312.

Turekian, K.K. & J.L. Kulp, 1956. The Geochemistry of Strontium. *Geochimica et Cosmochimica Acta* 10: 245-296.

Turner, C., 1970. Middle Pleistocene deposits at Marks Tey, Essex. *Philosophical Transactions of the Royal Society of London* B 257: 373-440.

Turner, C., 1973. Eastern England. In G.F. Mitchell, L.F. Penny, F.W. Shotton & R.G. West (eds.), A correlation of Quaternary deposits in the British Isles. *Geological Society of London Special Report* 4: 8-9.

Urban, B., H. Thieme & H. Elsner, 1988. Biostratigraphische, quartärgeologische und urgeschichtliche Befunde aus dem Tagebau 'Schöningen', Ldkr. Helmstedt. *Zeitschrift der Deutschen Geologischen Gesellschaft* 139: 123-154.

Veenstra, H.J., 1965. Geology of the Dogger Bank area, North Sea. *Marine Geology* 3: 245-262.

Velichko, A.A. (ed.), 1984. *Late Quaternary Environments of the Soviet Union*. 327 pp. London: Longman.

Ventris, P.A., 1984. Comments on pre-Devensian glaciation of Lincolnshire (eastern England) and adjacent areas. *Quaternary Science Reviews* 3: vii-viii.

Ventris, P.A., 1985. Pleistocene environmental history of the Nar Valley, Norfolk. Unpublished Ph.D. Thesis, University of Cambridge.

Ventris, P.A., 1986. The Nar Valley. In R.G. West & C.A. Whiteman (eds.), *The Nar Valley & North Norfolk*: 6-55. Coventry: Quaternary Research Association.

Vernon, P., 1966. Drumlins and Pleistocene ice flow over the Ards Peninsula/Strangford Lough area, County Down, Ireland. *Journal of Glaciology* 6: 401-409.

Versey, H.C., 1938a. The Speeton pre-glacial shell bed. *The Naturalist*: 227-229.

Versey, H.C., 1938b. The Tertiary history of East Yorkshire. *Proceedings of the Yorkshire Geological Society* 23: 302-314.

Vincent, P., 1985. Quaternary geomorphology of the southern Lake District and Morecambe Bay area. In R.H. Johnson (ed.), *The geomorphology of North-West England*: 159-177. Manchester: Manchester University Press.

Vinogradow, A.P., 1959. *The geochemistry of rare and dispersed chemical elements in soils*: 209 pp. New York: Consultants Bureau Inc. (translation).

Visser, J.N.J., 1983a. The problem of recognizing ancient subaqueous debris flow deposits in glacial sequences. *Transactions of the Geological Society of South Africa* 86: 127-135.

Visser, J.N.J., 1983b. Submarine debris flow deposits from the Upper Carboniferous Dwyka Tillite Formation in the Kalahari Basin, South Africa. *Sedimentology* 30: 511-524.

Visser, J.N.J., 1983c. Glacial-marine sedimentation in the Late Paleozoic Karoo Basin, Southern Africa. In B.F. Molnia (ed.), *Glacial-Marine Sedimentation*: 667-702. New York: Plenum Press.

von Engeln, O.D., 1911. Phenomena associated with glacier drainage and wastage, with especial reference to observations in the Yakutat Bay region, Alaska. *Zeitschrift für Gletscherkunde* 6: 104-150.

von Weymarn, J., 1974. Coastal development in Lewis and Harris, Outer Hebrides, with particular reference to the effects of glaciation. Unpublished Ph.D. Thesis, University of Aberdeen.

von Weymarn, J., 1979. A new concept of glaciation in Lewis and Harris. *Proceedings of the Royal Society of Edinburgh* 77 B: 97-105.

von Weymarn, J. & K.J. Edwards, 1973. Interstadial site on the island of Lewis, Scotland. *Nature* 246: 473-474.

Vorren, T.O., M. Hald, M. Edvardsen & O.-W. Lind-Hansen, 1983. Glacigenic sediments and sedimentary environments on continental shelves: General principles with a case study from the Norwegian shelf. In J. Ehlers (ed.), *Glacial Deposits in North-West Europe*: 61-73. Rotterdam: Balkema.

Walkden, G.M., 1972. The mineralogy and origin of interbedded clay wayboards in the Lower Carboniferous of the Derbyshire dome. *Geological Journal* 8 (1): 143-160.

Walker, D., 1966. Late Quaternary history of the Cumberland lowland. *Philosophical Transactions of the Royal Society of London* B 251: 1-210.

Walker, M.J.C., 1980. Late-Glacial history of the Brecon Beacons, South Wales. *Nature* 287: 133-135.

Walker, M.J.C., 1982. The Late glacial and Early Flandrian deposits at Treath Mawr, Brecon Beacons, South Wales. *New Phytologist* 90: 177-194.

Walker, M.J.C. & J.J. Lowe, 1982. Lateglacial and early Flandrian chronology of the Isle of Mull, Scotland. *Nature* 296: 558-561.

Ward, J.C., 1875. The glaciation of the southern part of the Lake District and the glacial origin of the Lakes of Cumberland and Westmoreland. *Quarterly Journal of the Geological Society* 31: 152-166.

Ward, J.C., 1876. The geology of the northern part of the Lake District. *Memoir of the Geological Survey*.

Warren, W.P., 1970. Cirque glaciation in the Wicklow Mountains. Unpublished B.A. Dissertation, Dublin, University College.

Warren, W.P., 1977. North East Iveragh. In D.Q. Bowen (ed.), *South and South West Ireland*, Guidebook for Excursion A 15, X INQUA Congress: 37-45. Norwich: Geo Abstracts.

Warren, W.P., 1978. The Glacial History of the MacGillycuddy's Reeks and the adjoining area: 312 pp. Unpublished Ph.D. thesis, National University of Ireland: 312 pp.

Warren, W.P., 1979a. The stratigraphic position and age of the Gortian interglacial deposits. *Geological Survey of Ireland Bulletin* 2: 315-332.

Warren, W.P., 1979b. *Interim report on the Quaternary geology of the Tullow Lowland area*. Internal report, Geological Survey of Ireland: 15 pp.

Warren, W.P., 1979c. Moraines on the northern slopes and foothills of the MacGillycuddy's Reeks, south-west Ireland. In Ch. Schlüchter (ed.), *Moraines and Varves*: 223-236. Rotterdam: Balkema.

Warren, W.P., 1980. Ice movement in southwest Ireland: comments on the supposed Connachtian Glaciation. *Quaternary Newsletter* 31: 12-18.

Warren, W.P., 1981. Features indicative of prolonged and severe periglacial activity in Ireland, with particular reference to the south-west. *Biuletyn peryglacjalny* 28: 241-248.

Warren, W.P., 1985. Stratigraphy. In K.J. Edwards & W.P. Warren (eds.), *The Quaternary history of Ireland*: 39-65. London: Academic Press.

Warren, W.P., 1987a. Glaciodiagenesis in gelifluced deposits on the south coast of County Cork, Ireland. In J.J.M. van der Meer (ed.), *Tills and Glaciotectonics*: 105-115. Rotterdam: Balkema.

Warren, W.P., 1987b. Site 8: Slieve Bloom. In R.F. Hammond, W.P. Warren & D. Daly, *Offaly and West Kildare*:

49-56. Field Guide No. 10, Irish Association for Quaternary Studies.

Warren, W.P., C.A. Lewis, I.M. Quinn, P. Woodman, R.J. Devoy, J. Shaw, J.D. Orford & R.G.W. Carter, 1986. *Corca Dhuibhne*: 59 pp. Field guide No.9: Irish Association for Quaternary Studies.

Warwick, G.T., 1956. Caves and Glaciation 1: Central and Southern Pennines. *Transactions of the Cave Research Group of Great Britain* 4: 125-160.

Waters, R.S. & R.H. Johnson, 1958. The terraces of the Derby-shire Derwent. *East Midlands Geographer* (2) June: 3-15.

Watson, E., 1971. Remains of pingos in Wales and the Isle of Man. *Geological Journal* 7: 381-387.

Watts, W.A., 1959a. Interglacial deposits at Kilbeg and Newtown, Co. Waterford. *Proceedings of the Royal Irish Academy* 60 B: 79-134.

Watts, W.A., 1959b. Pollen spectra from the interglacial deposits at Kirmington, Lincolnshire. *Proceedings of the Yorkshire Geological Society* 32: 145-151.

Watts, W.A., 1963. Late glacial pollen zones in Western Ireland. *Irish Geography* 4: 367-376.

Watts, W.A., 1964. Interglacial deposits at Baggotstown, near Bruff, Co. Limerick. *Proceedings of the Royal Irish Academy* 63 B: 167-189.

Watts, W.A., 1967. Interglacial deposits in Kildromin Townland, near Herbertstown, Co. Limerick. *Proceedings of the Royal Irish Academy* 65 B: 339-348.

Watts, W.A., 1970. Tertiary and interglacial floras in Ireland. In N. Stephens & R.E. Glasscock (eds.), *Irish Geographical Studies*: 17-33. Belfast: The Queen's University.

Watts, W.A., 1977. The Late Devensian vegetation of Ireland. In G.F. Mitchell & R.G. West (eds.), The changing environmental conditions in Great Britain and Ireland during the Devensian (Last) Cold Stage. *Philosophical Transactions of the Royal Society of London* B 280: 273-293.

Watts, W.A., 1985. Quaternary vegetation cycles. In K.J. Edwards & W.P. Warren (eds.), *The Quaternary History of Ireland*: 155-185. London: Academic Press.

Wedepohl, K.H., 1970. *Handbook of geochemistry*, 4 vols. Berlin: Springer Verlag.

Wee, M.W. ter, 1983 a. The Saalian Glaciation in the northern Netherlands. In J. Ehlers (ed.), *Glacial Deposits in North-West Europe*: 405-412. Rotterdam: Balkema.

Wee, M.W. ter, 1983 b. The Elsterian Glaciation in the Netherlands. In J. Ehlers (ed.), *Glacial Deposits in North-West Europe*: 413-415. Rotterdam: Balkema.

Wells, A.J., 1954. The glaciation of the Teesdale-Swaledale watershed. *Proceedings of the University of Durham Philosophical Society* 12: 82-93.

West, R.G., 1956. The Quaternary deposits at Hoxne, Suffolk. *Philosophical Transactions of the Royal Society of London* B 239: 265-356.

West, R.G., 1957. Interglacial deposits at Bobbitshole, Ipswich. *Philosophical Transactions of the Royal Society of London* B 246: 1-31.

West, R.G., 1961. Vegetational history of the Early Pleistocene of the Royal Society borehole at Ludham, Norfolk. *Proceedings of the Royal Society of London* B 155: 437-453.

West, R.G., 1969. A note on pollen analyses from the Speeton

Shell Bed. *Proceedings of the Geologists' Association* 80: 217-218.

West, R.G. (ed.), 1977a: *East Anglia*. INQUA X Congress 1977, Guidebook for Excursions A1 and C1: 64 pp. Norwich: Geo Abstracts.

West, R.G., 1977b. *Pleistocene Geology and Biology*: 440 pp. London: Longman.

West, R.G., 1977c. Early and Middle Devensian flora and vegetation. *Philosophical Transactions of the Royal Society* B 280: 229-246.

West, R.G., 1980a. *The pre-glacial Pleistocene of the Norfolk and Suffolk Coasts*: 203 pp. Cambridge: Cambridge University Press.

West, R.G., 1980b. Pleistocene forest history in East Anglia. *New Phytologist* 85: 571-622.

West, R.G., 1984. The future of Quaternary Research. *Quaternary Newsletter* 43: 30-33.

West, R.G., C.A. Dickson, J.A. Catt, A.H. Weir & B.W. Sparks, 1974. Late Pleistocene deposits at Wretton, Norfolk. II Devensian deposits. *Philosophical Transactions of the Royal Society of London* B 267: 337-420.

West, R.G. & J.J. Donner, 1956. The glaciations of East Anglia and the East Midlands: a differentiation based on stone orientation measurements of the tills. *Quarterly Journal of the Geological Society of London* 112: 69-91.

West, R.G., B.M. Funnell & P.E.P. Norton, 1980. An Early Pleistocene cold marine episode in the North Sea: pollen and faunal assemblages at Covehithe, Suffolk, England. *Boreas* 9: 1-10.

West, R.G. & D.G. Wilson, 1968. Plant remains from the Corton Beds at Lowestoft, Suffolk. *Geological Magazine* 105: 116-123.

Whatley, R.C. & D.G. Masson, 1979. The ostracod genus *Cytheropteron* from the Quarternary (*sic*) and Recent of Great Britain. *Revista Espanola de Micropaleontologia* XI: 223-277.

Whitaker, W., 1889. The geology of London and of part of the Thames Valley. *Memoir of the Geological Survey of the United Kingdom*. London: Her Majesty's Stationery Office. 2 volumes: 556 + 352 pp.

Whitaker, W. & W.H. Dalton, 1882. *Old Series geological sheet 48 NE*. Geological Survey of England and Wales.

Whitaker, W., W.H. Penning, W.H. Dalton & F.J. Bennett, 1878. The geology of part of north-west Essex, with parts of Cambridgeshire and Suffolk and the north-eastern part of Hertfordshire. *Memoir of the Geological Survey of the United Kingdom* London: Her Majesty's Stationery Office. 42 + vi pp.

Whiteman, C.A., 1983. Great Waltham. In J. Rose (ed.), *The Diversion of the Thames*, Field Guide: 163-169. Cambridge: Quaternary Research Association.

Whiteman, C.A., 1987. Till lithology and genesis near the southern margin of the Anglian ice sheet in Essex, England. In J.J.M. van der Meer (ed.), *Tills and Glaciotectonics*: 55-66. Rotterdam: Balkema.

Whiteman, C.A., R. Kemp & D. Wilson, 1983. Stebbing. In J. Rose (ed.), *Diversion of the Thames*, Field Guide: 149-161, Cambridge: Quaternary Research Association.

Whitley, N., 1882. The evidence of glacial action in Devon and Cornwall. *Transactions of the Royal Geological Society of Cornwall* 10: 132-141.

Whittington, R.J., 1977. A late-glacial drainage pattern in the Kish Bank area and post-glacial sediments in the Central Irish Sea. In C. Kidson & M.J. Tooley (eds.), *The Quaternary History of the Irish Sea*: 55-68. Liverpool: Seel House Press.

Whittow, J.B., 1974. *Geology and scenery in Ireland*: 301 pp. Harmondsworth: Penguin.

Wilkenson, S.B., A. McHenry, J.R. Kilroe & H.J. Seymour, 1908. The geology of the country around Londonderry. *Memoir of the Geological Survey of Ireland*: 105 pp.

Wilkenson, S.B., 1907. The Geology of Islay. *Memoir of the Geological Survey of Scotland*. Glasgow: Her Majesty's Stationery Office.

Williams, D.M., 1981. The Maumtrasna Formation, western Ireland. In M.J. Hambrey & W.B. Harland (eds.), *Earth's pre-Pleistocene glacial record*: 576-578. Cambridge: Cambridge University Press.

Williams, P.F. & B.R. Rust, 1969. The sedimentology of a braided river. *Journal of Sedimentary Petrology* 39: 649-679.

Wills, L.J., 1924. The development of the Severn Valley in the neighbourhood of Iron-Bridge and Bridgnorth. *Quarterly Journal of the Geological Society of London* 80: 274-314.

Wills, L.J., 1937. The Pleistocene history of the West Midlands. *Report of the British Association for the Advancement of Science* 1937: 71-94.

Wilson, G.V., W. Edwards, J. Knox, R.C.B. Jones & J.V. Stephens, 1935. The geology of the Orkneys. *Memoir of the Geological Survey of Scotland*.

Wintle, A.G., 1981. Thermoluminescence dating of Late Devensian loesses in southern England. *Nature* 289: 479-480.

Wintle, A.G. & J.A. Catt, 1985. Thermoluminescence dating of Dimlington Stadial deposits in eastern England. *Boreas* 14: 231-234.

Wirtz, D., 1953. Zur Stratigraphie des Pleistocäns im Westen der Britischen Inseln. *Neues Jahrbuch für Geologie und Paläontologie*, Abhandlungen 96 (2): 267-303.

Wiseman, C.R., 1978. A Palaeoenvironmental Reconstruction of Part of the Lower Thames Terrace Sequence Based on Sedimentological Studies from Aveley, Essex. Unpublished M.Sc. Thesis, Council for National Academic Awards, City of London Polytechnic.

Woldstedt, P., 1958. *Das Eiszeitalter*, Vol. 2, 2nd edition: 438 pp. Stuttgart: Enke.

Wood, S.V. Jr., 1870. Observations on the sequence of the glacial beds. *Geological Magazine* 7: 17-22.

Wood, S.V. & F.W. Harmer, 1868. The Glacial and Post-glacial Structure of Norfolk and Suffolk. *Geological Magazine* V (52): 452-456.

Woodland, A.W., 1970. The buried tunnel valleys of East Anglia. *Proceedings of the Yorkshire Geological Society* 37: 521-578.

Woodward, H.B., 1884. The geology of the country around Fakenham, Wells and Holt. *Memoirs of the Geological Survey*: 57 pp. London: Her Majesty's Stationery Office.

Woodward, H.B., 1885. Glacial drifts of Norfolk. *Proceedings of the Geologists' Association* 9: 111-129.

Woodward, H.B., 1907. *The history of the Geological Society of London*. London: Geological Society of London.

Woolacott, D., 1907. The origin and influence of the chief physical features of Northumberland and Durham. *Geographical Journal* 30: 36-54.

Woolacott, D., 1921. The interglacial problem and the glacial and postglacial sequence in Northumberland and Durham. *Geological Magazine* 58: 26-32.

Wooldridge, S.W., 1938. The glaciation of the London Basin and the evolution of the Lower Thames drainage system. *Quarterly Journal of the Geological Society of London* 94: 627-667.

Wooldridge, S.W., 1960. The Pleistocene succession in the London Basin. *Proceedings of the Geologists' Association* 71: 113-129.

Wooldridge, S.W. & I.W. Cornwall, 1964. A contribution to a new datum for the pre-history of the Thames Valley. *Bulletin of the Institute of Archaeology*, University of London, 4: 223-232.

Wooldridge, S.W. & D.L. Linton, 1939. Structure, surface and drainage in south-east England. *Transactions of the Institute of British Geographers* 10: 124 pp.

Wooldridge, S.W. & D.L. Linton, 1955. *Structure, Surface and Drainage in South-east England*: 176 pp. London: George Philip.

Worsley, P., 1967. Problems in naming the Pleistocene deposits of the north east Cheshire Plain. *Mercian Geologist* 2: 51-55.

Worsley, P., 1970. The Cheshire-Shropshire lowlands. In C.A. Lewis (ed.), *The glaciations of Wales and adjoining regions*: 83-106. London: Longman.

Worsley, P., 1980. Problems in radiocarbon dating the Chelford Interstadial of England. In R.A. Cullingford, D.A. Davidson & J. Lewin (eds.), *Timescales in Geomorphology*: 289-304. Chichester: John Wiley.

Worsley, P., 1985. Pleistocene history of the Cheshire-Shropshire Plain. In R.H. Johnson (ed.), *Geomorphology of north-west England*: 201-221. Manchester: Manchester University Press.

Worsley, P., 1986. On the age of wood in till at Broughton Bay. *Quaternary Newsletter* 49: 17-19.

Worsley,. P., 1988. Evidence for Early Devensian/Midlandian glaciation in Britain and Ireland. *Quaternary Science Reviews* 7: in press.

Worsley, P., G.R. Coope, T.R. Good, D.T. Holyoak & J.E. Robinson, 1983. A Pleistocene succession from beneath Chelford Sands at Oakwood Quarry, Chelford, Cheshire. *Geological Journal* 18: 307-324.

Wray, D.A., J.V. Stephens, W.N. Edwards & C.E.N. Bromehead, 1930. The geology of the country around Huddersfield and Halifax. *Memoir of the Geological Survey*. London: Her Majesty's Stationery Office.

Wright, W.B., 1912. The drumlin topography of south Donegal. *Geological Magazine* 9: 153-159.

Wright, W.B., 1914. *The Quaternary Ice Age*: 464 pp. London: MacMillan.

Wright, W.B., 1922. *Drift geology: Killarney and Kenmare District (parts of Sheets 173, 184)*. Geological Survey of Ireland.

Wright, W.B., 1937. *The Quaternary Ice Age* 2nd Edition: 478 pp. London: Macmillan.

Wright, W.B. & H.B. Muff, 1904. The Pre-Glacial Raised Beach of the South Coast of Ireland. *Scientific Proceedings of the Royal Dublin Society* 10: 250-324.

Wyatt, R.J., 1971. New evidence for drift-filled valleys in north-east Leicestershire and south Lincolnshire. *Bulletin of the Geological Survey of Great Britain* 37: 29-55.

Wyatt, R.J., A. Horton & R.J. Kenna, 1971. Drift-filled channels on the Leicestershire-Lincolnshire border. *Bulletin of the Geological Survey of Great Britain* 37: 57-79.

Wymer, J.J., 1983. The Lower Palaeolithic site at Hoxne. *Suffolk Institute of Archaeology and History* 25: 169-189.

Wymer, J.J. & A. Straw, 1977. Hand-axes from beneath glacial till at Welton-le-Wold, Lincolnshire and the distribution of palaeoliths in Britain. *Proceedings of the Prehistoric Society* 43: 355-360.

Yaalon, D.H., I. Brenner & H. Koyomdjinsky, 1974. Weathering and mobility sequence of minor elements on a basaltic pedomorphic surface, Galilee, Israel. *Geoderma* 12: 233-244.

Yates, E.M. & F. Moseley, 1967. A contribution to the glacial geomorphology of the Cheshire plain. *Transactions of the Institute of British Geographers* 42: 107-125.

Yorke, C., 1954. *The pocket deposits of Derbyshire*. Birkenhead: Private Publication.

Young, J.A.T., 1974. Ice wastage in Glenmore, upper Spey Valley, Inverness-shire. *Scottish Journal of Geology* 10: 147-158.

Young, J.A.T., 1975. Ice wastage in Glen Feshie, Inverness-shire. *Scottish Geographical Magazine* 91: 91-101.

Zagwijn, W.H., 1974. The palaeogeographic evolution of the Netherlands during the Quaternary. *Geologie en Mijnbouw* 53: 369-385.

Zagwijn, W.H., 1985. An outline of the Quaternary stratigraphy of the Netherlands. *Geologie en Mijnbouw* 64: 17-24.

Zagwijn, W.H., 1986. The Pleistocene of the Netherlands with special reference to glaciation and terrace formation. *Quaternary Science Reviews* 5: 341-345.

Zagwijn, W.H., H.M. van Montfrans & J.G. Zandstra, 1971. Subdivision of the 'Cromerian' in the Netherlands; pollen-analysis, palaeomagnetism and sedimentary petrology. *Geologie en Mijnbouw* 50: 41-58.

Zandstra, J.G., 1983. Fine gravel, heavy mineral and grain-size analyses of Pleistocene, mainly glacigenic deposits in the Netherlands. In J. Ehlers (ed.), *Glacial Deposits in North-West Europe*: 361-377. Rotterdam: Balkema.

Zalasiewicz, J.A. & S.J. Mathers, 1985. Lithostratigraphy of the Red and Norwich Crags of the Aldeburgh-Orford area, south-east Suffolk. *Geological Magazine* 122: 287-296.

Zalasiewicz, J.A., S.J. Mathers & J.D. Cornwell, 1985. The application of ground conductivity measurements to geological mapping. *Quarterly Journal of Engineering Geology* 18: 139-148.

Zalasiewicz, J.A., S.J. Mathers, M.J. Hughes, P.L. Gibbard, S.M. Peglar, R. Harland, G.S. Boulton, R.A. Nicholson, P. Cambridge & G.P. Wealthall, 1988. Stratigraphy and palaeoenvironments of the Red Crag and Norwich Crag formations between Aldeburgh and Sizewell, Suffolk, England. *Philosphical Transactions of the Royal Society of London* B.

List of authors

Peter Allen, Geography Section, City of London Polytechnic, Calcutta House, Old Castle Street, London E1 7NT, United Kingdom

Peter S. Balson, Marine Earth Sciences Research Programme, British Geological Survey, Keyworth, Nottingham NG12 5GG, United Kingdom

John Boardman, Department of Humanities and Countryside Research Unit, Brighton Polytechnic, Palmer, Brighton, Sussex BN1 9PH, United Kingdom

Geoffrey S. Boulton, Department of Geology, University of Edinburgh, West Mains Road, Edinburgh EH9 3JW, United Kingdom

D.Q. Bowen, Institute of Earth Studies, University College of Wales, Penglais Campus, Aberystwyth SY23 2DB, United Kingdom

Philip Browne, Department of Geography, Trinity College, Dublin 3, Ireland

Ian D. Bryant, Koninklijke Shell Exploratie en Productie Laboratorium, Postbus 60, NL-2280 AB Rijswijk (ZH), Netherlands

Cynthia V. Burek, Earth Sciences Information, Welsh Office, Newhaven House, Church Street, Holt, Wrexham, Clwyd LL13 9JP, United Kingdom

John A. Catt, Soils and Plant Nutrition Department, Rothamsted Experimental Station, Harpenden, Hertfordshire, AL5 2JQ, United Kingdom

D. Allan Cheshire, School of Natural Sciences, Hatfield Polytechnic, College Lane, Hatfield, Hertfordshire AL10 9AB, United Kingdom

E. Rodger Connell, Marathon Oil (UK) Ltd., Anderson Drive, Aberdeen AB2 4AZ, United Kingdom

Peter Coxon, Department of Geography, Trinity College, Dublin 3, Ireland

John M. Cubitt, Poroperm Geochem Laboratories Ltd., Chester Street, Saltney, Chester CH4 8RD, United Kingdom

Roger V. Dackombe, School of Applied Sciences, The Polytechnic, Wolverhampton WV1 1SB, United Kingdom

Robert Donnelly, Department of Geology, University College Cardiff, P.O. Box 78, Cardiff CF1 1XL, United Kingdom

Terry Douglas, Department of Environment, Lipman Building, Newcastle Polytechnic, Newcastle-upon-Tyne NE1 8ST, United Kingdom

Jürgen Ehlers, Geologisches Landesamt, Oberstr. 88, D-2000 Hamburg 13, Germany

Christopher Evans, British Geological Survey, Keyworth, Nottingham NG12 5GG, United Kingdom

Nicholas Eyles, Glaciated Basin Research Group, Department of Geology, University of Toronto, Scarborough Campus, 1265 Military Trail, Scarborough, Ontario M1C 1A4, Canada

Philip L. Gibbard, University of Cambridge, Subdepartment of Quaternary Research, Downing Street, Cambridge CB2 3EA, United Kingdom

J. Murray Gray, Department of Geography and Earth Science, Queen Mary and Westfield College, University of London, Mile End Road, London E1 4NS, United Kingdom

Adrian M. Hall, Fettes College, Carrington Road, Edinburgh EH4 1QX, United Kingdom

Charles Harris, Department of Geology, University College Cardiff, P.O. Box 78, Cardiff CF1 1XL, United Kingdom

Jane K. Hart, Department of Geography, University of Southampton, The University Road, Southampton SO9 5NH, United Kingdom

Richard W. Hey, Hartleton Lodge, Bromsash, Ross-on-Wye, Herefordshire HR9 7SB, United Kingdom

Peter G. Hoare, Sedimentology and Palaeobiology Laboratory, Anglia Higher Education College, East Road, Cambridge CB1 1PT, United Kingdom

Albert Horton, British Geological Survey, Keyworth, Nottingham NG12 5GG, United Kingdom

David Huddart, Liverpool Polytechnic, Science and Outdoor Education, I M Marsh Campus, Barkhill Road, Liverpool L17 6BD, United Kingdom

Dennis H. Jeffery, Marine Earth Sciences Research

Programme, British Geological Survey, Keyworth, Nottingham NG12 5GG, United Kingdom

John Anthony Little, Department of Civil Engineering, Heriot-Watt University, Edinburgh EH14 4AS, United Kingdom

Stephen John Mathers, British Geological Survey, Keyworth, Nottinghamshire NG12 5GG, United Kingdom

A. Marshall McCabe, Department of Environmental Science, The University of Ulster, Jordanstown Campus, Shore Road, Newtown Abbey, Co. Antrim BT37 04B, Northern Ireland

Michael Anthony Paul, Department of Civil Engineering, Heriot-Watt University, Edinburgh EH14 4AS, United Kingdom

J. Douglas Peacock, 18 McLaren Road, Edinburgh EH9 2BN, United Kingdom

Roger John Rice, Geography Department, The University, Leicester LE1 7RH, United Kingdom

Eric Robinson, Department of Geology, University College, Gower Street, London WC1E 6BT, United Kingdom

Jim Rose, University of London, Royal Holloway and Bedford New College, Department of Geography, Egham Hill, Egham, Surrey TW20 0EX, United Kingdom

James Scourse, School of Ocean Sciences, University College of North Wales, Menai Bridge, Gwynedd LL59 5EY, United Kingdom

Allan Straw, University of Exeter, Department of Geography, Armory Building, Rennes Drive, Exeter EX4 4RJ, United Kingdom

Donald G. Sutherland, 2 London Street, Edinburgh EH3 6NA, United Kingdom

Geoffrey S.P. Thomas, Department of Geography, The University of Liverpool, P.O. Box 49, Liverpool L69 3BX, United Kingdom

Peter W. Thorp, 10 Marks Avenue, Ongar, Essex CM5 9AY, United Kingdom

William P. Warren, Geological Survey of Ireland, Beggars Bush, Haddington Road, Dublin 4, Ireland

Gary Paul Wealthall, British Geological Survey, Keyworth, Nottinghamshire NG12 5GG, United Kingdom

Richard West, University of Cambridge, Botany School, Downing Street, Cambridge CB2 3EA, United Kingdom

Colin A. Whiteman, University of Cambridge, Subdepartment of Quaternary Research, Free School Lane, Cambridge CB2 3RS, United Kingdom

Peter Worsley, Postgraduate Research Institute for Sedimentology, University of Reading, Box 227, Reading RG6 2AB, United Kingdom

Jan Antoni Zalasiewicz, British Geological Survey, Keyworth, Nottinghamshire NG12 5GG, United Kingdom

Index

The following abbreviations have been used in this index: f - figure, m - map, p - plate, t - table.

545

Onecote 194m
Onich 138m, 147m
Ontario, Canada 438t
Oolite limestones 221
orbital rhythms 3
Orchy Glacier 142m
Ordovician
    glaciofluvial gravel 37
    Manx Slates 333
    rocks 175
Orford 407m
    Ness 407m
    Till 471, 475t
organic
    matter 479t
    weathering 483
Orkney Islands 4, 8, 10, 55, 121, 122m, 123, 125-127
    last glaciation of the 126m
Ormsby 234m, 242f
Oronsay 111m
Orrisdale
    facies 339
    Formation 335t
    Gravel 335t
    Head 328m, 334m, 335t, 336f, 342
        Isle of Man, stratified diamict 316f
    Sands 335t, 341, 341f, 342f
    Till 335t, 337, 338
        at Jurby Head, Isle of Man 337f, 338f
orthoclase 198, 200, 472, 472t, 474t, 476
orthosilicates 472
oscillations, glacial 371
Oslo erratics 18, 499
Ossian
    end and lateral moraines 143
    glacier 139, 142m
Ostend Sands 236f, 238t
ostracods 44, 75, 185, 304, 305, 306f, 309, 370
Oulton
    Beds 235t
    Moss 153
Ouse 214m
    Great 234m
    Valley 64
Out Skerries 123m, 124m
outlet glaciers 138, 139, 143, 144
outliers, limestone 417
outwash 4, 172, 196, 198, 221, 234, 265, 280, 286, 290, 297, 333, 390, 392, 396, 437, 440, 450
    deposits 235, 242f
    fan 146, 147, 438t, 446
        ice-contact subaqueous 313
        ice-front 437
        of Glen Forsa 446f
        remnants on North Sea floor 251
        subaqueous 342p
    gravel
        distal 258t
        proximal 258, 258t
        trains 63
    ice-marginal 326f
    plain 451
        tiered 439
    proglacial 163, 166, 286
    rivers 438t
    sand and gravel 243

sandar 405
sands 240, 358
    subaquatic 309
    subaqueous 324, 326f
    supraglacial 440
    terraces 373, 375, 446-453
        at Kilninver 451f
        of the Midlothian Esk, long profiles 452f
    train 448, 450, 453f
        braided 447f
        dissected 446
    valley train 146
Over Haddon 197
overbank-flood sequences, sand-dominated 342
Overby 154m
overconsolidated
    diamict 285, 286, 287
    till 279
overconsolidation 400
    ratio 396
overflow channels 156, 191
overfolds 455, 462
oversteepening, depositional 327
Overstrand 234m, 236f, 237f, 238, 242f, 460, 461f
overthrusts 343, 459
Ox Mountains 357m, 358
Oxford 14m, 15
    Clay 219
oxidation 474
    of till 67
oxides 485
oxygen 472
    isotope stages 3, 7, 10, 11, 12, 24, 42, 78, 181, 279, 493, 494, 499

pack ice 42
    cover 309
Palaeoargillic Soil, grain-size distribution 267f
palaeochannel 447f, 448
    system 208
palaeoclimate 140
palaeoclimatic
    control 149
    reconstruction 102
palaeocurrent 330
    analyses 19, 442
    data
        Black Combe Coast 162f
        Ffyndaff 289f
    determinations 384
    direction 288, 329
    measurements 286
palaeoenvironment 204, 207, 292, 314, 343, 384, 431
Palaeogene 302m, 304f
palaeogeographic reconstruction of the Devensian North Sea 253m
Palaeoloxodon antiquus 28
palaeomagnetic
    boundary 7
    stratigraphy 493
palaeosol 265t, 457f
    at Docking Common 214p, 218, 218f
    at Oakwood, Chelford 207f
    Laddray Wood 177t

Troutbeck 175, 177t, 181
palaeovalleys 208, 246, 250, 252
Palaeozoic
    chert 19
    greywacke, striated erratic of 297f
    greywackes, Lower 415
    Lower 66
    of the Midlands 15
    rocks 63
        Lower 418
Palliolum groenlandicum 119
Pallis 418
palynology 38, 292, 497
palynomorphs in Skipsea Till 188
Papa Stour 124m, 125
para-glacial sediments 4
parabolic-shaped drumlins 425
parallel roads 137, 147
    of Glen Roy 94
parallel-laminated sands 280, 326, 343
Parson's Bank 302m
particle-size
    analyses 57, 384, 387, 407
    characteristics 271, 382
    composition of tills from Holwell Hyde 403t
    curves
        of chalky tills from Vale of St. Albans 400f
        for laminated clays from Cheshire 397f
        for tills from west-central Scotland 395f
        of ice-marginal deposits from Cheshire 398f
        of tills from Cowden 399f
        of tills from Northumbria 394f
    data 419
    distribution 78, 268, 272, 338, 382t, 383t
        of hummocky moraine sediments 95
        of Stortford Till 273f
        of Ugley Till 275f
        of Westmill Till 276f
    frequency 415
    properties 278
Parton 156m
Partry Mountains 101, 357m
Pass
    of Llanberis 99m
    of Ryvoan 449, 449m
Pastonian 185, 494, 494t, 495
patterned ground 17
    periglacial 469
Paviland
    Glaciation 5t, 9
    Moraine 9
Paxford
    Gravels 455
    Oolite Gravel 468f
Peak
    Dale 197
    District 193-202
        ice flow direction and erratics 198m
        principal geological units 194m
        till distribution 195m
        X-ray diffraction trace from younger till 200f
        X-ray diffraction trace of loess and

T - #0056 - 101024 - C24 - 279/216/34 [36] - CB - 9789061918752 - Gloss Lamination